ALTERNATING CURRENT FUNDAMENTALS

ALTERNATING CURRENT FUNDAMENTALS

EIGHTH EDITION

Stephen L. Herman

DELMAR
CENGAGE Learning™

Australia • Brazil • Japan • Korea • Mexico • Singapore • Spain • United Kingdom • United States

Alternating Current Fundamentals, 8th Edition
Stephen L. Herman

Vice President, Career and Professional Editorial: Dave Garza

Director of Learning Solutions: Sandy Clark

Acquisitions Editor: Stacy Masucci

Managing Editor: Larry Main

Senior Product Manager: John Fisher

Senior Editorial Assistant: Dawn Daugherty

Vice President, Career and Professional Marketing: Jennifer Baker

Marketing Director: Deborah Yarnell

Associate Marketing Manager: Scott A. Chrysler

Production Director: Wendy Troeger

Production Manager: Mark Bernard

Senior Content Project Manager: Christopher Chien

Senior Art Director: David Arsenault

Technology Project Manager: Christopher Catalina

Production Technology Analyst: Thomas Stover

Library of Congress Control Number: 2010922546

ISBN-13: 978-1-111-12527-1

ISBN-10: 1-111-12527-9

Delmar
Executive Woods
5 Maxwell Drive
Clifton Park, NY 12065
USA

Cengage Learning is a leading provider of customized learning solutions with office locations around the globe, including Singapore, the United Kingdom, Australia, Mexico, Brazil and Japan. Locate your local office at **www.cengage.com/global**

Cengage Learning products are represented in Canada by Nelson Education, Ltd.

To learn more about Delmar, visit **www.cengage.com/delmar**

Purchase any of our products at your local bookstore or at our preferred online store **www.cengagebrain.com**

Printed in the United States of America
4 5 6 7 8 20 19 18 17 16

Contents

Preface

The eighth edition of *Alternating Current Fundamentals* remains a leader in textbooks that provide the information needed for technicians in the electrical field. *Alternating Current Fundamentals* is a companion text to *Direct Current Fundamentals* and builds on that platform. An understanding of alternating current theory is an absolute must for anyone desiring employment in the electrical field in almost any part of the world. *If industry runs, it runs on electricity!*

Alternating Current Fundamentals Eighth Edition begins with an introduction to alternating current and compares it to direct current. AC voltage, frequency, waveforms, and the differences between peak, RMS, and average values are discussed. Series and parallel circuits containing resistance, inductance, and capacitance with mathematical solutions to problems are given. Practice problems at the end of applicable units aid students in understanding the concepts of RL, RC, and RLC circuits.

Alternating Current Fundamentals Eighth Edition contains information on single-phase and three-phase transformers. This coverage includes examples of how to calculate values of voltage, current, and turns for different types of transformers. Alternators, three-phase motors, and single-phase motors are also covered.

NEW FOR EIGHTH EDITION

- Improved graphics.
- Determining the values of different types of capacitors.
- Expanded coverage on three-phase motors.
- Expanded coverage on single-phase motors.
- A Green Technology Icon within the text indicates new information about "Green Technology" relating to conservation and energy efficiency. This text contains information about *power factor correction.* Keeping the power factor in ac circuits as close to unity as possible is desirable because it reduces power losses in line wires and increases efficiency of transmission.

ACKNOWLEDGMENTS

The author and Delmar Cengage Learning, would like to extend their thanks to those who provided detailed reviews of the manuscript. The contributions and suggestions of the following individuals are greatly appreciated.

Michael Sullivan, South Shore Technical High School
Robert Blakely, Mississippi Gulf Coast Community College
Robert Morris, State University of New York at Delhi
Ronald Langley, Williamsburg Technical College
John Wieczerza, Macomb Community College

SUPPLEMENTS TO THIS TEXT

An *Instructor Resources* CD is available for this text. It contains tools and instructional resources that enrich your classroom and make your preparation time shorter. The elements of the *Instructor Resources* link directly to the text and tie together to provide a unified instructional system. Features contained in the *Instructor Resources* include the following:

- An *Instructor Manual* as a PDF file that contains answers to the end of unit Achievement Review questions and answers to the practice problems.
- Unit presentations created in PowerPoint: These slides provide the basis for a lecture outline that helps you to present concepts and material.
- Twenty-four customizable topical support slide presentations in PowerPoint format focus on key topic areas for electricity and electric motor control; these can also include video clips, animations, and photos.
- Test questions: More than 200 questions of varying levels of difficulty are provided in true/false and multiple choice formats. These questions can be used to assess student comprehension or can be made available to the student for self-evaluation.

ISBN: 1111125287

To access additional course materials and companion resources, please visit www.cengagebrain.com. At the CengageBrain.com home page, search for the ISBN of your title (from the back cover of your book) using the search box at the top of the page. This will take you to the product page where free companion resources can be found.

1

Introduction to Alternating Current

Objectives

After studying this unit, the student should be able to

- list the reasons why alternating current is preferred to direct current for large generating, transmission, and distribution systems.
- use the functions sine, cosine, and tangent, which define the relationships between right triangles and angles in the quadrants of a coordinate system.
- demonstrate the graphical method of generating sine and cosine waves (giving the formula for each).
- describe how a sine wave of voltage is obtained as a coil rotates in a uniform magnetic field (simple ac generator), giving the equation for instantaneous voltage.
- list the factors affecting the frequency of the voltage from ac generators and give the equation expressing the frequency.
- list the advantages of a 60-hertz (Hz) service over a 25-Hz service.
- define the following terms: coordinate system, quadrants, sine wave, cosine wave, ac generator, alternating voltage, alternating current, frequency, electrical time degrees, mechanical degrees, fundamental wave, harmonic wave.

INTRODUCTION

Much of the electrical energy used worldwide is produced by alternating-current (ac) generators. Such widespread use of alternating current means that students in electrical trades must understand the principles of electricity and magnetism and their application to alternating-current circuits, components, instruments, transformers, alternators, ac motors, and control equipment.

Uses for Direct Current

Although alternating current is more commonly used, there are a number of applications where direct current (dc) either must be used or will do the job better than alternating current. Several of these applications are described in the following list:

- Direct current is used for various electrochemical processes, including electroplating, refining of copper and aluminum, electrotyping, production of industrial gases by electrolysis, and charging of storage batteries.
- Direct current is used to excite the field windings of alternating-current generators.
- Direct current applied to variable speed motors results in stepless, precise speed adjustments. Such motors are used in metal rolling mills, papermaking machines, high-speed gearless elevators, automated machine tools, and high-speed printing presses.
- Traction motors require direct current. Such motors are used on locomotives, subway cars, trolley buses, and large construction machinery that will not be driven on highways. Using a dc motor in these applications eliminates the need for clutches, gear shifting transmissions, drive shafts, universal joints, and differential gearing. Thus, almost all large locomotives have diesel engines that drive direct-current generators to supply the power for dc traction motors installed in each locomotive truck.

Under normal conditions, the electrical energy produced by alternating-current generators is transmitted to the areas where it is to be used by alternating-current loads. If direct current is required, the alternating current is changed to direct current by rectifiers or motor–generator sets. Fortunately, alternating current is suited for use with heating equipment, lighting loads, and constant-speed motors. Loads of this type are the most common users of electrical energy. Thus, the costly conversion to direct current is needed only for certain load requirements.

Advantages of Alternating Current

Alternating current is preferred to direct current for large generating, transmission, and distribution systems for the following reasons:

- AC generators can be built with much larger power and voltage ratings than dc generators. An ac generator does not require a commutator. The armature or output winding of the ac generator can be mounted on the stator (stationary part) of the machine. The output connections are made with cables or bus bars bolted directly to the stator windings (stationary armature windings). Armature voltages of 13,800 volts or more are common. Currents of any desired value can be obtained with the proper machine design. The rotating member of the alternator is the field. This field is supplied with direct current by means of slip rings or by means of a brushless exciter from an external dc source. The voltage of the source is in the range of 100 to 250 V. In contrast to the ac generator, the armature or output winding of a dc generator must be the rotating part of the machine. The connection of the armature to the external load is made through a commutator and brushes. These components restrict the maximum voltages and currents that can be obtained from dc machines to practical levels. Large dc machines rated at 600 to 750 V are common. Occasionally, a machine rated at 1500 V is required for certain applications. The commutators of dc generators are usually rated at less than 8000 amperes (A). These large current ratings are practical only on slow-speed machines. For these reasons, dc generator ratings are limited to relatively low voltage and power values as compared to ac generators.

- With the use of alternating current, the voltage can be stepped up or stepped down efficiently by means of transformers. A transformer has no moving parts, and its losses are relatively low. The efficiency of most transformers at the rated load is high, from 95% to more than 99%. Transformers cannot be used with direct current. DC voltage changes are obtained by using series resistors, which give rise to I^2R losses, or motor–generator sets, which have relatively low overall efficiencies. However, the reduction or increase of dc voltages in dc systems is inefficient.
- Large ac generators having very high power ratings (Figure 1–1), plus efficient transformers to step up or step down the alternating voltage, make it possible to conduct

FIGURE 1–1 A 44,000 kW power company installation (*Courtesy of General Electric Company*)

ac energy economically over long distances from generating stations to the various load centers by way of high-voltage transmission systems. Thus, huge amounts of electrical energy can be generated at one location. For example, a large hydroelectric generating station may be located near a waterfall. Here, the energy can be generated at a relatively low cost per kilowatt-hour. Large steam-generating stations are also located where fuel is easy to obtain and abundant water is available. Steam-generating stations use very large-capacity alternators having efficiency ratings as high as 97%. Large high-speed turbines operating at very high steam pressures are used to turn the ac generators. The efficiency of these steam turbines, operating at speeds of 1200, 1800, or 3600 revolutions per minute (r/min), is much greater than that of steam turbines used in smaller generating plants. Completely automated control systems are used in modern generating stations to increase the total operating efficiency even more. As a result, large steam-generating plants and hydroelectric stations operate efficiently to produce electrical energy at a low generating cost per kilowatt-hour.

- The ac induction motor (Figure 1–2) has no commutator or brushes. This type of motor has a relatively constant speed. It is rugged and simple in construction. The initial purchase price and the maintenance and repair costs for the ac induction motor are considerably less than the costs for a dc motor of comparable horsepower, voltage, and speed. Further, the starting equipment used with a typical induction motor is also lower in cost initially when compared to the starting equipment used with dc motors having similar horsepower ratings. Because ac induction motors do not contain a commutator or brushes, they generally have a longer life span and require less maintenance than dc machines.

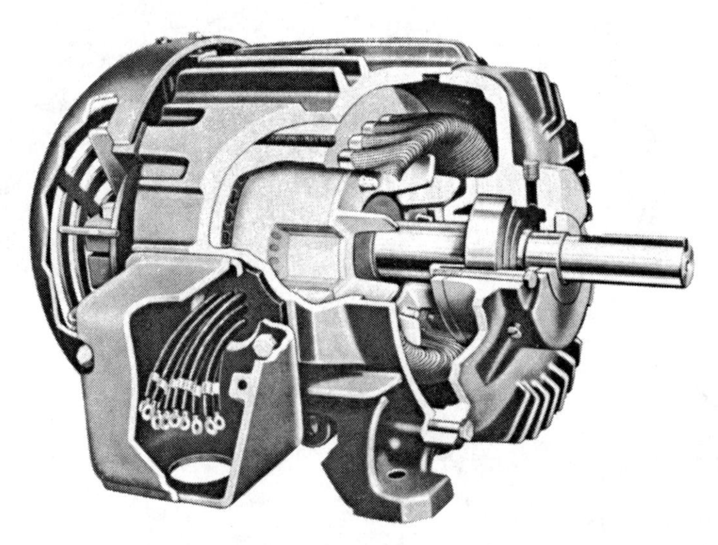

FIGURE 1–2 Cutaway view of a squirrel cage induction motor (*Courtesy of General Electric Company*)

ANGULAR RELATIONSHIPS

A basic knowledge of trigonometry is essential to an understanding of alternating-current concepts. That is, the student must know the basic mathematical relationships between right triangles and angles in the quadrants of a coordinate system.

Coordinate System and Angular Relationships

Figure 1–3 shows a *coordinate system* consisting of an X axis and a Y axis. These axes are mutually perpendicular lines that form four 90° angles called *quadrants*. Quadrants 1 through 4 in the figure are numbered *counterclockwise*.

Figure 1–4 represents a given X–Y coordinate system. The indicated angles are measured from the positive X axis to a given line. Lines OA and OB (quadrant 1) (Figure 1–4A) form a 90° angle at 0. Line OC (quadrant 1) and line OD (quadrant 2) (Figure 1–4B) form a 120° angle at 0. In Figure 1–4C, lines OE (quadrant 1) and OF (quadrant 3) form a 240°

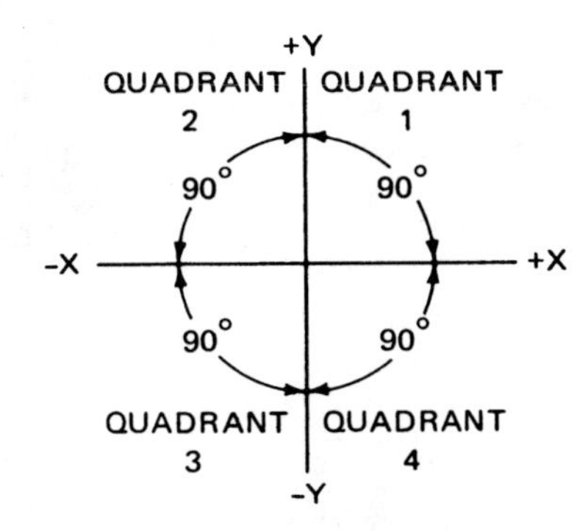

FIGURE 1–3 Coordinate system (X–Y axes) (*Delmar/Cengage Learning*)

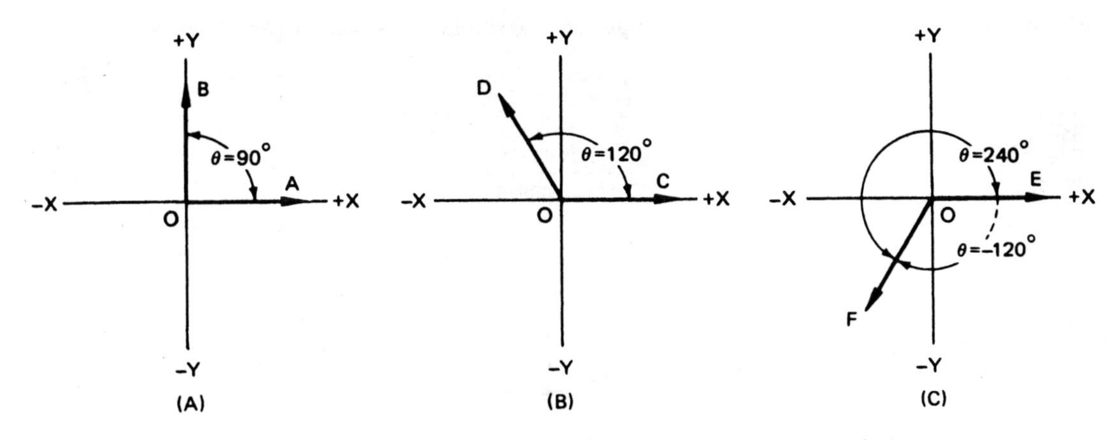

FIGURE 1–4 Angular relationships (*Delmar/Cengage Learning*)

angle at 0. In these examples, all of the angles are measured from the positive (+) X axis to the indicated line in the counterclockwise direction. If the angle is measured in the *clockwise* direction from the positive X axis, the angle is negative because the direction of measurement has changed. (See line OF in Figure 1–4C.) A simple saying can be used to help remember the relationships of sine, cosine, and tangent. Use the first letter of each word in the saying, "Oscar Had A Heap Of Apples" in Figure 1-5B.

Figures 1–5A and 1–6 summarize the angular relationships for the various quadrants. These relationships will be used throughout this text in vector problems.

For angle A:

1. The side opposite the right angle is called the ***hypotenuse*** (AB). The hypotenuse is always the longest side of a right triangle.
2. The side opposite angle A is called the *opposite side* (BC).
3. The side of angle A which is not the hypotenuse is called the *adjacent side* (AC).

For angle B:

4. The opposite side of angle B is (AC).
5. The adjacent side of angle B is (BC).

Sine (sin) of angle $= \dfrac{\text{opposite side}}{\text{hypotenuse}}$; $\sin A = \dfrac{a}{c}$

Cosine (cos) of angle $= \dfrac{\text{adjacent side}}{\text{hypotenuse}}$; $\cos A = \dfrac{b}{c}$

Tangent (tan) of angle $= \dfrac{\text{opposite side}}{\text{adjacent side}}$; $\tan A = \dfrac{a}{b} = \dfrac{\sin A}{\cos A}$

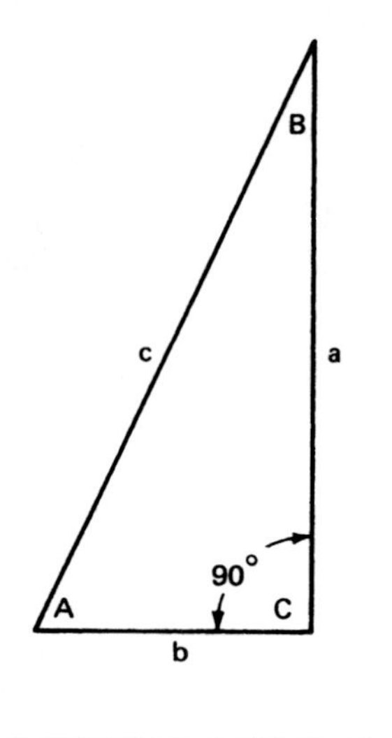

FIGURE 1–5A Rotating angular relationships—sine wave and cosine wave (*Delmar/Cengage Learning*)

$\dfrac{\text{Oscar}}{\text{had}}$	sin	$\dfrac{\text{opposite side}}{\text{hypotenuse}}$
$\dfrac{\text{a}}{\text{heap}}$	cos	$\dfrac{\text{adjacent side}}{\text{hypotenuse}}$
$\dfrac{\text{of}}{\text{apples}}$	tan	$\dfrac{\text{opposite side}}{\text{adjacent side}}$

FIGURE 1–5B Memory aid for trig functions (*Delmar/Cengage Learning*)

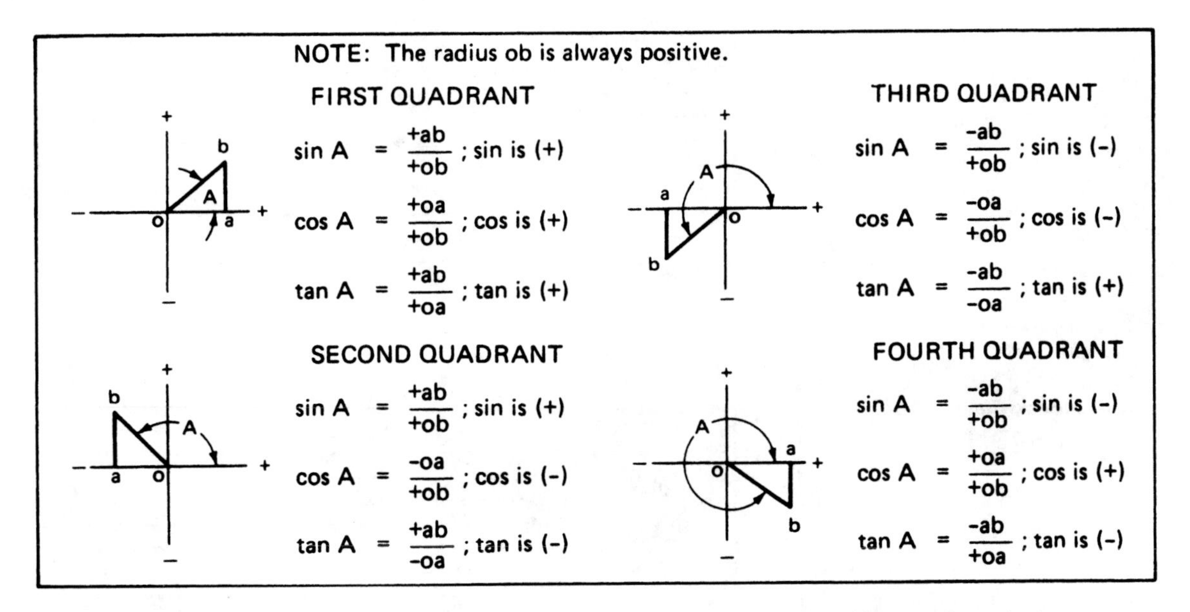

FIGURE 1–6 Angular relationships in the four quadrants (*Delmar/Cengage Learning*)

Generation of Sine and Cosine Waves

Figures 1–3 and 1–4 showed the static positions of a line in different quadrants. If the line is allowed to rotate *counterclockwise,* an analysis can be made of the projections (or shadows) of the line on the X and Y axes (Figure 1–7). A wave called a *sine wave* and one called a *cosine wave* will be generated by the rotating line.

The student should be able to visualize the shadow of the line as it rotates. As the angle theta (θ) increases, the shadow of the line on the Y axis increases and the shadow of the line on the X axis decreases.

Figure 1–8 shows the pattern obtained by rotating a line of magnitude R about the 0 point. The projections of R on the Y axis are plotted against the angle theta (θ) made by the line as it moves from the positive X axis. The wave pattern formed is called a *sine wave* and is expressed by the formula $y = R \sin \theta$, where R is the radius of the circle, θ is the angle moved (traversed) by the line from the positive X axis, and Y is the projection or shadow of the line on the Y axis.

In a similar manner, the projection of the rotating line on the X axis can be plotted against the angle θ made by the line as it rotates. Figure 1–9 shows the resulting wave pattern.

In Figure 1–9, the projection of the radius (R) on the X axis is zero at angles of 90° and 270°. The resulting wave pattern is called a *cosine wave*. This waveform is expressed by the formula $y = R \cos \theta$, where R is the radius of the circle.

Comparing the two wave patterns, it can be seen that when one has a magnitude of zero, the other has the maximum magnitude R, and vice versa. Note that the cosine wave has the same pattern as the sine wave, but reaches its maximum value 90° before the sine wave.

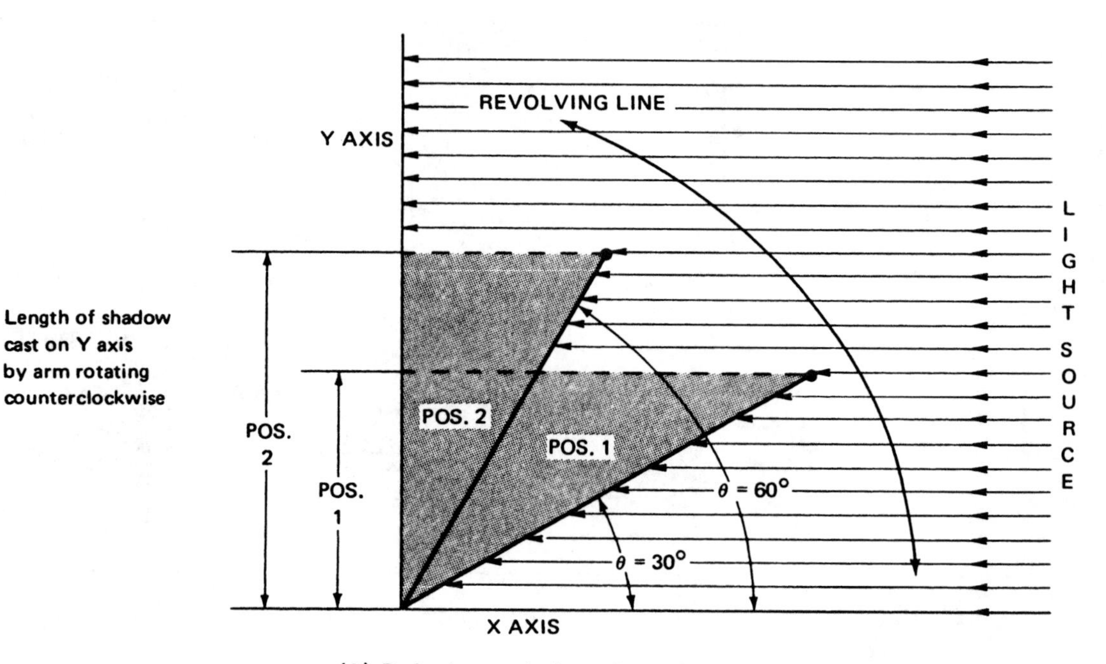

(A) Projection, or shadow, of a revolving line on the Y axis.

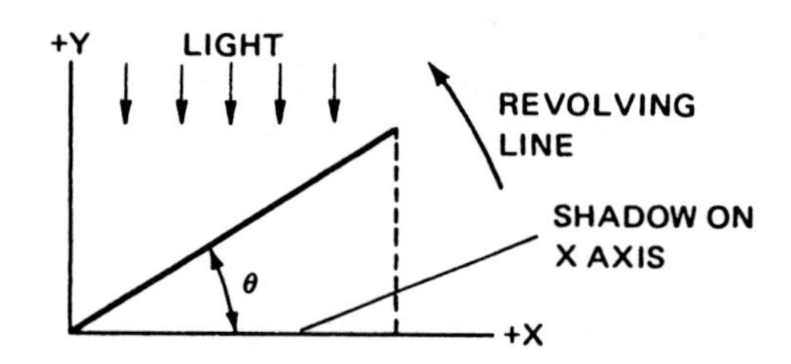

(B) Projection, or shadow, of a revolving line on the X axis.

FIGURE 1–7 Projections of a revolving line on the X and Y axes (*Delmar/Cengage Learning*)

Sine and cosine waves can be generated at the same time by rotating two lines that are 90° out of step or out of phase with each other. The projections of these two lines on the Y axis can be plotted against the common angle of movement (θ). Figure 1–10 shows the waveforms generated in this manner. The intermediate points have been deleted to reduce confusion in the drawing.

There may be some confusion because of the two angles marked θ in Figure 1–10. Recall that all angles are measured from the positive X axis. In this case, the angle θ represents not only the movement of line A to A' but also the movement of the A–B structure from the A–B position to the A'–B' position. The structure has moved an angle θ, which is measured from the positive X axis. Line OB generates a cosine wave, and line OA generates a sine wave.

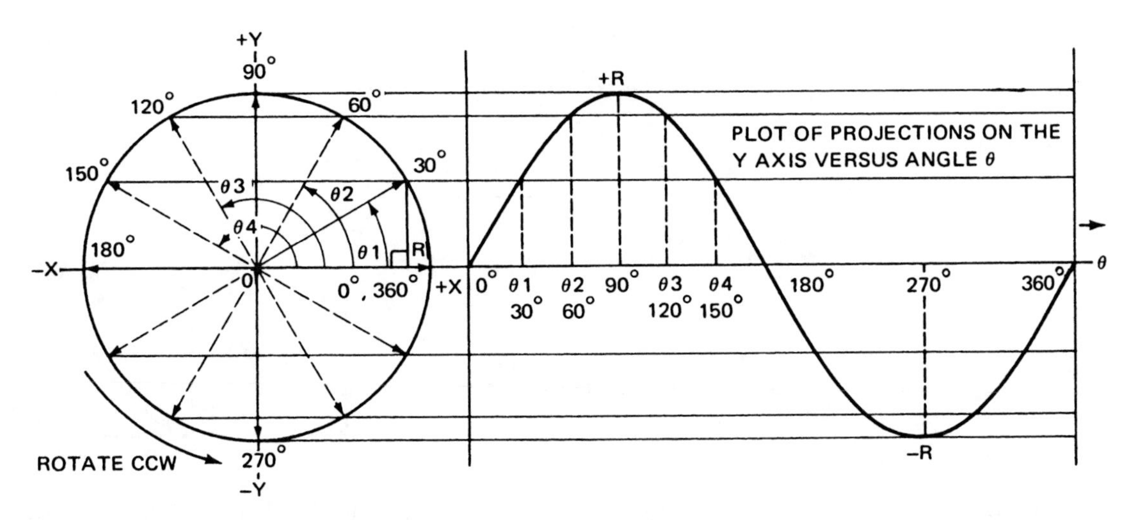

FIGURE 1–8 Projection obtained by rotating a line (having uniform circular motion) on the Y axis ***(Delmar/Cengage Learning)***

A discussion follows of how alternating voltages are generated using the principles learned in the study of direct current. This discussion will show that the sine wave is not generated by means of projections of a moving line on the Y axis. Rather, the sine wave is a function of the position of a coil in a magnetic field. One important point should be kept in mind: the mathematical relationships of all sine waves are the same, regardless of the method by which they are generated. All sine waves have the same form as expressed by the equation $Y = R \sin \theta$.

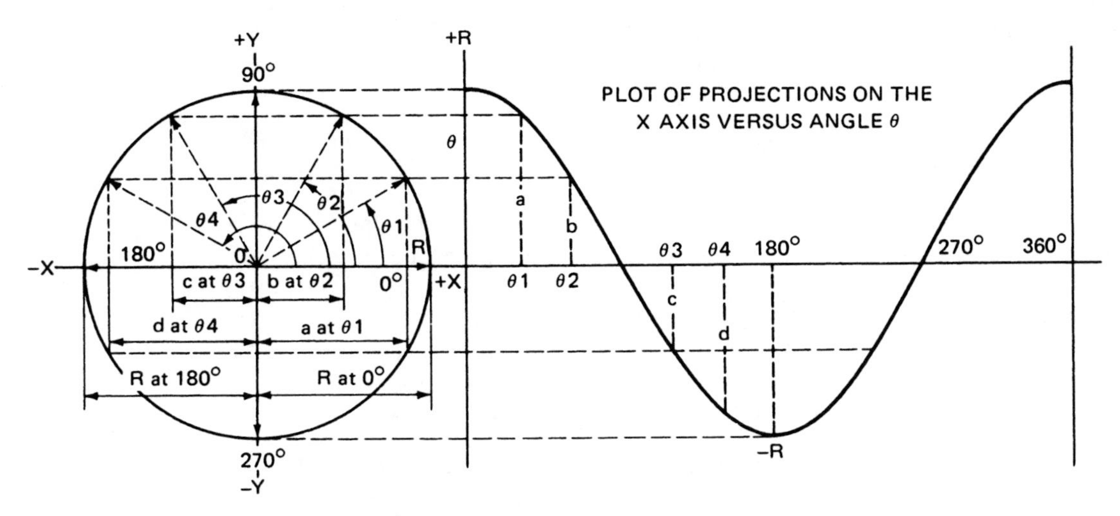

FIGURE 1–9 Projection of a rotating line (having uniform circular motion) on the X axis ***(Delmar/Cengage Learning)***

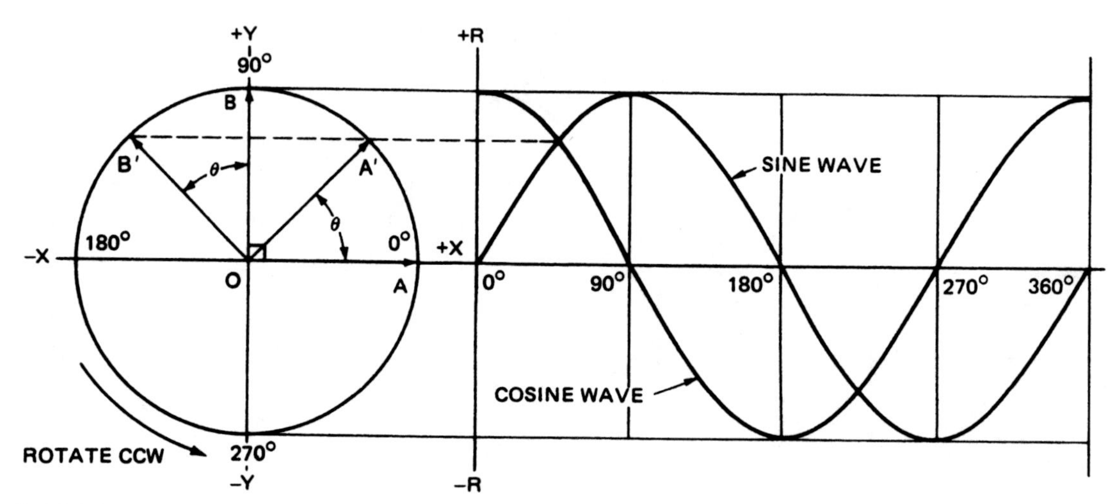

FIGURE 1–10 Projection of two rotating lines (which are 90° out of step or out of phase) on the Y axis (*Delmar/Cengage Learning*)

ALTERNATING-VOLTAGE GENERATION

A simple alternating-voltage generator consisting of a single coil rotating in a uniform magnetic field is shown in Figure 1–11.

The use of Fleming's generator rule shows that an alternating voltage is generated in the coil as it rotates. If the ends of the coil are connected to two slip rings, the alternating voltage can be observed on an oscilloscope. This voltage pattern is a typical sine wave, as shown in Figure 1–12.

The generated voltage in an armature conductor is expressed by the formula

$$\mathbf{V_{generated}} = \frac{\mathbf{BLv}}{\mathbf{10^8}}$$
$$= \mathbf{BLv \times 10^{-8}}$$

where $V_{generated}$ is the generated voltage in the armature conductor in volts, B is the magnetic flux of the field, L is the length of the armature conductor in inches, v is the velocity of rotation of the coil in inches per second, and 10^8 represents 100,000,000 lines of force that must be cut per second to cause one volt to be induced.

One of the magnetic measurements in the English system is the weber. One weber represents an amount of magnetic flux equal to 100,000,000 lines. Therefore, it can be stated that voltage is induced at a rate of one weber per second (1/Wb/s). The amount of voltage induced in a conductor is proportional to three factors:

1. The strength of the magnetic field (flux density)
2. The length of the conductor (often expressed as the number of turns of wire)
3. The speed of the cutting action

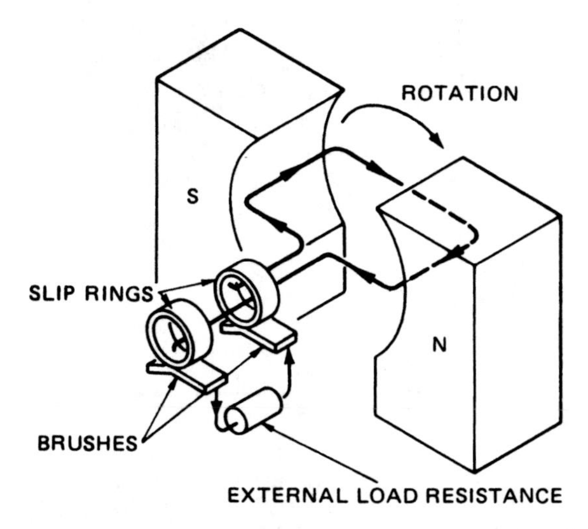

FIGURE 1–11 Elementary ac generator ***(Delmar/ Cengage Learning)***

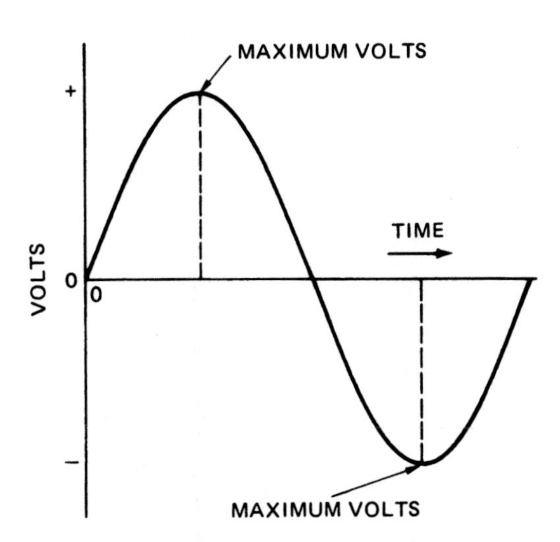

FIGURE 1–12 Sine wave of voltage ***(Delmar/ Cengage Learning)***

Most ac generators have stationary coil windings and rotating field windings. However, in Figure 1–11, the coil rotates and the field is stationary. In either case, the induced voltage in the coil windings depends upon the number of lines of force cut per second.

Development of an AC Sine Wave

To illustrate the development of the alternating-voltage sine-wave pattern shown in Figure 1–12, a more convenient form of the simple ac generator is needed.

A simple ac generator is shown in Figure 1–13. The conductors of the coil are moving parallel to the lines of force. At this instant, almost no lines of force are being cut and the generated voltage is zero.

In Figure 1–14 the conductors of the coil have moved counterclockwise to a point 30° from the starting position. The conductors of the coil are now cutting across the field flux. As a result, a voltage is induced in the coil. The instantaneous voltage in this position is determined by

$$v_{instantaneous} = V_{maximum} \times \sin \angle$$

Assuming that the maximum voltage is 141.4 V, the induced voltage at 30° is

$$v_i = V_{max} \times \sin 30° = 141.4 \times 0.5000 = 70.7 \text{ V}$$

Movement across a Magnetic Field. By examining the triangle in Figure 1–14, it can be seen that the total velocity of the conductor (V_T) has two components. There is a useless vertical component (v_p) parallel to the magnetic lines of force. The other component (v_c) is a useful horizontal component that *crosses*, or is perpendicular to, the magnetic lines of

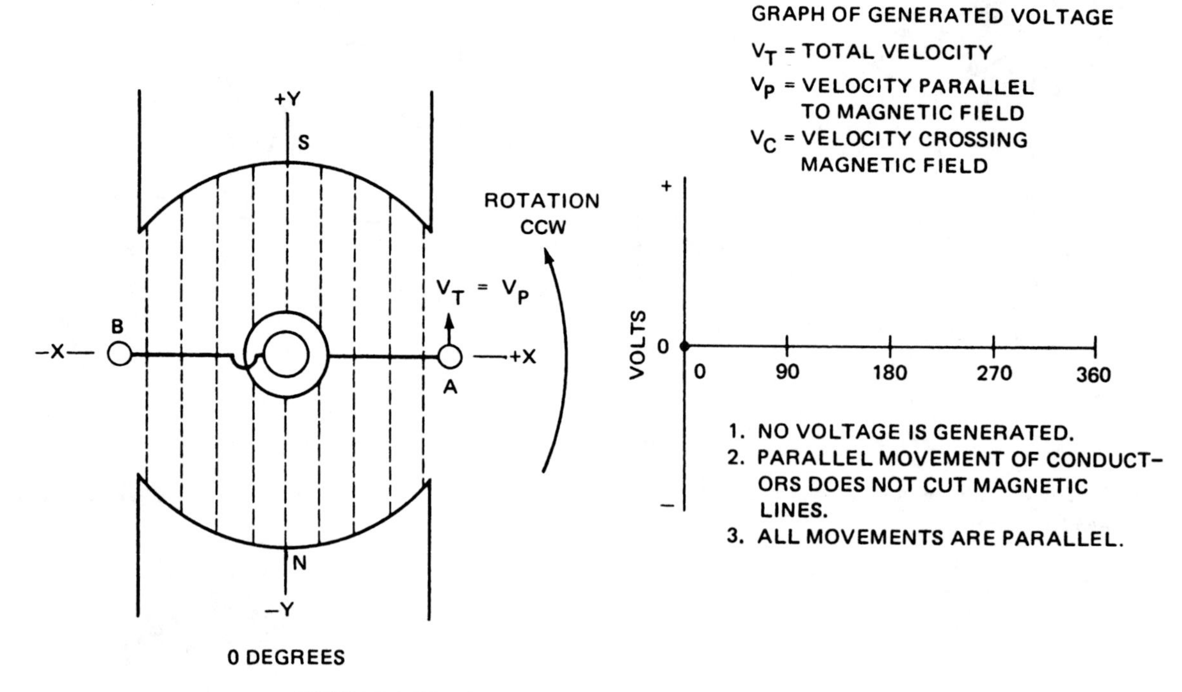

FIGURE 1–13 Start of cycle (*Delmar/Cengage Learning*)

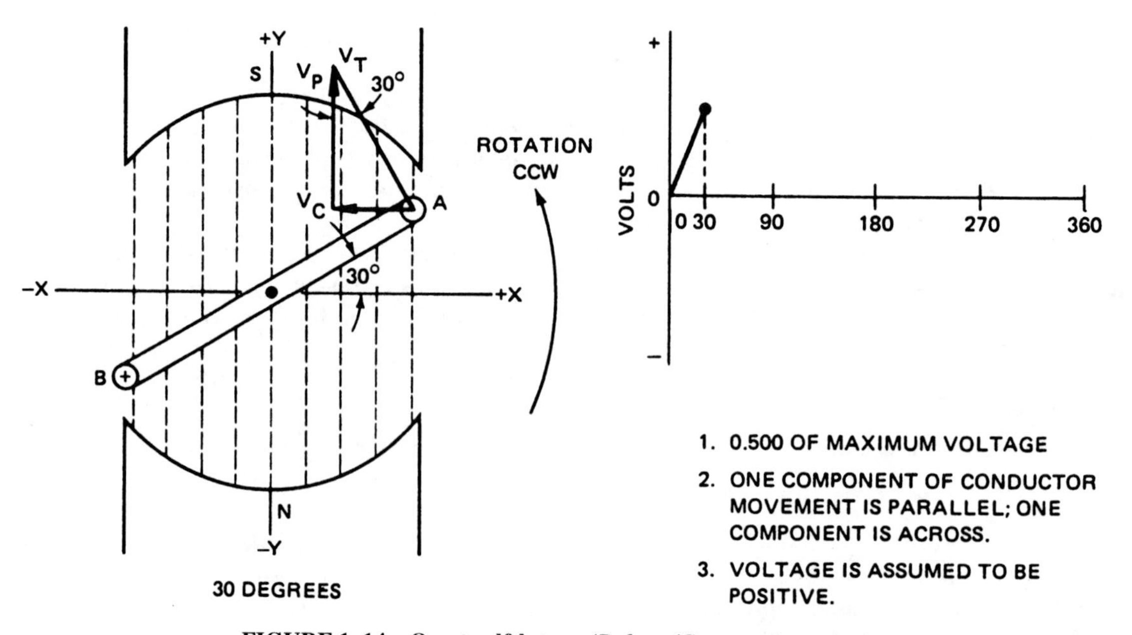

FIGURE 1–14 One-twelfth turn (*Delmar/Cengage Learning*)

force. Because voltage is generated only by the movement that cuts the magnetic field, the right triangle is solved for the v_c component:

$$v_c = V_T \times \sin 30°$$

Note the similarity between the equations expressing velocity, voltage, and projections on the Y axis:

1. $\mathbf{v_c = V_T \times \sin 30°}$
2. $\mathbf{v = V_{max} \times \sin 30°}$
3. $\mathbf{y = R \times \sin \theta}$

- Equation 1 was derived from a coil revolving in a magnetic field where the velocity components are resolved perpendicular to the magnetic field.
- Equation 2 was derived from a coil revolving in a magnetic field where the maximum voltage components are resolved into instantaneous voltage.
- Equation 3 was derived from rotating a line counterclockwise and taking the magnitude of its projection on the Y axis. In other words, R is resolved into its Y shadow.

The values v_c, v, and y in equations 1 through 3 are called *instantaneous values* of the sine wave. V_T, V_{max}, and R are called the *maximum values* of the sine wave. The general forms of a voltage sine wave and a current sine wave are

$$\mathbf{v = V_{max} \times \sin \theta}$$
$$\mathbf{i = I_{max} \times \sin \theta}$$

The coil at 45°. In Figure 1–15 the coil is at a new position 45° from the starting position. Refer to the right triangle construction in the figure. The component of the angular velocity has increased slightly (as compared to Figure 1–14). There is a proportional increase in the induced voltage to an instantaneous value determined as follows:

$$v = V_{max} \times \sin 45° = 141.4 \times 0.7071 = 100.0 \text{ V}$$

The coil at 90°. In Figure 1–16 the coil has rotated to an angle of 90° from the starting position. The sine of 90° is 1.0; therefore, the generated voltage has a maximum value of 141.4 V. The conductors of the coil are perpendicular to the flux field. Because the greatest number of lines of flux are cut in a given time period in this position, the induced voltage must be a maximum value. As the armature coil continues to rotate counterclockwise, the direction and the instantaneous value of voltage can be determined for any angle through 360° (one complete revolution). The resulting waveform is a sine wave of voltage. For each angular position of the coil in the magnetic field, the direction of the generated voltage can be obtained by Fleming's generator rule. The value of the instantaneous voltage generated in the coil for each angular position can be found from the sine of the angle times V_{max}.

During the design and construction of ac generators, an attempt is made to ensure a nearly perfect sine-wave voltage output. Motors, transformers, and other electrical equipment have better operating characteristics when they receive electrical energy from such ac generators.

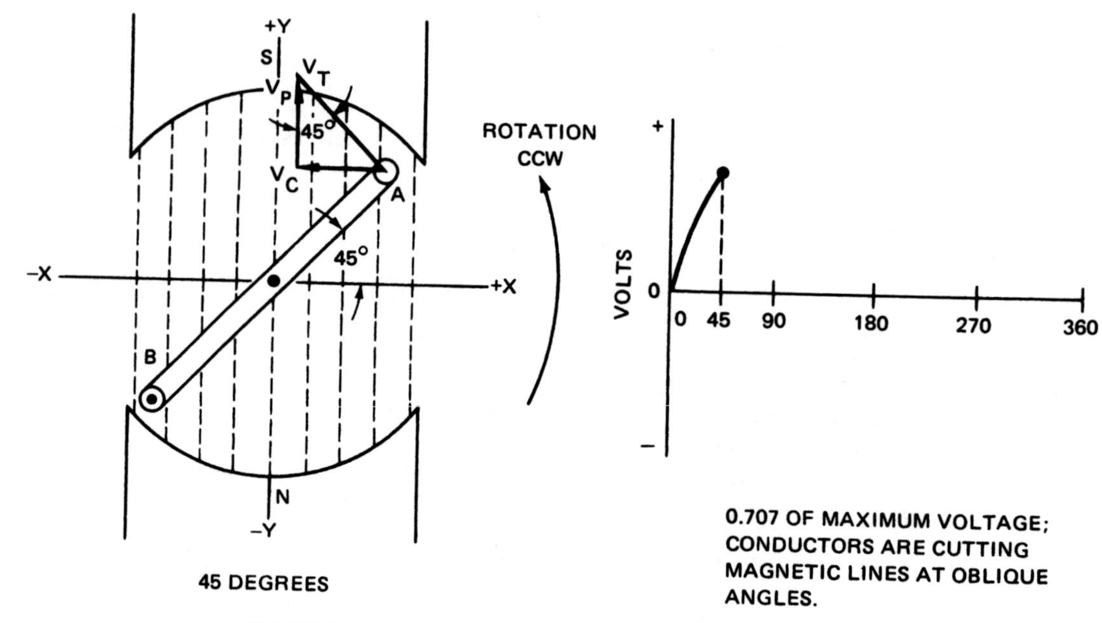

FIGURE 1–15 One-eighth turn (*Delmar/Cengage Learning*)

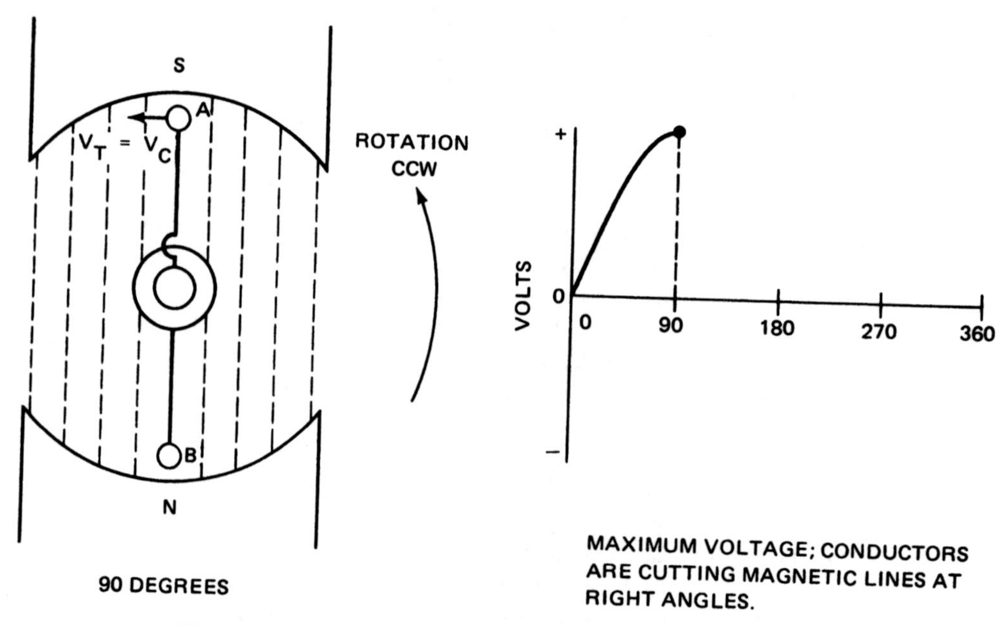

FIGURE 1–16 One-quarter turn (*Delmar/Cengage Learning*)

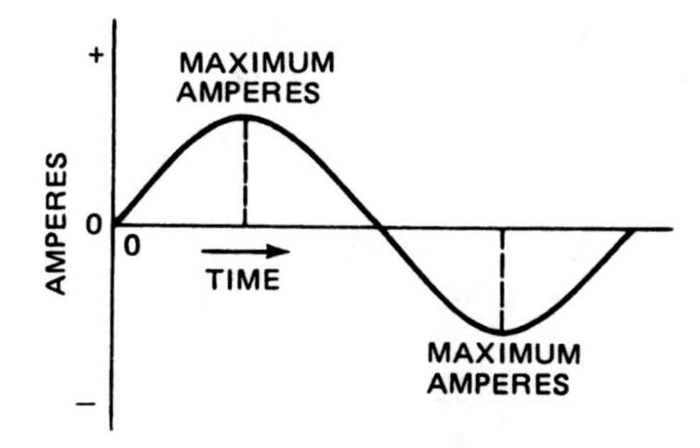

FIGURE 1–17 Sine wave of current (*Delmar/ Cengage Learning*)

Defining Alternating Voltage and Current

Alternating voltage may be defined as an electromotive force that changes continuously with time. It rises from zero to a maximum value in one direction and decreases back to zero. It then rises to the same maximum value in the opposite direction and again decreases to zero. These values are repeated again and again at equal intervals of time.

The alternator shown in Figure 1–11 is connected to a resistor, which is the external load. The alternating voltage of this generator causes an alternating current to be supplied to the load. As the alternating voltage varies in magnitude and direction, the current (in amperes) varies proportionally. Alternating current may be defined in a manner that is similar to the definition of alternating voltage. Refer to Figure 1–17.

Alternating current is a current that changes continuously with time. It rises from zero to a maximum value in one direction and decreases back to zero. It then rises to the same maximum value in the opposite direction and again decreases to zero. These values are repeated again and again at equal intervals of time.

FREQUENCY

The number of complete events or cycles per second is the frequency, measured in hertz. Sixty cycles per second equals 60 hertz, or 60 Hz.

In the United States and Canada, 60 Hz is used almost exclusively, with the exception of a few areas that use 25-Hz service. The advantage to using a higher-frequency service is that less iron and copper are required in the transformers. Therefore, they are lighter and lower in cost. Also, incandescent lamps operating at 60 Hz have no noticeable flicker. At 25 Hz, the flicker of incandescent lamps can be annoying.

The speed of a generator and the number of poles determine the frequency of the generated voltage. If a generator has two poles (north and south), and the coils rotate at a speed of one revolution per second, the frequency is one cycle per second. If the generator has two pairs of poles, then a cycle is generated every half-revolution, or 2 hertz per second (2 Hz/s).

Frequency of an AC Generator

In the simple alternator, one cycle of voltage is produced each time the coil makes one revolution between the two poles. If this coil makes 60 revolutions per second, the alternating voltage generated will have a frequency of 60 cycles per second (60 Hz). The frequency of an ac generator is expressed by the following formula

$$\mathbf{f = \frac{P \times S}{60}}$$

where

f = frequency, in hertz
P = number of *pairs* of poles
S = speed, in revolutions per minute (r/min)
60 = number of seconds in one minute

For example, if a two-pole ac generator turns at 3600 r/min, the frequency in hertz is

$$f = \frac{P \times S}{60} = \frac{1 \times 3600}{60} = 60 \text{ Hz}$$

If a four-pole ac generator is turned by a waterwheel at a speed of 750 r/min, the frequency of this generator is

$$f = \frac{P \times S}{60} = \frac{2 \times 750}{60} = 25 \text{ Hz}$$

Because there may be some confusion in using pairs of poles in the frequency formula, it is common practice to use the total number of poles of the alternator. In this case, the time constant of 60 s is doubled. For example, if a four-pole alternator turns at 1800 r/min, the frequency of the voltage output of the machine is

$$f = \frac{P \times S}{120} = \frac{4 \times 1800}{120} = 60 \text{ Hz}$$

where

P = number of *single* poles
$120 = 2 \times 60$ s

ELECTRICAL TIME DEGREES AND MECHANICAL DEGREES

When a coil makes one revolution in a generator with two poles, one cycle of voltage is generated. However, when a coil makes one revolution in a generator with four poles (Figure 1–18), two cycles of voltage are generated. Thus, a distinction must be made between mechanical and electrical degrees.

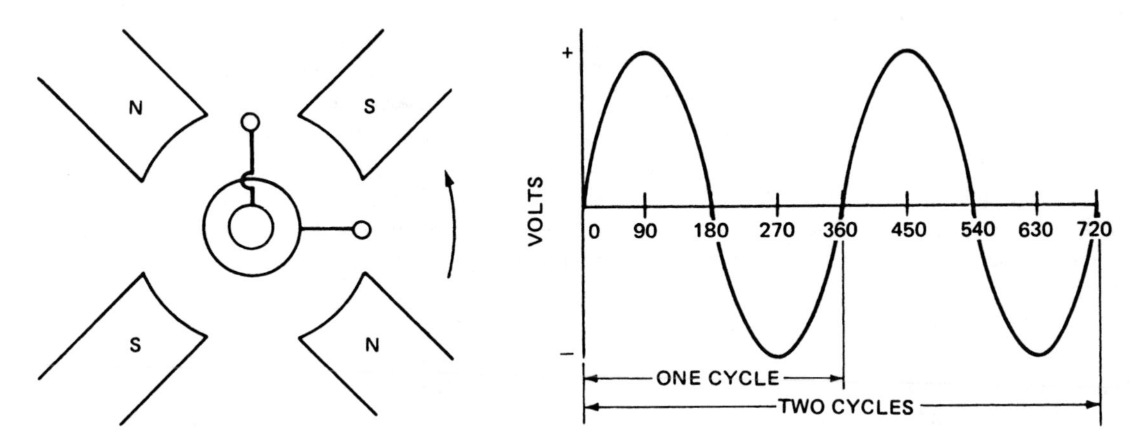

FIGURE 1–18 Four-pole generator, two cycles per revolution (*Delmar/Cengage Learning*)

When a coil or armature conductor makes one complete revolution, it passes through 360 *mechanical degrees*.

When an electromotive force or an alternating current passes through one cycle, it passes through 360 *electrical time degrees*.

As the number of poles in an ac generator increases, the actual required driven speed in r/min decreases proportionally for a given frequency. The relationship between speed, number of poles, and frequency is shown in Table 1–1. The frequency values may be checked using the following frequency equation:

$$f = P \times S \div 120$$

The values for the number of poles and the speed in r/min may be substituted in the formula for each frequency value.

Poles	Speed in r/min	
	60 Hz	25 Hz
2	3600	1500
4	1800	750
6	1200	500
8	900	375

TABLE 1–1 The relationship between speed, number of poles, and frequency

OTHER WAVEFORMS

Alternating-voltage waveforms are not all sine waves. For example, a square-wave output or a rectangular output can be generated by electronic equipment, such as a signal generator. One type of electronic oscillator has a voltage output pattern that resembles a sawtooth (Figure 1–19).

For electrical energy transmitted at frequencies of 60 Hz and 25 Hz, the voltage wave pattern may be distorted so that it is not a true sine wave. Such distortion is due to conditions that may exist in ac generators, transformers, and other equipment. A distorted wave pattern consists of a fundamental wave (which is the frequency of the circuit) and other waves having higher frequencies. These waves are called *harmonics* and are superimposed on the fundamental wave. The exact appearance of the distorted wave will depend on the frequencies, magnitudes, and phase relationships of the voltage waves superimposed on the fundamental wave. For example, assume that a harmonic wave having a frequency three times that of the fundamental wave is superimposed on the fundamental wave (Figure 1–20). The resulting distorted wave pattern depends on the phase relationship between the harmonic wave and the fundamental wave.

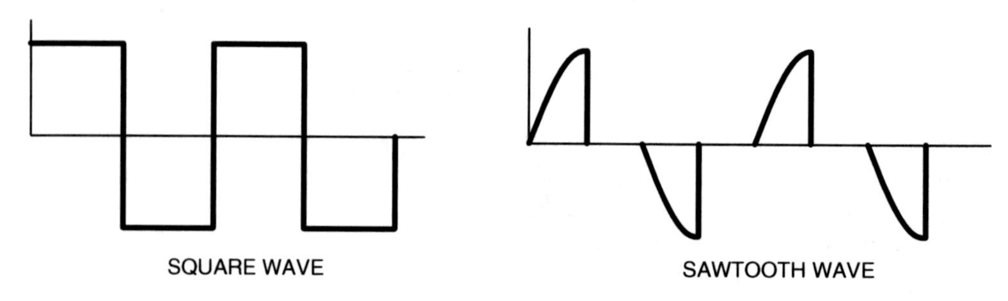

FIGURE 1–19 Not all waveforms are a sine wave (*Delmar/Cengage Learning*)

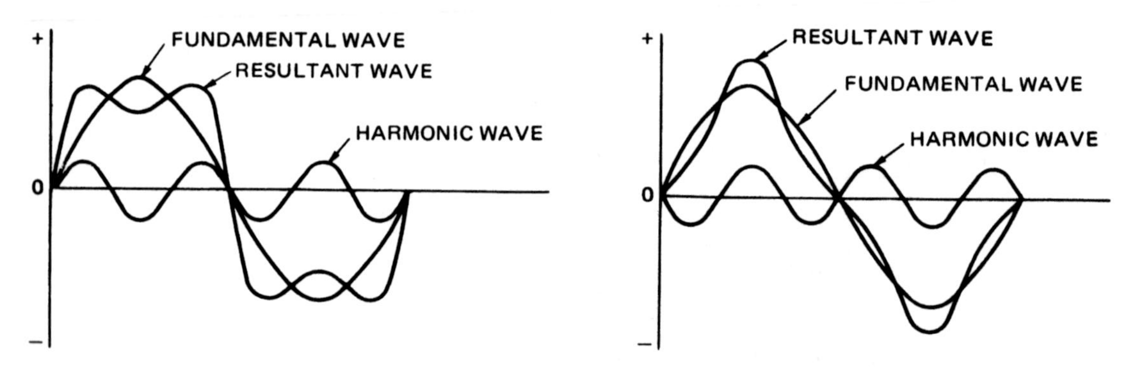

FIGURE 1–20 Sine-wave patterns formed with a fundamental frequency and its third harmonic (*Delmar/Cengage Learning*)

In Figure 1–20 the harmonic wave is shown referred to the zero axis. It has a frequency three times that of the fundamental and is superimposed on the fundamental wave. Note that the resultant pattern of the fundamental wave is different in the two diagrams. The difference arises because the phase relationship of the harmonic wave with the fundamental wave in the two illustrations is different.

This text cannot cover the various circuit problems involving unique ac voltage wave patterns. Therefore, it will be assumed that sine-wave voltage and current values are used throughout this text, unless otherwise noted.

HIGHER FREQUENCIES

It was noted earlier that the most common frequency used for the transmission of electrical energy is 60 Hz. Another value commonly used in aircraft and in other mobile equipment is 400 Hz.

In electronics, the frequencies used cover a very wide range. For example, audio frequencies between 20 and 16,000 Hz are used to operate speakers in amplifier units and radio receivers. Transmitted frequencies above 15,000 Hz are called *radio frequencies*. These higher frequencies are expressed in units of kilohertz (1 kHz = 1000 Hz), megahertz (1 MHz = 1,000,000 Hz or 1000 kHz), and gigahertz (1 GHz = 1,000,000,000 Hz or 1000 MHz).

SUMMARY

- Alternating current is more commonly used, but there are a number of applications where direct-current systems must be used or will do the job more efficiently than ac.
- AC alternators operate economically at relatively high voltages and heavy current ratings. DC generators are limited in both high voltages and large current ratings.
- The generation of large amounts of ac energy in large central stations is a more efficient and economical operation than in smaller local units.
- AC electrical energy can be transmitted at very high voltages over long distances (lowering I^2R losses). Transformers raise or lower voltages as needed at generating stations or distribution points. Transformers cannot be used on dc systems.
- The ac induction motor is simple and rugged in construction. It has excellent operating characteristics and is far more economical in initial costs, replacement, and maintenance than are dc motors.
- The *sine wave* is the function of the position of a coil in a magnetic field.
- The *cosine* wave has the same pattern as the sine wave, but reaches its maximum value 90° before the sine wave.
- Quadrants in a coordinate system are numbered *counterclockwise*.
- Angles are measured from the positive X axis to the indicated line in the *counterclockwise* direction.

- Angles measured in the *clockwise* direction from the positive X axis to the indicated line are *negative* because the direction of measurement has changed.
- The formula for the induced voltage in an armature conductor is

$$V_{induced} = BLv \times 10^{-8}$$

- The induced voltage is directly proportional to the velocity component, $V_T = v_c/(\sin \angle\theta)$, which is perpendicular to the magnetic field.
- The instantaneous value of alternating voltage is given by

$$V_{instantaneous} = V_{maximum} \times \sin \angle$$

- *Alternating voltage* is a voltage that changes continuously with time. It rises from zero to a maximum value in one direction, decreases to zero, rises to the same maximum value in the opposite direction, again decreases to zero, and then repeats these values at equal intervals of time.
- *Alternating current* is a current that changes continuously with time. It rises from zero to a maximum value in one direction, decreases to zero, rises to the same maximum value in the opposite direction, again decreases to zero, and then repeats these values at equal intervals of time.
- A *cycle* of alternating voltage or alternating current can be defined as that voltage or current that rises from zero to a positive maximum value, returns to zero, then rises to a negative maximum value, again returns to zero, and repeats these values at equal intervals of time.
- *Frequency* is the number of complete events or cycles per second (Hz) of alternating voltage or alternating current.
- Each cycle is divided into two alternations, with each alternation equal to 180 electrical time degrees.
- The relationship between the number of poles, speed, and frequency is expressed by

$$f = \frac{P \times S}{120}$$

- As the number of poles in an ac generator increases, the actual required driven speed in r/min decreases proportionally for a given frequency.
- When using a higher frequency, less iron and copper are required in transformers, motors, and other electrical equipment.
- Alternating-voltage waveforms are not all sine waves. They may be *distorted waveforms* caused by *harmonics* superimposed on the fundamental wave.

Achievement Review

1. State four reasons why most electrical energy produced is generated by alternators rather than by direct-current generators.
2. List five applications in which direct current is preferred over alternating current.
3. Name several ways by which alternating current is changed or rectified into direct current.
4. Explain the difference between the rotating line method and the rotating coil method of generating a sine wave.
5. Assuming that the rotating line of Figure 1–8 has a length of one unit, determine its projection or shadow on the X and Y axes at 30°, 45°, 120°, and 240°. (Refer to Figure 1–5 and Appendices 4 and 5.)
6. Prove that $\tan \theta = (\sin \theta)/(\cos \theta)$. (Refer to Figure 1–5.)
7. Using the answers to question 5, determine the tangents for 30°, 45°, 120°, and 240°. (Refer to Figure 1–5.) Check the answers with Appendices 4 and 5.
8. A sine-wave voltage produced by an ac generator has a maximum value of 170 V. Determine the instantaneous voltage at 45 electrical degrees after crossing the zero axis in a positive direction.
9. Determine the instantaneous voltage of the generator in question 8 at 240 electrical degrees.
10. The speed of a six-pole alternator is 1200 r/min. Determine the frequency of the output of the generator.
11. A 25-Hz alternator has two poles. Determine the speed of the alternator in r/min.
12. Two ac generators are to be operated in parallel at the same frequency. Alternator 1 has four poles and turns at a speed of 1800 r/min. Alternator 2 has 10 poles.
    a. What is the frequency of alternator 1?
    b. What speed must alternator 2 have so that it can operate in parallel with alternator 1?
13. Explain the difference between electrical time degrees and mechanical degrees.
14. Define (a) cycle, (b) alternation, (c) frequency.
15. Why is 60-Hz alternating-current service preferred to a frequency of 25 Hz in most areas of the United States and Canada?
16. Plot a sine wave of voltage for 360° or one cycle. The voltage has an instantaneous maximum value of 300 V.

17. Explain what is meant by a fundamental sine wave with a triple-frequency harmonic.

18. What is the advantage in using a frequency higher than 60 Hz for the electrical systems of various types of aircraft?

19. a. 1500 kilohertz = ___?___ hertz (Hz)
    b. 15,000 hertz = ___?___ kilohertz (kHz)
    c. 18 megahertz = ___?___ hertz (Hz)
    d. 18 megahertz = ___?___ kilohertz (kHz)

20. What is a megahertz? What is a kilomegahertz (normally called a *gigahertz*)?

PRACTICE PROBLEMS FOR UNIT 1

Sine-Wave Values

Find the missing values.

PEAK VOLTS	INSTANTANEOUS VOLTS	PHASE ANGLE
347	208	
780		43.5°
	24.3	17.6°
224	5.65	
48.7		64.6°
	240	45°
87.2	23.7	
156.9		82.3°
	62.7	34.6°
1,256	400	
15,720		12°
	72.4	34.8°

$V_{INST} = V_{MAX} \times \sin \angle\theta$ $\qquad V_{MAX} = \frac{V_{INST}}{\sin \angle\theta}$ $\qquad \sin \angle\theta = \frac{V_{INST}}{V_{MAX}}$

2

Alternating-Current Circuits Containing Resistance

Objectives

After studying this unit, the student should be able to

- discuss the phase relationship between voltage and current in an ac circuit containing resistance.
- compare the heating effect of ac current and dc current.
- determine the effective values of ac voltages and currents by making use of root-mean-square formulas.
- compute the instantaneous, average, and peak-to-peak values of voltages and currents by formulas.
- explain why the average power of an ac circuit is not always the product of the effective values of the voltage and current.
- plot a power curve, with current and voltage in phase.
- compute electrical energy in kilowatt-hours.
- define joule, calorie, and British thermal unit.
- explain why average values are used with rectifiers instead of effective values.
- discuss how dc instruments may be modified to measure ac voltages and currents.

CURRENT AND VOLTAGE IN PHASE

A simple alternating-current circuit consists of resistance only. Either an incandescent lighting load or a heating load, such as a heater element, is a noninductive resistive load.

Operation of an AC Generator

In Figure 2–1, an ac generator supplies current to a 100-ohm (Ω) heater element. The output of this generator is a voltage sine wave with a maximum or peak value of 141.4 V. When the voltage in this circuit is zero, the current is zero. When the voltage is

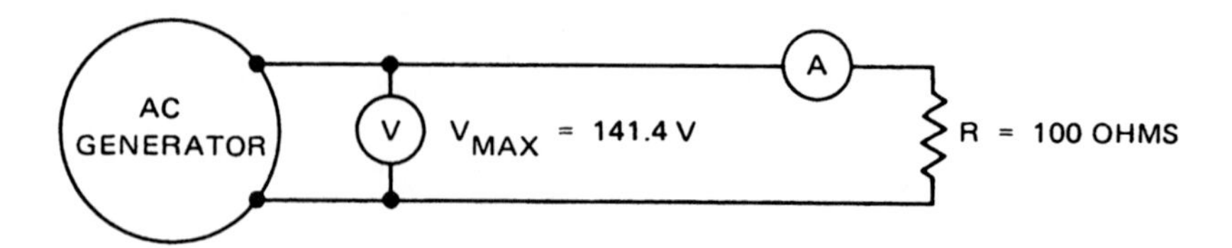

FIGURE 2–1 Alternating-current circuit with a resistance load (*Delmar/Cengage Learning*)

at its maximum value, the current is also at maximum. When the voltage reverses direction, the current also reverses direction. When the current and the voltage waveforms of a circuit are zero at the same time and reach their maximum values at the same time and in the same direction, these waves are said to be *in phase*.

Figure 2–2 shows the voltage and current sine waves in phase for the circuit of Figure 2–1. Ohm's law states that the current in a resistor is directly proportional to the voltage and inversely proportional to the magnitude of the resistance of the circuit. Note in Figure 2–2 that as the voltage increases from zero in either direction, the current increases proportionally in the same direction as required by Ohm's law. Thus, Ohm's law may be applied to ac circuits having a resistive load:

$$\mathbf{I} = \frac{\mathbf{V}}{\mathbf{R}}$$

When the voltage is at its maximum value of 141.4 V, the current is also at its maximum value. This value is

$$I_{maximum} = \frac{V_{maximum}}{R} = \frac{141.4\ V_{maximum}}{100\ \Omega} = 1.414 \text{ amperes (A)}$$

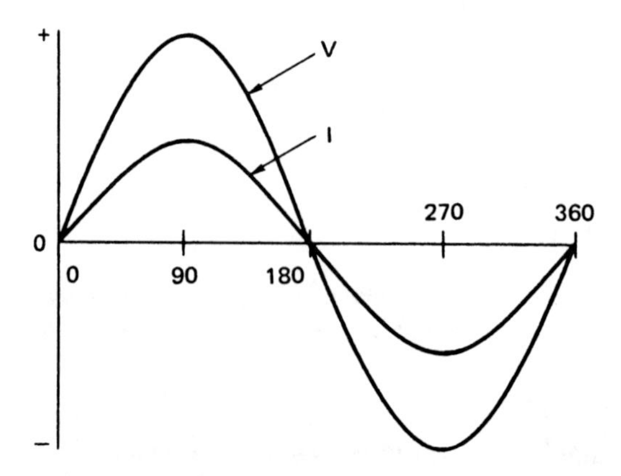

FIGURE 2–2 Current and voltage waveforms in phase (*Delmar/Cengage Learning*)

HEATING EFFECT OF AN ALTERNATING CURRENT

The alternating current in the circuit of Figure 2–1 is shown by the sine wave in Figure 2–3. The instantaneous maximum value of the current is 1.414 A.

If the current wave is considered for one complete cycle, then its average value is zero. This value is due to the fact that the negative alternation is equal to the positive alternation. If a dc ammeter is used to measure the current of this circuit, it will indicate zero. Thus, alternating current must be measured using an ac ammeter. Such an instrument measures the effective value of the current.

The *effective value* of alternating current is based on its heating effect and not on the average value of a sine-wave pattern. An alternating current with an effective value of one ampere is that current that will produce heat in a given resistance at the same rate as one ampere of direct current.

Plotting Sine Waves of Current

Direct Current Fundamentals showed that the heating effect of ac varies as the square of the current (watts = I^2R). For alternating current with a maximum value of 1.414 A, it is possible to plot a curve of the squared values of the current. If the horizontal scale of the curve is graduated in electrical time degrees, it is a relatively simple matter to obtain the instantaneous current values at regular intervals, such as 15° increments. The instantaneous values of current can then be squared and plotted to give a curve of current squared values for one cycle. For example, the instantaneous value of current at 30° is determined as follows:

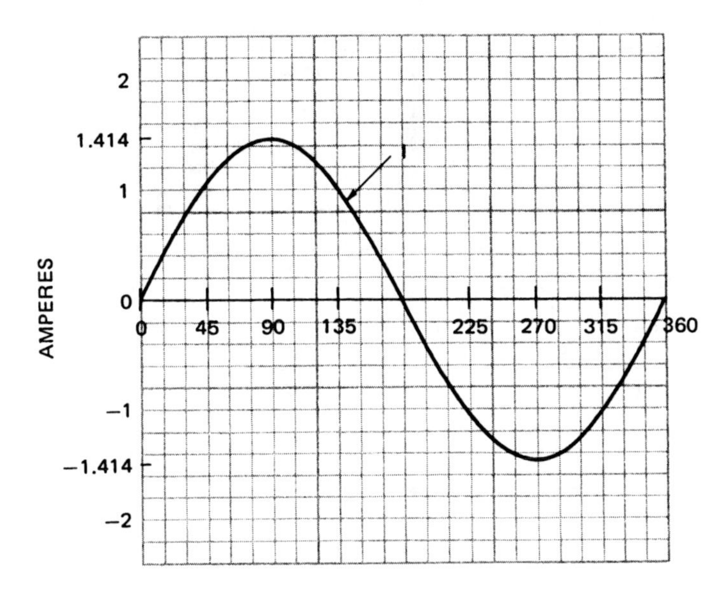

FIGURE 2–3 Sine wave of current (*Delmar/Cengage Learning*)

$$\begin{aligned} I_{instantaneous} &= I_{maximum} \times \sin \angle \\ &= 1.414 \times \sin 30^\circ \\ &= 1.414 \times 0.5000 \\ &= 0.707 \text{ A} \end{aligned}$$

The squared value of current at 30° is obtained by squaring the instantaneous current:

$$I^2_{instantaneous} = 0.707^2 = 0.5 \text{ A}$$

Table 2–1 gives values of the instantaneous current and the squared current for each 15° interval from 0° to 360°. Note that the instantaneous current values are negative as the

Angle in Degrees	Instantaneous Current	Current Squared
0	0	0
15	0.366	0.134
30	0.707	0.500
45	1.000	1.000
60	1.225	1.500
75	1.366	1.866
90	1.414	2.000
105	1.366	1.866
120	1.225	1.500
135	1.000	1.000
150	0.707	0.500
165	0.366	0.134
180	0	0
195	−0.366	0.134
210	−0.707	0.500
225	−1.000	1.000
240	−1.225	1.500
255	−1.366	1.866
270	−1.414	2.000
285	−1.366	1.866
300	−1.225	1.500
315	−1.000	1.000
330	−0.707	0.500
345	−0.366	0.134
360	0	0

TABLE 2–1 Instantaneous and squared current values

direction of current is reversed in the negative alternation of the cycle. The current squared values are all positive. (Recall that when any two negative numbers are multiplied, the product is positive.)

Graph of Sine Waves of Current. Figure 2–4 shows a sine wave of current that was plotted using the instantaneous current values given in Table 2–1. In addition, a current squared wave is also shown. This wave was plotted using the squared current values from Table 2–1.

The current squared wave in Figure 2–4 has a minimum value of 1.414^2 or 2 A. Note that the entire current squared wave is positive because the square of a negative value is positive. The graph shows that the current squared wave has a frequency that is twice that of the sine wave of current. The average value of the current squared wave is 1.0 A and is indicated by the dashed line on the graph.

The two areas of the current squared wave above the dashed line are the same as the areas of the two shaded valleys below the dashed line. This average value of one ampere over a period of one cycle gives the same heating effect as one dc ampere. The rectangular area formed by the dashed line and the zero reference line represents the heating effect of this alternating current. It also shows the heating effect of one direct-current ampere over a period of one cycle.

Root-Mean-Square Value of AC Current

An alternating current with a maximum value of 1.414 A and a steady direct current of one ampere both produce the same average heating through a resistance in a period of one cycle. In other words, both currents have the same average squared value. This value is called the *effective value* or the *root-mean-square* (*RMS*) *value* of current. Root-mean-square current is the abbreviated form of "the square root of the mean of

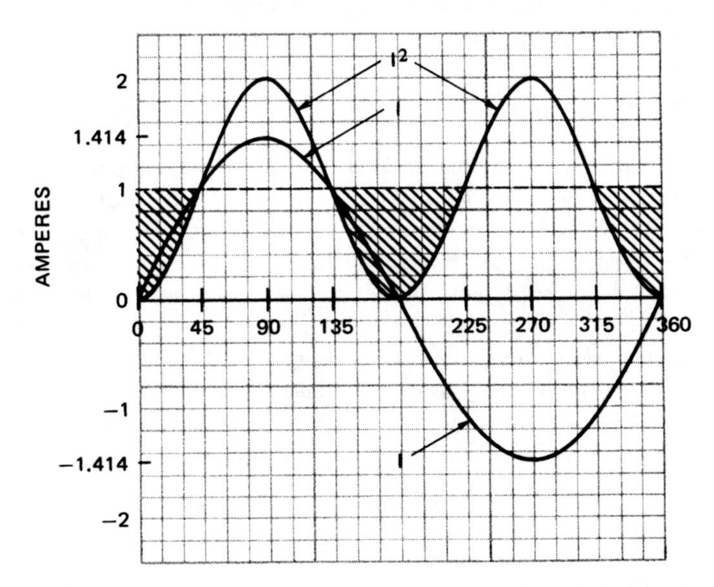

FIGURE 2–4 Root-mean-square values for a sine wave of alternating current (*Delmar/Cengage Learning*)

the square of the instantaneous currents." The RMS value of current is the current indicated by the typical ac ammeter. The relationship between the effective (RMS) current and the maximum current is

$$\mathbf{I_{RMS} = \frac{I_{max}}{\sqrt{2}} = \frac{I_{max}}{1.414} = 0.707\ I_{max}}$$

This means that the typical ac ammeter indicates 0.707 of the maximum value of a sine wave of current. This relationship can also be expressed as a ratio between the maximum value and the RMS value, or $\sqrt{2}$, which is 1.414. For example, if an alternating current has an instantaneous maximum value of 20 A, an ac ammeter will read

$$I_{RMS} = I_{max} \times 0.707 = 20 \times 0.707 = 14.1\ A$$

or

$$I_{RMS} = \frac{I_{max}}{1.414} = \frac{20}{1.414} = 14.1\ A$$

Either of these two ratios can be used to find the maximum value of current when the RMS or effective value is known, For example, an ac ammeter indicates a value of 15 A. The maximum current is

$$I_{max} = 1.414\ I_{RMS} = 1.414 \times 15 = 21.2\ A$$

or

$$I_{max} = \frac{I_{RMS}}{0.707} = \frac{15}{0.707} = 21.2\ A$$

In ac calculations, generally the effective value of current is used. This current is indicated by the letter I, which stands for *intensity* of current. The maximum current is shown as I_{max}.

EFFECTIVE (RMS) VOLTAGE

If one RMS ampere of alternating current passes through a resistance of one ohm, the RMS voltage drop across the resistor is one RMS ac volt. The RMS ac volt is 0.707 of the instantaneous maximum voltage. This RMS voltage is called the *effective voltage*. The typical ac voltmeter reads the effective value of voltage. The relationship between the maximum voltage and the effective voltage is the same as the one between the maximum current and the effective current.

The effective value of voltage is generally used in ac calculations. The effective voltage is indicated by the letter V. The maximum voltage is usually indicated by V_{max}.

For example, an ac voltmeter is connected across a lighting circuit and indicates a value of 120 V. What is the maximum instantaneous voltage across the circuit?

$$V_{max} = V \times 1.414 = 120 \times 1.414 = 169.7\ V$$

or

$$V_{max} = \frac{V}{0.707} = \frac{120}{0.707} = 169.7\ V$$

Unless otherwise specified, the voltage and current in an alternating-current circuit are always given as effective values. All standard ac ammeters and voltmeters indicate effective or RMS values.

Defining Terms

The following terms all relate to the sine wave: average, instantaneous, effective or RMS, peak, maximum, and peak-to-peak. These terms tend to confuse the student because they are closely related to each other. Figure 2–5 illustrates the various terms and their relationships.

RESISTANCE

In alternating-current circuits, the resistance is due to incandescent lighting loads and heating loads, just as in dc circuits. For these loads, inductance, hysteresis effects, and eddy current effects may be neglected. In a later unit of this text, the discussion will cover those factors that can change the ac impedance (resistance) of various types of loads. The term

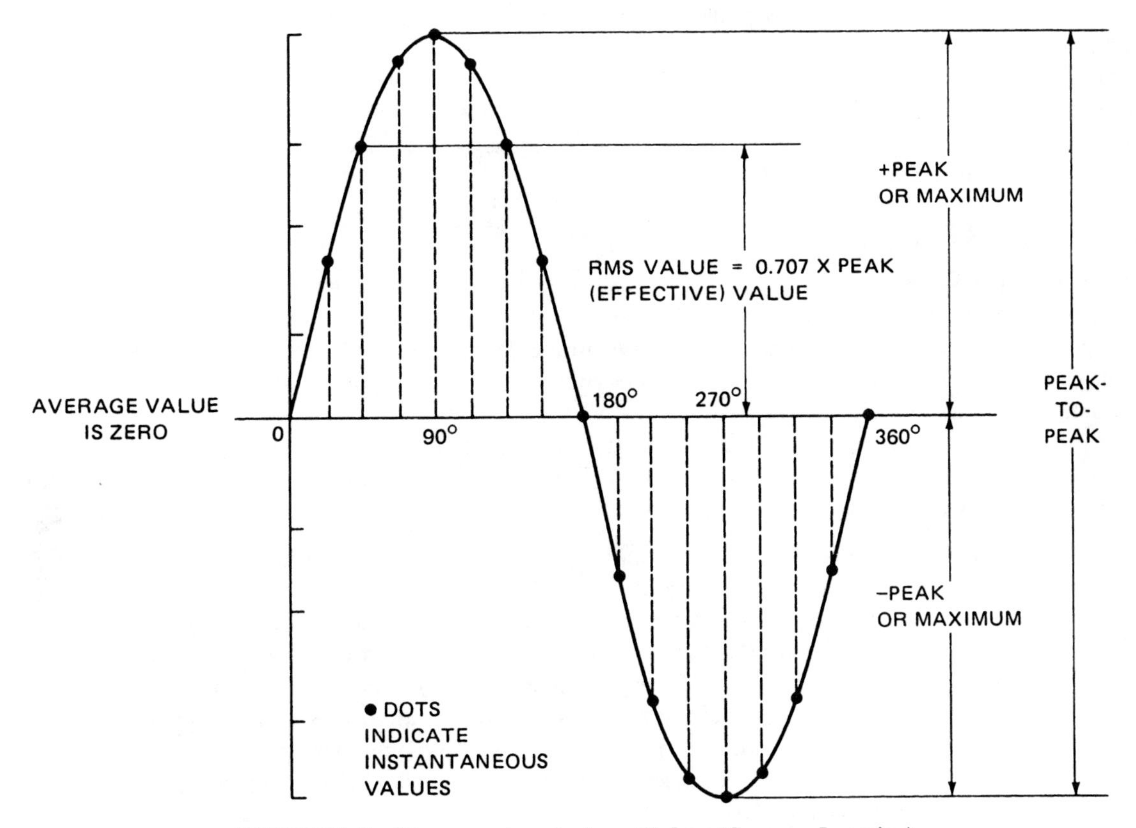

FIGURE 2–5 Sine-wave terminology (*Delmar/Cengage Learning*)

impedance is generally used to describe the total current limiting effect in alternating current circuits. Impedance is a combination of all current limiting properties such as resistance, inductance, and capacitance. Impedance will be discussed in later chapters of this text.

POWER IN WATTS

In dc circuits, the power in watts is equal to the product of volts and amperes. For ac circuits, the power in watts at any instant is equal to the product of the volts and amperes at that same instant. However, the average power of ac circuits is not always the product of the effective values of the voltage and current. Because many of the loads supplied by ac circuits have inductive effects, such as motors, transformers, and similar equipment, the current is out of phase with the voltage. As a result, the actual power in watts is *less* than the product of the voltage and current.

When the current and voltage are in phase in an ac circuit, the average power for a complete cycle is equal to the product of the RMS voltage and the RMS current. In the noninductive circuit shown in Figure 2–1, an ac generator supplies a sine wave of voltage. This voltage has a maximum value of 141.4 V and is applied across the terminals of the 100-Ω noninductive heating element. The sine wave of current for this circuit is in phase with the voltage sine wave. The maximum current is 1.414 A.

Watts are often called true power in ac circuits. Electricity is a form of pure energy, and in accord with physical laws, energy cannot be created or destroyed, but its form can be changed. Watts is a measure of the amount of electrical energy converted into some other form. In the case of a heating element, it measures the amount of electrical energy converted into thermal energy. In the case of a motor, it is a measure of the amount of energy converted into kinetic energy.

Plotting a Power Curve

It was stated previously that the power at any instant is equal to the product of the volts and amperes at that instant. If the product of the instantaneous values of voltage and current is obtained at fixed increments of electrical time degrees, a power curve can be plotted. Table 2–2 lists the instantaneous values of voltage, current, and power in watts for 15° intervals from 0° to 360°, or one cycle.

Figure 2–6 shows the voltage and current sine waves and the power curve plotted from the data given in Table 2–2. The power curve, indicated by "W" in Figure 2–6, gives the instantaneous power in the circuit at any point in the 360° time period of one cycle.

The Power Curve. It can be seen that all points on the power curve in Figure 2–6 are positive for this ac circuit. During the first alternation of the cycle, both the current and the voltage are positive. As a result, the power curve is also positive. During the second alternation of the cycle, both the current and the voltage are negative. However, the power curve is still positive because the power is the product of a negative current and a negative voltage. The power curve is represented above zero in a positive direction. This means that the load is taking power (in watts) from the source of supply. In this circuit, the voltage and current are acting together at all times. In other words, they are in phase. Thus, the power is positive in both alternations of the cycle.

Degrees	Instantaneous Voltage in Volts	Instantaneous Current in Amperes	Watts
0	0.0	0.0	0
15	36.6	0.366	13.4
30	70.7	0.707	50.0
45	100.0	1.000	100.0
60	122.5	1.225	150.0
75	136.6	1.366	186.6
90	141.4	1.414	200.0
105	136.6	1.366	186.6
120	122.5	1.225	150.0
135	100.0	1.000	100.0
150	70.7	0.707	50.0
165	36.6	0.366	13.4
180	0.0	0.0	0
195	−36.6	−0.366	13.4
210	−70.7	−0.707	50.0
225	−100.0	−1.000	100.0
240	−122.5	−1.225	150.0
255	−136.6	−1.366	186.6
270	−141.4	−1.414	200.0
285	−136.6	−1.366	186.6
300	−122.5	−1.225	150.0
315	−100.0	−1.000	100.0
330	−70.7	−0.707	50.0
345	−36.6	−0.366	13.4
360	0.0	0.0	0

TABLE 2–2 Instantaneous values of voltage, current, and power

The peak value of the power curve is the product of the maximum voltage and the maximum current:

$$W = 1.414\ V \times 1.414\ I = \sqrt{2}\ V \times \sqrt{2}\ I = 2\ VI$$

Now, if a dashed line is drawn across the power curve at the 100-watt (W) level on the vertical axis, the areas of the curve above the dashed line will just fit the shaded valleys of the curve below this line. In other words, the average power for the time period of one cycle is 100 W. This value is the product of the effective volts and the effective amperes.

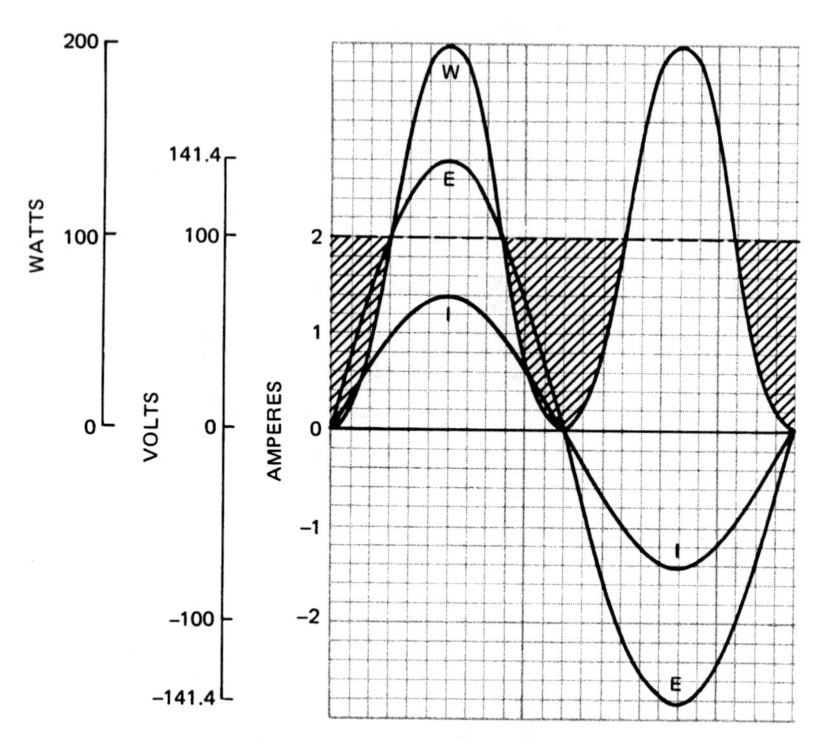

FIGURE 2–6 Power curve, with current and voltage in phase (Delmar/Cengage Learning)

The average power is W = V × I = 100 × 1 = 100 W. (Note the resemblance between the power curve in Figure 2–6 and the current-squared curve in Figure 2–4. They are the same curve and have the same kind of average, because power = current squared times the resistance, $P = I^2R$.)

ELECTRICAL ENERGY

The product of effective volts and effective amperes equals the power in watts in any ac circuit having a noninductive resistance load where the current and the voltage are in phase. To obtain a value for the electrical energy in watt-hours, the average power in watts is multiplied by the time in hours. The energy value in kilowatt-hours is obtained by dividing the value in watt-hours by 1000. This procedure was also given in *Direct Current Fundamentals* as a way of computing the energy in watt-hours and kilowatt-hours. The formulas for electrical energy are

$$\textbf{Watt-hours} = \mathbf{V \times I \times hours}$$

$$\textbf{Kilowatt-hours} = \frac{\mathbf{V \times I \times hours}}{\mathbf{1000}}$$

In ac circuits containing other than pure resistance, the energy formulas must be modified. Later units of this text will cover this situation.

MEASUREMENT OF ENERGY

Direct Current Fundamentals defined a unit of energy measurement known as the *joule*. One joule is the energy expended by one ampere at one volt, in one second. Repeated experiments showed that when one ampere was passed through a resistance of one ohm for one second, 0.2389 or 0.24 calorie of heat was liberated. One calorie is the quantity of heat energy required to raise the temperature of one gram of water one degree Celsius. One calorie is equal to 4.186 joules (J). One joule, which is a unit of energy, is equal to one watt-second. For a dc circuit or an ac circuit with a heater unit, the heat in calories is found using the formula

$$\mathbf{H = \frac{1}{4.186} \times I^2Rt}$$

where

H = heat energy, in calories
I = current, in amperes
R = resistance, in ohms
t = time, in seconds

In this formula, the I^2R term equals watts and I^2Rt equals the total energy in watt-seconds or joules. If 0.24 calorie (cal) is liberated for each joule of energy expended, the formula becomes

$$\mathbf{H = 0.24\ I^2Rt}$$

The calorie unit of the metric system is widely used. However, it is important that the student be familiar with another unit called the British thermal unit. The *British thermal unit* (Btu) is the amount of heat required to raise one pound of water one degree Fahrenheit. Because 1050 J is also required to raise one pound of water one degree Fahrenheit, 1050 J = 1 Btu.

For a dc circuit or an ac circuit consisting only of a resistance load, such as a heater unit, the Btu developed in the circuit can be found using the formula

$$\mathbf{H = \frac{I^2Rt}{1050}}$$

or

$$\mathbf{H = 0.000952\ I^2Rt}$$

where

H = heat energy, in British thermal units
I = current, in amperes
R = resistance, in ohms
t = time, in seconds

If the product in joules or watt-seconds in this formula is divided by 1050, the result is the heat energy in Btu. In the second form of the formula, the value in watt seconds is multiplied by the constant 0.000952. This constant is the decimal part of a Btu represented by one joule.

AVERAGE CURRENT AND VOLTAGE

Almost all alternating-current circuits and calculations use the effective or RMS values of the current and voltage. For example, ac voltmeters and ac ammeters indicate effective values. These values are 0.707 of the instantaneous maximum values.

There are some applications in which the *average* values of current and voltage are necessary. Some of these include rectifier units that use solid state devices such as diodes and silicon controlled rectifiers (SCRs) to convert alternating current into direct current.

Determining the Average Value

It was pointed out earlier that if an attempt is made to determine the average value of either a sine wave of voltage or current, the obvious procedure is to determine the average value of one alternation of a sine wave. This can be done by taking the average of the ordinates (values on the Y axis) at fixed increments in electrical degrees for one alternation. Figure 2–7 shows the ordinates at 10° intervals for 180° or one alternation.

Another method of determining the average value of a sine wave is to measure the area of the alternation between the wave and the zero reference line. A device called a *planimeter* is used to measure this area. If the area is then divided by the length of the baseline and multiplied by the ordinate scale, the result will be the average value of the alternation.

As shown in Figure 2–7, the average value of the maximum instantaneous value of a sine wave is 0.637. Because the effective or RMS value is 0.707 of the maximum value, a ratio can be made between the effective value and the average value. This ratio is $0.707 \div 0.637 = 1.11$. This value is called the *form factor* and is equal to the effective value divided by the average value.

Full-Wave Rectifier

Figure 2–8 shows a resistance-type load connected across the output of a full-wave rectifier. The full-wave rectifier causes both alternations of the cycle to be above the zero reference line in a positive direction. Even though the voltage and the resultant current at the resistance load are pulsating, they do not reverse direction. If a dc voltmeter is

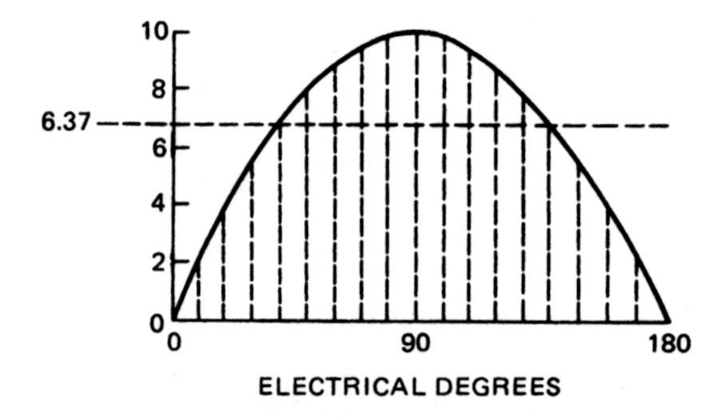

FIGURE 2–7 Determining the average value of current (*Delmar/Cengage Learning*)

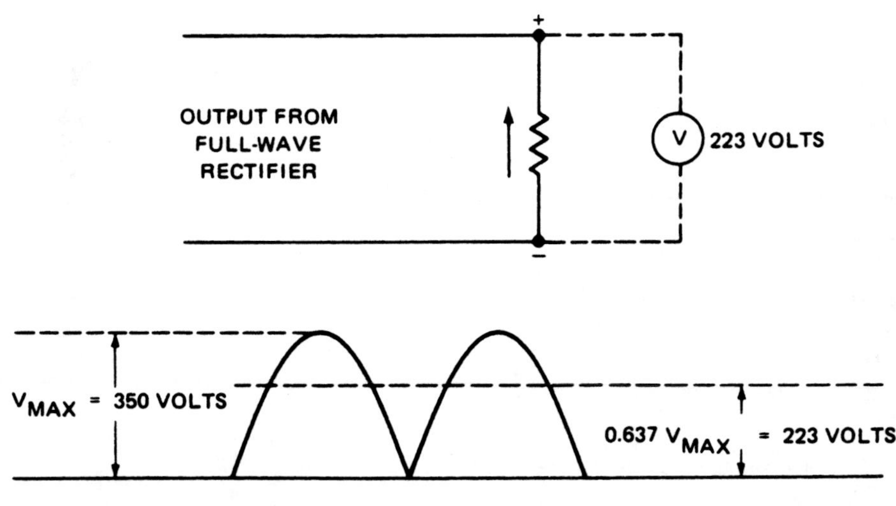

FIGURE 2–8 Average voltage for a full-wave rectifier (*Delmar/Cengage Learning*)

connected across the resistance load, as shown in Figure 2–8, the meter will indicate a value of 0.637 of the instantaneous maximum voltage. The dc voltmeter has a d'Arsonval movement, which operates on the same principle as a dc motor. This means that the deflection of the needle is determined by the average torque exerted on the movement for a time period of 360 electrical degrees. The resulting indication is the average value for the two alternations.

If the instantaneous maximum value of both alternations in Figure 2–8 is 350 V, then the average value indicated by the voltmeter is

$$V_{average} = V_{max} \times 0.637 = 350 \times 0.637 = 223 \text{ V}$$

Half-Wave Rectifier

Half-wave rectifiers (Figure 2–9) eliminate one half of the waveform and retain the other. Depending on the rectifier, it could eliminate the negative half and retain the positive half or eliminate the positive half and retain the negative half. Regardless, the output of a half-wave rectifier consists of only one half of the 360° cycle. Compare this with the full-wave rectifier, which inverts both halves of the waveform. The example shown in Figure 2–9 assumes that the negative half of the ac waveform has been eliminated.

The resistance load shown in Figure 2–9 is connected to the output of a half-wave rectifier. For half of each time period of 360 electrical degrees, there is no voltage or current. Therefore, the voltmeter will indicate the average of the one alternation over the 360° time period. This average is half of 0.637, or 0.318.

If the circuit in Figure 2–9 has a maximum instantaneous voltage of 350 V, the voltmeter indication will be

$$V_{average} = V_{max} \times 0.318 = 350 \times 0.318 = 111.5 \text{ V}$$

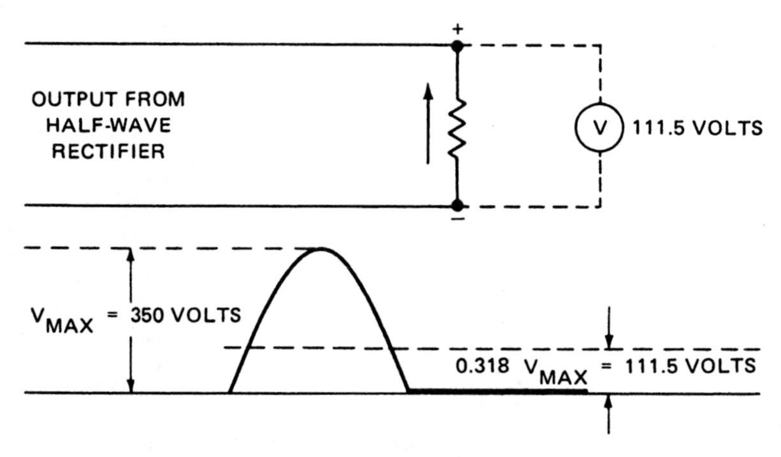

FIGURE 2–9 Average voltage for a half-wave rectifier *(Delmar/Cengage Learning)*

Measuring AC Voltages

Some manufacturers of dc instruments modify the circuit connections and the scale calibrations to measure ac voltages and currents. Direct-current instruments have uniform scale graduations and markings for the entire scale range. AC voltmeters and ac ammeters lack this uniformity. Thus, it is sometimes difficult to obtain accurate readings. This is the case near the lower end of the scale because of the nonlinear scale graduations.

Figure 2–10 shows how ac voltages can be measured using a dc voltmeter with a full-wave bridge rectifier. The small rectifier section is contained within the instrument. In general, silicon rectifiers are used in meters. The rectifiers permit electron flow in one direction only. The full-wave dc output of the rectifier is impressed directly across the terminals of the dc voltmeter.

The dc voltmeter with a full-wave rectifier, shown in Figure 2–10, is connected across an ac voltage source. The instrument indicates 108 V. This value is an average value equal

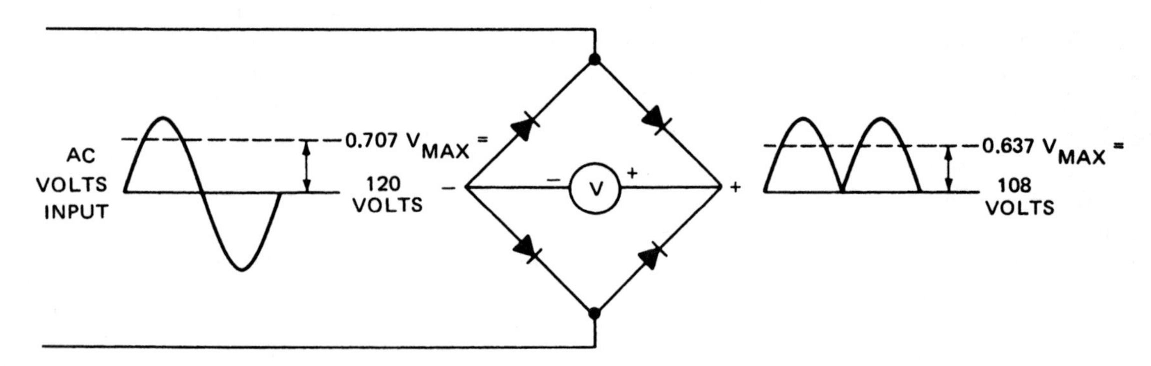

FIGURE 2–10 Use of a dc voltmeter with a full-wave (bridge-type) rectifier for measuring ac input voltage *(Delmar/Cengage Learning)*

to 0.637 of the instantaneous maximum value. If the losses in the rectifier can be neglected, then the actual effective value of ac voltage is

$$\begin{aligned} V &= V_{average} \times \text{form factor} \\ &= 108 \times 1.11 \\ &= 120 \text{ V} \end{aligned}$$

In other words, the form factor of 1.11 is applied as a multiplier to the dc voltmeter reading to obtain the effective value of ac volts. To eliminate the need to multiply the average voltage value by the form factor to obtain the effective voltage, the instrument is rescaled to read effective values. This is done by multiplying the original scale values by 1.11 and remarking the scale.

It is sometimes necessary to determine different values of voltage. The voltage ratings of solid state devices, for example, are given as PIV (peak inverse voltage) or PRV (peak reverse voltage). This is the value of voltage the device can withstand without being damaged. Both of these ratings list the peak value. AC voltmeters generally indicate the RMS value, not the peak value. If a solid state component is to be connected into an ac circuit, it is generally necessary to determine the peak value to make certain the component will not be damaged.

PROBLEM

Statement of the Problem

A diode is used as a half-wave rectifier. The diode has a PIV rating of 150 volts. Can it be connected to a 120-V RMS circuit without damage?

Solution

Determine the peak value of the ac circuit.

$$\begin{aligned} V_{peak} &= V_{RMS} \times 1.414 \\ V_{peak} &= 120 \times 1.414 \\ V_{peak} &= 169.7 \text{ volts} \end{aligned}$$

The diode will be damaged if it is connected to the 120-V ac line.

The chart in Figure 2–10 can be used to determine different voltage values.

SUMMARY

- An incandescent lighting load and a heating load, such as a heater element, are noninductive resistive loads. For a circuit with such a load, the current waveform is in phase with the voltage waveform.
- *In phase* means that the current and the voltage waveforms of a circuit are zero at the same time and reach their maximum values at the same time and in the same direction.
- Ohm's law may be applied directly to ac circuits having a resistive load.

- In a resistive ac circuit, inductance, hysteresis effects, and eddy current effects may be neglected.
- Alternating current must be measured using an ac ammeter. All standard ac ammeters and voltmeters indicate effective or RMS values.
- The relationship between maximum and effective values is

$$I_{RMS} = \frac{I_{max}}{\sqrt{2}} = \frac{I_{max}}{1.414} = 0.707\, I_{max}$$

$$V_{max} = V \times 1.414$$

or

$$V_{max} = \frac{V}{0.707}$$

- The effective value of ac is based on its heating effect and not on the average value of a sine-wave pattern.
- An effective ac current of one ampere will produce heat in a given resistance at the same rate as one ampere of direct current.
- The heating effect of ac varies as the square of the current (watts $= I^2R$).
- The product of the effective voltage and the effective current in amperes gives the power in watts in ac circuits when the current and voltage are in phase.
- Instantaneous values of current can be squared and plotted to give a curve of current-squared values for one cycle. The resulting curve is the power curve. All current-squared values are positive. (Recall that when two negative numbers are multiplied, the product is positive.)
- The power curve gives the instantaneous power in the circuit at any point in the 360° time period of one cycle.
- Because the current squared value is always positive and indicates the instantaneous power available in the circuit, the load takes power (in watts) from the source of supply during the complete cycle.
- The average power in watts of an ac circuit may be less than the product of the voltage and current. (In circuits containing other than pure resistance, the energy formulas must be modified.)
- Watt-hours $= V \times I \times$ hours (h)

  $$\text{Kilowatt-hours} = \frac{V \times I \times h}{1000}$$

  (In circuits containing other than pure resistance, the energy formulas must be modified.)
- One joule is the energy expended by one ampere at one volt, in one second.

- One calorie is the quantity of heat energy required to raise the temperature of one gram of water one degree Celsius.
- Relationships between joules, calories, and watt-seconds are as follows:

one joule = 0.2389 or 0.24 calorie (cal)
one calorie = 4.186 joules (J)
one joule = one watt-second (Ws)

$$H = \frac{1}{4.186} \times I^2Rt$$

or

$$H = 0.24\ I^2Rt$$

where
- H = heat energy, in calories
- I = current, in amperes
- R = resistance, in ohms
- t = time, in seconds

- The British thermal unit (Btu) is the amount of heat required to raise the temperature of one pound of water one degree Fahrenheit:

$$1050\ J = 1\ Btu$$

Therefore,

$$H(Btu) = \frac{I^2Rt}{1050}$$

or

$$H(Btu) = 0.000952\ I^2Rt$$

- Average values of current are used by rectifiers, vacuum-tube units, and dc instruments used with silicon rectifiers to measure ac values.
- A d'Arsonval movement operates on the same principle as a dc motor. The deflection of the needle is determined by the average torque exerted on the movement for a period of 360 electrical time degrees.
- DC instruments have uniform scale graduations and markings for the entire scale. AC instruments have nonlinear scale graduations with inaccuracies toward the lower end of the scale.
- An ac instrument indicates the effective value on the scale (0.707 × maximum value). A dc instrument with a silicon rectifier section, used to measure ac, will indicate the average value (0.637 × maximum value). Therefore, the dc instrument must be rescaled using a form factor multiplier.

- Form factor is a ratio between the effective value and the average value:

$$\text{Form factor} = \frac{0.707}{0.637} = 1.11$$

- A dc instrument with a silicon rectifier section, used to measure ac voltages, will be rescaled as follows:

$$V = V_{\text{average}} \times \text{form factor}$$

or

$$V = V_{\text{average}} \times 1.11$$

- The scale is marked to show this effective value.

Achievement Review

1. Explain what is meant by the term *effective current.*
2. Show how the term *root-mean-square current* was derived.
3. An ac voltmeter connected across the terminals of a heating element of an electric stove indicates a value of 240 V. What is the maximum instantaneous voltage impressed across the heating element?
4. An ac sine wave has an instantaneous maximum value of 7 A. What is the indication of an ac ammeter connected in this circuit?
5. A noninductive heater element with a resistance of 60 Ω is supplied by a voltage that has a pure sine-wave shape. The instantaneous maximum value of the voltage is 120 V.
   a. What is the instantaneous value of the voltage 30° after the voltage is zero and increasing in a positive direction?
   b. What is the effective value of the current?
   c. Show the relationship between the voltage and current sine waves for one complete cycle.
6. In question 5, what is the power in watts taken by the heater unit?
7. What is meant by the term *in phase*?
8. In question 5, what is the instantaneous value of current in amperes at 270 electrical degrees?
9. Twenty-four incandescent lamps are connected in parallel across a 120-V, 60-Hz supply. Twenty of the lamps are rated at 60 W, 120 V. Each lamp has a hot resistance of 240 Ω. The remaining four lamps are rated at 300 W, 120 V. Each

of these lamps has a hot resistance of 48 Ω. (Assume that the incandescent lamps are pure resistance.)
   a. Find the total current in amperes.
   b. What is the total power in watts?

10. a. If the load in question 9 is operated 5 h each day during a 30-day billing period, what is the total energy consumed in kilowatt-hours?
    b. What is the total cost at $.04 per kilowatt-hour to operate the lighting load for the 30-day billing period?

11. A noninductive heater element with a resistance of 5 Ω is connected across a 60-Hz source. The supply has a sine-wave voltage with an instantaneous maximum value of 141.4 V.
    a. Determine the power in watts taken by the heater unit.
    b. What is the energy, in kilowatt-hours, taken by the heater unit in one month if it is operated 5 h per day for a period of 20 days?

12. The nameplate rating of an electric iron is 120 V, 10 A. The heating element of this appliance is almost pure resistance with the current and voltage in phase.
    a. Determine the power in watts taken by the appliance.
    b. Calculate the resistance of the heater element.

13. Define (a) calorie; (b) British thermal unit (Btu).

14. It is desired to raise the temperature of one quart of water in a coffee percolator from 18°C to 100°C in 9 min. The supply voltage is 120 V.
    a. What is the wattage required by the heater unit to bring the water to a boiling temperature in 9 min?
    b. What is the resistance of the heater unit?
    c. Determine the current taken by the heater unit.
    [*Note*: one quart of water = 2.08 pounds (lb); 453.6 grams (g) = 1 lb.]

15. An electric hot-water heater is used to heat a 20-gallon (gal) tank of water from 60°F to 130°F in 100 min.
    a. Determine the wattage rating of the heater unit
    b. Assuming that the heater unit is used on a 230-V, 60-Hz service, find
       (1) the current in amperes taken by the heater unit.
       (2) the resistance of the unit.

16. When does the product of effective volts and effective amperes equal the power in watts in an ac circuit?

17. A full-wave rectifier supplies a pulsating dc voltage similar to the one shown in Figure 2–8. A voltmeter connected across the noninductive resistance load reads 250 V. What is the instantaneous maximum voltage impressed across the load?

18. Explain what is meant by the term *form factor*.

19. In Figure 2–11, a half-wave rectifier with a negligible resistance supplies a noninductive resistance load. A dc voltmeter connected across the load reads 54 V.
    a. Determine the instantaneous maximum voltage.
    b. What is the effective ac input voltage?

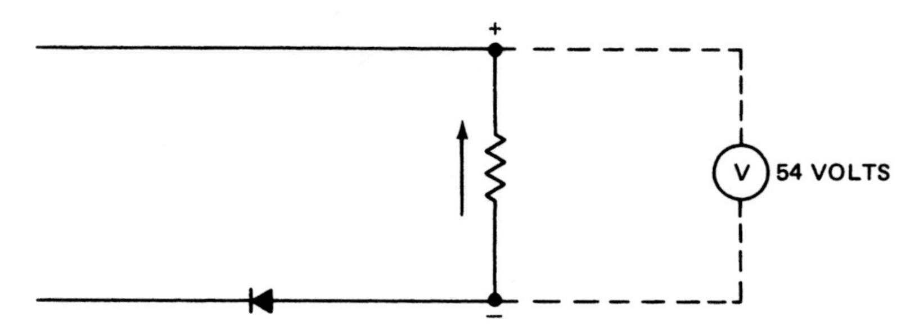

FIGURE 2–11 ***(Delmar/Cengage Learning)***

20. Show the type of rectified dc voltage that would be impressed across the load resistance in question 19.

PRACTICE PROBLEMS FOR UNIT 2

Peak, RMS, and Average Values

Find the missing values.

PEAK	RMS	AVERAGE
12.7		
	53.8	
		164.2
1235		
	240	
		16.6
339.7		
	12.6	
		9
123.7		
	74.8	
		108

3

Inductance in Alternating-Current Circuits

Objectives

After studying this unit, the student should be able to

- compare induction in an ac circuit to magnetic induction in a dc circuit.
- define a reactor and explain how it works in an ac circuit.
- demonstrate how counterelectromotive voltage opposes current in a lamp and inductor series circuit.
- calculate the impedance of a coil in an ac circuit by the use of the X_L formula.
- describe how an alternating-current circuit changes in magnitude and direction.
- using the X_L formula, explain how changing the frequency in an inductive ac circuit changes its characteristics.
- define a radian and show the number of mechanical degrees that each radian represents.
- describe the relationship between apparent power, reactive power, and true power in an inductive circuit.
- use the applications of vectors in an ac circuit to show how vectors are added and subtracted in both magnitude and direction.
- prove that current lags voltage in an ac inductive circuit.
- explain what the "angle theta" means and how it is used in an ac circuit.
- calculate the power factor of an ac series circuit containing both inductive reactance and resistance, by the use of trigonometric functions.

REVIEW OF ELECTROMAGNETISM

Direct Current Fundamentals discussed the following two basic principles of electromagnetism: (1) a magnetic field surrounds every current-carrying conductor or coil winding, and (2) an increase or decrease in the current causes an increase or decrease in the number of lines of force of this magnetic field.

A changing magnetic field induces a voltage in the conductor, coil, or circuit. This voltage is proportional to the rate of change of the lines of force cutting across the conductor, coil, or circuit. In a direct-current circuit, there will be no inductive effect once the current reaches the value defined by Ohm's law and remains constant. However, if the current changes in value, inductance does have a momentary effect. For example, if the current increases, more lines of flux will link the turns of wire in the coil winding. This change in flux linkage will cause a momentary induced voltage that opposes the increase in current. If the current decreases, there will be a decrease in the lines of flux linking the coil, resulting in a momentary induced voltage that attempts to maintain the current. These inductive effects are explained by Lenz's law, which states:

In all cases of electromagnetic induction, the induced voltage and the resulting current are in such a direction as to oppose the effect producing them.

Inductance in a DC Circuit

The effect of inductance in a direct-current circuit can be shown using the circuit in Figure 3–1.

This circuit consists of a group of lamps connected in series with a coil of wire mounted on a laminated steel core. (This coil has a relatively low resistance.) When the circuit is energized from a 120-V dc supply, a short amount of time passes before the lamps shine at full brightness. Also, the needle of the dc ammeter moves upscale to the Ohm's law value of current at a rate that is much slower than in a circuit containing only resistance. As the current increases to its Ohm's law value, more and more lines of force link the turns of the coil to form a magnetic field. This increasing flux causes an induced voltage that opposes the impressed voltage (120-V supply) and limits the current in amperes. This circuit situation is a typical example of Lenz's law in operation.

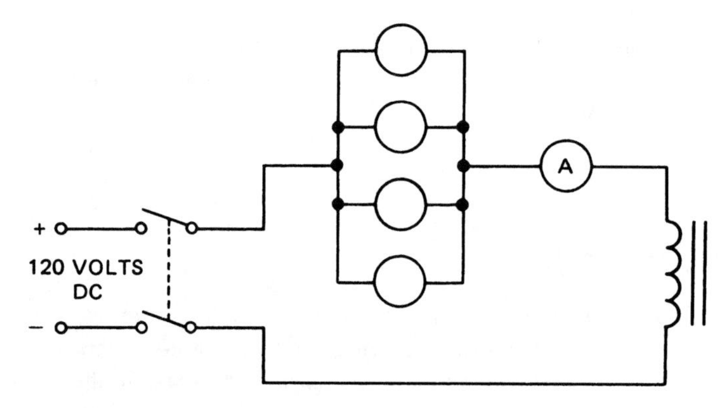

FIGURE 3–1 Reactor coil connected to a dc source (*Delmar/ Cengage Learning*)

Current in an Inductive DC Circuit

Figure 3–2 shows the current increasing to its Ohm's law value. Also shown is the momentary induced voltage for the circuit of Figure 3–1. At the instant the switch is closed, the current rises rapidly. Note, however, that the rate of current rise decreases with time. When the current reaches its Ohm's law value, the rate of increase becomes zero. At the same instant when the switch is closed and the circuit is energized, the induced voltage has its maximum value. At this time, there is the greatest rate of increase in the number of lines of force linking the turns of the coil. Thus, the maximum voltage is induced. As the rate at which the current rises decreases with time, the induced voltage also decreases. Finally, when the current reaches a constant (Ohm's law) value, the lines of force of the magnetic field will reach a maximum, resulting in a maximum field value. The induced voltage will decrease to zero when the current reaches its constant value.

When the circuit switch is opened, there will be a noticeable arc at the switch contacts. As the current decreases to zero (Figure 3–3), the lines of force collapse back into the turns of the coil. The cutting action of the collapsing lines of force is in a direction opposite that of the increasing field (when its switch is closed). As a result, the induced voltage will be in the same direction as the decaying current. This voltage will attempt to maintain the current, resulting in an arc at the switch contacts.

The decay of current in an inductive circuit is shown in Figure 3–3. The curve profiles for the decay of current and voltage may vary considerably from those shown. The actual curve profiles will depend on the time required for the line switch contacts to open.

A special switching arrangement must be used in the series circuit to observe the decay of the current and voltage. For example, assume that a single-pole switch is connected across the line wire of the series circuits in Figure 3–1. This switch closes at the instant the line switch contacts open. The length of time required for the current to decay to zero will equal the time required for the current to rise to its Ohm's law value when the circuit is energized.

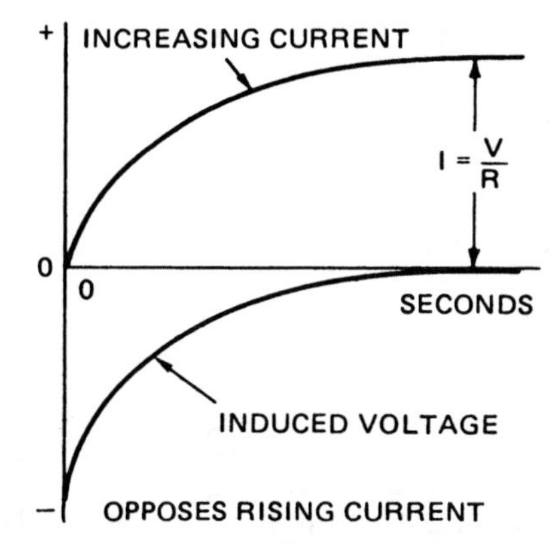

FIGURE 3–2 Current rises at an exponential rate (*Delmar/Cengage Learning*)

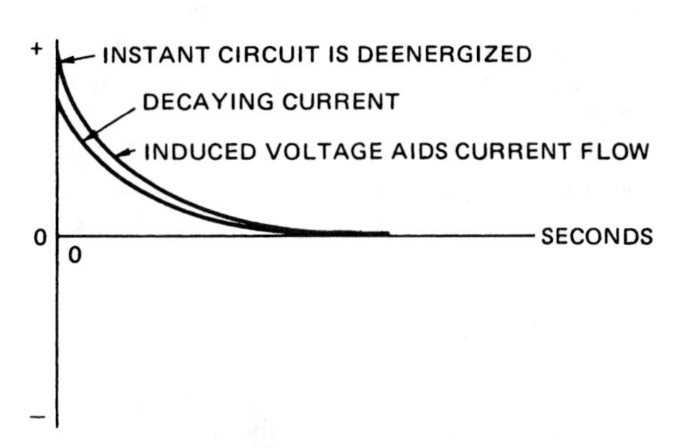

FIGURE 3–3 Current decreases at an exponential rate (*Delmar/Cengage Learning*)

In *Direct Current Fundamentals,* it was shown that inductance is really a form of electrical inertia. *Inductance* is the ability of a circuit component, such as a reactor coil, to store energy in the electromagnetic field. The unit of inductance is the *henry*. The symbol for the henry is "H." A henry is defined as follows:

> **A henry (H) is the inductance of a circuit, or a circuit component, when a current change of one ampere per second induces a voltage of one volt.**

The Series Circuit Connected to an AC Supply

If the series circuit shown in Figure 3–1, consisting of a reactor coil and lamps, is connected to a 120-V, 60-Hz alternating-current supply, the lamps will be very dim. An ac ammeter connected into the circuit will show an effective current value that is lower than that of the current in the dc circuit. This condition applies even though the effective value of the ac voltage is the same as that of the dc source. Also, the resistance of the lamps is almost the same as in the dc circuit. The current reduction indicated by the ammeter reading and the dim lamps is caused by the "choking" effect of inductance in an ac circuit.

The alternating current supplied by the 60-Hz source is changing in magnitude and direction continually. As a result, a countervoltage is induced in the coil. This voltage opposes the impressed voltage and thus limits the current in the series circuit.

To demonstrate that the inductive effect shown in the dc circuit is the cause of the current reduction, the laminated core is slowly removed from the coil. As the core is withdrawn from the coil, the lamps will increase in brightness and the ammeter reading will increase. When the core is completely removed, the reluctance of the magnetic circuit of the coil will increase. This means that there is less flux rising and collapsing around the turns of the coil. Thus, the induced voltage decreases. If the core is replaced in the coil, the lamps will dim again and the ac ammeter reading will decrease. If the frequency of the ac source is reduced from 60 Hz to 25 Hz and the effective value of the line voltage is the same, the current will increase. As a result, the ammeter reading will increase and the lamps will be brighter. The lower frequency means that there are fewer cycles per second. Thus, there are fewer changes of current and lines of force per second. The induced or counter voltage in the coil will be less, resulting in an increase in the current.

It can be seen from this discussion that inductance in an alternating-current circuit is just as effective in limiting the current as resistance. This means that both inductance and resistance must be considered in any calculations for ac circuits.

INDUCTIVE REACTANCE

The repeated changes in the direction and magnitude of alternating current give rise to an induced voltage, which limits the current in an inductive circuit. This opposition due to the inductance is called inductive reactance. *Inductive reactance* is indicated by the symbol X_L and is measured in ohms.

The inductive reactance in ohms can be found by the use of the formula

$$\mathbf{X_L = 2\pi fL}$$

where X_L = inductive reactance, in ohms
π = 3.14
f = frequency, in hertz
L = inductance, in henrys

Problem in Inductive Reactance

To illustrate the use of the formula for inductive reactance, assume that a coil has an inductance of 0.2652 henry (H) and negligible resistance. This coil is connected across a 60-Hz supply with an effective voltage of 100 V, as shown in Figure 3–4. Determine the inductive reactance of the coil in ohms and also determine the current.

The inductive reactance of the reactor coil in the figure is as follows:

$$\begin{aligned} X_L &= 2\pi fL \\ &= 2 \times 3.14 \times 60 \times 0.2652 \\ &= 100\ \Omega \end{aligned}$$

For a frequency of 60 Hz, it is an accepted practice to assume that the product of $2\pi f$ is 377 rather than 376.8. For a frequency of 25 Hz, the product of $2\pi f$ is 157. These values are convenient and may be used in the inductive reactance formula.

The current in the reactor coil of Figure 3–4 is found using the expression $I = \frac{V}{X_L}$.

This formula is really Ohm's law with the inductive reactance (X_L) in ohms used instead of the resistance (R) in ohms. The actual effective current in the reactor coil is

$$I = \frac{V}{X_L} = \frac{100}{100} = 1\ A$$

The formula for inductive reactance may also be written in the form

$$X_L = \omega L$$

The symbol ω is the lowercase Greek letter *omega*. It represents angular velocity. Before angular velocity is defined, it is important to review the meaning of the term *radian*.

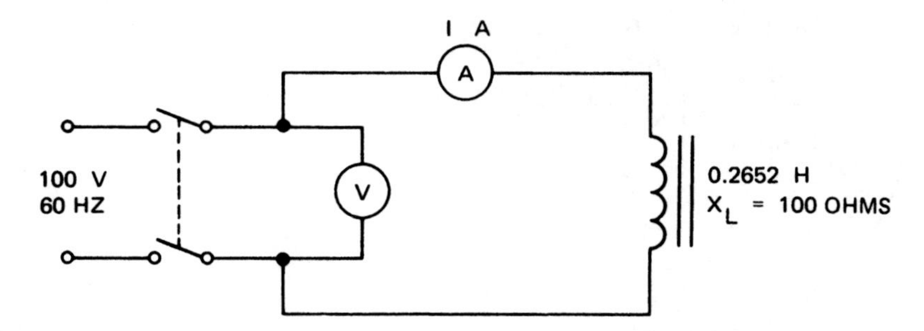

FIGURE 3–4 AC circuit containing pure inductance (*Delmar/Cengage Learning*)

Measurement in Radians

The *radian* is a unit of angular measurement. This unit is sometimes used in place of electrical time degrees. The definition of the radian is as follows:

One radian is equal to the angle at the center of a circle, subtended by an arc whose length is equal to the radius.

In Figure 3–5, the length of the arc AB is equal to the radius R, or OA. By definition, the angle AOB is equal to one radian and is slightly less than 60°. (Actually, one radian is 57.296°.) The diameter of a circle multiplied by π equals the circumference of the circle. The circumference of a circle can also be obtained by multiplying the radius of the circle by 2π. Thus, the actual angle represented by a radian is $360 \div 2\pi = 57.296°$.

One cycle is equal to 360 electrical time degrees. The circle shown in Figure 3–5, while illustrating angular measurement in radians, also represents 360°. Recall that the number of cycles per second is the frequency (f). Therefore, 2π radians is the angular velocity per cycle of the rotating radius line. The angular velocity per second is represented by $2\pi f$. The Y projection of the rotating radius line is the sine wave of frequency, f (see Figure 1–8). If ω represents the angular velocity per second, then the formula for inductive reactance may be stated in either of the following forms:

$$\mathbf{X_L = 2\,\pi\,fL}$$

or

$$\mathbf{X_L = \omega\,L}$$

where $\omega = 2\pi f$, the angular velocity per second.

Current in an Inductive Circuit

The reactor coil in Figure 3–4 consists of a 100-Ω inductive reactance and negligible resistance. If this coil is connected across an ac potential of 100 V, the current will be one ampere. However, it will be seen that this one-ampere current lags

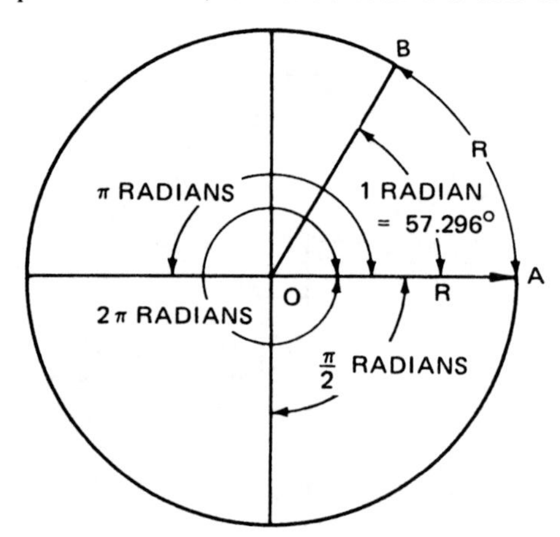

FIGURE 3–5 Angular measurement in radians (*Delmar/Cengage Learning*)

behind the voltage by 90 electrical degrees. Although this condition may not seem possible, a study of Lenz's law and the behavior of the induced voltage with an increase of current in a coil connected to a dc source (Figure 3–2) shows that the current may well lag behind the voltage.

In an ac circuit containing only inductance, the current is opposed by the induced voltage and thus cannot increase immediately with the impressed voltage. In Figure 3–6, the line voltage was omitted to make it easier to see the relationship between the alternating current and the induced voltage. The current is shown with its negative maximum value at the start of the cycle. At this point, and for a brief instant, there is no change in the current. Hence, there is no induced voltage. As the current decreases toward zero, the rate of change of current is increasing and the induced voltage increases. This voltage is represented by the dashed line.

The greatest rate of change of current for a given time occurs as it passes through zero. This means that the induced voltage has a maximum negative value at this same time. When the current passes through zero (at 90°) and increases to its positive maximum value, the rate of change of current is decreasing. As a result, the change in the lines of force in a given time period decreases proportionally and the induced voltage is reduced.

When the current reaches its positive maximum value at 180°, there will be a very brief period of time when there is no change in the current. For this period, the induced voltage is again zero. As the current decreases and passes through zero at 270° in a negative direction, the induced voltage reaches its positive maximum value. By developing the induced wave pattern for the remainder of the cycle, a sine wave of induced voltage is formed that lags behind the sine wave of current by 90 electrical degrees. To maintain the current in this coil, the line voltage applied must be equal and opposite to the induced voltage. Obviously, if the applied line voltage is directly opposite to the induced voltage, then the two voltages are 180° out of phase with each other.

Figure 3–7 is similar to Figure 3–6, except that the impressed line voltage is also shown. It can be seen that the line voltage and the induced voltage are 180° out of phase. Note also that the current lags the line voltage by 90°.

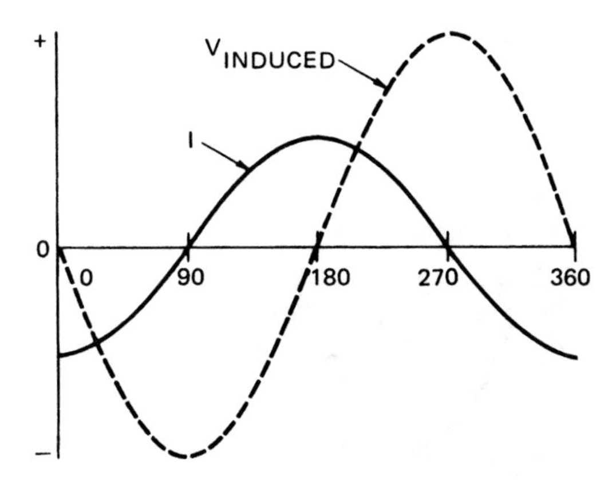

FIGURE 3–6 Current and induced voltage in an inductive circuit (supply voltage is not shown) (*Delmar/Cengage Learning*)

When the current and the line voltage are both positive, energy is being stored temporarily in the magnetic field of the coil. Energy is also being stored in the magnetic field when both the current and the line voltage are negative. However, when the current is positive and the line voltage is negative, energy is returned from the inductor coil to the supply. This return of energy occurs because the current and voltage are acting in opposition. In those parts of the cycle where the line voltage is positive and the current is negative, energy is also released from the magnetic field of the inductor coil and returned to the supply. This energy released from the electromagnetic field of an inductor coil maintains the current when the voltage and current are in opposition.

The inductor coil circuit in Figure 3–4 has an inductive reactance of 100 Ω. The line voltage is 100 V and the current is one ampere. The line voltage can be plotted as a sine wave with a maximum value of 141.4 V. The line current lags the line voltage by 90 electrical degrees. The RMS value of current is one ampere, and the maximum current value is 1.414 A.

The power in watts at any instant is equal to the product of the volts and amperes at the same instant. Table 3–1 lists the instantaneous values of voltage, current, and power at 15° intervals for one complete cycle (360°) for this inductive circuit.

The wave patterns in Figure 3–8 were plotted using the values given in Table 3–1. Note that the current lags the line voltage by 90 electrical degrees. It is assumed that the reactor coil has no resistance and the circuit consists of pure inductance. Between 0° and 90°, the current is negative and the voltage is positive. Therefore, the product of volts and amperes for this part of the cycle gives negative power.

Between 90° and 180°, both the current and the voltage are positive. This means that there is a pulse of power above the zero reference line. This positive power indicates that the supply is feeding energy to the electromagnetic field of the coil.

In the period between 180° and 270°, there is a second pulse of negative power. This is due to the negative voltage and the positive current. The negative power pulse indicates that

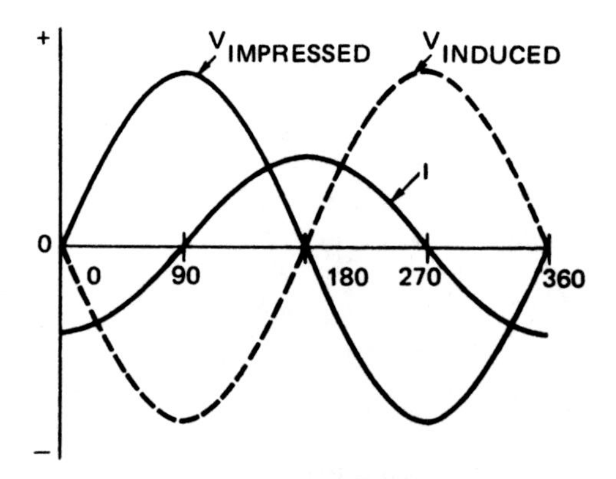

FIGURE 3–7 Impressed voltage, induced voltage, and current (*Delmar/Cengage Learning*)

Degrees	Instantaneous Voltage (Volts)	Instantaneous Current (Amperes)	Instantaneous Power (Watts)
0	0.0	−1.414	0
15	36.6	−1.366	−50.0
30	70.7	−1.225	86.6
45	100.0	−1.000	−100.0
60	122.5	−0.707	−86.6
75	136.6	−0.366	−50.0
90	141.4	0.0	0
105	136.6	0.366	50.0
120	122.5	0.707	86.6
135	100.0	1.000	100.0
150	70.7	1.225	86.6
165	36.6	1.366	50.0
180	0.0	1.414	0
195	−36.6	1.366	−50.0
210	−70.7	1.225	−86.6
225	−100.0	1.000	−100.0
240	−122.5	0.707	−86.6
255	−136.6	0.366	−50.0
270	−141.4	0.0	0
285	−136.6	−0.366	50.0
300	−122.5	−0.707	86.6
315	−100.0	−1.000	100.0
330	−70.7	−1.225	86.6
345	−36.6	−1.366	50.0
360	0.0	−1.414	0

TABLE 3–1 Instantaneous voltage, current, and power values

energy stored in the magnetic field of the coil is released to the supply. This energy, which is stored momentarily in the coil, maintains the current in opposition to the line voltage.

Between 270° and 360°, both the current and the voltage are negative and are acting together. The product of these negative values results in positive power, as shown by the last power pulse for the cycle in Figure 3–8. Because the power pulse is above zero, energy is being supplied by the source to the reactor coil.

The power waveform in Figure 3–8 shows that the areas of the two positive pulses of power are equal to the areas of the two negative pulses of power. Keep in mind that the pulses of power above the zero reference line are positive power that is fed by the source to the load. The pulses of power below the zero reference line are called *negative power*. These pulses represent power that is returned from the load to the source. The assumption is made that the areas of the two positive pulses of power are

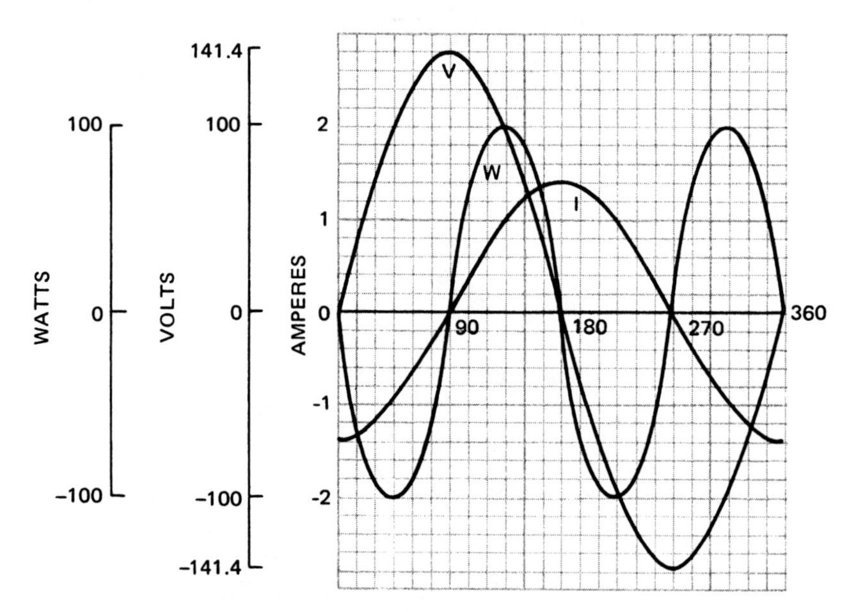

FIGURE 3–8 Power in a circuit with pure inductance (*Delmar/Cengage Learning*)

equal to the areas of the two negative pulses of power. This means that the net power taken by the inductor coil at the end of one complete cycle, or any number of cycles, is zero.

The actual power in watts taken by this circuit is zero. It should be noted that the product of the effective voltage and the effective current in amperes may not equal the power in watts in any circuit containing inductive reactance.

VECTOR ADDITION AND SUBTRACTION

The use of vectors simplifies the analysis of circuit conditions. The student should have a basic working understanding of vector diagrams and how they can aid in the solution of circuit problems.

Many physical quantities, such as cubic inches, pounds, minutes, and degrees, are expressed in specific units. These quantities do not contain a direction. Such units of measurement are known as *scalar* quantities when using vector terminology. A scalar unit is defined in terms of its magnitude only. Many quantities relating to electric circuits do not have magnitude alone, but also have direction. Any quantity having both magnitude and direction is called a *vector*. Examples of vectors include feet per second in a northeast direction, volts at 0°, and miles in a southerly direction. In electrical problems, rotating vectors are used to represent effective values of sinusoidal current and voltage in angular relationship to each other. According to convention, these vectors are rotated counterclockwise.

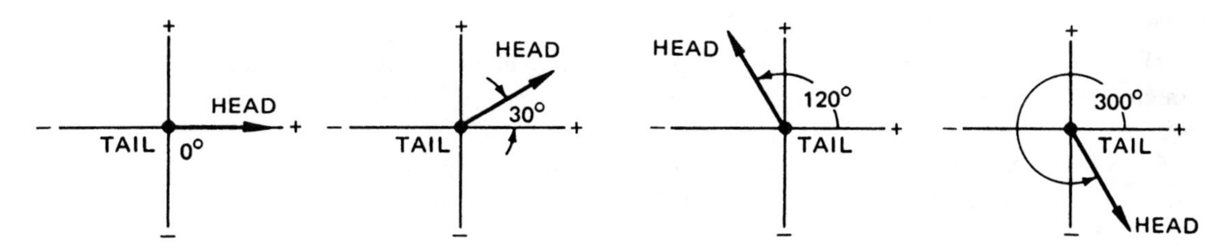

FIGURE 3–9 Vectors in different quadrants (*Delmar/Cengage Learning*)

Basic Rules of Vectors

There are some basic rules of vector addition and subtraction that are used throughout the text. The student must understand these and other basic vector concepts before any attempt can be made to solve vector diagrams.

Figure 3–9 shows vectors of a unit length in different quadrants. Each vector is labeled to indicate the head and the tail. This notation will not be carried throughout the text.

The following simple explanation of vector addition yields some basic rules that can be applied to more complicated problems. It is assumed that two people, A and B, exert a force of 6 lb and 4 lb, respectively, on a pole in the positive x direction. The total force exerted is 10 lb in the positive x direction. Figure 3–10 shows how these forces are arranged in a vector diagram. The relationship of these vectors is shown by the expression

$$\vec{V}_A + \vec{V}_B = \vec{R}$$

or

$$\vec{V}_B + \vec{V}_A = \vec{R}$$

The arrow above the V indicates a *vector* quantity. R is called the *resultant* of the two vectors. For this example, the resultant is 10 lb.

To add vectors, place the head of either vector to the tail of the other vector and measure or calculate the resultant from start to finish.

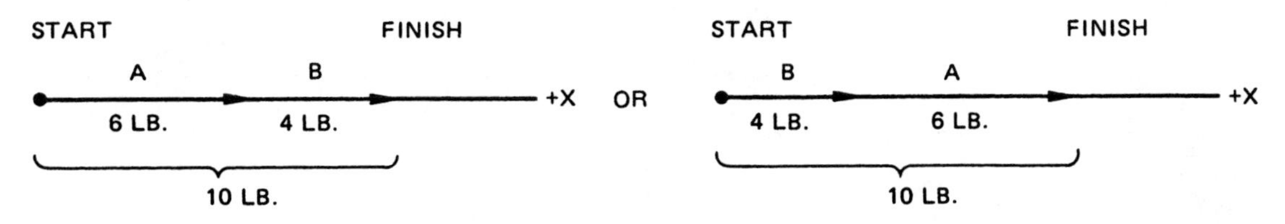

FIGURE 3–10 Vectors in a straight line (*Delmar/Cengage Learning*)

Applications of Vector Addition

The basic rule of vector addition can be applied to vector combinations that are not in a straight line (Figure 3–11). For each part of Figure 3–11, vector V_A is added to vector V_B, or vector V_B is added to vector V_A.

For each of the diagrams, the angles are measured from the positive X axis. Angles α and β in Figure 3–11C make the figure look more complicated than it really is.

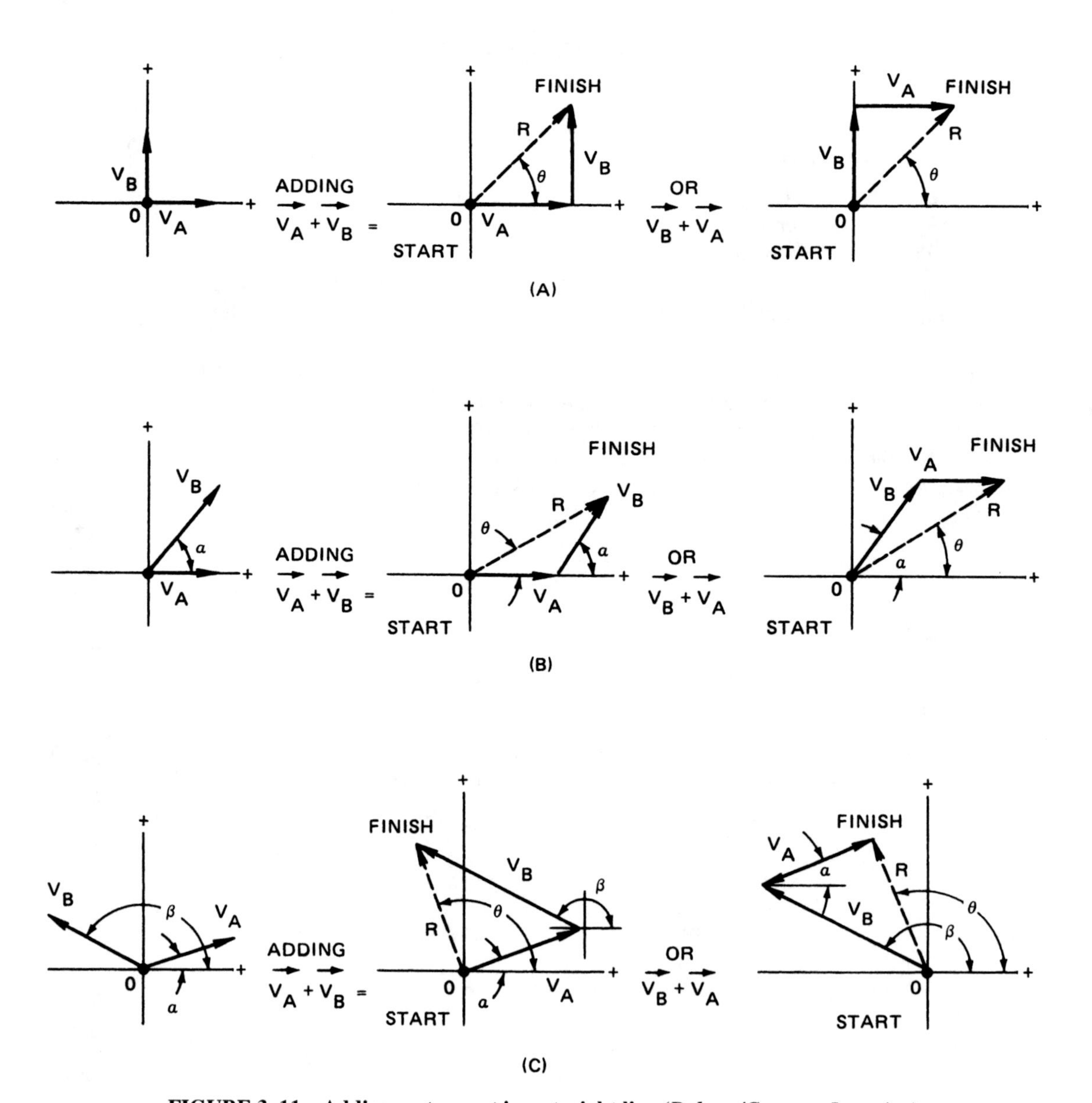

FIGURE 3–11 Adding vectors not in a straight line (*Delmar/Cengage Learning*)

The measurement of these angles ensures that vectors $\vec{V}_A$ and $\vec{V}_B$ are moved parallel to their original positions.

> **The addition of vectors in a straight line is the same as the arithmetic sum of their scalar values. Vectors that are not in a straight line require a geometric solution. Such vectors are *not* the sum of their scalar values.**

Subtraction of Vectors

Examples of vector subtraction are shown in Figure 3–12. The vector to be subtracted is reversed and is then added head to tail to the remaining vector.

The resultants in Figure 3–12 are all equal in magnitude. However, the resultants in parts B and C of the figure differ in direction by 180° from the resultants in parts D and E. It can be said that their scalar values are equal but their directions are different.

The information presented in Unit 1 on angular relationships could just as well have used the term *rotating vector* (or *phasor*) in place of *rotating line.*

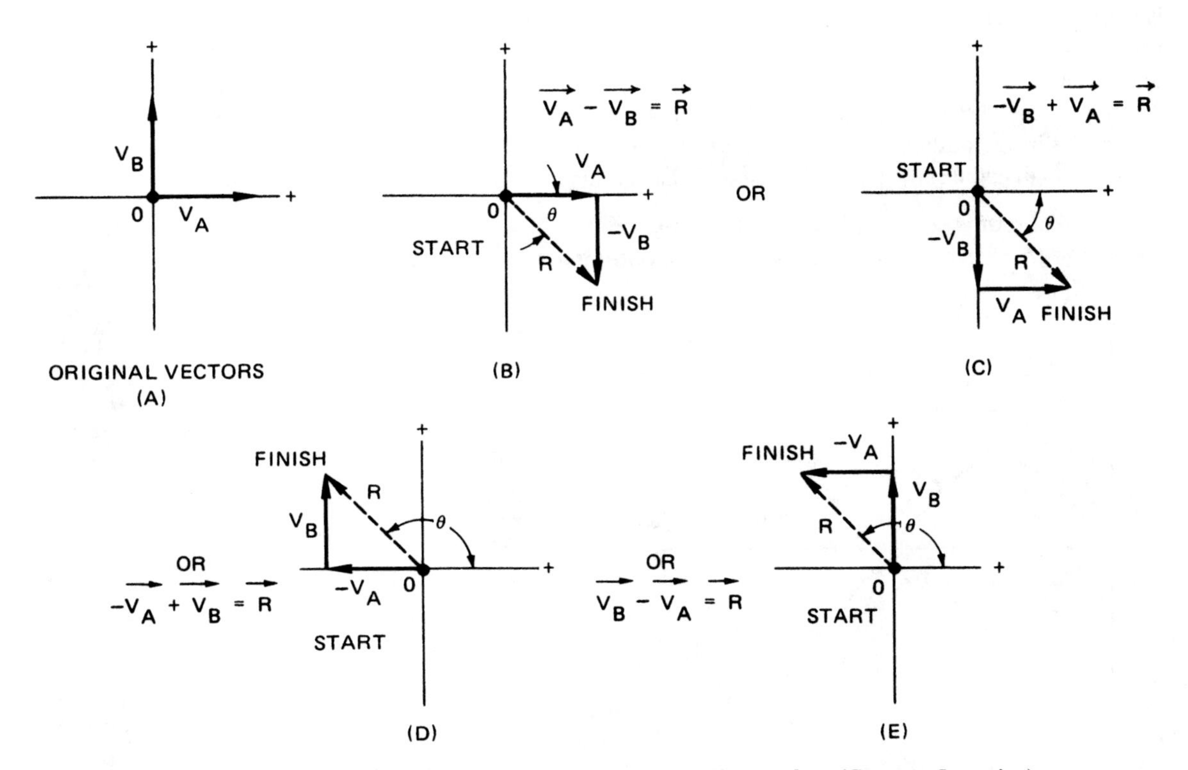

FIGURE 3–12 Subtracting vectors not in a straight line (*Delmar/Cengage Learning*)

RESISTANCE AND INDUCTIVE REACTANCE IN SERIES

Unit 2 described circuits containing resistance only. The first part of this unit covered inductance, inductive reactance, and the voltage and current relationships in ac circuits containing inductance only. This background makes it possible to analyze more practical circuits containing resistance and inductive reactance in series.

To show the application of vector diagrams to ac circuits, two simple examples will be discussed. The first example is an ac circuit with a noninductive load, where the current and voltage are in phase. The second example is a circuit containing only inductance, with the current lagging behind the impressed voltage by 90°.

The circuit shown in Figure 3–13 was used in Unit 2. This circuit consists of a 100-Ω noninductive heater unit with an effective line potential of 100 V. An effective current of one ampere is in phase with the voltage. The sine-wave patterns for the voltage and current for this circuit were developed in Unit 2. The student should recall that the first step was to determine the instantaneous voltage and current values at fixed increments in electrical time degrees. Then the sinusoidal wave patterns were plotted.

Diagrams for the Heater Current

The noninductive heater circuit has the sine-wave patterns shown in Figure 3–13. The vector diagram for this same circuit is also shown. The starting point of the vector diagram is marked "0." The V and I vectors are assumed to rotate counterclockwise around point 0. Line V represents the effective line voltage of 100 V. This line is drawn to a convenient scale. Line I is the vector representing the current of one ampere. This vector is placed directly on the voltage vector to indicate that it is in phase with the voltage. The lengths of the vectors in vector diagrams represent effective values of current and voltage. (Note that $\vec{I}$ and $\vec{V}$ both begin at 0; $\vec{V}$ is *not* added to $\vec{I}$ and is shown as having a greater magnitude.)

Current is constant in a series circuit. Thus, all angles are measured with respect to the current vector, making it a reference vector.

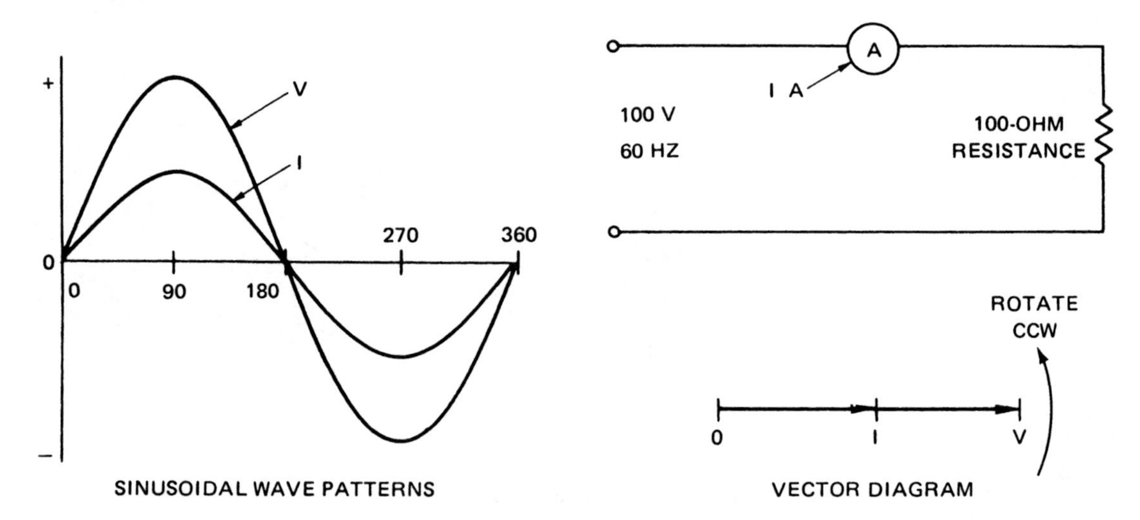

FIGURE 3–13 ***(Delmar/Cengage Learning)***

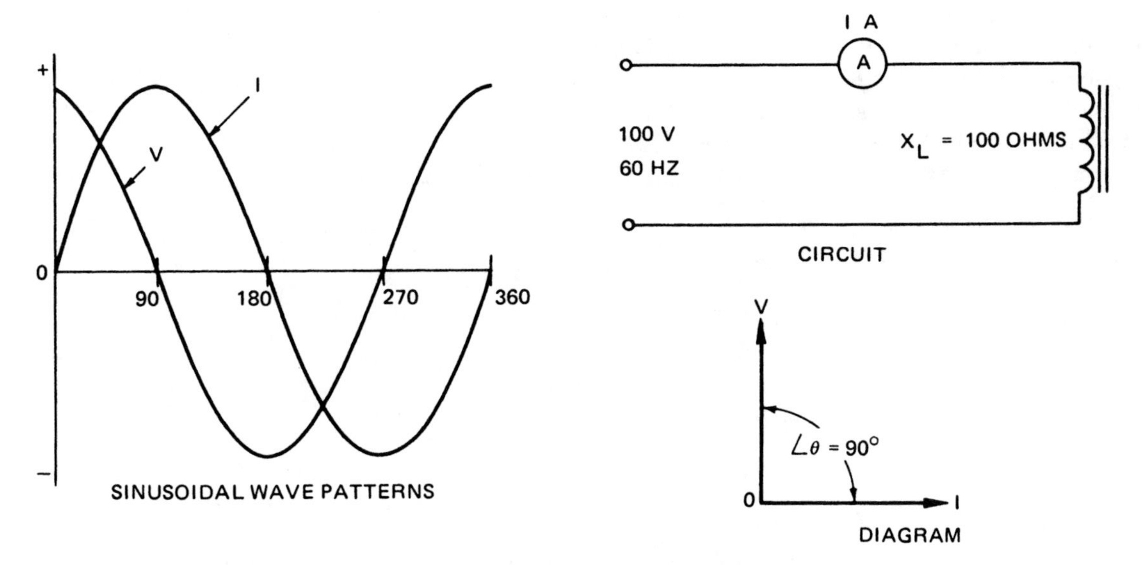

FIGURE 3–14 Current lagging voltage by 90° (*Delmar/Cengage Learning*)

Earlier in this unit, an inductor coil was used to show the effect of inductive reactance (X_L) of 100 Ω and was connected across an effective potential of 100 V. The effective current for this circuit was one ampere. The current lagged the voltage by 90°. For this circuit, the inductive reactive circuit, the sinusoidal wave patterns for the voltage and current, and the vector diagram are shown in Figure 3–14.

Refer to the vector diagram in Figure 3–14. The line current is drawn horizontally from point 0. Because the voltage leads the current by 90 electrical time degrees, the voltage vector must be drawn so that it is ahead of I by 90°. Vectors V and I are assumed to rotate counterclockwise. Thus, the voltage vector must point upward from 0. The angle between it and the current vector is 90°.

The angle by which the current lags behind the line voltage is called the *phase angle* or *angle theta*, abbreviated $\angle\theta$.

The series circuit shown in Figure 3–15 consists of a noninductive heater unit with a resistance of 20 Ω. The circuit also has an inductor coil with an inductive reactance (X_L)

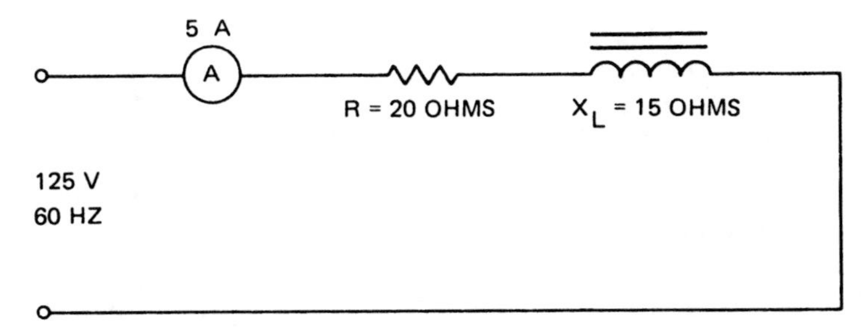

FIGURE 3–15 Resistance and inductive reactance in series (*Delmar/Cengage Learning*)

of 15 Ω. The coil is connected across a line potential of 125 V at a frequency of 60 Hz. The current in this circuit is limited in value by the resistance in ohms and the inductive reactance in ohms. Previously, it was shown that in a resistive circuit, the current is in phase with the voltage. In a circuit with inductive reactance, the current lags behind the voltage by 90°. There is both resistance and inductive reactance in the series circuit in Figure 3–15. For this circuit, the following quantities are to be found:

1. Voltage loss across the resistor
2. Voltage loss across the inductor coil
3. Line voltage
4. Impedance in ohms
5. True power in watts
6. Reactive power in VARs (volt-amperes-reactive)
7. Apparent power in volt-amperes
8. Power factor

Voltage Loss across R and X_L

The current in the circuit of Figure 3–15 is 5 A. Because this is a series circuit, the current is 5 A at all points in the series path. The voltage loss across the noninductive resistor will be in phase with the current. This loss is determined by using Ohm's law as follows:

$$\begin{aligned} V_R &= I \times R \\ &= 5 \times 20 \\ &= 100\text{ V across the resistor} \end{aligned}$$

The voltage loss across the inductor coil leads the current by 90°. In other words, the current lags the voltage across the inductor coil by 90°. Ohm's law is also used in this instance to determine the voltage loss across the inductor coil:

$$\begin{aligned} V_L &= I \times X_L \\ &= 5 \times 15 \\ &= 75\text{-V loss across the inductor} \end{aligned}$$

If the two voltage drops across the R and X_L components of the series circuit are added algebraically, the sum is 175 V. This value obviously does not represent the line voltage. The reason for this error is that the two voltage drops are 90 electrical degrees out of phase with each other. This means that the line voltage is the *vector sum* of the two voltage drops, or

$$\vec{V} = \vec{V}_R + \vec{V}_L$$

Vector Diagram of a Series Circuit

When a vector diagram is developed for a series circuit, it must be remembered that the current is the same at all points in the circuit. Therefore, the current is used as a reference line. In Figure 3–16, the 5-A current is drawn as a horizontal reference vector, using a workable scale.

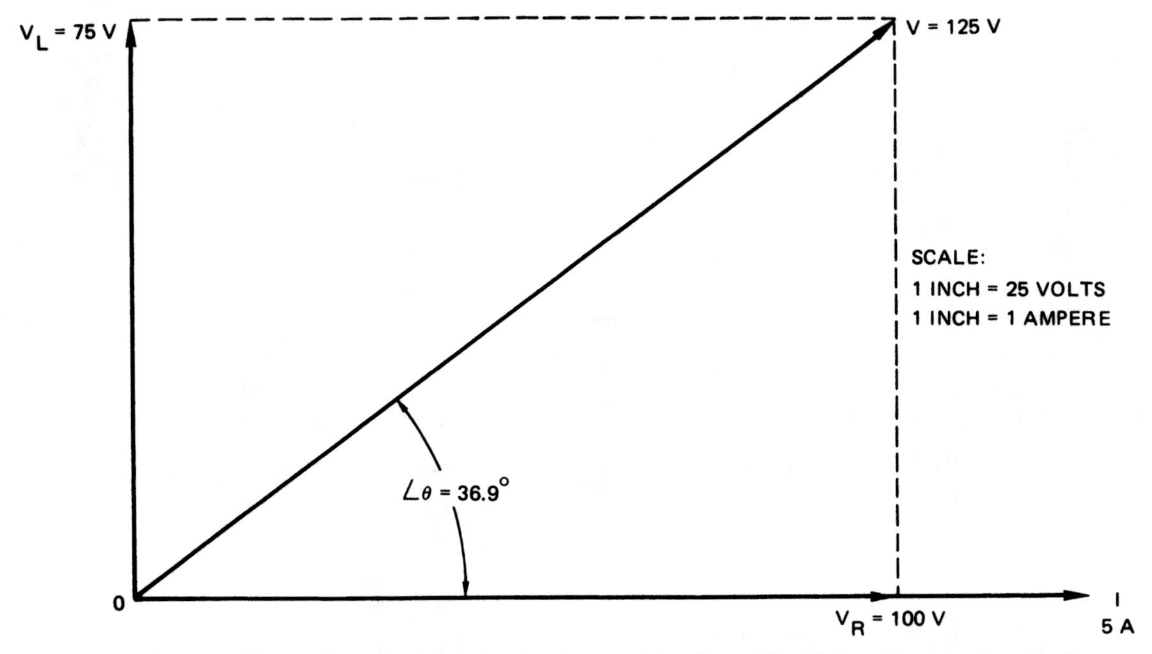

FIGURE 3–16 Vector diagram for series circuit containing R and X_L ***(Delmar/Cengage Learning)***

The voltage loss across the resistance (V_R) is placed on the current vector because both vectors are in phase. The voltage loss across the inductor coil is drawn in a vertical direction from point 0. This voltage (V_L) leads the current by 90°. The resultant voltage is the vector sum of the two voltage drops, which are 90° out of phase.

The vector diagram shows that the line voltage is the hypotenuse of a right triangle. The hypotenuse of a right triangle is equal to the square root of the sum of the squares of the other two sides. Therefore, the line voltage for the series circuit is

$$V = \sqrt{V_R^2 + V_L^2} = \sqrt{100^2 + 75^2} = 125 \text{ V}$$

This voltage value can be checked against Figure 3–15 to ensure that it is equal to the line voltage. The 125-V line voltage causes a current of 5 A in the series circuit, which consists of resistance and inductive reactance. Each ohmic value limits the current. The combined effect of resistance and inductive reactance is called *impedance* and is represented by the letter Z.

Impedance is the resulting value in ohms of the combination of resistance and reactance.

The Impedance Triangle

A triangle known as an *impedance triangle* is obtained by dividing the voltage triangle by a constant reference vector I (Figure 3–17). Several equations can be written by inspecting the right triangles in Figure 3–17:

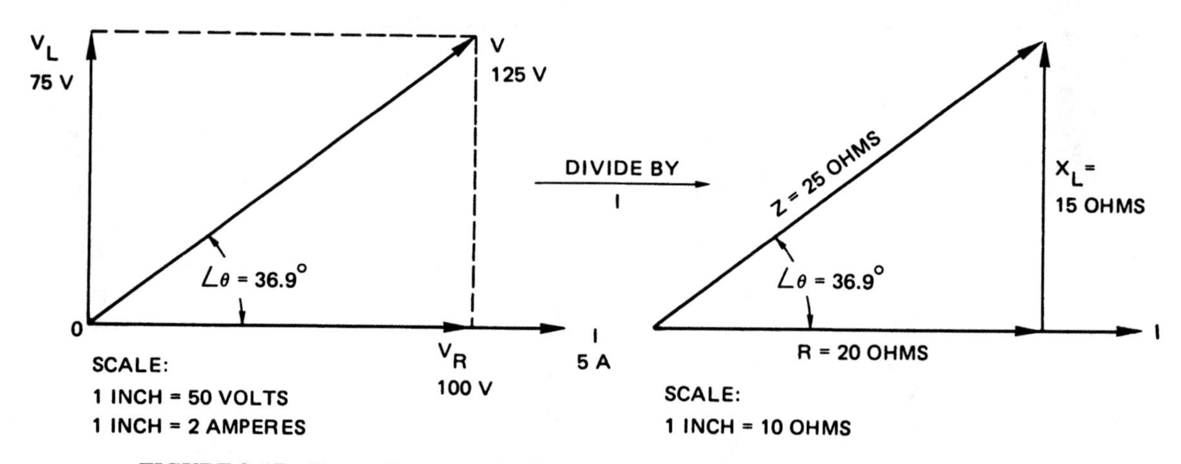

FIGURE 3–17 Vector diagrams of voltage and impedance (*Delmar/Cengage Learning*)

$$\vec{V} = \vec{V}_R + \vec{V}_L \text{ or } V = \sqrt{V_R^2 + V_L^2}$$

$$\vec{Z} = \vec{R} + \vec{X}_L \text{ or } Z = \sqrt{R^2 + X_L^2}$$

Obtained from the Pythagorean theorem

$$Z = \frac{V}{I};\ R = \frac{V_R}{I};\ X_L = \frac{V_L}{I}$$

Obtained by dividing the voltage triangle by the constant current (I)

(Note that five equations were derived from two triangles.)

Impedance of the Series Circuit. To determine the impedance of the series circuit in Figure 3–15, either of the following formulas may be used:

$$Z = \frac{V}{I} = \frac{125}{5} = 25\ \Omega$$

$$Z = \sqrt{R^2 + X_L^2} = \sqrt{20^2 + 15^2} = \sqrt{625} = 25\ \Omega$$

In the analysis of dc circuits:

$$I = \frac{V}{R_T}$$

In the analysis of ac circuits:

$$I = \frac{V}{Z}$$

To find the line current in ac series circuits, Ohm's law is used as follows:

$$I = \frac{V_R}{R}, \text{ or } I = \frac{V}{Z}, \text{ or } I = \frac{V_L}{X_L}, \text{ or } I = \frac{V_{X_L}}{X_L}, \text{ or } I = \frac{V_{coil}}{Z_{coil}}$$

V_L and V_{X_L} are used interchangeably when the coil does not have resistance. The term V_{coil} is used when the coil does have resistance. V_R and V_L equal V when either R or L is the only component in the circuit.

POWER AND POWER FACTOR

Several formulas for power can be written by inspecting the triangles in Figure 3–18. The impedance and voltage formulas were covered earlier in this unit.

1. $V \times I = VA$
   $V_R \times I = \text{watts}$
   $V_L \times I = \text{VARs}$
   } Derived by multiplying the voltage triangle by I

2. $I^2 Z = VA$
   $I^2 R = \text{watts}$
   $I^2 X_L = \text{VARs}$
   } Derived by multiplying the impedance triangle by I^2

3. $\vec{VA} = \vec{\text{watts}} + \vec{\text{VAR}}\text{s}$
   or
   $VA = \sqrt{\text{watts}^2 + \text{VARs}^2}$
   } Derived from the Pythagorean theorem

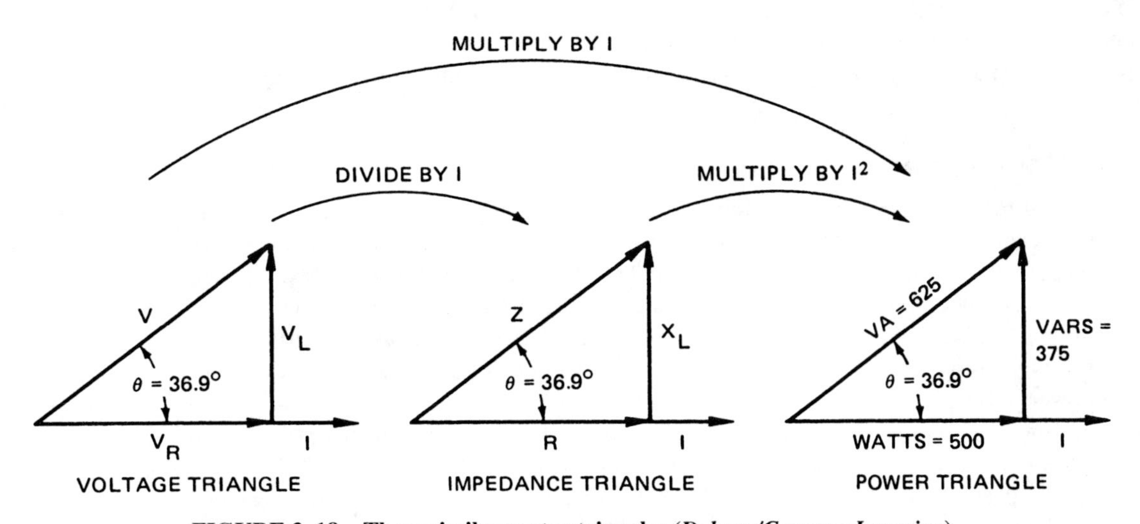

FIGURE 3–18 Three similar vector triangles (*Delmar/Cengage Learning*)

By applying the basic principles of similar triangles and the Pythagorean theorem to the triangles in Figure 3–18, *seven* more formulas are derived. For the circuit of Figure 3–15, the apparent power, in volt-amperes, is

$$V \times I = 125 \times 5 = 625 \text{ volt-amperes}$$

Determining True Power

In the noninductive circuit analyzed in Unit 2, the effective voltage and the effective current in amperes were in phase. The product of these values gave the power in watts taken by the circuit. The first part of the present unit deals with a circuit consisting of pure inductive reactance. In this circuit, the current lags the voltage by 90°. In a circuit with pure inductive reactance, the actual true power in watts is zero and is not the product of the volts and amperes. In a series circuit containing both inductive reactance and resistance (Figure 3–15), the current is neither in phase with the line voltage nor lagging 90° behind the line voltage. The angle of lag for a circuit containing both R and X_L will be somewhere between these two limits. Power will be taken by the circuit.

How is the true power in watts determined? One way is to use the formula $W = I^2R$. The actual power used in the resistance of any ac circuit is equal to the square of the effective current multiplied by the resistance in ohms.

$$\text{Actual watts} = I^2 \times R = 5^2 \times 20 = 500 \text{ W used in the resistance of the circuit in Figure 3–15}$$

The power taken by this circuit can also be found by multiplying the voltage loss across the resistance section of the circuit by the current. Because the current and voltage are in phase, their product is the true average power:

$$\text{Watts} = V_R \times I = 100 \times 5 = 500 \text{ W used in the resistance of the circuit}$$

Note that the true power in watts is the same using either method. Recall from *Direct Current Fundamentals* that electricity is a form of pure energy. Watts is a measure of the amount of electrical energy converted to some other form. When current flows through the resistive part of any circuit, electrical energy is converted into thermal energy in the form of heat.

Reactive Power

In addition to the true power used in the resistance, there is another power component to be considered. This component is called *reactive power* and is expressed in a unit called *volt-amperes-reactive*. This unit is usually shown in the abbreviated form "VARs." It represents the product of volts and amperes that are 90° out of phase with each other.

Figure 3–8 shows a graph of reactive power over one cycle. The total net true power in watts is zero for one cycle because the two negative power pulses cancel the two positive power pulses. As a result, the descriptive term *wattless power* is sometimes applied to this power component in an ac circuit. The correct terminology for this power component is *quadrature power* or *reactive power*. In the series circuit in Figure 3–15, the voltage drop across the component with pure inductive reactance is 75 V. Therefore, the quadrature power in VARs is the product of the voltage across the reactance and the current. These values are 90° out of phase. This quadrature, reactive, or wattless power is

$$\text{VARs} = V_L \times I = 75 \times 5 = 375 \text{ VARs}$$

To better understand wattless power or VARs, consider that true power or watts can be produced only when electrical energy is converted to some other form. When the alternating-current waveform flowing through an inductor increases in value, it causes a magnetic field to be produced around the inductor (Figure 3–19). Energy is required to create or establish the magnetic field. When the alternating-current waveform begins to decrease in value, the field collapses and returns the energy to the circuit. In the case of an inductor, electrical energy is stored in the form of a magnetic field and then returned to the circuit. The electrical energy is not changed to some other form.

The true power in watts is used in the resistance of the circuit. The quadrature power is identified with the reactance of the circuit. The combination of the two power components yields a resultant called the *apparent power* in volt-amperes. This resultant in volt-amperes can be regarded as the hypotenuse of a power triangle, as shown in Figure 3–18.

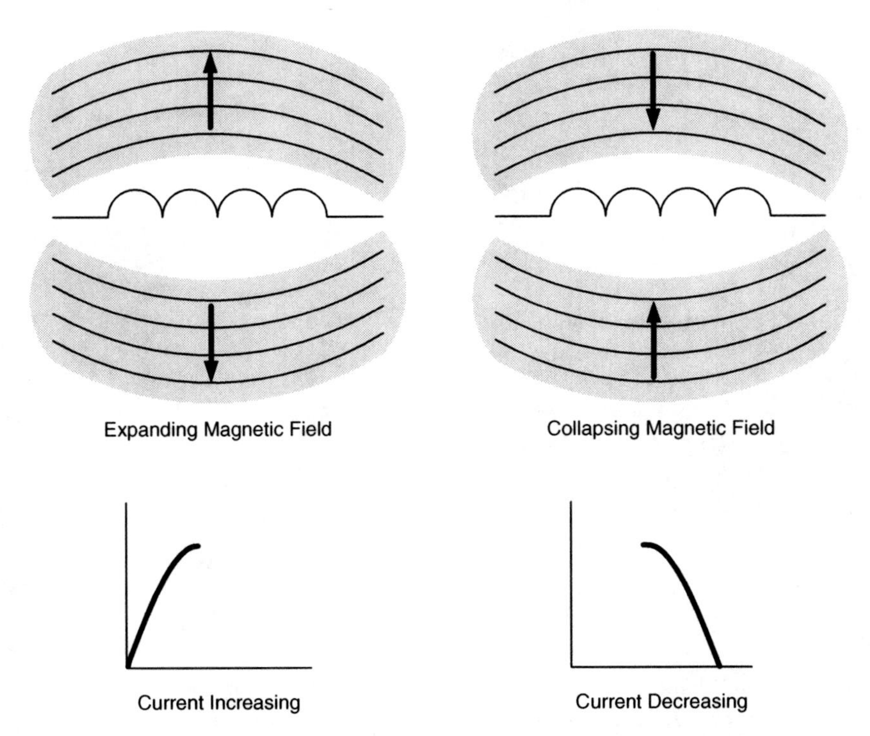

FIGURE 3–19 Energy is stored in the form of a magnetic field and then returned to the circuit. (*Delmar/Cengage Learning*)

Power Factor

Power factor (PF) is defined as the ratio between the true power in watts and the apparent power in volt-amperes. The power factor may be expressed as a decimal value or as a percentage:

$$PF = \frac{\text{true power (in watts)}}{\text{apparent power (in volt-amperes)}}$$

The power factor is the cosine of the angle θ (Figure 3–18). Three expressions for θ can be written by inspecting the three triangles in the figure:

$$\cos\theta = \frac{V_R}{V}$$ **From the voltage triangle**

$$\cos\theta = \frac{R}{Z}$$ **From the impedance triangle**

$$\cos\theta = \frac{P}{VA}$$ **From the power triangle**

Thus, the power vector of the circuit of Figure 3–15 is

$$\text{Power factor} = \frac{P}{VA} = \frac{500}{625} = 0.8 \text{ or } 80\% \text{ lag}$$

or

$$\text{Power factor} = \frac{R}{Z} = \frac{20}{25} = 0.8 \text{ lag}$$

or

$$\text{Power factor} = \frac{V_R}{V} = \frac{100}{125} = 0.8$$

Of the three forms given for the power factor, the one most commonly used is

$$\text{Power factor} = \frac{P}{VA}$$

Thus

$$P = V \times A \times \cos\theta$$

or

$$P = V \times I \times \cos\theta$$

where $\cos\theta$ = power factor

Unit 4 will present a thorough study of alternating-current resistance. This quantity is more properly called *effective resistance*. More realistic series circuits consisting of resistance and impedance components will be analyzed using the formulas derived in this unit.

SUMMARY

- The induced voltage of a conductor, coil, or circuit is caused by a changing magnetic field cutting across the circuit components. This voltage is proportional to the rate of change of the lines of force.
- In a dc circuit, the inductive effect stops once the current reaches the value defined by Ohm's law and remains constant.
- Lenz's law states that in all cases of electromagnetic induction, the induced voltage and the resultant current are in a direction that opposes the effect that produces them.
- Induction in an electrical circuit can be called *electrical inertia*.
- In a circuit containing inductance, current will increase at a much slower rate toward a maximum value than it does in a circuit containing resistance only.
- The length of time required for the current to decay to zero, once the circuit is deenergized, equals the time required for the current to rise to its Ohm's law value when the inductive circuit is energized.
- Inductance is the ability of a circuit component, such as a reactor coil, to store energy in the electromagnetic field.
- The unit of inductance is the *henry*. A circuit or circuit component is said to have an inductance of one henry when a current change of one ampere per second induces a voltage of one volt.
- In an ac circuit with the same effective voltage as an equivalent dc circuit, the circuit current will be less than in the dc circuit because of the choking effect of inductance in the ac circuit.
- The induced voltage or counterelectromagnetic voltage in a series inductive circuit opposes the impressed voltage and thus limits the circuit current.
- Removing the laminated iron core from a coil in an ac circuit will decrease the inductive reactance in the circuit. In effect, the circuit current is increased.
- Increasing the frequency of an ac inductive series circuit decreases the circuit current.
- Inductance in an ac circuit is just as effective in limiting current as resistance.
- *Inductive reactance* is the opposition in an ac circuit due to inductance.
    1. The symbol for inductive reactance is X_L.
    2. X_L opposition is measured in ohms.
    3. The formula for inductive reactance is

$$X_L = 2\pi fL$$

or

$$X_L = \omega L$$

where ω is the Greek letter *omega* and represents angular velocity; $2\pi f$ also is equal to angular velocity.

4. An accepted value for $2\pi f$ is 377, when the frequency is 60 Hz. For 25 Hz, $2\pi f = 157$.
5. Ohm's law for inductive reactance is

$$I = \frac{V_L}{X_L}$$

when the coil does not have resistance, or

$$I = \frac{V_{coil}}{Z_{coil}}$$

when the coil does have resistance.

- Radians
    1. One radian is equal to the angle at the center of a circle, subtended by an arc whose length is equal to the radius.
    2. One radian is equal in degrees to

$$\frac{360}{2\pi} = 57.296°$$

3. $\frac{\pi}{2}$ radians = 90°

π radians = 180°

2π radians = 360°

- The angle by which the current lags behind the line voltage is called the *phase angle* or *angle theta,* abbreviated $\angle\theta$.
    1. In a pure inductive circuit, the current lags the voltage by 90°.
    2. In a circuit containing both resistance and inductive reactance, the angle of lag is less than 90° and is equal to the vectorial sum of both the X_L and R components.
- A *scalar* is a measure of a specific unit of quantity [such as 6 inches (in.) or 5 pounds (lb)]. It has magnitude but does not show any direction.
- A *vector* is any quantity having both magnitude and direction.
- According to convention, vectors are rotated counterclockwise.
- To add vectors, place the head of either vector to the tail of the other vector and measure or calculate the resultant from start to finish.
    1. Vectors in a straight line can be added by taking the arithmetic sum of their scalar values.
    2. Vectors not in a straight line require a geometric solution and are not the sum of their scalar values.

- To subtract vectors, the vector to be subtracted is reversed and is then added head to tail to the remaining vector.
- All angles are measured with respect to the current vector as the reference vector, because current is constant in a series circuit.
- The actual power (true power) in watts taken by a pure inductive circuit is zero.
  1. The two positive pulses of power (when power is taken from the supply) are equal to the two negative pulses of power (when power is returned to the supply from the magnetic field). The net power taken by the inductor coil in one complete cycle is zero.
  2. True power is equal to I^2R.
     [*Note:* The product of the effective voltage and the effective current may not equal the power in watts in any circuit containing inductive reactance.]
- Ohm's law for an ac circuit containing both X_L and R is

$$V_R = I \times R$$
$$V_L = I \times X_L$$
$$V = I \times Z$$

  where impedance (Z) is the resulting value in ohms of the combination of R and X_L.
- Vector relationships:

$$\vec{V} = \vec{V}_R + \vec{V}_L \quad \text{or} \quad V = \sqrt{V_R^2 + V_L^2}$$
$$\vec{Z} = \vec{R} + \vec{X}_L \quad \text{or} \quad Z = \sqrt{R^2 + X_L^2}$$

- Power:
  1. Apparent power $= V \times I = VA$ or $I^2Z = VA$ or $V^2/Z = VA$
  2. True power (watts) $= V_R \times I$ or $I^2R = \text{Watts}$ or $V_R^2/R = \text{Watts}$
  3. Reactive power $= V_L \times I$ or $I^2X_L = \text{VARs}$ (wattless power) or $V_L^2/X_L = \text{VARs}$
  4. $VA^2 = \text{Watts}^2 + \text{VARs}^2$ or $VA = \sqrt{\text{watts}^2 + \text{VARs}^2}$
- Power factor:

$$PF = \frac{\text{true power (in watts)}}{\text{apparent power (in volt-amperes)}}$$

or

$$PF = \frac{\text{watts}}{VA}$$

or

$$PF = \cos\theta$$

or

$$\cos\theta = \frac{P}{VA}$$

or

$$\cos\theta = \frac{V_R}{V}$$

or

$$\cos\theta = \frac{R}{Z}$$

The most commonly used expressions for power factor are

$$PF = \frac{P}{VA}$$

or

$$\text{watts} = V \times I \times \cos\theta$$

or

$$W = V \times I \times PF$$

Achievement Review

1. State Lenz's law.
2. Define the standard unit of measurement of inductance.
3. What is meant by inductive reactance?
4. Add the following vectors (show the vector diagram for each addition):
   a. 208 V at 90° and 120 V at 180°
   b. 50 V at 0° and 42.1 V at 270°
   c. 20.3 V at 0° and 8.2 V at 0°
   d. 46.2 V at 0°, 71.4 V at 90°, and 38 V at 0°
5. Subtract the following vectors (show the vector diagram for each subtraction):
   a. 208 V at 90° from 120 V at 180°
   b. 50 V at 0° from 42.1 V at 270°
   c. 20.3 V at 0° from 8.2 V at 0°
   d. 84.2 V at 0° from 71.4 V at 90°
6. Resolve a voltage of 48 V into two coordinates, one of which is 23.2 V.
7. A line voltage (V) of 100 V is at an angle of 45° with respect to the line current. Find the horizontal component (V_R) and the vertical component (V_L) of the applied voltage.
8. An inductor coil with an inductance of 0.5 H and negligible resistance is connected across the terminals of a 230-V, 60-Hz supply.
   a. Determine the current in amperes taken by the inductor coil.
   b. What power in watts is taken by this inductor coil if the current lags the impressed voltage by 90 electrical degrees?

9. a. If the inductor coil in question 8 is connected across the terminals of a 230-V, 25-Hz supply, what is the current in amperes?
   b. Explain the reason for any change in current as the frequency changes.

10. An inductor coil connected across a 120-V, 60-Hz supply takes 10 A. The resistance of the inductor coil can be neglected. Determine
    a. the inductive reactance of the coil.
    b. the inductance of the coil.

11. An ac motor takes 10 A when connected across a 240-V, 60-Hz source. A wattmeter connected in the circuit reads 1500 W. Determine
    a. the power factor of the motor.
    b. the angle of lag of the current behind the voltage for the motor circuit.

12. An industrial load connected to a 440-V, 60-Hz line takes 50 A. The current lags the voltage by 30 electrical degrees.
    a. What is the power factor of this circuit?
    b. Determine
       1. the apparent power, in volt-amperes.
       2. the true power, in watts.

13. A series circuit (Figure 3–20) consists of resistance and inductive reactance connected in series across a 150-V, 60-Hz source. The resistance component of the series circuit consists of six 60-W, 120-V lamps, connected in parallel. Each lamp has a hot resistance of 240 Ω. The inductive reactive part of the series circuit consists of an inductor coil with an inductive reactance of 30 Ω and negligible resistance. Determine
    a. the impedance of the series circuit.
    b. the current, in amperes.
    c. the inductance of the inductor coil, in henrys.

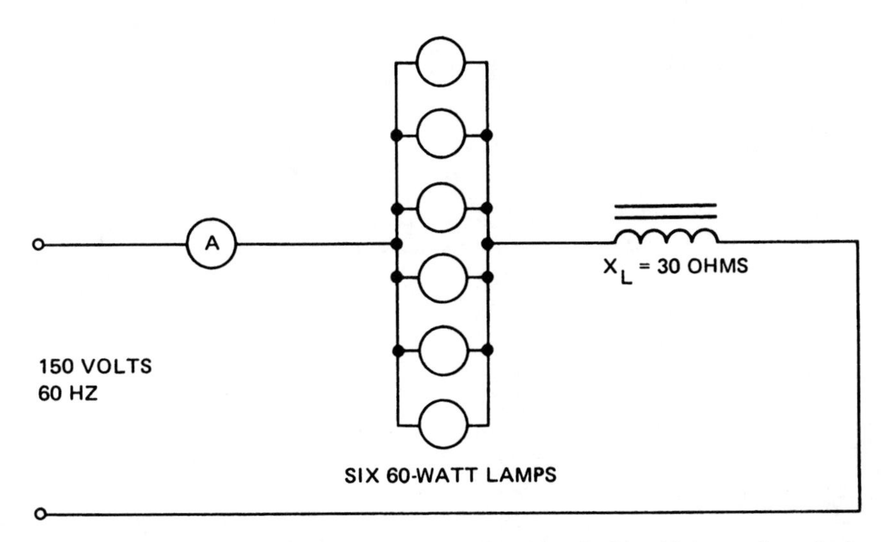

FIGURE 3–20 Resistance and inductance in series (*Delmar/Cengage Learning*)

14. For the circuit in question 13, determine
    a. the voltage loss across the bank of lamps.
    b. the voltage drop across the inductor coil.
    c. the true power in watts taken by the circuit.
    d. the reactive power in VARs taken by the circuit.

15. a. For the circuit in question 13, determine the power factor.
    b. What is the angle of lag or the power factor angle?

16. The following diagrams are to be drawn to scale for the circuit in question 13:
    a. A voltage vector diagram
    b. An impedance triangle
    c. A triangle of power components

17. A 30-Ω resistor and a reactor coil (inductor coil) with a 0.2-H inductance and negligible resistance are connected in series across a 120-V, 25-Hz supply. Determine
    a. the impedance of the circuit.
    b. the current.
    c. the true power in watts taken by the circuit.
    d. the power factor.
    e. the phase angle.

18. Draw a vector diagram for the circuit in question 17. The vectors must be drawn accurately to a convenient scale. Mark each vector with the proper letter identification.

19. A reactor coil that has negligible resistance is connected in series with a group of incandescent lamps connected in parallel. The purpose of the reactor coil is to reduce the voltage across the lamps. When the circuit is energized from a 120-V, 60-Hz source, the current in the reactor coil is 25 A. The voltage across the lamps is 30 V. Determine
    a. the circuit impedance, in ohms.
    b. the voltage loss across the reactor coil.
    c. the inductive reactance of the inductor coil, in ohms.
    d. the inductance of the reactor coil, in henrys.
    e. the power factor and the angle of lag.

20. Using the circuit given in question 19
    a. determine the power in watts used in the circuit.
    b. determine the quadrature, or vertical power component, in VARs.
    c. determine the apparent power, in volt-amperes.
    d. draw an impedance triangle and a power triangle to scale.
    e. draw a voltage vector diagram to scale.

PRACTICE PROBLEMS FOR UNIT 3

Inductive Circuits

Find the missing values.

INDUCTANCE	FREQUENCY	INDUCTIVE REACTANCE
1.2 H	60 Hz	
0.085 H		213.628 Ω
	1000 Hz	4712.389 Ω
0.65 H	600 Hz	
3.6 H		678.584 Ω
	25 Hz	411.459 Ω
0.5 H	60 Hz	
0.85 H		6408.849 Ω
	20 Hz	201.062 Ω
0.45 H	400 Hz	
4.8 H		2412.743 Ω
	1000 Hz	40.841 Ω

$$X_L = 2\pi fL \qquad L = \frac{X_L}{2\pi f} \qquad f = \frac{X_L}{2\pi L}$$

4

Series Circuits—Resistance and Impedance

Objectives

After studying this unit, the student should be able to

- explain effective resistance in an ac circuit and show how eddy current losses and hysteresis losses cause the effective resistance to be greater than the true ohmic resistance.
- calculate and solve ac series circuit problems containing resistive and impedance components, using the procedures outlined in this unit.
- analyze and show the solution to ac series circuits containing R and Z components by using vector diagrams.
- explain the relationship between impedance triangles, triangles of power values, and vector diagrams.
- discuss and use the units millihenrys and microhenrys by applying the X_L formula.
- define the figure of merit, Q, of an inductor coil and describe what high and low Q values mean to the circuit performance.

EFFECTIVE RESISTANCE

The effective resistance is the resistance that a circuit or component offers to alternating current. Also known as ac resistance, it may vary with the frequency, current, or voltage of the circuit. Effective resistance must not be confused with impedance. The quantity resulting from the combination of effective resistance and reactance is impedance.

SKIN EFFECT

Alternating current tends to flow along the surface of a conductor. Direct current acts through the entire cross-sectional area of the conductor in a uniform manner. The name *skin effect* is given to the action whereby alternating current is forced toward the surface of a conductor. Because of the skin effect, there is less useful copper conducting area with alternating current. As a result, there is an increase in resistance.

EDDY CURRENT LOSSES

Alternating current in a conductor or circuit sets up an alternating magnetic field. Eddy currents are induced by this field in any metal near the conductor. For example, eddy current losses occur in the iron cores of reactors, transformers, and stator windings in ac generators and motors. A reduction in eddy currents is achieved by laminating the cores used in ac equipment. However, there are small I^2R losses in each lamination of the core.

HYSTERESIS LOSSES

In alternating current, the direction of current is constantly changing. This means that the lines of force of the magnetic field are also changing direction repeatedly. In other words, millions of molecules reverse direction in the process of magnetizing, demagnetizing, and remagnetizing the structure of any iron core or other metallic material adjacent to the conductors of an ac circuit.

As these molecules reverse their direction with each change of magnetization, molecular friction results. Power is required to overcome this molecular friction. This loss occurs as heat in the metallic structure and is known as *hysteresis loss*. The ac circuit adjacent to the metallic material must supply the power in watts to overcome this hysteresis loss. All ac generators, ac motors, transformers, and other ac equipment experience hysteresis loss. To reduce this loss, special steels are used for the core structure. For example, silicon steel may be used because it has a relatively low molecular friction loss.

DIELECTRIC LOSSES

As the impressed voltage rises to a maximum value twice in each cycle, a voltage stress is placed on the insulation of the conductor. Such a stress occurs first in one direction and then in the other. As a result, there is a small heat loss in the insulation. This loss is called the *dielectric loss* and is very small when compared to the other losses in the circuit due to the skin effect, eddy currents, and hysteresis. Usually the dielectric loss can be neglected.

The losses just described all require the use of power supplied by the electrical circuit. The power in watts is expressed by the formula $P = I^2R$. This formula can be rearranged to find the resistance: $R = P \div I^2$. Thus, if the power increases while the current remains the same, the effective resistance increases.

For ac circuits, "R" represents the effective ac resistance. The dc resistance, or the true ohmic resistance, in an ac circuit may be designated as "R_O."

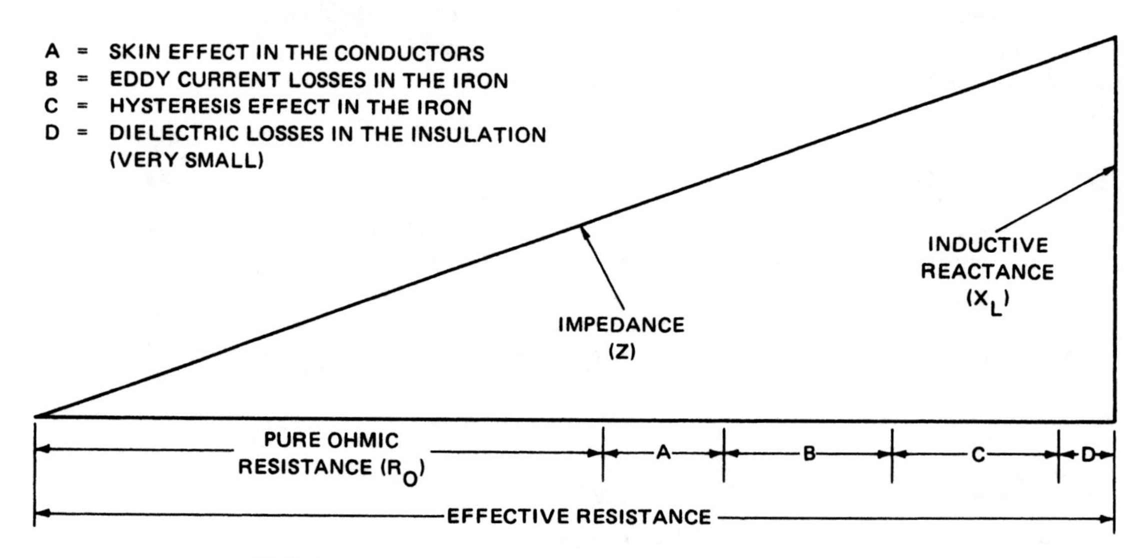

FIGURE 4–1 Impedance triangle (*Delmar/Cengage Learning*)

Representing DC and Effective (AC) Resistance

Figure 4–1 shows an impedance triangle. The base of the triangle represents the effective resistance. This resistance is divided into five parts representing the true dc resistance and the four losses just described.

Problem in DC and AC Resistance

A practical circuit problem can be used to compare the meanings of the true ohmic (dc) resistance and the effective (ac) resistance. Figure 4–2 shows a reactor coil (reactor), with a laminated silicon steel core, connected to a dc source.

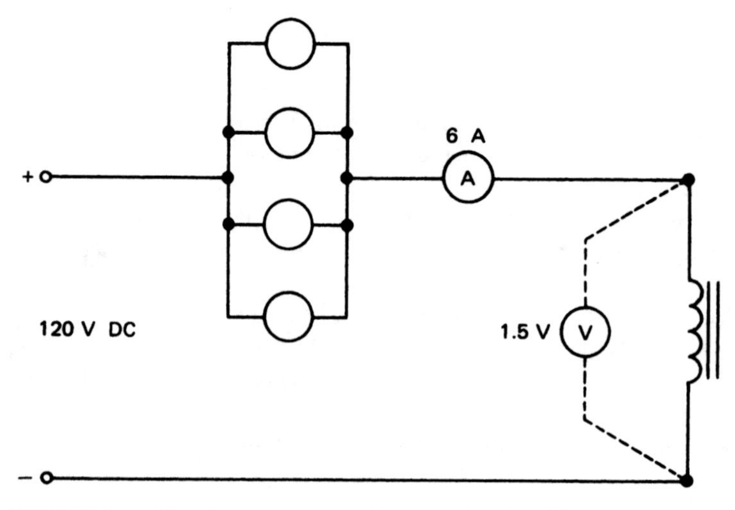

FIGURE 4–2 Reactor connected to dc source (*Delmar/Cengage Learning*)

The pure ohmic resistance of the reactor can be measured using the ammeter–voltmeter method. A bank of lamps is inserted in series with the reactor to limit the current to a safe value. The true ohmic resistance of the reactor in Figure 4–2 is

$$R_O = \frac{V_{coil}}{I}$$

$$= \frac{1.5}{6} = 0.25\ \Omega$$

$$P = V_{R_{coil}} \times I = 6 \times 1.5 = 9\ W$$

The iron core is now removed from the reactor, which is connected to a 120-V, 60-Hz source, as shown in Figure 4–3. This circuit does not require a current-limiting resistor connected in series with the reactor. Because the reactor is energized from an ac source, there is enough inductive reactance in the reactor to limit the current to a safe value. As a result, the bank of lamps is removed from the circuit.

Using the values given in Figure 4–3, the effective (ac) resistance of the air core reactor is

$$R = \frac{P}{I^2}$$

$$= \frac{8.25}{5^2} = \frac{8.25}{25} = 0.33\ \Omega$$

It has been determined that the dc (true ohmic) resistance is 0.25 Ω. The effective ac resistance is 0.33 Ω. The slight increase in effective resistance compared to the dc resistance is due to the skin effect and dielectric losses.

The reactor shown in Figure 4–4 has a laminated iron core. The power increases from 8.25 to 16 W, even though the current decreases from 5 to 4 A. This means that the effective resistance increased.

$$R = \frac{P}{I^2}$$

$$R = \frac{16}{4^2} = \frac{16}{16} = 1\ \Omega$$

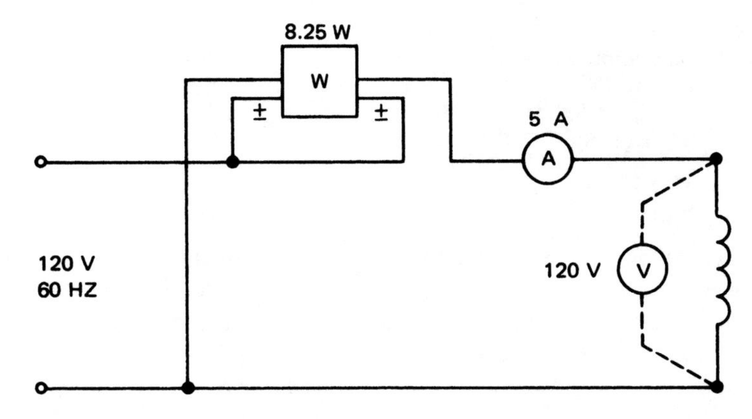

FIGURE 4–3 Reactor (with air core) connected to ac source (*Delmar/ Cengage Learning*)

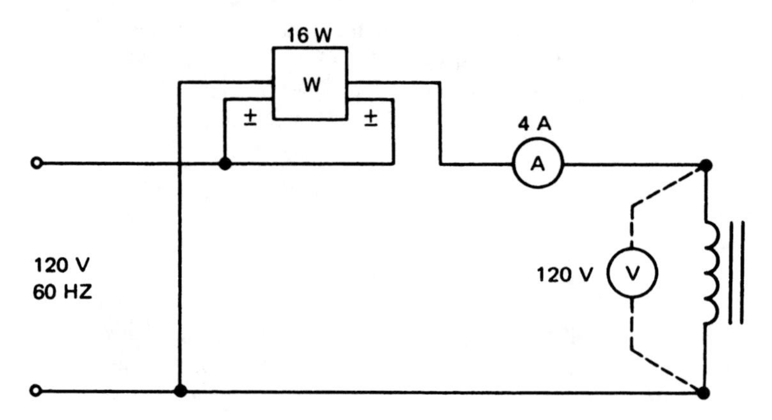

FIGURE 4–4 Reactor (with laminated core) connected to ac source (*Delmar/Cengage Learning*)

Increase of Effective (AC) Resistance. The increase in the effective (ac) resistance from 0.33 to 1 Ω is due to the eddy current and hysteresis losses in the steel core. The addition of the laminated core to the reactor results in a greater power loss in the reactor because of the skin effect and the dielectric, eddy current, and hysteresis losses. The increased power loss means there is also an increase in the effective resistance. Recall that the true ohmic resistance of the reactor is 0.25 Ω.

When the laminated core of the reactor is replaced with a solid cast-iron core (Figure 4–5), the voltage is still 120 volts and the current is 4.5 A. However, the wattmeter shows that there is a great increase in the power expended in the coil. Why does the power increase with a solid cast-iron core?

The constantly changing field induces voltages in the solid core. The resulting eddy currents have a low-resistance circuit path and are higher in value. Thus, the I^2R losses are greater. In addition, the hysteresis loss (molecular friction loss) is greater in the cast-iron core than in the silicon steel core. These increases in the eddy current and the hysteresis losses mean that more true power in watts is delivered to the reactor. As a result, the effective resistance of the reactor for this circuit increases:

$$R = \frac{P}{I^2}$$

$$= \frac{60.75}{4.5^2} = \frac{60.75}{20.25} = 3\ \Omega$$

Effect of Higher Frequencies on AC Resistance. The effective resistance of alternating-current equipment at low frequencies, such as 60 Hz, can be several times greater than the true ohmic resistance. In the high-frequency range, the effective resistance can be many times higher than the dc resistance. This increased resistance is due to the fact that the skin effect, dielectric losses, eddy current losses, and hysteresis losses all increase with an increase in frequency.

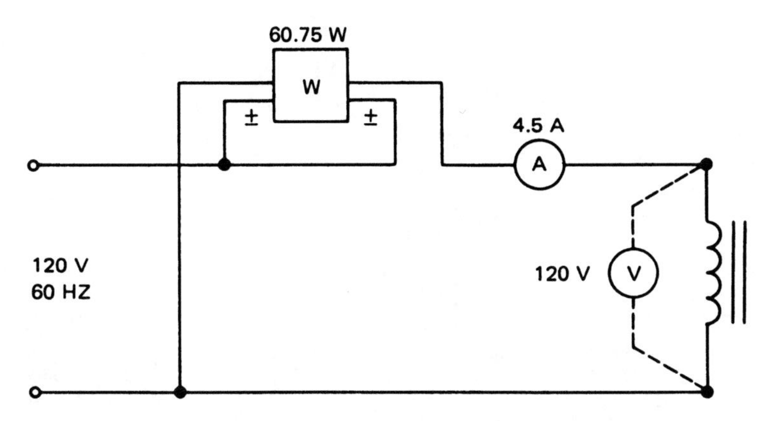

FIGURE 4–5 Reactor coil (with solid cast-iron core) connected to ac source (*Delmar/Cengage Learning*)

Table 4–1 summarizes the resistance, reactance, and current ranges for various circuits.

RESISTANCE AND INDUCTANCE IN SERIES

Unit 3 discussed series circuits containing only resistance and pure inductance. It is not possible for a circuit to have pure inductance because the coils of inductive equipment all have resistance. To obtain high values of inductance, iron cores are usually required. However, eddy currents and hysteresis losses in iron cores increase the ac resistance considerably. This effective resistance must be considered in all problems involving inductance. The discussion in this section will be confined to the study of circuits with resistance and inductance in series. The impedance section of each circuit considered consists of a reactor having a relatively large inductive reactance compared to the effective resistance.

Vector Resolution

The solution of most of the following problems uses a method called *vector resolution*. This method is widely used and its application should be clear to the student.

	DC	AC		
		Air Core	Laminated Core	Solid Core
Resistance	Lowest	Low	Average	High
Reactance	Zero	Low	High	Average
Current	Highest	High	Low	Average

TABLE 4–1 Comparing resistance, reactance, and current in dc circuits and ac circuits containing reactors

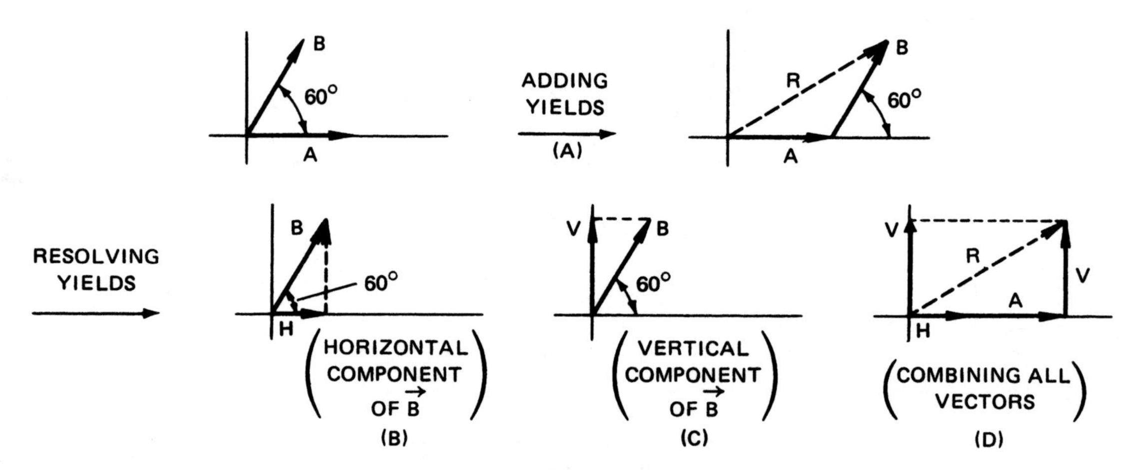

FIGURE 4–6 Resolving vectors (*Delmar/Cengage Learning*)

As an example of the method, consider two vectors, $\vec{A}$ and $\vec{B}$, that are out of phase by 60°.

Figure 4–6 shows both conventional vector addition and the resolution method. Both methods yield the same answer. Note, however, that the resolution method yields a *right triangle,* which can be solved using simple algebra.

Inspection of the resolution method shows the following relationships:

H (horizontal component of B) $= B \times \cos 60°$ (Figure 4–6B)

V (vertical component of B) $= B \times \sin 60°$ (Figure 4–6C)

If vector B is of unit length, then the horizontal component of B is the *cosine* side of the triangle and the vertical component of B is the *sine* of the triangle.

The vectors are combined as follows:

$$R = \vec{A} + \vec{B} \cos 60° + B \sin 60° \text{ (vector equation)}$$

$$= \text{X-axis vector} + \text{Y-axis vector}$$

or

$$R = \sqrt{(A + B \cos 60°)^2 + (B \sin 60°)^2} \text{ (Figure 4–6D)}$$

The final equation is the result of (1) algebraically adding the vectors on the X axis (horizontal components), (2) squaring the result, (3) algebraically adding the vectors on the Y axis (vertical components), and (4) squaring that result.

The final equation has the form $C = \sqrt{A^2 + B^2}$.

The vertical components should *not* be added algebraically to the horizontal components. Students often make this mistake when solving vector resolution problems.

Several sample problems dealing with vectors are given on the following pages, with their solutions fully explained.

PROBLEM 1

Statement of the Problem

Figure 4–7 shows a reactor connected directly across a 120-V, 60-Hz source. Meter readings for the line voltage, current, and power are shown on the circuit diagram. The impedance of the coil is the result of the combination of the effective resistance and inductive reactance. The impedance can be determined by considering the coil as a series circuit with a resistance component and an inductive reactance component.

The ohmic values of the reactor coil are determined as follows:

- Impedance of coil:

$$Z = \frac{V}{I} = \frac{120}{4} = 30\ \Omega$$

- Effective resistance of coil:

$$R = \frac{P}{I^2} = \frac{64}{4^2} = 4\ \Omega$$

- Inductive reactance of coil:

$$X_L = \sqrt{Z^2 - R^2} = \sqrt{30^2 - 4^2} = 29.8\ \Omega$$

The reactor is now connected in series with a 36-Ω resistor across a 250-V, 60-Hz source, as shown in Figure 4–8. The ohmic values for the coil ($Z = 30\ \Omega$, $R = 4\ \Omega$, and $X_L = 29.8\ \Omega$) are still the same in this circuit because the frequency is unchanged (60 Hz).

Solve the problem for the following information:

1. The total circuit resistance
2. The impedance of the entire series circuit

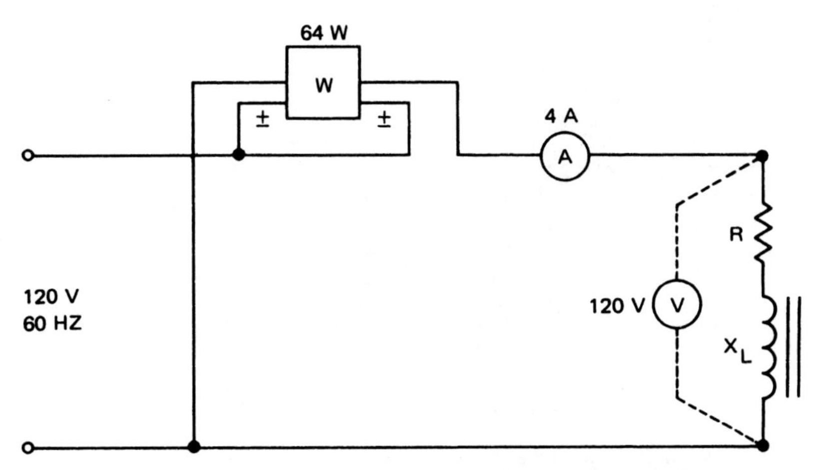

FIGURE 4–7 Reactor coil (*Delmar/Cengage Learning*)

3. The current, in amperes
4. The true power, in watts, taken by
   a. the resistor
   b. the coil
   c. the entire series circuit
5. The apparent power in volt-amperes taken by the entire series circuit
6. The power factor of
   a. the coil
   b. the entire series circuit

Then construct a vector diagram for the series circuit.

Solution

1. The total resistance of the series circuit is obtained by adding the resistance values. The resistor value is 36 Ω. The effective resistance, determined earlier, is 4 Ω. Therefore, the total resistance is

   $$R = 36 + 4 = 40\ \Omega$$

2. The inductive reactance of the entire series circuit is contained in the reactor coil. When the reactor was connected across a 60-Hz source, its inductive reactance was 29.8 Ω. The total resistance of the series circuit is 40 Ω. Thus, the total series circuit impedance is

   $$Z = \sqrt{R^2 + X_L{}^2} = \sqrt{40^2 + 29.8^2} = \sqrt{2488.04} = 49.88\ \Omega$$

3. The line voltage is 250 V, 60 Hz. This voltage is impressed across the total impedance of the circuit. The current, in amperes, in this series circuit path is

   $$I = \frac{V}{Z} = \frac{250}{49.88} = 5.02\ A = 5\ A$$

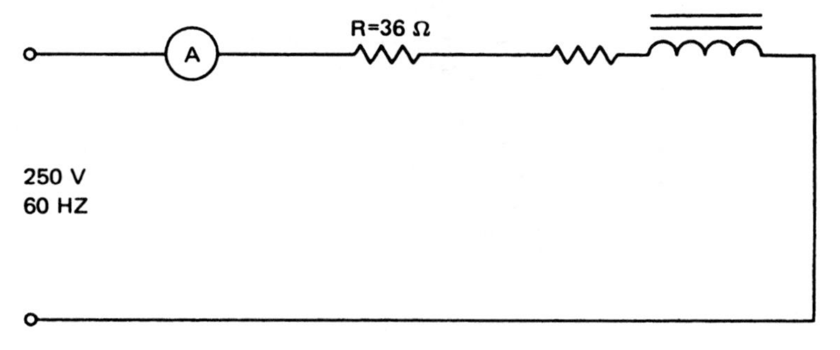

FIGURE 4–8 Resistance and impedance in series ***(Delmar/Cengage Learning)***

4. a. The true power taken by the resistor rated at 36 Ω is

$$P = I^2R = 5^2 \times 36 = 25 \times 36 = 900 \text{ W}$$

b. The true power expended in the coil is

$$P = I^2R = 5^2 \times 4 = 25 \times 4 = 100 \text{ W}$$

c. The total true power for the entire series circuit can be obtained by adding the power expended in the resistor and the reactor. The formula I^2R can also be used to find the total true power. R is the total effective resistance of the series circuit:

$$\begin{aligned} P_{total} &= P_{resister} + P_{coil} \\ &= 900 + 100 \\ &= 1000 \text{ W} \end{aligned}$$

or

$$\begin{aligned} P_{total} &= I^2R_{total} \\ &= 5^2 \times 40 \\ &= 1000 \text{ W} \end{aligned}$$

5. The total apparent power in volt-amperes for the entire series circuit is the product of the line potential and the series circuit current:

$$VA = V \times I = 250 \times 5 = 1250 \text{ volt-amperes (VA)}$$

6. a. The ohmic components for the impedance coil were determined earlier. The power factor of the coil is found using R and Z from the impedance triangle developed for the reactor.

$$PF = \frac{R}{Z} = \frac{4}{30} = 0.1333$$

The value 0.1333 is the cosine of an angle that is equal to 82.3°. Thus, the reactor current lags the impressed voltage across the reactor by 82.3°.

b. The power factor of the entire series circuit is the ratio of the total true power in watts to the total apparent power in volt-amperes. The power factor of the circuit is also the ratio of the total series circuit resistance to the total circuit impedance:

$$\begin{aligned} PF &= \frac{P}{VA} \\ &= \frac{1000}{1250} \\ &= 0.80 \end{aligned}$$

or

$$PF = \frac{R}{Z}$$

$$= \frac{40}{49.88}$$

$$= 0.80$$

The angle for a cosine value of 0.8000 is 36.9°. This means that the line current lags the line voltage by 36.9°.

7. Figure 4–9 shows one method of constructing the vector diagram for this series circuit. The line current, 5 A, is drawn as a horizontal line. A convenient scale is used. V_R, the component of the reactor voltage caused by the effective resistance of the coil, is placed directly on the current line. V_R is drawn from point 0 along the current line and has a magnitude of 20 V. The voltage drop across the 36-Ω resistor is 180 V. This value is added to the 20-V drop on the current vector. The total voltage drop caused by resistance in the series circuit is 200 V in phase with the current. The voltage component caused by inductive reactance in the reactor is drawn in a vertical direction from point 0. This component, V_{X_L}, is at an angle of 90° with the current. The current lags the voltage loss, caused by the inductive reactance, by an angle of 90°. The two voltage components of the reactor can be added vectorially to obtain the reactor voltage.

In Figure 4–9, the voltage across the reactor is shown as the vector sum of the two voltage components of the coil. These components are caused by the effective resistance and the inductive reactance. The phase angle between the voltage across the reactor and the current is 82.3°. This angle is indicated by the small Greek letter alpha (α). Note the right triangle formed by V_{X_L} and V_{coil}.

Vector addition is used to find the sum of the total voltage drop due to the resistance in the circuit and the voltage drop due to inductive reactance. This vector sum is the line voltage. The angle θ between the line voltage and the current is 36.9°, lagging.

Figure 4–10 shows a second method of constructing a vector diagram for the same series circuit. In this figure, the voltage drops are arranged differently because of the two resistance components of the series circuit. This change in placement on the current vector of the two resistance voltage losses means that the voltage components for the reactor must also be shown in a location different from that in Figure 4–9.

Although the two vector diagrams are drawn differently, the magnitude and phase relationships of all voltages, in reference to the current, are exactly the same. Therefore, either vector diagram may be used to analyze a series circuit containing R and Z components.

PROBLEM 2

Statement of the Problem

Figure 4–11 shows a noninductive heater unit connected in series with a reactor. The reactor controls the current in the series circuit. This means that the reactor also controls the temperature of the heater unit. The wattmeter indicates the total true power in watts taken by the circuit.

1. Determine the series circuit power factor and the phase angle.
2. Determine the true power, in watts, taken by the noninductive heating load.

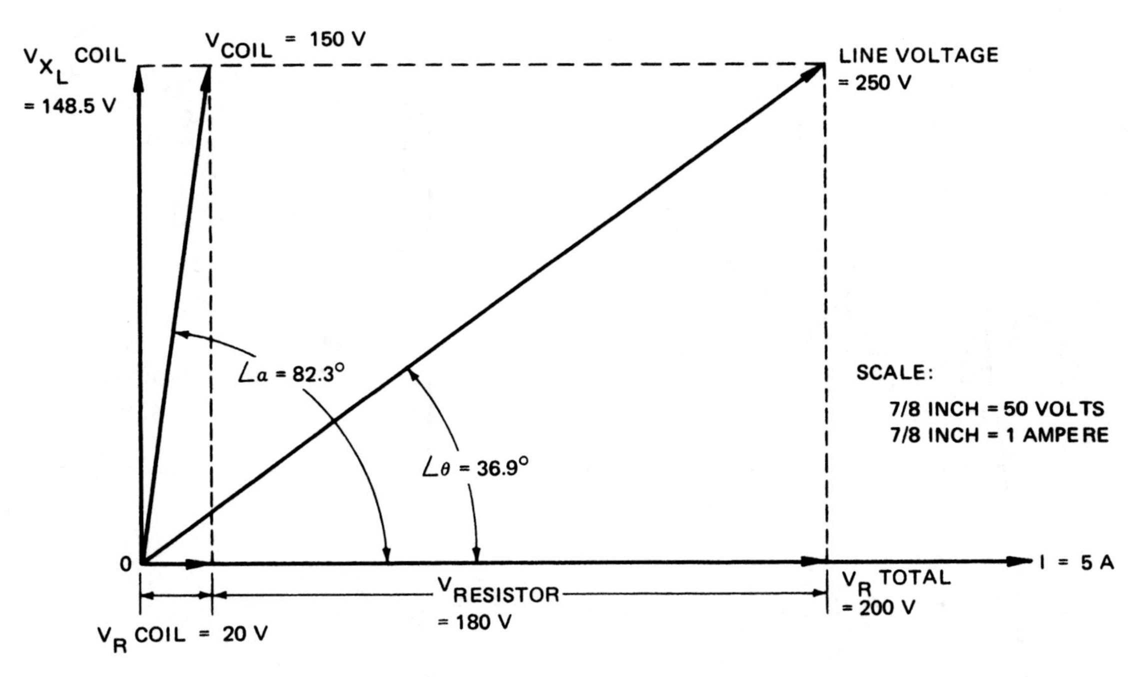

FIGURE 4–9 Vector diagram for R and Z in series (*Delmar/Cengage Learning*)

3. Determine the loss in watts in the coil.
4. Determine the effective resistance of the coil.
5. What is the power factor and angle of lag for the coil?
6. Determine the inductance, in henrys, of the coil at the circuit frequency of 25 Hz.
7. Draw a vector diagram, an impedance triangle, and a triangle of power values. Each diagram is to be drawn to scale and properly labeled.

Solution

1. The total true power taken by the circuit is 1650 W. The power factor is the ratio of the total true power in watts to the total apparent power in volt-amperes:

$$PF = \frac{P}{V \times I} = \frac{1650}{220 \times 10} = 0.75 \text{ lag}$$

The angle of lag of the line current behind the line voltage is 41.4°.

2. The true power in watts taken by the heater unit is the product of the voltage across the heater unit and the current. These two values are in phase:

$$P = V_R \times I = 150 \times 10 = 1500 \text{ W}$$

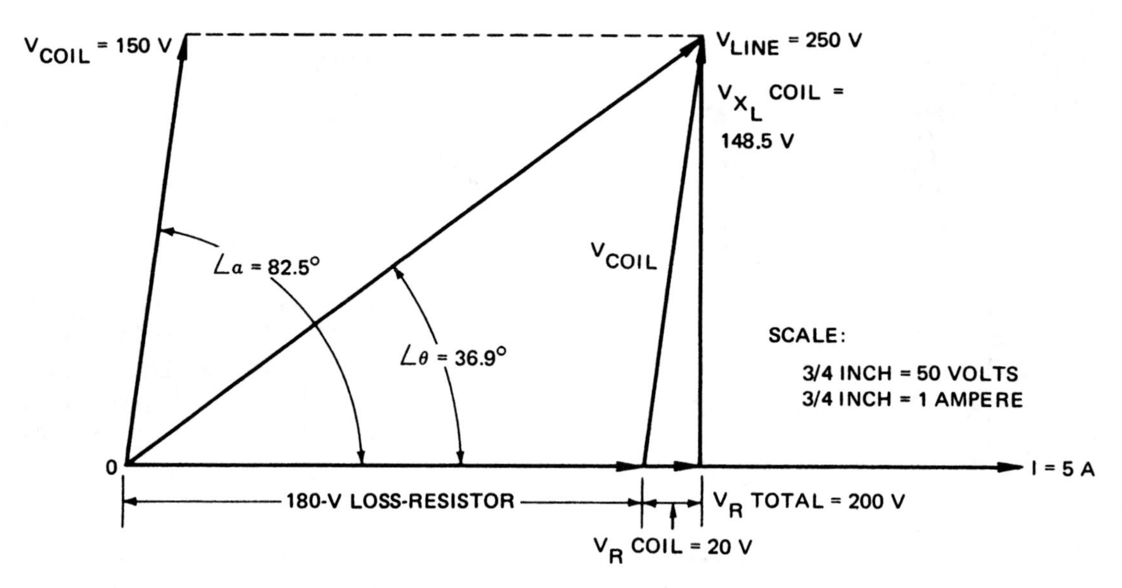

FIGURE 4–10 Vector diagram for R and Z in series (alternate method) (*Delmar/Cengage Learning*)

3. The true power lost in the reactor due to the effective resistance is

   $$P = 1650 - 1500 = 150 \text{ W}$$

4. If the current and the loss in watts are known, the effective resistance of the reactor can be determined from $W = I^2R$.

   $$150 = 10^2R$$
   $$R = 1.5\ \Omega$$

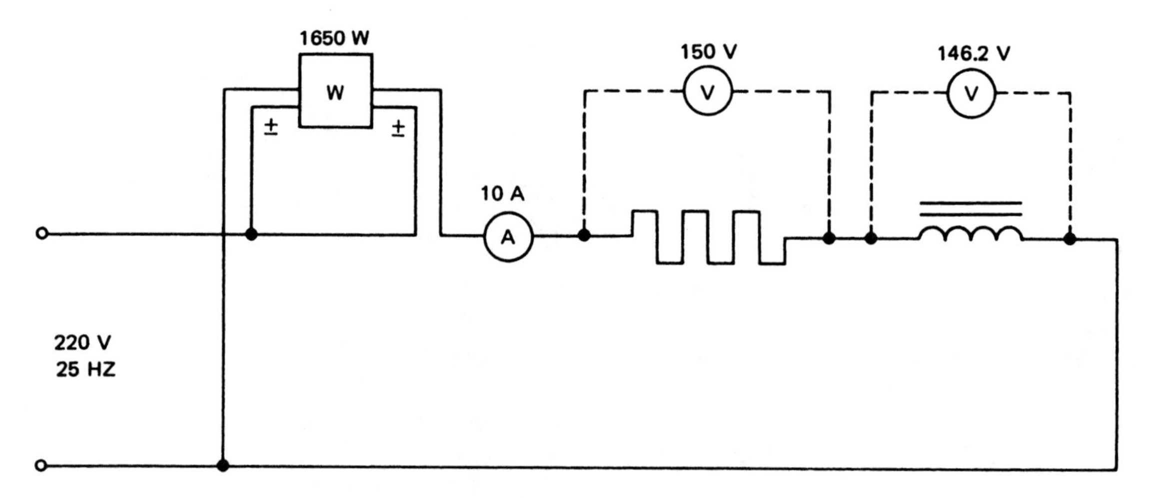

FIGURE 4–11 (*Delmar/Cengage Learning*)

5. The impedance of the reactor must be determined before the power factor for the reactor can be found. The effective resistance of the reactor was determined in step 4. Thus, the power factor for the reactor can be determined by the ratio R/Z:

$$Z_{coil} = \frac{V_{coil}}{I}$$

$$= \frac{146.2}{10} = 14.62\text{-}\Omega \text{ impedance}$$

$$PF = \frac{R}{Z}$$

$$= \frac{1.5}{14.62} = 0.1026 \text{ lag}$$

The angle of lag of the current behind the reactor voltage is 84.1°.

6. The inductance of the reactor can be obtained once the inductive reactance is known. The effective resistance and the impedance of the coil were calculated in steps 4 and 5, respectively. The inductive reactance is given by the expression

$$X_L = \sqrt{Z^2 - R^2} = \sqrt{14.62^2 - 1.5^2} = 14.54\ \Omega$$

Knowing the inductive reactance, the inductance can be found using

$$X_L = 2\pi fL$$

$$L = \frac{X_L}{2\pi f} = \frac{14.54}{157} = 0.093\ \text{H}$$

Figure 4–12 shows the impedance triangle, the triangle of power values, and the vector diagram. A study of the three diagrams shows that they are similar right triangles.

FIGURE OF MERIT

It has been pointed out that inductor coils have some resistance. For commercial coils, the inductive reactance is the useful component. The effective resistance is the unwanted component, which must be held to as low a value as possible. The ratio of the inductive reactance in ohms to the effective resistance in ohms is called the *figure of merit.* The symbol for the figure of merit is the letter Q.

The expression for the figure of merit is

$$Q = \frac{X_L}{R} = \frac{2\pi fL}{R} = \frac{\omega L}{R}$$

A large Q (high figure-of-merit value) is desirable because it means that the useful inductive reactance is high and the effective resistance is low. A low value of Q means that the resistance component is relatively high. As a result, there is a large power loss.

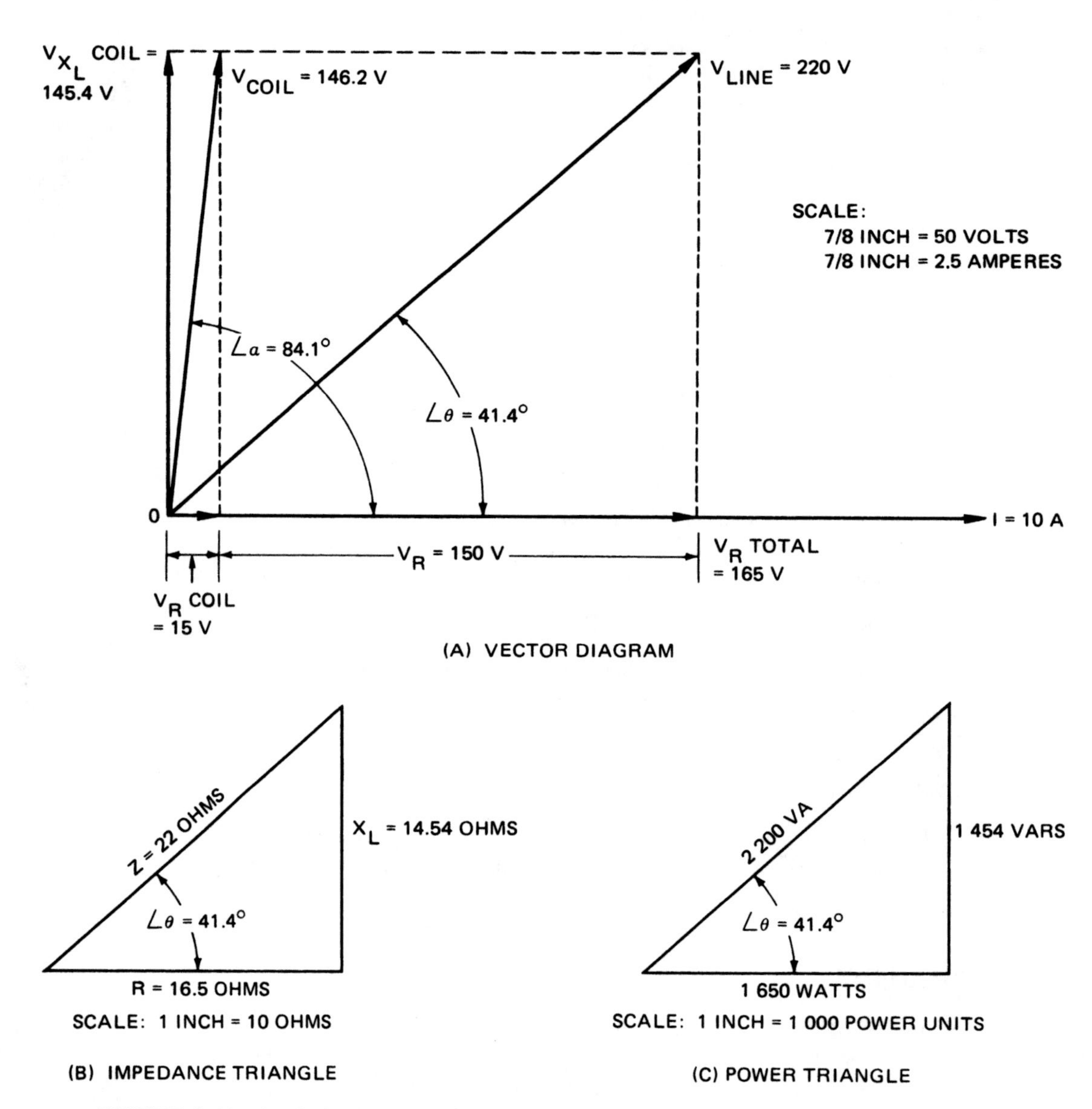

FIGURE 4–12 Analysis of an ac series circuit containing R and Z (*Delmar/Cengage Learning*)

By referring to the expressions for the figure of merit (Q), it can be seen that an increase in frequency will cause an increase in the inductive reactance and in the effective resistance. However, the inductive reactance and the effective resistance do not increase in the same ratio with an increase in frequency. Thus, the figure of merit (the Q of an inductor coil) must be determined for the frequency, or band frequencies, at which the coil is to be used.

For example, a 0.2-H inductor coil has a resistance of 100 Ω and is used at a frequency of 10,000 Hz. What is the Q of this coil?

$$Q = \frac{2\pi fL}{R}$$

$$= \frac{6.28 \times 10{,}000 \times 0.2}{100}$$

$$= 125.6 \text{ or } 126$$

Reactors that have a high Q value are often treated as pure inductors, and the amount of wire resistance is ignored because it is such a small part of the total impedance of the coil. For example, assume that an inductor has a Q of 10. This indicates that the amount of inductive reactance is 10 times greater than the wire resistance. In the following example, it will be assumed that an inductor has a wire resistance of 10 Ω and an inductive reactance of 100 Ω. When the total impedance of the coil is computed, it will be seen that the actual amount of wire resistance is only a small part of the coil's total impedance:

$$Z = \sqrt{R^2 + X_L^2}$$

$$= \sqrt{10^2 + 100^2}$$

$$= \sqrt{100 + 10{,}000}$$

$$= \sqrt{10{,}100}$$

$$= 100.5\ \Omega$$

PROBLEM 3

Statement of the Problem

Refer to Figure 4–13. It will be assumed that the inductor has a high Q value and contains zero resistance. The circuit contains 30 Ω of resistance and 40 Ω of inductive reactance and is connected to a 240-V, 60-Hz line. Refer to the formulas in the Resistive Inductive (Series) section of Appendix 15 to find the following unknown values:

1. Total circuit impedance (Z)
2. Current flow (I)
3. Voltage drop across the resistor (V_R)
4. Watts, true power (P)
5. Inductance of the inductor (L)
6. Voltage drop across the inductor (V_L)
7. Volt-amperes-reactive, reactive power (VARs)
8. Volt-amperes, apparent power (VA)
9. Power factor (PF)
10. Phase angle, which indicates how many degrees the voltage and current are out of phase with each other ($\angle\theta$)

Solution

1. The impedance (Z) can be computed using the following formula:

$$Z = \sqrt{R^2 + X_L^2}$$
$$= \sqrt{30^2 + 40^2}$$
$$= \sqrt{900 + 1600}$$
$$= \sqrt{2500}$$
$$= 50\ \Omega$$

2. Now that the total impedance of the circuit is known, the circuit current can be computed using the following formula:

$$I = \frac{V}{Z}$$
$$= \frac{240}{50}$$
$$= 4.8\ \text{A}$$

3. In a series circuit, the current is the same at any point in the circuit. Therefore, 4.8 A of current flows through both the resistor and the inductor. The amount of voltage dropped across the resistor can be computed by using the following formula:

$$V_R = I \times R$$
$$= 4.8 \times 30$$
$$= 144\ \text{V}$$

4. True power for the circuit can be computed by using any of the power formulas. The following formula is used in this example:

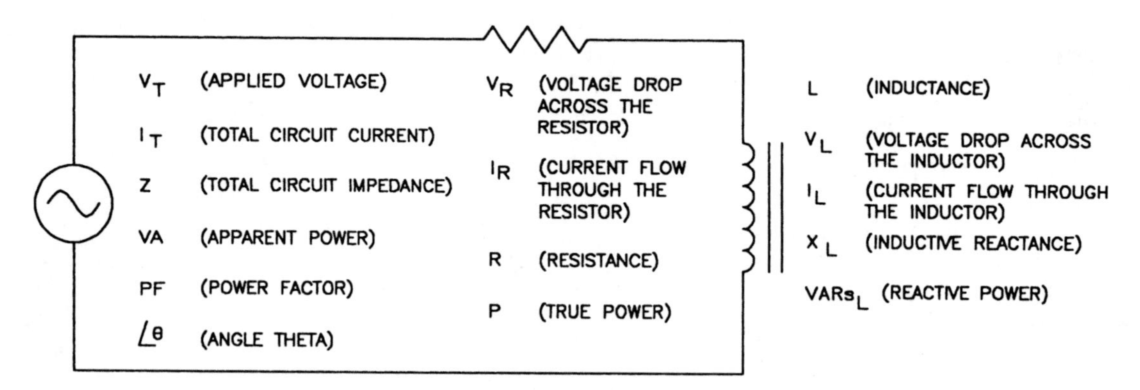

FIGURE 4–13 RL series circuit (*Delmar/Cengage Learning*)

$$P = V \times I$$
$$= 144 \times 4.8$$
$$= 691.2\ W$$

5. The amount of inductance can be computed using the following formula:

$$L = \frac{X_L}{2\pi f}$$
$$= \frac{40}{377}$$
$$= 0.106\ H$$

6. The amount of voltage dropped across the inductor can be computed using the following formula:

$$V_L = I \times X_L$$
$$= 4.8 \times 40$$
$$= 192\ V$$

7. *VARs* is the amount of reactive power in the circuit. It can be computed in a similar manner as watts except that reactive values of voltage and current are used instead of resistive values. In this example the following formula is used:

$$VARs = V_L \times I$$
$$= 192 \times 4.8$$
$$= 921.6 \text{ volt amperes-reactive}$$

8. *Volt-amperes* is the apparent power of the circuit. It can be computed in a similar manner as watts or VARs, except that total values of voltage and current are used. In this example the following formula is used:

$$VA = V_T \times I$$
$$= 240 \times 4.8$$
$$= 1152 \text{ volt amperes}$$

9. *Power factor* is the ratio of true power compared to apparent power. It can be computed by dividing any resistive value by its like total value. For example, the power factor can be computed by dividing the voltage drop across the resistor by the total circuit voltage, or resistance divided by impedance, or watts divided by volt-amperes. The power factor is generally expressed as a percentage. The decimal fraction computed from the division will, therefore, be changed to a percent by multiplying it by 100. In this circuit the following formula is used:

$$PF = \frac{P}{VA}$$
$$= \frac{691.2}{1152}$$
$$= 0.6 \times 100 \text{ or } 60\%$$

10. The power factor of a circuit is the cosine of the phase angle. The phase angle indicates how many degrees the current and voltage are out of phase with each other. Because the power factor of this circuit is 0.6, angle theta is as follows:

$$\cos \angle\theta = PF$$
$$= 0.6$$
$$\angle\theta = 53.13°$$

SUMMARY

- The resistance of a circuit or component to ac is the effective resistance.
- Impedance is the quantity resulting from the combination of effective resistance and reactance.
- Skin effect is the action whereby alternating current is forced toward the surface of a conductor.
- Eddy currents are induced by the alternating magnetic field around a current-carrying conductor in any metal near the conductor.
    1. Eddy current losses occur in the iron cores of reactors, transformers, and stator windings in ac generators and motors.
    2. Eddy currents are minimized by laminating the cores used in ac equipment.
- Hysteresis losses are caused by a constantly changing magnetic field due to ac.
    1. The molecules of the core structure or other metallic material near the conductors of an ac circuit also change direction as the field changes. The result of these changes is molecular friction.
    2. Power is required to overcome this molecular friction.
    3. Hysteresis loss occurs as heat in the metallic structure.
- Dielectric loss is a small heat loss in the insulation.
    1. The insulation of a conductor is subjected to a voltage stress as the impressed voltage rises to a maximum value twice in each cycle.
    2. Usually, this loss is so small it can be neglected.
- The effective resistance for an ac circuit is greater than the true ohmic resistance. Effective resistance consists of skin effect losses, eddy current losses, hysteresis losses, and

dielectric losses and is represented by the symbol "R." Power from the circuit is required to overcome the effective resistance.

- The effective (ac) resistance of a reactor can be found by using

$$R = \frac{P}{I^2}$$

- The symbol "R_O" represents the dc resistance or the true ohmic resistance in an ac circuit:

$$R_O = \frac{V_{coil}}{I}$$

$$W = V_{R_{coil}} \times I$$

- A silicon steel core placed in a coil instead of an iron core will greatly decrease the eddy current and hysteresis losses. As a result, the effective resistance of the reactor is decreased.
- In the high-frequency range, the effective resistance can be many times higher than the dc resistance.
- The vector resolution method yields a right triangle that can be solved using simple algebra. In this method, the vertical components should not be added algebraically to the horizontal components. The equation used with this method has the following form:

$$C^2 = A^2 + B^2$$

- The ohmic values of the reactor coil are determined as follows:
  1. Impedance of a coil:

$$Z = \frac{V}{I}$$

  2. Effective resistance of a coil:

$$R = \frac{P}{I^2}$$

  3. Inductive reactance of a coil:

$$X_L = \sqrt{Z^2 - R^2}$$

- Knowing the inductive reactance, the inductance can be found using

$$X_L = 2\pi fL$$

$$L = \frac{X_L}{2\pi f}$$

- The total circuit impedance is as follows

$$Z = \sqrt{R^2 + X_L^2}$$

or

$$Z = \frac{V}{I}$$

- The total true power for the entire series circuit can be obtained by adding the power expended in the resistor and the reactor

$$P_{total} = P_{resistor} + P_{coil}$$

or

$$P_{total} = I^2R_{total}$$

- The total apparent power in volt-amperes is the product of the line potential and the series circuit current:

$$VA = V \times I$$

- The power factor of the entire series circuit is determined as follows:
  1. The ratio of the total true power in watts to the total apparent power in volt-amperes:

$$PF = \frac{\text{watts}}{VA}$$

  2. The ratio of the total series circuit resistance to the total circuit impedance:

$$PF = \frac{R}{Z}$$

- The "law of cosines" formula is

$$a^2 = b^2 + c^2 - 2bc \cos \angle A$$

- The figure of merit (Q) of a coil is the ratio of the inductive reactance to the effective resistance of the coil

$$Q = \frac{X_L}{R}$$

or

$$Q = \frac{2\pi fL}{R}$$

or

$$Q = \frac{\omega L}{R}$$

Achievement Review

1. Determine the inductive reactance in ohms and the inductance in henrys for the coil shown in Figure 4–3.
2. Determine the inductive reactance in ohms and the inductance in henrys for the coil shown in Figure 4–4.
3. Explain why there are differences in the inductive reactance in ohms and the inductance in henrys in the results of questions 1 and 2, using the same coil.
4. Explain the difference between the true ohmic resistance and the effective resistance of ac components.
5. An electromagnet consists of a coil with a laminated core assembly. This device takes 6 A when connected to a 120-V dc source. When energized from a 120-V, 60-Hz ac source, the current is 2 A. A wattmeter indicates 100 W.
   a. Determine
      (1) the true ohmic resistance of the coil.
      (2) the effective resistance of the coil.
   b. Explain why there is a difference between the true ohmic resistance and the effective resistance values for the same coil winding.
6. Determine the following values for the electromagnet coil in question 5:
   a. The impedance
   b. The inductive reactance
   c. The inductance
   d. The power factor
7. Explain why the current taken by the electromagnet coil in question 5 decreased from 6 A at 120 V dc to a value of 2 A when connected to a 120-V, 60-Hz ac source.
8. A series circuit containing a noninductive resistance of 3 Ω and a coil is connected to a 120-V, 60-Hz supply. The current is 25 A. The true power, as indicated by a wattmeter, is 2500 W. Determine
   a. the power factor of the series circuit.
   b. the effective resistance of the coil.
   c. the impedance and inductive reactance for the coil.
   d. the power factor of the coil.
9. Determine the following values for the circuit given in question 8:
   a. The total circuit impedance
   b. The total resistance of the circuit

10. Determine the following values for the circuit given in question 8:
    a. The reactive power component for the coil
    b. The reactive power component for the entire series circuit
    c. The true power in watts taken by the coil
    d. The true power in watts taken by the noninductive 3-Ω resistance

11. Draw a vector diagram for the circuit in question 8. Select convenient scales for the current and voltage. Label all vectors properly.

12. The temperature of a noninductive heater unit is controlled by a reactor. The heater unit and the reactor are connected in series across a 250-V, 60-Hz source. The voltage across the heater unit is 200 V. The voltage across the reactor is 100 V. The current in the circuit is 10 A.
    a. Construct a vector diagram for the series circuit.
    b. Determine
        (1) the power factor of the series circuit.
        (2) the power factor of the reactor.
        (3) the impedance of the reactor.
        (4) the effective resistance of the reactor.
        (5) the inductive reactance of the reactor.

13. Determine the following data for the circuit given in question 12:
    a. The impedance of the entire series circuit
    b. The power expended in the heater unit, in watts
    c. The power expended in the coil, in watts
    d. The total power expended in the series circuit, in watts
    e. The total apparent power taken by the series circuit, in watts
    f. The reactive power component, in VARs, for the series circuit

14. Two reactors are connected in series across a 10-V, 10-kHz source. They are spaced so that there is no interaction between their electromagnetic fields. Coil 1 has an effective resistance of 200 Ω and an inductance of 200 millihenrys (mH). Coil 2 has an effective resistance of 300 Ω and an inductance of 10 mH. Determine
    a. the total resistance, in ohms.
    b. the total inductive reactance, in ohms.
    c. the total impedance, in ohms.
    d. the current, in milliamperes.
    e. the total true power expended in the series circuit, in watts.

15. Explain what is meant by the term *figure of merit*.

16. A coil has an inductance of 300 microhenrys (μH) and a figure of merit of 90 when operated at a frequency of 1500 kHz. Determine the effective resistance of the coil at this frequency.

PRACTICE PROBLEMS FOR UNIT 4

Resistive Inductive Series Circuits

Find the missing values for the following circuits. Refer to problem 3 and the circuit shown in Figure 4–13.

1. Assume that the circuit shown in Figure 4–13 is connected to a 480-V, 60-Hz line. The inductor has an inductance of 0.053 H, and the resistor has a resistance of 12 Ω.

V_T = 480 V	V_R = ________	V_L = ________
I_T = ________	I_R = ________	I_L = ________
Z = ________	R = 12 Ω	X_L = ________
VA = ________	P = ________	$VARs_L$ = ________
PF = ________	$\angle\theta$ = ________	L = 0.053 H

2. Assume that the voltage drop across the resistor (V_R) is 78 V, the voltage drop across the inductor (V_L) is 104 V, and the circuit has a total impedance (Z) of 20 Ω. The frequency of the ac voltage is 60 Hz.

V_T = ________	V_R = 78 V	V_L = 104 V
I_T = ________	I_R = ________	I_L = ________
Z = 20 Ω	R = ________	X_L = ________
VA = ________	P = ________	$VARs_L$ = ________
PF = ________	$\angle\theta$ = ________	L = ________

3. Assume that the circuit in Figure 4–13 has an apparent power of 144 volt-amperes and a true power of 115.2 W. The inductor has an inductance of 0.15915 H, and the frequency is 60 Hz.

V_T = ________	V_R = ________	V_L = ________
I_T = ________	I_R = ________	I_L = ________
Z = ________	R = ________	X_L = ________
VA = 144 volt-amperes	P = 115.2 W	$VARs_L$ = ________
PF = ________	$\angle\theta$ = ________	L = 0.15915 H

4. Assume that the circuit shown in Figure 4–13 has a power factor of 78%, an apparent power of 374.817 volt-amperes, and a frequency of 400 Hz. The inductor has an inductance of 0.0382 H.

V_T = ________	V_R = ________	V_L = ________
I_T = ________	I_R = ________	I_L = ________
Z = ________	R = ________	X_L = ________
VA = 374.817 volt-amperes	P = ________	$VARs_L$ = ________
PF = 78%	$\angle\theta$ = ________	L = 0.0382 H

5

Capacitors and RC Time Constants

Objectives

After studying this unit, the student should be able to

- explain how the atomic structure differs between materials that act as good conductors and those that act as good insulators.
- describe the elementary capacitor and the functions of each of its parts.
- explain how the insulating ability of the dielectric of a capacitor can break down, and indicate the extent of damage to the capacitor.
- define the following terms: capacitance, dielectric constant, dielectric strength rating, farad, microfarad, and picofarad.
- express as a formula the relationship between the capacitance, the charge in coulombs, and the voltage for a capacitor.
- explain how the total capacitance is determined for capacitors in series, in parallel, and in network systems.
- with the use of a voltmeter, demonstrate how a capacitor is charged and discharged.
- define the time constant for both an RC circuit and an RL circuit, and explain what is meant by "five time constants."
- plot an exponential curve used to determine the voltage across the plates of a capacitor; show the instantaneous values from the fully discharged state to the fully charged state of the capacitor.
- describe the construction of several common types of capacitors and give practical examples of the use of each.

INTRODUCTION

Direct Current Fundamentals explained how a material is a conductor or an insulator because of its atomic structure. It was stated that electrons are loosely held in the outer electron shells of each atom of a good conductor such as copper. Only a small force is required to dislodge these electrons. Good conductors have many free electrons in their structure.

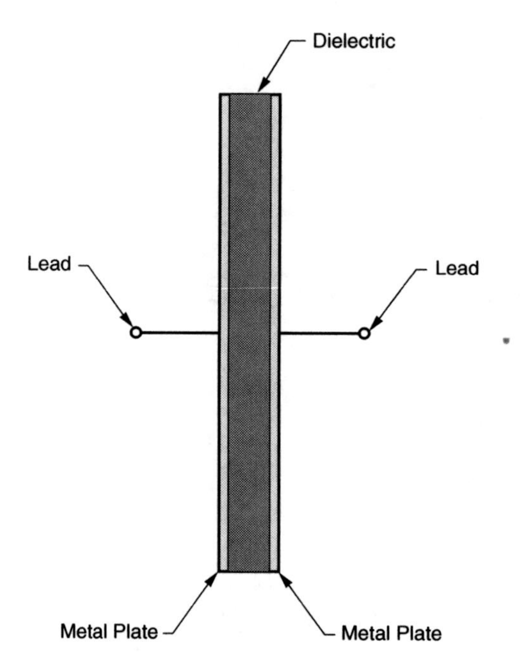

FIGURE 5–1 Simple capacitor (*Delmar/ Cengage Learning*)

A characteristic of insulating materials is that their electrons are firmly held in the electron shells of each atom of the material. A great deal of force is required to remove these electrons. Insulating materials have few free electrons in their structure.

The Elementary Capacitor

If an insulating material, known as a *dielectric*, is placed between two plates of a conducting material, the resulting assembly is an elementary capacitor.

In Figure 5–1, a simple capacitor is shown to consist of two metal plates separated from each other by a thickness of dielectric. Under normal conditions, with the capacitor deenergized, the electrons in the dielectric revolve in circular orbits around the nucleus of each atom.

Charging the Capacitor. The capacitor is shown connected to a dc voltage source in Figure 5–2. Electrons will flow from the negative side of the source to plate 2 and from plate 1 back to the source. This movement of electrons from negative to positive is the normal direction of electron flow.

Electrons will continue to move until the potential difference across the two metal plates is equal to the dc source voltage. When the potential and the voltage are the same, the flow will stop. There will be almost no flow of electrons through the dielectric (insulating) material between the plates. There will be a surplus of electrons on plate 2 and a deficiency of electrons on plate 1. The electrons in the atoms of the dielectric material will be attracted toward the positive plate and repelled from the negative plate. However, these electrons cannot flow from plate 1 to plate 2 because the electrons in a good dielectric material are firmly held in each atom.

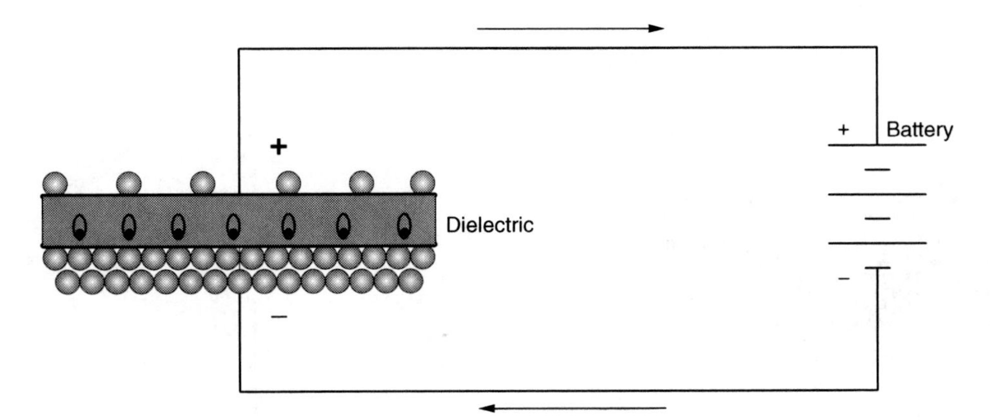

FIGURE 5–2 A capacitor can be charged by removing electrons from one plate and depositing electrons on the other. (*Delmar/Cengage Learning*)

Thus, the forces acting on the electrons cause their orbits in each atom in the dielectric to be distorted. As shown in Figure 5–3, the orbits form elliptical patterns.

The capacitor in Figure 5–3 is completely charged. The voltage across the capacitor plates is equal to the dc source voltage. Therefore, there is no electron flow. To simplify the figures, only three atoms are shown. In an actual capacitor, there are millions of atoms with distorted orbits in the dielectric.

If the capacitor is disconnected from the dc supply, the surplus electrons on the negative plate are held to this plate by the attraction of the positive charge of the other plate. In other words, an electrostatic field effect is created by the charged plates. This effect maintains the distortion of the electron orbits in the atoms of the dielectric. The atomic distortion of the dielectric is an indication of the electrical energy stored in the capacitor.

If the voltage applied to the capacitor is too large, the electrons in the atoms of the dielectric will be pulled from their orbits. The insulating ability of the dielectric breaks down, and the energy stored in the capacitor is released. In the case of a solid dielectric material, such a breakdown destroys its usefulness and a shorted capacitor results.

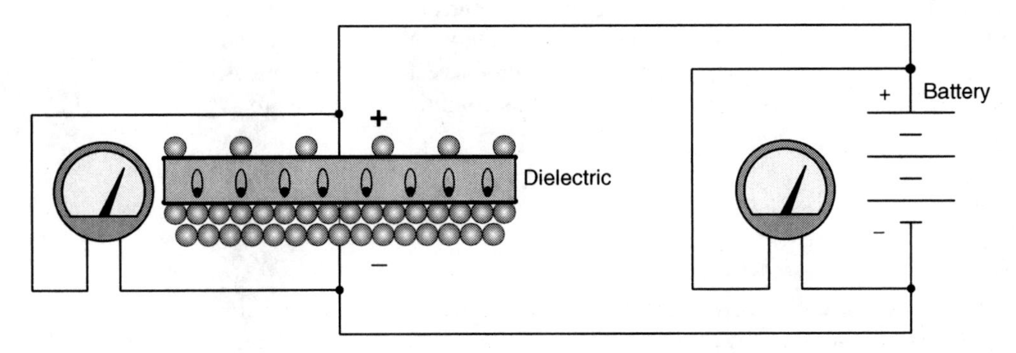

FIGURE 5–3 Current flows until the voltage across the capacitor is equal to the voltage across the source. (*Delmar/Cengage Learning*)

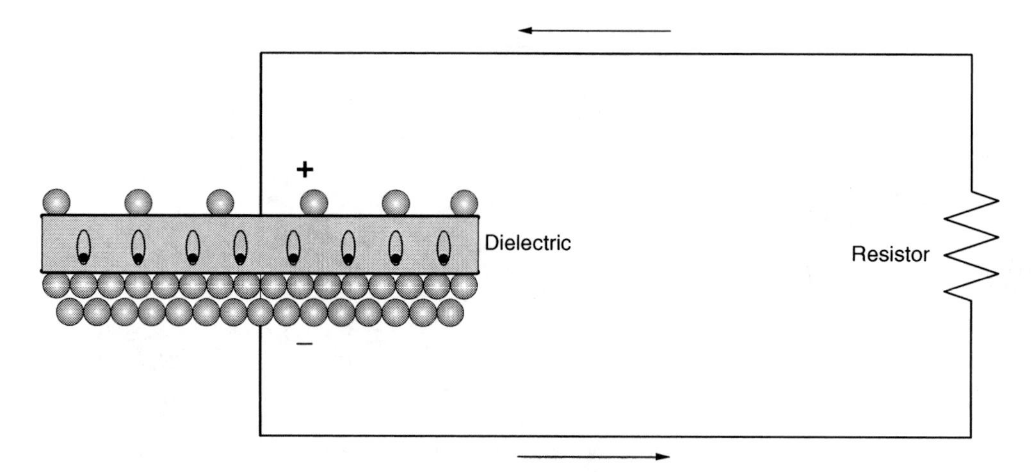

FIGURE 5–4 Capacitor being discharged (*Delmar/Cengage Learning*)

Discharging the Capacitor. A resistor is now connected across the capacitor as shown in Figure 5–4. The capacitor discharges through the conducting resistor. Electrons on the negatively charged plate (plate 2) will leave the plate and will flow toward plate 1 through the resistor. This electron flow continues until they are distributed equally in the circuit. Then the flow of electrons stops. While the electrons flow, the electrical energy stored in the electrostatic field is being released from the dielectric of the capacitor.

As the electrons flow from the negatively charged plate, a change takes place in the electron orbits of each atom of the dielectric. That is, the orbits change from the distorted elliptical pattern (Figure 5–4) to their normal circular pattern (Figure 5–5).

Leakage Current

In theory, it should be possible for a capacitor to remain charged forever. In actual practice, however, it cannot. No dielectric is a perfect insulator, and over a period of time electrons will eventually move through the dielectric from the negative plate to the positive plate, causing the capacitor to discharge (Figure 5–6). This current flow through the dielectric is called *leakage current* and is proportional to the resistance of the dielectric and the charge across the plates. If the dielectric of a capacitor becomes weak, it will permit an excessive amount of leakage current to flow. A capacitor in this condition is referred to as a *leaky capacitor.*

CAPACITANCE

The Meaning of Capacitance

A capacitor can store electrical energy. It can also return this energy to an electric circuit. It is important to understand what the term *capacitance* actually means. *Capacitance* is the property of a circuit, or circuit component, that allows it to store electrical energy in electrostatic form.

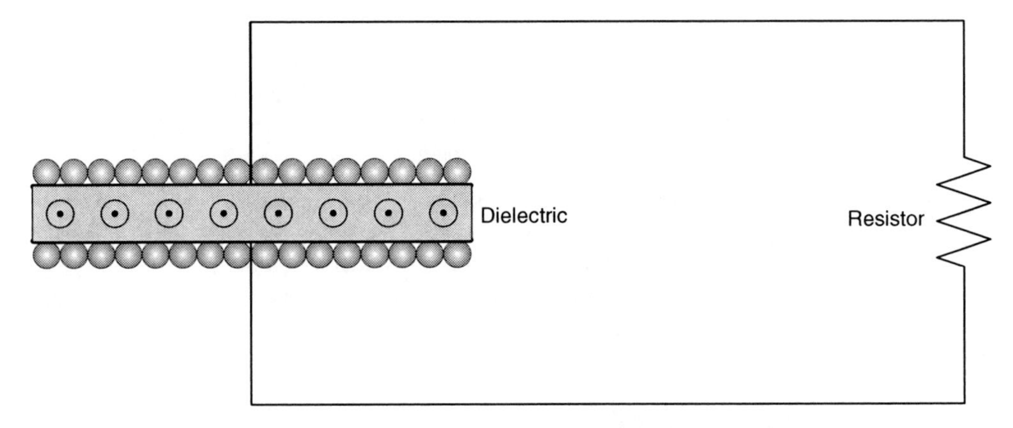

FIGURE 5–5 Capacitor completely discharged (*Delmar/Cengage Learning*)

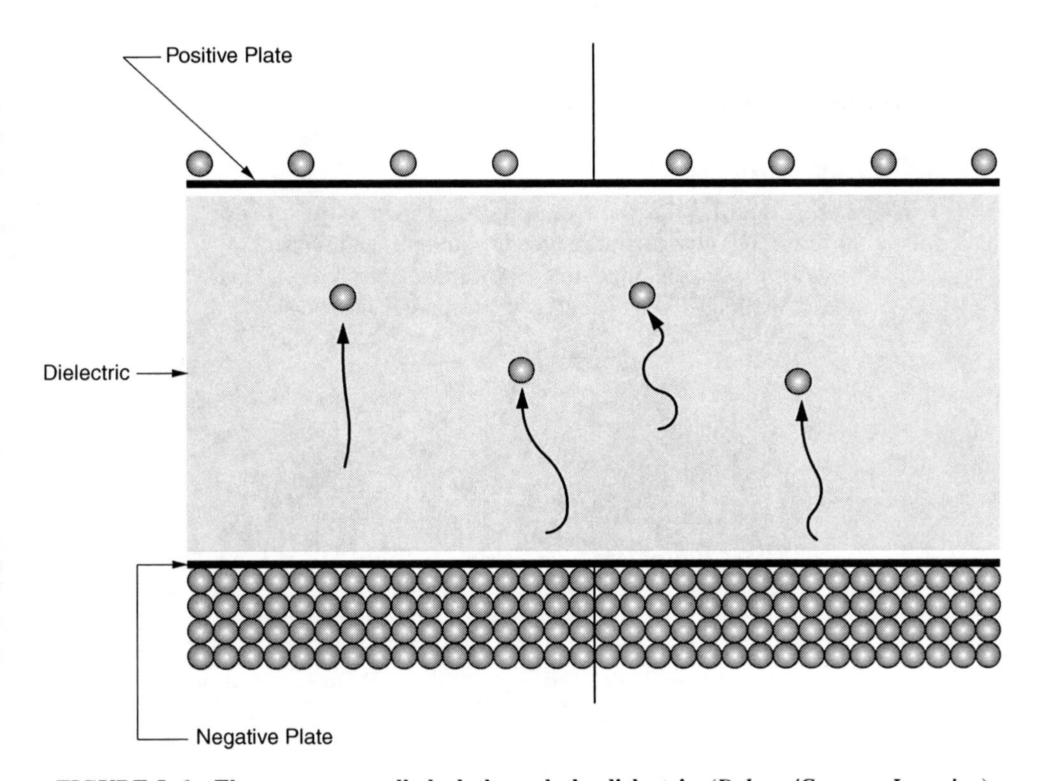

FIGURE 5–6 Electrons eventually leak through the dielectric. (*Delmar/Cengage Learning*)

The Measurement of Capacitance

Circuit components other than capacitors may create a capacitance effect. For example, when two wires of a circuit are separated by air, they will act as a capacitor. Also, adjacent turns of a coil winding, which are separated only by the insulation of the wire, will have some capacitance effect. The standard unit of measurement for capacitance is the farad. This unit of measurement is defined as follows:

A capacitor has a capacitance of one farad when a change of one volt across its plates results in a charge movement of one coulomb.

The farad is too large a unit for a typical capacitor. A smaller unit, known as a *microfarad*, is commonly used. The microfarad is equal to 1 ÷ 1,000,000, or 10^{-6} farad. The letter symbol for the farad is F. The microfarad is indicated by the symbol μF, where μ is the Greek letter *mu*.

Electronic circuits often require very small capacitors. In such cases, even the microfarad is too large. A smaller unit of capacitance measurement is required. This unit is the *picofarad*. One picofarad is equal to 1 ÷ 1,000,000th of a microfarad, or 10^{-12} farad. The letter symbol for the picofarad is pF.

The capacitance of a capacitor can be increased by any of the following factors:

- Increasing the plate area, resulting in an increase in the area of the dielectric under stress
- Moving the metal plates closer together, resulting in a decrease in the thickness of the dielectric
- Using a dielectric with a higher dielectric constant

The Charge of a Capacitor

For a given applied voltage, the charge on the plates of a capacitor is directly proportional to the capacitance of the capacitor. The charge is measured in coulombs and is directly proportional to the charging voltage. If the charge on the plates is directly proportional to both the capacitance and the impressed voltage, then the charge is expressed as follows:

$$Q = C \times V$$

where

Q = quantity of the charge, in coulombs
C = capacitance, in farads
V = charging potential, in volts

This expression may be written in three forms:

$$\mathbf{Q = C \times V}$$

$$\mathbf{C = \frac{Q}{V}}$$

$$\mathbf{V = \frac{Q}{C}}$$

As an example of the use of this formula, assume that a capacitor takes a charge of 0.005 coulomb (C) when connected across a 100-V dc source. Determine the capacitance of the capacitor in microfarads:

$$C = \frac{Q}{V}$$

$$= \frac{0.005}{100} = 0.00005 \text{ F} = 50\ \mu\text{F}$$

DIELECTRIC CHARACTERISTICS

Three factors were shown to affect the capacitance of a capacitor. One factor is the type of insulating material used for the dielectric. Most capacitors are constructed using a dielectric material having a higher dielectric constant than air.

The *dielectric constant* of an insulating material measures the effectiveness of the material when it is used as the dielectric of a capacitor. It is assumed that air has a dielectric constant of one. If a two-plate capacitor has a dielectric consisting of paper impregnated

Material	Dielectric Constant (K)
Air (1 atm) Air (100 atm)	1.00059 1.0548
Bakelite	4.0–10.0
Benzene	2.284
Castor oil	4.3–4.7
Cellulose acetate	7.0
Germanium	16
Hard rubber	2.0–4.2
Insulating oils	2.2–4.6
Lucite	2.4–3.0
Liquid ammonia (–78°C)	25
Mica	6.4–7
Mylar	3.1
Neoprene	6.7
Paper	2.0–2.6
Paraffin	1.9–2.2
Plexiglas	3.4
Polyethylene	2.25
Polyvinyl chloride	3.18
Pyrex glass	4.1–4.9
Rubber compounds	3.0–7.0
Strontium titanate	310
Vacuum	1
Water (distilled)	80.4

TABLE 5–1 Dielectric constants for some insulating materials

with paraffin, rather than a dielectric of air, the capacitance will increase. If the capacitance is doubled when using paper in place of air for the dielectric, the dielectric constant for paper is 2. This value indicates the degree of distortion of the electron orbits in the dielectric for a given applied voltage. Table 5–1 lists the dielectric constants of some insulating materials.

Dielectric Strength

If the voltage across the plates of a capacitor becomes too high, the dielectric may be burned or punctured by the high potential. That is, the high potential tears electrons from the orbits of the atoms of the dielectric material. As a result, the dielectric becomes a conducting material, resulting in permanent damage to the dielectric.

Dielectric materials are given a *dielectric strength rating*. This rating is stated as either "volts per centimeter" or "volts per mil" of thickness required to break down the dielectric. The dielectric strength rating is *not* the same as the dielectric constant rating. For example, the dielectric constant of paper is about 2 and that of Pyrex glass is approximately 4. However, the dielectric strength in volts per mil for some kinds of paper is about 1200 V. For Pyrex glass, the dielectric strength is only 325 V per mil.

CAPACITANCE FORMULAS

The capacitance of a two-plate capacitor is directly proportional to the area of one plate and inversely proportional to the distance of separation between the plates.

This statement can be expressed as a formula. Actually, there are two forms of the formula, depending on the units of measurement for the distances involved. To determine the capacitance in picofarads when the plate dimensions and the distance between the plates are given in inches, the formula is

$$C = \frac{K \times A}{4.45\ D}$$

where

C = capacitance, in pF
K = dielectric constant
A = area of one plate, in square inches
D = separation distance between the plates, in inches

When the plate dimensions and the distance between the plates are given in centimeters, the formula for the capacitance is

$$C = \frac{0.0885 \times K \times A}{D}$$

where

C = capacitance, in pF
K = dielectric constant
A = area of one plate, in cm^2
D = separation distance between plates, in cm

PROBLEM 1

Statement of the Problem

A paper capacitor consists of two tinfoil plates, each 8 feet (ft) long and 1 inch (in.) wide. The waxed paper that separates the two plates has a thickness of 0.05 in. Determine the capacitance of the capacitor. The dielectric constant of waxed paper is 2.

Solution

$$C = \frac{K \times A}{4.45\ D}$$

$$= \frac{2 \times 96}{4.45 \times 0.05} = \frac{192}{0.2225} = 863\ \text{pF}$$

If the dimensions of the plates and the distance between the plates are converted to centimeters, then the second formula can be used to find the same answer.

Converting inches to centimeters (cm) yields the following values:

- The length of one tinfoil plate is
  96 in. × 2.54 cm/in. = 243.84 cm
- The width of one tinfoil plate is
  1 in. × 2.54 cm/in. = 2.54 cm
- The area of one tinfoil plate is
  243.84 cm × 2.54 cm = 619.4 cm^2
- The distance between the plates is
  0.05 in. × 2.54 cm/in. = 0.127 cm

Substituting these values in the formula gives

$$C = \frac{0.0885 \times K \times A}{D}$$

$$= \frac{0.0885 \times 2 \times 619.4}{0.127}$$

$$= \frac{109.6338}{0.127}$$

$$= 863\ \text{pF}$$

CAPACITORS IN PARALLEL

In addition to increasing the plate area, an increase in capacitance can be obtained by increasing the number of plates in the capacitor.

The multiple plates of the capacitor shown in Figure 5–7 are placed so that a maximum plate area is obtained. Note that alternate plates are connected in parallel. To calculate

FIGURE 5–7 Multiplate capacitor (*Delmar/Cengage Learning*)

the capacitance of a multiplate capacitor, where the plates are identical and are separated by the same distance, the following expression is used:

$$C = \frac{0.0885\ KA\ (N - 1)}{D}$$

where

C = capacitance, in pF
K = dielectric constant
A = area of one plate, in cm^2
D = separation distance between plates, in cm
N = number of plates

When capacitors are connected in parallel, the effect is the same as increasing the number of plates. In other words, the total capacitance is the sum of the capacitance of each capacitor. For example, in Figure 5–8, three capacitors have the values 30 μF, 10 μF, and 15 μF. These capacitors are connected in parallel across the line voltage (V). The charge on each capacitor in coulombs is

$$Q_1 = C_1 \times V$$
$$Q_2 = C_2 \times V$$
$$Q_3 = C_3 \times V$$

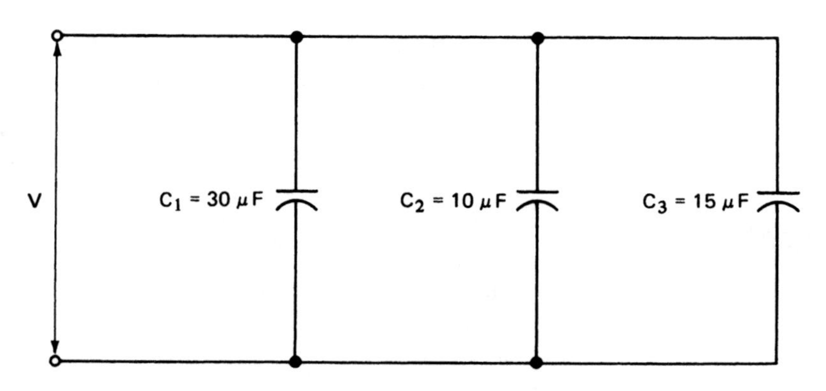

FIGURE 5–8 Capacitors in parallel (*Delmar/Cengage Learning*)

The total charge of the three capacitors in parallel is

$$Q_{total} = C_{total} \times V = Q_1 + Q_2 + Q_3$$

or

$$C_{total} \times V = C_1V + C_2V + C_3V$$
$$= V(C_1 + C_2 + C_3)$$

Therefore,

$$C_{total} = C_1 + C_2 + C_3$$

The total capacitance for the three capacitors in parallel in Figure 5–8 is

$$C_{total} = C_1 + C_2 + C_3 = 30 + 10 + 15 = 55\ \mu F$$

CAPACITORS IN SERIES

When capacitors are connected in series, the dielectrics of the individual capacitors are connected one after the other to form a single circuit path. This arrangement is equivalent to increasing the thickness of the dielectric of one capacitor. This means that the total capacitance of the circuit is less than the capacitance of any individual capacitor.

When capacitors are charged in a series circuit, the same number of electrons flow to each capacitor. As a result, each capacitor has the same charge (Q) in coulombs.

For example, Figure 5–9 shows three capacitors connected in series across a line voltage (V). The capacitors have the same ratings as in Figure 5–8 for a parallel circuit. It can be shown that the total capacitance of the capacitors connected in series is less than the capacitance of any one capacitor.

The voltage across the series circuit is

$$V = V_1 + V_2 + V_3$$

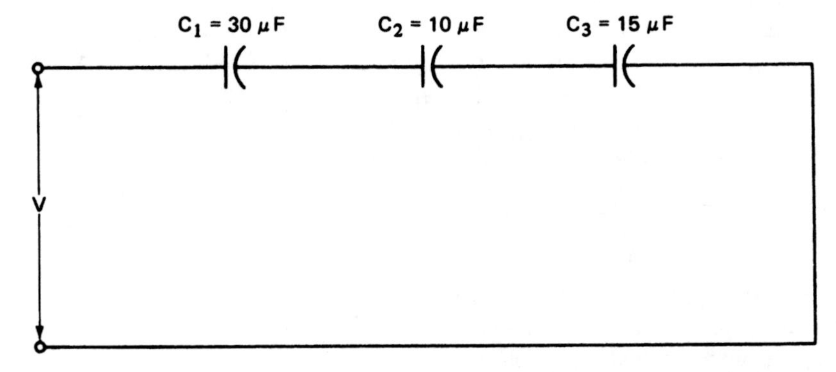

FIGURE 5–9 Capacitors in series (*Delmar/Cengage Learning*)

The charge in coulombs on each capacitor is the same:

$$Q_1 = Q_2 = Q_3 = Q$$

Therefore,

$$\frac{Q}{C_{total}} = \frac{Q}{C_1} + \frac{Q}{C_2} + \frac{Q}{C_3}$$

$$\frac{1}{C_{total}} = \frac{1}{C_1} + \frac{1}{C_2} + \frac{1}{C_3}$$

and

$$C_{total} = \frac{1}{\frac{1}{C_1} + \frac{1}{C_2} + \frac{1}{C_3}}$$

Substituting the values given in Figure 5–9 in the previous expression yields the total capacitance of the series capacitor bank:

$$C_T = \frac{1}{\frac{1}{C_1} + \frac{1}{C_2} + \frac{1}{C_3}}$$

$$= \frac{1}{\frac{1}{30} + \frac{1}{10} + \frac{1}{15}} = \frac{1}{\frac{6}{30}} = \frac{30}{6} = 5\ \mu F$$

In a parallel circuit, the total capacitance is

$$C_T = C_1 + C_2 + C_3 + \text{etc.}$$

In a series circuit, the total capacitance is

$$C_T = \frac{1}{\frac{1}{C_1} + \frac{1}{C_2} + \frac{1}{C_3} + \text{etc.}}$$

The formulas for total capacitance can be used only when all of the values are in the same unit of capacitance measurement, either microfarads or picofarads. If some of the capacitors in a circuit are rated in microfarads and the rest are rated in picofarads, then some values must be changed so that all of the values are expressed in the same unit of measurement before the total capacitance is found.

The total capacitance can also be determined using the product-over-sum formula:

$$C_T = \frac{C_1 \times C_2}{C_1 + C_2}$$

In the example above, three capacitors having values of 30 μF, 10 μF, and 15 μF are connected in series. The total capacitance can be determined by substituting two values at a time and then using that answer with the next capacitance value:

$$C_T = \frac{30 \times 10}{30 + 10} = \frac{300}{40} = 7.5\ \mu F$$

Use the total value of these two capacitors in the formula with the next capacitance value:

$$C_T = \frac{7.5 \times 15}{7.5 + 15} = \frac{112.5}{22.5} = 5\ \mu F$$

This procedure can be followed until all series capacitance values have been employed.

A third formula for determining the total capacitance of capacitors connected in series can be used only when all the capacitance values are the same. Assume that four capacitors, each having a value of 60 μF, are connected in series. The total capacitance can be determined by dividing the capacitance value of one capacitor by the number of capacitors.

$$C_T = \frac{C}{N} = \frac{60}{4} = 15\ \mu F$$

CAPACITORS IN NETWORK SYSTEMS

Capacitors can also be connected in network systems where the total capacitance must be determined. Figure 5–10 shows two capacitor network systems. To obtain the total capacitance of the capacitor network system in Figure 5–10A, the following procedure is used:

1. Total capacitance for branch 1:

$$\frac{1}{\frac{1}{C_1} + \frac{1}{C_2}} = \frac{1}{\frac{1}{100} + \frac{1}{100}} = \frac{100}{2} = 50\ \mu F$$

2. Total capacitance for branch 2:

$$\frac{1}{\frac{1}{C_1} + \frac{1}{C_4}} = \frac{1}{\frac{1}{20} + \frac{1}{30}} = \frac{60}{5} = 12\ \mu F$$

3. Total capacitance of both branches in parallel:

$$C_T = 50 + 12 = 62\ \mu F$$

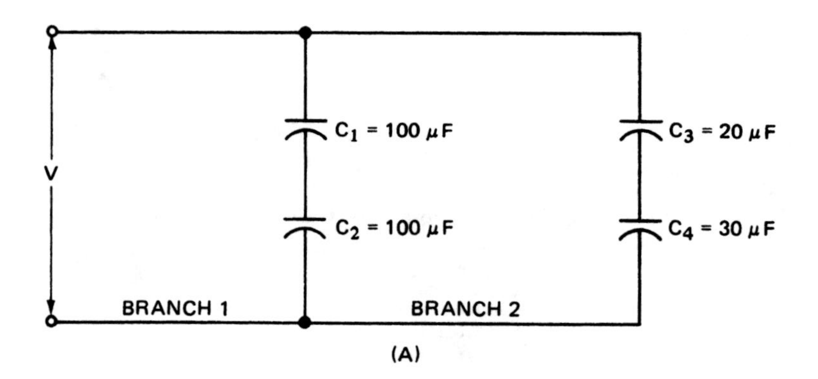

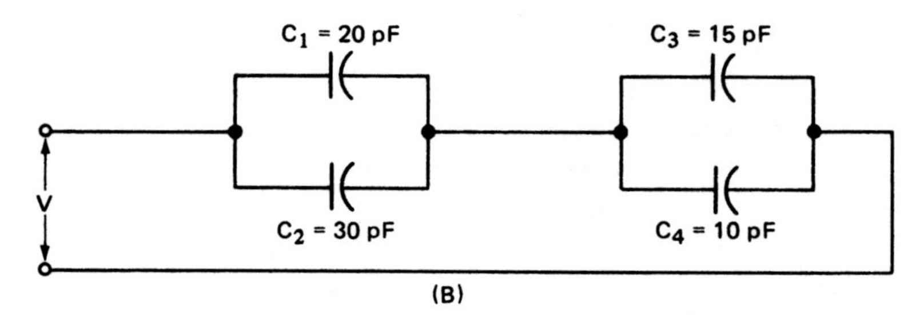

FIGURE 5–10 Capacitor network systems (*Delmar/Cengage Learning*)

A second capacitor network system is shown in Figure 5–10B. The ratings of the capacitors in this system are given in picofarads. For this system, the total capacitance is found by using the following procedure:

1. Total capacitance of C_1 and C_2 in parallel:

$$C_1 + C_2 = 20 + 30 = 50 \text{ pF}$$

2. Total capacitance of C_3 and C_4 in parallel:

$$C_3 + C_4 = 15 + 10 = 25 \text{ pF}$$

3. Total capacitance of the network system consisting of two parallel sections connected in series:

$$\frac{1}{\frac{1}{50} + \frac{1}{25}} = \frac{1}{\frac{1+2}{50}} = \frac{1}{\frac{3}{50}} = \frac{50}{3} = 16.67 \text{ pF}$$

ENERGY IN A CAPACITOR

It was shown at the beginning of this unit that a capacitor will store electrical energy. The amount of energy stored in a capacitor, in joules or watt-seconds, can be determined as follows:

$$\textbf{Watt-seconds} = \frac{1}{2}\mathbf{QV}$$

Thus, one-half the charge in coulombs times the charging potential in volts yields the energy in joules stored in the capacitor. Recall that $Q = C \times V$. Therefore, $C \times V$ may be substituted for Q in the previous formula with the result that

$$\textbf{Watt-seconds} = \frac{1}{2}\mathbf{C \times V \times V}$$

$$= \frac{1}{2}\mathbf{CV^2}$$

For example, a 50-μF capacitor is connected across a 300-V dc source. It is required to find the energy in watt-seconds stored in this capacitor:

$$\text{Watt-seconds} = \frac{1}{2}CV^2$$

$$= \frac{1}{2} \times 0.00005 \times 300^2 = 2.25 \text{ joules (J)}$$

RC TIME CONSTANTS

Direct Current Fundamentals showed that the intensity of electron flow per second in amperes is expressed by amperes = coulombs divided by seconds ($I = Q/t$). Also, the quantity of electrons in coulombs is given by coulombs = amperes times seconds ($Q = I \times t$). For a capacitor, the charge in coulombs on the plates is a function of the intensity of current (amperes) and the time (seconds) of charging. As the value of the current increases, less time is required to charge the capacitor fully.

Recall that the voltage across the plates of a capacitor equals the line voltage when the capacitor is fully charged. The charging time for a given capacitor depends on the charging current. Thus, the time required for a given voltage to build up across the capacitor can be controlled by increasing or decreasing the charging current. The charging current can be controlled by placing a series assembly of a resistor and a capacitor across the charging potential. In other words, this current controls the time that must elapse before a given voltage exists across the capacitor terminals.

Charging the Capacitor

The circuit shown in Figure 5–11A consists of a 50-μF capacitor connected in series with a resistor across a 300-V dc supply. In Figure 5–11B, the resistor consists of a dc voltmeter having a scale range of 300 V and a resistance of 1000 Ω/V. Thus, the resistance of this voltmeter is 300,000 Ω. At the instant the circuit is energized, the voltmeter needle swings to a full-scale reading. Such a value indicates that all of the line voltage is across the terminals of the voltmeter. (This voltmeter acts like a 300,000-Ω series resistor.) The voltage across the capacitor at this same instant is zero.

The equation for the voltage in the series circuit in Figure 5–11B is

$$\mathbf{V = V_C + V_V}$$

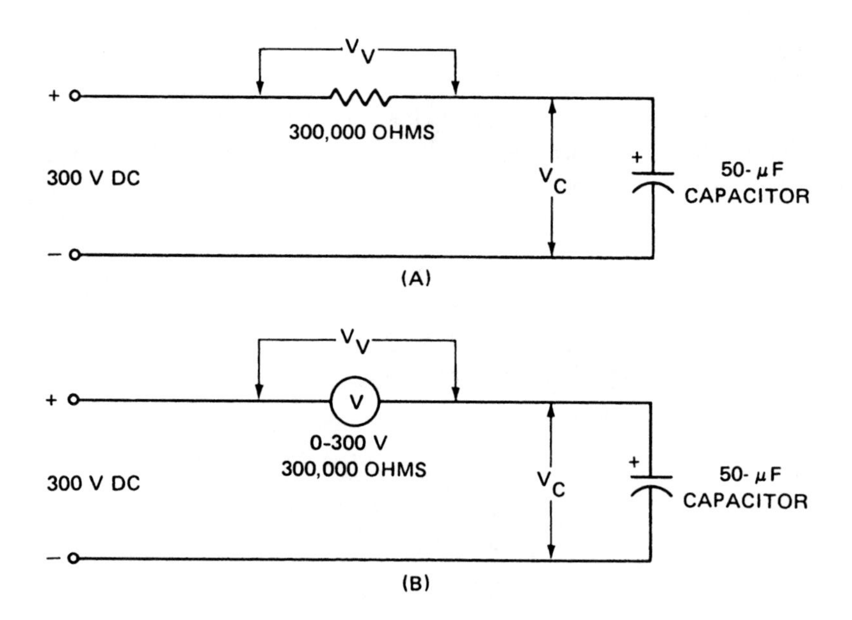

FIGURE 5–11 RC circuit with capacitor charging (*Delmar/Cengage Learning*)

where V_C is the voltage across the capacitor and V_V is the voltage across the voltmeter. The applied voltage (V) is constant. This means that as the voltage across the capacitor (V_C) increases, the voltage across the voltmeter (V_V) must decrease. As the voltage across the capacitor approaches the applied voltage (V), the current in the circuit approaches zero. In effect, the voltage across the capacitor is a backward voltage that reduces the current flow. As the current approaches zero, the voltage across the voltmeter approaches zero.

Plotting the Charging of the Capacitor. A curve can be plotted of the rising voltage across the capacitor plates. For the circuit of Figure 5–11B, voltmeter readings are taken at 15-s intervals. In each case, the reading is subtracted from the fixed line voltage (V). As a result, the voltage across the plates of the capacitor can be determined as the capacitor charges. Table 5–2 lists the voltage across the capacitor plates at 15-s intervals, as obtained by this method. At zero seconds, note that the capacitor is completely discharged. At the end of 75 s, the voltage across the capacitor plates has almost reached the value of the line voltage.

Drawing the Exponential Charging Curve. The data in Table 5–2 are plotted in Figure 5–12 to form a curve of the rising voltage across the capacitor plates for the interval of 75 s.

Note in the figure that the horizontal axis of the curve is marked both in time constants and time in seconds. The term *time constant* is defined as follows:

One time constant is the time, in seconds, required for a completely discharged capacitor to charge to 63.2% of the source voltage. This charging time is equal to the product of the resistance in ohms and the capacitance in farads:

Time in Seconds	Voltmeter Reading	Voltage across Capacitor Plates
0	300	300 − 300 = 0
15	110.4	300 − 110.4 = 189.6
30	40.5	300 − 40.5 = 259.5
45	15	300 − 15 = 285
60	5.4	300 − 5.4 = 294.6
75	2.1	300 − 2.1 = 297.9

TABLE 5–2 The charging voltage of a capacitor as a function of time

$$\tau = \mathbf{R} \times \mathbf{C}$$

The letter symbol for a time constant is the Greek letter tau (τ).

The time constant for the circuit shown in Figure 5–11 is

$$\begin{aligned} \tau &= R \times C \\ &= 300{,}000 \times 0.00005 = 15 \text{ seconds} \end{aligned}$$

Refer to Figure 5–12 and note that the 15-s mark on the horizontal scale corresponds to the mark for one time constant. From the definition of a time constant, the voltage across the capacitor plates in an RC series circuit is 63.2% of the impressed voltage at the end of one time constant. For this circuit, then, the voltage across the capacitor plates at the end of one time constant is 300 × 0.632 = 189 V.

Formula for an Exponential Curve. The curve shown in Figure 5–12 is known as an *exponential curve*. The voltage curve for a series RC circuit is expressed by the following mathematical formula

$$\mathbf{V_C = V(1 - \epsilon^{-\tau/RC})}$$

where

V_C = voltage across the capacitor plates, in volts
V = impressed or line potential, in volts
ϵ = Greek letter *epsilon*; represents the base of the Naperian system of logarithms, which is 2.718

The quantity $\mathbf{(1 - \epsilon^{-\tau/RC})}$ is called an *exponential operator* and is considered to be an operator on the voltage (V).

The formula for V_C may be used to find the percentage of the impressed voltage on the capacitor plates at the end of one time constant, two time constants, or any desired number of time constants. For example, the percentage of the impressed voltage on the capacitor

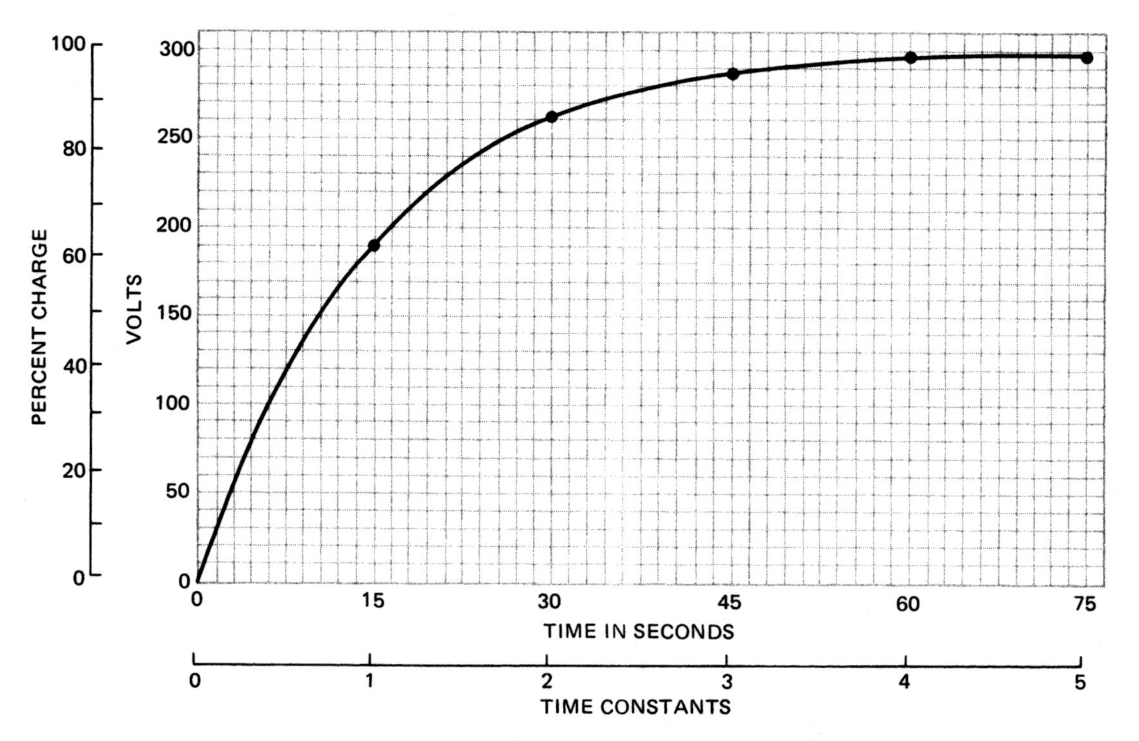

FIGURE 5–12 Exponential voltage curve for capacitor on charge (*Delmar/Cengage Learning*)

plates at the end of one time constant for the circuit illustrated graphically in Figure 5–12 is determined as follows:

$$\mathbf{V_C = V(1 - \epsilon^{-\tau/RC})}$$

When $\tau = \text{RC}$, then $\tau/\text{RC} = 1$.

Therefore,

$$V_C = V(1 - 2.718^{-1})$$

In algebra, a number with an exponent of minus one is the same as the reciprocal of that number. This means that the number 2.718^{-1} can be written as $1 \div 2.718$. Then

$$V_C = V\left(1 - \frac{1}{2.718}\right)$$

$$= V(1 - 0.37) = 0.63\ V$$

This calculation shows that the voltage across the capacitor plates at the end of one time constant is equal to 63.2% of the impressed voltage. For the RC series circuit of Figure 5–11, the actual voltage on the capacitor plates at the end of one time constant is

$$V_C = V\left(1 - \frac{1}{2.718}\right)$$
$$= 300(0.632) = 189 \text{ V}$$

Recall that this same value was determined graphically in Figure 5–12. To find the percentage of the impressed voltage across the capacitor plates at the end of two time constants, the calculations are as follows:

$$\mathbf{V_C = V(1 - \epsilon^{-\tau/RC})}$$
$$= V(1 - \epsilon^{-2})$$

In this problem, the exponent $\frac{\tau}{RC}$ is equal to two (two time constants are involved). A number with a negative exponent is the same as the reciprocal of that number to the given power. Because ϵ^{-2} is the same as $1 \div 2.718^2$, we obtain

$$V_C = V\left(1 - \frac{1}{2.718^2}\right)$$
$$= V(1 - 0.135) = V \times 0.865 = 0.865 \text{ V}$$

For the RC series circuit of Figure 5–11, the voltage across the capacitor plates at the end of two time constants is

$$V_C = 0.865 \text{ V}$$
$$= 0.865 \times 300 = 259.5 \text{ V}$$

These calculations can be repeated for any number of time constants. The resulting percentages are given in Table 5–3 for five time constants.

Based on the information given in Table 5–3, it is assumed that the capacitor is fully charged at the end of five time constants.

Discharging the Capacitor

When a capacitor is discharged through a resistor, the amount of time required depends on the value of the resistor. The 50-μF capacitor shown in Figure 5–13 is assumed to be completely charged at 300 V. This capacitor is to be discharged through the 300,000-Ω resistance of a voltmeter.

As a capacitor discharges through a resistor during one time constant, the voltage across the capacitor decreases to 37% of its value when the capacitor was fully charged. To determine capacitor voltages on discharge, the following exponential formula is used

$$\mathbf{v_C = V_C\,(\epsilon^{-\tau/RC})}$$

where

v_C = capacitor voltage for a given number of time constants, in volts
V_C = capacitor voltage when the capacitor is fully charged, in volts
ϵ = base of the Naperian system of logarithms, equal to 2.718

Number of Time Constants	Percentage of Impressed Voltage (%)
1	63.2
2	86.5
3	95.0
4	98.2
5	99.3

TABLE 5–3 Percentage of impressed voltage across a capacitor

When τ is one time constant, $\tau = R \times C$ and

$$v_C = V_C(2.718^{-1})$$
$$= V_C\left(\frac{1}{2.718}\right)$$
$$= V_C \times 0.37 = 0.37\ V_C$$

Plotting the Voltage Curve for a Capacitor on Discharge. Assume that the fully charged capacitor has a potential difference of 300 V across its terminals. When the capacitor is discharged, the voltage across the terminals at the end of one time constant is

$$v_C = 0.37\ V_C$$
$$= 0.37 \times 300 = 111\ V$$

This calculation can be repeated for any number of time constants. The results are summarized in Table 5–4, which lists the percentage of voltage on the plates of a discharging capacitor for each of five time constants.

By plotting the values given in Table 5–4, an exponential voltage curve is obtained for the discharge of the 50-μF capacitor connected across the 300,000-Ω voltmeter (Figure 5–14). The voltmeter serves as the discharge resistor. It also measures the decreasing voltage across the capacitor plates. The curve shows that the capacitor voltage decreases from 300 V

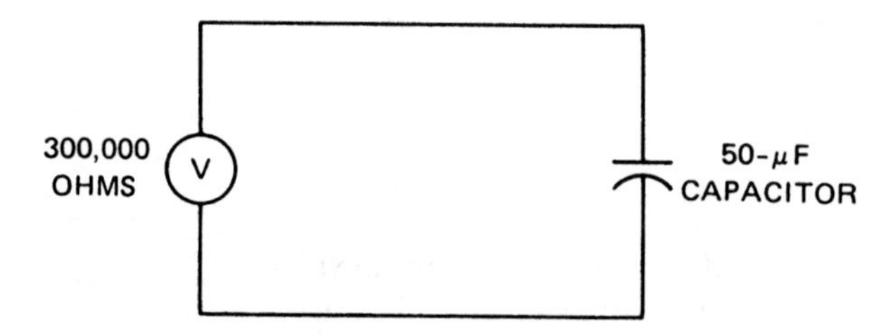

FIGURE 5–13 RC circuit with capacitor discharging (*Delmar/Cengage Learning*)

Number of Time Constants	Percentage of Fully Charged Potential (%)
1	37.0
2	13.5
3	5.0
4	1.8
5	0.7

TABLE 5–4 Percentage voltage across a discharging capacitor as a function of the number of time constants

to a value near zero. Table 5–5 shows the percentage of voltage and the actual voltage during the capacitor discharge for each of five time constants.

Current Relationships of a Capacitor

The study of RC circuits to this point has been based on the voltage relationships across the terminals of a capacitor. However, it is also important to study the *current* relationships for a capacitor when it is charging and discharging.

When a completely discharged capacitor is connected in series with a current-limiting resistor across a dc source, the initial current at the instant the circuit is energized is determined by $I = V \div R$. The capacitor dielectric offers no opposition to the current when the capacitor is discharged. As the charge on the capacitor plates increases, a voltage develops.

This voltage increases with time, as shown in Figure 5–12, until its value is equal to the line voltage. As the voltage increases, the current decreases following an exponential curve. When the voltage across the capacitor plates equals the line voltage, the charging current becomes zero.

If the same capacitor is discharged through a resistor, the direction of electron flow is reversed. At the instant the fully charged capacitor is connected to the resistor, the initial current is equal to $V \div R$. As the voltage across the capacitor terminals decreases following an exponential curve, the current also decreases in a similar manner.

Charge and Discharge Curves. Figure 5–15 shows the charge and discharge curves for current and voltage for an RC circuit. During charging, the charge current decreases as the potential across the capacitor plates increases. Both curves follow an exponential pattern. When the capacitor is discharged, the current decreases exponentially from its initial value to zero. The curve for the current on discharge is below the zero reference line defined by Ohm's law. This location means that the direction of current in the circuit is reversed when the capacitor is discharging. Note that the discharging voltage curve looks like the charging current curve. Likewise, the discharging current curve looks like the charging voltage curve.

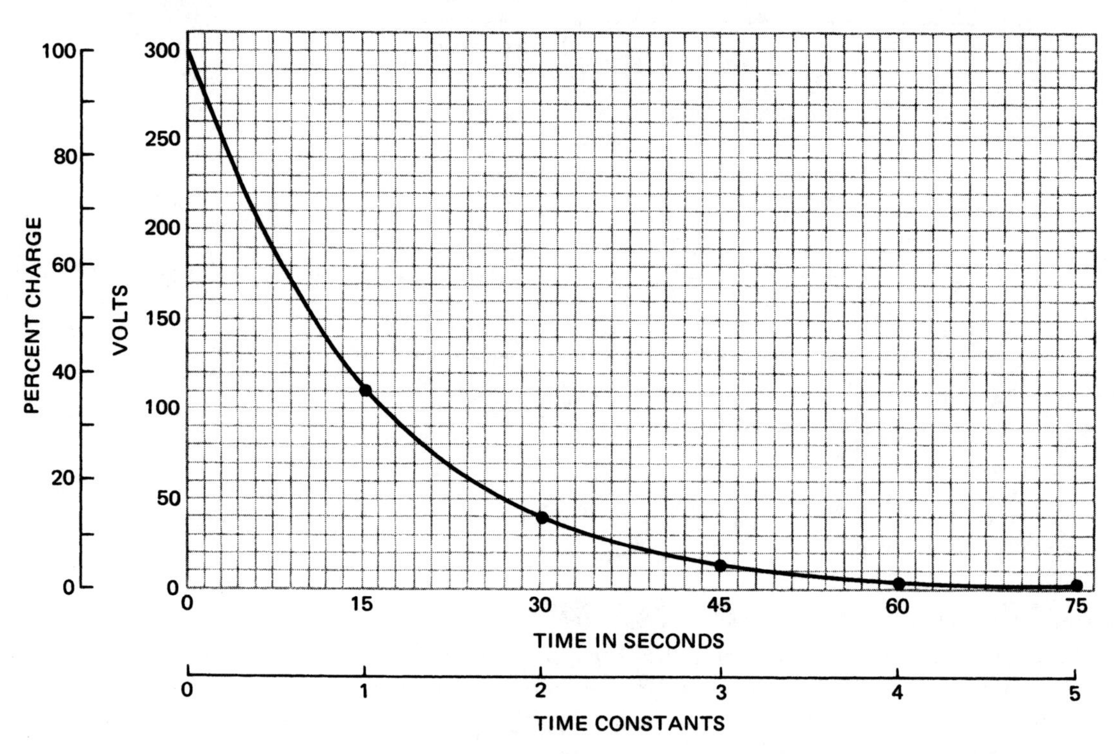

FIGURE 5–14 Exponential voltage curve for a capacitor on discharge *(Delmar/Cengage Learning)*

Number of Time Constants	Percentage of Impressed Voltage (%)	Capacitor Voltage (V)
1	36.8	111.0
2	13.5	40.5
3	5.0	15.0
4	1.8	5.4
5	0.7	2.1

TABLE 5–5 Capacitor voltage on discharge as a function of time constant

At the beginning of either a charge period or a discharge period, the initial current is given by Ohm's law. To find the current after a given number of time periods, for either charge or discharge, a modified form of Ohm's law is required.

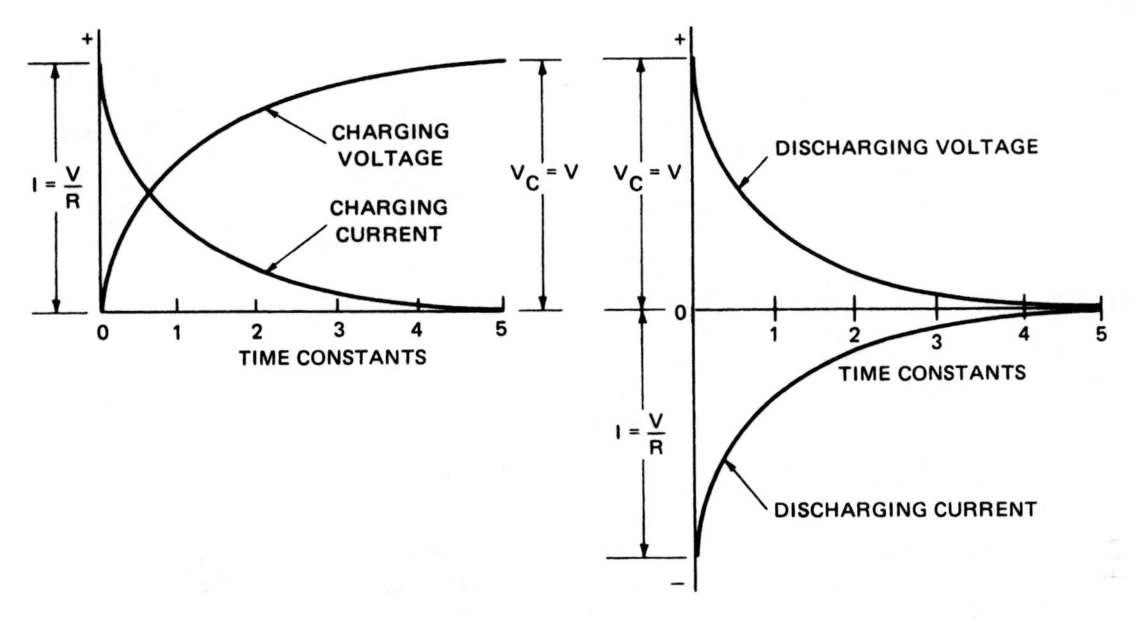

FIGURE 5–15 Charging and discharging currents and voltages in a series RC circuit (*Delmar/Cengage Learning*)

$$i = \frac{V^{(\epsilon^{-\tau/RC})}}{R}$$

The following problem illustrates the use of this equation in an RC circuit on charge.

PROBLEM 2

Statement of the Problem

Refer to the RC circuit shown in Figure 5–11. Determine

1. the initial charge current at the instant the circuit is energized.
2. the current at the end of one time constant.

Solution

1. At the instant the circuit is energized, the charging current is limited by the resistance of the voltmeter alone. The initial current is

$$I = \frac{V}{R}$$

$$= \frac{300}{300{,}000} = 0.001 \text{ A}$$

2. The charging current at the end of one time constant is

$$i = \frac{V(\epsilon^{-\tau/RC})}{R}$$

$$= \frac{300}{300{,}000} \times \frac{1}{2.718} = 0.001 \times 0.368 = 0.000368 \text{ A}$$

$$= 0.37 \text{ mA}$$

RL TIME CONSTANTS

A discussion of time constants for RL series circuits is given here because the exponential curves and calculations involved are similar to those for RC series circuits.

Current in an RL Circuit

Any inductance coil is really a combination of inductance and resistance in series. If the coil is connected to a dc supply, the final value of current is given by Ohm's law, $I = V/R$. However, the current does not rise to the Ohm's law value instantaneously. The current increases exponentially because the inductance opposes the change in current (Figure 5–16).

RL Exponential Current Curve. The exponential curve for current in an RL circuit is shown in Figure 5–16. When the circuit is energized, the induced voltage is at its maximum value. It then decreases as the rate of increase of current becomes less. The current reaches its Ohm's law value at the end of five time constants. At this point, the induced voltage is zero.

When the circuit is deenergized, the decay of current causes an induced voltage. This voltage attempts to maintain the current. Figure 5–17 shows the gradual decay of current as the induced voltage decreases. At the end of five time constants, the current and the induced voltage are close to zero.

The RL Time Constant

It was shown for RC circuits that one time constant is equal to the resistance in ohms times the capacitance in farads: $\tau = RC$.

In an RL circuit, the current at the end of one time constant is 63.2% of its final value as determined by Ohm's law. The time constant for an RL circuit is equal to the ratio of the inductance in henrys to the resistance in ohms: $\tau = L \div R$.

For example, assume that a coil has an inductance of 0.2 H and a resistance of 10 Ω. One time constant is found to be

$$\tau = \frac{L}{R} = \frac{0.2}{10} = 0.02 \text{ s}$$

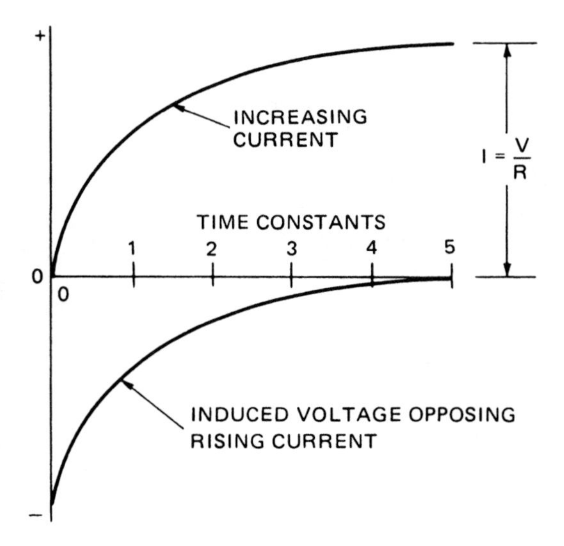

FIGURE 5–16 Current and induced voltage in an RL series circuit (*Delmar/Cengage Learning*)

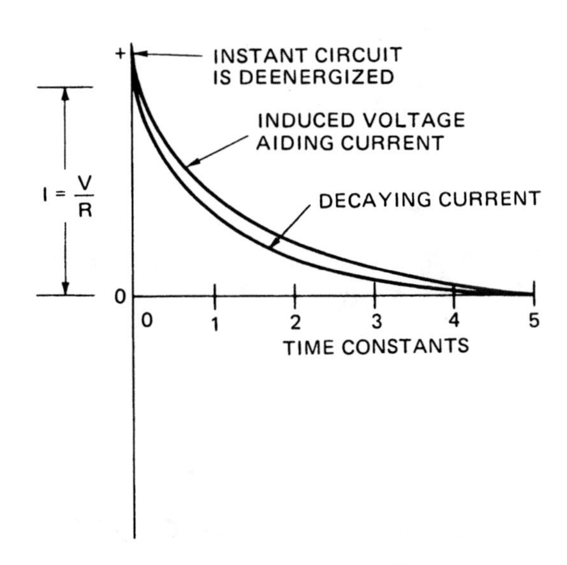

FIGURE 5–17 Decay of current and induced voltage in an RL series circuit (*Delmar/Cengage Learning*)

The coil is then connected across a 120-V dc source. The current at the end of a given length of time can be determined using the following formula:

$$\mathbf{i = \frac{V}{R}(1 - \epsilon^{-R\tau/L})}$$

$$\mathbf{= \frac{V}{R}\left(1 - \frac{1}{\epsilon^{\frac{R\tau}{L}}}\right)}$$

Using this expression to find the current at the end of two time constants after the coil is energized, the following calculations are required:

$$\mathbf{i = \frac{V}{R}\left(1 - \frac{1}{\epsilon^{\frac{R\tau}{L}}}\right)}$$

[*Note:* Two time constants for this circuit equal 0.04 s.]

$$i = \frac{120}{10}\left(1 - \frac{1}{2.718^{\frac{10 \times 0.04}{0.2}}}\right)$$

$$= \frac{120}{10}\left(1 - \frac{1}{2.718^{2}}\right) = 12 \times 0.865 = 10.38 \text{ A}$$

When the coil is deenergized, the coil current decreases toward zero according to the formula

$$i = \frac{V}{R} \epsilon^{\frac{-RT}{L}}$$

or

$$i = \frac{V}{R} \times \frac{1}{\epsilon^{\frac{RT}{L}}}$$

The current in the inductor coil can be determined as the current decreases to zero. At the end of two time constants, the current is

$$i = \frac{V}{R} \times \frac{1}{2.718^{\frac{10 \times 0.04}{0.2}}}$$

$$= \frac{120}{10} \times \frac{1}{2.718^2} = 12 \times 0.135 = 1.62 \text{ A}$$

APPLICATIONS OF CAPACITORS

There are many practical applications for capacitors in electricity and electronics. For example, in power applications, capacitors are used to improve the power factor of distribution circuits. The starting torque of certain types of single-phase induction motors can be improved by the addition of capacitors. Capacitors are used with rectifier units to change pulsating direct current to a constant dc. Nonpulsating direct current is used with electronic components.

Capacitors have many applications in communication electronics at audio, radio, and video frequencies. Capacitors are used in special phase shift circuits to control the plate current conduction of thyratron and ignitron tubes. Capacitors are also used in electronic time-delay systems and in phototube circuits, as well as in numerous other industrial applications. Examples of typical capacitors are shown in Figure 5–18.

TYPES OF CAPACITORS

The most basic form of fixed capacitor consists of two metal plates separated by a thin dielectric material. This material may be ceramic or mica. This simple capacitor is enclosed in a plastic case. Two leads pass through the case and connect to the plates. In some capacitors, a number of plates are placed between sheets of dielectric with alternate plates connected in parallel. This arrangement increases the plate area, resulting in an increase in the capacitance. Figure 5–19 shows one type of simple capacitor. Such a capacitor is usually rated in microfarads or picofarads.

Tubular Capacitors

A second type of capacitor is shown in Figure 5–20. The capacitance rating of this tubular paper capacitor is larger than that of a mica capacitor. The capacitor consists of a strip of waxed paper placed between two strips of tinfoil or aluminum foil. The paper and foil strips are about an inch wide and 8 to 12 ft in length. The strips are rolled and placed in

FIGURE 5–18 **Capacitors (*Delmar/Cengage Learning*)**

FIGURE 5–19 **Mica capacitor (*Delmar/ Cengage Learning*)**

a small plastic or metal cylinder. The resulting capacitor is physically small, but has a large plate area and a large value of capacitance.

The tinfoil or aluminum foil plates are often separated by a thin film of paper treated with an oil, or by a film of plastic. The paper capacitor (Figure 5–21) is enclosed in a flat case or an oil-filled can.

Ceramic Capacitors

Ceramic capacitors consist of a ceramic dielectric and silver plates. The dielectric is a ceramic compound such as barium titanate or titanium dioxide. The dielectric is usually in the form of a disc. The silver plates are secured to each side of this thin ceramic disc.

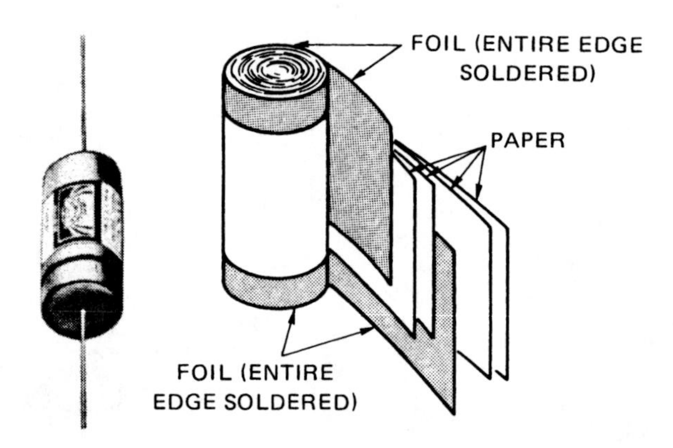

FIGURE 5–20 Paper and foil tubular capacitor *(Delmar/Cengage Learning)*

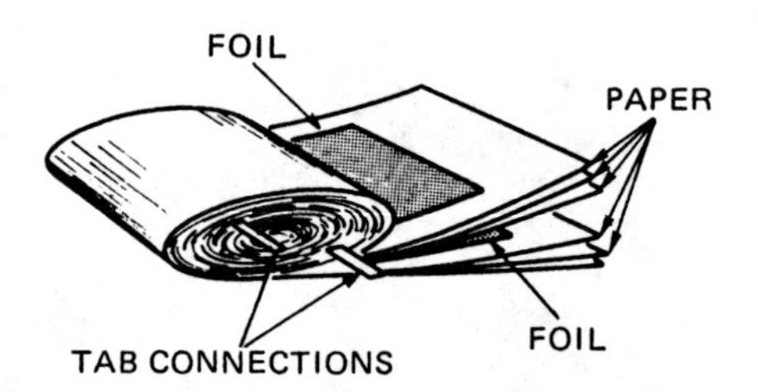

FIGURE 5–21 Flat paper and foil capacitor *(Delmar/Cengage Learning)*

Ceramic capacitors have capacitance ratings as high as 2.0 μF. This rating is large considering the relatively small size of the capacitor.

Electrolytic Capacitors

Another type of capacitor is used for direct-current applications generally. An electrolytic capacitor is shown in Figure 5–22. The positive plate is made from aluminum foil. This plate is immersed in an electrolyte consisting of a borax solution. The electrolyte serves as the negative plate of the capacitor. A second piece of aluminum foil is placed between the electrolyte and the negative terminal of the capacitor. When the capacitor is energized from a dc source, an insulating film of oxide develops on the positive aluminum foil. This film forms a very thin dielectric. The extremely thin dielectric means that the electrolytic capacitor has a very high capacitance rating for its physical size.

A more popular type of capacitor is the *dry electrolytic capacitor*. The electrolyte in this type of capacitor is gauze saturated with a borax solution. This dry capacitor has the advantage that there is no liquid electrolyte to leak from the case.

Connecting Electrolytic Capacitors. When connecting an electrolytic capacitor into a dc circuit, the positive aluminum plate must be connected to the positive side of the circuit. If this type of capacitor is installed with its connections reversed, the dielectric will be punctured. With the wet electrolytic capacitor, the dielectric will repair itself if the capacitor is reconnected properly. The dry capacitor, however, is permanently damaged if the connections are reversed.

AC Electrolytic Capacitor. The electrolytic capacitor described in the previous paragraphs cannot be used in an alternating-current circuit. Because the voltage is continually changing direction, the dielectric of the capacitor will be punctured repeatedly. However, a form of the electrolytic capacitor known as a *nonpolarized electrolytic capacitor* can be used in alternating-current circuits. This capacitor is really two wet, self-healing electrolytic capacitors assembled in a series back-to-back arrangement. The like plates of these

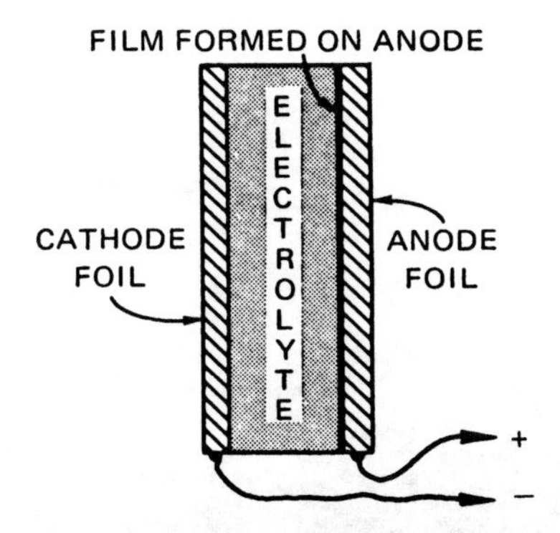

FIGURE 5–22 Electrolytic capacitor (*Delmar/Cengage Learning*)

capacitors are connected together. When an ac voltage is applied to the two outside terminals, one capacitor will be connected properly. At the same time, the dielectric of the second capacitor will be punctured. Then, in the other half-cycle, this process is reversed. Thus, there is always one capacitor connected properly, regardless of the direction of the ac voltage.

Air Capacitors

A common form of the *variable-air capacitor* is shown in Figure 5–23. One group of plates is mounted so they are movable with respect to a group of fixed plates. The movable plates are called the *rotor* and the fixed plates are called the *stator*. This type of capacitor is ideal for tuning radio receivers because the capacitance can be varied at low values, in the order of picofarads. The typical capacitance range of a variable air capacitor is between 500 pF and zero.

The *trimmer capacitor* (Figure 5–24) has two metal plates with an air dielectric. The spacing between the plates can be changed by means of an insulated setscrew. As the setscrew is tightened, the spacing is decreased. As a result, the capacitance is increased.

Capacitor Symbols

The symbol for any type of fixed capacitor is shown in Figure 5–25A. In schematic diagrams, the curved portion of the symbol is connected to ground, or to the low or negative voltage side of the circuit. The symbols for variable capacitors are given in Figure 5–25B.

CAPACITOR MARKINGS

Different types of capacitors are marked in different ways. Capacitance and voltage values are usually written directly on large ac oil-filled paper capacitors. The same is true for most electrolytic and small nonpolarized capacitors. Other types of capacitors, however, depend on color codes or code numbers and letters to indicate the capacitance value,

FIGURE 5–23 Variable air capacitor (*Delmar/Cengage Learning*)

FIGURE 5–24 Trimmer capacitor (*Delmar/Cengage Learning*)

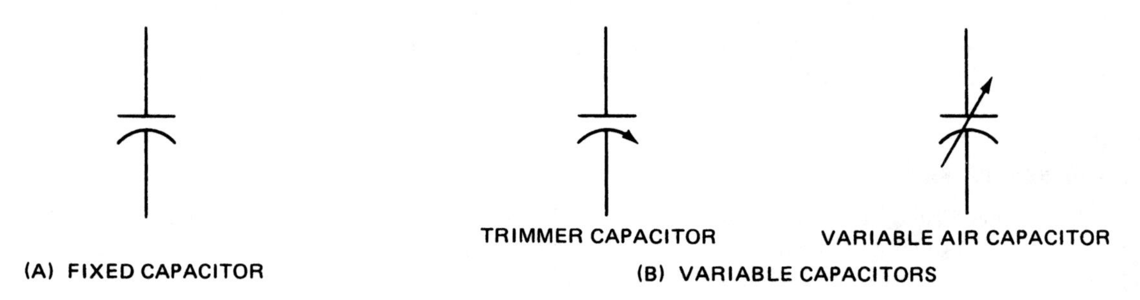

FIGURE 5–25 Capacitor symbols (*Delmar/Cengage Learning*)

tolerance, and voltage rating. Although color coding for capacitors has been abandoned in favor of direct marking by most manufacturers, it is still used by some. Also, many older capacitors with color codes are still in use. For this reason, this text will discuss color coding for several different types of capacitors.

Unfortunately, there is no actual set standard used by all manufacturers. The color codes presented are probably the most common. An identification chart for "postage stamp" mica capacitors and tubular paper or tubular mica capacitors is shown in Figure 5–26. It should be noted that most "postage stamp" mica capacitors use a five-dot color code. There are six-dot color codes, however. When a six-dot color code is used, the third color dot represents a third digit and the rest of the code is the same. The capacitance values given are in picofarads. Although these markings are typical, there is no actual standard, and it may be necessary to use manufacture's literature to determine the true values. A second method for

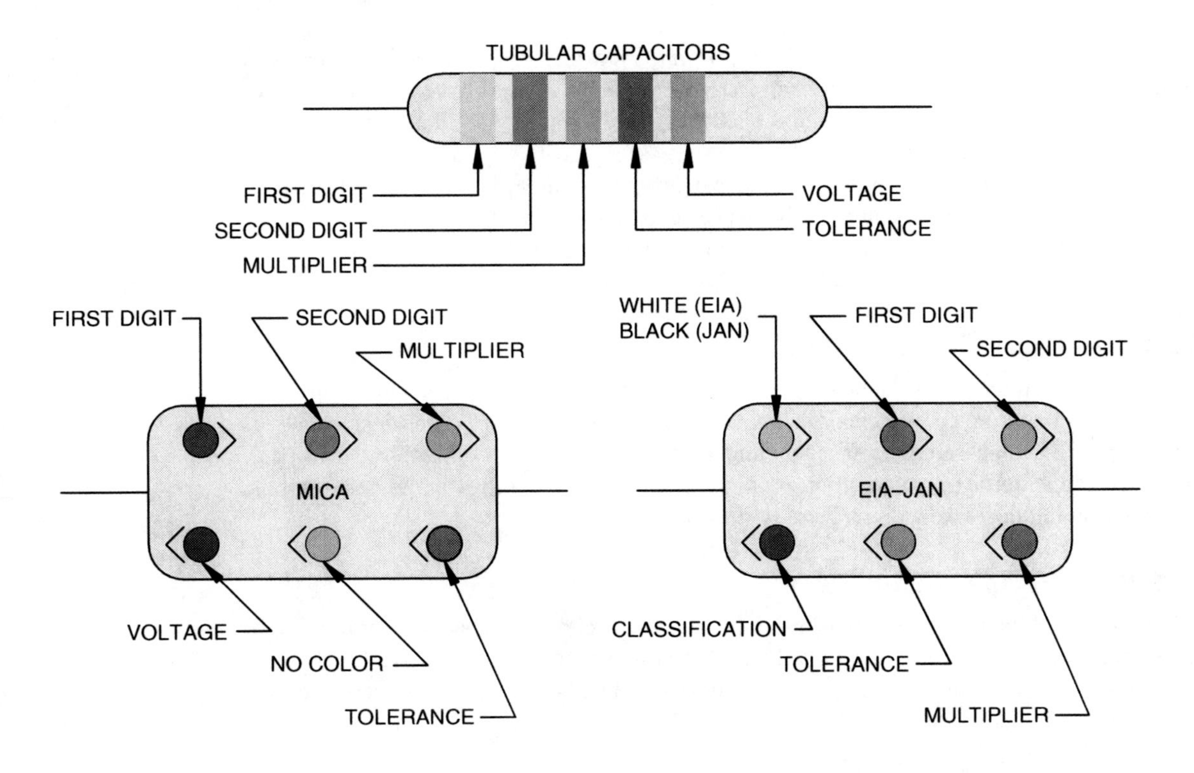

COLOR	NUMBER	MULTIPLIER	TOLERANCE	VOLTAGE
NO COLOR			20%	500
BLACK	0	1		
BROWN	1	10	1%	100
RED	2	100	2%	200
ORANGE	3	1,000	3%	300
YELLOW	4	10,000	4%	400
GREEN	5	100,000	5% (EIA)	500
BLUE	6	1,000,000	6%	600
VIOLET	7	10,000,000	7%	700
GRAY	8	100,000,000	8%	800
WHITE	9	1,000,000,000	9%	900
GOLD		0.1	5% (JAN)	1,000
SILVER		0.01	10%	2,000

FIGURE 5–26 Identification of mica and tubular capacitors (*Delmar/Cengage Learning*)

color coding mica capacitors is called the *Electronic Industries Association* (EIA) standard, or the *Joint Army-Navy* (JAN) standard. The JAN standard is used for electronic components intended for military use. When the EIA standard is employed, the first dot will be colored white. In some instances, the first dot may be colored silver instead of white. This indicates that the capacitor's dielectric is paper instead of mica. When the JAN standard is used, the first dot will be colored black. The second and third dots represent digits, the fourth dot is the multiplier, the fifth dot is the tolerance, and the sixth dot indicates classes A through E of temperature and leakage coefficients.

Temperature Coefficients

The temperature coefficient indicates the amount of capacitance change with temperature. Temperature coefficients are listed in parts per million (ppm) per degree Celsius (°C). A positive temperature coefficient indicates that the capacitor will increase its capacitance with an increase in temperature. A negative temperature coefficient indicates that the capacitance will decrease with an increase in temperature.

Ceramic Capacitors

Another capacitor that often uses color codes is the ceramic capacitor (Figure 5–27). This capacitor will generally have one band that is wider than the others. The wide band indicates the temperature coefficient, and the other bands are first and second digits, multiplier, and tolerance.

Dipped Tantalum Capacitors

A dipped tantalum capacitor is shown in Figure 5–28. This capacitor has the general shape of a match head but is somewhat larger in size. Color bands and dots determine the value, tolerance, and voltage. The capacitance value is given in picofarads.

Film Capacitors

Not all capacitors use color codes to indicate values. Some capacitors use numbers and letters. A film-type capacitor is shown in Figure 5–29. This capacitor is marked "105K." The value can be read as follows:

1. The first two numbers indicate the first two digits of the value.
2. The third number is the multiplier. Add the number of zeros to the first two numbers indicated by the multiplier. In this example, add five zeros to 10. The value is given in picofarads. This capacitor has a value of 1,000,000 pF, or 1 μF.
3. The K symbol indicates the tolerance. In this example, K indicates a tolerance of $\pm 10\%$.

SUMMARY

- The elementary capacitor consists of two metal plates separated from each other by a dielectric (insulating material).

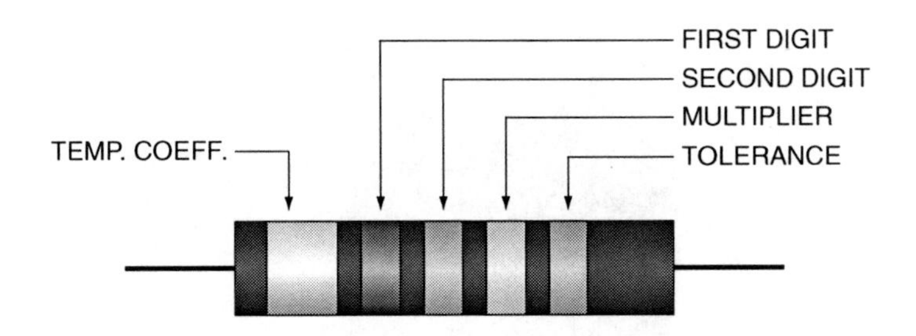

COLOR	NUMBER	MULTIPLIER	TOLERANCE OVER 10 pF	TOLERANCE 10 pF OR LESS	TEMP. COEFF.
BLACK	0	1	20%	2.0 pF	0
BROWN	1	10	1%		N30
RED	2	100	2%		N80
ORANGE	3	1,000			N150
YELLOW	4				N220
GREEN	5				N330
BLUE	6		5%	0.5 pF	N470
VIOLET	7				N750
GRAY	8	0.01		0.25 pF	P30
WHITE	9	0.1	10%	1.0 pF	P500

FIGURE 5–27 Color codes for ceramic capacitors (*Delmar/Cengage Learning*)

1. During the charge process, electrons flow from the negative side of the source to one plate and from the second plate back to the source.
2. There is almost no flow of electrons through the dielectric material.
3. The movement of electrons continues until the potential difference across the two metal plates is equal to the source voltage. At this point, the capacitor is said to be fully charged.
4. The charges on the plates of the capacitor act on the electrons of the atoms in the dielectric and cause the orbits of these electrons to be distorted.
5. An electrostatic field created by the charged plates maintains the distortion of the electron orbits in the dielectric atoms. This electrostatic field stores electrical energy.
6. The distortion dissipates once the capacitor is discharged. The electron orbits of the atoms of the dielectric return to their normal pattern.

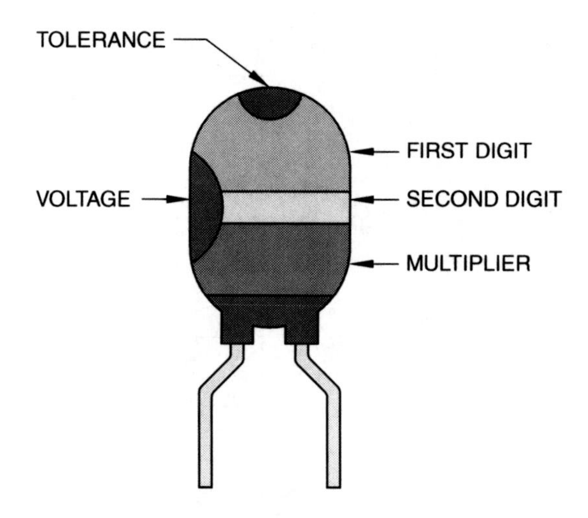

COLOR	NUMBER	MULTIPLIER	TOLERANCE	VOLTAGE
			NO DOT 20%	
BLACK	0			4
BROWN	1			6
RED	2			10
ORANGE	3			15
YELLOW	4	10,000		20
GREEN	5	100,000		25
BLUE	6	1,000,000		35
VIOLET	7	10,000,000		50
GRAY	8			
WHITE	9			3
GOLD			5%	
SILVER			10%	

FIGURE 5–28 Dipped tantalum capacitors (*Delmar/Cengage Learning*)

- Capacitance is the property of a circuit, or circuit component, that allows it to store electrical energy in electrostatic form.
    1. Capacitance is measured in farads. A capacitor has a capacitance of one farad when a change of one volt across its plates results in a charge movement of one coulomb:

$$1 \text{ microfarad } (1\ \mu F) = 1 \text{ farad } (1\ F) \div 1{,}000{,}000 = 10^{-6}\ F$$
$$1 \text{ picofarad } (1\ pF) = 1\ \mu F \div 1{,}000{,}000 = 10^{-12}\ F.$$

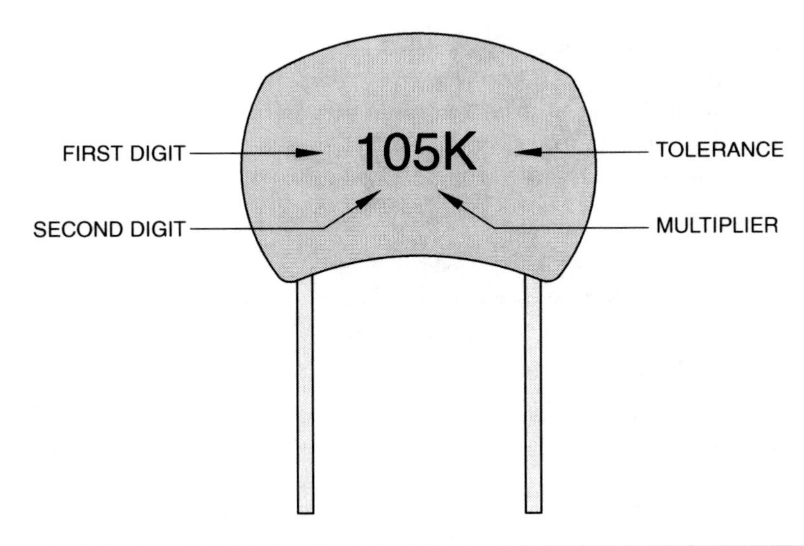

NUMBER	MULTIPLIER	TOLERANCE		
			10 pF OR LESS	OVER 10 pF
0	1	B	0.1 pF	
1	10	C	0.25 pF	
2	100	D	0.5 pF	
3	1,000	F	1.0 pF	1%
4	10,000	G	2.0 pF	2%
5	100,000	H		3%
6		J		5%
7		K		10%
8	0.01	M		20%
9	0.1			

FIGURE 5–29 Film-type capacitors (*Delmar/Cengage Learning*)

2. The capacitance of a capacitor can be increased by
   a. increasing the plate area to increase the area of dielectric under stress.
   b. moving the metal plates closer together to decrease the thickness of the dielectric, resulting in a greater stress.
   c. using a dielectric with a higher dielectric constant.

(1) The dielectric constant of an insulating material measures the effectiveness of the material when used as the dielectric of a capacitor. This gives a value indicating the degree of distortion of the electron orbits in the dielectric for a given applied voltage.

(2) The dielectric constants of all other materials are measured against air, which is assumed to be one.

- Circuit components other than capacitors may create a capacitance effect.
  1. When two wires of a circuit are separated by air, they will act as a capacitor.
  2. Adjacent turns of a coil winding, which are separated only by the insulation of the wire, will have some capacitance effect.

- The charge on the plates of a capacitor for a given applied voltage is directly proportional to the capacitance of the capacitor:

$$Q = C \times V \quad \text{or} \quad C = \frac{Q}{V} \quad \text{or} \quad V = \frac{Q}{C}$$

- Dielectric materials are given a dielectric strength rating:
  1. This rating is stated as either "volts per centimeter" or "volts per mil" of dielectric thickness required to break down the dielectric.
  2. It is *not* the same as the dielectric constant.

- The formulas used to determine the capacitance of a two-plate capacitor are

  Measurements in inches:

$$C = \frac{K \times A}{4.45\ D}$$

  Measurements in centimeters:

$$C = \frac{0.0885 \times K \times A}{D}$$

- The total capacitance when capacitors are connected in parallel is given by

$$C_T = C_1 + C_2 + C_3$$

  When capacitors are connected in parallel, the effect is the same as increasing the number of plates or the area of the plates.

- The total capacitance when capacitors are connected in series is given by

$$C_T = \frac{1}{\frac{1}{C_1} + \frac{1}{C_2} + \frac{1}{C_3}}$$

  When capacitors are connected in series, the effect is equivalent to increasing the thickness of the dielectric of one capacitor, resulting in less capacitance.

- The formulas for total capacitance can be used only when all of the values are in the same unit of capacitance measurement, either microfarads or picofarads.
- The amount of energy stored in a capacitor is measured in joules or watt-seconds:

  $$\text{Watt-seconds} = \frac{1}{2}QV$$

  Because $Q = C \times V$, we obtain

  $$\text{Watt-seconds} = \frac{1}{2}CV^2$$

- The amount of charge on the plates of a capacitor is a function of the current and time.
  1. As the value of the current increases, less time is required to charge the capacitor fully.
  2. The time required for a given voltage to build up across the capacitor can be controlled by increasing or decreasing the charging current

     $$\tau = R \times C \quad \text{or} \quad R = \frac{\tau}{C} \quad \text{or} \quad C = \frac{\tau}{R}$$

     where τ is the Greek letter *tau*, the letter symbol for a time constant expressed in seconds.
- One time constant is the time, in seconds, required for a completely discharged capacitor to charge to 63% of the source voltage.
  1. For all practical purposes, it is assumed that the capacitor is fully charged at the end of five time constants.
  2. For all practical purposes, it is assumed that a fully charged capacitor will be completely discharged at the end of five time constants of discharge.
- The relationship of the current to the supply voltage for a capacitor being charged can be found at any instant:

  $$I = \frac{(V - V_C)}{R}$$

- RL time constants are similar to those for RC series circuits:

  $$\tau = L \div R$$

- When connecting an electrolytic capacitor, the plates must be properly connected.
  1. For the wet electrolytic capacitor, the dielectric will repair itself after being punctured as a result of reversed connections.
  2. If the connections of the dry electrolytic capacitor are reversed, the dielectric is permanently damaged.
  3. The electrolytic capacitor cannot be used in an alternating-current circuit. However, a nonpolarized electrolytic capacitor can be used in ac circuits.

Achievement Review

1. List three factors that affect the capacitance of a capacitor.
2. Define the basic unit of measurement of capacitance.
3. A capacitor takes a charge of 0.05 coulomb (C) when connected across a 250-V source. Determine the capacitance of the capacitor in microfarads.
4. Explain what is meant by the term *dielectric constant.*
5. A paper capacitor consists of two aluminum foil plates, each 10 ft long and 1.5 in. wide. The waxed paper separating the two plates has a thickness of 0.04 in. The dielectric constant is assumed to be 2. Determine the capacitance of the capacitor, using both formulas:

$$C = \frac{K \times A}{4.45\ D}$$
$$= \frac{0.0885 \times K \times A}{D}$$

6. Three capacitors have ratings of 20 μF, 60 μF, and 30 μF, respectively. These capacitors are connected in parallel across a 220-V dc source. Determine
   a. the total capacitance of the three capacitors connected in parallel.
   b. the total charge, in coulombs, taken by the three capacitors in parallel when connected across a 220-V dc source.
7. The three capacitors in question 6 are connected in series across a 220-V dc source. Determine
   a. the total capacitance of the three capacitors connected in series.
   b. the total charge, in coulombs, taken by the three capacitors in series when connected across a 220-V dc source.
8. Explain what is meant by the term *dielectric strength rating.*
9. List five practical applications for capacitors.
10. Determine the energy, in watt-seconds, stored in a 100-μF capacitor when connected across a 400-V dc source.
11. A 100-μF capacitor is connected in series with a 500-V dc voltmeter across a 400-V dc supply. The voltmeter has a resistance of 1000 Ω/V.

a. Determine the time, in seconds, represented by one time constant for this RC series circuit.

b. Determine the voltage across the capacitor plates at the end of one time constant when the capacitor is on charge.

12. Using the circuit given in question 11, determine
    a. the voltage across the capacitor plates at the end of two time constants, when the capacitor is charging.
    b. the voltage across the capacitor plates, at the end of five time constants, when the capacitor is charging.

13. Using the circuit given in question 11, determine the voltage across the plates of the capacitor, at the end of one time constant, when the capacitor is discharging.

14. At the end of one time constant, what is the charging current in amperes for the RC series circuit given in question 11?

15. Determine the time constant, in seconds, for an inductor coil with an inductance of 0.5 H and a resistance of 10 Ω.

16. If the inductor coil in question 15 is connected across a 120-V source, determine
    a. the initial current, in amperes, at the instant the circuit is energized.
    b. the current, in amperes, at the end of one time constant after the circuit is energized.
    c. the current, in amperes, at the end of two time constants after the circuit is energized.

17. When the inductor coil in question 16 is deenergized, determine the value of the decaying current in amperes
    a. at the end of two time constants.
    b. at the end of five time constants.

18. a. Explain the difference between a wet and a dry electrolytic capacitor.
    b. Where are electrolytic capacitors used?

19. Name four other types of capacitors, in addition to the electrolytic capacitor, that were described in this unit. Briefly describe the material used for the plates and the dielectrics of each type.

20. In schematic wiring diagrams, the curved portion of the symbol for a fixed capacitor should be connected in a particular manner. Explain what is meant by this statement.

PRACTICE PROBLEMS FOR UNIT 5

RC Time Constants

Find the missing values.

RESISTANCE	CAPACITANCE	TIME CONSTANT	TOTAL TIME
150 kΩ	100 μF		
350 kΩ			35 s
	350 pF	0.05 s	
	0.05 μF		10 s
1.2 MΩ	0.47 μF		
	12 μF	0.05 s	
86 kΩ			1.5 s
120 kΩ	470 pF		
	250 nF		100 ms
	8 μF		150 μs
100 kΩ		150 ms	
33 kΩ	4 μF		

$\tau = RC \qquad R = \frac{\tau}{C} \qquad C = \frac{\tau}{R} \qquad \text{Total Time} = \tau \times 5$

where

τ = time in seconds
R = resistance in ohms
C = capacitance in farads

6

Capacitors in Alternating-Current Circuits

Objectives

After studying this unit, the student should be able to

- calculate the capacitive reactance in an ac circuit using the formula

$$X_C = \frac{1}{2\pi fC}$$

- show that the capacitive reactance of a circuit is inversely proportional to the capacitance and to the frequency of the impressed ac voltage.
- demonstrate that in an ac circuit with pure capacitance, the current leads the impressed voltage by 90 electrical degrees.
- define the terms: the angle of phase defect, power factor, and Q, as applied to capacitors.
- prove by the use of an exponential curve that the net power taken by a capacitor in one complete cycle is practically zero.
- define the term magnetizing VARs and state its relationship to capacitance in an ac circuit.
- solve series circuit problems involving resistive and capacitive reactive components, making use of the appropriate formulas.
- construct vector diagrams, impedance triangles, and voltage triangles and use these triangles to solve problems in series ac circuits having resistance and capacitive reactive components.

It was shown in Unit 5 that when a capacitor is connected to a dc source, an opposition voltage builds up across the capacitor plates. After a period of time, this voltage will equal the impressed voltage. The charging current has a maximum value when the capacitor is uncharged. The current decreases to zero as the capacitor voltage builds up to equal the source voltage. The current remains at zero amperes while the dc source voltage equals the opposition voltage of the capacitor.

CAPACITIVE REACTANCE

A capacitor connected to an ac supply repeatedly charges and discharges as the ac voltage changes direction. The charging current also alternates in direction as the capacitor charges and discharges.

In the circuit in Figure 6–1, a 26.52-μF capacitor is connected across a 100-V, 60-Hz source. The ammeter indicates a current of one ampere. The dielectric of the capacitor prevents electrons from being conducted through the capacitor. However, there is electron flow to and from the plates of the capacitor as it charges and discharges in each cycle. The opposition voltage increases as the electrostatic field builds up across the plates. This voltage opposes the line voltage and limits the current. The opposition voltage that develops with a capacitor is really a countervoltage.

Inductance in an ac circuit also causes a countervoltage. The current-limiting effect of this countervoltage is measured as ohms of inductive reactance. The current-limiting effect of capacitance in an ac circuit is called *capacitive reactance* and is measured in ohms.

Capacitive reactance (X_C) is given by the formula

$$\mathbf{X_C = 1 \div 2\pi fC}$$

where

$2\pi = 6.28$

f = frequency, in cycles per second (hertz)

C = capacitance, in farads

For the capacitor in Figure 6–1, the capacitive reactance is

$$X_C = \frac{1}{2\pi fC}$$

$$= \frac{1}{377 \times 0.00002652} = \frac{1}{0.010} = 100\Omega$$

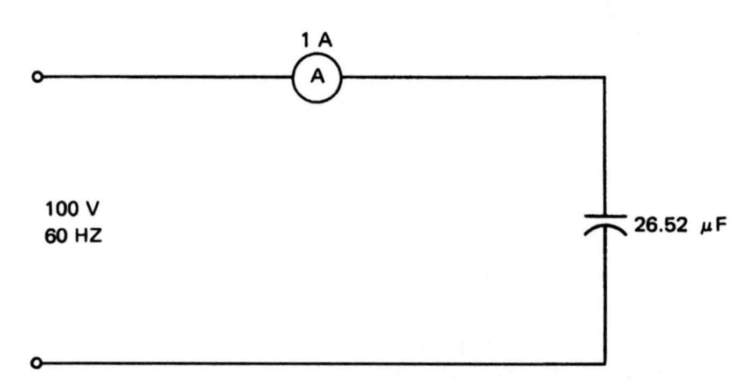

FIGURE 6–1 AC circuit containing pure capacitance only (*Delmar/Cengage Learning*)

Current in the Capacitive Circuit

The current, in amperes, in an ac circuit containing capacitance is determined using a form of Ohm's law. That is, capacitive reactance (X_C) is substituted for resistance (R) as follows:

$$I = V/X_C$$
$$= 100/100 = 1A$$

The formula for capacitive reactance shows that the value in ohms is inversely proportional to the capacitance in farads and the frequency of the impressed ac voltage. This relationship is logical because the charge on a capacitor is directly proportional to the capacitance for a given applied voltage. The pattern of this charge is an alternating current as the applied voltage alternates. If the capacitance increases for a given frequency and voltage, then the charge that flows in a given time must also increase. Thus, the current increases. If the frequency increases for a given capacitance and voltage, the same charge must flow per voltage alternation, but in a shorter time. Thus, an increase in frequency results in a larger current because current is a measure of the rate of flow of electrons.

An increase in either the frequency or the capacitance gives rise to an increase in current for a given applied voltage. It follows that an increase in the frequency or capacitance results in a decrease of the capacitive reactance (ohms).

LEADING CURRENT

The capacitor shown in Figure 6–1 has a capacitive reactance of 100 Ω and negligible resistance. The current in this circuit is one ampere. It was shown earlier in this text that the current in a pure resistive load is in phase with the voltage. In addition, it was shown that the current through a pure inductive reactance lags the impressed voltage by 90 electrical degrees. This unit will show that in a circuit with pure capacitive reactance, the current will *lead* the impressed voltage by 90 electrical degrees.

Alternating Voltage Applied to a Capacitor

A capacitor is charged when a dc voltage is applied to its terminals. The capacitor is discharged if a resistor is connected across its terminals. When an alternating voltage is applied to a capacitor, an alternating current of the same frequency repeatedly charges and discharges the capacitor. The charge in coulombs on the capacitor is proportional to the impressed voltage. As shown in Figure 6–2, the charge in coulombs is in phase with the impressed voltage ($Q = C \times V$). At zero voltage, the charge in coulombs is also zero. At the maximum positive and negative values of impressed voltage, the charge in coulombs is also at maximum.

Operation from 0° to 180°. The wave patterns for the impressed voltage and the charge are shown in Figure 6–2. The current is shown leading the line voltage by 90 electrical degrees. In the interval from 0° to 90°, the impressed voltage increases and the capacitor is charging. At 90° the capacitor voltage no longer increases. At this point, then, the current is zero. As the impressed voltage starts to decrease, the capacitor starts to discharge.

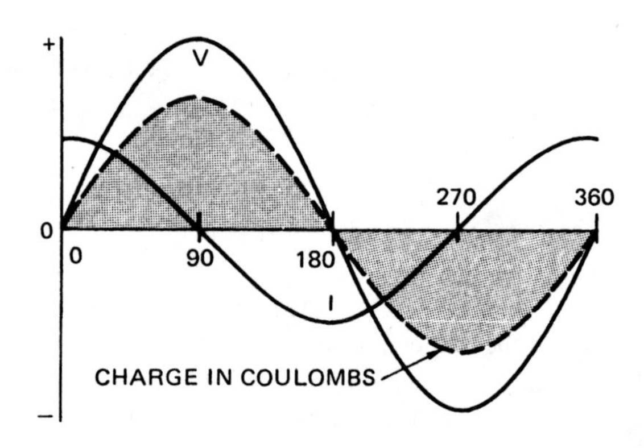

FIGURE 6–2 AC voltage, current, and charge in coulombs for a capacitor (*Delmar/Cengage Learning*)

The electrons now flow from the plates in a direction reversed from the initial charging direction. This is shown by the current as it increases in the negative (discharging) direction. At the same time, the impressed voltage decreases to zero during the interval from 90° to 180°.

Operation from 180° to 360°. At 180°, the line voltage is zero. The charge, in coulombs, is zero and the current is at its maximum value. The capacitor now begins to charge in the opposite direction, and the impressed voltage starts to rise toward its negative maximum value. The current is in the same direction as the impressed voltage in the range from 180° to 270°. The current decreases in magnitude as the rate of change of voltage decreases. At 270 electrical degrees, the capacitor voltage no longer increases. Therefore, the current is zero at 270°. Between 270° and 360°, the voltage again decreases to zero. As the negative charge is removed from the capacitor, the capacitor current increases in a positive direction. The capacitor voltage decreases to zero and the current rises in the positive direction to its maximum value at 360°. The capacitor is completely discharged at 360°.

ANGLE OF PHASE DEFECT

For circuits containing inductance coils, both the resistance and the inductive reactance must be considered. For circuits containing capacitors operating at commercial power frequencies, the resistance losses are usually assumed to be negligible. At such frequencies, the capacitor current leads the capacitor voltage by 90 electrical degrees.

Under actual operating conditions, the angle by which the capacitor current leads the voltage is slightly less than 90°. The *angle of phase defect* is the angle by which the phase angle between the capacitor voltage and current is less than 90°. The angles of phase defect for some capacitors having mica dielectrics are as small as three or four minutes. Capacitors having other types of dielectrics may have angles of phase defect greater than one degree. If the angle of phase defect becomes too large, the capacitor has a relatively high power loss. Such a loss causes an increase in the internal temperature of the capacitor. This temperature rise can shorten the useful life of the dielectric.

Power Factor of a Capacitor

The *power factor* of a capacitor relates the power losses of a capacitor to its volt-ampere rating. The power factor is the ratio of the power loss to the volt-ampere rating when the capacitor is operated at the rated voltage and frequency. The power factor is usually 0.01 [one percent (1%)] or less for power factor correction capacitors.

Q of a Capacitor

The term *Q* is used to describe capacitor losses. Q is given by the following expression:

$$Q = \frac{X_C}{R_e} = \frac{1}{2\pi f C R_e}$$

In this expression, the term R_e is the equivalent resistance that must be placed in series with a perfect capacitor to produce a loss in watts that is the same as the loss in the actual imperfect capacitor. This resistance includes the effects of dielectric losses and resistive losses. Capacitor Q may also be defined as follows:

$$Q = \frac{\text{VARs}}{\text{watts}}$$

As Q is a relatively large number, it is often simpler to use than the very small power factor values of capacitors.

VOLTAGE RATING OF A CAPACITOR

Capacitors that are used in either ac or dc circuits have a rating known as the *dc working voltage.* As an example, consider a paper capacitor made with Pyranol and enclosed in a metal container. Such a capacitor may have a rating specified as "600-V dc working voltage." It is of interest to determine whether this capacitor can be used on a 600-V ac circuit. A previous unit showed that ac voltages are given as effective values that are always 0.707 of the maximum value. The voltage of a 600-V ac circuit reaches a maximum instantaneous value twice in each cycle. This maximum value is

$$V_{\text{maximum}} = 1.414\ V_{(\text{RMS})} = 1.414 \times 600 = 848\ \text{V}$$

The dielectric of the capacitor is designed for 600 V. Because the ac voltage exceeds this value twice in each cycle, the dielectric can break down and the capacitor can be ruined. This maximum voltage has a short duration. However, the dc working voltage of a capacitor should be high enough to withstand the maximum or peak voltage of an ac circuit.

If the capacitor is not designed for ac service, it may have a short life when operated on ac, even if the applied maximum ac voltage does not exceed the dc working voltage. A capacitor operating on ac has an increase in the power losses. These greater losses may cause excessive internal operating temperatures. This and other factors result in the premature failure of dc capacitors operated on ac. It is recommended that a capacitor be used only in the type of service for which it is designed.

POWER IN A CAPACITOR

The circuit shown in Figure 6–1 has a maximum value of 141.4 V. The current leads the line voltage by 90°. Because the effective current is one ampere, the maximum value of the current is 1.414 A. The power, in watts, at any instant is equal to the product of the voltage and current values at the same instant. Table 6–1 summarizes

Degrees	Instantaneous Voltage (Volts)	Instantaneous Current (Amperes)	Instantaneous Power (Watts)
0	0.0	1.414	0.0
15	36.6	1.366	50.0
30	70.7	1.225	86.6
45	100.0	1.000	100.0
60	122.5	0.707	86.6
75	136.6	0.366	50.0
90	141.4	0.0	0.0
105	136.6	−0.366	−50.0
120	122.5	−0.707	−86.6
135	100.0	−1.000	−100.0
150	70.7	−1.225	−86.6
165	36.6	−1.366	−50.0
180	0.0	−1.414	0.0
195	−36.6	−1.366	50.0
210	−70.7	−1.225	86.6
225	−100.0	−1.000	100.0
240	−122.5	−0.707	86.6
255	−136.6	−0.366	50.0
270	−141.4	0.0	0.0
285	−136.6	0.366	−50.0
300	−122.5	0.707	−86.6
315	−100.0	1.000	−100.0
330	−70.7	1.225	−86.6
345	−36.6	1.366	−50.0
360	0.0	1.414	0.0

TABLE 6–1 Instantaneous voltage, current, and power for a circuit with capacitance

the instantaneous voltage, current, and power values at 15° intervals for the circuit in Figure 6–1.

Wave Patterns for Pure Capacitance

The values in Table 6–1 are plotted against electrical time degrees to obtain the voltage, current, and power waveforms shown in Figure 6–3. Between 0° and 90°, the voltage and the current are positive. The product of the instantaneous voltage and current values in this part of the cycle will give positive power. This means that the capacitor receives energy from the supply between 0° and 90°. This energy is stored in the electrostatic field of the capacitor.

In the period between 90° and 180°, the current rises to its negative maximum value. The impressed voltage is positive and decreases to zero. The product of the instantaneous negative current values and the positive voltage values from 90° to 180° gives negative power. Thus, the capacitor discharges in this period and returns its stored energy to the source.

From 180° to 270°, both the current and the impressed voltage are negative and act together. The product of these instantaneous values (with like signs) gives positive power for this part of the cycle. This means that the capacitor is charging again and storing electrical energy.

The voltage is negative from 270° to 360°. At the same time, the current rises in a positive direction. The product of a negative voltage value and a positive current value gives a negative power value. During this part of the cycle, the capacitor discharges, releasing electrical energy back into the circuit.

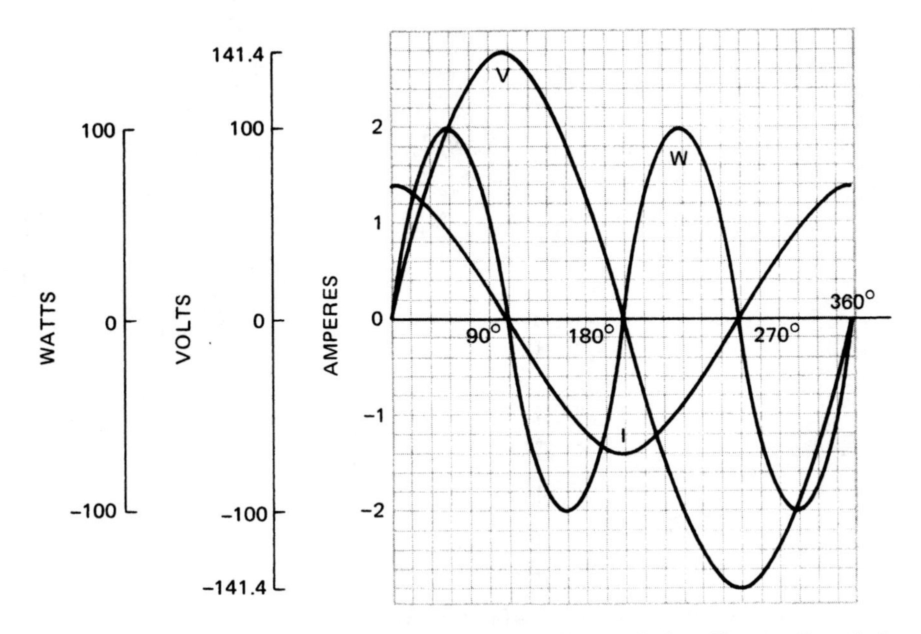

FIGURE 6–3 Power for a circuit with pure capacitance ***(Delmar/Cengage Learning)***

Power Waveform. Figure 6–3 shows that for the power waveform, the areas of the two positive pulses of power are equal to the areas of the two negative pulses of power. The positive power pulses represent power fed from the source to the load. The negative pulses of power represent the power returned to the source from the capacitor as it discharges. If the areas of positive power equal the areas of negative power, the net power taken by the capacitor at the end of one complete cycle, or at the end of any number of complete cycles, is zero.

CAPACITORS IN PARALLEL AND IN SERIES

Unit 5 explained the methods used to compute the total capacitance for capacitors connected in parallel and in series. Figure 6–4 shows three capacitors connected in parallel. The total capacitance is 55 μF. The total capacitive reactance, in ohms, of the three capacitors is less than the reactance of any single capacitor. This is due to the fact that the group has more plate area than a single capacitor.

The capacitive reactance of the three capacitors in parallel is

$$X_C = \frac{1}{2\pi fC}$$

$$= \frac{1}{377 \times 0.000055} = \frac{1}{0.020735} = 48.22\ \Omega$$

The total current taken by the three capacitors is

$$I = \frac{V}{X_C}$$

$$= \frac{120}{48.22} = 2.48\ \text{A}$$

This current leads the impressed ac voltage by nearly 90 electrical degrees. This means that the true power, in watts, taken by capacitors is zero.

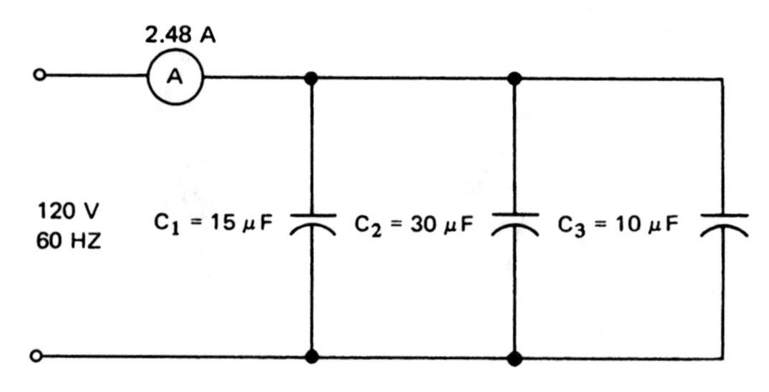

FIGURE 6–4 Capacitors in parallel (*Delmar/Cengage Learning*)

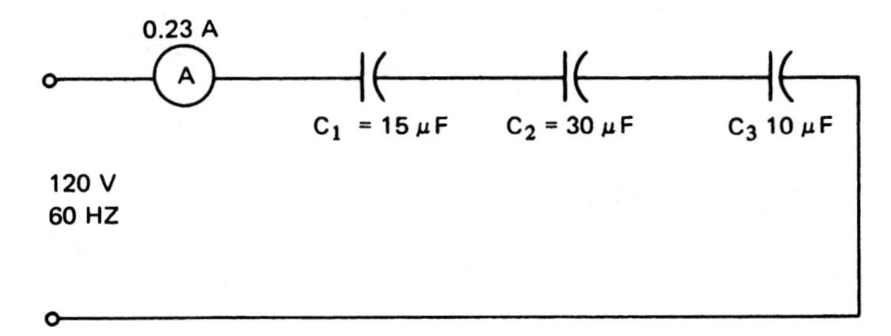

FIGURE 6–5 Capacitors in series ***(Delmar/Cengage Learning)***

In Figure 6–5, the same capacitors are connected in series. The total capacitance = 5 μF. When the series capacitors are connected to a 120-V, 60-Hz supply, the capacitive reactance is

$$X_C = \frac{1}{2\pi fC}$$

$$= \frac{1}{377 \times 0.000005} = 530.5\ \Omega$$

The low value of total capacitance for this series connection causes the high capacitive reactance. The current in the series circuit is

$$I = \frac{V}{X_C}$$

$$= \frac{120}{530.5} = 0.23\ \text{A}$$

This current is the same at all points in the series circuit. The current is 90° out of phase with the voltage.

MAGNETIZING VARs

When a pure resistance is connected to an ac source, the current is in phase with the applied voltage. The true power in this case is given by any of the following equations:

$$\mathbf{W = VI}$$

$$\mathbf{= I^2R}$$

$$\mathbf{= \frac{V^2}{R}}$$

If a pure inductance is connected to an ac source, the current lags the applied voltage by 90 electrical degrees. This means that no power is lost. The product of the applied voltage and current values is called *magnetizing VARs* (or *VARs*). This value does not represent a true power loss. If the pure inductance is a coil having a reactance of one ohm, and an ac voltage of one volt is applied, there will be a current of one ampere lagging by 90°. Under these conditions, the coil requires one magnetizing VAR.

The value of magnetizing VARs for a pure inductance can be calculated using any of the following formulas:

$$\mathbf{VARs} = \mathbf{VI}$$

$$= \mathbf{I^2X_L}$$

$$= \frac{\mathbf{V^2}}{\mathbf{X_L}}$$

This unit explained that alternating current in a pure capacitance leads the applied voltage by 90°. Thus, no true power is lost in the circuit. It was stated that the product of the applied ac voltage and the current is called *VARs*. Coil current lags its applied voltage by 90°, and capacitor current leads its applied voltage by 90°. Thus, it can be seen that the capacitor and coil currents act in opposite ways. Coils can be viewed as requiring or *consuming* magnetizing VARs. Capacitors may be considered as *sources* of magnetizing VARs.

The magnetizing VARs produced by a pure capacitance can be calculated by any of the following equations:

$$\mathbf{VARs} = -\mathbf{VI}$$

$$= -\mathbf{I^2X_C}$$

$$= -\frac{\mathbf{V^2}}{\mathbf{X_C}}$$

The minus sign means that the VARs for a capacitor are opposite in direction to those for an inductance.

The principle of the conservation of energy can be applied to steady-state ac circuit theory. This principle states that the total power supplied to a circuit is equal to the sum of the values of power consumed in each of the individual circuit components.

Another principle states that the value of the magnetizing VARs supplied to a circuit is equal to the sum of the magnetizing VAR values required by each of the inductive elements. This quantity is in addition to the sum of all of the magnetizing VARs produced by all of the capacitive elements. Note that the sign of the capacitive VARs is taken as negative.

It will be shown later that the inductive elements of certain circuits require a value of magnetizing VARs equal to that produced by capacitive elements. Such circuits require no external VARs. Thus, when the circuit is viewed from the input terminals, it appears as a pure resistance. This type of circuit is said to be *in resonance* because there is a balance between the magnetizing VAR supply and the demand.

It is important to understand the similar concepts of conservation of energy and conservation of magnetizing VARs. In ac power systems, an adequate supply of both watts and magnetizing VARs must be available. Otherwise, the system will not perform as expected. Inductive load equipment such as fluorescent light ballasts, power distribution lines, and induction motors all require capacitors to ensure the supply of magnetizing VARs.

PROBLEM 1

Statement of the Problem

The series circuit in Figure 6–6 consists of a noninductive lighting load with a resistance of 30 Ω and a capacitor with a capacitive reactance of 40 Ω. These components are connected across a 200-V, 60-Hz source. The current in the circuit is limited by resistance and by capacitive reactance. Determine the following quantities for this circuit:

1. The impedance, in ohms
2. The current, in amperes
3. The true power, in watts
4. The reactive power, in VARs
5. The apparent power, in volt-amperes
6. The power factor

Solution

1. The impedance in an ac series circuit is the result of combining the resistance in ohms and the reactance in ohms. The reactance may be inductive reactance or capacitive reactance. In the series circuit in Figure 6–6, the resistance and the capacitive

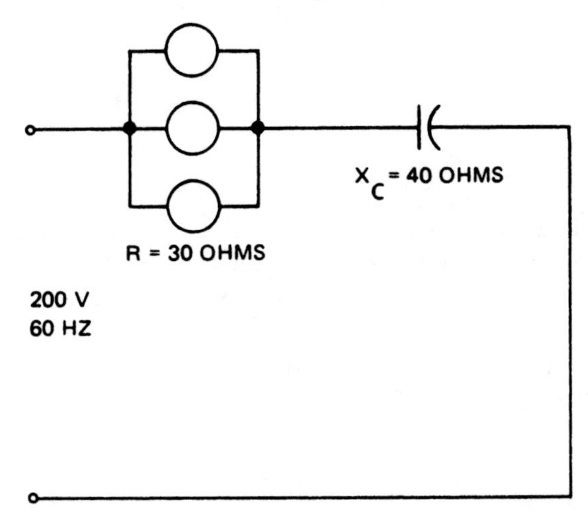

FIGURE 6–6 R and X_C in series ***(Delmar/Cengage Learning)***

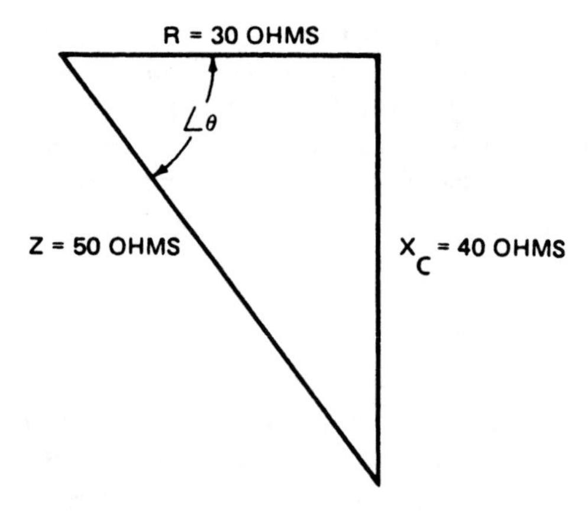

FIGURE 6–7 Impedance triangle for a series circuit containing R and X_C (*Delmar/Cengage Learning*)

reactance are combined to obtain the impedance. The impedance triangle for this circuit is shown in Figure 6–7. The triangle consists of a base leg of resistance (30Ω), an altitude leg of capacitive reactance (40 Ω), and a hypotenuse of impedance (50 Ω). Note that the impedance triangle for the circuit of Figure 6–6 is inverted. The reason for the position of the impedance triangle will be explained later in this problem. The hypotenuse of the impedance triangle is equal to the square root of the sum of the squares of the other two sides:

$$Z = \sqrt{R^2 + X_C^2}$$

$$= \sqrt{30^2 + 40^2}$$

$$= \sqrt{2500} = 50\ \Omega$$

2. The line voltage is 200 V. This voltage causes a current in the series circuit of

$$I = \frac{V}{Z} = \frac{200}{50} = 4\text{ A}$$

3. The true power in the circuit is used in the resistance of the circuit because the resistance of the capacitor is negligible. The true power can be found from the formula $W = I^2R$. The value of the lighting load is obtained by multiplying the voltage drop across the lighting load by the current. Another method of determining this load is to divide the resistance into the square of the voltage across the resistor. The actual true power can be computed by any one of the formulas given:

$$P = I^2R$$

$$= 4^2 \times 30$$

$$= 16 \times 30$$

$$= 480\text{ W}$$

or

$$P = V_R I$$
$$= 120 \times 4$$
$$= 480 \text{ W}$$

or

$$P = \frac{V_R^2}{R}$$
$$= \frac{120 \times 120}{30}$$
$$= 480 \text{ W}$$

4. The reactive power is known as the *magnetizing VARs*. This quantity is the product of the current and the voltage at the capacitor terminals. This voltage is 90° out of phase with the current.

$$V_C = IX_C = 4 \times 40 = 160 \text{ V across the capacitor}$$
$$\text{VARs} = -V_C I = -160 \times 4 = -640 \text{ VARs}$$

This value can also be obtained by using either of the remaining formulas:

$$\mathbf{VARs = -I^2X_C}$$

or

$$\mathbf{VARs = \frac{-V_C^2}{X_C}}$$

5. The apparent power input in volt-amperes for this circuit can be found by either of two methods:
   a. Line voltage × line current = volt-amperes (VA)
      VI = 200 × 4 = 800 VA
   b. The input power in volt-amperes is also the result of combining the true power and the magnetizing VARs. This input power is represented in Figure 6–8 as the hypotenuse of a right triangle. The base leg in this case is the true power in watts and the altitude leg is the magnetizing VARs. The power triangle for the circuit is inverted in the same manner as the impedance triangle in Figure 6–7. The input power is the square root of the sum of the squares of the other two sides of the triangle:

$$\text{Volt-amperes} = \sqrt{P^2 + \text{VARs}^2}$$
$$= \sqrt{480^2 + 640^2}$$
$$= 800 \text{ VA}$$

6. The input power factor for an ac series circuit is the ratio of the true power, in watts, to the input power, in volt-amperes. In other words, the input power factor is the ratio of the power leg of the power triangle to the hypotenuse. Thus, the power factor is the cosine of the included angle θ.

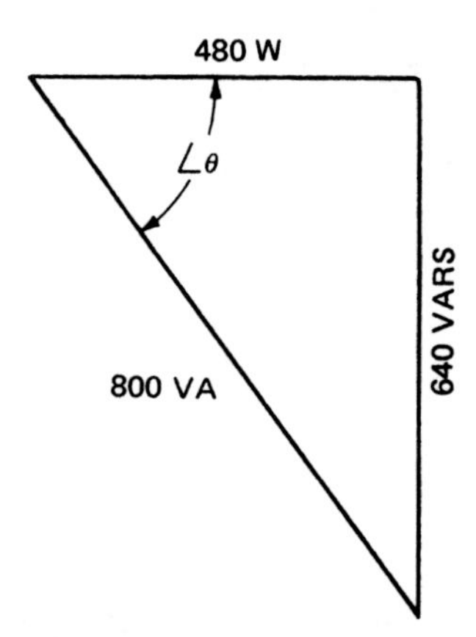

FIGURE 6–8 Power triangle for series circuit containing R and X_C (*Delmar/Cengage Learning*)

DIAGRAMS FOR THE SERIES CIRCUIT

Figure 6–9 shows the impedance triangle, the power triangle, and the vector diagram for this series circuit. In Figure 6–9C, the voltage across the capacitor is drawn downward from point 0 at a 90° angle with the current vector. This position of the capacitor voltage (V_C) indicates that the current leads the voltage by 90 electrical degrees.

Because the impedance triangle is drawn in an inverted position, Figure 6–9A is identical to the triangle of voltages that forms a part of the vector diagram. The capacitive reactance, in ohms, is drawn in a downward direction so that it can be compared to the voltage drop across the capacitor. This drop is shown in a similar direction in the vector diagram. The angle θ is in the same position in both the impedance triangle and in the triangle of voltages in the vector diagram.

The power triangle is also drawn in an inverted position. It is similar to the other diagrams. Because the capacitor produces magnetizing VARs, the VARs are shown in a downward direction. Thus, the phase angle θ is in the same position for all three diagrams. This means that the line current leads the line voltage.

The student should recognize that the formulas and triangles describing an RL circuit are the same as those describing an RC circuit. However, the V_C vector is 180° out of phase with the V_L vector.

The power factor of a circuit containing resistance and capacitive reactance can be obtained using the values from any one of the three triangles given in Figure 6–9.

Using values from the impedance triangle, we obtain

$$\text{Cosine } \angle\theta = \frac{R}{Z}$$
$$= \frac{30}{50}$$
$$= 0.60 \text{ power factor, leading}$$
$$= 53.1^\circ \text{ leading}$$

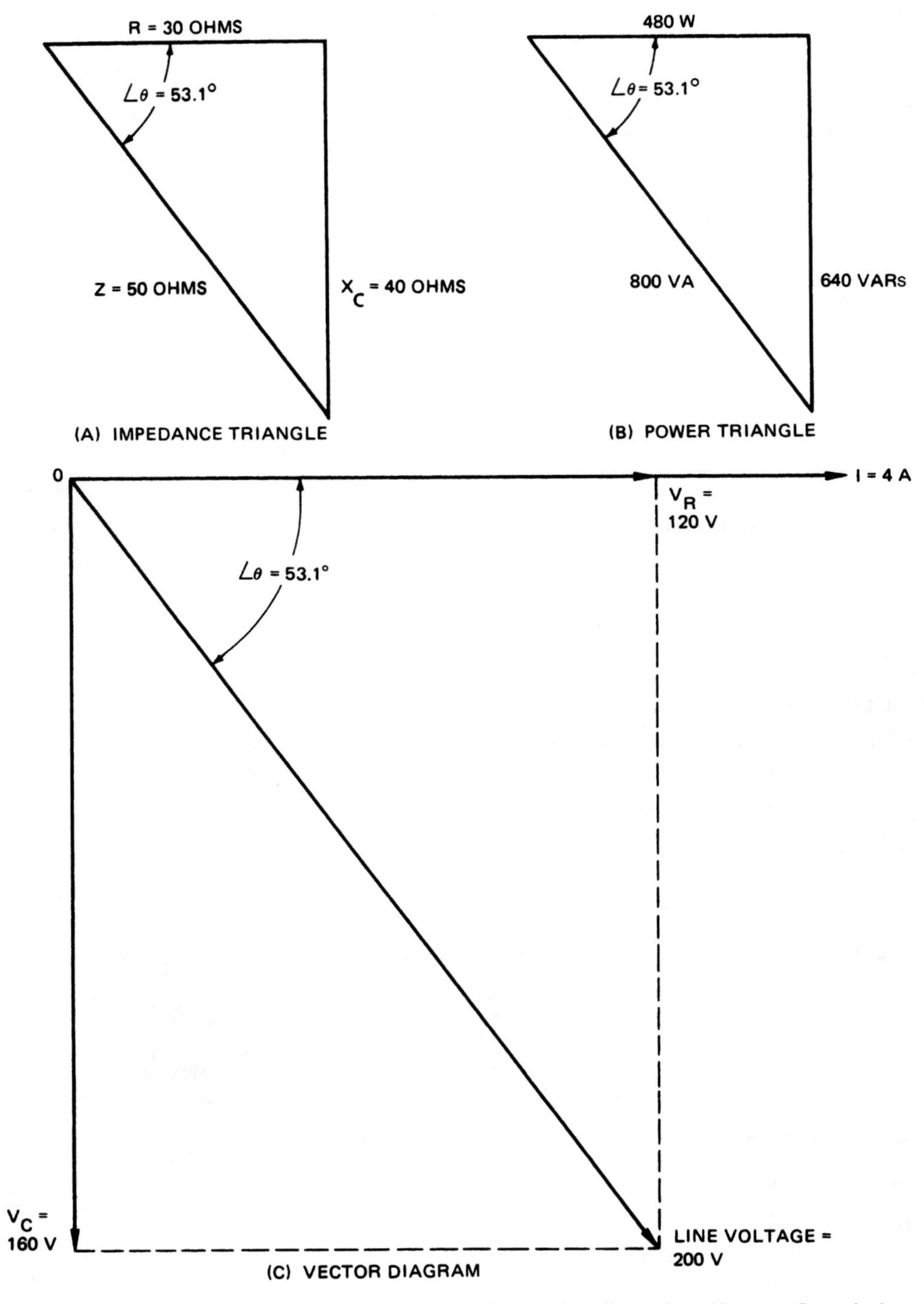

FIGURE 6–9 Analysis of an ac series circuit containing R and X_C ***(Delmar/Cengage Learning)***

Using values from the power triangle yields

$$\text{Cosine } \angle\theta = \frac{P}{VA}$$
$$= \frac{480}{800}$$
$$= 0.60 \text{ power factor, leading}$$
$$= 53.1^\circ \text{ leading}$$

Using values from the vector diagram of voltages, we obtain

$$\text{Cosine } \angle\theta = \frac{V_R}{V}$$
$$= \frac{120}{200}$$
$$= 0.60 \text{ power factor, leading}$$
$$= 53.1^\circ \text{ leading}$$

The power factor angle (θ) for this circuit is 53.1° leading. This means that the line current leads the line voltage by 53.1°.

PROBLEM 2

Statement of the Problem

In the circuit shown in Figure 6–10, a resistor and a capacitor are connected in series. The circuit is connected to a 240-V, 60-Hz line. An ammeter measures a total current of 18.75 A, and a wattmeter indicates a true power of 3375 W. Use the formulas

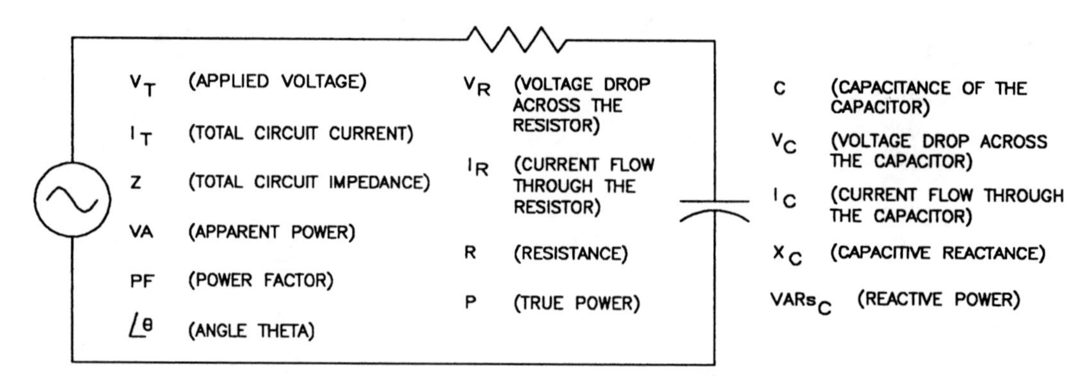

FIGURE 6–10 **RC series circuit (*Delmar/Cengage Learning*)**

shown in the Resistive Capacitive (Series) section of Appendix 15 to find the following values:

1. Volt-amperes, apparent power (VA)
2. Volt-amperes-reactive, reactive power (VARs)
3. Capacitive reactance (X_C)
4. Capacitance (C)
5. Voltage drop across the capacitor (V_C)
6. Resistance of the resistor (V_R)
7. Total circuit impedance (Z)
8. Power factor (PF)
9. Angle theta, which indicates the angle at which the voltage and current are out of phase with each other ($\angle\theta$)

Solution

1. Because the applied voltage and total current are known, the apparent power can be computed using the following formula:

$$VA = V_T \times I_T$$
$$= 240 \times 18.75$$
$$= 4500 \text{ volt-amperes}$$

2. Now that both the apparent power and the true power are known, the reactive power (VARs) can be computed using the following formula:

$$VARs = \sqrt{VA^2 - P^2}$$
$$= \sqrt{4500^2 - 3375^2}$$
$$= 2976.47 \text{ VARs}$$

3. In a series circuit, the current flow must be the same at any point in a circuit. Because the current flow and reactive power are both known, the capacitive reactance can now be computed using the following formula:

$$XC = \frac{VARs}{I_C^2}$$
$$= \frac{2976.47}{351.562}$$
$$= 8.466\ \Omega$$

4. The capacitance of the capacitor can be computed using the following formula:

$$C = \frac{1}{2\pi f X_C}$$

$$= \frac{1}{3191.68}$$

$$= 0.0003133 \text{ F} \quad \text{or} \quad 313.3\ \mu\text{F}$$

5. The voltage drop across the capacitor can be found using the following formula:

$$V_C = I_C \times X_C$$

$$= 18.75 \times 8.466$$

$$= 158.74 \text{ V}$$

6. The resistance of the resistor can be computed using the following formula:

$$R = \frac{P}{I_R^2}$$

$$= \frac{3375}{351.562}$$

$$= 9.6\ \Omega$$

7. The total circuit impedance can be computed using the following formula:

$$Z = \frac{V_T}{I_T}$$

$$= \frac{240}{18.75}$$

$$= 12.8\ \Omega$$

8. The circuit power factor can be determined by using the following formula:

$$PF = \frac{R}{Z}$$

$$= \frac{9.6\ \Omega}{12.8\Omega}$$

$$= 0.75 \quad \text{or} \quad 75\%$$

9. The phase angle difference between the current and voltage can be computed using the following formula:

$$\cos \angle\theta = PF$$

$$\angle\theta = 41.41°$$

SUMMARY

- Charging a capacitor
  1. An opposition voltage builds up across the capacitor plates.
  2. The charging current has a maximum value when the capacitor is uncharged.
  3. The current decreases to zero as the capacitor voltage builds up to equal the source voltage.
  4. The current remains at zero amperes while the source voltage equals the opposition voltage of the capacitor.

- Reactance
  1. The opposition voltage that develops as a capacitor charges is really a countervoltage because it opposes the line voltage and limits the current.
  2. The current-limiting effect of this countervoltage (capacitance) in an ac circuit is expressed in a unit called *capacitive reactance* and is measured in ohms

     $$X_C = \frac{1}{2\pi fC}$$

     where

     $2\pi = 6.28$
     f = frequency, in cycles per second (hertz)
     C = capacitance, in farads
  3. The value of X_C in ohms is inversely proportional to the capacitance in farads and the frequency of the impressed ac voltage.

- Ohm's law for determining the current, in amperes, in an ac circuit containing capacitance is

  $$I = \frac{V}{X_C}$$

- Lead/lag currents
  1. Current in a pure resistor is in phase with the voltage.
  2. Current in a pure inductive reactance lags the impressed voltage by 90 electrical degrees.
  3. Current in a circuit with pure capacitive reactance leads the impressed voltage by 90 electrical degrees.

- Angle of phase defect
  1. The phase angle between the capacitor voltage and current is slightly less than 90°.
  2. In some capacitors having mica dielectrics, the angle of phase defect is as small as three or four minutes.
  3. Capacitors having other types of dielectrics may have angles of phase defect greater than one degree, causing high power loss.
  4. If the angle of phase defect is too large, the internal temperature of the capacitor may increase and shorten the useful life of the dielectric.

- The power factor of a capacitor relates the power losses of a capacitor to its volt-ampere rating.
    1. The power factor is the ratio of the power loss to the volt-ampere rating when the capacitor is operated at the rated voltage and frequency.
    2. The power factor is usually 0.01 (1%) or less for power factor correction capacitors.
- The term *Q* describes capacitor losses and is expressed as

$$Q = \frac{X_C}{R_e}$$

where R_e is the equivalent resistance that includes the effects of dielectric losses and resistive losses within the capacitor.

- Q is a relatively large number and is easier to use than the small values of the power factors for capacitors.
- The dc working voltage of a capacitor should be high enough to withstand the maximum or peak voltage of an ac circuit:

$$V_{maximam} = 1.414 \times V_{(RMS)}$$

- A capacitor should be used only for the type of service for which it is designed.
- The power in a capacitor, in watts, at any instant is equal to the product of the voltage and current values at the same instant.
- The true power or net power taken by the capacitor at the end of one complete cycle, or at the end of any number of complete cycles, is zero.
- Magnetizing VARs are present in a pure inductive circuit where the current lags the applied voltage by a value up to 90 electrical degrees.
- Magnetizing VARs are present in a pure capacitive circuit where the current leads the applied voltage by a value up to 90 electrical degrees.
- Magnetizing VARs are represented by the product of the applied voltage and the current. (This value does not represent a true power loss.)
- For a pure inductance:

$$\text{VARs} = \frac{V^2}{X_L} = VI = I^2X_L$$

- For a pure capacitance:

$$\text{VARs} = -\frac{V^2}{X_C} = -VI = -I^2X_C$$

(The minus sign means that the VARs for a capacitor are opposite to those for an inductance.)

- Coils consume magnetizing VARs, and capacitors supply magnetizing VARs.

- A circuit is in resonance when
  1. the circuit is viewed from the input terminals and it appears to be pure resistance.
  2. there is a balance between the magnetizing VAR supply and the demand.
  3. the inductive elements of certain circuits receive a value of magnetizing VARs equal to that produced by the capacitive elements.
- The conservation of energy principle states that the total power supplied to a circuit is equal to the sum of the values of power consumed in each of the individual circuit components.
- In an ac power system, an adequate supply of watts and magnetizing VARs must be available if the system is to perform as expected.
- Inductive load equipment, such as fluorescent light ballasts, power distribution lines, and induction motors, all require capacitors to ensure an adequate supply of magnetizing VARs.

Achievement Review

1. What is capacitive reactance?
2. Three capacitors are connected in parallel across a 230-V, 60-Hz supply. These capacitors have values of 10 μF, 30 μF, and 60 μF.
   a. A single capacitor can replace the three capacitors. What value of capacitance is required to do this?
   b. Determine the total current taken by the three capacitors.
   c. What is the current in the 10-μF capacitor?
3. The three capacitors in question 2 are reconnected in series across the same ac supply.
   a. What is the total capacitance, in microfarads, of these capacitors?
   b. What is the capacitive reactance, in ohms, of these capacitors connected in series?
   c. What is the current in the 10-μF capacitor?
4. a. A 100-μF capacitor is connected across a 240-V, 60-Hz source. Determine
   (1) the capacitive reactance, in ohms.
   (2) the current, in amperes.

b. Assuming that this same capacitor is connected across a 240-V, 25-Hz source, determine
   (1) the capacitive reactance, in ohms.
   (2) the current, in amperes.

c. Give reasons for any differences in the capacitive reactance and the current in parts (a) and (b) of this problem.

5. Explain what is meant by the term *angle of phase defect.*

6. A capacitor has an angle of phase defect of 1.2°.
   a. What is the actual angle by which the current leads the impressed voltage for this capacitor?
   b. What is its power factor?

7. A noninductive load with a resistance of 30 Ω is connected in series with a capacitor. The capacitor has a capacitive reactance of 25 Ω and negligible resistance. The series circuit is energized from a 210-V, 60-Hz source. Determine
   a. the impedance of the series circuit.
   b. the current, in amperes.
   c. the loss in volts across the noninductive resistance load.
   d. the loss in volts across the capacitor.

8. For question 7, determine
   a. the power in watts expended in the series circuit.
   b. the volt-amperes-reactive component for the series circuit, in VARs.
   c. the apparent power, in volt-amperes.
   d. the circuit power factor.
   e. the power factor angle for this circuit.

9. For the series circuit in question 7, draw the following diagrams using convenient scales:
   a. An impedance triangle
   b. A power triangle
   c. A vector diagram

10. What is the rating, in microfarads, of the capacitor used in the series circuit in question 7?

11. An experimental series circuit consists of a noninductive variable resistor connected in series with a 1-μF capacitor of negligible resistance. The variable resistor has a maximum resistance of 5000 Ω. This series circuit is connected across a 50-V, 60-Hz source.
    a. If the current is to lead the line voltage by a phase angle of 45°, what is the resistance in ohms that must be inserted in the series circuit by adjusting the variable resistor?

b. What is the current in this series circuit?

c. What is the power factor of this circuit?

d. What is the power expended in the circuit?

12. In question 11, determine
    a. the loss in volts across the variable resistor.
    b. the loss in volts across the capacitor.
    c. the volt-ampere reactive component (VARs) for the capacitor.
    d. the apparent power for the entire series circuit.

13. For the circuit in question 11, draw the impedance triangle and the vector diagram, using convenient scales.

14. A paper capacitor can be used in ac and dc circuits. It has a dc working voltage of 500 V. Can this capacitor be connected directly across a 440-V, 60-Hz source?

15. A capacitor is connected across a 120-V, 60-Hz supply. An ammeter indicates 5 A and a wattmeter indicates 10 W. Determine
    a. the resistance of the capacitor.
    b. the power factor.
    c. the phase angle to the nearest tenth of a degree.
    d. the angle of phase defect to the nearest tenth of a degree.

PRACTICE PROBLEMS FOR UNIT 6

Resistive Capacitive Series Circuits

Find the missing values in the following circuits. Refer to problem 2 and Figure 6–10.

1. Assume that the circuit shown in Figure 6–10 is connected to a 480-V, 60-Hz line. The capacitor has a capacitance of 165.782 μF, and the resistor has a resistance of 12 Ω.

V_T = 480 volts	V_R = ________	V_C = ________
I_T = ________	I_R = ________	I_C = ________
Z = ________	R = 12 Ω	X_C = ________
VA = ________	P = ________	$VARs_C$ = ________
PF = ________	$\angle\theta$ = ________	C = 165.782 μF

2. Assume that the voltage drop across the resistor (V_R) is 78 V, the voltage drop across the capacitor (V_C) is 104 V, and the circuit has a total impedance (Z) of 20 Ω. The frequency of the ac voltage is 60 Hz.

V_T = ________	V_R = 78 V	V_C = 104 V
I_T = ________	I_R = ________	I_C = ________
Z = 20 Ω	R = ________	X_C = ________
VA = ________	P = ________	$VARs_C$ = ________
PF = ________	$\angle\theta$ = ________	C = ________

3. Assume the circuit in Figure 6–10 has an apparent power of 144 VA and a true power of 115.2 W. The capacitor has a capacitance of 6.2833 μF, and the frequency is 60 Hz.

V_T = ________	V_R = ________	V_C = ________
I_T = ________	I_R = ________	I_C = ________
Z = ________	R = ________	X_C = ________
VA = 144 VA	P = 115.2 W	$VARs_C$ = ________
PF = ________	$\angle\theta$ = ________	C = 6.2833 μF

4. The circuit shown in Figure 6–10 has a power factor of 68%, an apparent power of 300 VA, and a frequency of 400 Hz. The capacitor has a capacitance of 4.7125 μF.

V_T = ________	V_R = ________	V_C = ________
I_T = ________	I_R = ________	I_C = ________
Z = ________	R = ________	X_C = ________
VA = 300 VA	P = ________	$VARs_C$ = ________
PF = 68 %	$\angle\theta$ = ________	C = 4.7125 μF

7

Series Circuits: Resistance, Inductive Reactance, and Capacitive Reactance

Objectives

After studying this unit, the student should be able to

- analyze a series ac circuit containing R, X_L, and X_C components.
- calculate the net reactance of an RLC series circuit.
- determine the impedance for a series ac circuit containing R, X_L, and X_C components, by the use of formulas.
- calculate the power factor for a series RLC circuit.
- perform vector analysis of a series ac circuit containing R, X_L, and X_C components.
- develop vector diagrams showing the relationship between the voltage drops across the R, X_L, and X_C components and the applied line voltage.
- plot the phase relationships between the voltage and current in each resistive, inductive, and capacitive component in a series ac circuit.
- define a resonant series circuit.
- describe and show graphically the effects of frequency at resonance, frequencies above resonance, and frequencies below resonance on R, X_L, and X_C components.

Many alternating-current series circuits contain resistance, inductance, and capacitance. Inductance and capacitance have opposite effects in an ac circuit. The student must be able to analyze this type of circuit and perform calculations to obtain the various circuit values.

SERIES RLC CIRCUIT

Figure 7–1 shows a series circuit containing a resistor, an inductance coil with negligible resistance, and a capacitor. For this type of circuit, it is important to remember that (1) the voltage (V) across an inductor (L) leads the current (I), and (2) the current (I) in a capacitor (C) leads the voltage (V).

This information can be expressed by means of a vector diagram, as in Figure 7–2.

The voltage component across the inductor (V_L) is rotated counterclockwise so that it leads the current (I) by 90°. The current in the capacitor leads the voltage (V_C) by 90°. The diagram shows that the voltage drops across the inductor and the capacitor are 180° out of phase (Figure 7–2A). This means that they oppose each other. When the components in the voltage vector diagram are divided by I, the resulting reactance vectors also oppose each other (Figure 7–2B).

Net Reactance of an RLC Series Circuit

The net reactance, in ohms, for an RLC series circuit is found by subtracting the capacitive reactance from the inductive reactance. For the circuit in Figure 7–1:

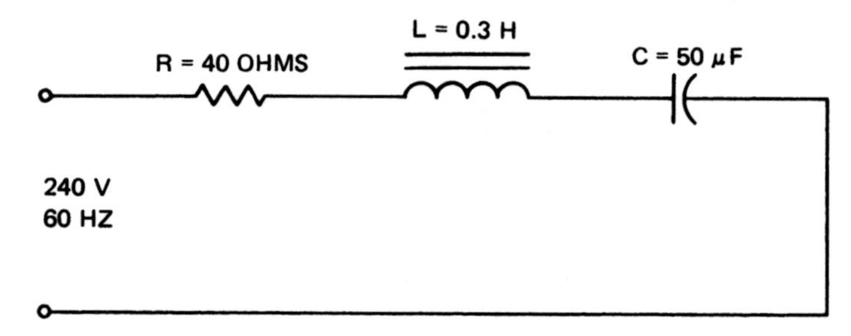

FIGURE 7–1 Series circuit containing R, X_L, and X_C ***(Delmar/Cengage Learning)***

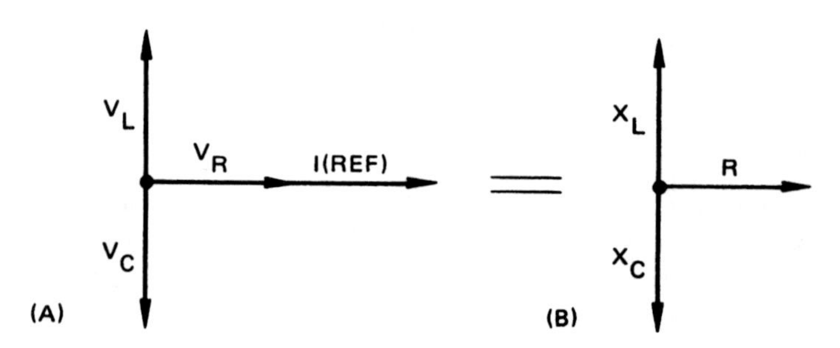

FIGURE 7–2 Vector diagram of V_R, V_L, and V_C ***(Delmar/Cengage Learning)***

Inductive reactance:

$$X_L = 2\pi fL = 2 \times 3.14 \times 60 \times 0.3 = 113.1\ \Omega$$

Capacitive reactance:

$$X_C = \frac{1}{2\pi fc} = \frac{1}{2 \times 3.14 \times 60 \times 0.00005}$$

$$= \frac{1}{0.01885}$$

$$= 53.1\ \Omega$$

Net inductive reactance:

$$X = X_L - X_C = 113.1 - 53.1 = 60\ \Omega$$

This example shows that 53.1 Ω of inductive reactance is canceled by the 53.1 Ω of capacitive reactance. The difference of 60 Ω represents the inductive reactance that affects the operation of the series circuit.

Impedance of the RLC Series Circuit

Impedance is the result of the combination of resistance and reactance. In Figure 7–1, the impedance is the combination of the resistance of 40 Ω and the difference between the inductive reactance and the capacitive reactance. The impedance for this circuit is

$$Z = \sqrt{R^2 + (X_L - X_C)^2}$$

$$= \sqrt{40^2 + (113.1 - 53.1)^2}$$

$$= \sqrt{40^2 + 60^2}$$

$$= 72.1\ \Omega$$

The current in this series circuit is

$$I = \frac{V}{Z} = \frac{240}{72.1} = 3.33\ A$$

Voltage Drop across the Elements

The voltage drops across the resistor, the inductor coil, and the capacitor are determined as follows:

Voltage drop across R:

$$V_R = IR$$

$$= 3.33 \times 40$$

$$= 133.2\ V$$

Voltage drop across L:

$$V_L = IX_L$$
$$= 3.33 \times 113.1$$
$$= 376.6 \text{ V}$$

Voltage drop across C:

$$V_C = IX_C$$
$$= 3.33 \times 53.1$$
$$= 176.82 \text{ V}$$

Vector Diagram of the Series RLC Circuit

The vector diagram for the series RLC circuit of Figure 7–1 is shown in Figure 7–3. V_L is shown in the positive Y direction, and V_C is shown in the negative Y direction. The magnitude of $V_L - V_C$ is found by means of the vector addition rule of placing vectors head to tail. Thus, the V_C vector is placed on top of the V_L vector (head to tail) to find the difference: $V_L - V_C = 200$ V. It is important that the student understand that the vectors were not subtracted to equal 200 V, but were added.

The vector sum of the voltage across the resistor, 133.2 V, and the voltage across the net inductive reactance, 200 V, is equal to the line voltage, 240 V:

$$V = \sqrt{V_R^2 + V_L^2}$$
$$= \sqrt{133.2^2 + (376.6 - 176.6)^2} = 240 \text{ V}$$

Power Factor of the Series RLC Circuit

The power factor of this type of circuit can be determined by any of the methods used in previous series circuit problems. The power factor of this circuit is

$$PF = \frac{V_R}{V} = \frac{133.2}{240} = 0.555 \text{ lag}$$
$$= \frac{R}{Z} = \frac{40}{72.1} = 0.555 \text{ lag}$$

Because the inductive reactance is greater than the capacitive reactance, this series circuit has a lagging power factor. The power factor has a lagging phase angle of 56.3°. This means that the current lags the line voltage by 56.3°, as shown in Figure 7–3.

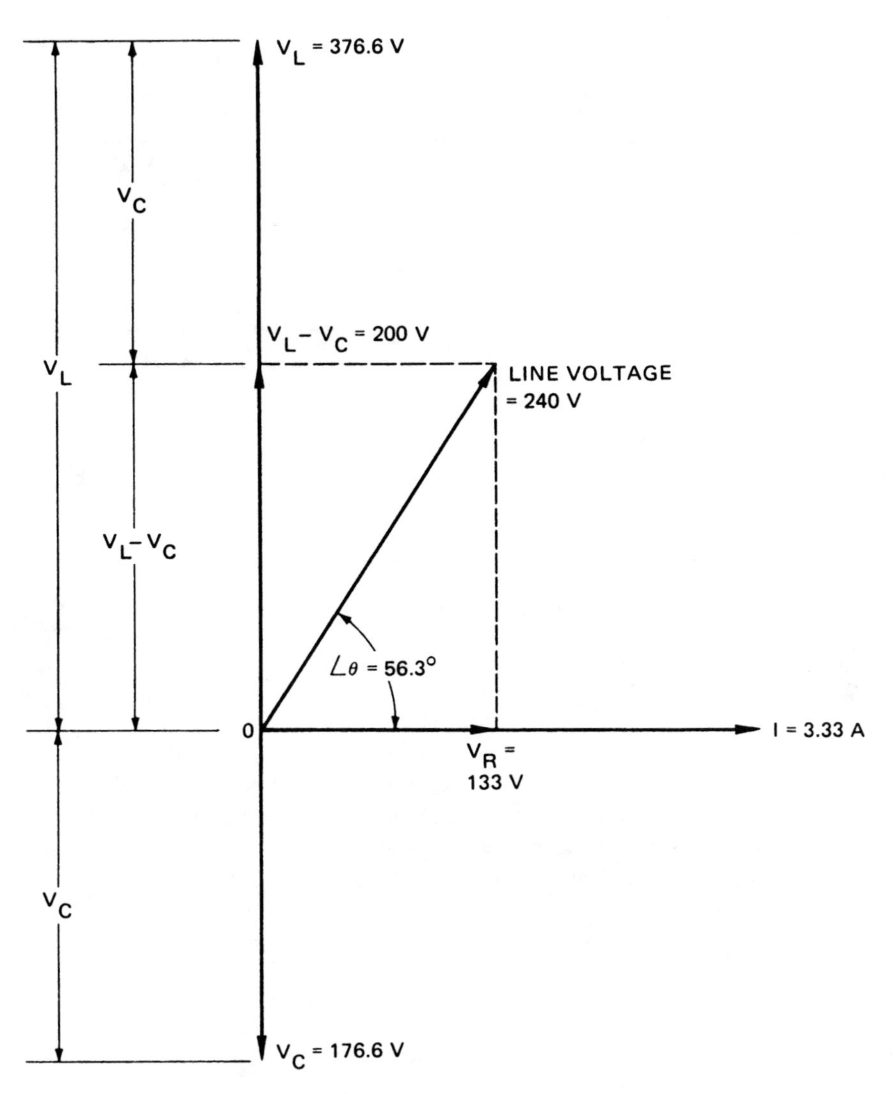

FIGURE 7–3 ***(Delmar/Cengage Learning)***

PROBLEM 1

Statement of the Problem

For the circuit in Figure 7–1, the inductive reactance is greater than the capacitive reactance. For the circuit in Figure 7–4, the capacitive reactance is more than the inductive reactance.

This series circuit consists of three components: a noninductive 20-Ω resistor, an inductance coil having an inductance of 0.1 H and negligible resistance, and a 50-μF

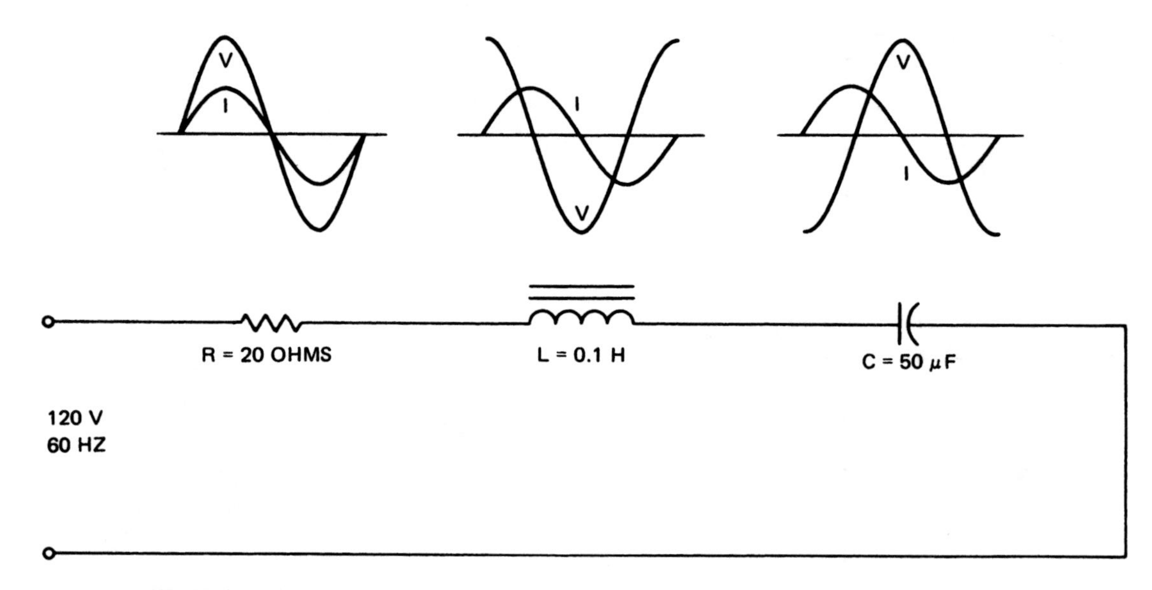

FIGURE 7–4 Series circuit containing R, X_L, and X_C (*Delmar/Cengage Learning*)

capacitor. These components are connected to a 120-V, 60-Hz source. Figure 7–4 also shows the relationship between the line current and voltage waveforms across each component. The voltage across the resistance is in phase with the current, the voltage across the inductor coil leads the current by 90°, and the voltage across the capacitor lags the current by 90°.

The following circuit values are to be determined for this series RLC circuit:

1. The impedance
2. The current
3. The voltage across (a) the resistor, (b) the coil, and (c) the capacitor
4. The line voltage (check)
5. The true power, in watts, taken by the series circuit
6. The magnetizing VARs required by the coil
7. The magnetizing VARs supplied by the capacitor
8. The net magnetizing VARs supplied to the power source
9. The input volt-amperes for the series circuit
10. The circuit power factor and the phase angle

Solution

1. The inductive reactance equals $2\pi fL$, and the capacitive reactance equals $\frac{1}{2\pi fC}$. Therefore, the impedance formula can be stated in either of these forms:

$$Z = \sqrt{R^2 + (X_L - X_C)^2}$$

or

$$\mathbf{Z = \sqrt{R^2 + \left(2\pi fL - \frac{1}{2\pi fC}\right)^2}}$$

If the capacitive reactance is greater than the inductive reactance, the quantity in parentheses in the formulas is negative. Because the square of a negative number is positive, $(X_L - X_C)^2$ is always positive regardless of the term that is greater, X_L or X_C. The impedance of the series circuit is

$$Z = \sqrt{R^2 + \left(2\pi fL - \frac{1}{2\pi fC}\right)^2}$$

$$= \sqrt{20^2 + \left(2 \times 3.14 \times 60 \times 0.1 - \frac{1}{2 \times 3.14 \times 60 \times 0.00005}\right)^2}$$

$$= \sqrt{400 + 235.6225}$$

$$= 25.21\ \Omega$$

2. The current in the circuit is

$$I = \frac{V}{Z} = \frac{120}{25.21} = 4.76\ \text{A}$$

3. The voltage across each of the three circuit components is

Resistor:	$V_R = IR$	$= 4.76 \times 20$	$= 95.2$ V
Coil:	$V_L = IX_L$	$= 4.76 \times 37.7$	$= 179.45$ V
Capacitor:	$V_C = IX_C$	$= 4.76 \times 53.05$	$= 252.52$ V

4. The voltage across the capacitor is greater than the voltage across the coil. These two voltages are 180° out of phase and oppose each other. Subtracting these voltages yields

   $$252.52 - 179.45 = 73.07\ \text{V}$$

   This means that the part of the capacitor voltage that is not canceled by the coil voltage is equal to 73 V. The vectorial addition of 73 V and the loss across the resistor, 95.2 V, gives the line voltage, or 120 V. The line voltage can be determined using the right-triangle method, in which

$$V = \sqrt{V_R^2 + V_C^2}$$

$$= \sqrt{95.2^2 + 73^2} = 120.0\ \text{V}$$

5. The true power, in watts, taken by the series circuit can be determined by

$$P = \frac{V_R^2}{R}$$

$$= \frac{95.2^2}{20}$$

$$= 453.2 \text{ W}$$

or

$$P = V_R \times I$$

$$= 95.2 \times 4.76$$

$$= 453.2 \text{ W}$$

or

$$P = I^2R$$

$$= 4.76^2 \times 20$$

$$= 453.2 \text{ W}$$

6. The magnetizing VARs required by the coil is given by

$$\text{VARs} = \frac{V_L^2}{X_L}$$

$$= \frac{179.45^2}{37.7}$$

$$= 854.17 \text{ VARs}$$

or

$$\text{VARs} = V_L \times I$$

$$= 179.45 \times 4.76$$

$$= 854.18 \text{ VARs}$$

or

$$\text{VARs} = I^2X_L$$

$$= 4.76^2 \times 37.7$$

$$= 854.19 \text{ VARs}$$

7. The magnetizing VARs supplied by the capacitor is given by

$$\text{VARs} = \frac{X_L^2}{X_C}$$

$$= \frac{252.52^2}{53.05}$$

$$= 1202 \text{ VARs}$$

or

$$\begin{aligned} \text{VARs} &= V_C \times I \\ &= 252.52 \times 4.76 \\ &= 1202 \text{ VARs} \end{aligned}$$

or

$$\begin{aligned} \text{VARs} &= I^2 X_C \\ &= 4.76^2 \times 53.05 \\ &= 1202 \text{ VARs} \end{aligned}$$

8. The net magnetizing VARs supplied to the source is the difference between the VARs supplied by the capacitor and the VARs required by the coil: 1202 − 854 = 348 VARs. The net magnetizing VARs can also be calculated using the net reactance (15.35 Ω capacitive) and the net reactive voltage (73.07 V):

$$\begin{aligned} \text{VARs} &= \frac{(V_C - V_L)^2}{(X_C - X_L)} \\ &= \frac{73.07^2}{15.35} \\ &= 348 \text{ VARs} \end{aligned}$$

or

$$\begin{aligned} \text{VARs} &= (V_C - V_L) \times I \\ &= 73.07 \times 4.76 \\ &= 348 \text{ VARs} \end{aligned}$$

or

$$\begin{aligned} \text{VARs} &= I^2 \times (X_C - X_L) \\ &= 4.76^2 \times 15.35 \\ &= 348 \text{ VARs} \end{aligned}$$

9. The input volt-amperes for the entire series circuit is

$$\begin{aligned} \text{VA} &= VI \\ &= 120 \times 4.76 \\ &= 571.2 \text{ volt-amperes} \end{aligned}$$

or

$$\begin{aligned} \text{VA} &= \sqrt{(\text{watts})^2 + (\text{net VARs})^2} \\ &= \sqrt{453.2^2 + 348^2} \\ &= 571.4 \text{ volt-amperes} \end{aligned}$$

10. There are three methods of calculating the power factor of the series circuit:

$$PF = \frac{R}{Z}$$
$$= \frac{20}{25.2}$$
$$= 0.7936 \text{ lead}$$

or

$$PF = \frac{P}{VA}$$
$$= \frac{453.2}{571.2}$$
$$= 0.7934 \text{ lead}$$

or

$$PF = \frac{V_R}{V}$$
$$= \frac{95.2}{120}$$
$$= 0.7933 \text{ lead}$$

The capacitive reactance is greater than the inductive reactance of this circuit. This means that all of the inductive reactance is canceled. The circuit power factor and the phase angle are determined by the resistance and the capacitive reactance that is not canceled by the inductive reactance. As a result, the series circuit has a leading power factor. The current leads the line voltage by a phase angle of 37.5°.

Voltage and Current Vector Diagram

A voltage and current vector diagram is shown in Figure 7–5 for the series circuit of Figure 7–4. The two reactance voltages are 180 electrical degrees out of phase. Because the voltage across the capacitor is larger, the voltage across the coil is subtracted from V_C. The difference of $V_C - V_L$ is combined by vector addition with the voltage across the resistor to give the line voltage. The current leads the line voltage in a counterclockwise direction by 37.5°.

RESONANCE IN SERIES CIRCUITS

In the solution to problem 1 in this unit, the inductive reactance was greater than the capacitive reactance. As a result, there was a lagging power factor. In problem 2, the capacitive reactance is larger than the inductive reactance and the power factor is leading.

If the inductive reactance equals the capacitive reactance, then V_L equals V_C. Because these voltages are 180° out of phase with each other, they will cancel exactly. The effects of both the inductive reactance and the capacitive reactance are now removed from the

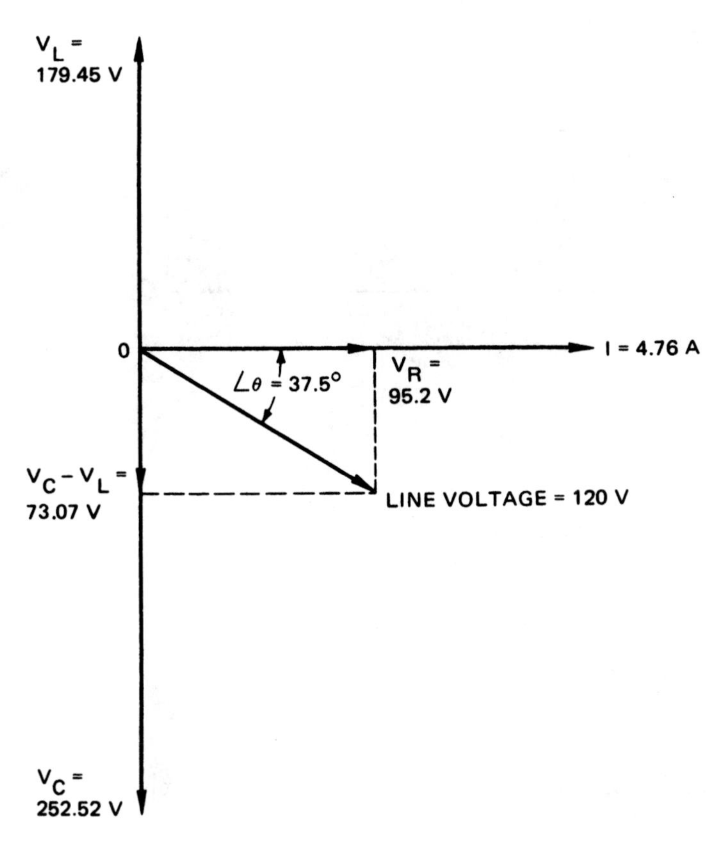

FIGURE 7–5 ***(Delmar/Cengage Learning)***

series circuit. Thus, only the resistance of the circuit remains to limit the current. A circuit for which these conditions are true is called a *resonant* circuit. The full line voltage appears across the resistance component.

PROBLEM 2

Statement of the Problem

A resonant series circuit is shown in Figure 7–6. The phase angle between the current and the line voltage for this circuit is zero. The power factor is 1.00.

Another way of looking at this circuit is to recognize that the value of the magnetizing VARs required by the coil is equal to the magnetizing VARs supplied by the capacitor. These values cancel each other so that no VARs appear at the input terminals.

Determine the following quantities for the resonant series circuit in Figure 7–6:

1. The impedance
2. The current
3. The voltage across the resistor
4. The voltage across the coil

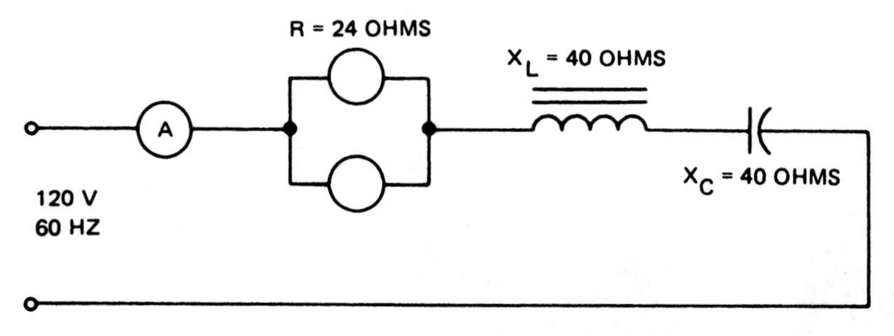

FIGURE 7–6 Resonant series circuit (*Delmar/Cengage Learning*)

5. The voltage across the capacitor
6. The loss in the resistor, in watts
7. The magnetizing VARs required by the coil
8. The magnetizing VARs supplied by the capacitor
9. The input in volt-amperes
10. The power factor and the phase angle for the series circuit

Then develop the vector diagram for the circuit.

Solution

1. The impedance of a resonant circuit is the same as the resistance of the circuit. The formula for impedance is

$$\mathbf{Z = \sqrt{R^2 + \left(2\pi fL - \frac{1}{2\pi fC}\right)^2}}$$

or

$$\mathbf{Z = \sqrt{R^2 + (X_L - X_C)^2}}$$

At resonance, the inductive reactance equals the capacitive reactance. Thus, when these values are subtracted, the reactance of the resonant circuit is zero:

$$2\pi fL - \frac{1}{2\pi fC} = 0$$

or

$$(X_L - X_C) = 0$$

$$(40 - 40) = 0$$

The impedance in a resonant series circuit becomes

$$Z = \sqrt{R^2 + 0^2} = R$$

For the circuit in Figure 7–6, the impedance is

$$Z = \sqrt{R^2 + (X_L - X_C)^2}$$
$$= \sqrt{24^2 + (40 - 40)^2} = 24\ \Omega$$

2. The impedance and resistance are the same in a resonant series circuit. Therefore, Ohm's law defines the current:

$$\mathbf{I} = \frac{\mathbf{V}}{\sqrt{\mathbf{R}^2 + (\mathbf{X}_L - \mathbf{X}_C)^2}}$$
$$= \frac{\mathbf{V}}{\sqrt{\mathbf{R}^2 + \mathbf{0}^2}} = \frac{\mathbf{V}}{\sqrt{\mathbf{R}^2}} = \frac{\mathbf{V}}{\mathbf{R}}$$

The current for the resonant circuit is

$$I = \frac{V}{R} = \frac{120}{24} = 5\ A$$

3. The voltage across the total resistance of a resonant series circuit equals the line voltage. It is assumed that there is no resistance in either the coil or the capacitor. The voltage across the lamp load is

$$V_R = IR$$
$$= 5 \times 24 = 120\ V = V$$

4. The voltage across the coil is

$$V_L = IX_L = 5 \times 40 = 200\ V$$

5. The voltage across the capacitor is

$$V_C = IX_C = 5 \times 40 = 200\ V$$

6. The power loss in the resistor is 600 W. This value is determined as follows:

$$P = \frac{V_R^2}{R}$$
$$= \frac{120^2}{24}$$
$$= 600\ W$$

or

$$P = V_R \times I$$
$$= 120 \times 5$$
$$= 600\ W$$

or

$$P = I^2R$$
$$= 5^2 \times 24$$
$$= 600 \text{ W}$$

7. The magnetizing VARs required by the coil can be calculated using any one of the following expressions:

$$\text{VARs} = \frac{V_L^2}{X_L}$$
$$= \frac{200^2}{40}$$
$$= 1000 \text{ VARs}$$

or

$$\text{VARs} = V_L \times I$$
$$= 200 \times 5$$
$$= 1000 \text{ VARs}$$

or

$$\text{VARs} = I^2X_L$$
$$= 5^2 \times 40$$
$$= 1000 \text{ VARs}$$

8. The magnetizing VARs supplied by the capacitor is determined using one of these expressions:

$$\text{VARs} = \frac{V_C^2}{X_C}$$
$$= \frac{200^2}{40}$$
$$= 1000 \text{ VARs}$$

or

$$\text{VARs} = V_C \times I$$
$$= 200 \times 5$$
$$= 1000 \text{ VARs}$$

or

$$\text{VARs} = I^2X_C$$
$$= 5^2 \times 40$$
$$= 1000 \text{ VARs}$$

9. The input volt-amperes for the series circuit is the product of the line voltage and the current. The input volt-amperes and the true power are the same in a series resonant circuit:

$$VA = VI = 120 \times 5 = 600 \text{ VA}$$

10. Because the true power and the input volt-amperes are the same for this circuit, the power factor is unity. A unity power factor is also obtained from the ratio of the resistance of the resonant series circuit and the input impedance:

$$PF = \frac{P}{VA} = \frac{600}{600}$$
$$= 1.00 \text{ (unity)}$$

or

$$PF = \frac{R}{Z} = \frac{24}{24}$$
$$= 1.00 \text{ (unity)}$$

11. The phase angle for a power factor of unity is 0°. This means that the line current and the line voltage are in phase, as shown in Figure 7–7.

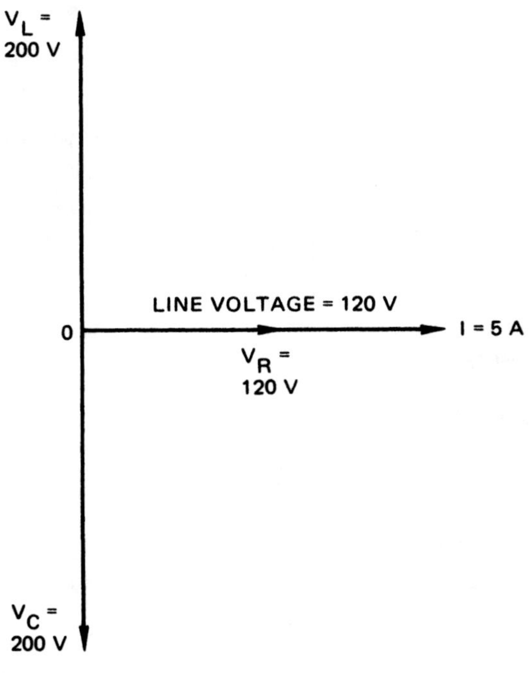

FIGURE 7–7 Vector diagram for resonant series circuit
(Delmar/Cengage Learning)

THE PROPERTIES OF SERIES RESONANCE

Figure 7–8 summarizes the properties of series resonance. Each curve is based on the mathematics of series resonance.

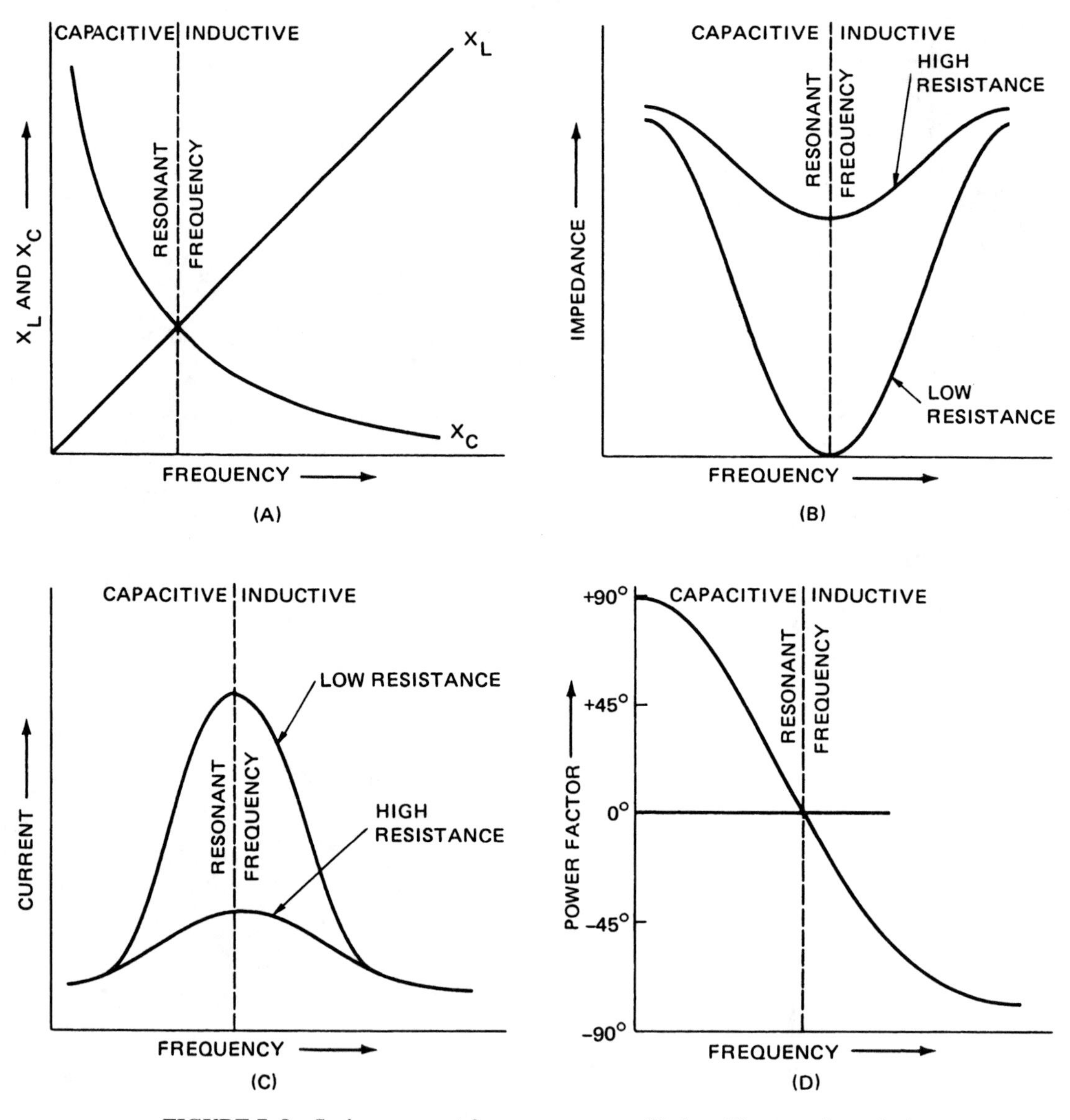

FIGURE 7–8 **Series resonant frequency curves** (*Delmar/Cengage Learning*)

X_L and X_C versus Frequency (Figure 7–8A)

The equation $X_L = 2\pi fL$ shows that as the frequency increases, X_L increases. Also, as the frequency decreases, X_L decreases. When frequencies above resonance are applied to RLC circuits, they become inductive.

The equation $X_C = 1/(2\pi fC)$ shows that as the frequency reaches very high values, X_C approaches zero. As the frequency approaches zero, X_C approaches infinity. When frequencies below resonance are applied to RLC circuits, they become capacitive.

Resonance is defined by the point at which X_L and X_C cross ($X_L = X_C$). Mathematically, this is expressed as follows:

$$X_L = X_C$$

$$2\pi fL = \frac{1}{2\pi fC}$$

$$f^2 = \frac{1}{(2\pi)^2 LC}$$

$$f = \frac{1}{2\pi\sqrt{LC}}$$

The resulting frequency is the natural, or resonant frequency of *any* L + C combination in a circuit. This equation is known as a general equation and the curves are known as *general curves*.

Impedance versus Frequency (Figure 7–8B)

The equation $Z = \sqrt{R^2 + (X_L - X_C)^2}$ is more difficult to plot. However, a general idea of the curve can be obtained by examining Figure 7–8A and the impedance formula.

At frequencies below resonance, the impedance formula takes the form $Z = \sqrt{R^2 + X_C^{\,2}}$ because X_L is negligible. A high value of impedance results at these frequencies.

For frequencies at resonance, $X_L = X_C$ and $X_L - X_C = 0$. As a result, the impedance formula takes the form $Z = \sqrt{R^2}$. At resonance, low values of impedance are obtained.

At frequencies above resonance, the impedance formula has the form $Z = \sqrt{R^2 + X_L^2}$. In this case, X_C is negligible. Again, high values of impedance are obtained. This means that the curve goes from a high value to a low value and back to a high value. These methods of determining the curves yield approximate curves, not exact curves.

Current versus Frequency (Figure 7–8C)

The impedance/frequency waveform of Figure 7–8B is useful in interpreting the current versus the frequency waveform in Figure 7–8C.

According to the equation $I = V/Z$, as the impedance rises, the current falls. The reverse is also true. This means that the current curve is the reciprocal or inverse of the impedance curve.

Graphically, it was convenient to use the impedance curve to explain the current curve. In the laboratory, however, it is easier to vary the frequency and note ammeter readings than it is to vary the frequency and calculate impedance values.

Power Factor versus Frequency (Figure 7-8D)

The formula that defines the power factor curve is PF = R/Z. This formula was selected because it lends itself to the impedance curve.

At frequencies below resonance, the equation PF = R/Z shows the power factor to be inversely proportional to impedance. For high values of impedance, where the frequencies are below resonance, the power factor becomes small. Therefore, for small values of R, it is possible to obtain a power factor near zero. The cosine of the phase angle is zero. Thus, the phase angle equals 90°. Because the frequency is below resonance, the circuit is capacitive, and the power factor is leading or positive.

For frequencies at resonance, Z = R. Thus, the equation PF = R/Z becomes unity. The angle whose cosine is unity is 0°.

For frequencies above resonance, the equation PF = R/Z has the same values as it does when the frequencies are below resonance. In this case, the circuit is now inductive and the power factor is lagging or negative.

SUMMARY

- In an inductive circuit, the current lags the applied voltage across the inductor.
- In a capacitive circuit, the current leads the applied voltage across the capacitor.
- The net reactance of an RLC series circuit is

$$X = X_L - X_C$$

where

$$X_L = 2\pi fL$$

$$X_C = \frac{1}{2\pi fC}$$

[*Note:* The square of a negative number is positive; thus, $(X_L - X_C)^2$ is always positive.]

- The impedance of the RLC series circuit is

$$Z = \sqrt{R^2 + (X_L - X_C)^2}$$

[*Note:* (1) If X_L is greater than X_C, the circuit has a lagging power factor; (2) If X_C is greater than X_L, the circuit has a leading power factor.]

- The power factor of the series RLC circuit is

$$PF = \frac{V_R}{V}$$

or

$$PF = \frac{R}{Z}$$

or

$$PF = \frac{P}{VA}$$

- The true power, in watts, taken by the series RLC circuit is

$$P = \frac{V_R^2}{R}$$

or

$$P = V_R \times I$$

or

$$P = I^2R$$

- The vector sum of the voltage across the resistor and the voltage across the net inductive reactance gives the line voltage

$$V = \sqrt{V_R^2 + V_X^2}$$

where $V_X = V_L - V_C$.

- The magnetizing VARs *required* by the coil is

$$VARs = \frac{V_L^2}{X_L}$$

or

$$VARs = V_L \times I$$

or

$$VARs = I^2X_L$$

- The magnetizing VARs *supplied* by the capacitor is

$$VARs = \frac{V_C^2}{X_C}$$

or

$$VARs = V_C \times I$$

or

$$VARs = I^2X_C$$

- The net magnetizing VARs can be found as follows:

$$\text{VARs} = \frac{(V_C - V_L)^2}{(X_C - X_L)^2}$$

or

$$\text{VARs} = (V_C - V_L) \times I$$

or

$$\text{VARs} = I^2 \times (X_C - X_L)$$

- The input volt-amperes for the entire series circuit is

$$\text{VA} = \text{VI}$$

or

$$\text{VA} = \sqrt{(\text{watts})^2 + (\text{net VARs})^2}$$

- The circuit power factor (PF) and the phase angle ($\angle\theta$) are determined by the resistance and the capacitive or inductive reactance that is *not* canceled.
- At the resonant frequency
  1. The phase angle between the current and the line voltage is zero.
  2. The power factor is 1.00, and the input volt-amperes and the true power are the same: VA = VI. Also,

$$\text{PF} = \frac{P}{\text{VA}}$$

or

$$\text{PF} = \frac{R}{Z}$$

  3. The magnetizing VARs required by the inductors is *equal* to the magnetizing VARs supplied by the capacitor:

$$\frac{V_L^2}{X_L} = \frac{V_C^2}{X_C}$$

or

$$V_L \times I = V_C \times I$$

or

$$I^2X_L = I^2X_C$$

  4. The inductive reactance equals the capacitive reactance.
  5. The reactance is zero.

6. The impedance of the circuit is

$$Z = \sqrt{R^2 + (X_L - X_C = 0)^2} = R$$

7. The current is defined by Ohm's law for the series resonant circuit:

$$I = \frac{V}{R}$$

$$P = \frac{V_R^2}{R}$$

or

$$P = V_R \times I$$

or

$$P = I^2R$$

8. The resonant frequency is

$$f = \frac{1}{2\pi\sqrt{LC}}$$

- At frequencies above resonance:
  1. $Z = \sqrt{R^2 + X_L^2}$ (X_C is negligible)
  2. The circuit is inductive, and the phase angle is such that the current lags the line voltage.
  3. The power factor (PF) is less than 1.00 because

  $$PF = \frac{R}{Z}$$

  4. I = V/Z; thus, as the impedance rises, the current falls.
- At frequencies below resonance:
  1. $Z = \sqrt{R^2 + X_C^2}$ (X_L is negligible)
  2. Impedance values are high and the power factor becomes small.
  3. For small values of R, it is possible to obtain a power factor near zero:

  $$PF = \frac{R}{Z}$$

  4. The circuit is capacitive and the power factor is leading or positive.
- To determine the total impedance of the circuit, both the resistance of the resistor and that of the coil must be represented in R of the formula

$$Z = \sqrt{R^2 + (X_C - X_L)^2}$$

where

$$R_{coil} = \frac{P_{coil}}{I^2}$$

The inductive reactance of the coil is

$$XL = \sqrt{Z^2_{coil} - R^2_{coil}}$$

- The power factor and phase angle of the coil are determined as follows:

$$PF_{coil} = \frac{R_{coil}}{Z_{coil}}$$

Achievement Review

1. A series circuit consists of a 100-Ω resistor, a coil with an inductance of 0.5 H and negligible resistance, and a 40-μF capacitor. These components are connected to a 115-V, 60-Hz source.
   a. Determine
      1. the impedance of the circuit.
      2. the current in the circuit.
      3. the power factor and the phase angle of the circuit (indicating whether the power factor and the phase angle are leading or lagging).
   b. Draw a vector diagram for the circuit.

2. A coil, having a resistance of 100 Ω and an inductance of 0.2 H, is connected in series with a 20-μF capacitor across a 120-V, 60-Hz supply. Determine
   a. the impedance of the circuit.
   b. the current.
   c. the power factor of the circuit.
   d. the voltage across the capacitor.
   e. the instantaneous maximum voltage across the terminals of the capacitor.

3. In the series circuit shown in Figure 7–9, determine
   a. the total impedance.
   b. the voltage drop across the coil.
   c. the capacitance required to obtain resonance.

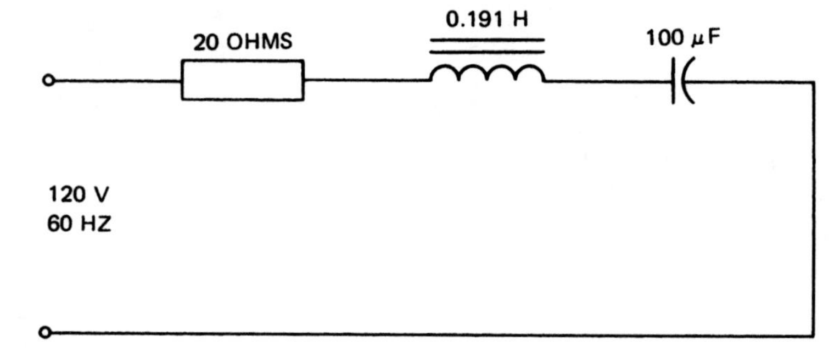

FIGURE 7–9 ***(Delmar/Cengage Learning)***

4. a Explain what is meant by the term *resonance* when used with ac series circuits.
   b. What precaution must be observed when working with series circuits having inductive and capacitive circuit components?

5. For the series circuit shown in Figure 7–10, determine
   a. the frequency at which this circuit will resonate.
   b. the value of the impedance of the circuit at resonance.
   c. the power factor at resonance.
   d. the voltage across the capacitor at the resonant frequency.

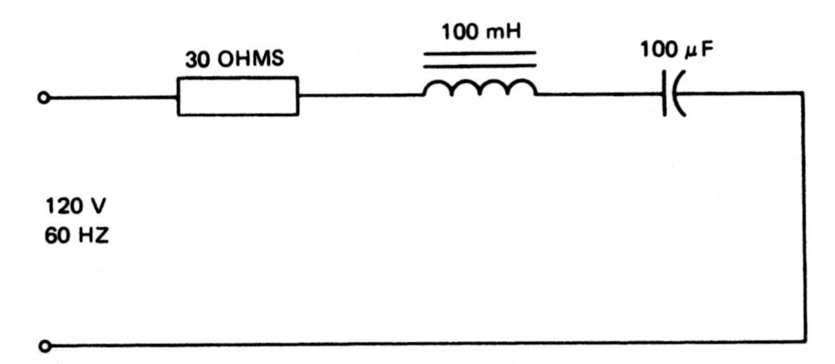

FIGURE 7–10 ***(Delmar/Cengage Learning)***

6. The starting winding circuit of a capacitor-start, induction-run motor consists of a 15-Ω resistance and a 20-Ω inductive reactance. These components are in series with a 50-μF capacitor. The circuit is connected to a 120-V, 60-Hz source. Determine
   a. the impedance of the series circuit.
   b. the current.
   c. the true power.
   d. the power factor.

7. A simple tuning circuit consists of a 100-μH inductance, a 200-pF capacitance, and a 20-Ω resistance connected in series. The voltage from the antenna to ground is 100 μV.
   a. Determine the resonant (natural) frequency of the circuit.
   b. Determine the current, in microamperes, at the resonant frequency.

8. For the circuit given in question 7, determine the voltage across the coil at the resonant frequency.

9. In the circuit shown in Figure 7–11, determine the voltmeter reading for each of the following conditions:
   a. When point A is grounded
   b. When the ground is removed from A and placed at B
   c. When there is no ground at either point A or point B, but there is a "break" in the coil

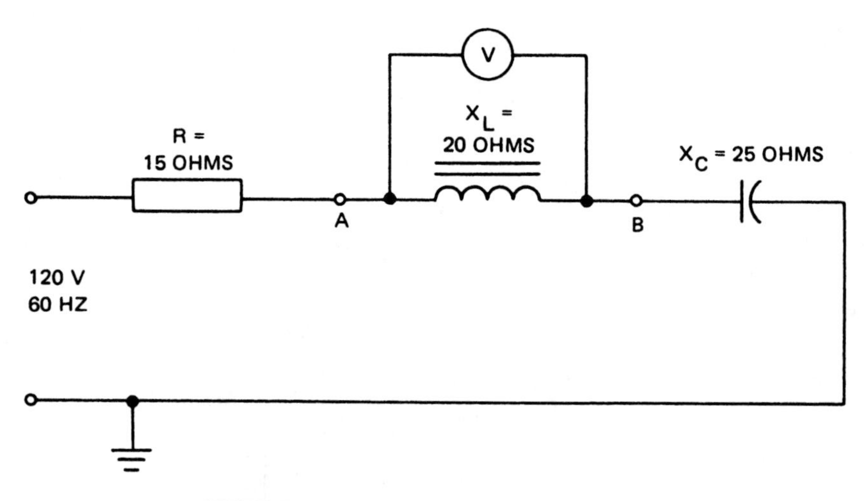

FIGURE 7–11 *(Delmar/Cengage Learning)*

10. For question 9, determine the voltmeter reading for each of the following conditions:
    a. There is no ground at either point A or point B, and the coil and capacitor are in good condition, but the resistor is completely "shorted."
    b. There are no faults in the circuit.

11. A single-phase, 115-V, 60-Hz motor uses a 100-μF capacitor in series with the starting winding. The starting winding has an effective resistance of 5 Ω and an inductance of 0.01 H. Determine
    a. the total impedance, in ohms, of the starting winding circuit, including the series-connected capacitor.
    b. the current.
    c. the voltage across the capacitor.

12. Draw a labeled vector diagram for the circuit in question 11.

13. A coil has a resistance of 100 Ω and an inductance of 0.2 H. This coil is connected in series with a 20-μF capacitor across a 120-V, 60-Hz supply. Determine
    a. the total impedance of the series circuit.
    b. the current.
    c. the power factor and the power factor angle for the series circuit.
    d. the impedance of the coil.
    e. the power factor and the power factor angle for the coil.

14. Draw a labeled vector diagram for the series circuit in question 13.

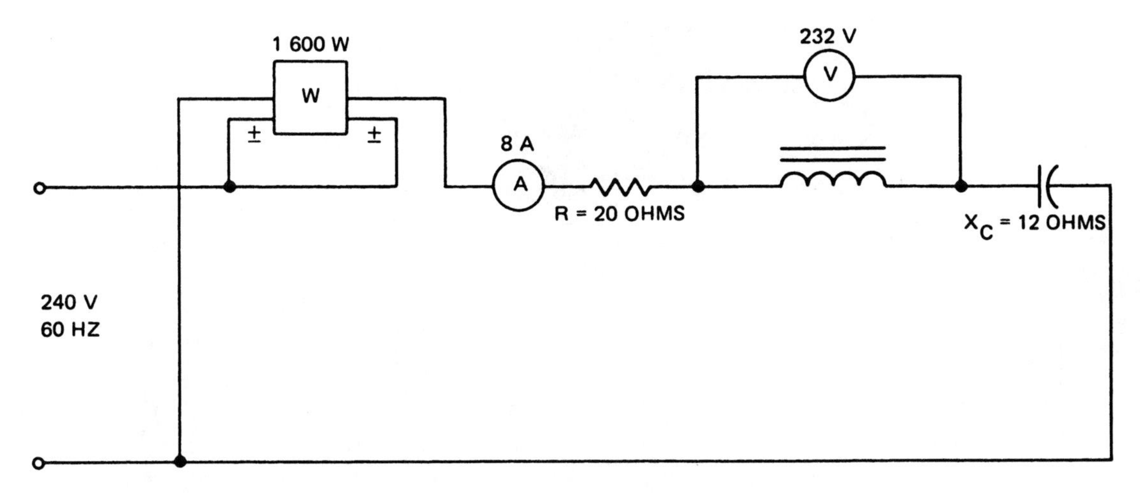

FIGURE 7–12 (*Delmar/Cengage Learning*)

15. Using the values given on the circuit diagram in Figure 7–12, determine
    a. the impedance of the coil.
    b. the resistance of the coil.
    c. the inductive reactance of the coil.
    d. the power factor and the phase angle for the coil.
16. Using the circuit given in question 15, determine
    a. the impedance of the entire series circuit.
    b. the power factor and the power factor angle for the series circuit.
    c. the loss in volts across the resistor and across the capacitor.
    d. the inductance of the coil, in henrys.
    e. the capacitance of the capacitor, in microfarads.
17. Construct a vector diagram for the circuit in question 16.

PRACTICE PROBLEMS FOR UNIT 7

Resistive, Inductive, Capacitive Series Circuits

Find the missing values in the following circuits. Refer to Figure 7–13 and the formulas listed in the Resistance, Inductive, Capacitive (Series) section of Appendix 15.

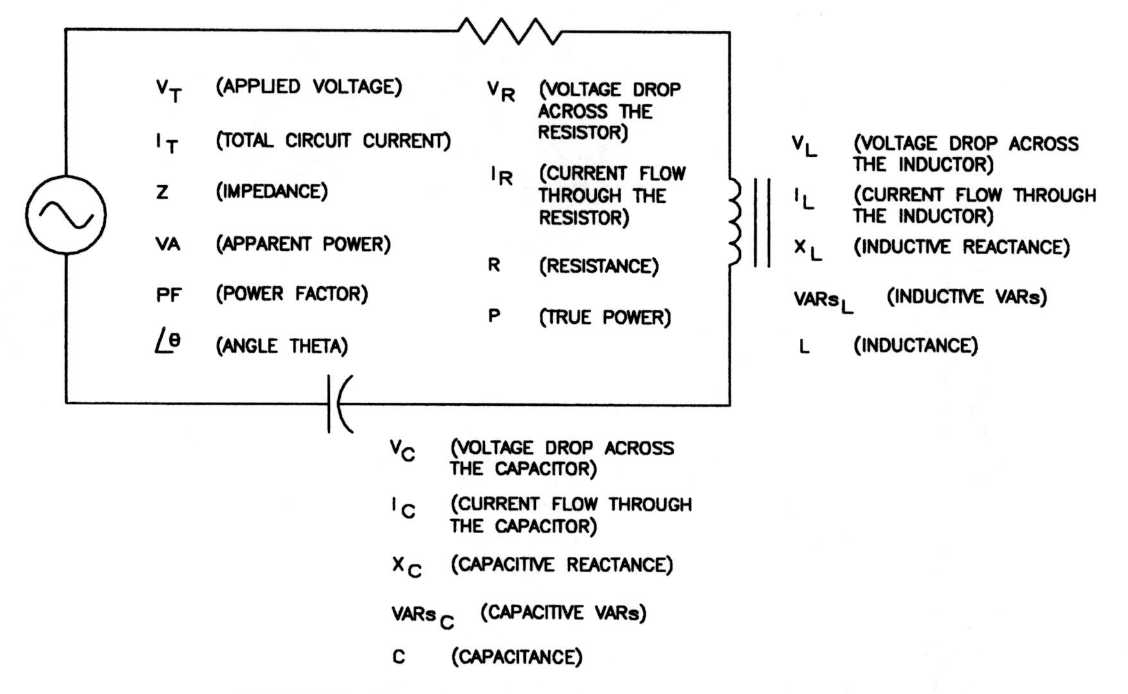

FIGURE 7–13 RLC series circuit (*Delmar/Cengage Learning*)

1. The circuit in Figure 7–13 is connected to a 120-V, 60-Hz line. The resistor has a resistance of 36 Ω, the inductor has an inductive reactance of 100 Ω, and the capacitor has a capacitive reactance of 52 Ω.

V_T = 120 V	V_R = ________	V_L = ________
I_T = ________	I_R = ________	I_L = ________
Z = ________	R = 36 Ω	X_L = 100 Ω
VA = ________	P = ________	$VARs_L$ = ________
PF = ________	$\angle\theta$ = ________	L = ________

V_C = ________

I_C = ________

X_C = 52 Ω

$VARs_C$ = ________

C = ________

2. The circuit shown in Figure 7–13 is connected to a 400-Hz line with an applied voltage of 35.678 V. The resistor has a true power of 14.4 W. There are 12.96 inductive VARs and 28.8 capacitive VARs.

V_T = 35.678 V	V_R = ________	V_L = 104 V
I_T = ________	I_R = ________	I_L = ________
Z = ________	R = ________	X_L = ________
VA = ________	P = 14.4 W	$VARs_L$ = 12.96 VARs
PF = ________	$\angle\theta$ = ________	L = ________
	V_C = ________	
	I_C = ________	
	X_C = ________	
	$VARs_C$ = 28.8 VARs	
	C = ________	

3. The circuit in Figure 7–13 is connected to a 60-Hz line. The apparent power in the circuit is 29.985 (VA), and the power factor is 62.5%. The resistor has a voltage drop of 14.993 V. The inductor has an inductive reactance of 60 Ω, and the capacitor has a capacitive reactance of 45 Ω.

V_T = ________	V_R = 14.993 V	V_L = 104 V
I_T = ________	I_R = ________	I_L = ________
Z = ________	R = ________	X_L = 60 Ω
VA = 29.985 VA	P = ________	$VARs_L$ = ________
PF = 62.5%	$\angle\theta$ = ________	L = ________
	V_C = ________	
	I_C = ________	
	X_C = 45 Ω	
	$VARs_C$ = ________	
	C = ________	

4. The circuit shown in Figure 7–13 is connected to a 1000-Hz line. The resistor has a voltage drop of 185 V. The inductor has a voltage drop of 740 V, and the capacitor has a voltage drop of 444 V. The circuit has an apparent power of 51.8 VA.

V_T = ________	V_R = 185 V	V_L = 740 V
I_T = ________	I_R = ________	I_L = ________
Z = ________	R = ________	X_L = ________
VA = 51.8 VA	P = ________	$VARs_L$ = ________
PF = ________	$\angle\theta$ = ________	L = ________

V_C = 444 V

I_C = ________

X_C = ________

$VARs_C$ = ________

C = ________

8

AC Parallel Circuits

Objectives

After studying this unit, the student should be able to

- demonstrate the ability to analyze ac parallel circuits by solving problems for parallel circuits containing
  1. noninductive resistive branches only.
  2. noninductive resistive and pure inductive branches.
  3. noninductive resistive and capacitive branches.
  4. noninductive resistive and impedance branches.
  5. R, X_L, and X_C branches.
  6. R, Z, and X_C branches.
- construct vector diagrams for ac parallel circuits, showing the relationship between current and voltage for circuits containing R, X_L, and X_C components.
- construct impedance diagrams and power triangles for ac parallel circuits containing R, X_L, and X_C components.
- list the properties of a parallel circuit at resonance.
- discuss the similarities and differences between series ac circuits and parallel ac circuits at resonance.
- solve equations to determine the size of the capacitor or capacitor bank required to correct the power factor of an ac induction motor.
- list the advantages of a unity power factor for industrial use and explain the results gained.

INTRODUCTION

There are more applications in alternating-current work for parallel circuits than for series circuits. Nearly all commercial, industrial, and residential power circuits are connected in parallel. The voltage across any branch circuit in a parallel arrangement is the same as the line voltage. This is true if the voltage drop in the line wires is neglected. Recall from earlier studies that the total (line) current for an ac parallel circuit may not be equal to the arithmetic sum of the current in each of the branch circuits. The currents in the branch circuits may be out of phase. Thus, the total (line) current must be calculated using vectors. The approximate value of the line current can also be found by drawing a vector diagram accurately to scale.

This unit develops methods of analyzing ac parallel circuits. These methods are based on concepts learned in the study of series circuits.

The voltage is constant across each branch of a parallel circuit. Thus, all angles are measured with respect to the voltage vector. This means that the voltage vector is a reference vector.

PARALLEL CIRCUIT WITH RESISTIVE LOAD

The first parallel circuit to be studied has a noninductive resistive load in each branch (Figure 8–1). All of the branch currents for this circuit are in phase with the line voltage. In this case, both the vector sum and the arithmetic sum of the branch currents give the total line current.

This parallel circuit is connected to a 120-V, 60-Hz source. The calculations are the same for an ac parallel circuit with noninductive resistance loads only and for a direct-current parallel circuit.

PROBLEM 1

Statement of the Problem

For the parallel circuit shown in Figure 8–1, find the following quantities:

1. The combined resistance
2. The current taken by each branch
3. The line current
4. The total power taken by the parallel circuit
5. The power factor and the phase angle
6. The vector diagram

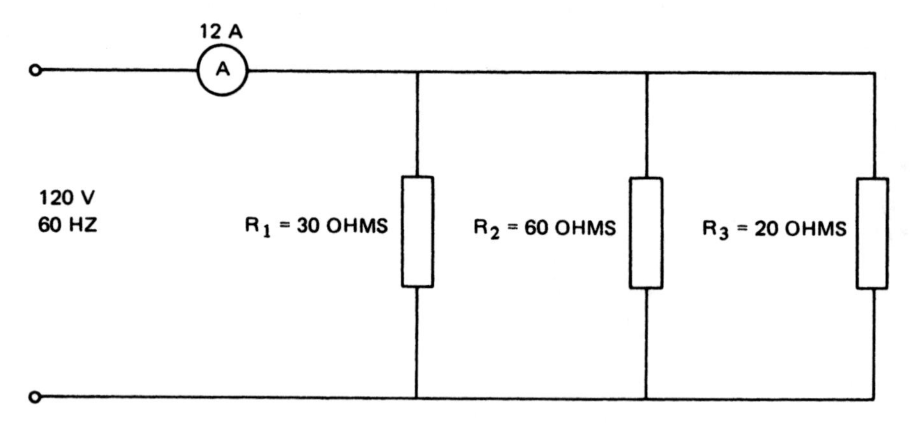

FIGURE 8–1 Parallel circuit with noninductive resistance in the branches *(Delmar/ Cengage Learning)*

Solution

1. The combined resistance is obtained by using the reciprocal resistance formula (discussed in *Direct Current Fundamentals*):

$$R_C = \frac{1}{\frac{1}{R_1} + \frac{1}{R_2} + \frac{1}{R_3}}$$

$$= \frac{1}{\frac{1}{30} + \frac{1}{60} + \frac{1}{20}} = \frac{1}{\frac{6}{60}} = \frac{60}{6} = 10\ \Omega$$

2. The current taken by each branch is given by Ohm's law:

$$I_1 = \frac{V}{R_1}$$

$$= \frac{120}{30}$$

$$= 4\ \text{A}$$

or

$$I_2 = \frac{V}{R_2}$$

$$= \frac{120}{60}$$

$$= 2\ \text{A}$$

or

$$I_3 = \frac{V}{R_3}$$

$$= \frac{120}{20}$$

$$= 6\ \text{A}$$

3. The total current can be found by either of two methods. The individual branch current values can be added, or Ohm's law can be applied to the entire parallel circuit. Either method gives a total current of 12 A.

4. The total power taken by the entire parallel circuit is the product of the line voltage and the line current. (These values are in phase.)

$$P = VI_{\text{total}}$$

$$= 120 \times 12$$

$$= 1440\ \text{W}$$

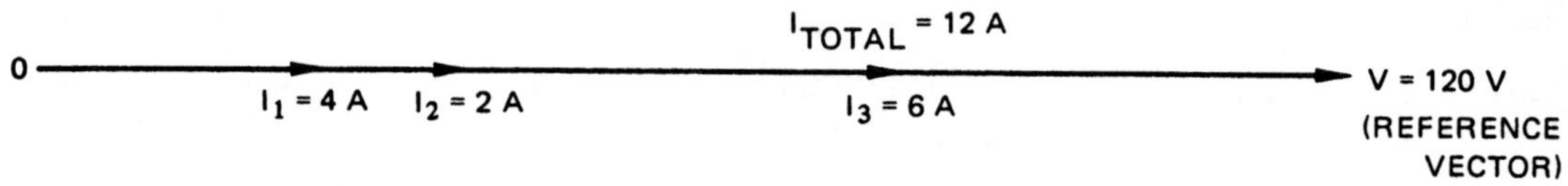

0 $I_{TOTAL} = 12$ A V = 120 V (REFERENCE VECTOR)

$I_1 = 4$ A $I_2 = 2$ A $I_3 = 6$ A

FIGURE 8–2 **Vector diagram for a parallel circuit with noninductive resistive loads (*Delmar/Cengage Learning*)**

or

$$
\begin{aligned}
P &= I^2R \\
&= 12^2 \times 10 \\
&= 144 \times 10 \\
&= 1440 \text{ W}
\end{aligned}
$$

or

$$
\begin{aligned}
P &= \frac{V^2}{R} \\
&= \frac{120^2}{10} \\
&= \frac{14{,}400}{10} \\
&= 1440 \text{ W}
\end{aligned}
$$

5. The power factor is a ratio of the power in watts to the input volt-amperes:

   $$PF = \frac{1440}{1440} = 1.00 \text{ (unity)}$$

   The angle whose cosine is 1.00 is 0°. In other words, the line current is in phase with the line voltage.

6. The vector diagram for this parallel circuit is shown in Figure 8–2. The individual branch current values are placed on the voltage vector. Both the arithmetic sum and the vector sum of the branch currents equal the total line current.

PARALLEL CIRCUIT WITH BRANCHES CONTAINING R AND X_L

Another type of parallel circuit has one branch containing a noninductive resistance load and a second branch containing pure inductance (Figure 8–3).

The two branch currents for this parallel circuit are 90 electrical degrees out of phase with each other. The current in the noninductive resistance branch is in phase with the line voltage. The current in the pure inductive branch lags the line voltage by 90°. Thus, the total line current can be found using vector methods.

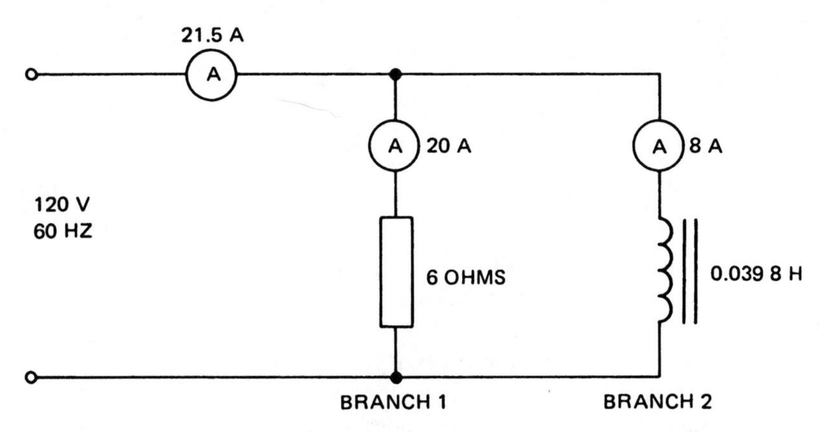

FIGURE 8–3 Parallel circuit with R and X_L branches ***(Delmar/Cengage Learning)***

PROBLEM 2

Statement of the Problem

For the parallel circuit shown in Figure 8–3, determine the following:

1. The current taken by each of the two branches
2. The vector diagram for the parallel circuit
3. The total current
4. The power taken by the parallel circuit
5. The power factor for the parallel circuit
6. The phase angle
7. The combined impedance

Solution

1. The voltage across each branch of the circuit is 120 V. The current in each branch of this circuit is as follows:

Branch 1:

$$I_1 = \frac{V}{R}$$

$$= \frac{120}{6}$$

$$= 20 \text{ A}$$

Branch 2:

$$X_L = 2\pi fL$$

$$= 377 \times 0.0398$$

$$= 15\ \Omega$$

$$I_2 = \frac{V}{X_L}$$

$$= \frac{120}{15}$$

$$= 8\ \text{A}$$

2. The current and voltage vector diagram for an RL parallel circuit is shown in Figure 8–4. The vector I_L is in the negative Y direction. Recall that in series circuits, the vector V_L was in the positive Y direction. The change in direction does not involve new concepts, but occurs because the reference vector has changed from I to V. The line current is the vector sum of the branch currents and lags behind the line voltage by an angle of 21.5°. (The calculation of this angle is given in steps 5 and 6.)

3. The total line current is really the hypotenuse of the right triangle in Figure 8–4, where

$$I_{total} = \sqrt{I_R^2 + I_L^2}$$

$$= \sqrt{20^2 + 8^2}$$

$$= \sqrt{400 + 64} = 21.5\ \text{A}$$

4. The power taken by the parallel circuit is used in the first branch containing resistance. No power is taken by the pure inductive load in the second branch. The power for the parallel circuit is

$$P = VI_R$$

$$= 120 \times 20 = 2400\ \text{W}$$

5. The power factor of any ac circuit is the cosine of the angle θ. This value equals the ratio of the power to the input volt-amperes. The power factor is also the ratio of

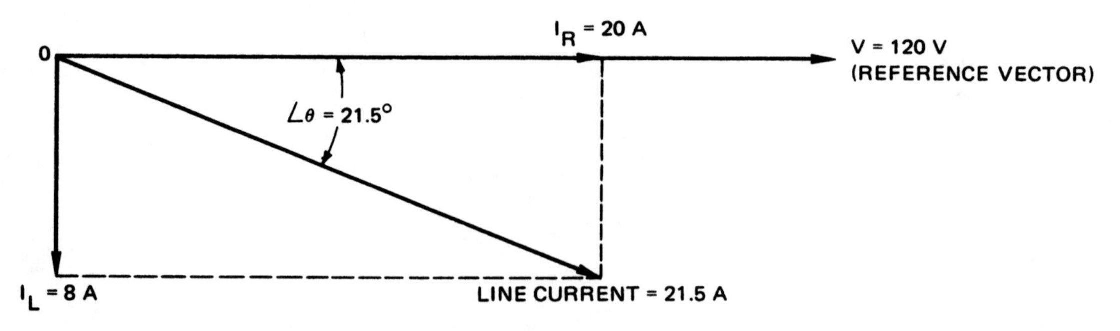

FIGURE 8–4 Vector diagram for parallel circuit with R and X_L branches (*Delmar/Cengage Learning*)

the in-phase current to the total line current. For the circuit in Figure 8–4, the power factor is

$$\text{Power factor} = \frac{\text{watts}}{V \times I}$$

$$= \frac{2400}{120 \times 21.5} = 0.9302 \text{ lagging}$$

or

$$\text{Power factor} = \frac{I_{\text{in phase}}}{I_{\text{total}}}$$

$$= \frac{20}{21.5} = 0.9302 \text{ lagging}$$

6. The power factor angle (θ) is 21.5° lagging.
7. The combined impedance of a parallel circuit is found using Ohm's law. The impedance formula for a series circuit cannot be used because each parallel branch connected across the line wires tends to reduce the total parallel circuit impedance:

$$Z_{\text{combined}} = \frac{V_{\text{line}}}{I_{\text{line}}}$$

$$= \frac{120}{21.5} = 5.58\ \Omega$$

Impedance Triangle for Parallel Circuits

In series circuits, the voltage triangle is divided and multiplied by current to obtain the impedance and power triangles, respectively. In parallel circuits, the current triangle is divided by the reference vector V to obtain the impedance triangle (Figure 8–5).

The impedance triangle shown in Figure 8–5 for a parallel circuit does not resemble the series circuit impedance triangle. The impedance formula for a parallel circuit is

$$\frac{1}{Z} = \sqrt{\left(\frac{1}{R}\right)^2 + \left(\frac{1}{X_L}\right)^2}$$

or

$$Z = \frac{1}{\sqrt{\left(\frac{1}{R}\right)^2 + \left(\frac{1}{X_L}\right)^2}}$$

Compare this formula with the series circuit impedance formula:

$$Z = \sqrt{R^2 + X_L^2}$$

Another formula that can be used to determine the impedance of resistance and inductive reactance connected in parallel is

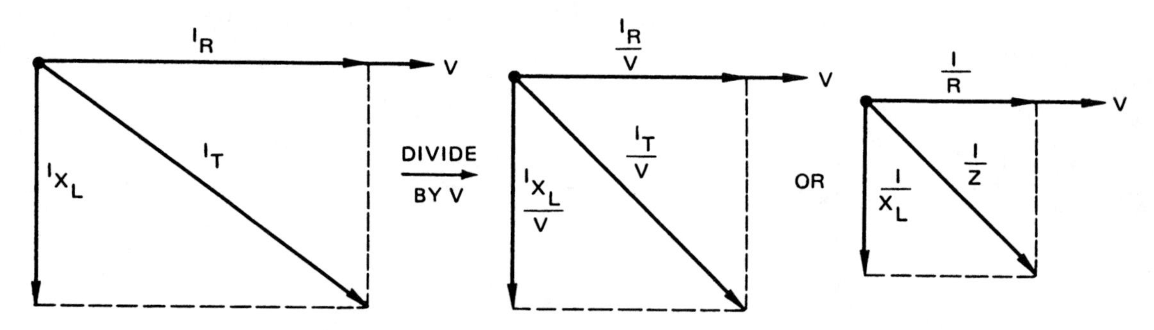

FIGURE 8–5 Impedance triangle (*Delmar/Cengage Learning*)

$$Z = \frac{R \times X_L}{\sqrt{R^2 + X_L^2}}$$

Assume that a parallel circuit contains a 15-Ω resistor connected in parallel with an inductor that has an inductive reactance of 20 Ω. To determine circuit impedance, substitute the values in the following formula:

$$Z = \frac{15 \times 20}{\sqrt{15^2 + 20^2}}$$

$$= \frac{300}{\sqrt{625}}$$

$$= \frac{300}{25}$$

$$= 12\ \Omega$$

PARALLEL CIRCUIT WITH BRANCHES CONTAINING R AND X_C

Another type of parallel circuit is shown in Figure 8–6. In this circuit, two branches contain noninductive resistance loads and a third branch contains a capacitor. Assuming that the capacitor has negligible resistance, the current in the capacitive branch leads the line voltage by 90 electrical degrees. This means that the line current is the vector sum of the currents in the branches.

PROBLEM 3

Statement of the Problem

The circuit shown in Figure 8–6 contains noninductive resistance and capacitance. For this circuit:

1. Find the current taken by each branch.
2. Construct a vector diagram.

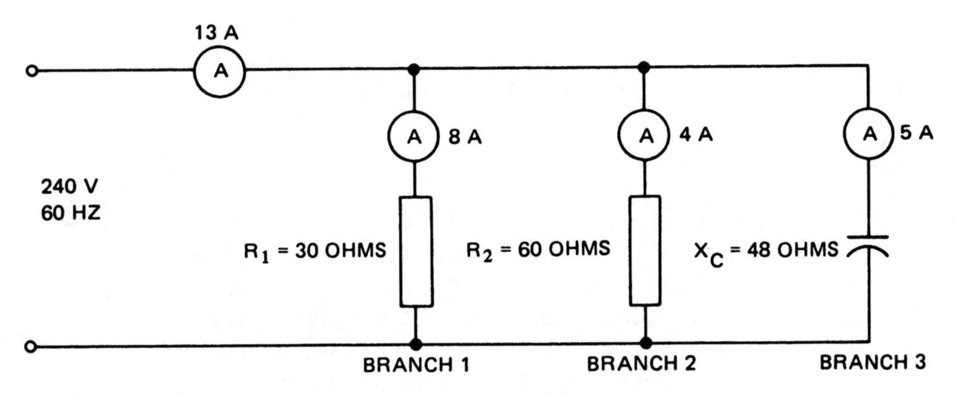

FIGURE 8–6 Parallel circuit with branches containing R and X_C *(Delmar/Cengage Learning)*

3. Find the total or line current.
4. Find the power taken by the parallel circuit.
5. Find the power factor for the parallel circuit.
6. Find the phase angle.
7. Find the combined impedance.

Solution

1. Branches 1 and 2 contain pure resistance only. The current in each of these branches is in phase with the line voltage.

Branch 1:

$$I_1 = \frac{V}{R_1}$$

$$= \frac{240}{30} = 8 \text{ A}$$

Branch 2:

$$I_2 = \frac{V}{R_2}$$

$$= \frac{240}{60} = 4 \text{ A}$$

Branch 3:

$$I_3 = \frac{V}{X_C}$$

$$= \frac{240}{48} = 5 \text{ A}$$

The total in-phase current for the parallel circuit is the sum of the currents in the first two branches: 8 + 4 = 12 A. The quadrature (out-of-phase) component of current is supplied to the capacitor load in branch 3. This current leads the line voltage by 90° and is given by

$$\frac{V}{X_C} = \frac{240}{48} = 5\ A$$

2. The vector diagram for the circuit is given in Figure 8–7. The two in-phase current values are placed on the line voltage vector. The total in-phase current is 12 A. The current in the capacitive branch is drawn on the vector diagram in a vertical direction from point 0. Thus, the quadrature current leads the line voltage by 90°. The line current is the vector sum of the total in-phase current and the leading quadrature current. Note that the line current leads the line voltage by the angle θ. As a result, this parallel circuit operates with a leading power factor.

3. The line current is the hypotenuse of the right triangle shown in Figure 8–7:

$$\mathbf{I_{line} = \sqrt{I^2_{total\ in\ phase} + I^2_{quadrant}}}$$

$$\mathbf{= \sqrt{(8 + 4)^2 + 5^2}}$$

$$\mathbf{= \sqrt{144 + 25} = 13\ A}$$

4. The total power taken by the parallel circuit is the product of the total in-phase current and the line voltage.

$$P = V \times I_{in\ phase}$$

$$= 240 \times 12 = 2880\ W$$

5. The power factor is

$$PF = \frac{P}{V \times I}$$

$$= \frac{2880}{240 \times 13} = 0.9231 \text{ leading}$$

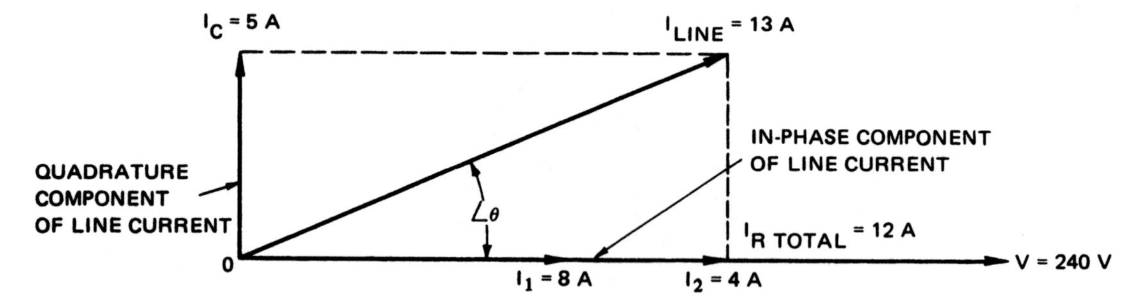

FIGURE 8–7 Vector diagram for parallel circuit with branches containing R and X_C *(Delmar/Cengage Learning)*

or

$$PF = \frac{I_{in\ phase}}{I_{line}}$$

$$= \frac{12}{13}$$

$$= 0.9231 \text{ leading}$$

6. The line current leads the line voltage by a phase angle (θ) of 22.6°.
7. The combined impedance (in ohms) of the parallel circuit is

$$Z_{combined} = \frac{V}{I_{total}}$$

$$= \frac{240}{13} = 18.46\ \Omega$$

The total impedance of the circuit can also be found using the formula

$$Z = \frac{1}{\sqrt{\left(\frac{1}{R}\right)^2 + \left(\frac{1}{X_C}\right)^2}}$$

$$Z = \frac{1}{\sqrt{\left(\frac{1}{20}\right)^2 + \left(\frac{1}{48}\right)^2}}$$

$$= \frac{1}{\sqrt{(0.05)^2 + (0.02083)^2}}$$

$$= \frac{1}{\sqrt{0.0025 + 0.000434}}$$

$$= \frac{1}{\sqrt{0.002934}}$$

$$= \frac{1}{0.054167}$$

$$= 18.46\ \Omega$$

Another formula that can be used to determine the impedance in a circuit that contains both resistance and capacitive reactance connected in parallel is

$$Z = \frac{R \times X_C}{\sqrt{R^2 + X_C^2}}$$

PARALLEL CIRCUIT WITH BRANCHES CONTAINING R AND Z_{coil}

The parallel circuit shown in Figure 8–8 contains two branches. One branch contains a pure resistance of 30 Ω. The second branch contains a load that has both resistance and inductance. This load is similar to a motor or transformer. The symbol Z_{coil} is used instead of X_L because of the coil resistance. It should be noted, however, that the value of Z_{coil} applies only to the impedance of the coil and should not be confused with the value for the total impedance of the circuit, which is Z_{total}.

In this circuit, a wattmeter is used to measure the true power of the circuit, and ammeters are used to measure the total current flow and the current flow through each branch. In this circuit, the current will lag the voltage by some value less than 90°.

PROBLEM 4

Statement of the Problem

For the circuit in Figure 8–8, the following items are to be determined:

1. The current taken by each branch
2. The resistance of the coil
3. The power factor of the coil
4. The in-phase component of current in the coil
5. The quadrature component of current in the coil
6. Total impedance of the circuit

Draw a vector diagram for the circuit, and determine (1) the line current and (2) the circuit power factor.

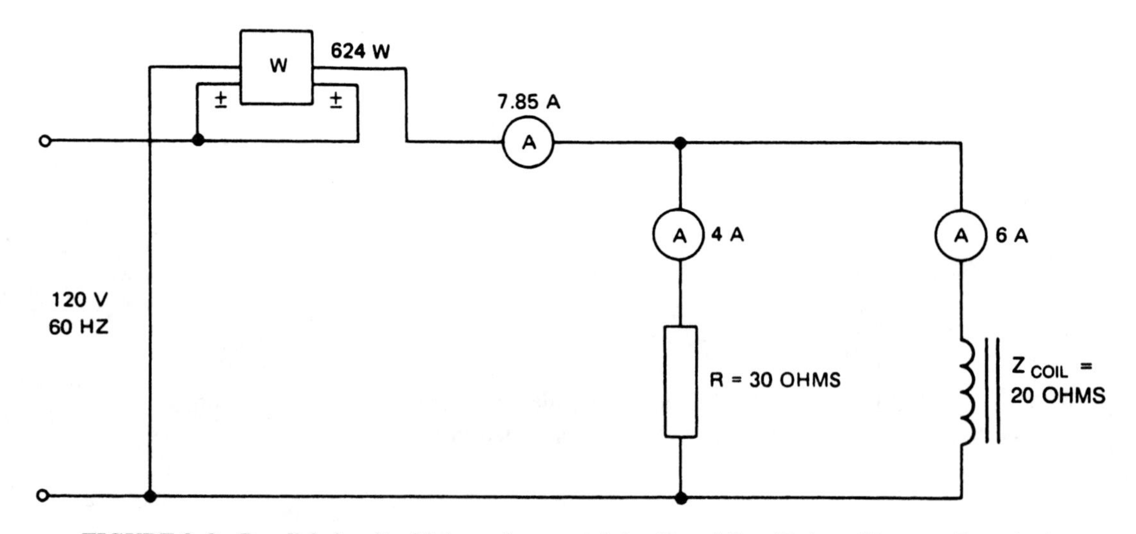

FIGURE 8–8 Parallel circuit with branches containing R and Z_{coil} *(Delmar/Cengage Learning)*

Solution

1. The current in each branch of the circuit is given by Ohm's law:

$$I_R = \frac{V}{R} = \frac{120}{30} = 4A \qquad \text{(resistance branch)}$$

$$I_{Zcoil} = \frac{V}{Z_{coil}} = \frac{120}{20} = 6\ A \qquad \text{(impedence branch)}$$

2. The resistance of the coil can be found from the expression for the power taken by the coil:

$$\mathbf{P = I^2R}$$

The true power taken by the resistance branch is

$$\begin{aligned} P &= V \times I_R \\ &= 120 \times 4 \\ &= 480\ W \end{aligned}$$

or

$$\begin{aligned} P &= I^2R \\ &= 4^2 \times 30 \\ &= 480\ W \end{aligned}$$

or

$$\begin{aligned} P &= \frac{V^2}{R} \\ &= \frac{120^2}{30} \\ &= 480\ W \end{aligned}$$

The true power taken by the resistance branch is subtracted from the total power taken by the parallel circuit to find the power taken by the coil. Thus, the actual power used in this coil is

$$P_{coil} = 624 - 480 = 144\ W$$

Therefore, the resistance of the coil is

$$\begin{aligned} R &= \frac{P}{I^2} \\ &= \frac{144}{6^2} = \frac{144}{36} = 4\ \Omega \end{aligned}$$

3. To find the power factor of the coil, it is assumed that the coil is a series circuit containing resistance and inductive reactance. For this case, the power factor is expressed as follows:

$$\text{Power factor} = \frac{R}{Z} = \frac{4}{20} = 0.20 \text{ lagging}$$

For a power factor of 0.20, the coil current lags behind the line voltage by an angle of 78.5°.

4. The 6-A current in the coil can be resolved into two components. One component is the in-phase current. The second component, or quadrature current, lags the line voltage by 90° because of the inductive reactance in this branch.

$$PF = \frac{I_{\text{in phase}}}{I_Z}$$

$$0.2 = \frac{I_{\text{in phase}}}{6} = 1.2 \text{ A, in phase component of the coil current}$$

5. To find the quadrature current, refer to the vector diagram in Figure 8–9. The sine of the angle alpha is given as follows:

$$\sin \angle a = \frac{I_{\text{quadrature}}}{I_Z}$$

$$0.9799 = \frac{I_{\text{quadrature}}}{6} = 5.88 \text{ A}$$

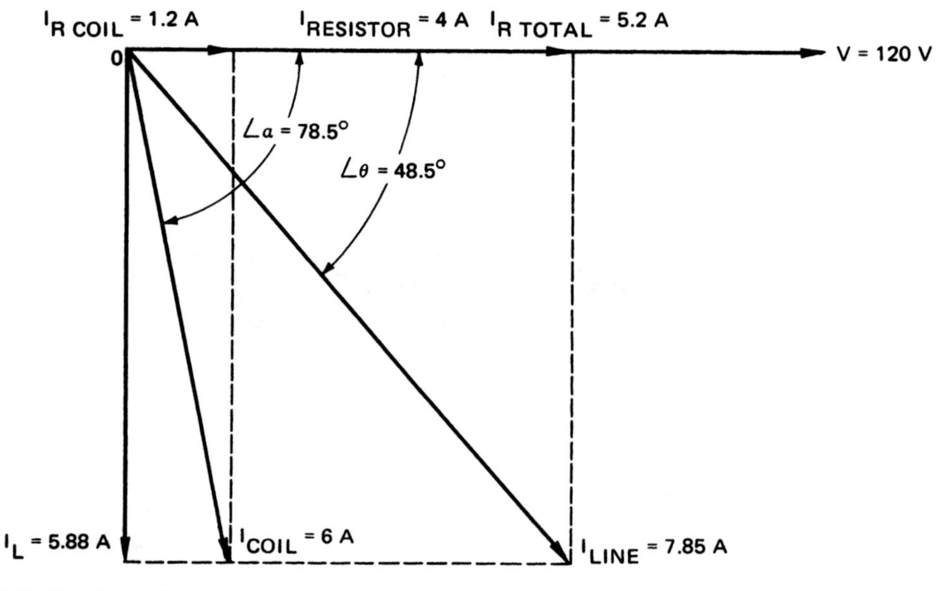

FIGURE 8–9 Vector diagram for a parallel circuit with branches containing R and Z (*Delmar/Cengage Learning*)

6. The vector diagram for the circuit is shown in Figure 8–9. The in-phase current for the coil (1.2 A) is placed on the line voltage vector. The quadrature component of the current (5.88 A) is in a downward direction at an angle of 90° to the line voltage. The vector sum of these components is the current taken by the coil (6 A). The phase angle between the coil current and the line voltage is 78.5°. The current taken by the resistance branch (4 A) is added to the in-phase current of the coil circuit. Thus, the total in-phase current is 5.2 A. The circuit must be supplied with 5.2 A in phase with the line voltage and a quadrature current of 5.88 A lagging the line voltage by 90°. The vector sum of these two components is the line current (7.85 A). The line current lags the line voltage by 48.5°. The total line current is the hypotenuse of a right triangle in the vector diagram of Figure 8–9:

$$I_{line} = \sqrt{I^2_{in\ phase} + I^2_{quadrature}}$$

$$= \sqrt{(1.2 + 4)^2 + 5.88^2} = \sqrt{5.2^2 + 5.88^2}$$

$$= \sqrt{27.04 + 34.5744} = \sqrt{61.6144} = 7.85\ A$$

The power factor for the circuit can be found by either of two methods:

$$PF = \frac{W}{VI}$$

$$= \frac{624}{120 \times 7.85} = 0.6624 \text{ lagging}$$

or

$$PF = \frac{I_{in\ phase}}{I_{line}}$$

$$= \frac{5.2}{7.85} = 0.6624 \text{ lagging}$$

The power factor angle whose cosine is 0.662 is 48.5° lagging.

7. The total impedance of the circuit can be found by using the formula

$$Z_{total} = \frac{V_{total}}{I_{total}}$$

$$= \frac{120}{7.85} = 15.3\ \Omega$$

PARALLEL CIRCUIT WITH BRANCHES CONTAINING R, X_L, AND X_C

Figure 8–10 illustrates a parallel circuit consisting of three branch circuits. One branch contains resistance, a second branch contains pure inductance, and the third branch contains capacitance.

PROBLEM 5

Statement of the Problem

For the circuit shown in Figure 8–10, determine the following items:

1. The current taken by each branch
2. The line current
3. The power
4. The magnetizing VARs required by the coil
5. The magnetizing VARs supplied by the capacitor
6. The net magnetizing VARs supplied by the line
7. The power factor and the phase angle for the circuit
8. Total impedance of the circuit

Construct a vector diagram for the parallel circuit.

Solution

1. The current taken by each branch of the circuit is as follows:

Resistance branch:

$$I_R = \frac{V}{R} = \frac{240}{20} = 12\text{A}$$

Coil branch:

$$I_L = \frac{V}{X_L} = \frac{240}{16} = 15\text{ A}$$

Capacitor branch:

$$I_C = \frac{V}{X_C} = \frac{240}{24} = 10\text{ A}$$

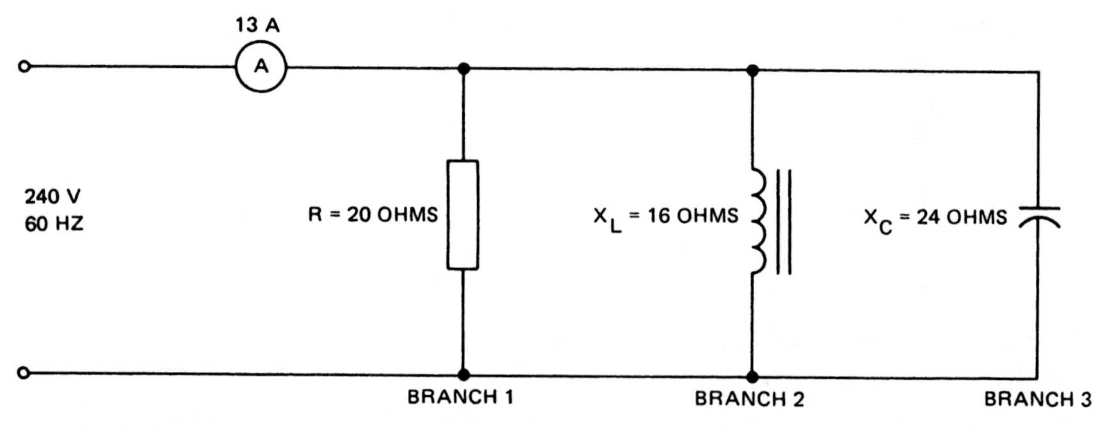

FIGURE 8–10 **Parallel circuit with branches containing R, X_L, and X_C** ***(Delmar/Cengage Learning)***

2. The current in the capacitor branch leads the line voltage by 90°. The coil current lags the same line voltage by 90°. In other words, I_C and I_L are 180° out of phase with each other. $I_L - I_C$ (15 A − 10 A) yields a net quadrature current of 5 A inductive. This means that the line current consists of an in-phase component of 12 A and a lagging quadrature component of 5 A. The line current is

$$I_{line} = \sqrt{I_R^2 + (I_L - I_C)^2}$$

$$= \sqrt{12^2 + 5^2} = 13 \text{ A}$$

3. All of the power taken by this circuit is used in the resistance branch. The currents taken by the coil branch and the capacitor branch are 90° out of phase with the line voltage. As a result, the power taken by either of these branches is zero. The power for the resistance branch is the power for the entire parallel circuit. This value, in watts, can be determined by any one of the following expressions:

$$P = VI_R$$

$$= 240 \times 12$$

$$= 2880 \text{ W}$$

or

$$P = I^2R$$

$$= 12^2 \times 20$$

$$= 144 \times 20$$

$$= 2880 \text{ W}$$

or

$$P = \frac{V^2}{R}$$

$$= \frac{240^2}{20}$$

$$= 2880 \text{ W}$$

4. The magnetizing VARs required by the coil is determined as follows:

$$\text{VARs} = VI_L$$

$$= 240 \times 15$$

$$= 3600 \text{ VARs}$$

or

$$\text{VARs} = I_L^2 \times X_L$$

$$= 15^2 \times 16$$
$$= 3600 \text{ VARs}$$

or

$$\text{VARs} = \frac{V^2}{X_L}$$
$$= \frac{240^2}{16}$$
$$= 3600 \text{ VARs}$$

5. The value of magnetizing VARs supplied by the capacitor is

$$\text{VARs} = VI_C$$
$$= 240 \times 10$$
$$= 2400 \text{ VARs}$$

or

$$\text{VARs} = I_C^2 X_C$$
$$= 10^2 \times 24$$
$$= 2400 \text{ VARs}$$

or

$$\text{VARs} = \frac{V^2}{X_C}$$
$$= \frac{240^2}{24}$$
$$= 2400 \text{ VARs}$$

6. The net magnetizing VARs supplied by the line is the difference between the VARs required by the coil and the VARs supplied by the capacitor:

$$\text{Net VARs} = 3600 - 2400 = 1200 \text{ VARs}$$

7. The power factor for this parallel circuit may be found using either of two ratios: (a) the ratio of the power in watts to the input volt-amperes, or (b) the ratio of the in-phase current to the total line current. The resulting power factor is lagging. This lag is due to the fact that the quadrature current taken by the coil branch is greater than the current taken by the capacitor branch. The net quadrature current supplied by the ac source lags the line voltage by 90°. The power factor is

$$PF = \frac{P}{VI}$$
$$= \frac{2880}{240 \times 13}$$

$$= \frac{2880}{3120}$$

$$= 0.9231$$

or

$$PF = \frac{I_{in\ phase}}{I_{line}}$$

$$= \frac{12}{13}$$

$$= 0.9231 \text{ lagging}$$

8. The angle θ for a lagging power factor of 0.9231 is 22.6°. The vector diagram for the parallel circuit with branches containing R, X_L, and X_C is shown in Figure 8–11.

9. The total impedance of the circuit can be found using either of the following formulas:

$$Z = \frac{V}{I_{total}}$$

$$= \frac{240}{13} = 18.48\ \Omega$$

or

$$Z = \frac{1}{\sqrt{\left(\frac{1}{R}\right)^2 + \left(\frac{1}{X_L} - \frac{1}{X_C}\right)^2}}$$

$$= \frac{1}{\sqrt{\left(\frac{1}{20}\right)^2 + \left(\frac{1}{16} - \frac{1}{24}\right)^2}}$$

$$= \frac{1}{\sqrt{(0.05)^2 + (0.0625 - 0.04167)^2}}$$

$$= \frac{1}{\sqrt{(0.05)^2 + (0.02083)^2}}$$

$$= \frac{1}{\sqrt{0.0025 + 0.000434}}$$

$$= \frac{1}{\sqrt{0.002934}}$$

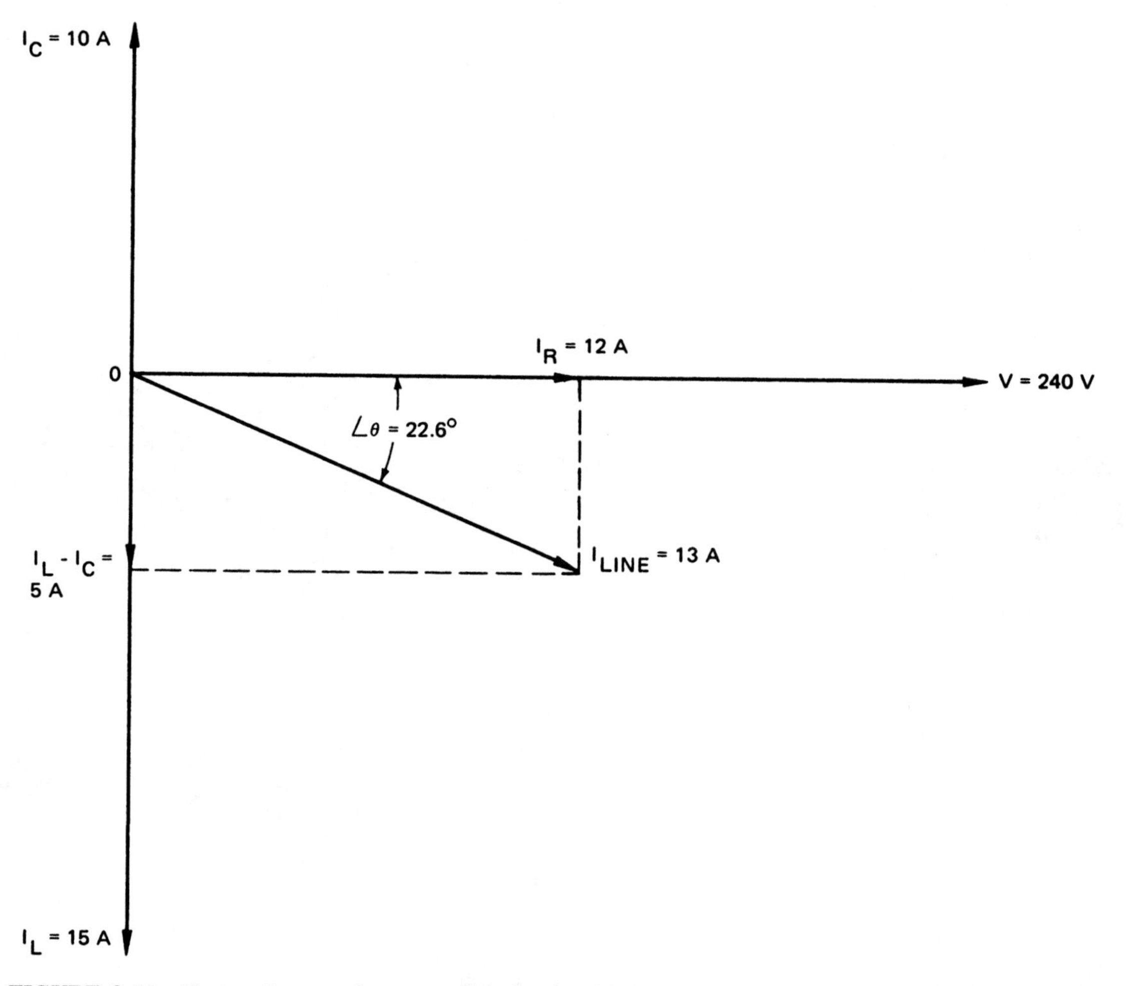

FIGURE 8–11 Vector diagram for a parallel circuit with branches containing R, X_L, and X_C (*Delmar/ Cengage Learning*)

$$= \frac{1}{0.05416}$$

$$= 18.46\ \Omega$$

Another formula that can be used to determine the total impedance of a circuit containing resistance, inductive reactance, and capacitive reactance is

$$Z = \frac{R \times X}{\sqrt{R^2 + X^2}}$$

where

$$X = \frac{X_L \times X_C}{X_L + X_C}$$

In this formula, X_L is a positive number and X_C is a negative number. Therefore, Z will be either positive or negative depending on whether the circuit is more inductive (positive) or capacitive (negative).

Assume a parallel circuit contains a 12-Ω resistor connected in parallel with an inductor that has an inductive reactance of 8 Ω and a capacitor that has a capacitive reactance of 16 Ω. To find the total impedance of this sample circuit using this formula, first determine the value of X:

$$X = \frac{X_L \times X_C}{X_L + X_C}$$

$$= \frac{8 \times (-16)}{8 + (-16)}$$

$$= \frac{-128}{-8}$$

$$= 16$$

Now that the value of X has been determined, the impedance can be computed using the following formula:

$$Z = \frac{R \times X}{\sqrt{R^2 + X^2}}$$

$$= \frac{12 \times 16}{\sqrt{12^2 + 16^2}}$$

$$= \frac{192}{\sqrt{400}}$$

$$= \frac{192}{20}$$

$$= 9.6\ \Omega$$

PARALLEL CIRCUIT RESONANCE

The current in a series circuit reaches its maximum value at resonance. At this point, the effects of inductive reactance and capacitive reactance cancel each other. Also, the full line voltage is impressed across the resistance of the circuit. Thus, for the series circuit, the current and line voltage are in phase.

An analysis of a series and a parallel resonant circuit having identical values of R, L, and C will show the following similarities:

1. The line currents and the line voltages are in phase.
2. The power factors are unity.
3. The values of X_L and X_C are equal.
4. The impedances are purely resistive.

These similarities may give the impression that both circuits are identical at resonance. However, there are two important differences between series and parallel circuits, which can be seen by looking at the magnitudes of the currents and impedances:

1. The current is at a maximum in a series resonant circuit and at a minimum in a parallel resonant circuit.
2. The impedance is at a minimum in a series resonant circuit and at a maximum in a parallel resonant circuit.

These factors make parallel resonance a useful feature in tuned circuits.

Current and Impedance Discussion

Figure 8–12 shows the current and impedance curves for a parallel resonant circuit. These curves are opposite to those shown for a series resonant circuit. The current equation for the parallel circuit is

$$I_L = \sqrt{I_R^2 + (I_{XL} - I_{XC})^2}$$

At frequencies below resonance, the current equation takes the form

$$I_L = \sqrt{I_R^2 + I_{XL}^2}$$

where I_{XC} is negligible at low frequencies. The capacitive reactance has high values at frequencies below resonance. This means that only very small currents flow. The opposite is true for the inductive branch, where I_L becomes a maximum value.

For frequencies at resonance, the current equation takes the following form:

$$I_L = \sqrt{I_R^2} = I_R$$

At resonance, X_L and X_C are equal and 180° out of phase. Their currents are also equal and 180° out of phase and thus cancel. The current remaining is the resistive component, I_R. (I_L is at a minimum.) At resonance, it is possible in theory to have the following values: an inductive current of 10 A, a capacitive current of 10 A, and a line current of 0 A.

At frequencies above resonance, the current equation takes the following form:

$$I_L = \sqrt{I_R^2 + I_{XC}^2}$$

By examining what happens at frequencies below resonance, the student should be able to analyze the previous resonance case where I_L becomes a maximum value again.

The impedance curve is the reciprocal of the current curve. The impedance curve is defined by the equation Z = V/I.

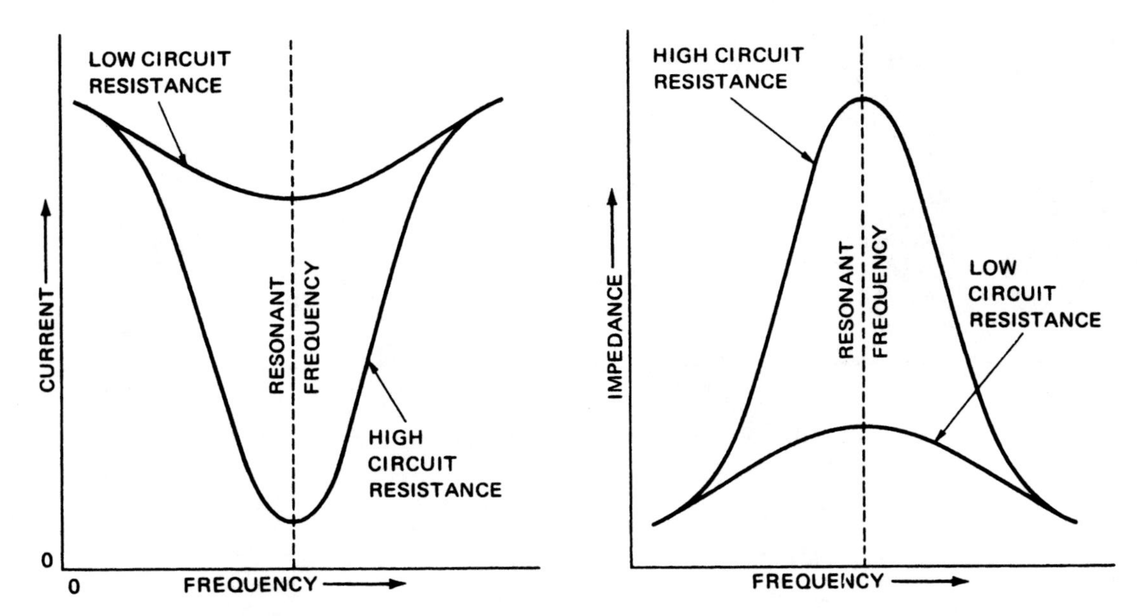

FIGURE 8–12 Parallel resonant frequency curves (*Delmar/Cengage Learning*)

The parallel circuit shown in Figure 8–10 has a lagging power factor of 0.9231. A larger capacitor may be used in the capacitive branch of the circuit. This increased value causes an increase in the leading current. For example, for a capacitor with $X_C = 16\ \Omega$, the quadrature current for this capacitor is 15 A leading. The coil in branch 2 of the circuit also requires 15 A. The 15 A taken by the capacitor leads the line voltage by 90°. The 15 A supplied to the coil lags the line voltage by 90°.

Resonant Parallel Circuit

Figure 8–13 shows a parallel circuit in which the coil and capacitor branches each take 15 A. This circuit is a resonant parallel circuit. The 15 A in the coil branch and the 15 A in the capacitor branch are 180° out of phase. As a result, the currents cancel each other. The source supplies the in-phase current only for the resistance load in the circuit. Therefore, the line current is the same as the current taken by the resistor.

PROBLEM 6

Statement of the Problem

For the circuit given in Figure 8–13, determine the following items:

1. The line current
2. The power
3. The power factor and the phase angle

Construct a vector diagram for this parallel resonant circuit.

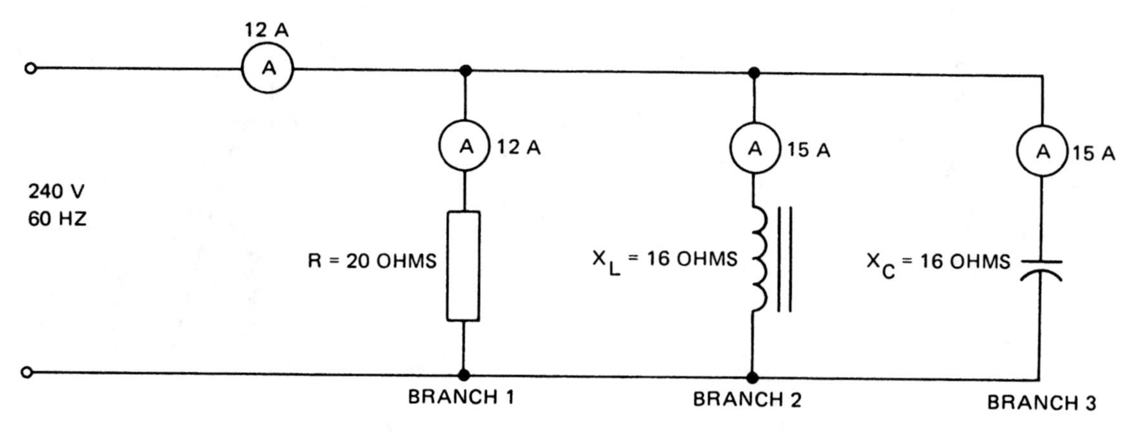

FIGURE 8–13 Resonant parallel circuit (*Delmar/Cengage Learning*)

Solution

1. $I_{line} = I_R = 12$ A
2. $P = I^2R = 12^2 \times 20 = 2880$ W; or
   $VI = 240 \times 12 = 2880$ W
3. PF = watts ÷ VI = 2880 ÷ 2880 = 1.00 (unity). The phase angle θ is 0°.
4. The vector diagram (Figure 8–14). shows that the line current and the in-phase current are the same. In addition, it can be seen that the two quadrature currents are equal and 180° out of phase with each other.

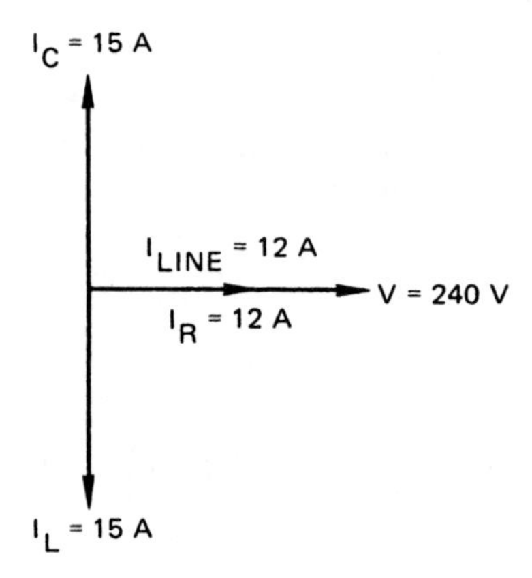

FIGURE 8–14 Vector diagram for resonant parallel circuit (*Delmar/Cengage Learning*)

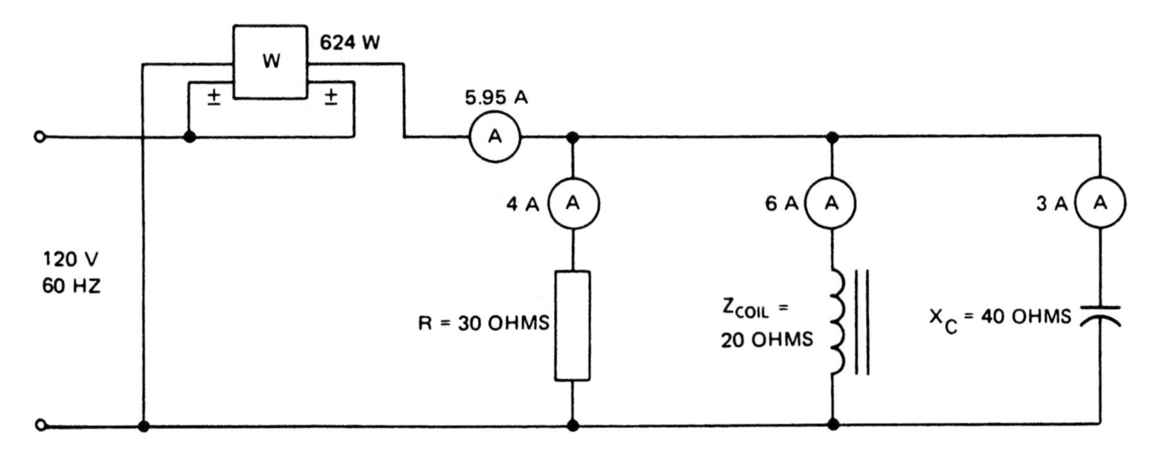

FIGURE 8–15 Parallel circuit with branches containing R, Z_{coil}, and X_C *(Delmar/Cengage Learning)*

PARALLEL CIRCUIT WITH BRANCHES CONTAINING R, Z_{coil}, AND X_C

The circuit shown in Figure 8–15 contains three branches: a wattmeter, which indicates the true power in the circuit; and ammeters, which indicate the line current and the current in each branch.

One branch contains a pure resistance of 30 Ω. A second branch contains a load that has both resistance and inductance. This is similar to the load shown in Figure 8–8. As before, the value of Z_{coil} is not to be confused with the value of Z_{total}. Z_{coil} is used to represent the value of impedance for the coil only, and Z_{total} is used to represent the value of impedance for the entire circuit. Finally, the third branch contains a capacitor with a capacitive reactance of 40 Ω.

Power Factor for the Circuit

The power factor of the circuit is found by the use of the following expression:

$$PF = \frac{P}{VI}$$

$$= \frac{624}{120 \times 5.95} = 0.874 \text{ lag}$$

The entire circuit has a lagging power factor because the inductive reactive current is more than the current in X_C. The phase angle is 29° lagging.

Power Factor for the Impedance Coil

To find the power factor of the impedance branch, the power loss in the coil must be determined first:

$$\text{Coil power loss} = 624 - (120 \times 4)$$
$$= 624 - 480$$
$$= 144 \text{ W}$$
$$R_{coil} = \frac{P_{coil}}{I^2_{coil}}$$
$$= \frac{144}{6^2}$$
$$= \frac{144}{36} = 4\ \Omega$$
$$\text{Power factor} = \frac{R}{Z}$$
$$= \frac{4}{20} = 0.20 \text{ lagging}$$

The current in the impedance coil lags the line voltage. The cosine of this angle of lag is 0.20. Thus, the angle itself is 78.5°.

The Vector Diagram. The angles and current values calculated for this problem are used to obtain the vector diagram shown in Figure 8–16. A graphical solution to the parallel circuit is shown in the following procedure. Steps 1 through 3 require a compass, protractor, and ruler.

1. Lay out the current vectors as shown:

   $\vec{I}_R = 4$ A, $\vec{I}_C = 3$ A, and $\vec{I}_{coil} = 6$ A
2. Combine the two vectors $\vec{I}_R$ and $\vec{I}_C$. These vectors are 90° out of phase. Their resultant is: $I_C + I_R = 5$ A.
3. Because $\vec{I}_T = \vec{I}_C + \vec{I}_R + \vec{I}_{coil}$, the I_{coil} vector should be added to the resultant $(\vec{I}_C + \vec{I}_R)$. The total current, I_T or I_{line}, has a value of 5.95 A. Note that the magnitude of I_{line} is less than that of the branch current (I_{coil}), or 6 A.

This problem can be solved mathematically by resolving the I_{coil} vector (6 A) into a horizontal (in-phase) component and a vertical (quadrature) component. This method is used in the following section on power factor correction.

CORRECTING MOTOR POWER FACTOR

It is sometimes necessary to correct the power factor of a motor. The following procedure can be used to perform this task. Before the power factor of a motor can be corrected, it must first be determined how much out of phase current and voltage are with each other. In the circuit shown in Figure 8–17, a wattmeter, an ammeter, and a voltmeter are used to measure circuit values. The voltmeter indicates a voltage of 480 V connected to the motor. The ammeter shows a total current draw of 250 A, and the wattmeter shows the true power of the circuit to be 96 kW.

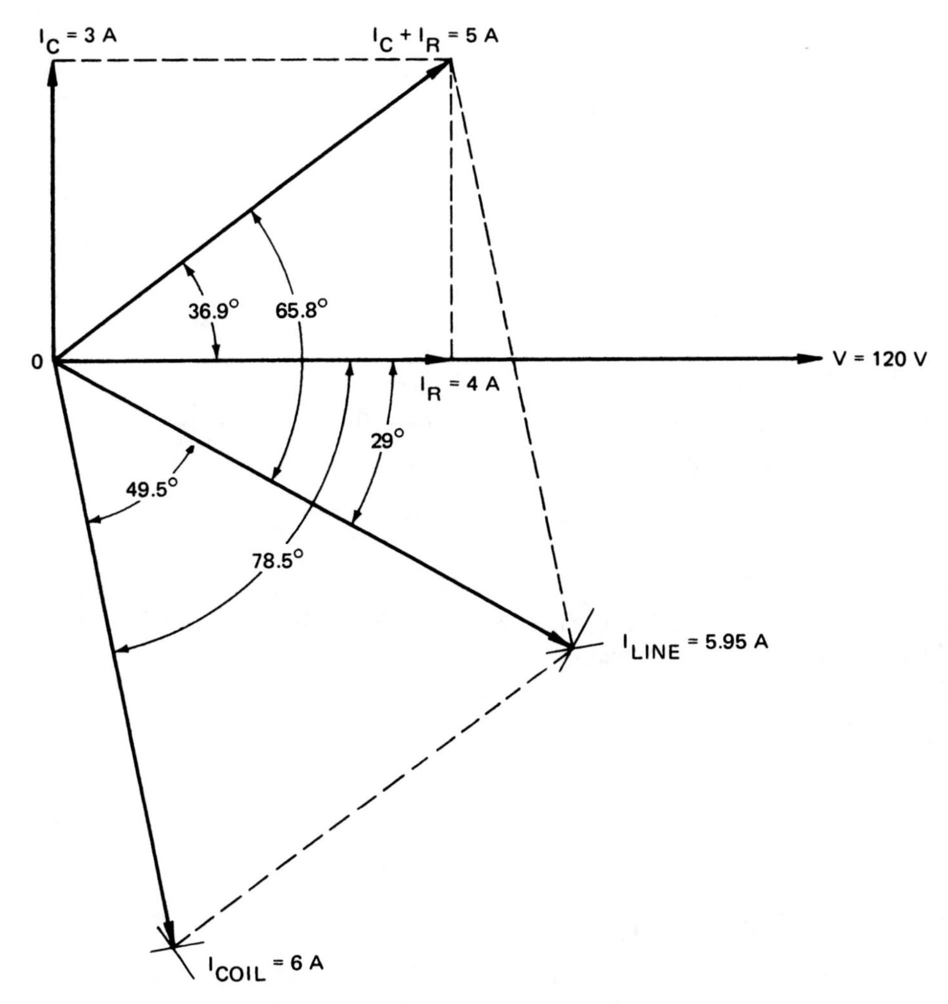

FIGURE 8–16 Vector diagram for a parallel circuit with branches containing R, Z_{coil}, and X_C *(Delmar/Cengage Learning)*

Computing Apparent Power

The apparent power or volt-amperes of the circuit can be determined by multiplying the applied voltage by the total current:

$$\begin{aligned} VA &= V \times I \\ &= 480 \times 250 \\ &= 120{,}000 \text{ VA} = 120 \text{ kVA} \end{aligned}$$

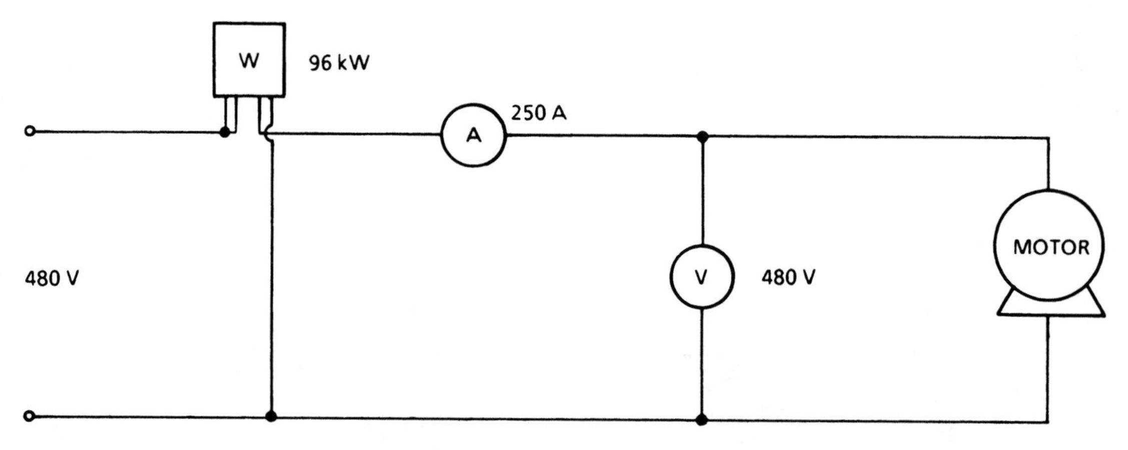

FIGURE 8–17 **AC motor circuit (*Delmar/Cengage Learning*)**

Computing Power Factor

Because both the true power and the apparent power are known, the power factor of the circuit can be found using the following formula:

$$PF = \frac{P}{VA}$$

$$= \frac{96\text{ kW}}{120\text{ kVA}}$$

$$= 0.8 \quad \text{or} \quad 80\%$$

Calculating the Reactive Power

The amount of reactive power or VARs (volt-amperes-reactive) can be determined using the formula

$$VARs = \sqrt{VA^2 - P^2}$$

In this example, the value of kilovolt-amperes-reactive (kVARs) will be computed by using the value of kilovolt-amperes and kilowatts:

$$kVARs = \sqrt{120^2 - 96^2}$$

$$= \sqrt{14{,}400 - 9216}$$

$$= \sqrt{5184}$$

$$= 72\text{ kVARs}$$

The 72 kVARs represents the amount of reactive power in the circuit. Because this reactive component is caused by a motor, it is inductive. If the power factor is to be corrected, the same amount of capacitive VARs must be added to the circuit. The capacitive VARs will cancel the inductive VARs, and the current and voltage will be in phase with each other.

Calculating the Capacitance Needed

To calculate the amount of capacitance needed, first determine the amount of current that must flow to produce 72 capacitive kVARs. This can be found by using the following formula:

$$I = \frac{VARs}{V}$$

$$= \frac{72{,}000}{480}$$

$$= 150\ A$$

The amount of capacitive reactance needed to produce a flow of 150 A through a capacitor can now be computed using the following formula:

$$X_C = \frac{V}{I}$$

$$= \frac{480}{150}$$

$$= 3.2\ \Omega$$

The amount of capacitance needed to produce a capacitive reactance of 3.2 Ω at 60 Hz can now be computed using the formula

$$C = \frac{1}{2\pi f X_C} \qquad C = \frac{1}{377 \times 3.2} \qquad C = \frac{1}{1206.4}$$

where $C = 0.0008289\ F$ or $828.9\ \mu F$

If a capacitance of 828.9 μF is connected across the motor as shown in Figure 8–18, the power factor will be corrected to unity or 100%.

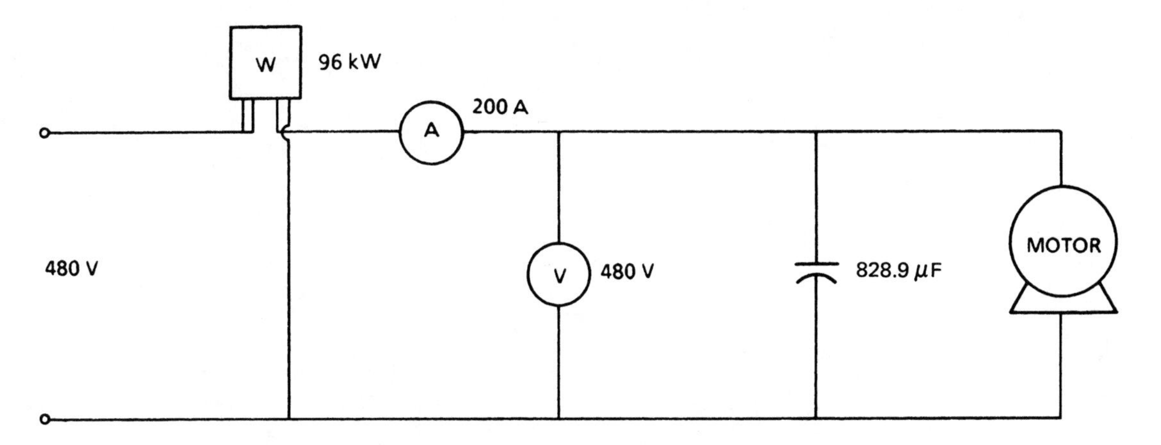

FIGURE 8–18 A capacitor corrects power factor for the motor (*Delmar/Cengage Learning*)

Advantages of Unity Power Factor. Utility companies and industrial plants try to keep the power factor of their ac circuits as close to unity as possible. The unity power factor is desirable because of the following:

1. The quadrature current in the line wires causes I^2R losses, just as the in-phase current causes power losses in the circuit conductors. Therefore, if the power factor is raised to a value near unity, the power losses in the line wires are reduced and the efficiency of transmission is increased.
2. If the power factor is corrected to a value near unity, there is less voltage drop in the line. As a result, the voltage at the load is more constant.
3. AC circuits operated at a high power factor improve the efficiency and operating performance of ac generators and transformers supplying these circuits.

SUMMARY

- The voltage across any branch circuit in a parallel arrangement, neglecting line voltage drops, is the same as the line voltage.
- In the analysis of series circuits, the current is used as the reference vector.
- In the analysis of parallel circuits, the voltage is used as the reference vector. Thus, all angles in the analysis of parallel circuits are measured with respect to the voltage vector.
- For an ac parallel circuit with branches containing only pure resistance, the calculations are the same as those for a direct-current parallel circuit:

$$R_C = \frac{1}{\frac{1}{R_1} + \frac{1}{R_2} + \frac{1}{R_3}}$$

$$I_{line} = I_1 + I_2 + I_3$$

or

$$I_{line} = \frac{V}{R_C}$$

$$P_{total} = V \times I_{total}$$

or

$$P_{total} = I^2_{total} \times R_C$$

or

$$P_{total} = \frac{V^2}{R_C}$$

For an ac parallel circuit, the current in each branch circuit containing a pure resistive component is in phase with the line voltage. Thus, the power factor is

$$PF = \frac{P}{VA} = 1.00 \text{ (resistance only)}$$

The angle whose cosine is 1.00 is 0°.

- In a parallel ac circuit containing R, X_C, and X_L components, the total line current may be out of phase with the line voltage. In this case, the line current must be calculated using vectors:

$$\vec{I}_{total} = \vec{I}_R + \vec{I}_{XL} = \vec{I}_{XC}$$

or

$$I_{line} = \sqrt{I_R^2 + (I_L - I_C)^2}$$

- A parallel circuit with branches containing R, X_C, and X_L has the following properties as determined by calculation:

1. **Current**

   a. Resistive branch:

   $$I_R = \frac{V}{R}$$

   The current in the resistive branch is in phase with the line voltage.

   b. Coil branch:

   $$I_L = \frac{V}{X_L}$$

   The current in the capacitive branch leads the line voltage by 90°.

   c. Capacitive branch:

   $$I_C = \frac{V}{X_C}$$

   The current in the inductive branch lags the same line voltage by 90°.

   I_C and I_L are 180° out of phase with each other, and $(I_L - I_C)$ yields a net quadrature current. The line current consists of an in-phase component and a leading or lagging quadrature component:

   $$I_{line} = \sqrt{I_R^2 + (I_L - I_C)^2}$$

2. **Power.** All of the power taken by this circuit is used in the resistive branch. Because the currents taken by the coil branch and the capacitor branch are 90° out of phase with the line voltage, no power is taken in either of these branches. The power for the resistance branch is the power for the entire parallel circuit:

   $$P = VI_R \quad \text{or} \quad P = I^2R \quad \text{or} \quad P = \frac{V^2}{R}$$

3. **Power factor (PF).** The power factor for an ac parallel circuit is determined using either of the following ratios:

$$PF = \frac{P}{VA}$$

or

$$PF = \frac{I_{in\ phase}}{I_{line}}$$

4. **Magnetizing VARs.** Magnetizing VARs required by the coil:

$$VARs = V \times I_L \quad \text{or} \quad I_L^2 \times X_L \quad \text{or} \quad \frac{V^2}{X_L}$$

Magnetizing VARs supplied by the capacitor:

$$VARs = V \times I_C \quad \text{or} \quad I_C^2 \times X_C \quad \text{or} \quad \frac{V^2}{X_C}$$

The net magnetizing VARs supplied by the line is the difference between the VARs required by the coil and the VARs supplied by the capacitor.

- Resonance in ac circuits:
  1. Series and parallel resonant circuits having identical values of R, L, and C have the following similarities:
     a. The line currents and the line voltages are in phase.
     b. The power factors are unity.
     c. The values of X_L and X_C are equal.
     d. The impedances are purely resistive.
  2. Series and parallel resonant circuits having the same values of R, L, and C are different in the following respects:
     a. The current is at a maximum in a series resonant circuit and at a minimum in a parallel resonant circuit.
     b. The impedance is at a minimum in a series resonant circuit and at a maximum in a parallel resonant circuit.

  These factors make parallel resonance a useful feature in tuned circuits.

- The current and impedance curves for a parallel resonant circuit are opposite to those for a series resonant circuit. The current equation for the parallel resonant circuit is

$$I_L = \sqrt{I_R^2 + (I_{XL} - I_{XC})^2}$$

  1. At frequencies below resonance:

$$I_L = \sqrt{I_R^2 + I_{XL}^2}$$

  where I_{XC} is negligible at low frequencies (X_C has high values) and a small current flows. I_L is at a maximum.

  2. For frequencies at resonance:

$$I_L = \sqrt{I_R^2} = I_R$$

  where X_L and X_C are equal and 180° out of phase. Their currents are equal and opposite and thus cancel. I_L is at a minimum.

3. At frequencies above resonance:

$$I_L = \sqrt{I_R^2 + I_{XC}^2}$$

where I_{X_L} is negligible and X_L has high values. I_L is at a maximum again.

4. The impedance curve is the reciprocal of the current curve:

$$Z = \frac{V}{I}$$

- For a parallel ac circuit containing impedance, the power lost in the coil, the resistance of the coil, and the power factor of the coil are determined as follows:
  1. Power loss, coil $= P - (V \times I_L)$
  2. $R_{coil} = \frac{P_{coil}}{I^2_{coil}}$
  3. Power factor $= \frac{R_{coil}}{Z_{coil}}$
- Power factor correction for an ac induction motor:
  1. Use instruments to measure values of voltage, current, and power.
  2. Compute apparent power of the circuit using the formula

     $$VA = V \times I$$

  3. Compute the power factor using the formula

     $$PF = \frac{P}{VA}$$

  4. Compute the amount of reactive power in the circuit using the formula

     $$VARs = \sqrt{VA^2 - P^2}$$

  5. Compute the amount of capacitance needed using the formulas

     $$I = \frac{VARs}{V}$$

     $$X_C = \frac{V}{I}$$

     $$C = \frac{X_C}{2\pi f}$$

- Advantages of a unity power factor:
  1. The quadrature current causes I^2R losses. Correcting the power factor to a value near unity reduces the power losses in the line wires and the efficiency of transmission is increased.
  2. Correcting the power factor to a value near unity decreases the voltage drop. As a result, the voltage at the load is more constant.
  3. If ac circuits are operated at a high power factor, there is an improvement in the efficiency and operating performance of ac generators and transformers supplying these circuits.

Achievement Review

1. A parallel circuit consists of two branches connected to a 120-V, 60-Hz source. The first branch has a noninductive resistance load of 50 Ω. The second branch consists of a coil with an inductance of 0.2 H and negligible resistance.
   a. Find the current in each branch.
   b. Determine the line current.
   c. Determine the power factor of the parallel circuit.
   d. Draw a labeled vector diagram for this circuit.

2. A resistance of 40 Ω is connected in parallel with a pure inductance of 0.24 H across a 120-V, 25-Hz supply.
   a. Find the current taken by each branch.
   b. Determine the line current.
   c. Determine the power factor of the parallel circuit.
   d. Determine the combined impedance of the parallel circuit.
   e. Draw a labeled vector diagram for the parallel circuit.

3. A 120-V, 60-Hz source supplies a parallel circuit consisting of two branches. Branch 1 is a noninductive resistance load of 5 Ω. Branch 2 feeds a 1000-μF capacitor with negligible resistance.
   a. Determine
      (1) the current taken by each branch.
      (2) the line current.
      (3) the combined impedance of the parallel circuit.
      (4) the power factor and phase angle for the parallel circuit.
   b. Draw a labeled vector diagram for the parallel circuit.

4. A 240-V, 60-Hz single-phase source supplies a parallel circuit consisting of two branches. Branch 1 has a resistance of 20 Ω in series with an inductance of 0.04 H. Branch 2 feeds a 50-μF capacitor with negligible resistance.
   a. Determine
      (1) the current in each branch.
      (2) the total current.
      (3) the power factor of branch 1 (containing R and X_L components in series).
      (4) the power factor of the entire parallel circuit.
   b. Draw a labeled vector diagram for the parallel circuit.

5. A 240-V, 60-Hz parallel circuit supplies three branches. Branch 1 consists of a noninductive heating load with a resistance of 12 Ω, branch 2 feeds a pure inductive reactance load of 8 Ω, and branch 3 is a capacitor having a capacitive reactance of 16 Ω.
   a. Determine
      (1) the current taken by each of the three branches.
      (2) the total current.

(3) the combined impedance of the parallel circuit.
(4) the circuit power factor.
(5) the total power in kilowatts taken by the parallel circuit.
b. Draw a labeled vector diagram for this circuit.

6. Using the circuit shown in Figure 8–19, determine
   a. the current taken by each branch.
   b. the power expended in the resistance branch, in watts.
   c. the power expended in the coil branch, in watts.
   d. the impedance of the coil.
   e. the effective resistance and the inductive reactance components of the coil.

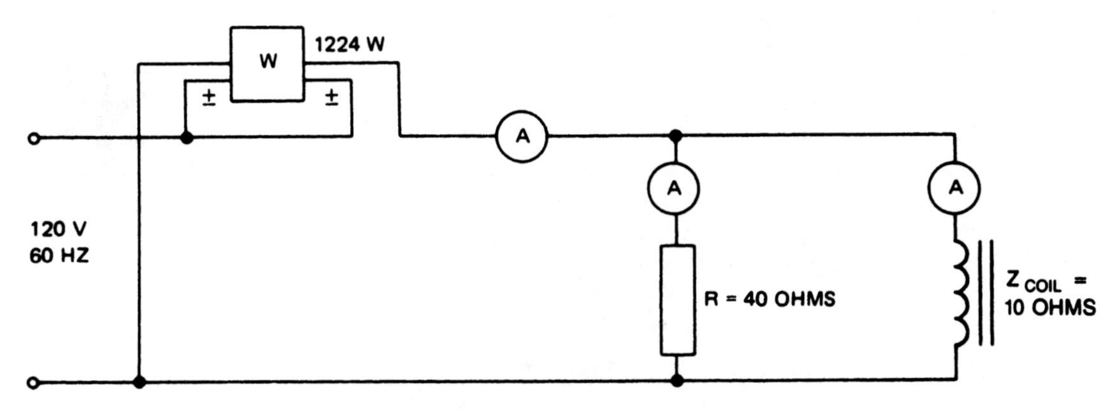

FIGURE 8–19 ***(Delmar/Cengage Learning)***

7. a. Using the circuit in the preceding question, determine
      (1) the total current.
      (2) the power factor and the phase angle for the inductive branch circuit.
      (3) the power factor and the phase angle for the entire parallel circuit.
      (4) the combined impedance of the parallel circuit.
   b. Draw a vector diagram and label it properly for this circuit.

8. A 125-V, 60-Hz parallel circuit has three branches. Branch 1 has a resistance of 20 Ω, branch 2 feeds a coil with an inductance of 0.0211 H, and branch 3 supplies a 200-μF capacitor. The resistances of both the coil and the capacitor are negligible. Determine
   a. the current in each branch.
   b. the total current.
   c. the power factor for the parallel circuit.

9. Explain the meaning of the term *parallel resonance*.

10. An ac parallel circuit consists of three branches connected to a 120-V, 60-Hz supply. Branch 1 consists of a noninductive resistance load, branch 2 feeds a coil, and branch 3 supplies a capacitor. The resistances of the coil and the capacitor are negligible. The total line current is 10 A. The power factor of the entire parallel circuit is 90% lagging. The capacitor has a capacitive reactance of 12.5 Ω.
    a. Draw a labeled vector diagram of the circuit.
    b. Determine
        (1) the current taken by each of the three branches.
        (2) the true power in watts taken by the entire parallel circuit.

11. The capacitor in the circuit described in question 5 is changed to one having the proper rating to give a circuit power factor of unity. Determine the following data for the capacitor value that causes the circuit to be in parallel resonance:
    a. Rating in VARs
    b. Capacitive reactance
    c. Rating in microfarads

12. In the circuit given in question 6, there is to be a third branch feeding a capacitor. This capacitor has the proper rating to correct the power factor to unity. Using the information provided in question 6, determine
    a. the rating of the capacitor, in VARs.
    b. the capacitive reactance of the capacitor (assume the resistance to be negligible).
    c. the rating of the capacitor, in microfarads.

    Construct a vector diagram for this circuit.

13. A noninductive heater and an ac motor are operated in parallel across a 120-V, 25-Hz source. The heater takes 600 W. The motor takes 360 W at a lagging power factor of 60%. Determine
    a. the current taken by each of the two branches.
    b. the total current.
    c. the circuit power factor.

14. A 120-V induction motor requires 24 A at a lagging power factor of 75%. If a lamp load of 30 A is connected in parallel with the motor, what are the power factor and the phase angle for the entire parallel circuit?

15. Why is it important to maintain a high power factor with alternating-current systems?

16. At full load, a 220-V, 2-horsepower (hp), single-phase induction motor takes 12 A when operated on a 60-Hz service. The full-load power factor is 80% lagging. At full load, determine
    a. the phase current.
    b. the power, in watts.
    c. the quadrature current.
    d. the magnetizing VARs required.
    e. the input volt-amperes.

17. The power factor of the circuit in question 16 is to be corrected to a value of unity by connecting a capacitor in parallel with the motor. Determine
    a. the rating of the capacitor, in VARs.
    b. the capacitive reactance of the capacitor.
    c. the rating of the capacitor, in microfarads.
    d. the line current after the power factor is increased to 100% or unity.

18. An industrial plant has a load of 50 kW at a power factor of 70% lag, feeding from a 240-V, 60-Hz system. Determine
    a. the line current.
    b. the capacitor rating, in kilovars, required to raise the power factor to unity.
    c. the line current after the capacitor is added to the circuit.

PRACTICE PROBLEMS FOR UNIT 8

Resistive Inductive Parallel Circuits

Find the missing values for the following circuits. Refer to Figure 8–20 and the formulas under the Resistive Inductive (Parallel) section of Appendix 15.

1. Assume that the circuit shown in Figure 8–20 is connected to a 60-Hz line and has a total current flow of 34.553 A. The inductor has an inductance of 0.02122 H, and the resistor has a resistance of 14 Ω.

V_T = ________	V_R = ________	V_L = ________
I_T = 34.553 A	I_R = ________	I_L = ________
Z = ________	R = 14 Ω	X_L = ________
VA = ________	P = ________	$VARs_L$ = ________
PF = ________	$\angle\theta$ = ________	L = 0.02122 H

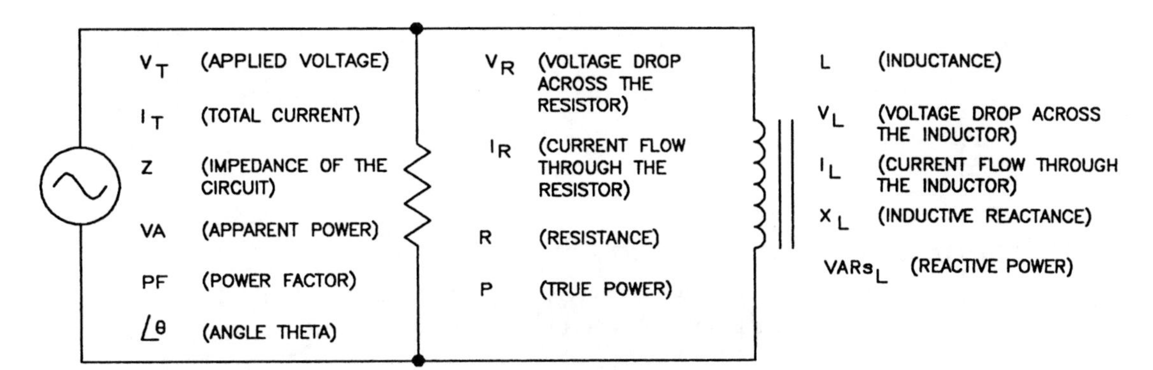

FIGURE 8–20 RL parallel circuit (*Delmar/Cengage Learning*)

2. Assume that the circuit is connected to a 60-Hz line and has a total impedance of 21.6 Ω. The resistor has a resistance of 36 Ω and the inductor has a current flow of 2 amperes through it.

V_T = ________	V_R = ________	V_L = ________
I_T = ________	I_R = ________	I_L = 2 A
Z = 21.6 Ω	R = 36 Ω	X_L = ________
VA = ________	P = ________	$VARs_L$ = ________
PF = ________	$\angle\theta$ = ________	L = ________

3. Assume that the circuit in Figure 8–20 is connected to a 60-Hz line and has a current flow through the resistor of 82 A and a current flow through the inductor of 156 A. The total impedance of the circuit is 2.7236 Ω.

V_T = ________	V_R = ________	V_L = ________
I_T = ________	I_R = 82 A	I_L = 156 A
Z = 2.7236 Ω	R = ________	X_L = ________
VA = ________	P = ________	$VARs_L$ = ________
PF = ________	$\angle\theta$ = ________	L = ________

4. Assume that the circuit in Figure 8–20 is connected to a 60-Hz line and has a true power of 375 Watts and a reactive power of 150 VARs. The total current flow in the circuit is 6 A.

V_T = ________	V_R = ________	V_L = ________
I_T = 6 A	I_R = ________	I_L = ________
Z = ________	R = ________	X_L = ________
VA = ________	P = 375 W	$VARs_L$ = 150 VARs
PF = ________	$\angle\theta$ = ________	L = ________

Resistive Capacitive Parallel Circuits

Find the missing values for the following circuits. Refer to Figure 8–21 and the formulas listed under the Resistive Capacitive (Parallel) section of Appendix 15.

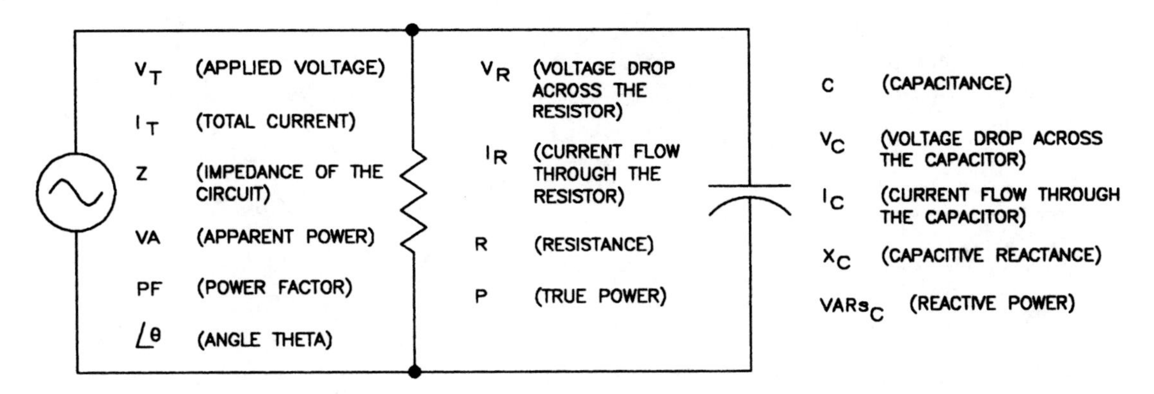

FIGURE 8–21 RC parallel circuit (*Delmar/Cengage Learning*)

5. Assume that the circuit shown in Figure 8–21 is connected to a 60-Hz line and has a total current flow of 10.463 A. The capacitor has a capacitance of 132.626 μF, and the resistor has a resistance of 14 Ω.

V_T = ________	V_R = ________	V_C = ________
I_T = 10.463 A	I_R = ________	I_C = ________
Z = ________	R = 14 Ω	X_C = ________
VA = ________	P = ________	$VARs_C$ = ________
PF = ________	$\angle\theta$ = ________	C = 132.626 μF

6. Assume that the circuit is connected to a 400-Hz line and has a total impedance of 64.8 Ω. The resistor has a resistance of 108 Ω, and the capacitor has a current flow of 2 A through it.

V_T = ________	V_R = ________	V_C = ________
I_T = ________	I_R = ________	I_C = 2 A
Z = 64.8 Ω	R = 108 Ω	X_C = ________
VA = ________	P = ________	$VARs_C$ = ________
PF = ________	$\angle\theta$ = ________	C = ________

7. Assume that the circuit in Figure 8–21 is connected to a 600-Hz line and has a current flow through the resistor of 65.6 A and a current flow through the capacitor of 124.8 A. The total impedance of the circuit is 2.17888 Ω.

V_T = ________	V_R = ________	V_C = ________
I_T = ________	I_R = 65.6 A	I_C = 124.8 A
Z = 2.17888 Ω	R = ________	X_C = ________
VA = ________	P = ________	$VARs_C$ = ________
PF = ________	$\angle\theta$ = ________	C = ________

8. Assume the circuit is connected to a 1000-Hz line and has a true power of 486.75 W and a reactive power of 187.5 VARs. The total current flow in the circuit is 7.5 A.

V_T = ________	V_R = ________	V_C = ________
I_T = 7.5 A	I_R = ________	I_C = ________
Z = ________	R = ________	X_C = ________
VA = ________	P = 486.75 W	$VARs_C$ = 187.5 VARs
PF = ________	$\angle\theta$ = ________	C = ________

Resistive-Inductive-Capacitive Parallel Circuits

Find the missing values in the following circuits. Refer to Figure 8–22 and the formulas under the Resistive-Inductive-Capacitive (Parallel) section of Appendix 15.

9. The circuit shown in Figure 8–22 is connected to a 120-volt, 60-Hz line. The resistor has a resistance of 36 Ω, the inductor has an inductive reactance of 40 Ω and the capacitor has a capacitive reactance of 50 Ω.

V_T = 120 V	V_R = ________	V_L = ________
I_T = ________	I_R = ________	I_L = ________
Z = ________	R = 36 Ω	X_L = 40 Ω
VA = ________	P = ________	$VARs_L$ = ________
PF = ________	$\angle\theta$ = ________	L = ________

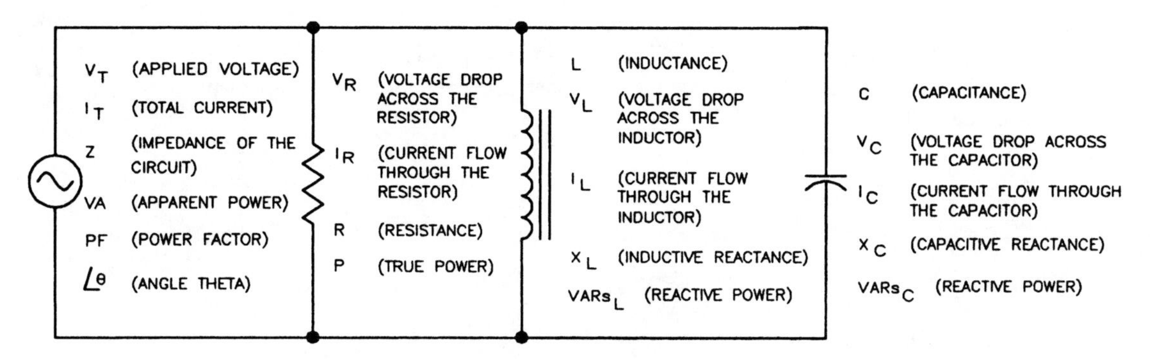

FIGURE 8–22 RLC parallel circuit (*Delmar/Cengage Learning*)

V_C = ________

I_C = ________

X_C = 50 Ω

$VARs_C$ = ________

C = ________

10. The circuit in Figure 8–22 is connected to a 400-Hz line with a total current flow of 22.627 A. There is a true power of 3840 W. The inductor has a reactive power of 1920 VARs, and the capacitor has a reactive power of 5760 VARs.

V_T = ________	V_R = ________	V_L = ________
I_T = 22.267 A	I_R = ________	I_L = ________
Z = ________	R = ________	X_L = ________
VA = ________	P = 3840 W	$VARs_L$ = 1920 VARs
PF = ________	$\angle\theta$ = ________	L = ________

V_C = ________

I_C = ________

X_C = ________

$VARs_C$ = 5760 VARs

C = ________

11. The circuit in Figure 8–22 is connected to a 60-Hz line. The apparent power in the circuit is 48.106 VA. The resistor has a resistance of 12 Ω. The inductor has an inductive reactance of 60 Ω, and the capacitor has a capacitive reactance of 45 Ω.

V_T = ________	V_R = ________	V_L = ________
I_T = ________	I_R = ________	I_L = ________
Z = ________	R = 12 Ω	X_L = 60 Ω
VA = 48.106 VA	P = ________	$VARs_L$ = ________
PF = ________	$\angle\theta$ = ________	L = ________

V_C = ________

I_C = ________

X_C = 45 Ω

$VARs_C$ = ________

C = ________

12. The circuit shown in Figure 8–22 is connected to a 1000-Hz line. The resistor has a current flow of 60 A, the inductor has a current flow of 150 A, and the capacitor has a current flow of 70 A. The circuit has a total impedance of 4.8 Ω.

V_T = ________	V_R = ________	V_L = ________
I_T = ________	I_R = 60 A	I_L = 150 A
Z = 4.8 Ω	R = ________	X_L = ________
VA = ________	P = ________	$VARs_L$ = ________
PF = ________	$\angle\theta$ = ________	L = ________

V_C = ________

I_C = 70 A

X_C = ________

$VARs_C$ = ________

C = ________

9

Series–Parallel Circuits

Objectives

After studying this unit, the student should be able to

- use the GBY method to solve ac series–parallel circuits containing impedances with different resistive and reactive components.
- define and give the formula and the appropriate unit for each of the following terms: admittance, conductance, and susceptance.
- explain the relationship between admittance, conductance, and susceptance and compare this with the relationship between impedance, resistance, and reactance, using impedance and admittance triangles.
- calculate values for the admittance, conductance, and susceptance of an ac network system using the GBY method.

There are several methods that can be used to solve ac series–parallel circuits containing impedances with different resistive and reactive components. The method used in this unit introduces the terms *conductance*, *susceptance*, and *admittance*. The following abbreviations are used for these terms: G for conductance, B for susceptance, and Y for admittance. Known as the GBY method, it is used here because only basic mathematical processes are needed to solve circuit problems.

ADMITTANCE, CONDUCTANCE, AND SUSCEPTANCE

Admittance

Impedance is the measurement of the opposition to electron flow in a circuit containing resistance and reactance. The unit of *admittance* is a measurement of the ease of electron flow through a circuit or component containing resistance and reactance. This means that admittance is the inverse, or reciprocal, of impedance. The unit of measurement for admittance is the *siemens* (abbreviated as S). (Formerly the unit of admittance was known as the *mho,* ℧.) Admittance is indicated by the letter Y.

The relationship between admittance, in siemens, and the impedance, in ohms, is

$$Y = \frac{1}{Z}$$

Also

$$Z = \frac{V}{I}$$

Therefore,

$$Y = \frac{I}{V}$$

Conductance

Conductance is a measurement of the ease of electron flow through a resistance. Conductance, like admittance, is measured in siemens. For dc circuits and ac circuits containing only noninductive resistance, the conductance is the reciprocal of the resistance. For ac circuits containing resistance and reactance, the conductance (G) is equal to the resistance divided by the square of the impedance:

$$\mathbf{G = R \div Z^2}$$

Susceptance

Susceptance measures the ease of electron flow through the reactance of an ac circuit. Like admittance and conductance, the unit of susceptance is the siemens. Susceptance (B) is equal to the reactance, in ohms, divided by the square of the impedance:

$$\mathbf{B = X \div Z^2}$$

Susceptances (B) in parallel and conductances (G) in parallel are added.

Relationship between Admittance, Conductance, and Susceptance

The relationship between the admittance, the conductance, and the susceptance is the same as the relationship between the impedance, the resistance, and the reactance, as shown in Figure 9–1.

The admittance triangle in Figure 9–1 shows that the admittance (Y) is the hypotenuse of a right triangle:

$$\mathbf{Y = \sqrt{G^2 + B^2}}$$

PROBLEM 1

Statement of the Problem

The schematic diagram of a series–parallel circuit is shown in Figure 9–2. The series part of the circuit contains resistance and inductive reactance. The GBY method is to be used to find the following circuit values:

1. The total impedance, in ohms
2. The current for the series–parallel circuit
3. The power factor of the series–parallel circuit

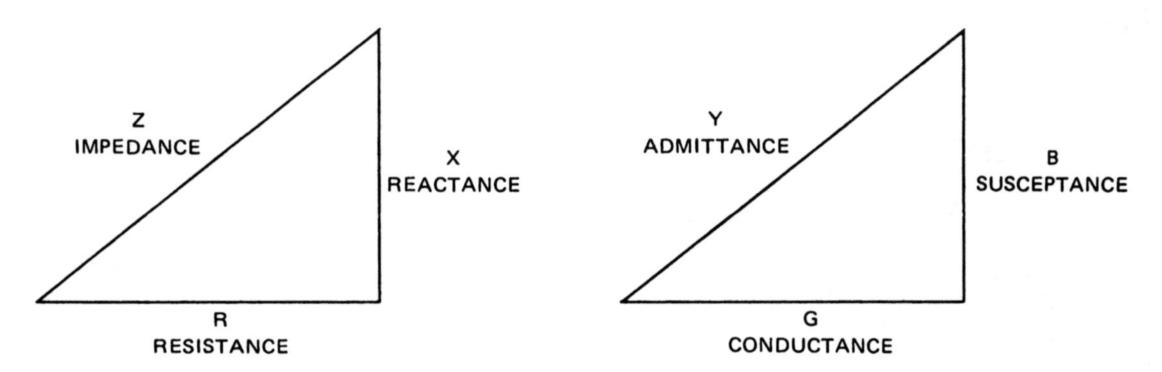

FIGURE 9–1 Impedance and admittance triangles (*Delmar/Cengage Learning*)

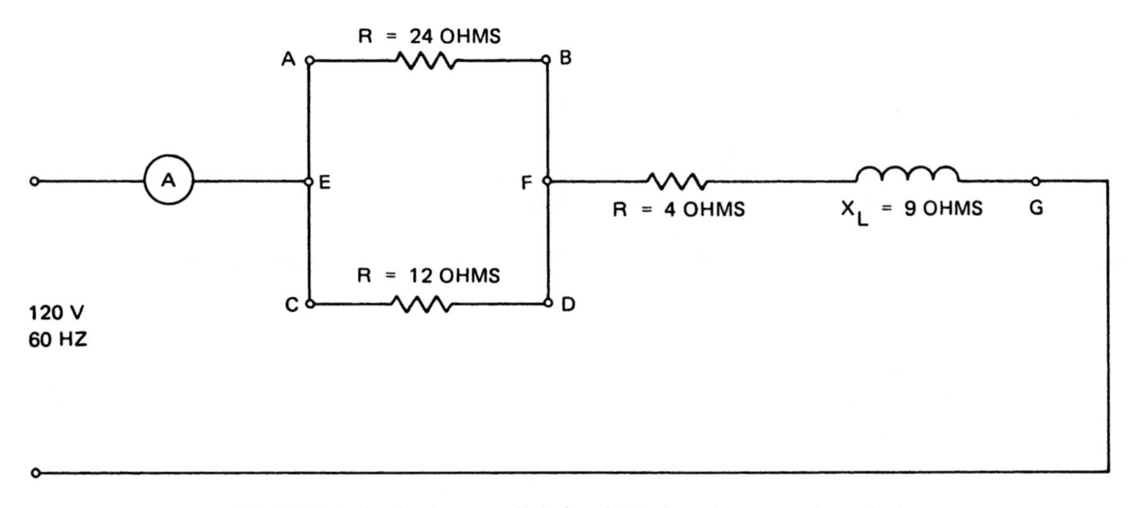

FIGURE 9–2 Series–parallel circuit (*Delmar/Cengage Learning*)

Solution

1. This problem is solved by considering the circuit in sections. The values of the various quantities for sections A–B and C–D of the circuit in Figure 9–2 are shown in this table. The values in each circuit section connected in parallel are placed in the table as a convenient reference.

Symbol	Circuit	
	A–B	**C–D**
R	24	12
X	0	0
Z	24	12
G	0.0417	0.0833
B	0	0

Circuit A–B:

$$R = 24\ \Omega$$
$$X = 0\ \Omega$$
$$Z = 24\ \Omega$$
$$G_{AB} = \frac{R_{AB}}{Z^2_{AB}} = \frac{24}{24^2} = 0.0417\ S$$
$$B_{AB} = \frac{X_{AB}}{Z^2_{AB}} = \frac{0}{24^2} = 0\ S$$

Circuit C–D:

$$R = 12\ \Omega$$
$$X = 0\ \Omega$$
$$Z = 12\ \Omega$$
$$G_{CD} = \frac{R_{CD}}{Z^2_{CD}} = \frac{12}{12^2} = 0.0833\ S$$
$$B_{CD} = \frac{X_{CD}}{Z^2_{CD}} = \frac{0}{12^2} = 0\ S$$

The admittance of the parallel portion of the circuit between E and F is determined as follows:

$$Y_{EF} = \sqrt{(G_{AB} + G_{CD})^2 + (B_{AB} + B_{CD})^2}$$
$$= \sqrt{(0.0417 + 0.0833)^2 + (0 + 0)^2}$$
$$= 0.125\ S$$

Impedance of E–F:

$$Z_{EF} = \frac{1}{Y}$$
$$= \frac{1}{0.125}$$
$$= 8\ \Omega$$

Resistance of E–F:

$$G_{EF} = \frac{R_{EF}}{Z^2_{EF}}$$
$$R_{EF} = G_{EF} \times Z^2_{EF}$$
$$= 0.125 \times 8^2$$
$$= 8\ \Omega$$

Reactance of E–F:

$$B_{EF} = \frac{X_{EF}}{Z^2_{EF}}$$

$$X_{EF} = B_{EF} \times Z^2_{EF}$$

$$= 0 \times 8^2$$

$$= 0\ \Omega$$

In the series part of the circuit between F and G, the resistance is 4 Ω and X_L is 9 Ω. For the entire circuit, the resistance of the parallel section (E–F) is added to the resistance of the series portion (F–G):

$$R_{EG} = R_{EF} + R_{FG} = 8 + 4 = 12\ \Omega$$

The total reactance (X_L) for the entire circuit is obtained by adding the reactance for the parallel section (E–F) to that of the series portion (E–F):

$$X_{EG} = X_{EF} + X_{FG} = 0 + 9 = 9\ \Omega$$

The impedance for the entire circuit is

$$Z_{EG} = \sqrt{R^2_{EG} + X^2_{EG}}$$

$$= \sqrt{12^2 + 9^2}$$

$$= 15\ \Omega$$

2. The current taken by this series–parallel circuit is

$$I = \frac{E}{Z_{EG}}$$

$$= \frac{120}{15} = 8\ A$$

3. The power factor of the entire circuit is

$$PF = \frac{R_{EG}}{Z_{EG}}$$

$$= \frac{12}{15} = 0.80 \text{ lagging}$$

PROBLEM 2

Statement of the Problem

In the series–parallel circuit shown in Figure 9–3, both the parallel and the series sections contain resistance and reactance.

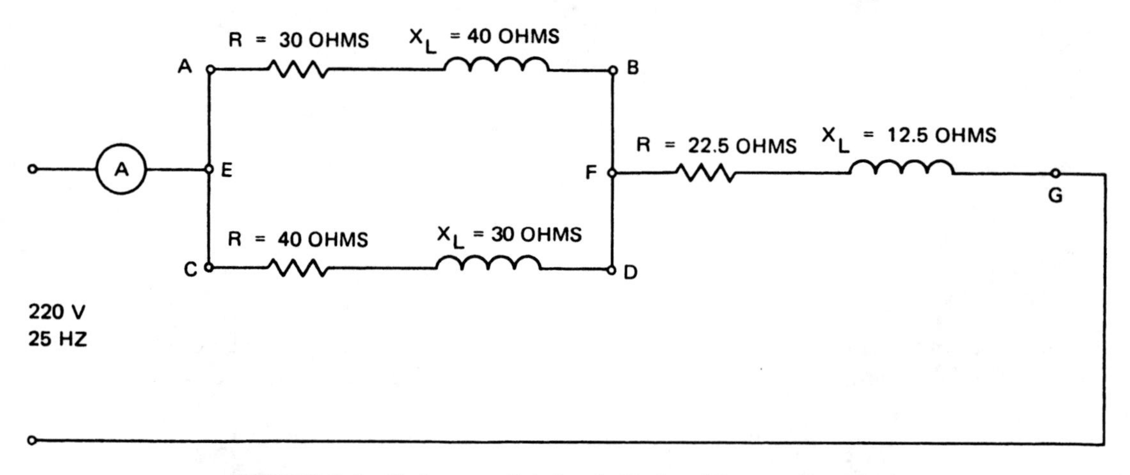

FIGURE 9–3 Series–parallel circuit (*Delmar/Cengage Learning*)

The GBY method is to be used to determine the following values for this circuit:

1. The total impedance, in ohms
2. The current indicated by the ammeter
3. The power factor of the series–parallel circuit

Solution

1. The following table lists the given values for the parallel branches (A–B and C–D) of the circuit.

Symbol	Circuit	
	A–B	C–D
R	30	12
X	40	30
Z	50	50
G	0.012	0.016
B	0.016	0.012

Branch A–B:

$$R_{AB} = 30\ \Omega$$

$$X_{AB} = 40\ \Omega$$

$$Z_{AB} = \sqrt{R_{AB}^2 + X_{AB}^2}$$

$$= \sqrt{30^2 + 40^2}$$

$$= \sqrt{2500} = 50\ \Omega$$

$$G_{AB} = \frac{R_{AB}}{Z^2_{AB}} = \frac{30}{50^2} = 0.012 \text{ S}$$

$$B_{AB} = \frac{X_{AB}}{Z^2_{AB}} = \frac{40}{50^2} = 0.016 \text{ S}$$

Branch C–D:

$$R_{CD} = 40\ \Omega$$

$$X_{CD} = 30\ \Omega$$

$$Z_{CD} = \sqrt{R^2_{CD} + X^2_{CD}}$$

$$= \sqrt{40^2 + 30^2}$$

$$= \sqrt{2500} = 50\ \Omega$$

$$G_{CD} = \frac{R_{CD}}{Z^2_{CD}} = \frac{40}{50^2} = 0.016 \text{ S}$$

$$B_{CD} = \frac{X_{CD}}{Z^2_{CD}} = \frac{30}{50^2} = 0.012 \text{ S}$$

The admittance for the series branch E–F is

$$Y_{EF} = \sqrt{(G_{AB} + G_{CD})^2 + (B_{AB} + B_{CD})^2}$$

$$= \sqrt{(0.012 + 0.016)^2 + (0.016 + 0.012)^2}$$

$$= 0.0396 = 0.04 \text{ S}$$

Additional values for the series branch E–F are determined as follows:

$$Z_{EF} = \frac{1}{Y_{EF}}$$

$$= \frac{1}{0.04}$$

$$= 25\ \Omega$$

$$R_{EF} = G_{EF} \times Z^2_{EF}$$

$$= 0.028 \times 25^2$$

$$= 17.5\ \Omega$$

$$X_{EF} = B_{EF} \times Z^2_{EF}$$

$$= 0.028 \times 25^2$$

$$= 17.5\ \Omega$$

For the entire circuit:

$$R_{EG} = 17.5 + 22.5 = 40\ \Omega$$

$$X_{EG} = 17.5 + 12.5 = 30\ \Omega$$

Thus, the total impedance of the entire circuit, in ohms, is

$$Z_{EG} = \sqrt{R_{EG}^2 + X_{EG}^2}$$

$$= \sqrt{40^2 + 30^2}$$

$$= 50\ \Omega$$

2. The current shown on the ammeter is

$$I_{EG} = \frac{V}{Z_{EG}}$$

$$= \frac{220}{50} = 4.4\ \text{A}$$

3. For the series–parallel circuit, the power factor is

$$PF = \frac{R_{EG}}{Z_{EG}}$$

$$= \frac{40}{50}$$

$$= 0.80 \text{ lagging}$$

SUMMARY

- The GBY method of solving ac network circuits is commonly used because only basic mathematical processes are needed.
  1. G = conductance. This unit is a measurement of the ease of electron flow through a circuit or component containing resistance.
     a. It is measured in siemens.
     b. It is the inverse or reciprocal of resistance in a dc circuit or an ac circuit containing only noninductive resistance:

     $$G = \frac{1}{R}$$

     c. For ac circuits containing resistance and reactance, the conductance is

     $$G = \frac{R}{Z^2}$$

2. B = susceptance. This unit is a measurement of the ease of electron flow through the reactance of an ac circuit.
   a. It is measured in siemens.
   b. It is expressed as the ratio of the reactance and the impedance squared:

$$B = \frac{X}{Z^2}$$

3. Y = admittance. This unit is a measurement of the ease of electron flow through a circuit or component containing resistance and reactance (impedance).
   a. It is measured in siemens.
   b. It is the inverse or reciprocal of impedance.
   c. The relationship between admittance, in siemens, and the impedance, in ohms, is

$$Y = \frac{1}{Z}$$

Also,

$$Z = \frac{V}{I}$$

Therefore,

$$Y = \frac{I}{V}$$

- The relationship between the admittance, the conductance, and the susceptance is the same as the relationship between impedance, resistance, and reactance.
- The admittance (Y) is the hypotenuse of a right triangle:

$$Y = \sqrt{G^2 + B^2}$$

- To determine the total susceptance, in siemens, for parallel branches in an ac circuit, the susceptance of the capacitor branch must be subtracted from the susceptance values of the other two branches.
  1. The inductive reactance (X_L) and the capacitive reactance (X_C) cause opposite effects in an ac circuit.
  2. The reciprocal values of X_L and X_C, expressed as susceptance, in siemens, also have opposite effects.
- The total admittance, in siemens, for a parallel circuit with a number of branches is

$$Y = \sqrt{(G_1 + G_2 + G_3)^2 + (B_1 + B_2 + B_3)^2}$$

- The combined resistance for an ac circuit containing series and parallel branches is the sum of the resistance in the series branches and the resistance in the parallel branches, where the resistance in the parallel branches is

$$R_{total} = GZ^2$$

- The combined reactance (XL) for a circuit with series and parallel branches is the sum of the reactance in the series branches and the reactance in the parallel branches, where the reactance in the parallel branches is

$$X_{total} = BZ^2$$

- The total impedance for a circuit with series and parallel branches is

$$Z = \sqrt{R_{total}^2 + X_{total}^2}$$

Achievement Review

1. Explain the meaning of the following terms:
   a. Admittance
   b. Conductance
   c. Susceptance
2. In the series–parallel circuit shown in Figure 9–4, determine
   a. the admittance, in siemens, of the parallel section of the circuit between points E and F.
   b. the total impedance of the series–parallel circuit.

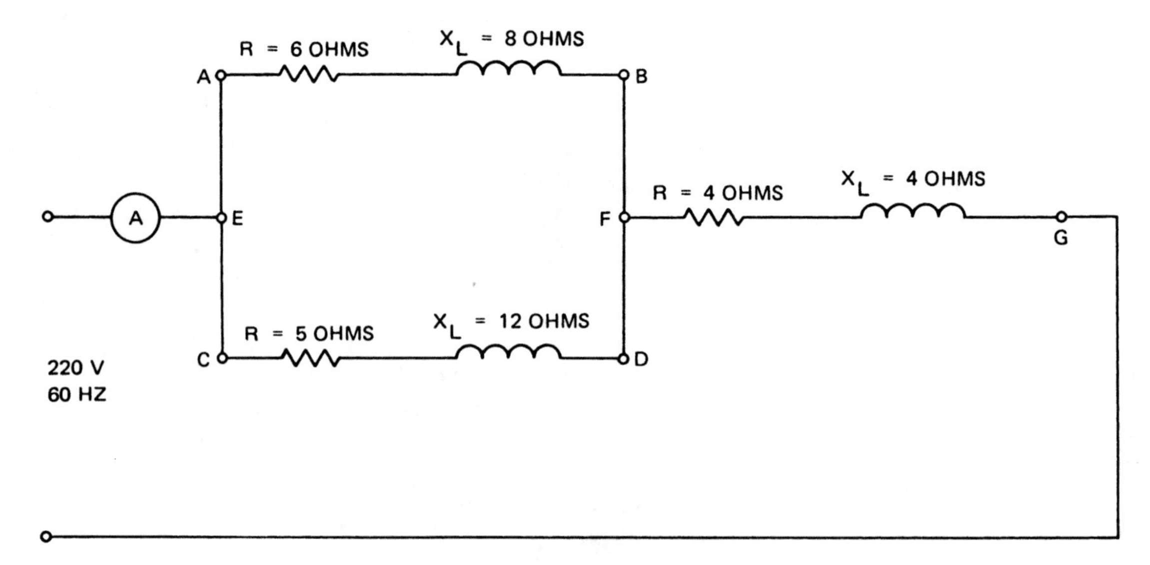

FIGURE 9–4 (*Delmar/Cengage Learning*)

3. Using the circuit given in question 2, determine
   a. the reading of the ammeter.
   b. the current in amperes in the A–B circuit branch.
   c. the power factor of the entire series–parallel circuit.
   d. the total power, in watts, taken by the entire series–parallel circuit.
4. In the series–parallel circuit shown in Figure 9–5, determine
   a. the admittance, in siemens, of the three branches connected in parallel.
   b. the impedance, in ohms, of the entire series–parallel circuit.

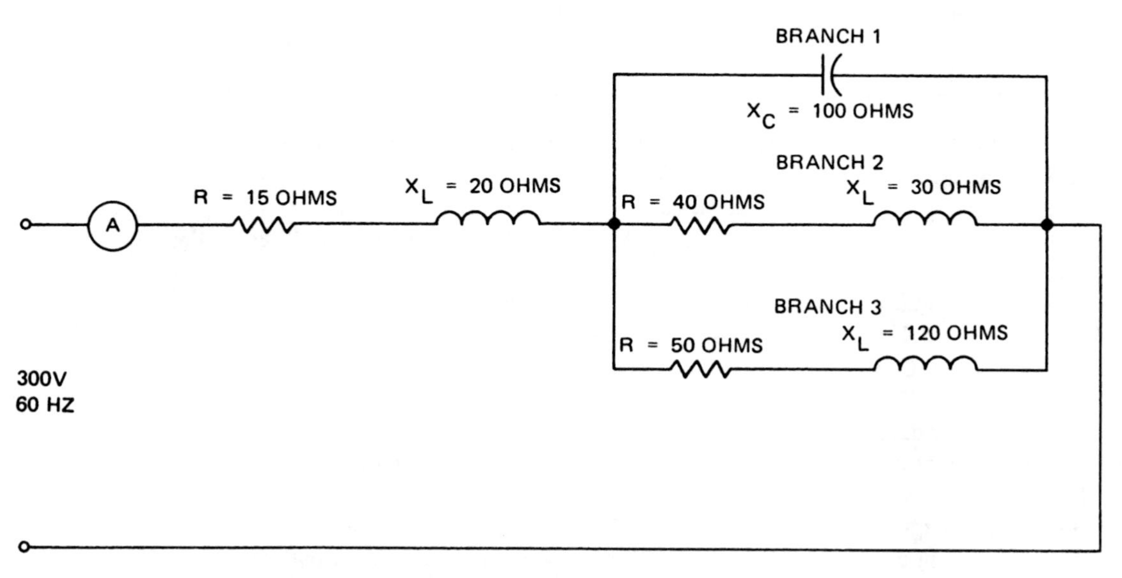

FIGURE 9–5 (*Delmar/Cengage Learning*)

5. Using the circuit given in question 4, determine
   a. the total current taken by the entire series–parallel circuit.
   b. the current taken by each of the three branches.
6. Using the circuit given in question 4, determine
   a. the power factor of the entire series–parallel circuit.
   b. the power factor of each branch circuit.
   c. the total power, in watts, taken by the series–parallel circuit.

10

Three-Phase Systems

Objectives

After studying this unit, the student should be able to

- describe the modern three-phase generation system, which produces energy and then transmits and distributes this energy to users.
- list the advantages of a three-phase power system over a single-phase power system.
- explain how the windings of the three coils of a three-phase generator are placed on the stator armature.
- define phase sequence and describe how it may be changed.
- connect the single-phase windings of generators, motors, transformers, and other electrical devices, using the wye connection and the delta connection, to form three-phase circuits.
- lay out a vector diagram of voltages in a three-phase wye circuit.
- show the current and voltage relationships for a wye-connected circuit using a vector diagram.
- plot a vector diagram of voltages in a three-phase delta circuit.
- show the current and voltage relationships for a delta-connected circuit using a vector diagram.
- apply the law of cosines to calculate the line current in a three-phase system.
- calculate the power (in volt-amperes), the true power (in watts), and the total power (in watts) in three-phase systems.
- draw the connections for the two-wattmeter method and the three-wattmeter method of measuring the power, in watts, taken by a three-phase, three-wire system or a three-phase, four-wire system.
- describe the difference between an isolated two-phase, four-wire system and a two-phase, three-wire system.

INTRODUCTION

Most alternating-current energy is generated by three-phase generators. This energy is then distributed over three-phase transmission systems. The three-phase circuits used are actually three single-phase circuits combined into one circuit having either three or four wires. Single-phase motors and other single-phase loads may be operated from a three-phase circuit.

Unit 1 gave several reasons why three-phase service is preferred to single-phase service for many applications. The rest of this section is a review of these reasons.

1. For given physical sizes of three-phase motors and generators, the horsepower ratings of the motors and the kVA ratings of the generators are larger. A three-phase generator or an induction motor has a capacity that is about 150% that of a single-phase machine having a comparable frame size.
2. The power delivered by a single-phase circuit is pulsating. Figure 2–7 showed the sinusoidal wave patterns of voltage, current, and power for a resistive load. At unity power factor, when the current and the voltage are in phase, the power is zero twice in each cycle. When the current and the voltage are out of phase, the power is zero four times in each cycle. In certain parts of each cycle, the power is negative. Pulsating power is supplied to each of the three single-phase circuits that make up the three-phase system. However, the total power delivered to the balanced three-phase circuit is the same at any instant. As a result, the operating characteristics of three-phase motors and other machines are superior to those of similar single-phase machines.
3. A balanced three-phase, three-wire circuit having the same voltage between the line wires uses only 75% of the copper required for a single-phase, two-wire circuit. Both circuits have the same kVA capacity, voltage rating, length of circuit, and efficiency of transmission.

THREE-PHASE VOLTAGE

Figure 10-1 illustrates a single-phase alternator. A coil of wire is cut by a rotating magnetic field. Because the magnet contains both north and south magnetic poles, the induced voltage will alternate positive and negative values. The frequency of the ac voltage is determined by the speed of the rotating magnet.

Figure 10-3 illustrates the construction of a three-phase alternator. In this example three separate phase windings are spaced 120 mechanical degrees apart. The rotating magnet will induce voltage into each winding. Due to the placement of the phase windings, the three induced voltages will be 120 electrical degrees apart. This example does not illustrate how the separate phase windings are connected together to form common three-phase connections.

In practice, the coil windings are connected so that only three or four conductors are required to supply a three-phase circuit. There are two standard methods of connecting the

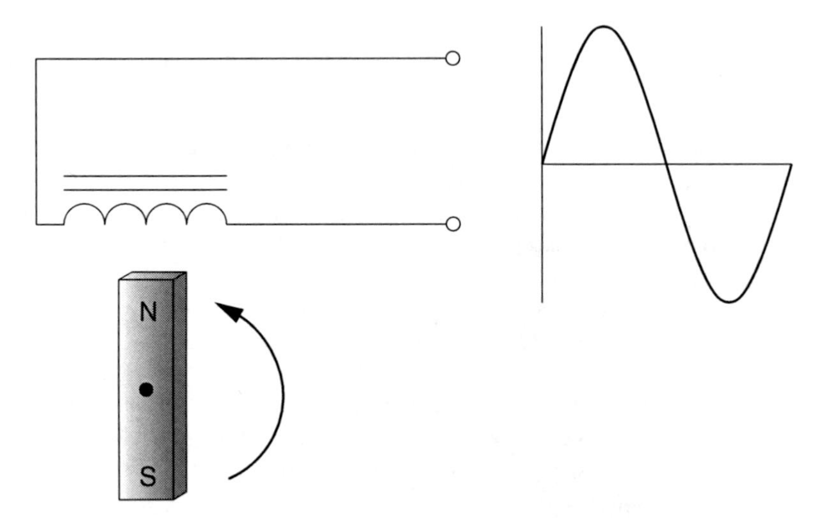

FIGURE 10–1 A single-phase voltage is produced when a rotating magnetic field cuts a coil of wire. (*Delmar/Cengage Learning*)

single-phase windings of generators, motors, transformers, and other devices to form three-phase circuits. These methods are known as the *wye connection* and the *delta connection.* See Figures 10–3 and 10–10.

PHASE SEQUENCE

The phase sequence, or the phase rotation, is the order in which the three voltages of a three-phase circuit follow one another. For example, phase A, in Figure 10–2, starts to rise in a positive direction at zero electrical degrees. At the same instance in time, phase B is positive and heading in the negative direction while phase C has almost reached its peak negative position. These three voltages are 120 electrical degrees apart. Phase sequence is also expressed as phase rotation. When three-phase voltage is connected to a three-phase motor, a rotating magnetic field is produced inside the motor. The direction of rotation is determined by the phase sequence. The direction of the rotating magnetic field can be reversed by changing any two of the three-phase lines connected to the motor. The phase sequence or phase rotation can also be determined by connecting an oscilloscope to the lines to observe the relative position of the different sine wave voltages or by connecting a phase rotation meter to the three lines.

THE WYE CONNECTION

The wye connection is the most commonly used way of connecting the three single-phase windings of three-phase generators. The three coil or phase windings are placed in the slots of the stationary armature. This armature is called the stator. The

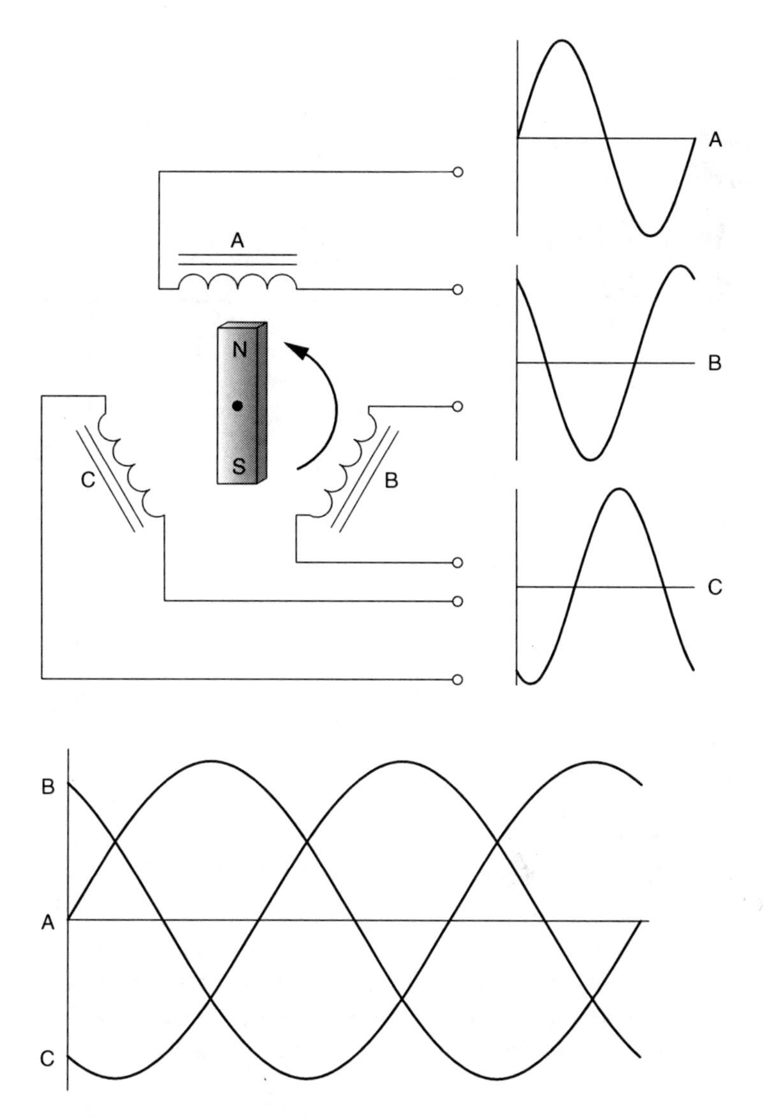

FIGURE 10–2 A three-phase alternator is constructed by placing three separate phase windings 120 mechanical degrees apart. (*Delmar/Cengage Learning*)

windings are placed so that the three induced voltages are 120 electrical degrees apart. If the ends (marked 0) of each of the phase windings are connected at a common point and the beginnings of the windings (marked A, B, and C) are brought out as the three line leads, the resulting arrangement is the wye connection. Figure 10–3 shows the schematic diagram of a three-phase wye connection.

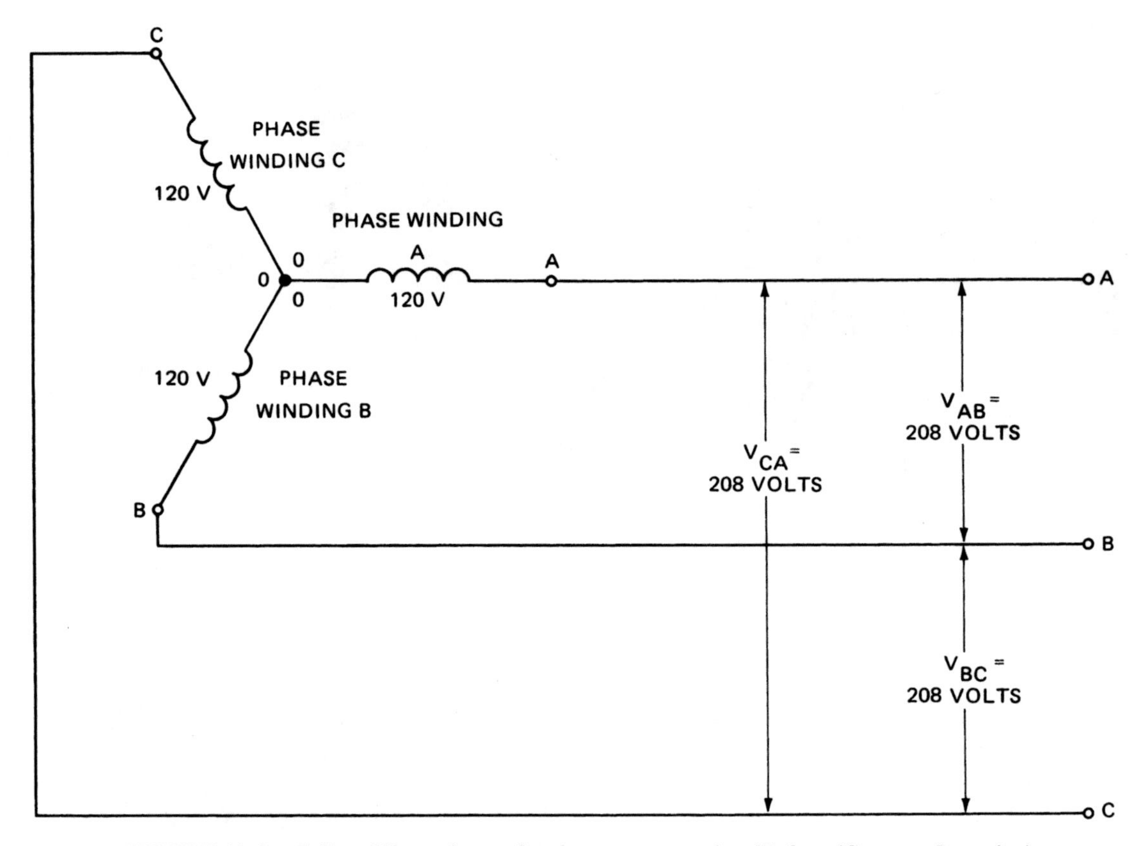

FIGURE 10–3 **Coil and line voltages for the wye connection (*Delmar/Cengage Learning*)**

The Phase Voltages

The induced voltage in each phase winding is called the *phase voltage*. The voltage across the line wires is called the *line-to-line voltage*. If the voltage induced in each phase winding is 120 V, then the voltage across each pair of line wires is 208 V. Thus, the voltages between A and B, B and C, and C and A are all 208 V.

Phase Voltages Are Out of Phase. The two-phase voltages connected together do not add up to 240 V because they are 120° out of phase. Refer to Figure 10–4 during the following discussion of this statement. The phase windings OA and OB are shown in this figure.

Unit 2 stated that the voltages given in ac problems are the effective (RMS) values of the sinusoidally varying voltages. Thus, the actual voltage of phase A (V_{OA}) is shown in Figure 10–4C. This sinusoidal waveform has an effective value of 120 V. V_{OB} is a similar sinusoidal waveform that reaches its maximum value 120 electrical degrees later than does V_{OA}. The student should realize that this 120° difference is due to the position of the winding in the generator.

The vector diagram in Figure 10–4B shows that the phase voltages V_{OA} and V_{OB} are separated by 120°. As explained in the section on vectors in Unit 3, each voltage vector

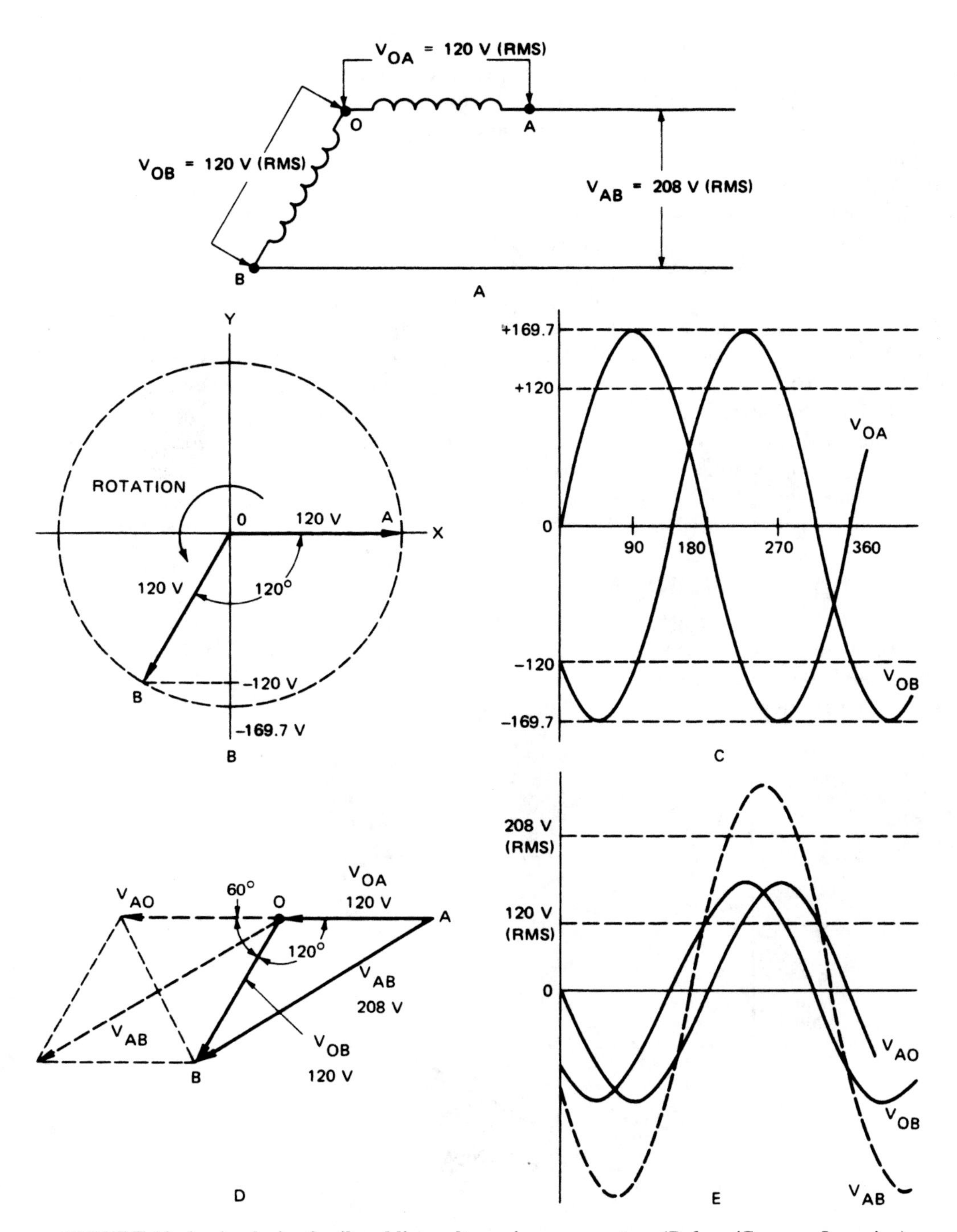

FIGURE 10–4 Analysis of coil and line voltages in a wye system (*Delmar/Cengage Learning*)

is assumed to be rotating counterclockwise at the given line frequency. Figure 10–4B represents only one instant in time. The length of the vectors represents the magnitude of the peak voltage: $1.414 \times 120 = 169.7$ V. As the vectors rotate counterclockwise, their projections on the Y axis produce the waveforms shown in Figure 10–4C.

Figure 10–4E shows V_{OA} inverted to give V_{AO}. The result is added to V_{OB} to produce V_{AB}. The two sinusoidal waveforms are similar, but they are not quite in phase. Therefore, their sum is not quite equal to (2×120 V) RMS. The actual value is 208 V RMS. Although $\overrightarrow{V_{OA}}$ and $\overrightarrow{V_{OB}}$ are 120° apart, $\overrightarrow{V_{AO}}$ and $\overrightarrow{V_{BO}}$ are only 60° apart. Thus, the total of 208 V is close to 240 V.

By vector addition, the line voltage V_{AB} equals V_{AO} (the reverse of V_{OA}) + V_{OB}. Figure 10–4D shows this sum as a vector diagram. The vector $\overrightarrow{V_{AO}}$ is moved to the left, and the other dashed lines are added. A close look at the triangles formed shows that $V_{AB} = \sqrt{3} \times V_{AO}$. Therefore, in wye-connected systems, the line-to-line voltage is equal to 1.73 ($\sqrt{3}$) times the line-to-line voltage.

In Figure 10–4D, vector V_{AB} appears twice. The solid-line position shows the correct location of this vector (between lines A and B). The dashed-line position, however, originates from point 0 with the other vectors. This line shows more clearly the relative phases. This type of diagram will be used again later in this unit.

Vector Diagram of Voltage

Figure 10–5 is a vector diagram for a three-phase, wye-connected circuit. The diagram shows the three coil voltages and the three line-to-line voltages. Only the voltage relationships for a three-phase circuit have been discussed so far.

What are the current relationships in a three-phase, wye-connected system?

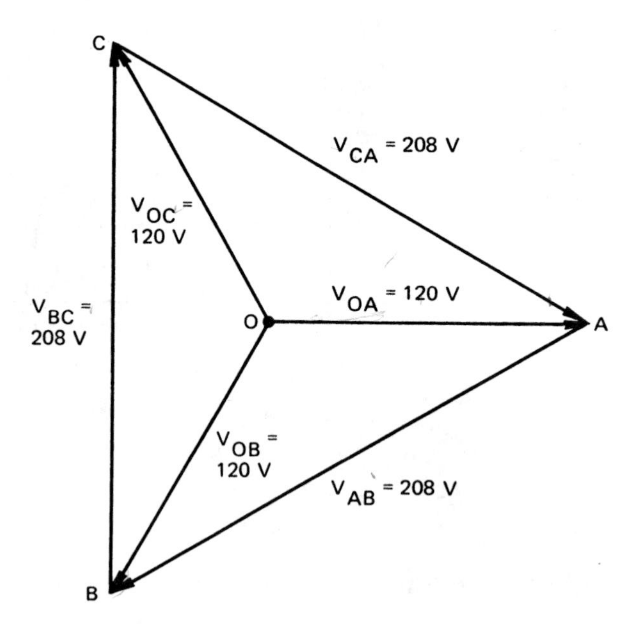

FIGURE 10–5 Vector diagram of voltages in a three-phase, wye-connected circuit (*Delmar/Cengage Learning*)

Current Relationships

The line current and the phase winding current are the same. These currents are equal because each phase winding is connected in series with one of the three line wires. The following statements give the relationships between the coil and line currents and the voltage in a balanced three-phase wye system.

- The line voltage is equal to $\sqrt{3}$ times the phase, or coil winding, voltage in a balanced three-phase, wye-connected system.
- The line current values and the phase winding current values are the same in a balanced three-phase, wye-connected system.

Three-Phase, Wye-Connected Circuit

The schematic diagram shown in Figure 10–6 is for a three-phase, wye-connected generator supplying current to a three-phase, noninductive heating load. Kirchhoff's current law states that the sum of the currents at a junction point in a circuit network is always zero. Therefore, at the source junction (O), the vector sum of the current is zero. Similarly, the vector sum of the three currents at the junction (O) of the heating loads is zero. The currents in the three-phase, three-wire, wye-connected system shown in Figure 10–6 may be unequal because of an unbalanced load. However, the vector sum of the currents at either junction is still zero.

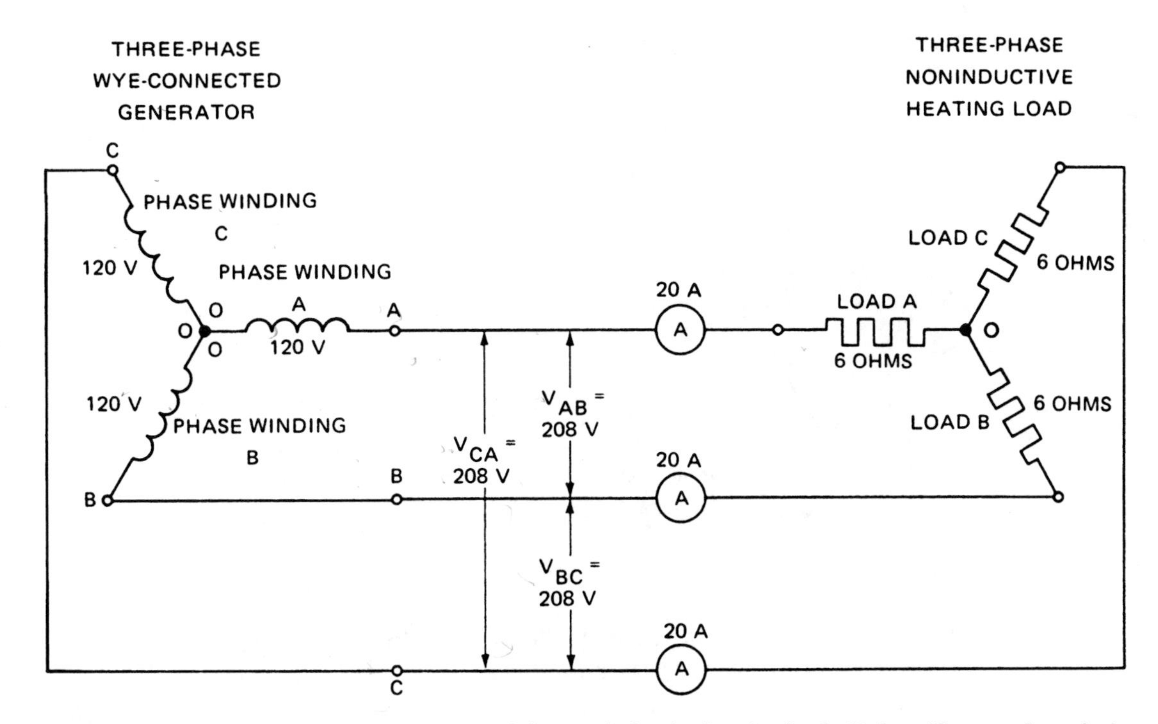

FIGURE 10–6 Wye-connected system supplying noninductive heating loads (*Delmar/Cengage Learning*)

Currents and Voltages in a Wye System

Figure 10–7 is a vector diagram of the currents and voltages for the wye-connected circuit. This circuit operates at a unity power factor. The load shown in Figure 10–6 consists of three noninductive heating elements. Each element has a resistance of 6 Ω. For a line voltage of 208 V, the voltage across each heater element is

$$V_{coil} = \frac{V_{line}}{\sqrt{3}} = \frac{208}{1.73} = 120\ \text{V}$$

The current taken by the heater coil is

$$I_{coil} = \frac{V_{coil}}{R_{coil}} = \frac{120}{6} = 20\ \text{A}$$

Each 120-V phase voltage in this three-phase generator is in phase with its own coil current of 20 A. However, for a unity power factor, there is a phase angle between the line voltage and the line current, which is also the coil current. This phase angle is 30°, as shown in the vector diagram. Note that the coil voltage (V_{OA}) is in phase with the coil current, I_A. Also, the coil voltage, V_{OB}, is in phase with the current I_B, and the coil voltage V_{OC} is in phase with the current I_C. There is a phase angle of 30 electrical degrees between each of

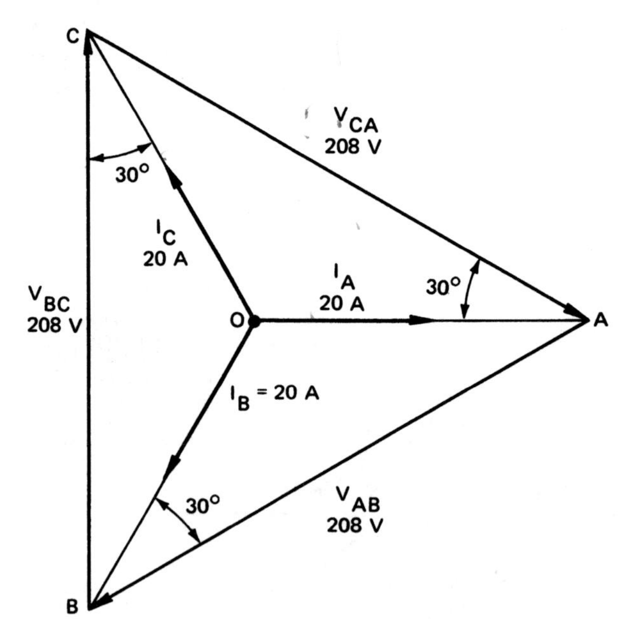

FIGURE 10–7 Current and voltage relationships for the wye-connected circuit (*Delmar/Cengage Learning*)

the following: the line voltage V_{CA} and the line current I_A, the line voltage V_{AB} and the line current I_B, and the line voltage V_{BC} and the line current I_C.

> **For resistive loads, the unity power factor causes the line current to lead the line voltage by 30°. Thus, the power factor angle is measured between the coil voltage and the coil current.**

Balanced Wye-Connected Motor

Figure 10–8 shows a balanced three-phase, wye-connected motor load supplied from a three-phase, wye-connected source. The angle of lag of each coil current behind its respective coil voltage is 40 electrical degrees. In other words, the phase angle (θ) of the power factor is 40°.

The vector diagram in Figure 10–9 shows the relationship between the current and voltage values for the three-phase motor circuit shown in Figure 10–8. The angle θ (the power factor angle) is the angle between the coil voltage and the coil current.

POWER IN THE WYE SYSTEM

The value of volt-amperes produced in each of the three single-phase windings of the three-phase generator is

Volt-amperes $= V_{coil} \times I_{coil}$

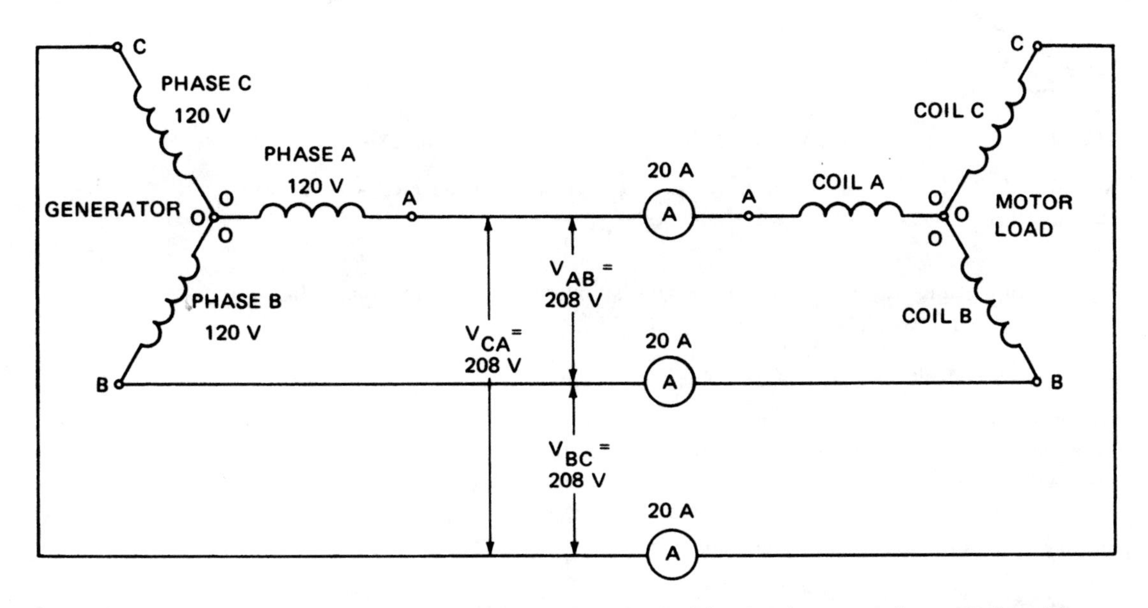

FIGURE 10–8 Wye-connected system supplying a motor load with a lagging power factor (*Delmar/Cengage Learning*)

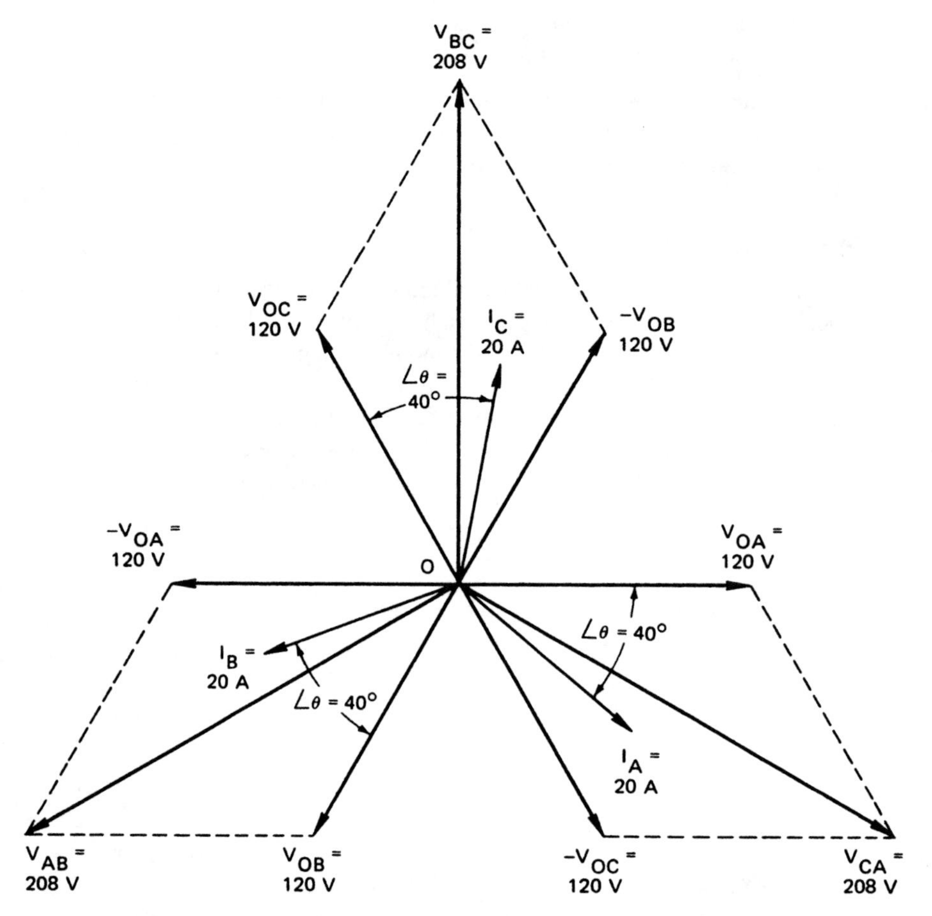

FIGURE 10–9 Current and voltage relationships in a wye-connected system with a lagging power factor (*Delmar/Cengage Learning*)

If the voltage and current values of the wye system are balanced, the total volt-amperes produced by all three windings is

$$\textbf{Total Volt-amperes} = 3 \times V_{coil} \times I_{coil}$$

In practice, it is easier to measure line voltages and line currents than it is to find the coil voltages and the coil currents. Therefore, I_{line} is substituted for I_{coil} in the equation for the total volt-amperes. (I_{line} and I_{coil} are equal in a wye-connected system.)

It was shown earlier that $V_{coil} = V_{line}/\sqrt{3}$. Making the substitutions for V_{coil} and I_{coil}, the equation for the total volt-amperes becomes

$$\textbf{Total Volt-amperes} = \frac{3 \times V_{line} \times I_{line}}{\sqrt{3}}$$

$$= \sqrt{3} \times V_{line} \times I_{line}$$

Each of the three coil windings of the wye-connected generator supplies power, in watts, equal to

$$\text{Watts} = V_{coil} \times I_{coil} \times \cos \angle\theta$$

Three-Phase Power

The total three-phase power, in watts, can be determined for these conditions:

1. The three coil currents are the same.
2. The three coil voltages are equal.
3. The power factor angle is the same for each coil winding.

The equations used to determine the total three-phase power are

$$\textbf{Watts} = 3 \times V_{coil} \times I_{coil} \times \cos \angle\theta$$

$$= 3 \times V_{coil} \times I_{line} \times \cos \angle\theta$$

$$= \frac{3 \times V_{line} \times I_{line}}{\sqrt{3}}$$

$$= \sqrt{3} \times V_{line} \times I_{line} \times \cos \angle\theta$$

The input in kilovolt-amperes and the power in kilowatts are determined using these formulas:

Input kilovolt-amperes:

$$\text{kVA} = \frac{\sqrt{3} \times V_{line} \times I_{line}}{1000}$$

Power in kilowatts:

$$\text{kW} = \frac{\sqrt{3} \times V_{line} \times I_{line} \times \cos \angle\theta}{1000}$$

Power Factor

The power factor of a balanced three-phase wye-connected system can be determined when the total power (true power or watts) and the total input (apparent power or volt-amperes) are given. The equation expressing the relationship is

$$\cos \angle\theta = \frac{\text{total power}}{\sqrt{3} \times V_{line} \times I_{line}}$$

In a balanced three-phase circuit, the power factor is always the cosine of the angle between the coil voltage and the coil current. If the current values are severely unbalanced, or if the three voltages differ greatly, then the three-phase power factor has almost no meaning. When the unbalance is minor, then average values of the line current and the line voltage are used in the power factor formula.

THE DELTA CONNECTION

There is a second standard connection method by which the three single-phase coil windings of a three-phase generator can be interconnected. This second method is called the *delta connection*. Loads connected to a three-phase system may be connected in delta. The name *delta* is used because the schematic diagram of this connection closely resembles the Greek letter *delta* (Δ).

The schematic diagram of a delta connection is shown in Figure 10–10. This figure represents a three-phase generator consisting of three coil windings. The end of each winding is identified by the letter O. The beginning of each phase winding is marked with the letter A, B, or C.

Making the Delta Connection

The delta connection is made by connecting the *beginnings* of the coil windings as follows: coil winding A to the end of coil winding B and to line A, coil winding B to the end of coil winding C and to line B, and coil winding C to the end of coil winding A and to line C, as in Figure 10–10.

Figure 10–10 shows line voltage V_{AB} connected across phase winding B at points OA and OB. Line voltages V_{BC} and V_{CA} are then connected across phase windings C and A, respectively. The phase and line voltages have common points; thus, these voltages must be equal.

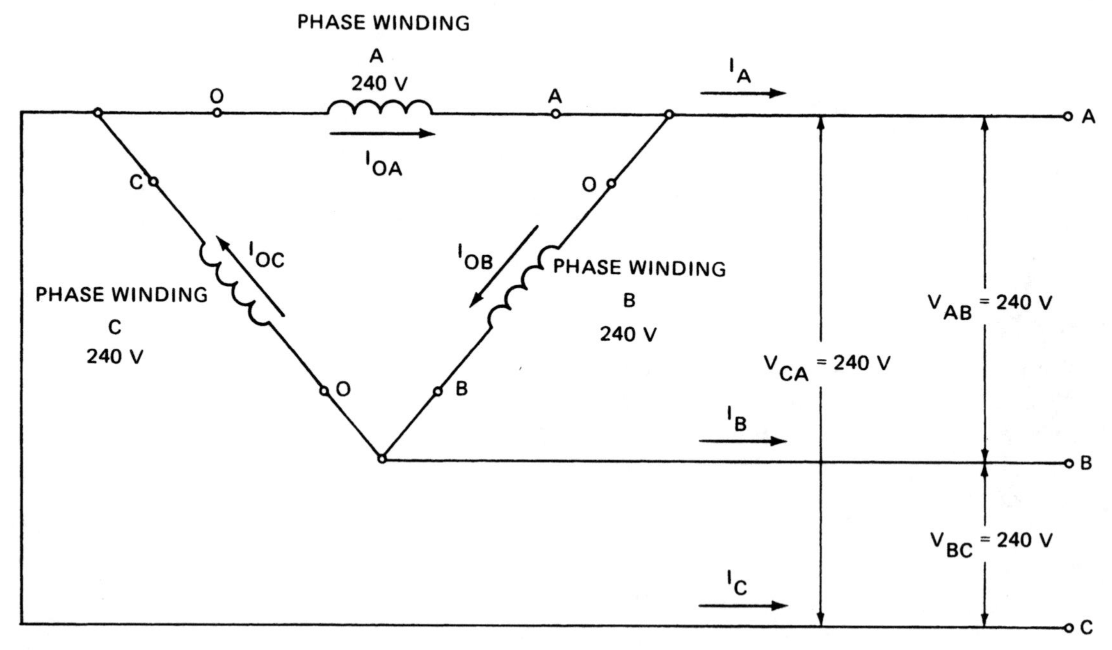

FIGURE 10–10 In a delta connection the line and phase voltage values are the same, but the line current is greater than the phase current by a factor of the square root of 3. (*Delmar/Cengage Learning*)

Vectors for the Delta Connection

The vector concepts presented in Unit 2 will be reviewed by solving the problem shown in Figure 10–11. The vector equation for the line voltage V_{AB} becomes $\overrightarrow{V_{AB}} = \overrightarrow{V_{AO}} + \overrightarrow{V_{CO}}$, where V_{AO} and V_{CO} are phase voltages. The equation was derived by passing from point OA through point OC to point OB. By moving in a direction opposite to that of the electromotive force arrow, the vectors V_{OA} and V_{OB} are reversed. The resulting line voltage vector V_{AB} matches coil vector V_{OC}.

Current Relationships

The current relationships for a delta connection are an interesting feature of this arrangement. Figure 10–10 shows the currents entering and leaving points OA, OB, and OC. Line currents I_A, I_B, and I_C cannot leave these points simultaneously. Thus, the current directions are shown at 120° intervals of time. At point OA, an analysis shows that currents I_A and I_{OB} are leaving and current I_{OA} is entering. Kirchhoff's current law states that the sum of the currents leaving a point must equal the sum of the currents entering that point. Because the currents in Figure 10–10 are out of phase, the vector equation becomes $\overrightarrow{I_{OA}} = \overrightarrow{I_A} + \overrightarrow{I_{OB}}$. Solving this equation for I_A yields

$$\overrightarrow{I_A} = \overrightarrow{I_{OA}} - \overrightarrow{I_{OB}}$$

The minus sign indicates that the vector I_{OB} is reversed.

Figure 10–12B shows the addition of I_{OA} to $-I_{OB}$ to give the line current I_A. Line currents I_B and I_C are obtained by similar methods. The student should practice these procedures.

The three line currents, I_A, I_B, and I_C, are 120 electrical degrees apart. This means that the three-phase system is balanced.

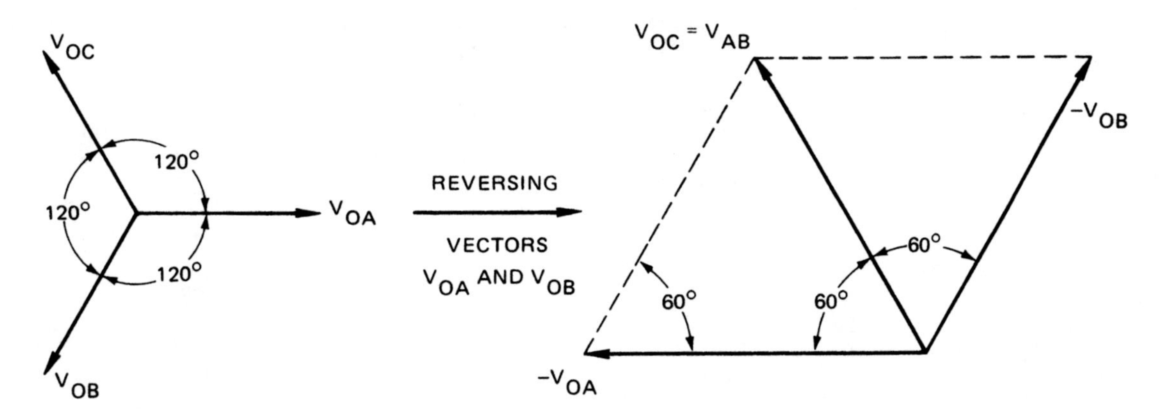

FIGURE 10–11 Vector diagram of voltages in a three-phase delta circuit (*Delmar/Cengage Learning*)

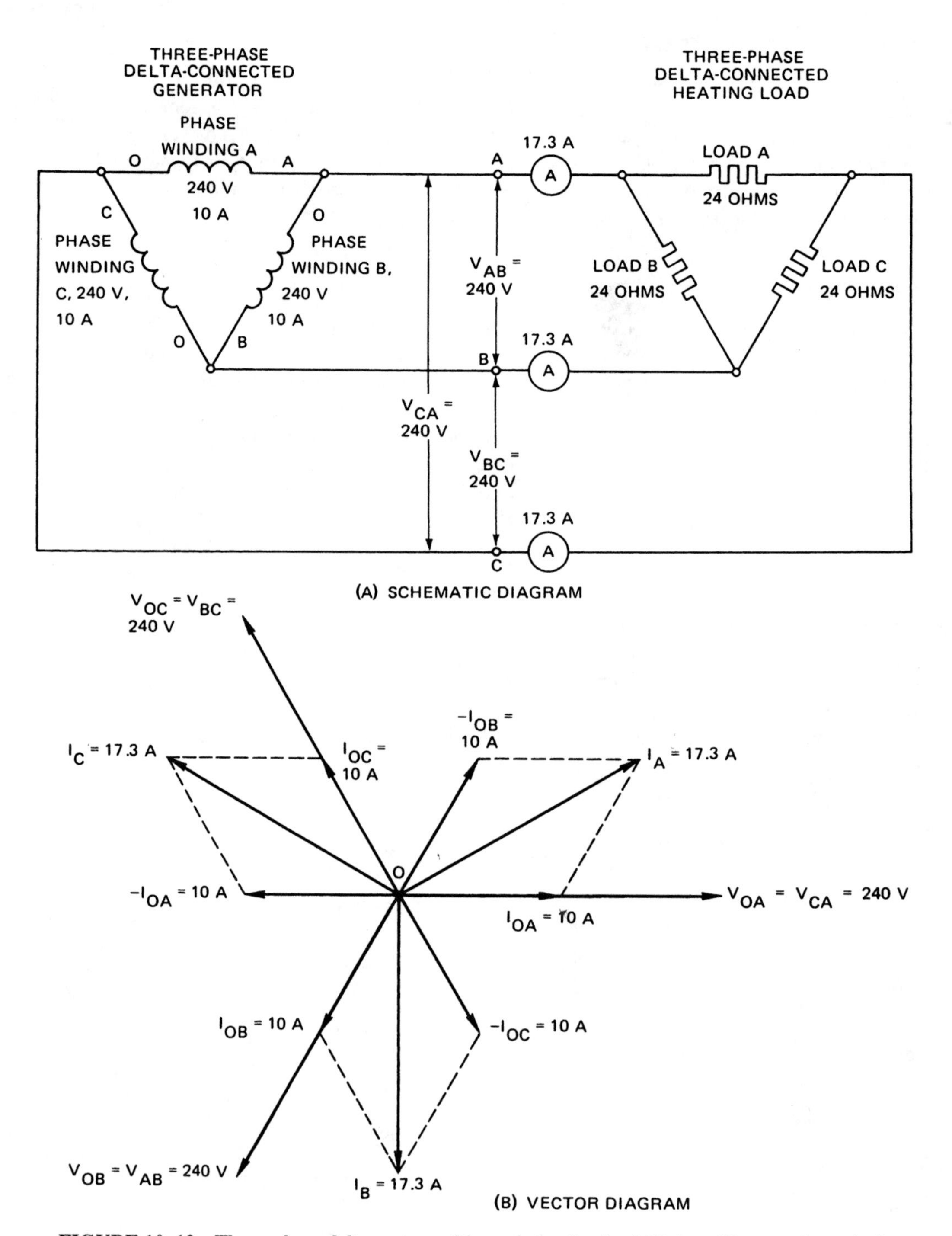

FIGURE 10–12 Three-phase delta system with noninductive load (*Delmar/Cengage Learning*)

Adding Coil Currents. The coil currents for the circuit of Figure 10–12 must be added using a method similar to that used in Figure 10–4. Using the vector triangle, it can be shown that

$$\begin{aligned} I_{line} &= 2 \times \cos 30^\circ \times I_{coil} \\ &= 2 \times 0.8660 \times I_{coil} \\ &= 1.73 \times I_{coil} \\ &= \sqrt{3} \times I_{coil} \end{aligned}$$

For Figure 10–12A, it is assumed that the current in each phase winding is 10 A. The load consists of three noninductive heater units connected in delta. Each heater unit has a resistance of 24 Ω. If the line-to-line voltage is 240 V, the voltage across each heater unit is also 240 V. The current in each load resistance is

$$I = \frac{V}{R} = \frac{240}{24} = 10\text{ A}$$

This means that the current in each of the three coil windings of the three-phase generator and in each load resistor is 10 A. The coil current is in phase with the coil voltage in each phase winding because the load consists of noninductive resistance.

Lagging Power Factor

A three-phase balanced system connected in delta will have a lagging power factor if the system supplies an inductive load, such as an induction motor. Figure 10–13 shows a three-phase generator connected in delta supplying a three-phase motor, also connected in delta. Figure 10–13A is the schematic diagram of this system. The vector diagram for the circuit is shown in Figure 10–13B. If the coil current lags the coil voltage by 40°, the power factor is 0.766 lagging. (The power factor equals the cosine of 40°, or 0.766.) The line current is a vector sum and is equal to 1.73 times the coil current.

Relationships between the Currents and Voltages

The relationships between the phase winding and the line values of the current and voltage for a balanced three-phase delta system are as follows:

- The phase winding voltage and the line voltage values are equal.
- The line current is equal to the $\sqrt{3}$, or 1.73, times the phase winding current.

POWER IN THE DELTA SYSTEM

The value of volt-amperes supplied by each phase winding of a three-phase generator is

$$\textbf{Volt-amperes} = \mathbf{V_{coil} \times I_{coil}}$$

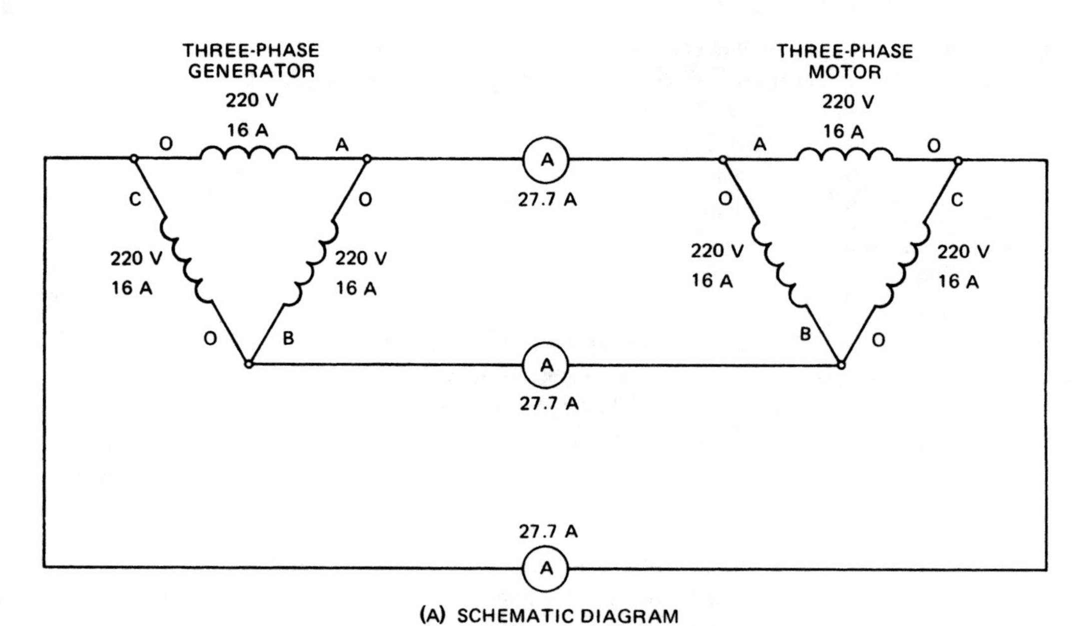

(A) SCHEMATIC DIAGRAM

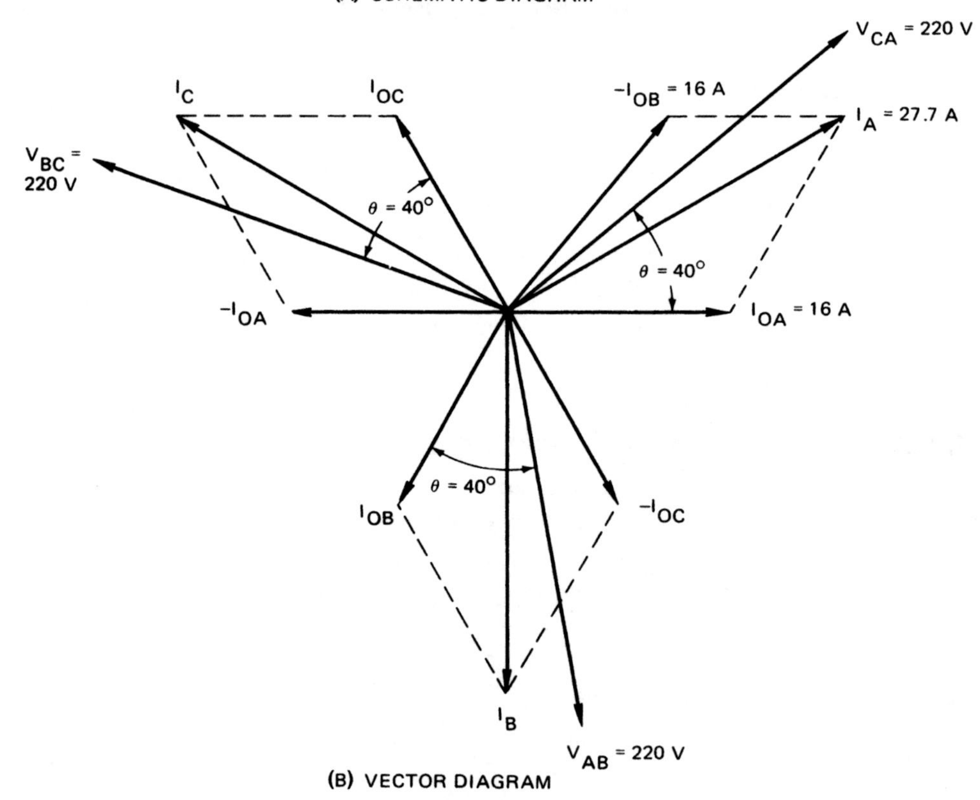

(B) VECTOR DIAGRAM

FIGURE 10–13 Three-phase delta system with an inductive load (*Delmar/Cengage Learning*)

If the voltages and currents of a delta-connected system are balanced, the total volt-amperes of all three windings is

$$\textbf{Total VA} = 3 \times V_{coil} \times I_{coil}$$

It is easier to measure the line voltage and the line current than it is to measure coil values. Therefore, the equation for the total volt-amperes of a balanced three-phase delta system is rewritten using the line voltage and the line current values. The line voltage may be substituted for the coil voltage because they are the same in a delta system. Thus, the equation becomes

$$\textbf{Total VA} = 3 \times V_{line} \times I_{coil}$$

For a balanced delta system, the coil current equals the line current $\div \sqrt{3}$. This means that $I_{line} \div \sqrt{3}$ can be substituted for I_{coil} in the equation for total VA.

$$\textbf{Total VA} = 3 \times V_{line} \times \frac{I_{line}}{\sqrt{3}}$$

or

$$\textbf{Total VA} = \sqrt{3} \times V_{line} \times I_{line}$$

The true power, in watts, supplied by each coil winding of a delta-connected generator is

$$\textbf{Watts} = V_{coil} \times I_{coil} \times \cos \angle\theta$$

The equation for the total three-phase power, in watts, can be written if it is assumed that the three coil voltages are equal, the three coil currents are the same, and the power factor angle is the same for each coil. Thus,

$$\begin{aligned}\textbf{Total power, in watts} &= 3 \times V_{coil} \times I_{coil} \times \cos \angle\theta \\ &= 3 \times V_{line} \times I_{coil} \times \cos \angle\theta \\ &= 3 \times V_{line} \times \frac{I_{line}}{\sqrt{3}} \times \cos \angle\theta \\ &= \sqrt{3} \times V_{line} \times I_{line} \times \cos \angle\theta\end{aligned}$$

The total volt-amperes and the total power for the balanced three-phase wye system and the balanced three-phase delta system are expressed by the same equations:

$$\textbf{Total Volt-amperes} = \sqrt{3} \times V_{line} \times I_{line}$$

$$\textbf{Total power(watts)} = \sqrt{3} \times V_{line} \times I_{line} \times \cos \angle\theta$$

The power factor of a balanced delta-connected, three-phase system is the ratio of the total power (in watts) to the total volt-amperes:

$$\cos \angle\theta = \frac{\text{total power}}{\sqrt{3} \times V_{line} \times I_{line}} = \text{PF}$$

If the currents and voltages in a delta system are severely unbalanced, the three-phase power factor has no real significance.

PROBLEM 1

Statement of the Problem

The wye-connected, three-phase generator shown in Figure 10–6 supplies power to a three-phase noninductive load. Determine

1. the line voltage.
2. the line current.
3. the input volt-amperes.
4. the power, in watts.

Solution

1. In a three-phase wye system, the line voltage is equal to $\sqrt{3}$ times the phase winding voltage:

   $$V_{line} = \sqrt{3} \times V_{coil} = 1.73 \times 120 = 208 \text{ V}$$

2. The line current equals the phase winding current. Each heater unit has a resistance of 6 Ω. If the voltage across each resistor is 120 V, then the coil current is $I = V \div R = 120 \div 6 = 20$ A. This value is the coil current at both the load and the source. Because each phase winding is in series with a line wire and a load element, the line current is also 20 A.

3. The value of the input volt-amperes is

   $$\begin{aligned}\text{Volt-amperes} &= \sqrt{3} \times V_{line} \times I_{line} \\ &= 1.73 \times 208 \times 20 \\ &= 7200 \text{ VA}\end{aligned}$$

4. The power factor is unity. This means that the power, in watts, is equal to the value of volt-amperes. The phase angle is zero because the coil current and the voltage are in phase for a noninductive heating load. The true power is

   $$\begin{aligned}\text{Total power(watts)} &= \sqrt{3} \times V_{line} \times I_{line} \times \cos \angle\theta \\ &= 1.73 \times 208 \times 20 \times 1.00 \\ &= 7200 \text{ W}\end{aligned}$$

PROBLEM 2

Statement of the Problem

A wye-connected, three-phase generator is shown in Figure 10–8. This generator supplies a motor load. The power factor angle is 40° lagging. Determine

1. the value of the input volt-amperes.
2. the true power, in watts.

Solution

1. $\text{Volt-amperes} = \sqrt{3} \times V_{line} \times I_{line}$

$= 1.73 \times 208 \times 20$

$= 7200 \text{ VA}$

2. $\text{Total power} = \sqrt{3} \times V_{line} \times I_{line} \times \cos \angle\theta$

$= 1.73 \times 208 \times 20 \times \cos 40°$

$= 1.73 \times 208 \times 20 \times 0.7660$

$= 55.15 \text{ W}$

PROBLEM 3

Statement of the Problem

A three-phase alternator is connected in wye. Each phase winding is rated at 8000 V and 418 A. The alternator is designed to operate at a full-load output with a power factor of 80% lag. Determine

1. the line voltage.
2. the line current.
3. the full load volt-ampere rating, in kilovolt-amperes.
4. the full load power, in kilowatts.

Solution

1. $V_{line} = \sqrt{3} \times V_{coil} = 1.73 \times 8000 = 13{,}840 \text{ V}$

2. $I_{line} = I_{coil} = 418 \text{ A}$

3. $$\text{kVA} = \frac{\sqrt{3} \times V_{line} \times I_{line}}{1000} = \frac{1.73 \times 13{,}840 \times 418}{1000}$$

$= 10{,}008 \text{ kVA}$

The generator rating is 10,000 kVA

4. $$\text{Full-load power} = \frac{\sqrt{3} \times V_{line} \times I_{line} \times \cos \angle\theta}{1000}$$

$$= \frac{1.73 \times 13{,}840 \times 418 \times 0.8}{1000}$$

$= 8006 \text{ kW}$

The generator rating at an 80% lagging power factor is 8000 kW

PROBLEM 4

Statement of the Problem

The delta-connected generator shown in Figure 10–13 supplies a delta-connected induction motor. The three-phase power factor of the motor is 0.7660 lagging. Find

1. the line voltage.
2. the line current.
3. the apparent power input to the motor, in volt-amperes.
4. the true power input to the motor, in watts.

Solution

1. For a circuit connected in delta, the line voltage and the coil voltage are the same:

$$V_{line} = V_{coil} = 220\ V$$

2. The line current is equal to $\sqrt{3}$ times the coil current:

$$I_{line} = \sqrt{3} \times I_{coil} = 1.73 \times 16 = 27.7\ A$$

3. The volt-amperes taken by the motor is

$$VA = \sqrt{3} \times V_{line} \times I_{line}$$

$$= 1.73 \times 220 \times 27.7$$

$$= 10{,}542.6\ VA$$

4. The power, in watts, taken by the motor is

$$P = \sqrt{3} \times V_{line} \times I_{line} \times \cos \angle\theta$$

$$= 1.73 \times 220 \times 27.7 \times 0.766$$

$$= 8075.6\ W$$

POWER MEASUREMENT IN THREE-PHASE SYSTEMS

The power, in watts, taken by a three-phase, three-wire system can be measured with two wattmeters. This method can be used to measure the power in a three-wire wye system or in a three-wire delta system.

The Two-Wattmeter Method

Figure 10–14 shows the standard connections for the two-wattmeter method. In this case, the method is used to measure the power supplied by a three-phase, three-wire system to a wye-connected load. The current coils of the two wattmeters are connected in series with two of the three line leads. The potential coil of each wattmeter is connected between the line wire from the current coil to the third line wire.

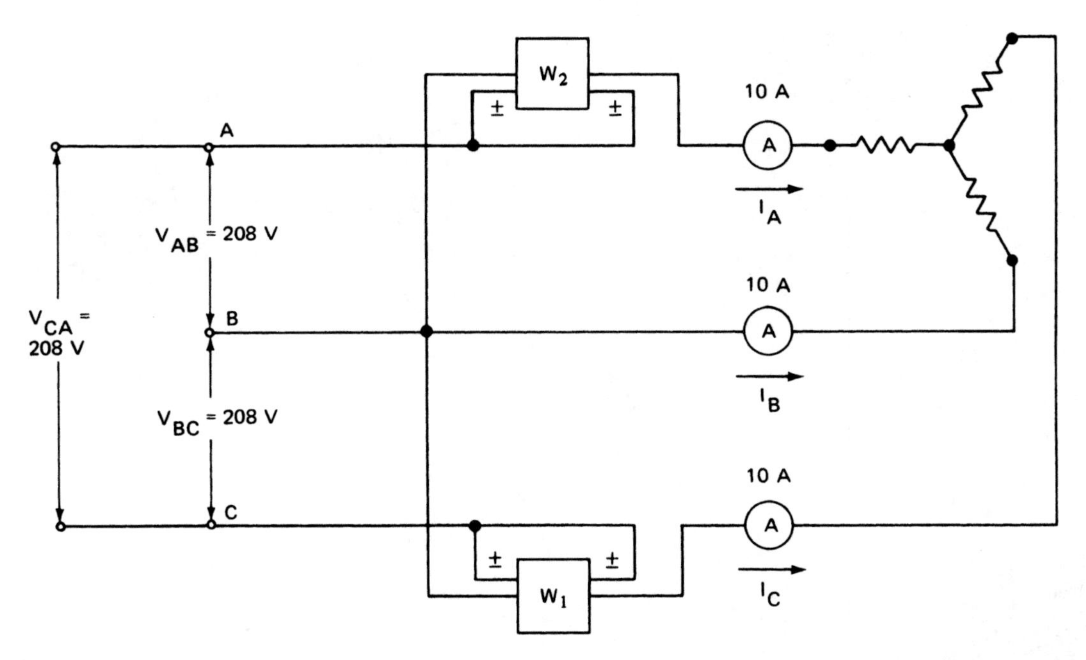

FIGURE 10–14 Two-wattmeter method of determining power in a three-phase system (*Delmar/Cengage Learning*)

Careful attention must be paid to the polarity marks (±) on the voltage and current coils of the wattmeters. The connections must be made exactly as shown in Figure 10–14. The total power for the three-phase system is

$$P_T = W_1 + W_2$$

The ± side of the voltage coil of W_2 is connected to line A. The other side of this coil is connected to line B. As a result, the voltage coil of the wattmeter reads the voltage at A with respect to B, or V_{AB}.

Similarly, the voltage coil of W_1 reads V_{BC}. Figure 10–15 shows the construction of these voltage vectors for a unity power factor.

A wattmeter will read the product of V_{line} and I_{line} multiplied by the cosine of the angle between the two values. Recall, however, that the power factor for a three-phase system is measured between V_{coil} and I_{coil}. The following two examples illustrate the use of the two-wattmeter method of determining the power.

Case I: Unity Power Factor. At a power factor of unity, the angle between V_{line} and I_{line} is 30°, as shown in Figure 10–15:

$$P = V_{line} \times I_{line} \times \cos 30°$$

or

$$P_1 = V_{BC} \times I_C \times \cos 30°$$

$$P_2 = V_{AB} \times I_A \times \cos 30°$$

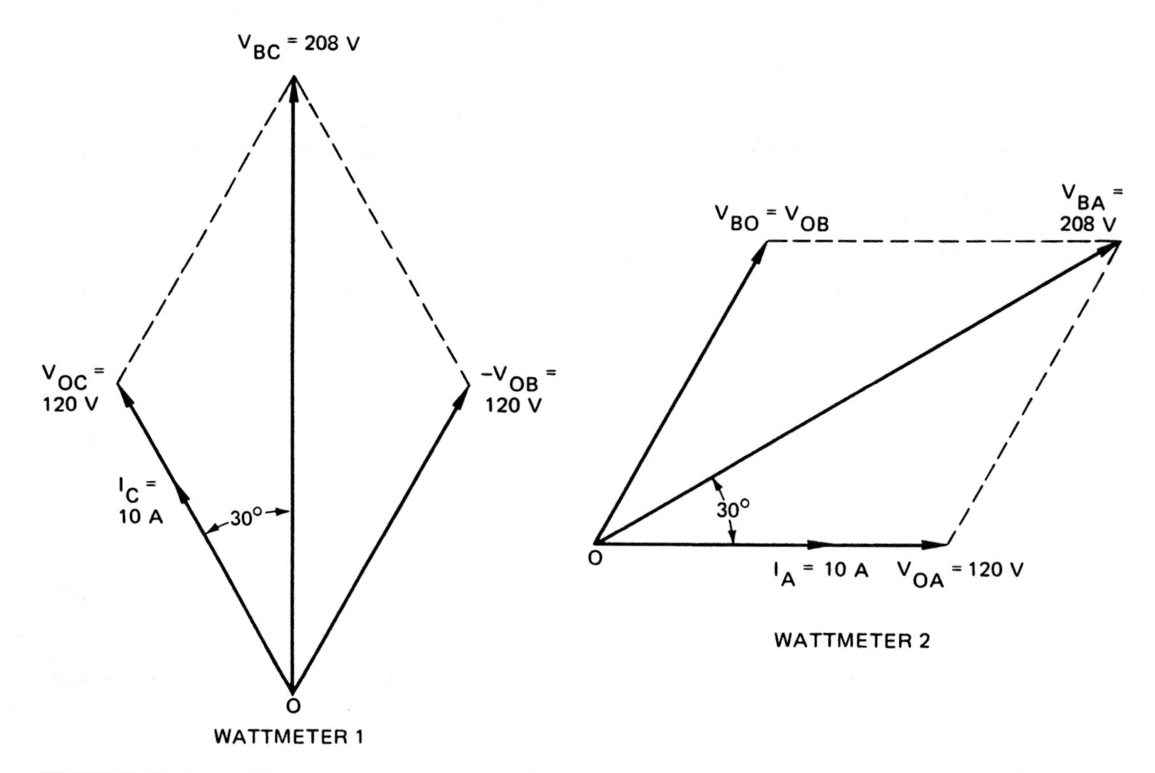

FIGURE 10–15 A 30 degree phase angle exists between the line voltage and the line current (*Delmar/ Cengage Learning*)

For balanced loads, where $I_A = I_C$:

$$\mathbf{W_1 = W_2}$$

Case II: Lagging Power Factor. Figure 10–16 shows that the power factor angle θ between V_{coil} and I_{coil} is greater than 30°. The wattmeter indication is based on the angle between V_{line} and I_{line}. Thus, the general form of the power equation is

$$\mathbf{P = V_{line} \times I_{line} \times \cos (30° \pm \theta)}$$

or

$$\mathbf{W_1 = V_{BC} \times I_C \times \cos (30° - \theta)}$$

$$\mathbf{W_2 = V_{AB} \times I_A \times \cos (30° + \theta)}$$

An analysis of the power values for different values of θ produces the following results:

1. At $\theta = 0°$, $W_1 = W_2$, and W_1 and W_2 are positive.
2. At θ = less than (<) 30°, W_1 is greater than (>) W_2, and W_1 and W_2 are positive.
3. At $\theta = 30°$, $W_2 = ½W_1$, and W_1 and W_2 are positive.
4. At $\theta < 60°$, $W_1 = W_{total}$; and $W_2 = 0$.

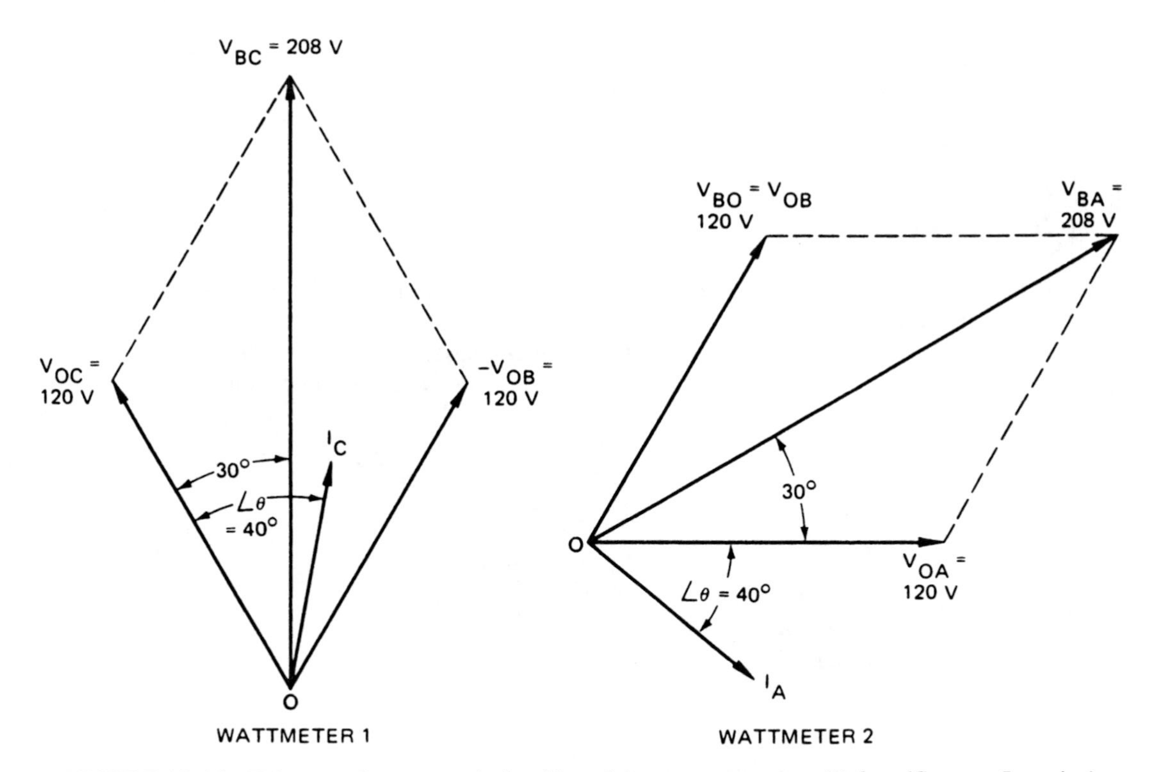

FIGURE 10–16 Voltage and current relationships of the two watt meters (*Delmar/Cengage Learning*)

5. At $\theta > 60°$, W_2 is the negative and W_1 is positive. In the case of $\theta > 60°$, W_2 is negative. Thus, it is necessary to reverse the voltage coil connections of wattmeter 2 so that it reads upscale. The reading must be recorded as a negative value and must be subtracted from W_1 to obtain the total power:

$$P_T = W_1 - P_2$$

Mathematical Proof of $W_T = W_1 \pm W_2$. The general forms of the wattmeter equations are

$$\mathbf{W_1 = V_{line} \times I_{line} \times \cos(30° - \theta)}$$

$$\mathbf{W_2 = V_{line} \times I_{line} \times \cos(30° + \theta)}$$

Trigonometric identities are used to obtain the following:

$$\cos(30° - \theta) = (\cos 30° \times \cos\theta - \sin\theta \times \sin\theta)$$

$$\cos(30° + \theta) = (\cos 30° \times \cos\theta + \sin\theta \times \sin\theta)$$

Substituting these identities in the wattmeter equation yields

$$W_1 = V_{line}\, I_{line}\, (\cos 30° \times \cos\theta - \sin\theta \times \sin\theta)$$

$$W_2 = V_{line}\, I_{line}\, (\cos 30° \times \cos\theta + \sin\theta \times \sin\theta)$$

Adding, we obtain

$$W_1 + W_2 = 2\, V_{line}\, I_{line}\, (\cos 30° \times \cos \theta)$$

$$W_1 + W_2 = 2\, V_{line}\, I_{line}\, \frac{\sqrt{3}}{2} \cos \theta$$

Thus

$$\mathbf{W_1 + W_2 = \sqrt{3}\, V_{line}\, I_{line}\, ¥ \cos \theta}$$

This equation for the sum of W_1 and W_2 is the same as the equation derived for the total power in a three-phase system.

Refer to the previous computations for a balanced three-phase, wye-connected load, operating at different values of lagging power factor. These calculations show that the two wattmeter readings are

Wattmeter 2:

$$W_2 = V_{line} \times I_{line} \times \cos (30° + \theta)$$

Wattmeter 1:

With lag $\leq 30°$:

$$W_1 = V_{line} \times \cos (30° - \theta)$$

With lag $> 30°$:

$$W_1 = V_{line} \times I_{line} \times \cos (\theta - 30°)$$

These formulas are also correct for a balanced three-phase, delta-connected load.

Using the curve shown in Figure 10–17, it is possible to obtain the power factor without finding the input volt-amperes. Power factor values form the vertical scale. The ratios of the smaller wattmeter reading to the larger reading form the horizontal scale. The curve in Figure 10–17 is obtain by substituting different values of the angle θ in the following ratio:

$$\mathbf{\frac{Watts\ 2}{Watts\ 1} = \frac{VI \cos (30° + \theta)}{VI \cos (30° - \theta)} = \frac{\cos (30° + \theta)}{\cos (30° - \theta)}}$$

For example, if wattmeter 1 reads 2049 W and wattmeter 2 reads 711 W, the ratio of the smaller reading to the larger reading is

$$W_2 \div W_1 = 711 \div 2049$$
$$= 0.347$$

When applied to the curve of Figure 10–17, this ratio gives a power factor of 0.76. As another example, wattmeter 1 has a phase angle of 75° lagging and a power factor of 0.2588. Wattmeter 1 reads 1471 W and wattmeter 2 reads −538 W. The ratio of these two values is $W_2 \div W_1 = -538 \div 1471 = -0.36$. When applied to the curve, the ratio gives a power factor of 0.26 lagging. This reading is close to the given value of 0.2588.

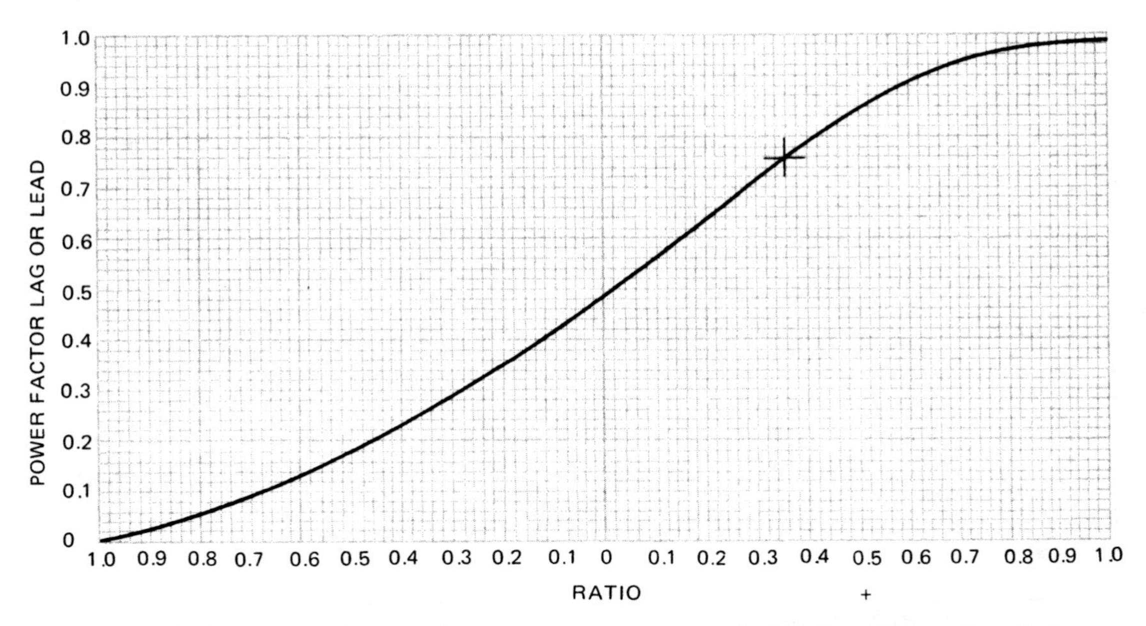

FIGURE 10–17 Power factor ratio curve—two-wattmeter method (*Delmar/Cengage Learning*)

THE THREE-WATTMETER METHOD

Many wye-connected systems have a neutral wire in addition to the three line wires. Such a system is called a three-phase, four-wire, wye-connected circuit (Figure 10–18). The neutral wire connects at the common point where the three coil windings terminate in the alternator. The neutral then runs directly to the common point of the wye-connected load. This type of system is used when 120-V, single-phase service is required for lighting loads. In addition, it is used when three-phase, 208-V service is required for three-phase motor loads. The neutral wire helps to maintain relatively constant voltages across the three sections of the wye-connected load when the currents are unbalanced.

For this type of system, the three-wattmeter method is used to measure the power in the circuit.

Connections for the Three-Wattmeter Method

The connections for the three-wattmeter method are shown in Figure 10–18. The current coil of each single-phase wattmeter is connected in series with one of the three line wires. The potential coil of each wattmeter is connected between the line wire to which its current coil is connected and the common neutral wire. Thus, each wattmeter indicates only the power taken by one of the three sections of the wye-connected load. In this type of circuit connection, the wattmeter will never read backward. However, different readings are obtained if the loads are unbalanced. If the currents are balanced and

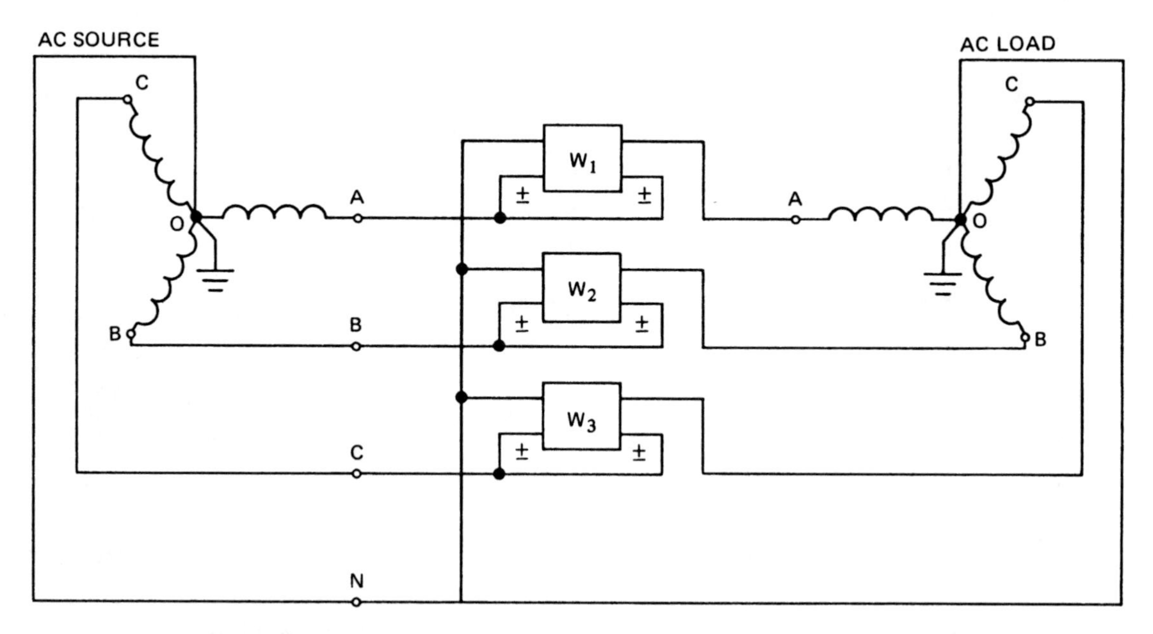

FIGURE 10–18 Three-phase, four-wire system (*Delmar/Cengage Learning*)

the voltages are equal, then all three wattmeters will read equal values. According to the three-wattmeter method, the total power taken by a three-phase, four-wire system is

$$\textbf{Total watts} = W_1 + W_2 + W_3$$

TWO-PHASE SYSTEM

Two-phase systems are rapidly being replaced by the three-phase system. The reasons for the popularity of three-phase systems were given at the beginning of this unit. However, the student may have to work with a two-phase system at some point. Thus, some basic information about such systems is presented here. Basically, a two-phase generator consists of two single-phase windings placed 90 mechanical degrees apart in the slots of the stator core. The output of this generator consists of two sine waves of voltage 90 electrical degrees apart.

One type of two-phase system is the two-phase, four-wire system (Figure 10–19A). This system consists of two separate single-phase circuits. These circuits are isolated electrically from each other. In the second type of two-phase system (Figure 10–19B), the two phases are interconnected. The result of this arrangement is a two-phase, three-wire system. The voltage across the outside wires of this system is equal to $\sqrt{2}$ times the coil voltage. It should be evident to the student that this voltage value results because the induced voltages in the phase windings are 90 electrical degrees apart and are of equal magnitude.

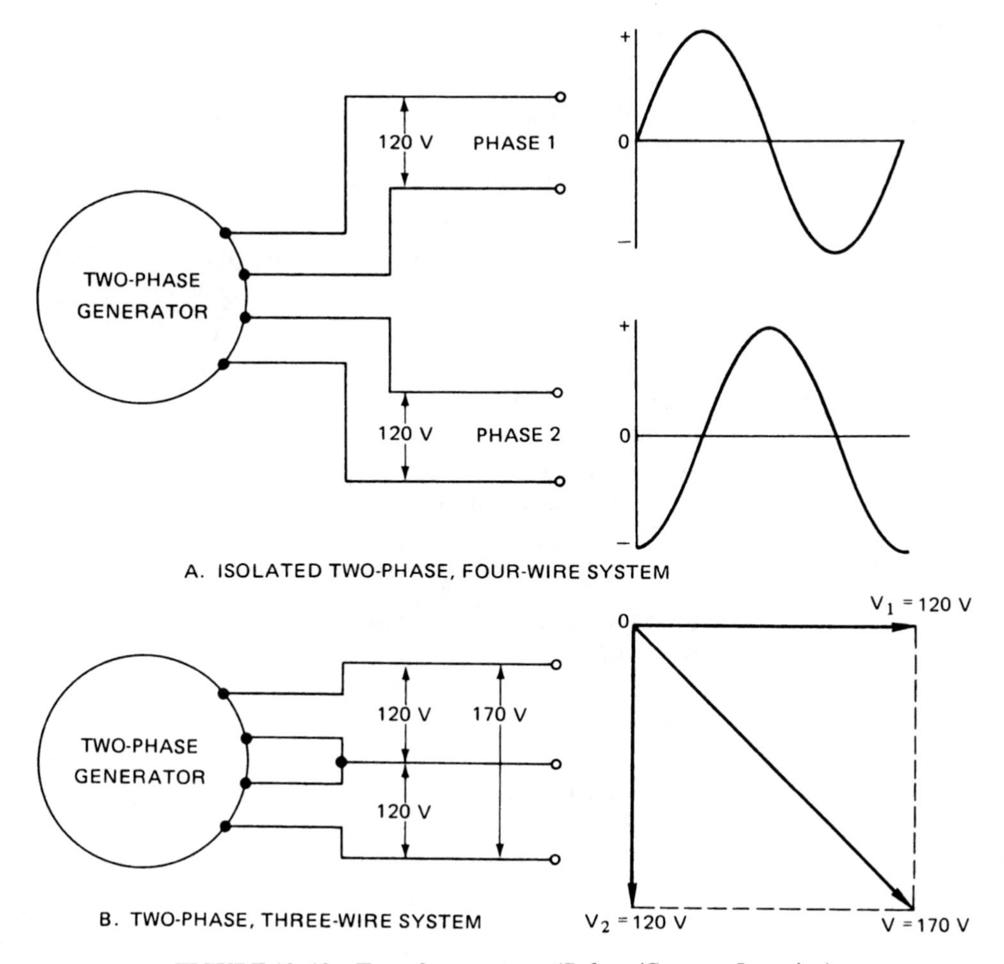

FIGURE 10–19 **Two-phase systems (*Delmar/Cengage Learning*)**

SUMMARY

- A three-phase system has the following advantages as compared to a single-phase system:
    1. Three-phase generators and motors have a capacity approximately 150% that of single-phase units of the same physical size.
    2. Three-phase power is constant and single-phase power is pulsating.
    3. Generators, motors, transformers, feeders, and other three-phase devices have a savings in copper of approximately 25% over single-phase devices.
    4. Three-phase devices are lower in initial cost and maintenance than are single-phase devices.
- A three-phase circuit consists of three single-phase circuits combined into one circuit having either three or four wires.

- Single-phase motors and other single-phase loads may be operated from a three-phase system.
- A simple three-phase generator consists of three coils or phase windings placed in the slots of the stationary armature (the stator). The windings are placed so that the three induced voltages are 120 electrical degrees apart.
  1. The induced voltage in each phase winding is called the *phase voltage.*
  2. The voltage across the line wires is called the *line-to-line voltage.*
  3. Three-phase generators, motors, and transformers can be connected in the wye or delta configuration.
- The phase sequence, or the phase rotation, is the order in which the three voltages of a three-phase circuit follow one another.
  1. Counterclockwise rotation of the generator produces the phase sequence ACB. Clockwise rotation produces the phase sequence ABC.
  2. The phase sequence may be changed by reversing the direction of rotation of the three-phase generator or by interchanging the connections of any two of the three line wires.
- Wye connection:
  1. The wye-connection is the most commonly used way of connecting the three single-phase windings of three-phase generators.
  2. A wye connection is made by connecting one end of each phase winding or coil to a common point. The other end of each coil is brought out and connected separately, one to each line lead.
  3. If the voltage induced in each phase winding is 120 V, the voltage across each pair of line wires is not equal to 240 V. The two phase voltages are 120° out of phase. The line-to-line voltage will be 208 V:

$$\begin{aligned} V_{line} &= \sqrt{3} \times V_{phase} \\ &= 1.73 \times 120 \\ &= 208 \text{ V} \end{aligned}$$

  4. The voltages between phases A and B, B and C, and C and A are all 208 V.
  5. The line current and the phase winding current are the same because each phase winding is connected in series with one of the three line wires.
- Kirchhoff's current law states that
  1. the sum of the currents at a junction point in a circuit network is always zero.
  2. the sum of the currents leaving a point must equal the sum of the currents entering that point.
- Power in the wye-connected system:
  1. $VA_{phase} = V_{coil} \times I_{coil}$
  2. If the V and I values of the wye system are balanced, then

$$VA_{total} = 3 \times V_{coil} \times I_{coil}$$

3. It is easier to measure line voltages and line current than it is to find the coil voltages and the coil currents ($I_{line} = I_{coil}$ in a wye-connected system). Also

$$V_{coil} = \frac{V_{line}}{\sqrt{3}}$$

Therefore,

$$\text{(Input) } VA_{total} = \frac{3 \times V_{line} \times I_{line}}{\sqrt{3}}$$

$$= \sqrt{3} \times V_{line} \times I_{line}$$

4. Power in kilowatts:

$$kW = \frac{\sqrt{3} \times V_{line} \times I_{line} \times \cos \angle\theta}{1000}$$

5. Power factor for a balanced three-phase, wye-connected system:

$$\cos \angle\theta = \frac{\text{total power}}{\sqrt{3} \times V_{line} \times I_{line}}$$

If the current values are severely unbalanced, or the three voltages differ greatly, then the three-phase power factor has almost no meaning.

- Delta connection:
  1. This is the second standard connection method by which the three single-phase coil windings of a three-phase generator can be interconnected.
  2. The term *delta* is used because the schematic diagram of this connection closely resembles the uppercase Greek letter *delta* (Δ).
  3. The phase and line voltages have common points; thus, these voltages are equal.
  4. The three line currents, I_A, I_B, and I_C are 120 electrical degrees apart in a balanced three-phase system:

$$I_{line} = \sqrt{3} \times I_{line} = 1.732 \times I_{coil}$$

  5. The total volt-amperes for both a balanced three-phase, three-wire, wye-connected system and a balanced three-phase, three-wire, delta-connected system is

$$VA = \sqrt{3} \times V_{line} \times I_{line}$$

  6. The total power for both a balanced three-phase, three-wire, wye-connected system and a balanced three-phase, three-wire, delta-connected system is

$$\text{Watts} = \sqrt{3} \times V_{line} \times I_{line} \times \cos \angle\theta$$

  7. The power factor of a balanced delta-connected, three-phase system is

$$\cos \angle\theta = \frac{\text{total power}}{\sqrt{3} \times V_{line} \times I_{line}} = PF$$

If the currents and voltages in a delta system are severely unbalanced, the three-phase power factor has no real significance.

- The power, in watts, taken by a three-phase, three-wire system can be measured with two wattmeters. This method can be used to measure the power in both a three-wire wye system and a three-wire delta system.
  1. The current coils of the two wattmeters are connected in series with two of the three line leads.
  2. The potential coil of each wattmeter is connected between the line wire connecting the current coil and the third line wire.
  3. Careful attention must be paid to the polarity marks (±) on the voltage and current coils of the wattmeters.
  4. $W_T = W_1 \pm W_2$
- Three-wattmeter method:
  1. Many wye-connected systems have a neutral wire in addition to the three line wires. This neutral wire is connected between the common point of the coils in the alternator and the common point of the wye-connected load. Such a system is called a *three-phase, four-wire, wye-connected circuit.*
  2. The three-wattmeter method is used to measure the power in this circuit.
     a. The current coil of each single-phase wattmeter is connected in series with one of the three line wires.
     b. The potential coil of each wattmeter is connected between the line wire to which its current coil is connected and the common neutral wire.
     c. Total watts $= W_1 + W_2 + W_3$.
- Two-phase system:
  1. Basically, a two-phase generator has two single-phase windings placed 90 mechanical degrees apart in the slots of the stator core.
  2. The two line voltages of a two-phase system are 90 electrical degrees apart.
  3. A two-phase, four-wire system consists of two separate single-phase circuits, isolated electrically from each other.
  4. In a two-phase, three-wire system, the two single-phase windings are electrically connected at one end of each coil and are brought out as one of the three line wires. The other two line wires each connect to the free end of the phase coil.
     a. The voltage across the outside wires of this system is

$$V_{phase} = \sqrt{2} \times V_{coil}$$

     b. The voltage from either one of the outside wires to the center wire is equal to the coil voltage.

Achievement Review

1. The three windings of a three-phase, 60-Hz ac generator are connected in wye. Each of the three coil windings is rated at 5000 VA and 120 V. Determine
   a. the line voltage.
   b. the line current when the generator is delivering its full-load output.
   c. the full-load rating of the three-phase generator, in kilovolt-amperes.

2. The three-phase generator in question 1 delivers the rated output to a three-phase noninductive heating load. As a result, the current in each coil of the generator is in phase with its respective voltage.
   a. Determine the full-load output of the generator in kilowatts.
   b. Draw a vector diagram to scale of the resulting voltages and currents when the three-phase generator delivers the full-load output to this noninductive load. All vectors must be properly labeled.

3. The three-phase generator in question 1 is connected to a balanced three-phase load. This connection causes each coil current of the generator to lag its respective coil voltage by 30°.
   a. Determine the output of the alternator, in kilowatts, when the full-load output is delivered to this type of load.
   b. Draw a vector diagram to scale of the voltages and currents for the alternator when it delivers the rated output with a phase angle of 30°. All vectors must be properly labeled.

4. Give several reasons why three-phase connections are preferred to single-phase connections for many alternating-current installations.

5. A heating load consists of three noninductive heating elements connected in delta. Each heating element has a resistance of 24 Ω. This heating load is supplied by a 240-V, three-phase, three-wire service. Determine
   a. the voltage across each heater element.
   b. the current in each heater element.
   c. the line current.
   d. the total power taken by this three-phase load.

6. Draw a vector diagram to scale of the currents and voltages for the circuit of question 5. All vectors must be properly labeled.

7. A three-phase, delta-connected alternator is rated at 720 kVA, 2400 V, 60 Hz. At the rated load, determine
   a. the output, in kilowatts, at an 80% lagging power factor.
   b. the coil current.
   c. the line current.
   d. the voltage rating of each of the three windings.

8. A three-phase, wye-connected alternator is rated at 720 kVA, 2400 V, 60 Hz. At the rated load, determine
   a. the output, in kilowatts, at an 80% lagging power factor.
   b. the full-load line current.
   c. the full-load current rating of each of the three windings.
   d. the voltage across each phase winding.
9. A 5-kVA, 208-V, three-phase ac generator is connected in wye.
   a. Determine
      (1) the voltage of each coil of the phase windings.
      (2) the coil current at full load.
   b. If the phase windings of this alternator are reconnected in delta, what are the new line voltage and the current values at full load?
10. Three coils are connected in delta across a 240-V, three-phase supply. The line current is 20 A. The total power delivered to the three coils is 6000 watts. Determine
    a. the total load, in volt-amperes.
    b. the three-phase power factor.
    c. the current in each coil and the voltage across each coil.
    d. the impedance, in ohms, of each coil.
11. Draw a vector diagram to scale of the voltages and currents for the three-phase, delta-connected circuit in question 10. All vectors must be properly labeled.
12. Using a circuit diagram, show how to obtain the test data required to determine the total power and the total load volt-amperes in a three-phase, three-wire circuit. (Assume that two single-phase wattmeters, three ammeters, and one voltmeter are to be used to determine the data.)
13. The following test data were obtained for a three-phase, 220-V, 5-hp motor operating at full load:

Line Voltage			Line Current			Wattmeter Readings	
Volts A–B	Volts B–C	Volts C–A	Amperes A	Amperes B	Amperes C	No. 1 Watts	No. 2 Watts
220	220	220	13.3	13.3	13.3	2920	1330

    At full load, determine
    a. the power taken by the three-phase motor.
    b. the power factor.
14. The following data were obtained by a technician for a 10-hp, 220-V, three-phase motor delivering the rated load output:

Line Voltage			Line Current			Wattmeter Readings	
Volts A–B	Volts B–C	Volts C–A	Amperes A	Amperes B	Amperes C	No. 1 Watts	No. 2 Watts
220	220	220	27	27	27	5940	2800

At the rated load, determine

a. the input volt-amperes.
b. the power input.
c. the power factor.
d. the efficiency of the motor.

15. Using the curve given in Figure 10–17, determine
    a. the power factor of the motor in question 13.
    b. the power factor of the motor in question 14.

16. The power input to a three-phase motor is measured by the two-wattmeter method. The three line voltages are 220 V and the current in each of the three line wires is 8 A. If the three-phase power factor is 0.866 lag, what are the values of power indicated by wattmeter 1 and wattmeter 2?

17. Using the two-wattmeter method, the following test data were obtained for a 440-V, three-phase motor:

Line Voltage			Line Current			Wattmeter Readings	
Volts A–B	Volts B–C	Volts C–A	Amperes A	Amperes B	Amperes C	No. 1 Watts	No. 2 Watts
440	440	440	14	14	14	4200	−1800

    a. Determine
       (1) the power input to the motor.
       (2) the input in volt-amperes.
       (3) the power factor.
    b. Using the curve given in Figure 10–17, check the power factor obtained in step a(3) of this question.

18. Show the connections for the three-wattmeter method used with a three-phase, four-wire, wye-connected system. Both 120-V, single-phase service and 208-V, three-phase service are to be available.

19. A three-phase, four-wire, wye-connected system supplies a noninductive lighting load only. The current in line A is 8 A, in line B the current is 10 A, and in line C the current is 6 A. The voltage from each line wire to the neutral wire is 120 V. Determine
    a. the power, in watts, indicated by each of the three wattmeters.
    b. the total power, in watts, taken by the entire lighting load.

20. With the aid of diagrams, explain the difference between a two-phase, four-wire system and a two-phase, three-wire system.

11

AC Instruments and Meters

Objectives

After studying this unit, the student should be able to

- explain the operation of the following types of movements used with ac ammeters and voltmeters: attraction, inclined coil, repulsion, repulsion–attraction, and dynamometer.
- diagram the connections for a bridge rectifier used with a dc permanent-magnet moving coil-type instrument.
- list the advantages and limitations of using a permanent-magnet moving coil instrument to measure ac voltage and current values.
- calculate the metered power of a load connected to a watt-hour meter by the use of formulas.
- describe the construction, operation, and use of each of the following special-purpose measuring instruments:
    - Dynamometer wattmeter
    - Two-element wattmeter
    - Varmeter
    - Power factor meters
    - Synchroscope
    - Resonant circuit frequency meter
- define register ratio and gear ratio as they apply to recording instruments.
- describe the construction and operation of a watt-hour meter.
- describe the connections and formulas to be used to test the accuracy of a watt-hour meter using a watt-hour meter standard.

MEASUREMENT OF AC QUANTITIES

The equipment used to measure ac quantities differs somewhat from the equipment used in dc measurements. This unit will describe the instruments generally used to measure voltage, power factor, VARs, current, watts, frequency, and phase angle. In addition, meters will be described for measuring watt-hours, Watt-hour demand, varhours, and varhour demand.

For the instruments used to measure ac voltage, current, watts, and VARs, each movement must have three basic components:

1. A spring mechanism to produce an opposing torque; the magnitude of this torque depends on the quantity being measured.
2. A restoring spring mechanism to restore the indicator or pointer of the meter to zero after the required measurement is made. In practice, one mechanism is used to produce the opposing torque and then return the indicator to zero.
3. A damping system to prevent the pointer from overshooting and excessive swinging. If too much damping is provided, a long time is required before the pointer can reach a new reading after a change occurs in the measured quantity. The proper damping means that the pointer moves quickly and stabilizes quickly when the quantity being measured is changed. AC instruments often use electromagnetic damping. As shown in Figure 11–1, a permanent magnet induces eddy currents in some part of the instrument movement. The magnetic effect of these eddy currents opposes the motion of the pointer. Air damping is another method of controlling the motion of the pointer. In this method, a vane retards the movement of the pointer.

MEASUREMENT OF VOLTAGE AND CURRENT

Alternating current and voltage can be measured using several different types of movements. These movements vary in their ability to meet the criteria expressed in the following questions:

1. Does the instrument scale indicate RMS or average values?
2. To what value of volts or amperes must the movement respond?
3. Is the calibration of the scale linear in the useful range, or are the numbers on the scale crowded together in some important region?
4. How accurate is the instrument?
5. How much does it cost?

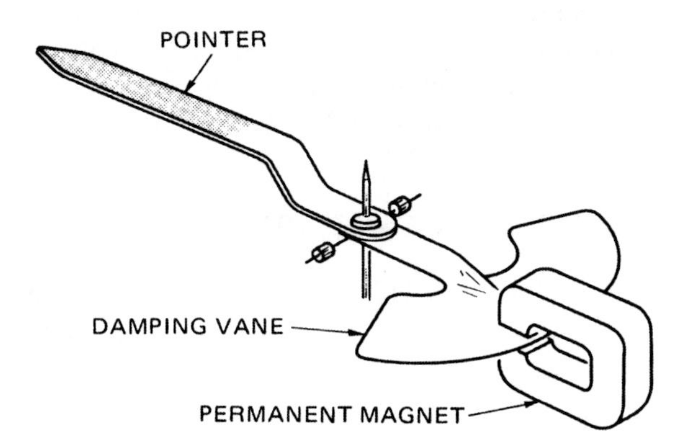

FIGURE 11–1 Magnetic damping mechanism
(Delmar/Cengage Learning)

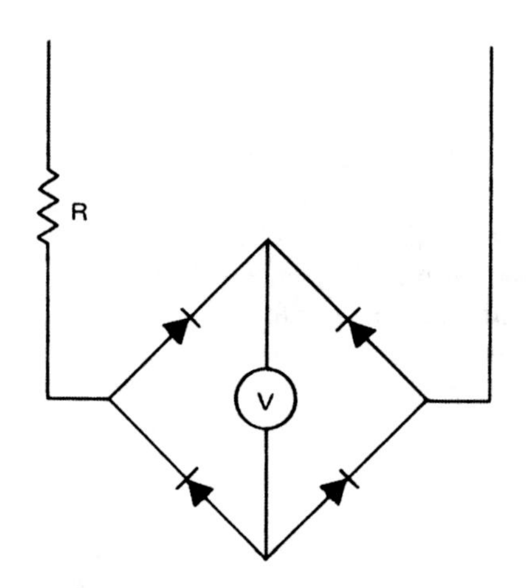

FIGURE 11–2 Alternating current is converted to direct current with a bridge rectifier (*Delmar/Cengage Learning*)

RECTIFIER INSTRUMENTS WITH D'ARSONVAL MOVEMENT

Direct Current Fundamentals described the d'Arsonval dc instrument movement. This movement can be used for ac measurement if a rectifier is also used.

Figure 11–2 shows a bridge-connected full-wave rectifier used with a voltmeter. (R is the series resistor normally required with a voltmeter.)

The dc d'Arsonval movement develops a torque that is proportional to the average value of the current in the moving coil. For an ac wave, only the RMS (effective) values are of interest. Thus, the ac scale is calibrated in RMS values. The RMS voltage is 1.11 times the average voltage value of a sine wave. This means that rectifier-type measuring instruments are accurate only when pure sine-wave quantities are involved. If the voltage to be measured has another type of wave shape, the instrument will give erroneous readings.

Multimeters are instruments that measure both ac and dc quantities. d'Arsonval movements are used in most multimeters. When used as a voltmeter, a multimeter is expected to have a high resistance linear scale and consume relatively little power. When used as an ammeter, most rectifier-type multimeters must be used in the range of microamperes or milliamperes. The full-scale readings will range from about 100 μA to 1000 mA. A multimeter with a d'Arsonval movement is shown in Figure 11–3.

OTHER TYPES OF AC VOLTMETER AND AMMETER MOVEMENTS

Several other types of ac movements can be used in measuring instruments. These movements include the following:

- Magnetic vane attraction movement
- Inclined coil movement
- Repulsion movement
- Repulsion–attraction movement
- Dynamometer movement

All of the movements listed respond to RMS values of voltage or current.

FIGURE 11–3 Analog-type multimeter
(Courtesy of Simpson Electric Co.)

The Magnetic Vane Attraction Movement

This type of movement has a soft iron plunger that projects into a stationary field coil (Figure 11–4). Current in the field coil produces a magnetic force that pulls the plunger deeper into the coil. The instantaneous value of this magnetic force is proportional to the square of the current in the coil. This means that the average torque turning the movement is proportional to the average, or mean, of the squares of the coil current (the RMS values).

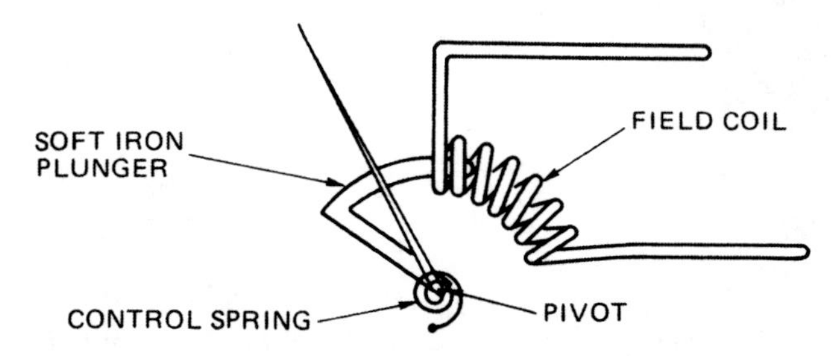

FIGURE 11–4 Simple magnetic vane attraction-type movement ***(Delmar/Cengage Learning)***

This torque is independent of current direction. Thus, the instrument can be used for either ac or dc measurements.

The magnetic force that attracts the plunger has a minimum value when the plunger is just entering the coil. The value of the force increases rapidly as more of the soft iron vane enters the coil. This means that the numbers are crowded together at the lower end of the scale and are expanded for the high end of the scale. This type of movement is commonly used in low-cost ammeters. When a coil of many turns and a series resistor are added, this movement can be used to make voltage measurements.

Inclined Coil Movement

The Thompson inclined coil movement (Figure 11–5) is used in portable and switchboard ammeters and voltmeters. The scale of this movement is long and reasonably linear. An iron vane is free to move in a magnetic field. The vane tends to take a position parallel to the flux. Figure 11–5 shows a pair of elliptical iron vanes attached to a shaft passing through the center of the stationary field coil. If the current in the field coil is increased, an increasing force is produced, which tends to align the vanes with (parallel to) the coil flux. As a result, the shaft turns and moves the attached pointer across the scale.

Repulsion Movement

The repulsion-type movement can be used for both current and voltage measurements. A repulsion force is developed between two soft iron vanes that are affected by the same magnetic field (Figure 11–6). One iron vane is attached to the instrument shaft.

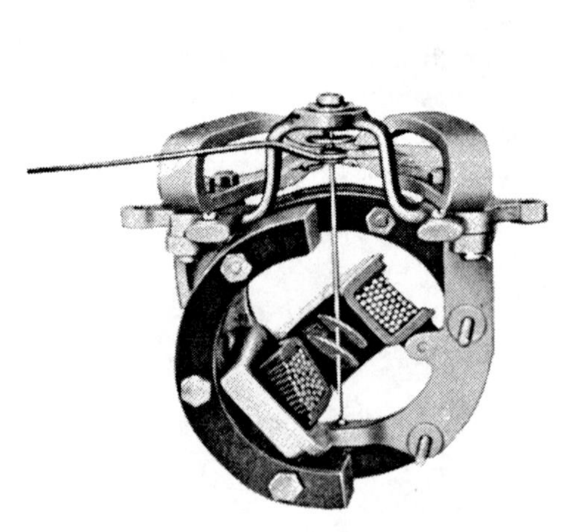

FIGURE 11–5 Cutaway view of an inclined-coil attraction-type movement (*Delmar/Cengage Learning*)

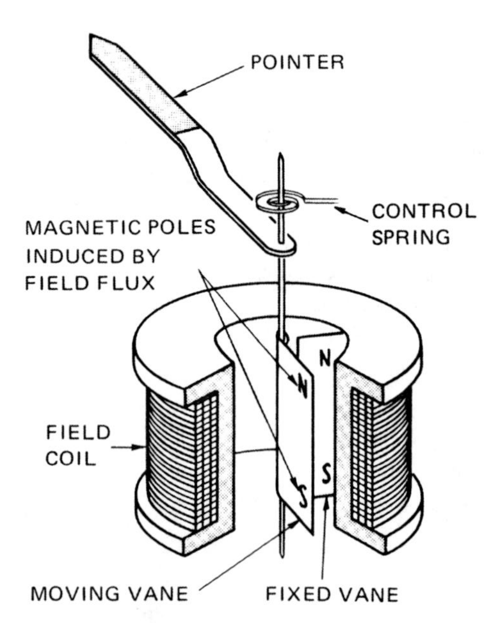

FIGURE 11–6 Repulsion-type movement (*Delmar/Cengage Learning*)

The other vane is mounted on the stationary field coil. When there is no current in the coil, the control spring holds the movable vane close to the fixed vane. Alternating current in the field coil magnetizes both vanes. The like poles of the vanes repel each other and create a torque. This torque turns the instrument shaft. The repulsion force between the two vanes varies according to the square of the current. This force also varies inversely as the square of the distance between the vanes. As a result, the movement has a reasonably uniform scale.

The Repulsion-Attraction Movement (Figure 11-7)

This type of movement is used for both ammeters and voltmeters. It produces more torque per watt than do any of the other ac movements included in this list. A cutaway view of the structure of the repulsion–attraction movement is shown in Figure 11–7.

The instantaneous polarities of the iron vanes are shown in Figure 11–8. The movable vane is repelled from the wide end of the middle fixed vane first. The repelling force decreases as the vane moves to the narrow end of the middle vane. An attracting force increases as the ends of the moving vane come closer to the upper and lower fixed attraction vanes. When these fixed attraction vanes have the correct size and spacing, a scale length representing 250° of angular deflection can be provided. The distribution of values along the scale is determined by the shape and separation of the vanes. Meters can be designed to broaden the scale at any point.

Dynamometer Movement

The dynamometer movement produces a torque by the interaction of magnetic fields. One field is caused by the current in a moving coil. A second field is due to the current in a stationary coil. This stationary magnetic field is not constant. It varies with the amount

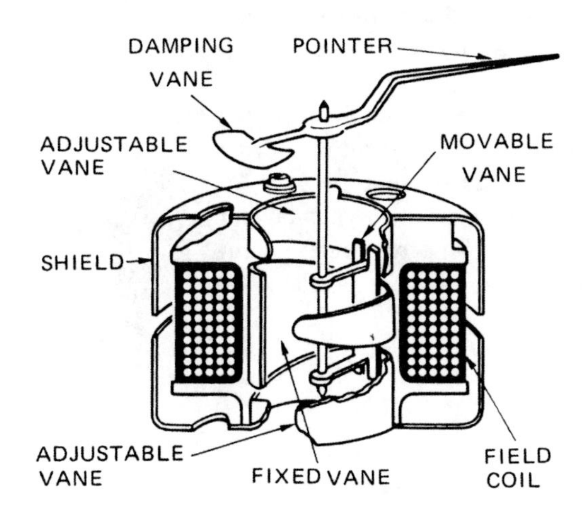

FIGURE 11–7 Cutaway view of repulsion–attraction movement (*Delmar/Cengage Learning*)

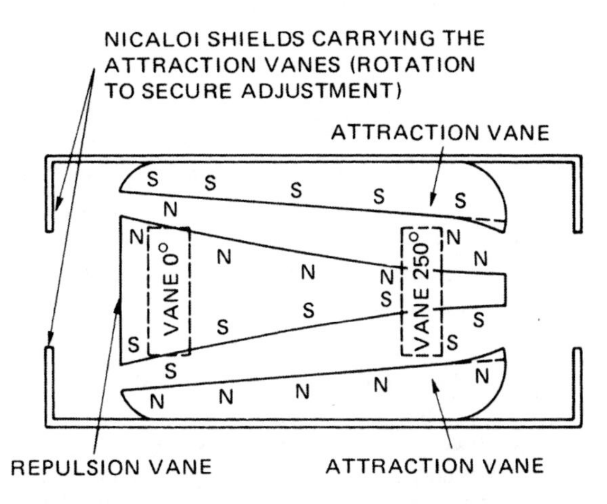

FIGURE 11–8 Development of repulsion–attraction magnetic system (*Delmar/Cengage Learning*)

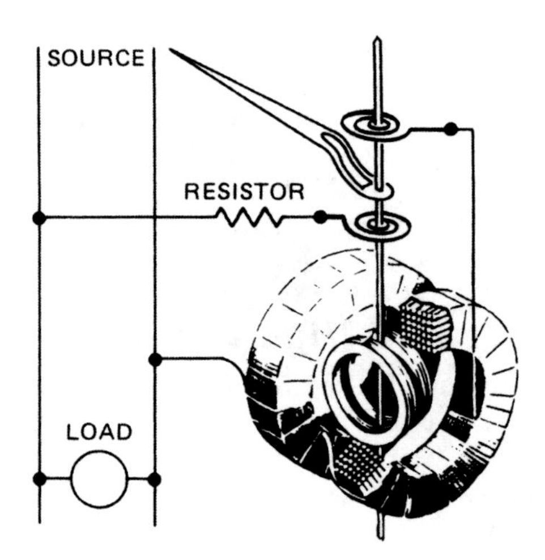

FIGURE 11–9 Dynamometer movement used as a voltmeter (*Delmar/Cengage Learning*)

of current in the stationary coils. Thus, the torque produced in this movement depends on the moving coil current and the stationary coil current. The fixed and moving coils are connected in series. The dynamometer movement can be adapted easily to make voltage measurements by adding the proper series resistor.

Figure 11–9 shows a dynamometer movement used as a voltmeter. When compared with d'Arsonval movements, the dynamometer movement is more efficient and accurate (to one-quarter of 1% or better). This type of movement is seldom used as an ammeter for the following reasons: (1) the lead-in spirals to the moving coil can carry a limited current only, (2) frequency variations influence the inductance of the coils and introduce error, and (3) the resistance of the two coils in series may produce an undesirably high voltage drop across the shunt.

The torque is determined by changes in the stationary coil current or the movable coil current. Because of this fact, the dynamometer movement is a very useful measuring device for several other applications. Although this type of movement may be used for dc, most of its practical uses are for ac. For example, instruments with dynamometer movements are used to measure power in watts and reactive volt-amperes. Such instruments are also used to measure power factor and frequency and to indicate synchronism in ac circuits.

AMMETERS AND VOLTMETERS

Ammeters and voltmeters of the same type operate on the same basic principle. The main difference is that ammeter movements have a few turns of heavy wire and voltmeter movements have many turns of fine wire. Voltmeters also have resistors connected in series with the movement to obtain the desired ranges.

PRACTICAL RANGES FOR MOVING IRON INSTRUMENTS

AC Ammeters

The physical size of ac ammeters, using any type of moving iron movement, determines the current rating of the instrument. The size of the instrument is influenced by the amount of heat to be dissipated and the size of the connection terminals to be supported. In small panel instruments, 100 A is the maximum practical current rating. For large portable instruments, 200 A is the maximum rating. In some large ammeters designed for switchboard use, the current rating may be as high as 600 A.

Larger Current Ratings. Other means must be used to obtain larger current ratings for moving iron ac ammeters. Permanent-magnet moving coil instruments commonly use shunts to obtain higher current ratings. However, shunts are not satisfactory for moving iron instruments. One reason is that the movement is less sensitive and requires a greater voltage drop across the shunt than in dc ammeters. Thus, there is more heat dissipation in the shunt. As a result, the resistance increases in the various parts of the instrument circuit, creating errors in accuracy.

The use of a shunt for moving iron instruments also introduces frequency errors. The inductive reactance of the shunt is low and the inductive reactance of the coil is relatively high. The impedance of the shunt remains nearly unchanged over a range of frequencies. The impedance of the coil, however, changes considerably as the frequency varies. Because the changes in frequency do not affect the coil and the shunt equally, there will be a large error if the instrument is used on a frequency other than the one at which it was calibrated.

Many moving iron instruments are used only for ac measurements. As a result, it is a standard practice to use an instrument current transformer to obtain an increase in the current range of a 5-A instrument. When the measuring instrument is connected to the secondary of the transformer, the current in the transformer primary will be indicated accurately. The instrument can be calibrated to indicate the primary current. The actual calibration depends on the ratio between the primary current and the secondary current. Detailed information on instrument current transformers is given in a later unit of this text.

AC Voltmeters

When moving iron instruments are used as ac voltmeters, series resistors are used to extend the scale range for voltages up to 750 V. The effect is the same as that obtained when a series resistor or a voltage multiplier is used with permanent-magnet moving coil movements in dc voltmeters.

When ac voltages greater than 750 V are to be measured, larger ohmic resistance values cannot be used. Because of the higher ohmic values, more power would be expended in the resistors. Also, there would be high-voltage insulation problems. Thus, large ac voltages are measured using an instrument potential transformer with the movement. The primary winding and insulation of the transformer are suitable for the higher voltage. The secondary winding is usually rated at 120 V. The ac voltmeter usually has a coil rating of 150 V. In many cases, the instrument scale is calibrated to indicate the primary voltage directly. Detailed information on instrument potential transformers is given in a later unit of this text.

THE DYNAMOMETER WATTMETER

To measure the power in watts with an instrument having a dynamometer movement, the stationary field coils are connected in series with the line. Thus, the field flux depends on the current. The moving coils are connected across the line so that the moving coil flux is proportional to the system voltage. Figure 11–10 shows a typical wattmeter circuit. The resistor is connected in series with the moving coil. It can be shown that the instantaneous torque is proportional to the product of the instantaneous field current and the instantaneous moving coil voltage. The average torque for a whole cycle is proportional to the average of the power pulses. This means that the pointer deflection is proportional to the power as expressed by the following equation:

$$\text{Watts} = VI \cos \theta$$

Operation of the Dynamometer Wattmeter

The operation of the dynamometer wattmeter is shown by the wave patterns in Figure 11–11. In Figure 11–11A, the current, voltage, and power waves are shown for one cycle when the current and voltage are in phase. Note that the power curve at any instant is positive. When the current and voltage are in phase, the field flux and the armature flux increase and decrease together. These quantities reach their maximum values at the same time. The deflection of the movement pointer represents the average of the product of the instantaneous voltage and current. This value is the true power, in watts, for the circuit.

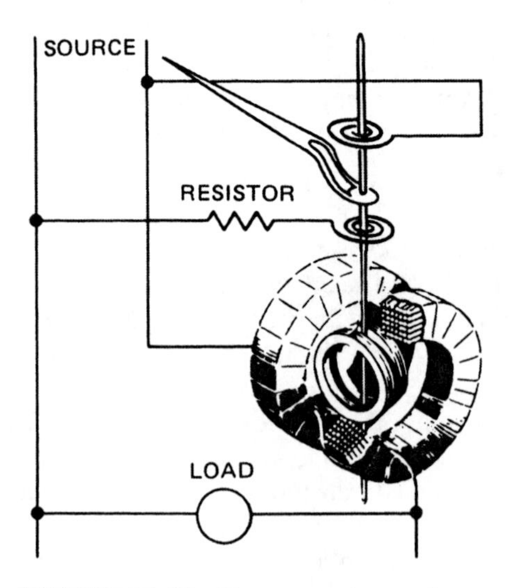

FIGURE 11–10 Dynamometer used as a wattmeter (*Delmar/Cengage Learning*)

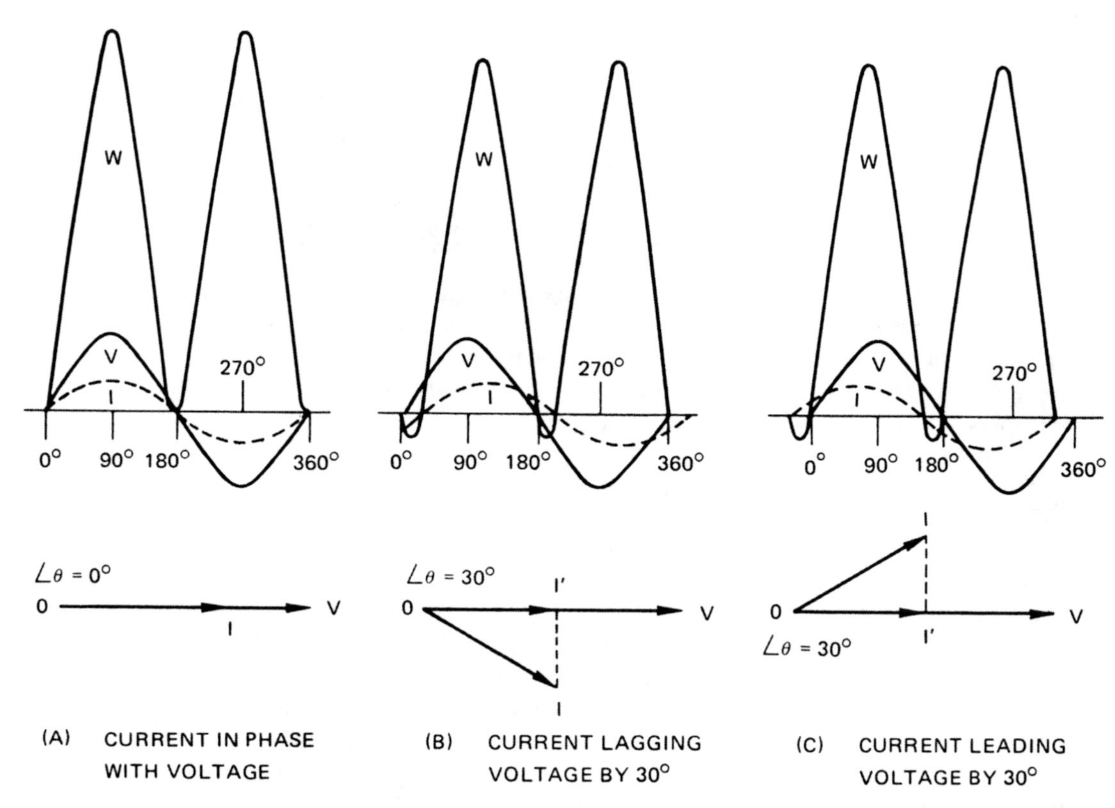

FIGURE 11–11 Curves for voltage, current, and power for a single-phase circuit (*Delmar/Cengage Learning*)

Figure 11–11B shows the current, voltage, and power relationships for a circuit where the current lags the voltage by 30°. In this case, the field flux and the armature flux do not reach their maximum values at the same time. The field flux reaches its maximum value 30° behind the maximum value of armature flux. This means that the torque never reaches as high a value as in the case where the current and voltage are in phase. Instead, the torque always has an average value corresponding to the product of the voltage, current, and power factor. Figure 11–11B shows that the average value of the power is less than the value shown in Figure 11–11A for the condition of the current and voltage in phase.

In Figure 11–11C, the current, voltage, and power are shown for a circuit where the current leads the voltage by 30°. The wattmeter indication is the same for this case as for the lagging current condition. Again, the average torque on the movement of the instrument is determined by the product of the instantaneous current and voltage values. Therefore, Figure 11–11C shows that the average value of the power is less than the value shown in Figure 11–11A for the condition of the current and voltage in phase.

Connecting a Dynamometer Wattmeter

When connecting an instrument having a dynamometer movement, the technician must consider the instantaneous direction of current in each of the coils. This direction determines the flux, which, in turn, specifies the direction of the deflecting torque.

The diagram in Figure 11–12 shows the marking (±) next to one of the terminals of the potential coil circuit (the armature) and also one of the terminals of the current coil (the field). These terminals are connected to the same side of the line to ensure that the deflection has the correct direction.

There are two different methods of connecting the potential coil of a wattmeter. In Figure 11–12, the potential circuit of the wattmeter is not connected directly across the load. Instead, it measures a voltage higher than the load voltage by an amount equal to the voltage drop in the current coil. In other words, the wattmeter indicates too high a value of watts. The extra power is that expended in the current coil. For the connections shown in Figure 11–12, the true power of this circuit is

$$\textbf{True power} = \textbf{wattmeter reading} - \mathbf{I^2R}\ \textbf{of current coil}$$

A second method of connecting the wattmeter is shown in Figure 11–13. In this diagram, the potential coil is connected directly across the load voltage. The current coil of the wattmeter now reads both the potential coil current and the load current. This reading is due to the fact that the potential coil is really a high-resistance load in parallel with the actual load. In summary, the wattmeter indicates a value that is higher than the actual power taken by the load. The power in excess of that taken by the load is equal to the power expended in the potential circuit of the wattmeter. Using the connections in Figure 11–13, the true power is

$$\textbf{True power} = \textbf{wattmeter reading} - \frac{\mathbf{V^2}\ \textbf{of load}}{\textbf{R of potential circuit}}$$

For either connection (Figures 11–12 and 11–13) the wattmeter indicates a value slightly larger than the true power. However, using the connections in Figure 11–13, the percentage error will be slightly less because the potential coil circuit is connected directly across the load.

Using a Wattmeter

To use a wattmeter, the rating of its potential coil and current coil must correspond to the current and voltage ratings of the circuit in which the instrument is to be used. For example, a wattmeter may be used in an ac circuit having a low power factor. The wattmeter can indicate a power value within the scale range of the instrument even though the current coil is greatly overloaded. Even if the circuit has a high power factor, the load current may be much larger than the current coil rating. However, because of a low voltage, the wattmeter pointer is on scale. Again, the voltage across the potential coil may be excessive. But the presence of a low current means that the power indication of the wattmeter is on scale. A wattmeter is always rated according to its potential and current coil ratings rather than in watts.

Figure 11–14 shows a voltmeter, an ammeter, and a wattmeter connected into a single-phase circuit. The voltmeter and ammeter readings will show whether the voltage or current rating of the wattmeter is exceeded. Dynamometer type wattmeters are very

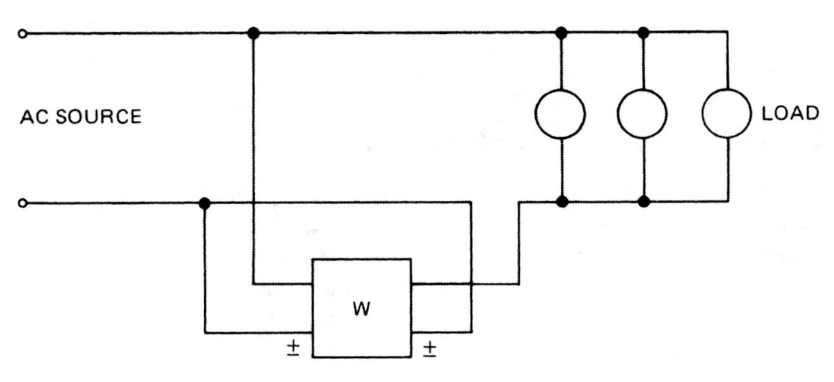

FIGURE 11–12 Wattmeter connections (*Delmar/Cengage Learning*)

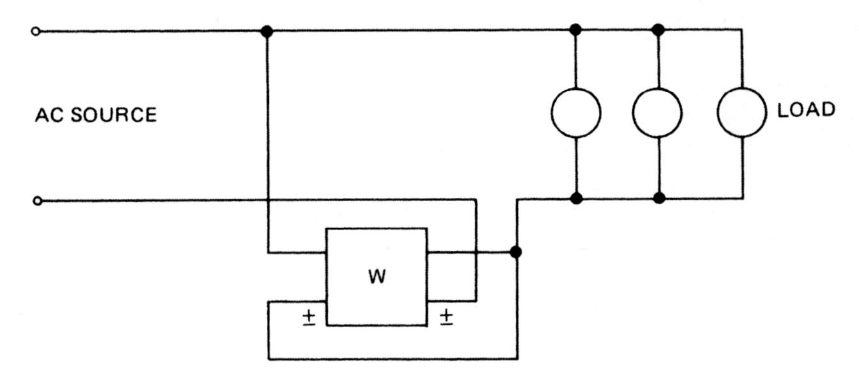

FIGURE 11–13 Wattmeter connections (*Delmar/Cengage Learning*)

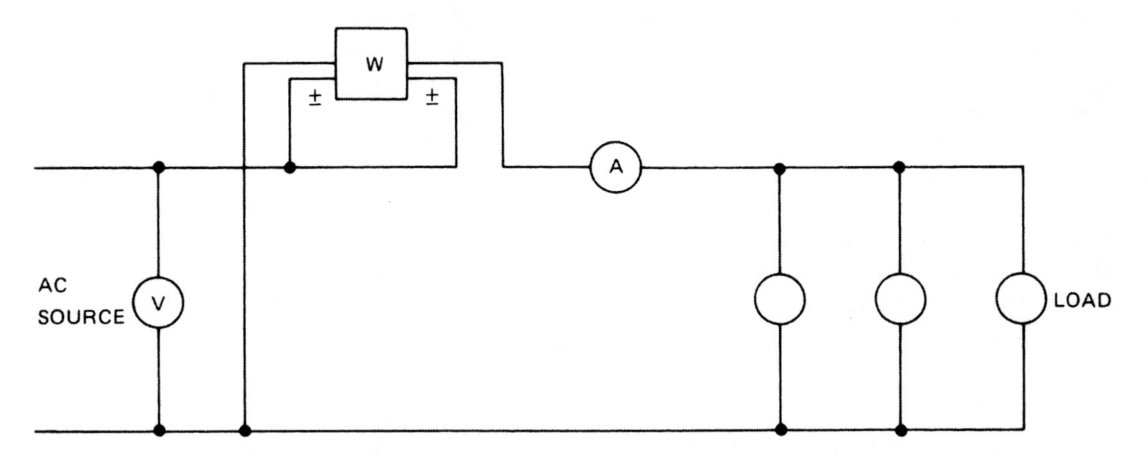

FIGURE 11–14 Wattmeter connections on a single-phase circuit (*Delmar/Cengage Learning*)

expensive to produce because they must contain a stationary coil for measuring current and a moving coil for measuring voltage. Electronic type wattmeters are rapidly replacing the dynamometer type. Electronic wattmeters employ an electronic circuit to measure the quantities of voltage and current, and supply this information to a common meter movement or digital readout. Electronic wattmeters are connected into the circuit in the same manner as a dynamometer type wattmeter.

TWO-ELEMENT WATTMETER FOR THREE-PHASE SYSTEM

Unit 10 described power measurement in three-phase circuit systems. Recall that two single-phase wattmeters were used to measure the power in a three-phase, three-wire system. The two single-phase wattmeters can be combined into a single instrument. The scale of this instrument indicates the sum or difference of the power values indicated by the separate meters. To make the single wattmeter, two sets of potential coils are mounted on a single shaft. Also, two sets of field coils are mounted on the instrument frame so that they have the proper relationship to the armature coils. In this way, each of two power measuring mechanisms develops a torque that is proportional to the power in the circuit to which it is connected. These torque values are added to obtain the total power in the three-phase, three-wire circuit.

If the power factor of the system is less than 0.5, the torque of one mechanism opposes that of the second mechanism. The difference between the torque values is the power indication.

A wattmeter containing two dynamometer mechanisms (Figure 11–15) is called a *two-element wattmeter.*

In Figure 11–16, a two-element wattmeter is shown connected into a three-phase circuit. The current and potential terminals are shown on the instrument case for both elements. Note that the connections shown in Figures 11–15 and 11–16 are the same as the connections given for the two-wattmeter method described in Unit 10.

VARMETERS

The previous section showed how the dynamometer-type mechanism is used to measure true power in watts. This same instrument can also be used to measure the reactive volt-amperes (VARs) in an ac circuit. In this use, the instrument is known as a *varmeter*. The wattmeter indicates the product of the circuit voltage and the in-phase component of the current. The varmeter indicates the product of the circuit voltage and the current component 90° out of phase with the voltage.

To measure VARs, the phase of the potential coil voltage must be shifted by 90°. The flux of the potential coil is then in phase with the flux due to the quadrature component of current in the stationary coil. The phase shift in a single-phase circuit is obtained by connecting an external impedance in series with the potential coil. In this way, the current in the movable coils lags the voltage by 90°.

Measurement of VARs in a Three-Phase Circuit

To find the VARs in a three-phase circuit, a two-element varmeter is used. This device has the same construction as a two-element wattmeter. However, as shown in Figure 11–17, an external phase-shifting autotransformer is added to the varmeter to shift the potential coil voltages 90 electrical degrees.

In Figures 11–17 and 11–18, voltages 1–2 and 3–2 are applied to the transformer from the three-phase line. Taps on the autotransformer are selected as follows:

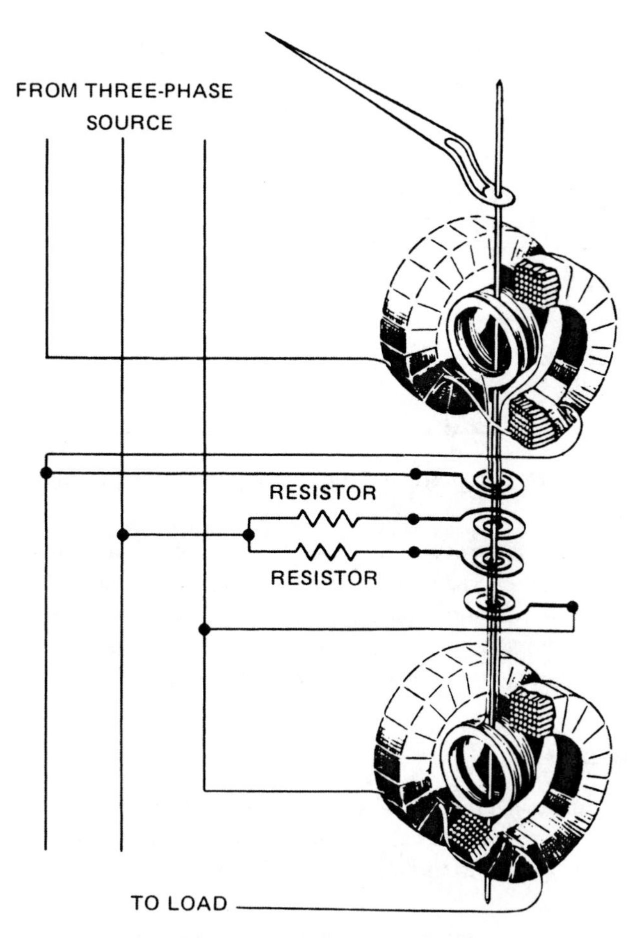

FIGURE 11–15 Diagram of two-element wattmeter connected in single-phase, three-wire circuit (*Delmar/Cengage Learning*)

1. Taps 5–4 are connected to the potential coil whose current coil is in line 1. Voltage $V_{5\text{-}4}$ is equal in magnitude to $V_{1\text{-}2}$, but is 90° out of phase with $V_{1\text{-}2}$.
2. Taps 7–6 are connected to the potential coil whose current coil is line 3. Voltage $V_{7\text{-}6}$ is equal in magnitude to $V_{3\text{-}2}$, but is 90° out of phase with $V_{3\text{-}2}$.

A wattmeter reads VI cos θ, and the varmeter reads VI cos $(\theta - 90°)$. However, cos $(\theta - 90°)$ is equal to sin θ. This means that the effect of shifting the voltages 90° is to make the instrument read VI sin θ (or VARs).

Figure 11–19 shows a diagram of a simpler and less expensive instrument that can be used to measure VARs in a three-phase system where the currents are always balanced. In a balanced three-phase circuit, the power = 3 × a line current × the voltage on the line to neutral × the cosine of the angle between the voltage and current. In the vector diagram in Figure 11–19, power is expressed as $3 \times V_{02} \times I_2 \cos \theta$. Because $V_{1\text{-}3}$ is 90° out of phase with V_{02}, the varmeter will read $V_{1\text{-}3}\, I_2 \cos (\theta - 90°)$. This indication equals $V_{1\text{-}3}\, I_2 \sin \theta$. But $V_{1\text{-}3} = V_{02}\sqrt{3}$.

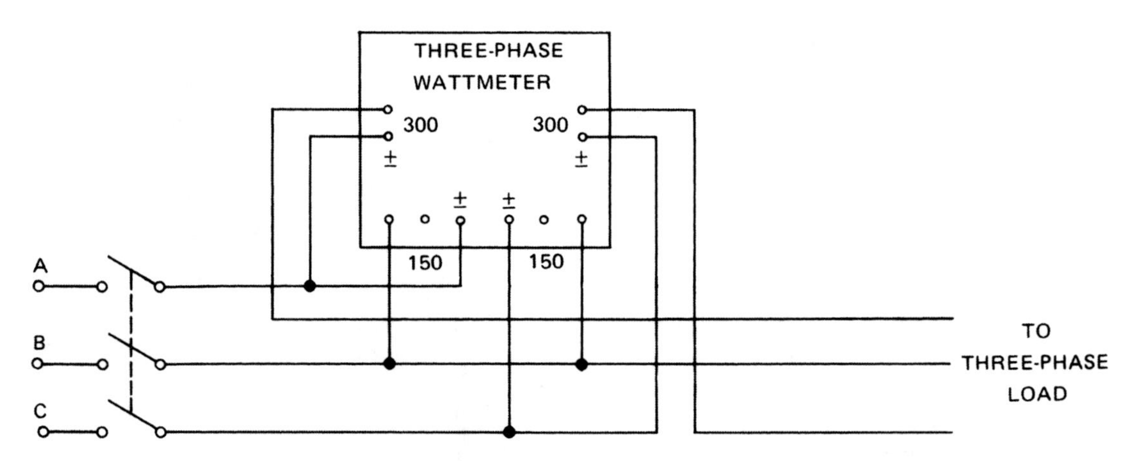

FIGURE 11–16 Connections for two-element wattmeter in a three-phase circuit *(Delmar/Cengage Learning)*

Thus, the instrument reads $\sqrt{3}\, V_{02}\, I_2 \sin\theta$. If the instrument scale reading is multiplied by $\sqrt{3}$, or if the scale is calibrated with a $\sqrt{3}$ multiplier, the instrument indicates $3\, V_{02}\, I_2 \sin\theta$. This reading is the system VARs. The circuit in Figure 11–19 is commonly used to measure the VARs in three-phase motor circuits with balanced currents. An auxiliary autotransformer is not required with this circuit, and a single-element dynamometer can be used.

POWER FACTOR METERS

The power factor of a circuit can be determined by taking simultaneous readings with a wattmeter, an ammeter, and a voltmeter. However, this method is too inconvenient to be used when the power factor of a system is to be determined repeatedly. The task is

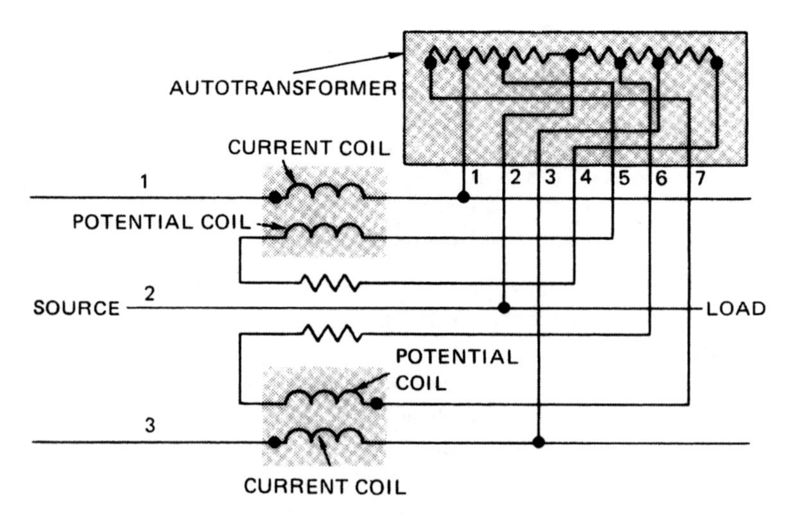

FIGURE 11–17 Connection of dynamometer mechanisms and autotransformer in circuit to measure VARs *(Delmar/Cengage Learning)*

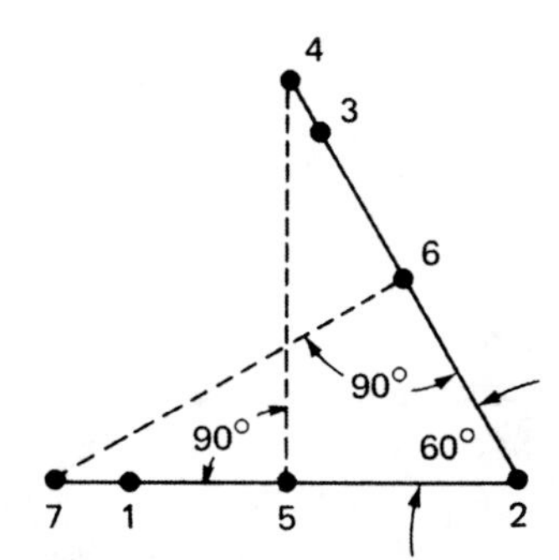

FIGURE 11–18 Vector diagram of autotransformer voltages ***(Delmar/Cengage Learning)***

simplified by the use of a power factor meter. Any deflection of the meter pointer indicates the ratio of the power in watts to the apparent power in volt-amperes. The instrument also shows whether the current is lagging, or leading, the voltage.

Single-Phase Power Factor Meter

A single-phase power factor meter resembles a single-phase wattmeter. The stationary field coils are connected in series with one side of the line. The field coils carry the line current and produce the field flux. The single-phase power factor meter differs from the wattmeter in that it has no control springs. The moving coil, or armature construction, is also different in the power factor meter. The moving mechanism has two armature coils. These coils are mounted on the same shaft. The axes of the coils are 90° apart. One moving coil is connected across the line with a noninductive resistance in series. The flux of this coil reacts with that of the field coil to produce a torque that is proportional to the in-phase component of current. The other moving coil is connected across the line and has an inductive reactance in series with it. The current in this coil lags the line voltage by almost 90°. Thus, the torque for this coil is proportional to the line current component, which is 90° out of phase with the line voltage.

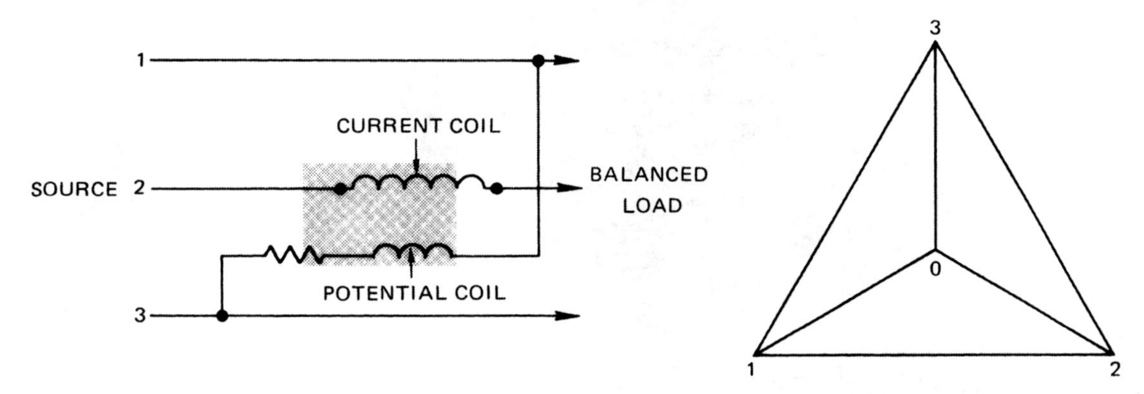

FIGURE 11–19 Connection of single-element dynamometer measuring VARs in balanced three-phase circuit ***(Delmar/Cengage Learning)***

Crossed-Coil Power Factor Meter

A schematic diagram of a single-phase power factor meter is shown in Figure 11–20. This meter is called a *crossed-coil power factor meter* because the two moving coils are crossed at a 90° angle. A cutaway view of the mechanism of a crossed-coil power factor meter is shown in Figure 11–21.

When this type of meter is connected into a circuit containing a noninductive unity power factor load, the entire line current is in phase with the voltage. This means that full torque is developed by the moving coil. The voltage of this coil is in phase with the line current. The quadrature component of current at unity power factor is zero. Thus, torque is not developed by the moving coil whose current lags the line voltage by 90°. As a result, the mechanism moves to a position where the flux alignment between the field current and the active moving coil is at a maximum. The pointer then indicates a power factor of unity, or 1.00.

Assume that the meter shown in Figure 11–21 is connected to a circuit having a power factor less than unity. The movable coil is in series with the inductive reactance. This coil develops torque in a direction determined by the lagging or leading quadrature current

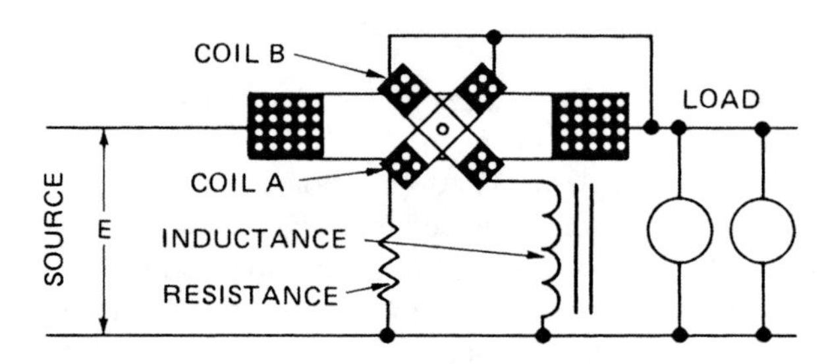

FIGURE 11–20 Diagram of single-phase, crossed-coil power factor meter (*Delmar/Cengage Learning*)

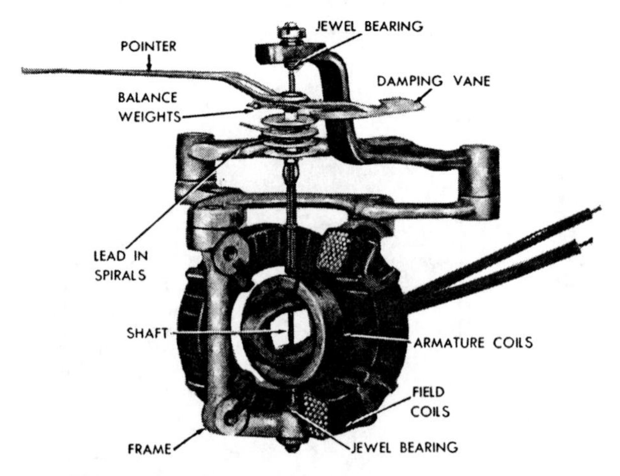

FIGURE 11–21 Cutaway view of mechanism of crossed-coil power factor meter (*Delmar/Cengage Learning*)

component in the fixed field coils. The torque magnitude is determined by the amount of quadrature current. A torque is also created by the in-phase component of current opposing the torque caused by the quadrature current. The resultant of these two torques determines the final position of the pointer. As a result, control springs are not required. The pointer position shows the lagging or leading power factor value.

Another form of the crossed-coil power factor meter is shown in Figure 11–22. In this meter, separate elements are used for the in-phase current and the quadrature potential current.

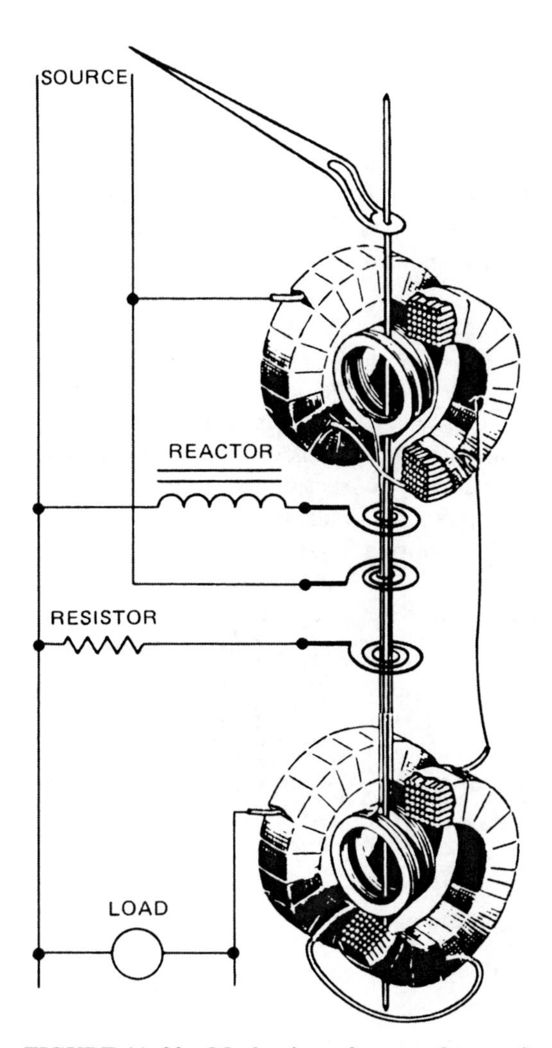

FIGURE 11–22 Mechanism of a second type of crossed-coil power factor meter (*Delmar/Cengage Learning*)

Three-Phase Meter

The crossed-coil power factor meter designed for three-phase service is similar in appearance and structure to the single-phase meter. However, the operating principle of the three-phase meter is different. In the three-phase meter, the crossed coils of the moving mechanism are connected in series with resistors across two phases. The connections are made so that the torques of the two coils oppose each other. The stationary field coil (current coil) is connected in series with one phase leg. The two potential coils are energized from two phases common to the line wire in which the field coil or current coil is connected.

Figure 11–23 is a schematic diagram of the connections for a three-phase, crossed-coil power factor meter. The potential coils are placed at an angle of 60° with each other. For a balanced load, power factor variations change the phase angle between the field coil current and the two potential coil currents. Thus, one phase angle increases and the other angle decreases. Therefore, the torque of one part of the element is proportional to the cosine of 30°, plus the circuit phase angle. The torque of the other part of the element is the cosine of 30°, minus the circuit phase angle. The deflection shown by the meter actually varies by the ratio of the readings that would be given by separate single-phase wattmeters connected in corresponding phases. This ratio can be calibrated directly as the power factor.

SYNCHROSCOPES

The synchroscope is an instrument that shows the relative phase angle and the frequency difference between two alternating voltages. This instrument indicates when alternators are in phase. It also indicates whether the frequency of the incoming generator is higher or lower than that of the generator already connected to the line.

There are a number of versions of synchroscopes available. Two commonly used synchroscopes are the polarized-vane type and the moving iron type. These synchroscopes are

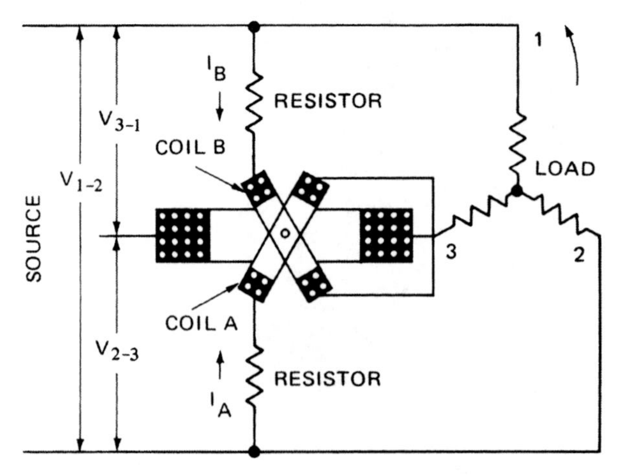

FIGURE 11–23 Three-wire, three-phase, crossed-coil power factor meter (*Delmar/Cengage Learning*)

designed for operation on single-phase circuits. Both devices show when the voltages of two single-phase ac generators are synchronized. They also can be used with three-phase generators if the phase sequences of the generators are known.

Polarized-Vane Synchroscope

The polarized-vane synchroscope (Figure 11–24) uses a mechanism that is similar in physical structure to that of the polarized-vane power factor meter. The basic difference in this mechanism is that the polarizing coil is wound as a potential coil rather than as a current coil. The stator winding has a phase-splitting network and is connected across one phase of the incoming generator. The polarizing winding for the vanes is connected across the corresponding phase of the generator already on the line.

The stator winding of a polarized-vane synchroscope is arranged so that a two-phase field effect is obtained by a phase-splitting network, as shown in Figure 11–24. In the stator network system, capacitor C_2 causes current I_D to lead the voltage (V) by a large phase angle of 75° to 80°. Current I_A lags V because of the inductance of coil A. The amount of lag is about 10° to 15°. As a result, the angle between currents I_A and I_D is 90°.

Operation of the Polarized-Vane Synchroscope

Figure 11–25 shows the relationships between the field and vane fluxes for one cycle with the voltages of the two generators in synchronism. If the rotating field is 61 Hz and the field of the vane polarization is 60 Hz, then the rotating field is 1/60th revolution ahead each time the vanes reach maximum magnetism in one polarity. The vanes line up with the position of the field at the instant of maximum vane magnetism. Thus, each complete cycle of vane polarization brings the vanes 1/60th of a turn further around. For this case, the vanes make one complete revolution in one second. If the stator field rotates at 62 Hz, the vanes will rotate twice as fast. That is, they will make two revolutions in one second.

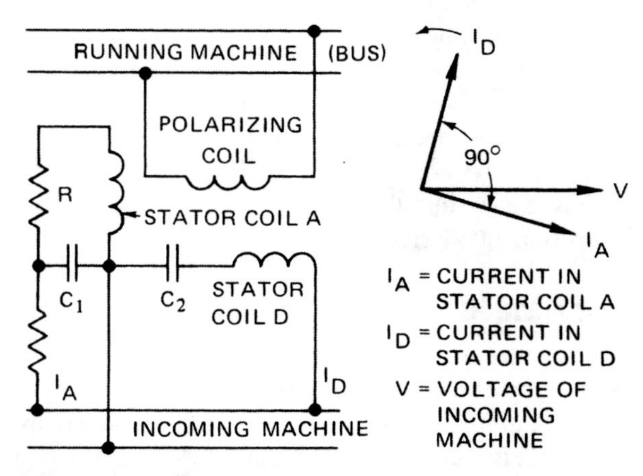

FIGURE 11–24 Connections and phase relationships for the polarized-vane synchroscope (*Delmar/Cengage Learning*)

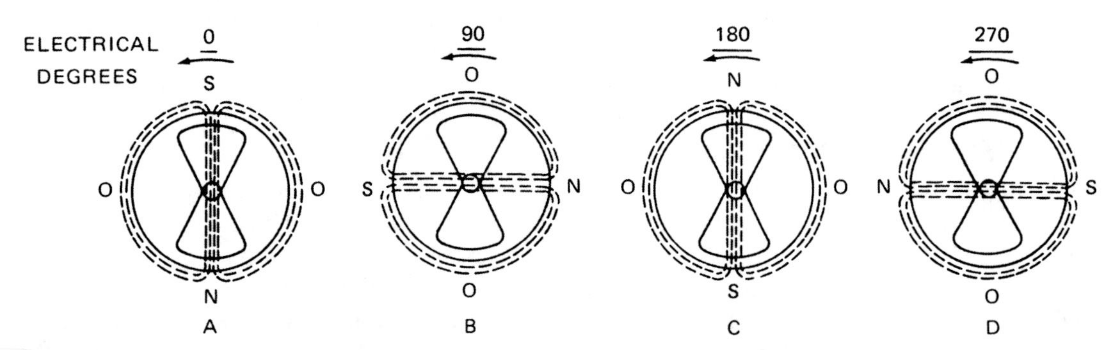

FIGURE 11–25 Relationship of field and vane fluxes during one cycle with running machine and incoming machine synchronized *(Delmar/Cengage Learning)*

Assume that a frequency of 59 Hz is applied to the stator windings and 60 Hz is applied in the polarizing coil. The vanes will then make one revolution per second in the opposite direction. In other words, the speed of the pointer is the difference in frequency of the incoming generator and the machine already connected to the line wires. The direction of rotation of the pointer shows whether the speed of the incoming generator is too fast or too slow. When the two generators are operating at the same frequency, the vanes do not rotate. The position of the pointer for this case indicates the phase relationship between the two voltages.

FREQUENCY METERS

Electric motors, transformers, and other types of machines require the correct voltage for efficient operation. These machines are also designed to operate at a definite frequency. In the case of electric clocks, it is very important that the frequency be accurately indicated. A variation of a small fraction of a cycle, continued through a long period of time, can result in serious errors in time indication on an electric clock. Therefore, electric power systems must operate at the correct frequency. It is obvious that frequency-indicating instruments must be used to show the frequency of a system.

Standard practice requires that ac systems operate at a single frequency. Thus, frequency indications are required to cover only a narrow band of frequency values on either side of the normal frequency. This means that the accuracy of the instrument is improved over what it would be if the instrument covered a wide frequency range.

Resonant Circuit Frequency Meter

One type of commonly used frequency meter is the resonant circuit meter. The physical structure of such a frequency meter resembles that of the dynamometer.

A schematic diagram of a resonant circuit frequency meter is shown in Figure 11–26. The use of two series resonant circuits provides a deflecting torque that has a definite relationship to the applied voltage regardless of its magnitude. The two field coils are alike and are connected so that their fluxes oppose each other. Each field coil is

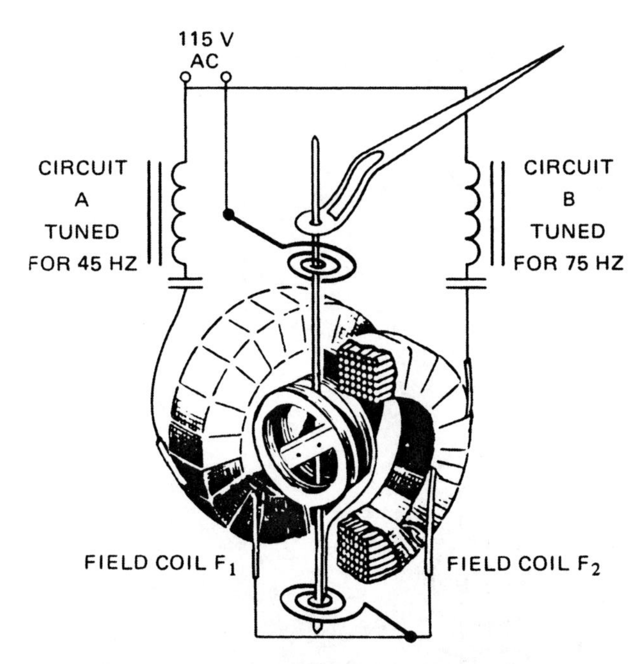

FIGURE 11–26 Schematic diagram of resonant circuit frequency meter ***(Delmar/Cengage Learning)***

connected in series with an inductor–capacitor combination. The constants of this combination permit series resonance below the normal operating frequency in one field coil and series resonance above the normal frequency in the other field coil. If a frequency meter is designed to operate on a normal frequency at 60 Hz, the field circuits are designed so that they are in resonance at 45 Hz and 75 Hz. The armature coil is connected through lead-in spirals. It has almost no countertorque effects and carries the total current of both field circuits.

Changing Circuit Impedance. The change in the circuit impedance of each field coil circuit with a change in frequency is shown in Figure 11–27. Figure 11–27B shows the currents in each of the field circuits and the total armature current. The magnitudes of these currents vary with frequency. Within the operating range of the frequency, the curves show that the impedance of the circuit resonating at 45 Hz is inductive. For the circuit resonating at 75 Hz, the impedance is capacitive. This means that the current in field coil F_1 in circuit A always lags the terminal voltage. Also, the current in field coil F_2 in circuit B always leads the voltage. The actual value of lag or lead and the magnitude of the current in each circuit all depend on the frequency. The current in the armature is the vector sum of the two field currents. Both the armature current and the armature flux lead or lag the terminal voltage, depending on which field current is greater. When the frequency is such that the leading and lagging currents are equal, the armature current is in phase with the terminal voltage.

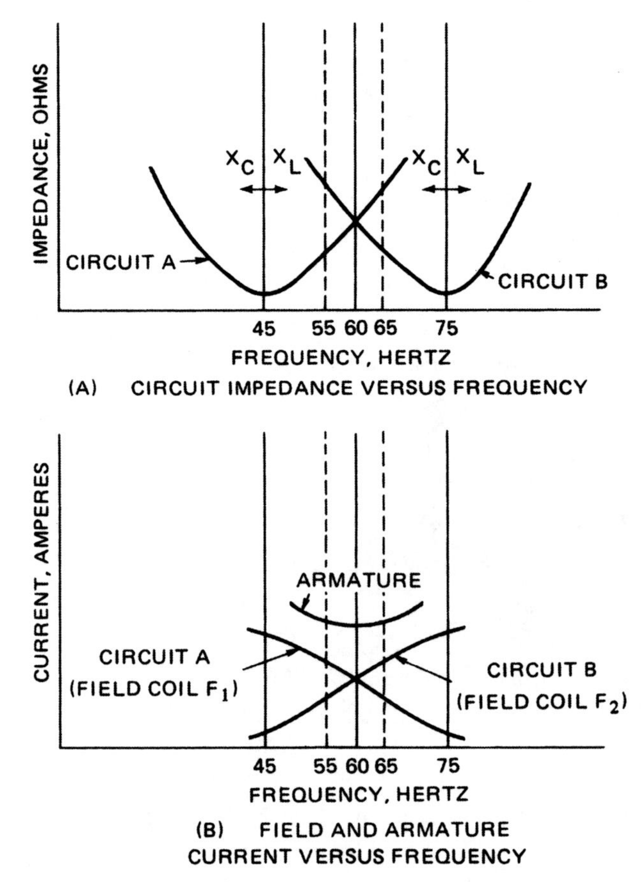

FIGURE 11–27 Electrical characteristics of resonant circuit frequency meter (*Delmar/Cengage Learning*)

Field Flux

The fluxes of the field coils oppose each other. Thus, the resultant field flux is the vector difference between the two fluxes. At 55 Hz, the resultant flux leads the armature flux by an angle that is slightly larger than 90°, as shown in Figure 11–28A.

The torque due to the resultant flux acting on the iron vane is proportional to the product of the armature flux and the in-phase component of the resultant field flux. The direction of the torque in Figure 11–28A causes the pointer to move downscale.

This deflection of the movement causes the iron vane to move out of alignment with the field flux. A countertorque is developed as the vane flux and the field flux align themselves to obtain the shortest possible flux path. When the countertorque equals the armature coil torque, the pointer comes to rest.

Increasing Frequency. With a higher frequency, the leading current (I_B) increases and its phase angle with the line voltage decreases. The lagging current (I_A) decreases and its phase angle with the line voltage increases. As a result, the phase angle between the

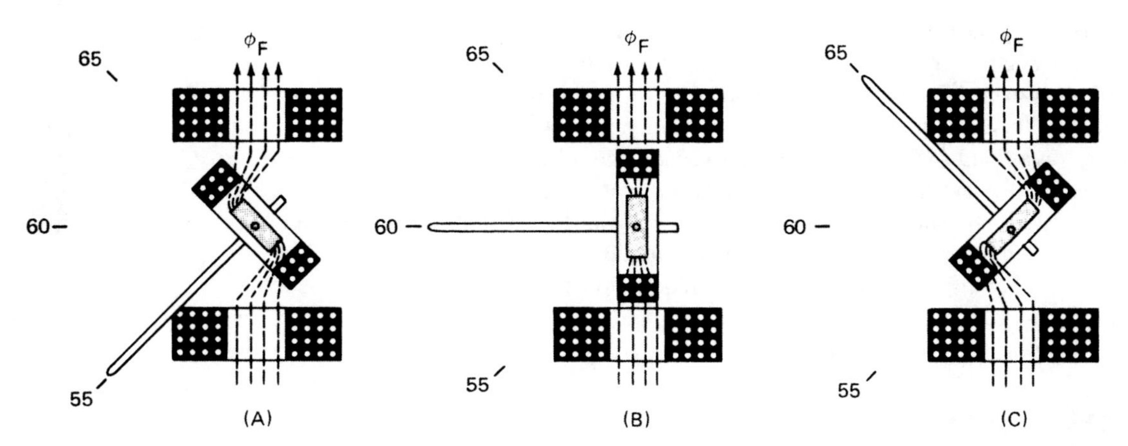

FIGURE 11–28 Resultant field flux acting on iron vane to produce restoring torque (*Delmar/Cengage Learning*)

terminal voltage and the resultant field flux also decreases, as does the phase angle between the armature current and the applied voltage.

Figure 11–28B represents a condition in which the two field currents are equal. These currents have equal and opposite phase angles with the applied voltage. The armature current is in phase with the applied voltage. The resultant of the two field fluxes is 90° out of phase with the voltage. Therefore, the in-phase component of the field flux is zero and there is no deflecting torque. The field flux aligns the iron vane so that the pointer is held at midscale.

If the frequency increases even more, the leading current increases and its phase angle with the applied voltage decreases. The lagging current in the other circuit decreases and has a larger phase angle with the applied voltage. The armature current now lags the applied voltage, as shown in Figure 11–28C. The in-phase component of the resultant field flux is now in the opposite direction. The resulting torque causes the pointer to move upscale.

Resonant-type frequency meters are usually designed to operate on single-phase 115- or 120-V circuits. If the meter scale has a range of 55–60–65 Hz, the meter will indicate frequency values to an accuracy of 0.15 Hz.

RECORDING INSTRUMENTS

Many applications require that the conditions existing in an electrical circuit be monitored constantly. However, it is uneconomical to assign a person to record instrument readings repeatedly. To overcome this problem, recording instruments are used. Such instruments provide a graphical record of the actual circuit conditions at any time. Recording instruments may be grouped into two broad categories:

1. Instruments to record electrical values, such as volts, amperes, watts, power factor, and frequency.
2. Instruments to record nonelectrical quantities. For example, a temperature recorder uses a potentiometer system to record the output of a thermocouple.

Recording instruments are similar to indicating instruments in many ways. They use a permanent-magnet, moving-coil-type construction for dc circuits. For ac circuits, recording instruments may use either the moving iron or the dynamometer-type construction. Whereas the pointer of an indicating instrument just shows the measured quantity on a fixed scale, a recording instrument provides a permanent graphical record. The measured quantities are drawn on a scaled paper chart as it moves past a pen at a constant speed. Because of the friction between the pen and the chart, the indicating movement must have a higher torque than is required in an indicating instrument. As a result, recording instruments are larger and require more power to operate than is required by indicating instruments of the same scale range. Recording instruments are also more highly damped than are indicating instruments so that the pen does not overshoot the chart.

Strip-Chart Recorder

The strip-chart recorder is the most commonly used graphical recording instrument. The permanent record of measurements is made on a strip of paper 4 to 6 in. wide and up to 60 ft long. Figure 11–29 shows an ac recording voltmeter. This strip-chart-type voltmeter has several advantages over nonrecording voltmeters. The long charts allow the record to cover a considerable amount of time. This means that a minimum of operator attention is required. The chart can be operated at a relatively high speed to provide a detailed graphical record.

The principal parts of a strip-chart recording instrument are

1. the frame supporting the various parts of the instrument.
2. the system that moves according to variations in the quantity being measured.
3. the chart carriage, consisting of the chart, the clock mechanism, the timing gears and drum, the chart spool, and the reroll mechanism.
4. the fixed scale on which the value of the quantity being recording is indicated.
5. the recording system, consisting of a special pen-and-ink reservoir or an inkless marking system.

The same basic types of moving systems are used in both graphical recording instruments and indicating instruments. A permanent-magnet moving coil mechanism is shown in Figure 11–30. This mechanism is used with a dc recording instrument to overcome the friction between the pen and the chart. The pen in this movement is carefully counterbalanced. A repulsion-vane moving system for an ac recording ammeter is shown in Figure 11–31A. Figure 11–31B illustrates a two-element electrodynamometer mechanism for a three-phase recording wattmeter.

Chart and Drum. The graphical record is drawn on a chart that is graduated (scaled) in two directions. One of these scales corresponds to the scale range of the instrument. The second scale represents hours, minutes, or seconds, depending on the clock mechanism and the timing gears of the chart carriage. The timescale is uniformly spaced for a constant chart speed, which is provided by the drive mechanism. The drive for the timing drum may be a synchronous electric clock, a conventional hand-wound spring clock, or a spring-type clock mechanism wound by a small electric motor. The last drive listed provides the convenience of an electric drive. Also, it guarantees that a power failure will not stop the chart motion until the spring runs down.

FIGURE 11–29 Typical switchboard-type recording voltmeter (*Courtesy of Amprobe Instruments*)

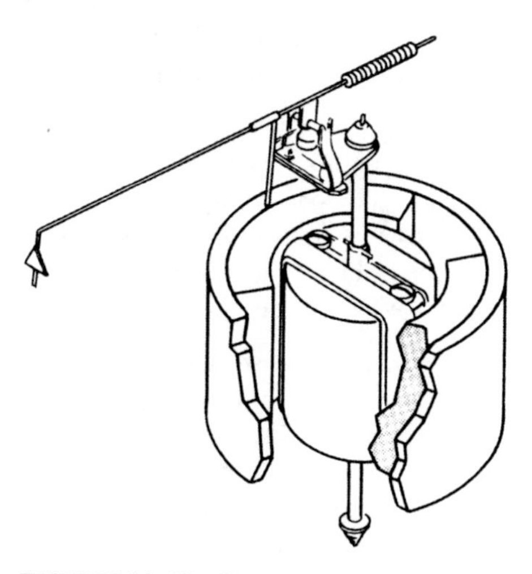

FIGURE 11–30 Permanent magnet moving coil mechanism for dc recorder (*Delmar/Cengage Learning*)

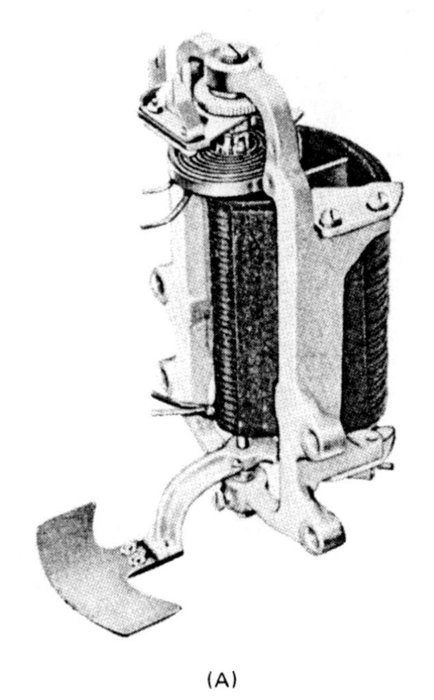

(A)

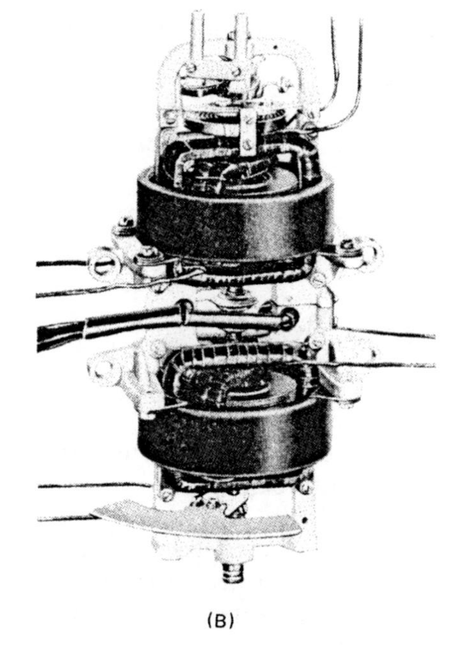

(B)

FIGURE 11–31 Moving mechanisms for ac recording instruments (*Delmar/Cengage Learning*)

The friction between the chart and the pen requires that the pen drive mechanism have a greater torque than in a simple indicating instrument. The larger instrument movements are costly and take a great deal of power from the circuit being measured. The amount of power required can be reduced by using electronic amplifiers. Such an amplifier can drive the recorder element using less power from the circuit to be measured. Voltage and current signals can be amplified directly; the measurement of power and VARs requires special amplifiers.

Amplifiers. One type of amplifier used for power and VARs measurements is a photoelectric device. This device uses conventional wattmeter elements mounted on an auxiliary shaft with a reflecting mirror. An optical system with photocells and electronic amplifiers develops a dc current proportional to the quantity to be measured. This type of amplifier has sufficient power to drive the recorder pen.

Pen Positioning. Many recorders use a motor-and-gear system to position the pen. A regulating system compares the input voltage to the voltage of a precision slide-wire potentiometer (which is driven from the recorder output shaft). If the input voltage changes, the servoamplifier causes the motor to run until the pen and its attached potentiometer are positioned so that the slider voltage is equal to the signal voltage.

A recorder using this system requires very little input power from the circuit being measured. The output drive is so powerful that accessories can be added, such as auxiliary slide-wire potentiometers, limit switches, and devices to code the measured quantities into digital computer language. These accessories can operate alarms and remote devices and can supply information to digital computers.

Other accessories permit recorders to graph many different quantities on one chart. A stepping switch connects the measuring circuit to a different input at regular intervals. As soon as the pointer is positioned, the printer places a dot with a number beside it on the chart to indicate the input being measured. The stepping switch then moves to the next input, and the process is repeated. As a result, the chart contains a series of dots for each of the input signals.

Most recorders operate from a dc signal having a magnitude in millivolts. Such recorders require transducers to convert quantities such as pressure, flow, strain, ac volts, ac amperes, watts, and VARs to dc millivolt signals that can be recorded.

THERMAL CONVERTERS

Thermal converters change ac voltage and current signals into a dc signal in millivolts. This signal is proportional to the product of VI cos θ. AC watts can be measured using a thermal converter with a dc recording or indicating device.

Figure 11–32 is a schematic diagram of a single-phase thermal converter. The currents in the resistors depend on the values of the voltage and current inputs and the phase angle θ between the inputs. Thermocouples are attached to each resistor. These thermocouples are connected in series so that their thermal electromotive forces (emfs) subtract. The total thermocouple circuit output is proportional to the temperature difference between the two resistors.

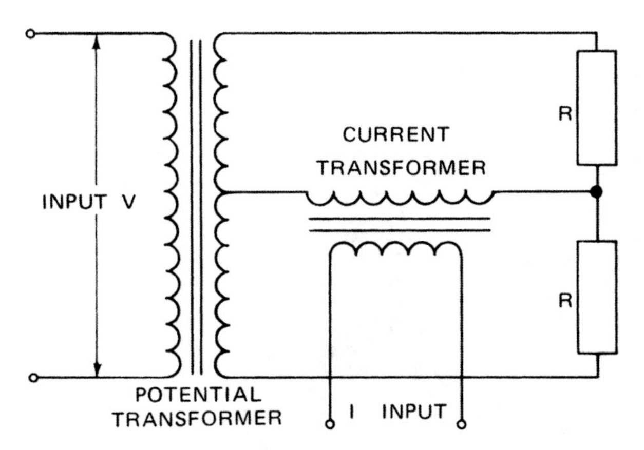

FIGURE 11–32 Single-phase thermal converter circuit
(*Delmar/Cengage Learning*)

This thermocouple output is also proportional to the difference in power dissipation in the resistors. Thus, the output is equal to a constant times VI cos θ. This means that the dc millivolt signal output of the converter indicates the ac power in the circuit to which the V and I inputs are attached.

The outputs of two of these converters can be connected in series. As a result, a dc voltage is developed that is proportional to the sum of two ac power inputs. In summary, three-phase ac power can be measured in two ways: (1) using the two-wattmeter method or (2) using one dc millivoltmeter or recorder.

External phase-shifting autotransformers can be added so that the same circuit will measure three-phase VARs.

THE WATT-HOUR METER

Direct Current Fundamentals explained that electrical work is the use of electrical energy over a period of time. Electric power is the rate at which electrical energy is used. The basic unit of measurement for electric power is the watt. In ac circuits, power in watts is the product of the potential in volts, the current in amperes, and the power factor. The basic unit of measurement for electric energy is the watt-hour. This value is found by multiplying the power (in watts) of a circuit by the total time in hours during which the power is used in the circuit. The watt-hour meter measures the electrical energy consumed in the circuit.

The connections for a typical single-phase watt-hour meter are shown in Figure 11–33. The watt-hour meter is similar to the wattmeter because it also has current coils connected in series with the load and potential coils connected across the line voltage. The interacting magnetic field of the current and voltage coils produces a torque in an armature. This torque is always proportional to the power in the circuit. In the watt-hour meter, the armature is a disc that rotates at a speed proportional to the power in the circuit. The rate at which the disc rotates corresponds to the power. The number of revolutions of the disc corresponds to the total energy used.

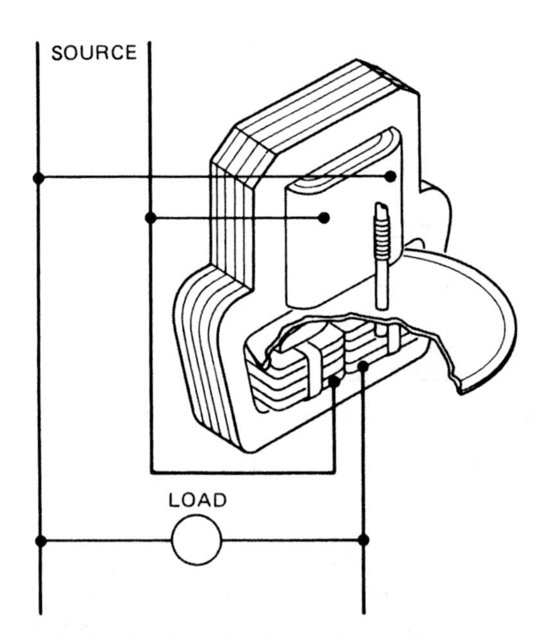

FIGURE 11–33 Diagram of single-phase watt-hour meter (*Delmar/Cengage Learning*)

Components of the Watt-hour Meter

A single-phase watt-hour meter consists of the following parts: an electromagnetic element, the magnetic braking system, the register, the frame, the base (including terminal connections), and the cover.

An induction-type motor is used in a watt-hour meter. The rotor is an aluminum disc mounted on a shaft. This shaft is free to turn in bearings held in the metal frame. A worm gear drives the gear register. Generally, this gear is cut directly into the shaft. In some instances, the rotor may be suspended magnetically. Guide pins are then used to maintain vertical alignment of the shaft. In all cases, the disc is mounted so that a portion of it rotates in the airgap of the stator assembly or electromagnet.

Operation

The electromagnet has two sets of windings assembled on a laminated soft iron core. The potential coil winding has many turns of fine wire and a high impedance. This winding is connected across the source voltage. The current coil winding consists of a few turns of heavy wire. This winding is connected in series with the metered circuit. The core laminations are riveted together to form a rigid mechanical structure. The permanent alignment that results maintains the correct magnetic flux distribution and so ensures a consistent performance.

A torque results from the interaction between the flux (produced by the current in one of the coils) and the eddy currents (induced in the disc by the flux created by the other coil). This torque causes the disc to turn at any given instant. The current coil conducts the load current. Because this coil has only a few turns of large-size wire, its inductance is very small. This means that the current coil flux is nearly in phase with the load current. The current coil flux produces eddy currents in the disc. These eddy currents lag the current coil flux by 90°. Recall that an induced emf always lags 90° behind the flux producing it.

The potential coil is highly inductive. Therefore, the current and flux of the potential coil lag the source voltage by nearly 90°. If a load has a unity power factor, then the potential coil flux is in phase with the eddy currents produced by the current coil. The potential coil poles are above that part of the disc where the eddy currents flow. These eddy currents react with the potential coil flux to develop a torque that is proportional to the line voltage and the load current.

The poles on which the current coil is wound are located beneath the part of the disc where the eddy currents from the potential coil flux flow. These eddy currents react with the flux of the current coil. This reaction produces an additional torque that is also proportional to the line voltage and the load current.

Assembly. An electromagnet assembly is shown in Figure 11–34 for a single-phase watt-hour meter. For a load with a power factor other than unity, the eddy currents lag or lead the fluxes with which they react. The amount of lag or lead corresponds to the phase difference between the line voltage and the load current. The torque developed in each of the eddy current and flux reactions is reduced by a proportional amount.

It is possible for the disc to turn at an excessive speed. To prevent this, a magnetic braking system is used. This system consists of two permanent magnets mounted so that the disc is located between the poles of the magnets. As the disc rotates, it cuts the flux of the two permanent magnets. Eddy currents are thus induced in the disc. The eddy currents react with the permanent-magnet flux to produce a damping torque. The torque opposes the meter torque, and the disc turns at the desired speed for a given load.

The gear register is located on the disc shaft and consists of a train of gears driven by a worm gear or a pinion gear. The gear register turns several dial pointers to show the number of times the disc has turned. This means that the watt-hour meter determines and adds together

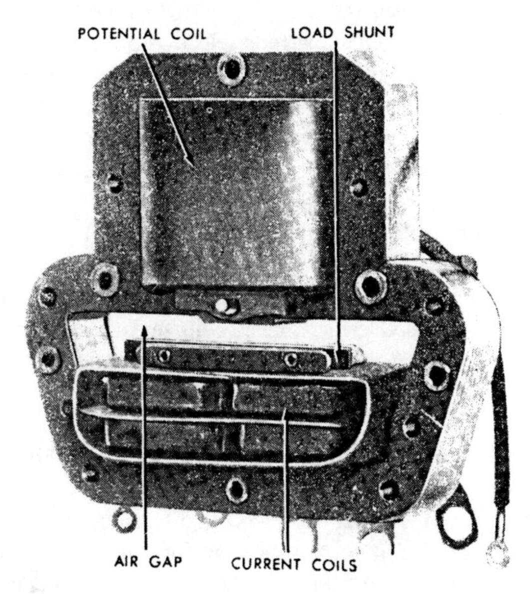

FIGURE 11–34 Electromagnet assembly for a single-phase watt-hour meter (*Delmar/Cengage Learning*)

(integrates) all of the instantaneous power values. As a result, there is an indication of the total energy used over a period of time. Compare this action with that of a wattmeter, which indicates only the instantaneous power or rate of energy use in a circuit.

Interpreting the Dial Readings of the Watt-hour Meter

The *watt-hour constant* of the meter is the number of watt-hours represented by one revolution of the disc. The pointer of the right-hand dial of the gear register indicates one kilowatt-hour after the disc makes the required number of revolutions. The gearing is arranged so that each division on the right-hand dial is one kilowatt-hour (kWh). On the second dial from the right, each division represents 10 kWh. The third dial from the right has divisions representing 100 kWh. For the dial on the left, each division represents 1000 kWh.

The *register ratio* is the number of revolutions made by the first gear wheel as it meshes with the worm or pinion gear on the disc shaft for one revolution of the right-hand dial pointer. The *gear ratio* is the number of revolutions made by the meter disc in causing one revolution of the right-hand dial pointer.

DIGITAL MULTIMETERS

Digital multimeters have become increasingly popular. The most apparent difference between digital meters and analog meters is the fact that digital meters display their readings in discrete digits rather than with a pointer and scale. A digital multimeter is shown in Figure 11–35. Some digital meters have a range switch similar to the range switch used with analog meters. This switch sets the full range value of the meter. Many digital meters have voltage range settings from 200 mV to 2000 V. The lower ranges are used for accuracy. For example, assume that it is necessary to measure a voltage of 16 V. The meter will be able to make a more accurate measurement when set on the 20-V range than it will when set on the 2000-V range.

Some digital meters do not contain a range setting control. These meters are known as *autoranging meters*. They contain a function control, which permits selection of the electrical quantity to be measured such as ac volts, dc volts, or ohms. When the meter probes are connected to the object to be tested, the meter automatically selects the proper range and displays the value. Appearance is not the only difference between digital meters and analog meters. Analog meters change scale value by inserting or removing resistance from the meter circuit (Figure 11–36). The typical resistance of an analog meter is 20,000 Ω/V for dc and 5000 Ω/V ac. This means that if the meter is set for a full-scale value of 60 V, there will be 1.2 MΩ of resistance connected in series with the meter if it is being used to measure dc ($60 \times 20{,}000 = 1{,}200{,}000$) and 300 kΩ if it is being used to measure ac ($60 \times 5000 = 300{,}000$). The impedance of the meter is of little concern if it is used to measure circuits that are connected to a high-current source. For example, assume that the voltage of a 480-V panel is to be measured with a multimeter having a resistance of 5000 Ω/V. If the meter is set on the 600-V range, the resistance connected in series with the meter is 3 MΩ ($600 \times 5000 = 3{,}000{,}000$). This will permit a current of 160 μA to flow in the meter circuit ($480/3{,}000{,}000 = 0.000160$). This 160 μA of current would not be enough to affect the circuit being tested.

Now assume that this meter is to be used to test a 24-V circuit that has a current flow of 100 μA. If the 60-V range is used, the meter circuit contains a resistance of 300 kilohms ($60 \times 5000 = 300{,}000$). This means that a current of 80 μA will flow when the meter is

FIGURE 11–35 Digital multimeter *(Courtesy of Simpson Electric Co.)*

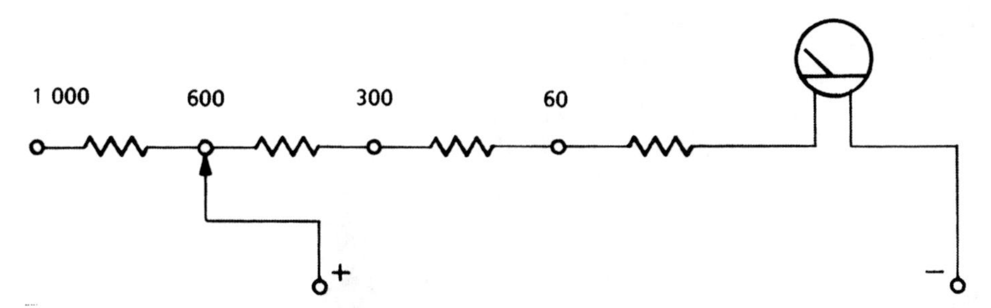

FIGURE 11–36 Analog voltmeters operate by inserting or removing resistance connected in series with the meter movement. *(Delmar/Cengage Learning)*

connected to the circuit (24/300,000 = 0.000080). The connection of the meter to the circuit has changed the entire circuit operation.

Digital meters do not have this problem. Most digital meters have an input impedance of about 10 MΩ on all ranges. This is accomplished by using field effect transistors (FETs) and a voltage divider circuit. A simple schematic for this circuit is shown in Figure 11–37. Notice in

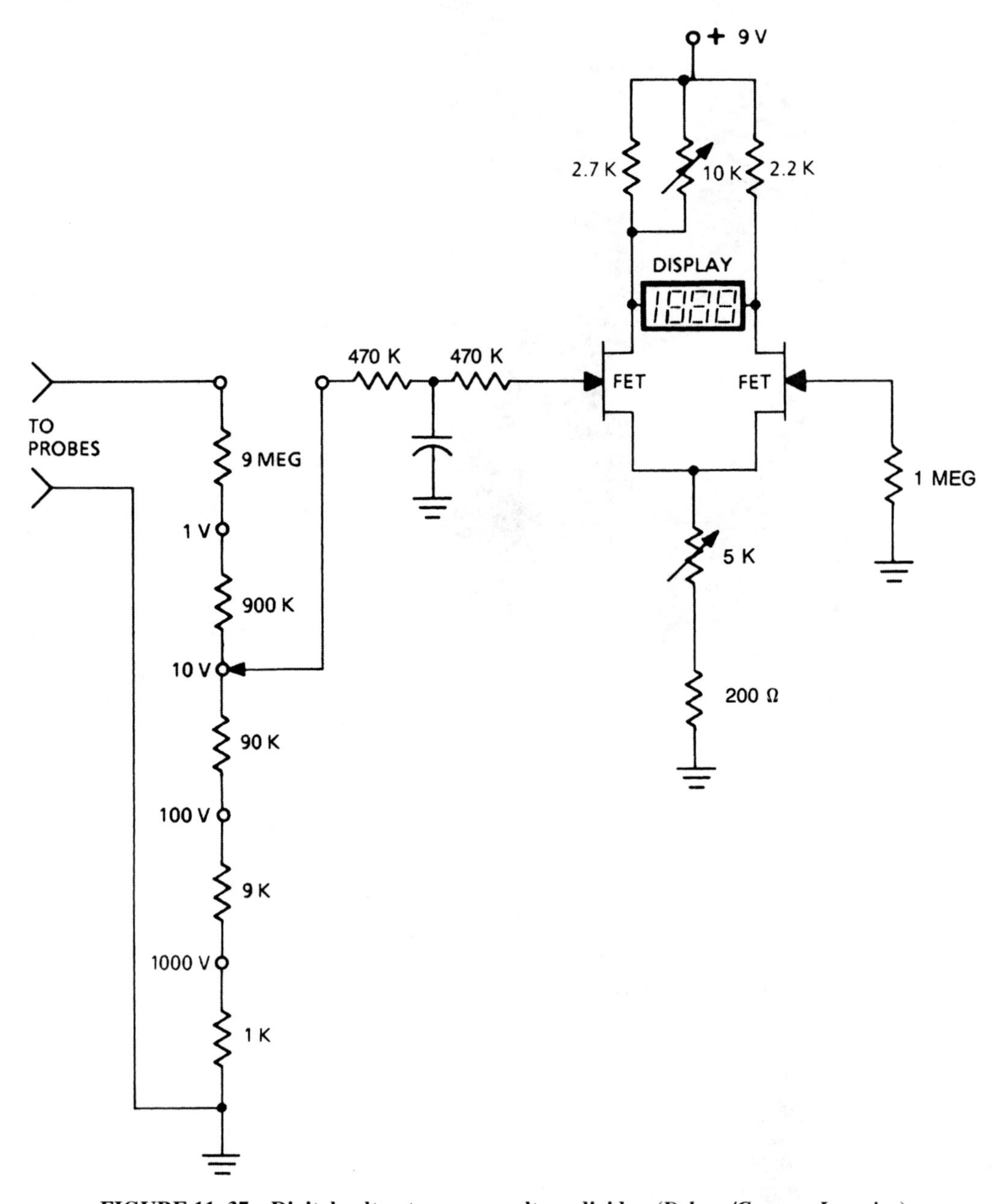

FIGURE 11–37 Digital voltmeters use a voltage divider. (*Delmar/Cengage Learning*)

this circuit that the meter input is connected across 10 MΩ of resistance regardless of the range setting of the meter. If this meter is used to measure the voltage of the 24-V circuit, a current of 2.4 μA will flow through the meter. This is not enough current to upset the rest of the circuit, and voltage measurements can be made accurately.

CLAMP-ON AMMETERS

Another meter frequently used is the clamp-on ammeter (Figure 11–38). Most of these meters operate on the current transformer principle. The movable jaw is the core of a current transformer. The range selection switch connects the meter to different taps on the transformer (Figure 11–39). The conductor around which the movable jaw is connected forms a one-turn primary for the transformer. The secondary is the winding around the iron core. The movable tap changes the turns ratio of the transformer.

If the current is too low to be measured easily, extra turns of wire can be wrapped around the movable jaw. Each turn of wire increases the scale factor of the meter. For example, if two turns of wire are wrapped around the movable jaw, the ammeter reading

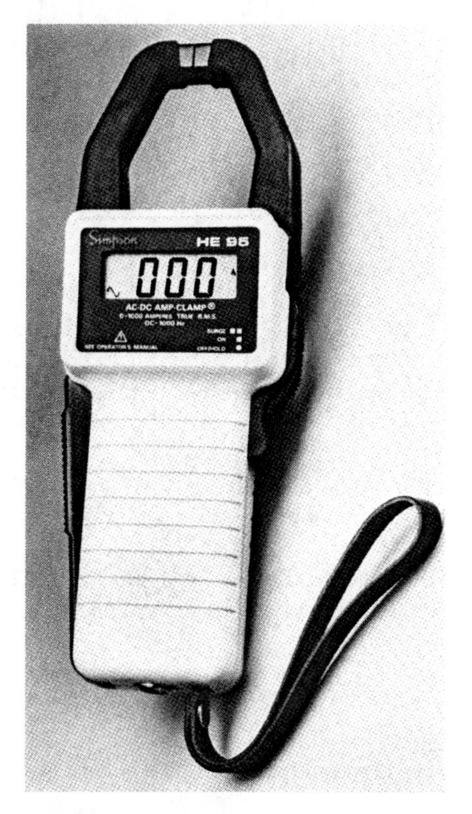

FIGURE 11–38 Clamp-on ammeter with a digital display (***Courtesy of Simpson Electric Co.***)

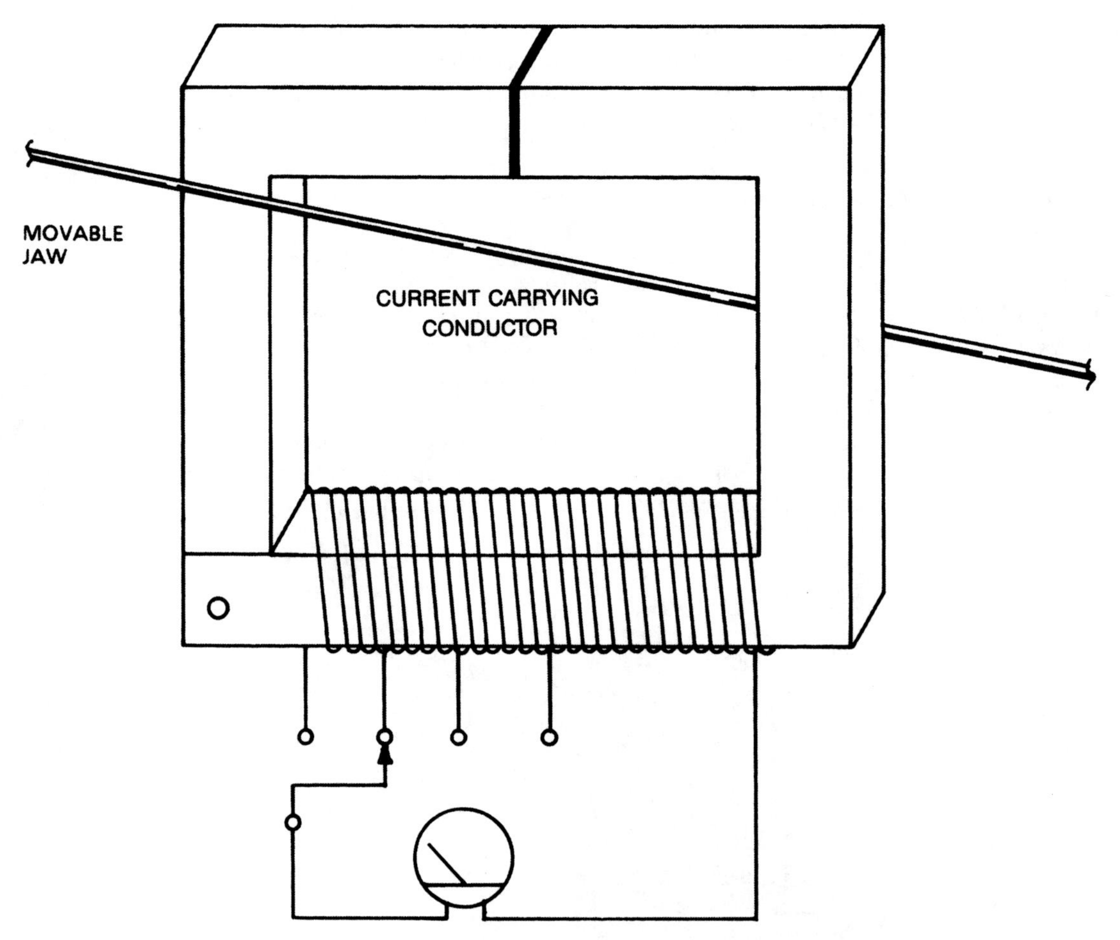

FIGURE 11–39 Clamp-on ammeters operate on the current transformer principle. (*Delmar/Cengage Learning*)

will double. To determine the actual amount of current in the circuit, it would be necessary to divide the reading by 2. If 10 turns of wire are wrapped around the movable jaw, the reading would have to be divided by 10.

The advantage of the clamp-on ammeter is that the circuit does not have to be broken to make measurements. This is a great advantage when it is necessary to check the current draw of a piece of operating equipment.

THE OSCILLOSCOPE

Many of the electronic control systems in today's industry produce voltage pulses that are meaningless to a volt-ohm-milliammeter (VOM). In many instances, it is necessary to know not only the amount of voltage present at a particular point but also the length or duration of the pulse and its frequency. Some pulses may be less than one volt and last for only a millisecond. A VOM would be useless for measuring many of these things. It is therefore necessary to use an oscilloscope to learn what is actually happening in the circuit.

The oscilloscope is a powerful tool and will perform many jobs that will not be discussed in this text. The first thing to understand about an oscilloscope is that it is a voltmeter. It does not measure current, resistance, or watts. The oscilloscope not only measures the voltage, it draws a picture of it. What the oscilloscope actually does is measure a voltage during a particular period of time, creating a two-dimensional image.

Voltage Range Selection

The oscilloscope is divided into two main sections. One section is the voltage section, and the other is the time base. The display of the oscilloscope is divided by vertical and horizontal lines (Figure 11–40). Voltage is measured on the vertical, or Y, axis of the display, and time is measured on the X axis. When using a VOM, a range selection switch is used to determine the full-scale value of the meter. Ranges of 600, 300, 60, and 12 V are common. Having the ability to change ranges permits more accurate measurements to be made. In addition, the oscilloscope has a voltage range selection switch (Figure 11–41). The voltage range-selection switch on an oscilloscope selects volts per division instead of volts full scale. The voltage range switch shown in Figure 11–41 is set for 10 m at the IX position. This means that each of the lines in the vertical direction or on the Y axis of the display has a value of 10 mV. Assume that the oscilloscope has been adjusted to permit 0 V to be

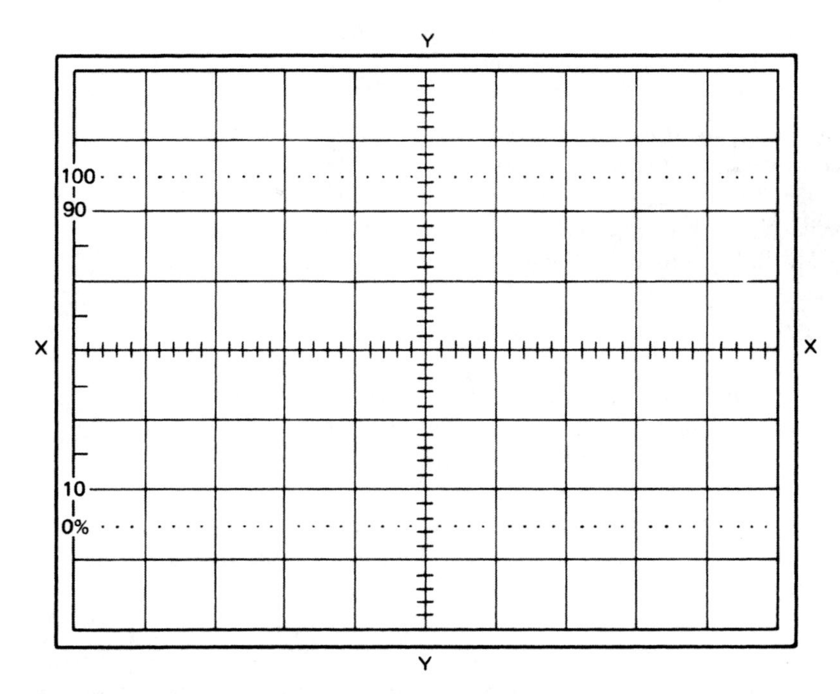

FIGURE 11–40 Oscilloscope display ***(Copyright Tektronix, Inc. All rights reserved. Reprinted with permission.)***

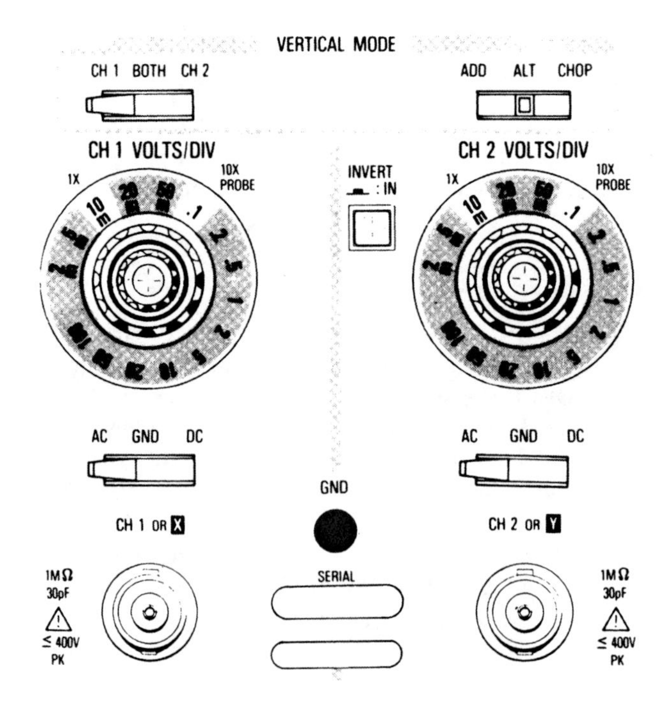

FIGURE 11–41 Voltage control (*Copyright Tektronix, Inc. All rights reserved. Reprinted with permission.*)

shown on the centerline of the display. If the oscilloscope probe is connected to a positive voltage of 30 mV, the trace would rise to the position shown in Figure 11–42A. If the probe is connected to a negative 30 mV, the trace will fall to the position shown in Figure 11–42B. Note that the oscilloscope has the ability to display both positive and negative voltages. If the range switch is changed to 20 V per division, Figure 11–42A would be displaying 60 V positive.

The Time Base

The next section of the oscilloscope to become familiar with is the time base (Figure 11–43). The time base is calibrated in seconds per division and has range values from seconds to microseconds. The time base controls the value of the division of the lines in the horizontal direction. If the time base is set for 5 ms per division, the trace will sweep from one division to the division beside it in 5 ms. With the time base set in this position, it will take 50 ms to sweep from one side of the display screen to the other. If the time base is set for 2 μs per division, the trace will sweep the screen in 20 μs.

Measuring Frequency

Because the oscilloscope has the ability to measure the voltage with respect to time, it is possible to compute the frequency of the waveform. The frequency of an ac waveform can be found by dividing the time it takes to complete one cycle into one (1/f). For example,

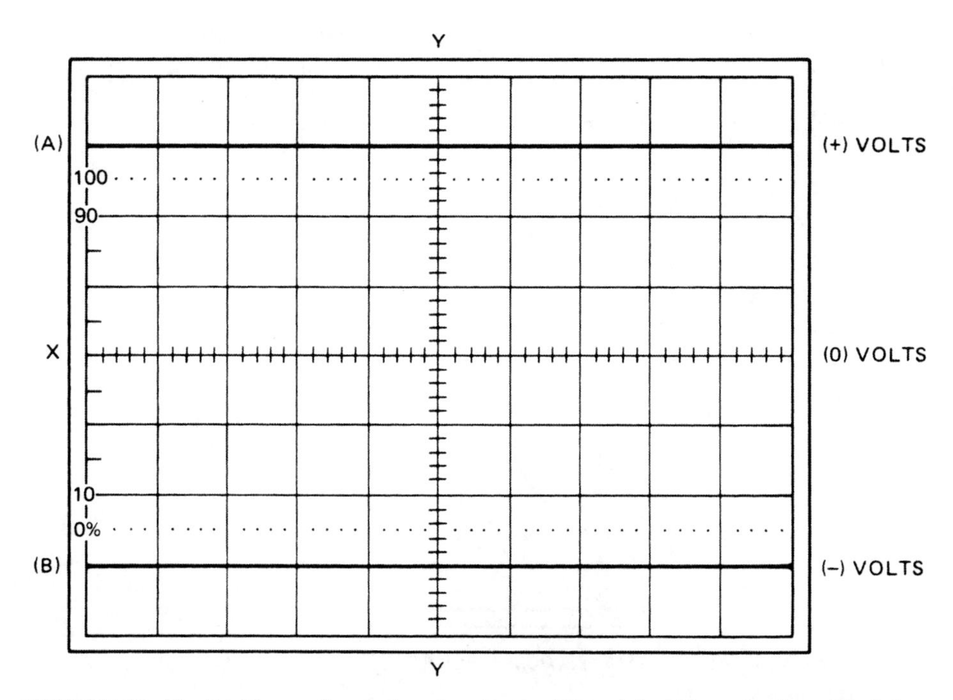

FIGURE 11–42 Positive and negative dc voltages (*Copyright Tektronix, Inc. All rights reserved. Reprinted with permission.*)

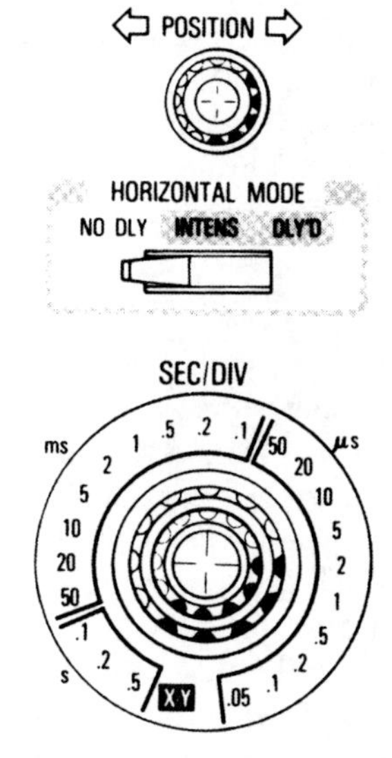

FIGURE 11–43 Time base (*Copyright Tektronix, Inc. All rights reserved. Reprinted with permission.*)

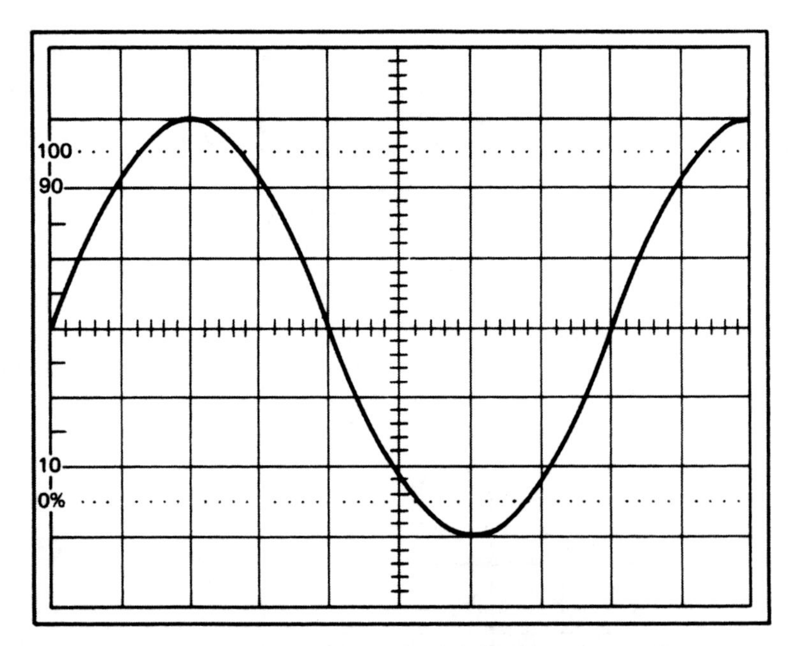

FIGURE 11–44 An ac sine wave (*Copyright Tektronix, Inc. All rights reserved. Reprinted with permission.*)

assume that the time base is set for 0.5 ms per division and the voltage range is set for 20 V per division. If the oscilloscope has been set so that the centerline of the display is 0 V, the ac waveform shown in Figure 11–44 has a peak value of 60 V. The oscilloscope displays the peak or peak-to-peak value of voltage and not the RMS or effective value. To measure the frequency, count the time it takes to complete one full cycle. The waveform shown in Figure 11–44 takes 4 ms to complete one full cycle. The frequency is, therefore, 250 Hz (1/0.004 = 250).

Attenuated Probes

Most oscilloscopes use a probe that acts as an attenuator. An *attenuator* is a device that divides or makes smaller the input signal (Figure 11–45). An attenuated probe is used to permit higher voltage readings than are normally possible. For example, most attenuated probes are 10 to 1. This means if the voltage range switch is set for 5 V per division, the display would actually indicate 50 V per division. If the voltage range switch is set for 2 V per division, each division on the display actually has a value of 20 V per division.

Probe attenuators are made in different styles by different manufacturers. In some probes the attenuator is located in the probe head itself, whereas in others the attenuator is located at the scope input. Regardless of the type of attenuated probe used, it may have to be compensated or adjusted. In fact, probe compensation should be checked frequently. Different manufacturers use different methods for compensating their probes, so it is generally necessary to follow the procedures given in the operator's manual for the oscilloscope being used.

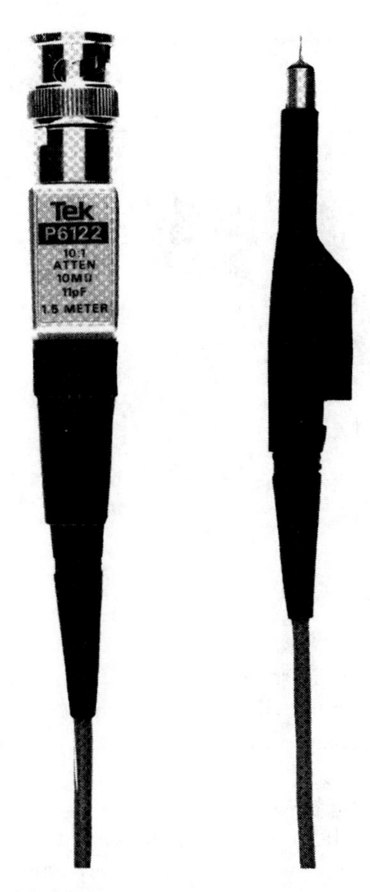

FIGURE 11–45 Oscilloscope probe and attenuator ***(Copyright Tektronix, Inc. All rights reserved. Reprinted with permission.)***

Oscilloscope Controls

The following is a list of common controls found on the oscilloscope. Refer to the oscilloscope shown in Figure 11–46.

1. **Power.** The power switch is used to turn the oscilloscope on or off.
2. **Beam find.** This control is used to locate the position of the trace if it is off the display. The beam finder button will indicate the approximate location of the trace. The position controls are then used to move the trace back on the display.
3. **Probe adjust.** This is a reference voltage point used to compensate the probe. Most probe adjust points produce a square-wave signal of about 0.5 V.
4. **Intensity and focus.** The intensity control adjusts the brightness of the trace. A bright spot should never be left on the display because it will burn a spot on the face of the cathode-ray tube (CRT). This burned spot results in permanent damage to the CRT. The focus control sharpens the image of the trace.

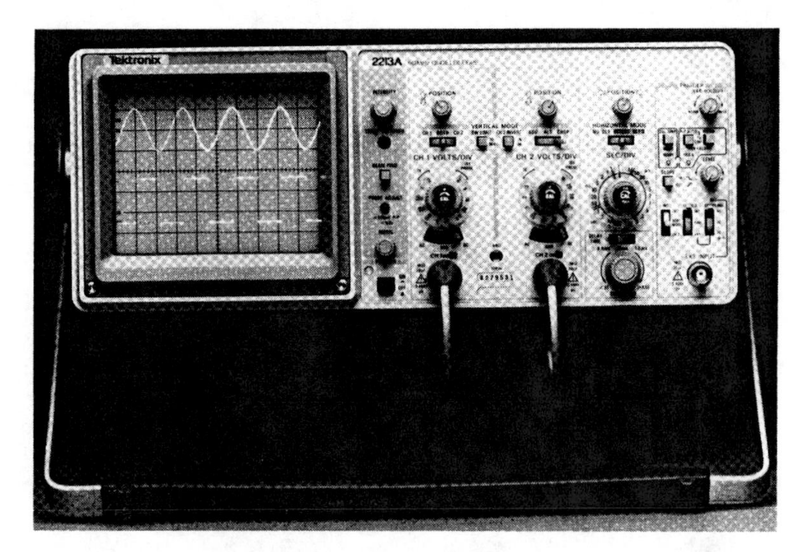

FIGURE 11–46 Tektronix 2213 oscilloscope *(Copyright Tektronix, Inc. All rights reserved. Reprinted with permission.)*

5. **Vertical position.** This control is used to adjust the trace up or down on the display. If a dual-trace oscilloscope is being used, there will be two vertical position controls. (A dual-trace oscilloscope contains two separate traces, which can be used separately or together.)
6. **Ch 1–both–Ch 2.** This control determines which channel of a dual-trace oscilloscope is to be used, or whether they are to both be used at the same time.
7. **Add–Alt–Chop.** This control is active only when both traces are being displayed at the same time. The add adds the two waves together. Alt stands for alternate. This alternates the sweep between channel 1 and channel 2. The chop mode alternates several times during one sweep. This makes the display appear more stable. The chop mode is generally used when displaying two traces at the same time.
8. **AC–Grd–DC.** The ac is used to block any dc voltage when only the ac portion of the voltage is to be seen. For instance, assume an ac voltage of a few millivolts to be riding on a dc voltage of several hundred volts. If the voltage range is set high enough so that 100 V dc can be seen on the display, the ac voltage could not be seen. The ac section of this switch inserts a capacitor in series with the probe. The capacitor blocks the dc voltage and permits the ac voltage to pass. Because the 100 V dc has been blocked, the voltage range can be adjusted for millivolts per division, which will permit the ac signal to be seen.

   The Grd section of the switch stands for ground. This section grounds the input so the sweep can be adjusted for 0 V at any position on the display. The ground switch grounds at the scope and does not ground the probe. This permits the ground switch to be used when the probe is connected to a live circuit. The dc section permits the oscilloscope to display all of the voltage, both ac and dc, connected to the probe.
9. **Horizontal position.** This control adjusts the position of the trace from left to right.

10. **Auto–normal.** This control determines whether the time base will be triggered automatically or whether it is to be operated in a free-running mode. If this control is operated in the normal setting, the trigger signal is taken from the line to which the probe is connected. The scope is generally operated with the trigger set in the automatic position.
11. **Level.** The level control determines the amplitude the signal must reach before the scope triggers.
12. **Slope.** The slope permits selection as to whether the trace is triggered by a negative or positive waveform.
13. **Int–Line–Ext.** The Int stands for internal. The scope is generally operated in this mode. In this setting, the trigger signal is provided by the scope. In the line mode, the trigger signal is provided from a sample of the line. The Ext, or external, mode permits the trigger pulse to be applied from an external source.

These are not all the controls shown on the oscilloscope in Figure 11–46, but they are the major controls. Most oscilloscopes contain these controls.

Interpreting Waveforms

Being able to interpret the waveforms on the display of the oscilloscope takes time and practice. When using the oscilloscope, it must be kept in mind that the display shows the voltage with respect to time.

In Figure 11–47, it is assumed that the voltage range has been set for 0.5 V per division, and the time base is set for 2 ms per division. It is also assumed that 0 V has been set

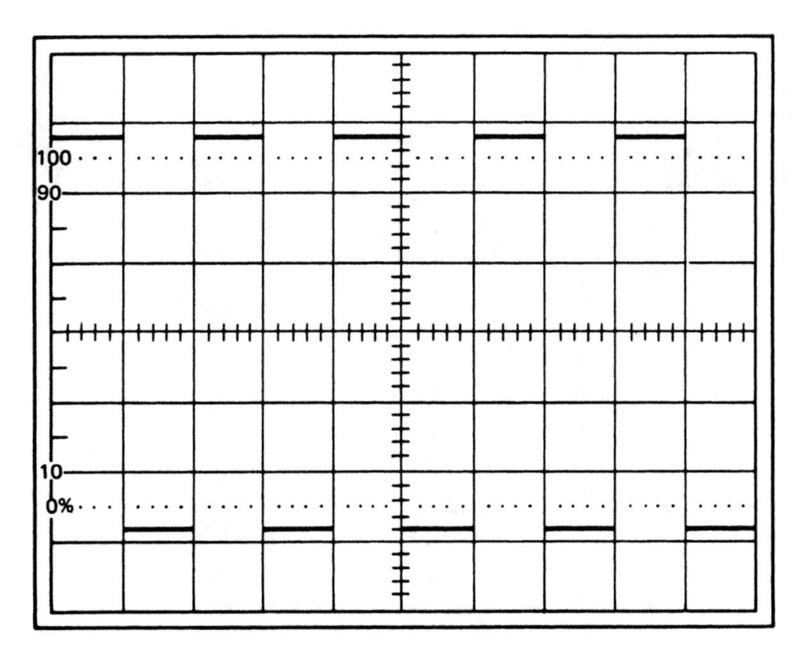

FIGURE 11–47 An ac square wave ***(Copyright Tektronix, Inc. All rights reserved. Reprinted with permission.)***

on the centerline of the display. The waveform shown is a square wave. The display shows the voltage rises in the positive direction to a value of 1.4 V and remains there for 2 ms. The voltage then drops to 1.4 V negative and remains there for 2 ms before going back to positive. Because the voltage changes between positive and negative, it is an ac voltage. The length of one cycle is 4 ms. The frequency is, therefore, 250 Hz (1/0.004 = 250).

In Figure 11–48, the oscilloscope has been set for 50 mV per division and 20 μs per division. The display shows a voltage that is negative to the probe's ground lead and has a peak value of 150 mV. The waveform lasts for 20 μs, which produces a frequency of 50 kHz (1/0.000020 = 50,000). The voltage is dc because it never crosses the zero reference and goes in the positive direction.

In Figure 11–49, assume that the scope is set for 50 V per division and 0.1 ms per division. The waveform shown rises to a value of 150 V in the positive direction and then drops to about 25 V. The voltage remains at 25 V for 0.15 ms and drops back to 0 V. The voltage remains at 0 for 0.3 ms before the cycle starts over again. The voltage shown is dc because it remains in the positive direction. To compute the frequency, measure from the beginning of one wave to the beginning of the next wave. This is the period of one complete cycle. In this case the length of one cycle is 0.6 ms. The frequency is, therefore, 1666 Hz (1/0.0006 = 1666).

Learning to interpret the waveforms seen on the display of an oscilloscope will take time and practice, but it is well worth the effort. The oscilloscope is the only means by which many of the waveforms and voltages found in electronic circuits can be understood.

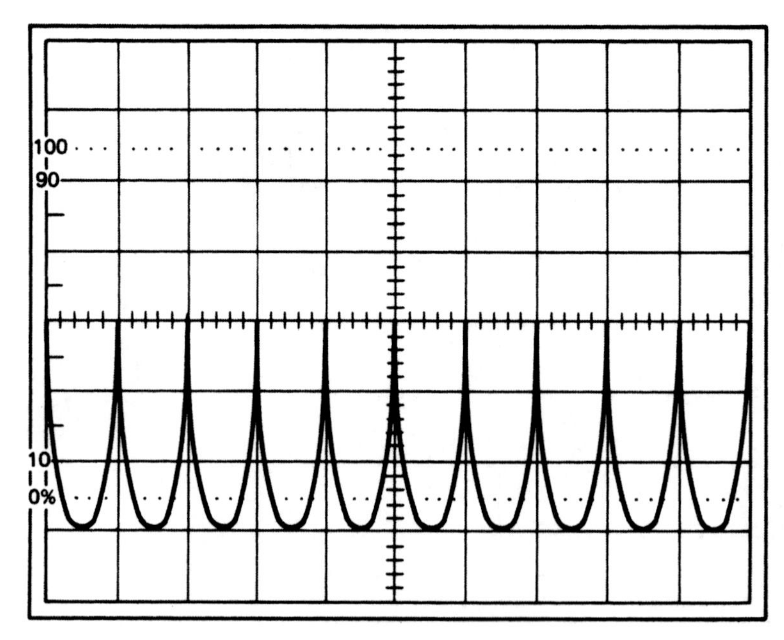

FIGURE 11–48 A dc waveform (***Copyright Tektronix, Inc. All rights reserved. Reprinted with permission.***)

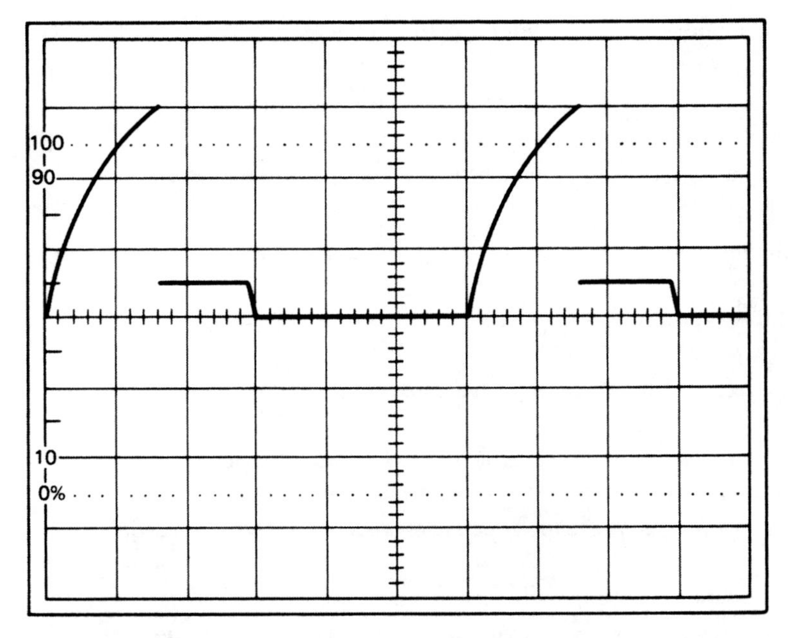

FIGURE 11–49 A dc waveform *(Copyright Tektronix, Inc. All rights reserved. Reprinted with permission.)*

SUMMARY

- Instruments used to make ac measurements must have three basic components:
  1. A spring mechanism to produce an opposing torque
  2. A restoring spring mechanism to restore the pointer to a zero reading
  3. A damping system to prevent excessive swinging of the pointer
- The selection of an instrument for a particular use is usually based on its ability to
  1. indicate RMS or average values, as required.
  2. respond within a certain range of volts or amperes.
  3. provide linear calibration in the useful range of the scale.
  4. indicate values with the desired accuracy.
  5. make measurements at a reasonable cost.
- *DC d'Arsonval movement* with a rectifier:
  1. can be used for ac measurements.
  2. develops a torque proportional to the average value of the current in the moving field; the rectifier makes use of the RMS voltage (1.11 times the average voltage value of a sine wave).

3. is accurate only when pure sine-wave quantities are involved.
4. is used in most multimeters.
   a. A multimeter used as a voltmeter has a high-resistance linear scale and uses relatively little power.
   b. A multimeter used as an ammeter measures in microamperes or milliamperes.

- *Magnetic vane attraction movement* has a soft iron plunger that projects into a stationary field coil. Also:
  1. Current in the field coil produces a magnetic force that pulls the plunger deeper into the coil.
  2. The instantaneous value of this force is proportional to the square of the RMS current value in the coil.
  3. The torque developed in turning the movement is independent of current direction; thus, the instrument can be used for ac or dc measurements.
  4. The scale numbers are crowded at the lower end and expanded for the high end of the scale.
  5. This instrument is commonly used in low-cost ammeters.
  6. This instrument can be used to make voltage measurements by adding a coil of many turns and a series resistor.
- *Inclined coil movement* is used in portable and switchboard ammeters and voltmeters. In addition
  1. The scale is long and reasonably linear.
  2. An iron vane is free to move in a magnetic field.
  3. The vane tends to take a position parallel to the flux.
  4. Increasing the current in the field coil increases the force aligning the vanes with the coil flux; thus, the shaft turns and moves the attached pointer across the scale.
- *Repulsion movement* employs two soft iron vanes; one vane is attached to the instrument shaft and the other is mounted on the stationary field coil. Also:
  1. A repulsion force is developed between the two soft iron vanes because both are affected by the same magnetic field.
  2. With no current through the field coil, a control spring holds the movable vane close to the fixed vane.
  3. As current increases through the field coil, the like poles of the vanes repel each other and create a torque that turns the shaft.
  4. The movement has a reasonably uniform scale and can be used for both current and voltage measurements.
- *Repulsion–attraction movement* produces a greater torque than do most other movements. The following features also apply:
  1. The movable vane is attached to the pointer and is repelled from the wide end of the middle fixed vane as the current through the field coils is increased.

2. The repelling force decreases as the movable vane moves to the narrow end of the middle fixed vane; an attraction force increases as the moving vane comes closer to the upper and lower fixed attraction vanes.
3. A scale length representing 250° of angular deflection can be provided.
4. The distribution of values (volts or amperes) along the scale is determined by the shape and separation of the vanes.
5. Meters can be designed to broaden the scale at any point.

- *Dynamometer movement* produces a torque by the interaction of magnetic fields; this torque varies with the amount of current in the stationary coils. Both the fixed and moving coils are connected in series. Dynamometer movement
  1. can be adapted easily to make voltage measurements by adding the proper series resistor.
  2. is more efficient and accurate than d'Arsonval movement.
  3. is seldom used as an ammeter for the following reasons:
     a. The lead-in spirals to the moving coil can carry only a limited current.
     b. Frequency variations introduce error.
     c. The resistance of the two coils in series may produce an undesirably high voltage drop across the shunt.
  4. is used for instruments measuring power, power factor, and frequency, and to indicate synchronism in ac circuits.

- Practical ranges for moving iron instruments:
  1. When the instrument is used for ammeters, its physical size is influenced by the amount of heat to be dissipated and the size of the connection terminals to be supported.
     a. For small panel instruments, 100 A is the maximum practical current rating.
     b. For large portable instruments, 200 A is the maximum rating.
     c. Some large ammeters for switchboard use are rated as high as 600 A.
     d. Shunts are not satisfactory for moving iron instruments.
     e. It is standard practice to use an instrument current transformer to increase the current range.
  2. When used as ac voltmeters
     a. series resistors are used to extend the scale range for voltages up to 750 V.
     b. instrument potential transformers are used with the movement for voltages greater than 750 V; in many cases, the instrument scale is calibrated to indicate the primary voltage directly.
  3. When used as a wattmeter
     a. the stationary field coils are connected in series with the line so that the flux depends on the current.
     b. the moving coils are connected across the line so that the moving coil flux is proportional to the system voltage.

c. the instantaneous torque is proportional to the instantaneous moving coil voltage.

d. pointer deflection is proportional to the power:

$$\text{Watts} = \text{VI} \cos \theta$$

e. the power curve at any instant is positive.

Also, when using an instrument with a dynamometer movement, the technician must consider the instantaneous direction in each of the coils. There are two different methods of connecting the potential coils of a wattmeter.

a. The potential coil *is not* connected directly across the load:

$$\text{True power} = \text{wattmeter reading} - \text{I}^2\text{R of current coil}$$

b. The potential coil *is* connected across the load voltage:

$$\text{True power} = \text{wattmeter reading} - \frac{\text{V}^2 \text{ of the load}}{\text{R of potential current}}$$

For this second method, the percentage of error is slightly less as compared to the first method.

A wattmeter is always rated according to its potential and current coil ratings rather than in watts.

4. When used as a varmeter,
   a. it indicates the product of the circuit voltage and the current component 90° out of phase with the voltage; that is, it measures the reactive volt-amperes.
   b. the varmeter reads VI cos (θ – 90) or VI sin θ; the wattmeter reads VI cos θ.
5. When used as a power factor meter,
   a. a single-phase power factor meter resembles a single-phase wattmeter.
   b. the stationary field coils are connected in series with one side of the line.
   c. the field coils carry the line current and produce the field flux.
   d. it differs from the wattmeter in that it has no control springs.
   e. the flux of the moving coils reacts with that of the field coil to produce a torque proportional to the in-phase component of current.
   f. the torque magnitude is determined by the amount of quadrature current.
   g. the resultant of the torque due to the in-phase component of current and that due to the quadrature component determines the pointer position and shows the lagging or leading power factor value.

- Synchroscope
  1. The synchroscope shows the relative phase angle and the frequency difference between two alternating voltages to indicate when two alternators are in phase.

2. This instrument also indicates whether the frequency of the incoming generator is higher or lower than that of the generator already connected to the line.
3. Both the polarized-vane type and the moving-iron type of synchroscope are commonly used.
4. These instruments are designed for operation on single-phase circuits. They may be used with three-phase generators if the phase sequences of the generators are known.

- A commonly used frequency meter is known as the *resonant circuit meter.*
  1. The structure of this meter resembles that of the dynamometer.
  2. The two field coils are alike and are connected so that their fluxes oppose each other.
  3. Each field coil is connected in series with an inductor–capacitor combination. Because of the constants of this combination,
     a. series resonance occurs below the normal operating frequency in one field coil and above the normal frequency in the other field coil.
     b. the current in the armature is the vector sum of the currents in the two field coils. The value of lead or lag depends on which field current is greater.
     c. when the frequency is such that the leading and lagging currents are equal, the armature current is in phase with the terminal voltage.
     d. the pointer movement is caused by the torque, due to the resultant flux acting on the iron vane; this torque is proportional to the product of the armature flux and the in-phase component of the resultant field flux.
- Recording instruments
  1. provide a graphical record of the actual circuit conditions at any time.
  2. are grouped into two broad categories:
     a. Instruments that record electrical values such as volts, amperes, watts, power factor, and frequency
     b. Instruments that record nonelectrical quantities such as temperature
  3. use a permanent-magnet, moving-coil-type construction for dc circuits.
  4. for ac circuits, may use either the moving-iron or the dynamometer-type construction.

A strip-chart recorder is the most commonly used graphical recording instrument:

1. A strip of paper 4 to 6 in. wide and up to 60 ft long is used for the permanent record.
2. The long chart means that the record can cover a considerable amount of time.
3. The chart can be operated at a relatively high speed to provide a detailed graphical record.

Some recording instruments have a spring-type clock mechanism wound by a small electric motor. This mechanism guarantees that a power failure will not stop the chart motion until the spring runs down. Some of the larger instruments require a large amount of power for operation. The amount of power required can be reduced by using electronic amplifiers.

- The thermal converter
  1. changes ac voltage and current signals into a dc signal in millivolts; this signal is proportional to the product of VI cos θ.
  2. can be used to measure ac watts using a dc recording or indicating device.

  The thermocouple output (dc millivolt signal) is proportional to ac power in the circuit to which the V and I inputs are attached.

- Watt-hour meter
  1. The watt-hour meter determines and adds together (integrates) all of the instantaneous power values to give an indication of the total energy used over a period of time.
  2. The watt-hour constant of the meter is the number of watt-hours represented by one revolution of the disc.
  3. Each division on the right-hand dial = one kilowatt-hour (kWh).
  4. Second dial from the right = 10 kWh.
  5. The third dial from the right = 100 kWh.
  6. The dial on the left = 1000 kWh.
  7. The register ratio is the number of revolutions made by the first gear wheel for one revolution of the right-hand dial pointer.
  8. The gear ratio is the number of revolutions made by the meter disc to cause one revolution of the right-hand dial pointer.
  9. A full-load adjustment of a meter means that the proper amount of magnetic braking is provided to give the correct speed at the rated voltage and current for a unity power factor.
  10. A light-load adjustment is made by moving a shading pole loop to produce a lag in the time phase of part of the potential flux; this adjustment overcomes errors due to friction and torque at light loads.
  11. Holes are drilled in the disc 180° apart to overcome the slow rotation (creeping) of the disc at light loads. The disc rotates until one of the hole positions open circuits the eddy currents in the disc. The resulting distortion of the eddy currents produces a locking torque to stop the rotation of the disc.

- Digital multimeter
  1. displays reading with individual digits instead of a pointer and scale.
  2. has a high input impedance.
  3. input impedance remains constant on all voltage ranges.

- The clamp-on ammeter:
    1. Most clamp-on ammeters operate on the principle of a current transformer.
    2. The range selection switch changes the ratio of the current transformer.
    3. The circuit does not have to be broken to measure the current flow.
- The oscilloscope
    1. measures voltage and time.
    2. can be used to measure frequency.
    3. displays a two-dimensional image of the voltage waveform.
    4. may use an attenuated probe to permit the measurement of higher voltages.

Achievement Review

1. a. Describe the construction of a magnetic vane, attraction-type instrument movement.

   b. Describe the operation of the magnetic vane, attraction-type instrument movement when used as
      1. an ac ammeter.
      2. an ac voltmeter.
2. Give one advantage and two disadvantages of the magnetic vane, attraction-type instrument movement.
3. Describe the operation of the inclined coil movement, when used as a voltmeter.
4. Explain the operation of a repulsion–attraction instrument movement.
5. A repulsion–attraction instrument movement has a full-scale deflection of 6 mA. The effective resistance of the coil is 3800 Ω. To use this instrument as a voltmeter with a full-scale deflection of 750 Ω, a series resistor is added. This resistor has a resistance of 120,000 Ω. With full-scale deflection, determine the power loss, in watts, in the

   a. instrument coil.

   b. series resistor.

   c. entire instrument.
6. Show the connections for a dc moving coil, permanent-magnet-type movement. Be sure to show the full-wave rectifier and series resistor used to measure the ac voltage.
7. a. If the losses in the rectifier shown in question 6 are negligible, what reading will the dc voltmeter indicate if the maximum value of the ac voltage measured is 340 V?

b. What factor is used to recalibrate the dc instrument scale to read the effective value of ac volts?

8. a. What are the advantages of using rectifier instruments?
   b. What is one limitation of using a rectifier instrument to measure current?
9. Describe the operation of a dynamometer-type movement when used as an ac voltmeter.
10. List several reasons why the dynamometer-type movement is seldom used in ac ammeters.
11. Describe the construction and operation of the dynamometer-type wattmeter.
12. Why is it often more satisfactory to use a two-element wattmeter to measure power in a three-phase, three-wire system than it is to use two single-phase wattmeters?
13. a. The current coil (or the potential coil) of a wattmeter can burn out even though the instrument reading is well below the full-scale deflection. Explain why this can happen.
    b. What precautions are taken when determining whether the current coil or voltage circuits are overloaded?
14. a. Show the connections for a two-element wattmeter used to measure power in a three-phase, three-wire system.
    b. Describe the operation of a two-element wattmeter used to measure three-phase power.
15. How can a dynamometer instrument be used to measure the reactive power in VARs?
16. Describe the operation of
    a. a single-phase power factor meter of the crossed-coil type.
    b. a three-phase power factor meter of the crossed-coil type.
17. Explain the operation of a polarized-vane synchroscope.
18. Explain the operation of a resonant-type frequency meter.
19. What is the value of a recording instrument? Explain.
20. a. List the parts of a typical single-phase watt-hour meter.
    b. Explain the meaning of each of these terms:
       1. Register ratio
       2. Gear ratio
       3. Watt-hour meter disc constant
21. An analog multimeter has a scale factor of 5000 Ω/V. If the meter is set on the 300-V range, how much current will flow through the meter if it is connected to 240 V?
22. What is a common input impedance for a digital multimeter?

23. What is the principle of operation of most clamp-on ammeters?
24. A clamp-on ammeter has five turns of wire wrapped around its movable jaw. If the meter is indicating a current of 15 A, how much current is actually flowing in the circuit?
25. An oscilloscope indicates an ac waveform. If one cycle is completed in 4 μs, what is the frequency of the waveform?

12

Alternating-Current Generators

Objectives

After studying this unit, the student should be able to

- classify alternators by the type of construction used and their principle of operation.
- list the advantages of the revolving field alternator as compared to the revolving armature alternator.
- explain the operation of a field discharge circuit used with the separately excited field in an alternator.
- list the methods of cooling and ventilation used in large alternators.
- apply Fleming's generator rule to a stationary armature conductor and a moving field to find the direction of the induced voltage.
- use formulas to calculate the induced voltage of an alternator at different load power factors.
- define synchronous impedance and synchronous reactance.
- explain the effect of synchronous impedance and synchronous reactance on the voltage regulation of ac generators, for different load power factors.
- demonstrate the standard test procedures used with three-phase wye- and delta-connected alternators to obtain the synchronous impedance, synchronous reactance, and effective resistance.
- interpret a saturation curve for an alternator and define the terms *knee* and *saturation*.
- list the data given on the nameplate of a typical ac generator.
- determine the efficiency of an alternator.
- explain the principles of synchronizing ac generators, including the conditions necessary for synchronization.
- describe two methods of using lamps to determine the phase sequence and synchronizing of alternators.
- connect a synchroscope to show synchronism and explain its operation.
- define load sharing and hunting.
- list the steps necessary to correct the problems of load sharing and hunting.

INTRODUCTION

It was shown in Unit 1 that an alternating voltage having a sine-wave pattern is induced in a single conductor, or armature coil, rotating in a uniform magnetic field with stationary field poles. Similarly, an emf is generated in stationary armature conductors when the field poles rotate past the conductors. A voltage will be induced in the armature conductors whenever there is relative motion between the armature conductors and the field.

DC generators have stationary field poles and rotating armature conductors. The *alternating* voltage induced in the armature conductors is changed to a *direct* voltage at the brushes by means of the commutator.

AC generators are also known as *alternators* because they supply electrical energy with an alternating voltage. These machines do not have commutators. Therefore, the armature is not required to be the rotating member.

Alternators are classified into two groups, depending on the type of construction. One group consists of the revolving armature type of ac generator. This machine has stationary field poles and a revolving armature. The second group consists of the revolving field type of ac generator, which has a stationary armature, or *stator*. The field poles rotate inside the stator for this type of ac generator.

REVOLVING ARMATURE-TYPE ALTERNATOR

Figure 12–1 shows a revolving armature-type alternator. The kilovolt-ampere capacity and the low-voltage rating of such an alternator are usually rather small. This

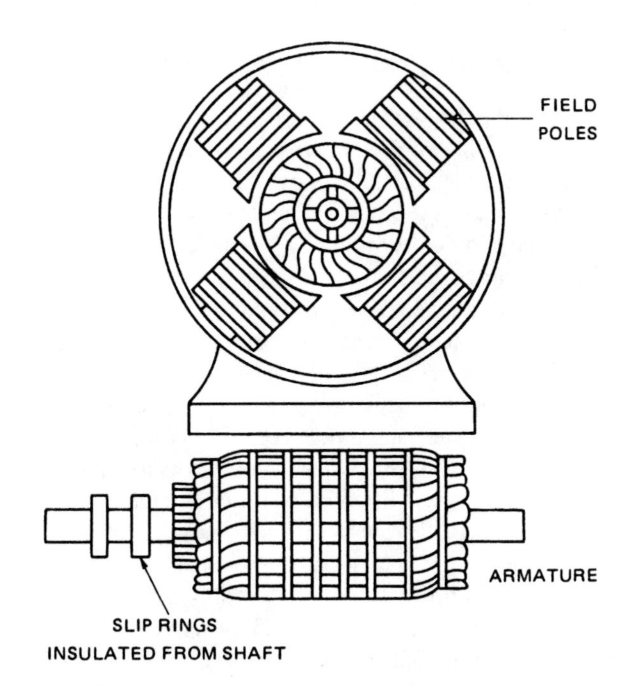

FIGURE 12–1 Revolving armature-type alternator (*Delmar/Cengage Learning*)

machine resembles a dc generator but has slip rings rather than a commutator. An ac generator cannot supply its own field current. Thus, the field excitation is direct current and is supplied from an external direct-current source.

REVOLVING FIELD-TYPE ALTERNATOR

Revolving field alternators are used for most applications. In this type of machine, the revolving field structure (rotor) uses slip rings and brushes to take the excitation current from an external dc source. The stationary part of the generator (stator) is a laminated core. This core consists of thin steel punchings, or laminations, securely clamped together and held in place in the steel frame of the generator. The armature coils are placed in slots in the stator. The field voltage is usually in the range between 100 and 250 V. The amount of power delivered to the field circuit is relatively small.

The rotating field alternator has two advantages over the rotating armature ac generator:

1. Voltages can be generated as high as 11,000 and 13,800 V. These values can be reached because the stationary armature windings do not undergo vibration and centrifugal stresses.
2. Alternators can have relatively high current ratings. Such ratings are possible because the output of the alternator is taken directly from the stator windings through heavy, well-insulated cables to the external circuit. Neither slip rings nor a commutator are used.

The Revolving Field

Two different types of revolving field structures are commonly used. The first type to be described is the *salient pole rotor*. The second type of structure is known as the *cylindrical rotor*. The salient pole rotor has projecting field poles. It is used with alternators operating at speeds below 1800 r/min. Prime movers for such slow-speed alternators include diesel units and waterwheels. (The prime mover supplies the mechanical energy input to the generator.)

Figure 12–2 shows a salient pole rotor, which is used on a slow-speed alternator. Each pole has a laminated steel core to reduce the eddy current losses. The field coil windings are mounted on the laminated poles. The windings are connected in series to give alternate north and south polarities. The field poles are magnetized by low-voltage dc taken from an external source by two slip rings. Each field pole is bolted to the fabricated steel spider. In some cases, poles are dovetailed to the spider. This type of construction is used as a safeguard against the centrifugal force of the rotating members causing the structure to fly apart. The spider is keyed to the generator shaft. The salient pole rotor in Figure 12–2 has slots on each of the pole faces. A damper winding, also known as an *amortisseur winding*, is placed in each of these slots. The purpose of this winding will be described later.

Large steam turbine-driven alternators normally operate at speeds of 1800 r/min and 3600 r/min. At these speeds, large salient pole rotors are impractical. As a result, most steam turbine-driven alternators have cylindrical rotors. If a steam turbine-driven alternator is rated at less than 5000 kW, then frequently this machine is a 1200-r/min salient pole type. A speed reduction gear is used with this machine, as well as 5000- or 6000-r/min turbines. The higher-speed turbine is more efficient than a 3600-r/min turbine. The entire unit costs less and is more efficient than a direct-driven, 3600-r/min motor–generator set using a cylindrical rotor generator.

FIGURE 12–2 Alternator armature of the salient pole type (*Courtesy of Cutler Hammer*)

The field coils of cylindrical rotors are embedded in slots. They are not wound on protruding or salient poles. Slip rings conduct the low-voltage dc excitation current to the revolving field circuit. Generally, the exciter is mounted directly on the generator shaft. A cylindrical rotor is shown in Figure 12–3.

FIGURE 12–3 Slots are cut into a cylindrical rotor (*Courtesy of CenterPoint Energy*)

Field Discharge Circuit

The separately excited field of an alternator can be disconnected from the dc supply by a two-pole switch. As the switch is closed, a momentary voltage is induced in the field windings. This voltage arises because the collapsing lines of flux cut the turns of the field windings. The induced voltage is large enough to damage equipment. To eliminate this voltage, a special field discharge switch is used.

Figures 12–4 and 12–5 show the connections for the field circuit of a separately excited alternator. In the closed position, the field discharge switch acts like a normal double-pole, single-throw switch. When the field discharge switch is in the closed position, an auxiliary switch blade is in an open position.

As the switch is opened, the auxiliary blade closes just before the main switch blades open. When the main switch blades are fully open, a circuit path still exists through the auxiliary switch blade. This path goes through the field discharge resistor and bypasses the field rheostat and the ammeter. As a result, the field discharge resistor is connected directly across the field windings.

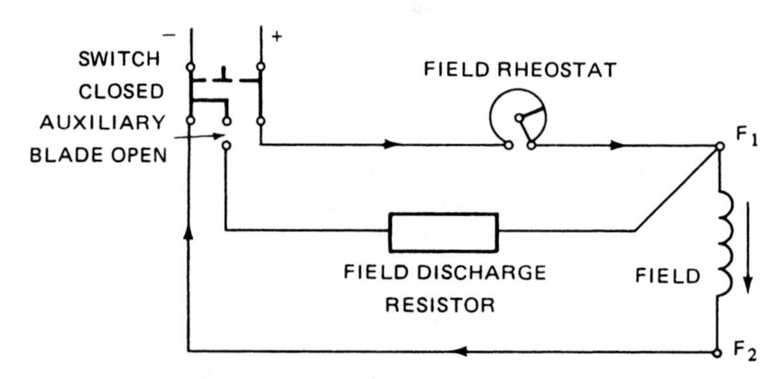

FIGURE 12–4 Field discharge circuit-switch closed (*Delmar/Cengage Learning*)

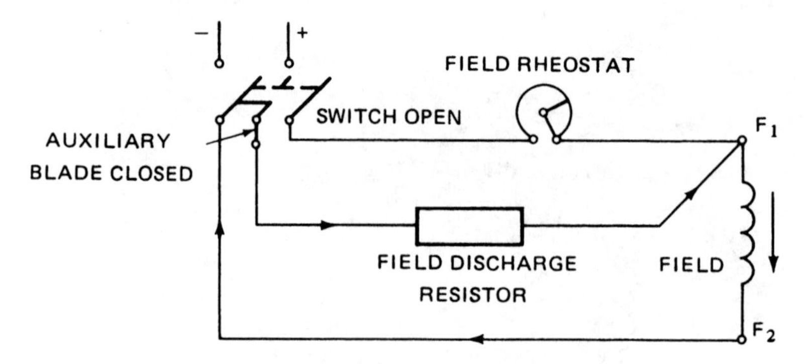

FIGURE 12–5 Field discharge circuit-switch open (*Delmar/Cengage Learning*)

The voltage induced in the field coils by the collapsing magnetic field dissipates quickly as a current through the field discharge resistor. This arrangement eliminates any danger to persons opening the circuit using a two-pole switch. In addition, the insulation of the field windings is protected from damage. All types of alternators use such a field circuit or one that is very similar. A larger machine may use a field contactor or field circuit breaker for the same purpose. Each of these devices will have two normally open main poles and one overlapping normally closed discharge pole.

The Brushless Exciter

Most large alternators use an exciter that contains no brushes. This is accomplished by adding a separate three-phase armature winding to the shaft of the large alternator rotor. The brushless exciter armature rotates between stationary wound electromagnets, as shown in Figure 12–6.

The dc excitation current is connected to the wound stationary magnets. The amount of voltage induced into the brushless exciter armature can be controlled by the amount of dc excitation current applied to the electromagnets. The output of the three-phase armature winding is connected to a three-phase bridge rectifier mounted on the same rotating shaft, as shown in Figure 12–7.

The rectifier converts the three-phase ac produced in the armature winding into direct current before it is applied to the main rotor windings. Because the brushless exciter armature winding, fuses, rectifier, and main rotor windings are all mounted on the same shaft, they rotate together, eliminating the need for brushes or sliprings. A rotor with a brushless exciter winding is shown in Figure 12–8.

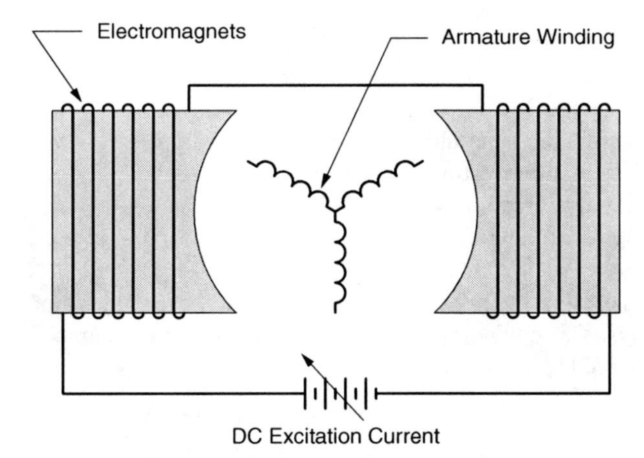

FIGURE 12–6 A brushless exciter is mounted on the main rotor shaft. Stationary electromagnets control the amount of induced voltage. (*Delmar/Cengage Learning*)

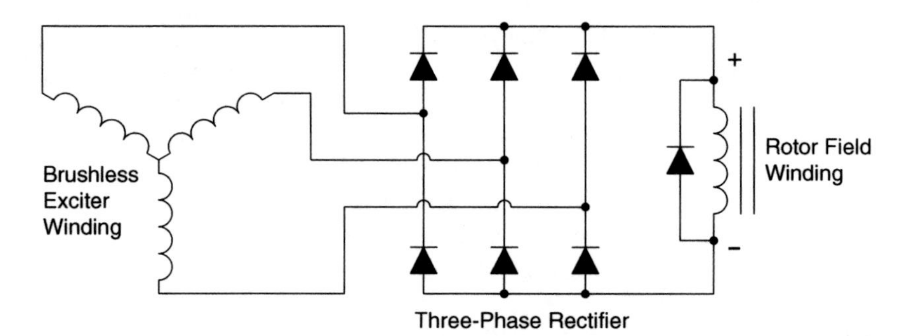

FIGURE 12–7 A three-phase bridge rectifier changes the three-phase voltage produced in the armature into direct current. (*Delmar/Cengage Learning*)

FIGURE 12–8 Rotor with a brushless exciter armature winding (*Courtesy of General Electric Company*)

THE STATOR WINDINGS

It was stated earlier that the dc armature generates alternating current. If properly connected slip rings are used, instead of a commutator, both alternating voltage and current can be obtained. For a revolving armature type of single-phase alternator having a small capacity and low voltage, the armature winding generally used is the same as that of a dc generator.

Large three-phase, revolving field-type ac generators can use one of several types of stationary armature (stator) windings. These windings generally consist of an even number of coils spaced around the perimeter of the stator core.

The Formed Coil

A typical formed coil is shown in Figure 12–9. Such a coil is machine wound and insulated before it is installed in the slots of the stator core. The coil is sized so that it will span the distance between the poles of opposite polarity. A coil having this dimension is said to have *full pitch*. A coil smaller than this dimension is said to have *fractional pitch*. All of the coils of one single-phase winding are collectively known as a *phase belt*.

Figure 12–10 shows the formed coils of a phase belt (single-phase winding) connected in series. The resulting voltage has the maximum value possible. In Figure 12–11, the same coils are reconnected in parallel to give the maximum current possible, at a lower voltage value.

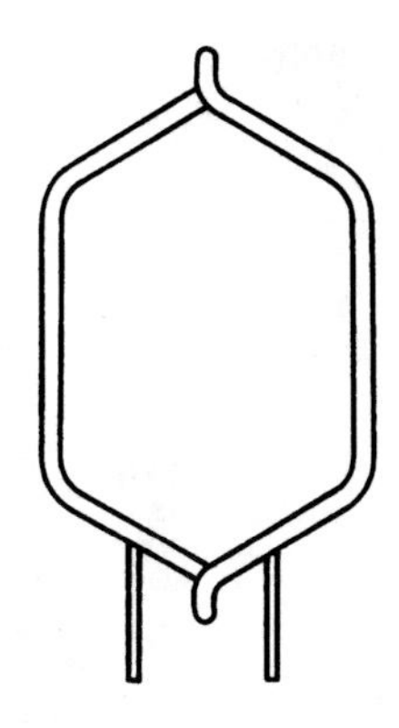

FIGURE 12–9 Formed coil (*Delmar/Cengage Learning*)

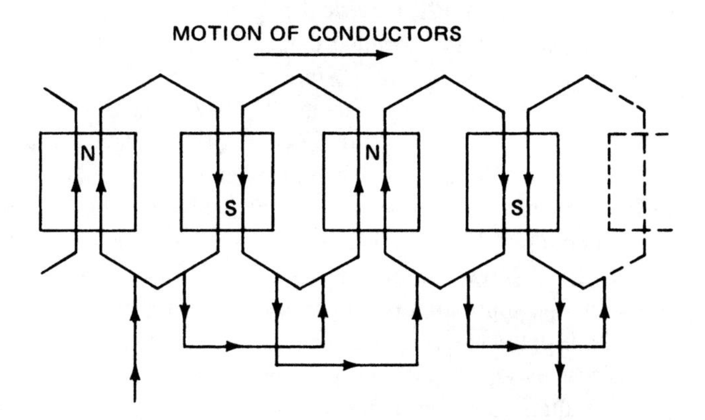

FIGURE 12–10 Formed coils connected for maximum voltage (*Delmar/Cengage Learning*)

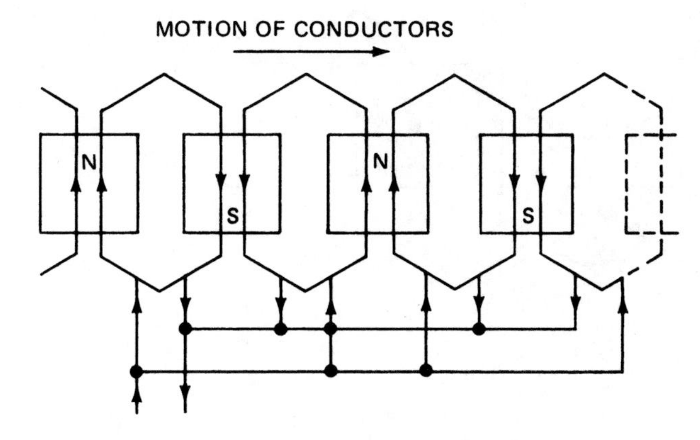

FIGURE 12–11 Formed coils connected for maximum current (*Delmar/Cengage Learning*)

Three-Phase Winding

A three-phase alternator has three separate windings that are placed in the slots of the stator core. The windings are arranged so that three voltages are produced that are 120 electrical degrees apart. Figure 12–12 represents a basic form of a three-phase winding. Three separate single-phase windings are connected in wye. (The three single-phase windings, or phase belts, may be connected in either delta or wye.) The wye connection is used more often than the delta connection because a higher terminal voltage is obtained.

Figure 12–13 shows the stator winding of a three-phase alternator, using a salient pole rotor. Formed coils, similar to the one shown in Figure 12–9, are placed in the slots of the laminated stator core. These coils are connected in three single-phase windings to give three voltages 120 electrical degrees apart. The three phase belts are connected in wye to give a terminal voltage that is 1.732 times the phase winding voltage. The three terminal leads from the stator windings are brought out at the bottom of the stator core. In this location, they can be connected directly to the external three-phase circuit. The flux of the revolving field poles continuously cuts the stator core when the alternator is operating. This action produces induced voltages and the resultant eddy currents in the stator core. These eddy currents can be reduced by the use of a laminated core consisting of thin strips of steel clamped together. The laminated stator is braced with steel channels in the frame. Figure 12–13 shows a typical stator assembly. Note the ventilating ducts in the core of the stator and the ventilating passages in the steel frame. This ventilation prevents the temperature of the stator windings from becoming too high.

The slow-speed alternator shown in Figure 12–13 has a relatively large diameter. Such an alternator is suitable for use with a salient field rotor, similar to the one shown in Figure 12–2.

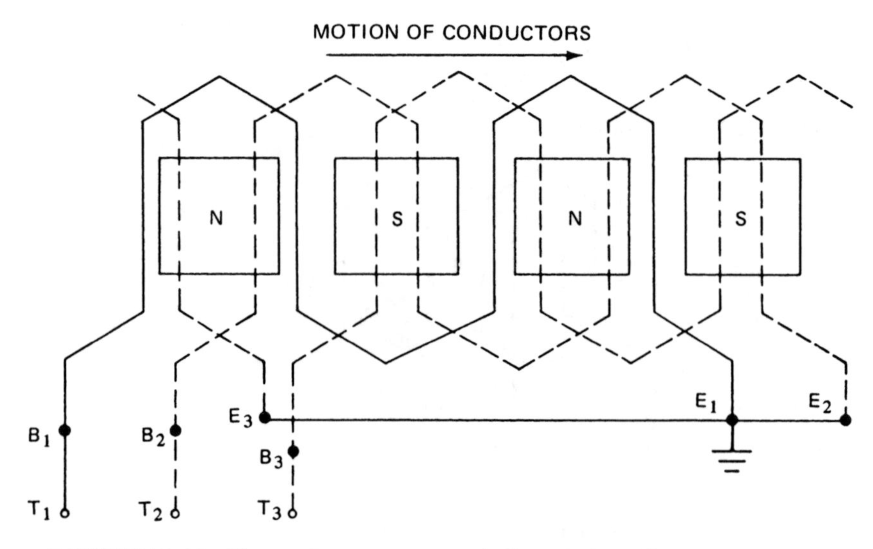

FIGURE 12–12 Three-phase armature winding (*Delmar/Cengage Learning*)

FIGURE 12–13 Stator winding of alternator (*Delmar/Cengage Learning*)

Stator Windings for Three-Phase Alternator

The stator core and windings shown in Figure 12–14 may be used with a turbine-driven, three-phase alternator. The stator in this case is also laminated to reduce eddy current losses. Rows of perforations are provided below the teeth of the slots on the core for the windings. These small holes serve as ventilating ducts to circulate air, or hydrogen, to reduce the heat produced in the stator core and windings.

If there is a short circuit, then the resulting short-circuit currents are great enough to stress the end connections. These current stresses can twist and tear the coil ends from the normal positions. To prevent this, special bracing is used for the coil ends, as shown in Figure 12–14.

The diameter of the stator core and windings for a turbine-driven, three-phase alternator is relatively small. The rotor that is used with the stator of Figure 12–14 is a cylindrical nonsalient type, as shown in Figure 12–1.

VENTILATION

Ventilation is not a problem in the operation of salient pole alternators. The salient poles act as fans to circulate the air. Also, large surface areas in this type of alternator are exposed to the surrounding air. Thus, heat is quickly dissipated.

FIGURE 12–14 Stator core and windings for turbine-driven, three-phase alternator (*Delmar/ Cengage Learning*)

However, high-speed alternators with cylindrical rotors do have ventilation problems. There is almost no fan effect in circulating air with the smooth cylindrical rotor. Only a limited surface area is exposed to the surrounding air in this type of alternator. The stationary armature of this alternator is relatively long, and its diameter is small. To handle the field flux, there must be enough thickness to the laminated stator core in back of the slots. This means that there must be another means of cooling such alternators. The usual method is to completely enclose the alternator with a cooling system using either air or hydrogen. Larger alternators also have hollow conductors through which oil or water can be circulated as a coolant.

Enclosed Cooling System

Most large, high-speed, turbine-driven alternators use a totally enclosed cooling system with hydrogen as the coolant. There are several reasons why hydrogen is preferred. The density of hydrogen is approximately 10% that of air. This means that the windage losses of the high-speed alternator are reduced. Hydrogen has almost seven times the heat conductivity of air. As a result, there is more effective cooling of the stator core and windings. Disadvantages to the use of hydrogen are that it is explosive and costly to replace. To overcome these problems, a gastight sealing system must be used for the alternator frame. The hydrogen is circulated by a blower system through ducts in the alternator. A water-cooling system is used to cool the machine.

OPERATION OF A ROTATING FIELD

The sinusoidal field voltage shown in Figure 12–15 is induced in the conductors of a stationary armature by the flux of two poles of a rotating field structure. The stationary armature is shown on a horizontal plane. The line of travel of the field poles is also shown on a horizontal plane, moving from right to left. The relative motion between the armature and the field remains unchanged if the armature conductors are viewed as moving from left to right, with the field poles stationary. Using Fleming's generator rule, it is assumed that the armature conductors are moving to the right in a stationary field. As a result, it can be determined that the induced voltage in armature conductor A tends to drive electrons away from the reader. At the same time, the induced voltage in conductor B causes electrons to move toward the reader. If the two conductors are connected together to make a closed coil loop, these voltages add constructively. The voltage produced depends on the flux density in gauss (B), the active length of the armature conductors (L), and the velocity (v) of the armature conductors:

$$V_{induced} = \frac{BLv}{10^8}$$

Note that this is the same formula given earlier in Unit 1. The voltage induced in the coil loop will have a sine-wave pattern if the field poles turn at a constant speed and have a sinusoidal flux distribution.

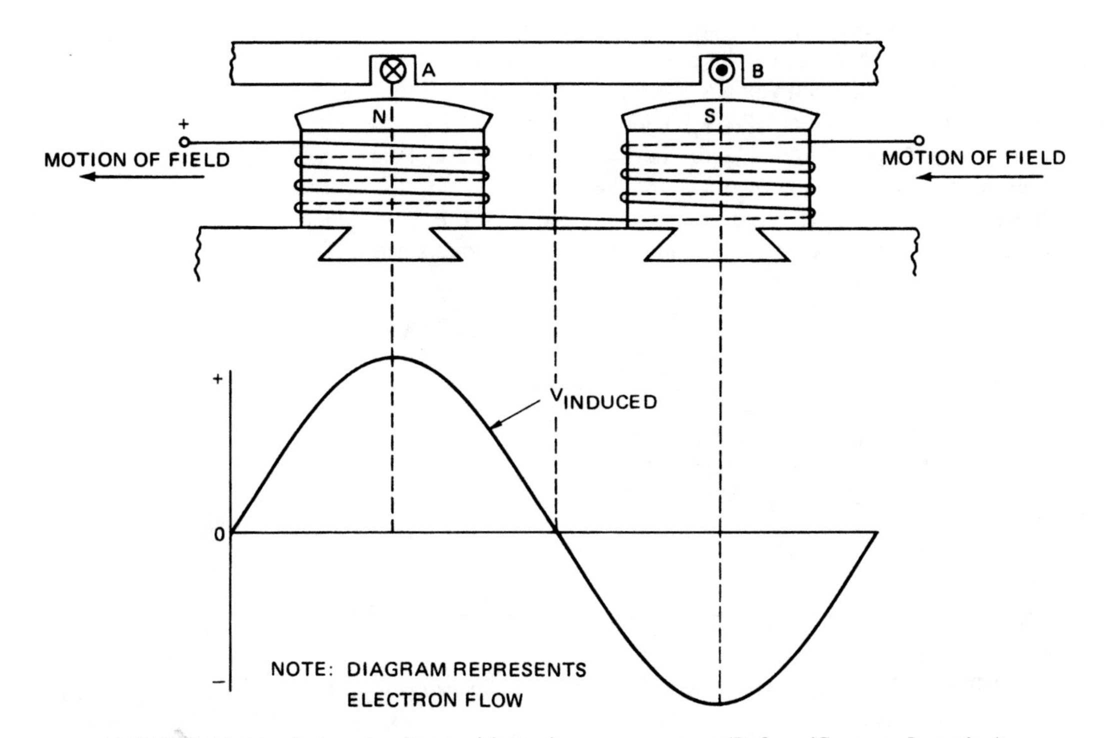

FIGURE 12–15 Induced voltage with stationary armature (*Delmar/Cengage Learning*)

As indicated in Unit 1, the frequency of the induced voltage depends on the number of poles and the speed of the ac generator. When a north pole and a south pole move past a coil, one cycle of voltage is generated. Therefore, the number of cycles of voltage generated in one revolution of the rotating field will be equal to the number of pairs of poles on the rotor. (A pair of poles means one north pole and one south pole.) The frequency of the induced voltage is expressed as follows:

$$f = \frac{P}{2} \times \frac{S}{60} = \frac{PS}{120}$$

where

f = frequency, in hertz
P = total number of poles
$\frac{P}{2}$ = number of pairs of poles
S = speed, in r/min
$\frac{S}{60}$ = speed, in r/s

The frequency table in Unit 1 lists the frequencies obtained for different rotor speeds and numbers of field poles.

Controlling the Field Current

The voltage induced in any alternator depends on the field strength and the speed of the rotor. To maintain a fixed frequency, an alternator must operate at a constant speed. Thus, the magnitude of the generated voltage depends on the dc field excitation. A method of changing or controlling the terminal voltage is shown in Figure 12–16.

A rheostat is connected in series with the separately excited field circuit. The three-phase voltage output of the alternator is controlled by adjusting the resistance of the rheostat. If an alternator is operated at a constant speed with a fixed field excitation current, the terminal voltage will change with an increase in the load current. The actual

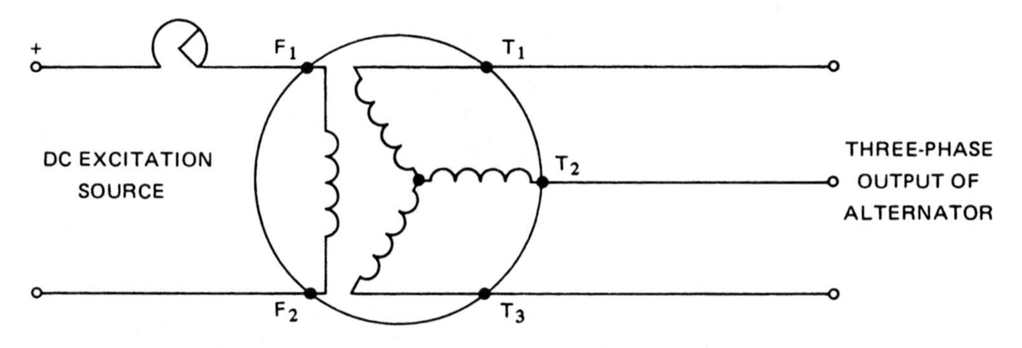

FIGURE 12–16 Three-phase alternator (*Delmar/Cengage Learning*)

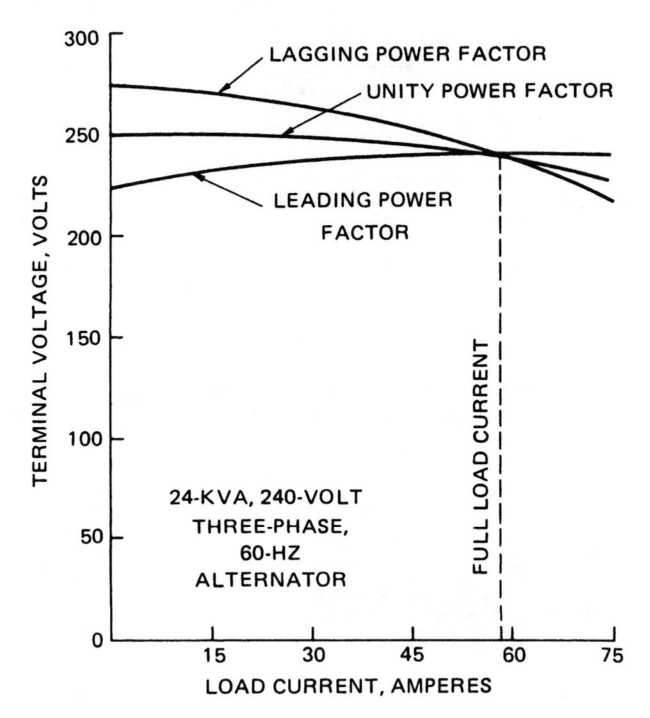

FIGURE 12–17 External load voltage characteristics ***(Delmar/Cengage Learning)***

change in voltage is influenced by the power factor of the load circuit and the impedance of the armature windings.

Effect of the Load Power Factor

In Figure 12–17, the graph shows the effect of the load power factor on the terminal voltage. For each load voltage characteristic curve in Figure 12–17, the alternator field rheostat was adjusted to give the same terminal voltage of 240 V when delivering the full-load current. The term *percent voltage regulation* is applied to the change in the terminal voltage from a full-load to a no-load condition at a constant speed and a fixed field excitation current.

Percent voltage regulation of an alternator is defined as the percentage change in terminal voltage as the load current is decreased from the full-load value to zero at constant values of speed and field excitation:

$$\textbf{Percent voltage regulation} = \frac{\textbf{no-load volts} - \textbf{full-load volts}}{\textbf{full-load volts}} \times 100$$

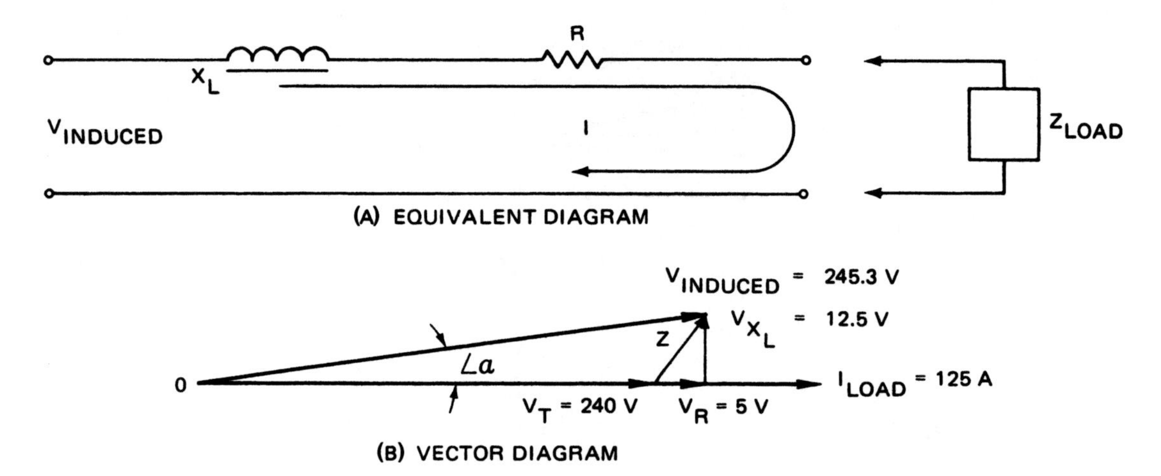

FIGURE 12–18 Alternator having a unity power factor load (*Delmar/Cengage Learning*)

ARMATURE VOLTAGE LOSSES (NEGLECTING ARMATURE REACTION)

All alternator armatures have coils and generate ac voltages. Thus, the equivalent diagrams of such alternators must show a resistor and an inductor (Figure 12–18A).

The equivalent diagram in this case is a series RL circuit. This means that a series circuit analysis can be made to solve for the required quantities.

The vector equation is

$$\mathbf{V_{induced}} = \overrightarrow{\mathbf{IX_L}} + \overrightarrow{\mathbf{IR}} + \overrightarrow{\mathbf{V_T}}$$

Unity Power Factor

For resistive loads, V_T and IR are in phase with the reference current I. IX_L is 90° out of phase with I. The resulting vector diagram is shown in Figure 12–18B. The effects that resistive loads have on the terminal alternator voltage are illustrated in the following example.

A single-phase, 60-Hz alternator has a full-load rating of 30 kVA at 240 V. The armature windings have an effective resistance of 0.04 Ω and an inductive reactance of 0.1 Ω. What is the induced voltage in the alternator windings when the full-load current is delivered at a load power factor of unity?

The current output of the alternator at rated load is

$$I = \frac{VA}{V} = \frac{30{,}000}{240} = 125 \text{ A}$$

The effective resistance and the inductive reactance cause voltage losses in the armature windings. At full load, these losses are

$$V_R = I \times R = 125 \times 0.04 = 5 \text{ V}$$
$$V_{X_L} = I \times X_L = 125 \times 0.1 = 12.5 \text{ V}$$

The vector diagram shows that the induced voltage is the hypotenuse of a right triangle. Thus

$$\begin{aligned} V_{induced} &= \sqrt{(V_T + IR)^2 + (IX_L)^2} \\ &= \sqrt{(240 + 5)^2 + 12.5^2} \\ &= \sqrt{60,180} \\ &= 245.3 \text{ V} \end{aligned}$$

The voltage at the terminals of the generator is 240 V and is in phase with the load current. At the same time, the induced voltage in the armature is 245.3 V. The load current lags the induced voltage by the angle α.

Lagging Power Factor

Figure 12–19 is a vector diagram for the same 30-kVA alternator described in the previous example. In this case, however, the power factor load is lagging. The alternator delivers the rated load current of 125 A at a terminal voltage of 240 V to a load with a 0.8 lagging power factor. The load current lags the terminal voltage by the angle θ. In this example, θ equals 37°. In the diagram, the 5-V IR drop in the armature windings is placed at the end of the terminal voltage. Its position is such that it is in phase with the load current. A voltage loss also occurs because of the inductive reactance of the armature windings. This loss leads the load current by 90°. The resultant of IR and IXL is the armature impedance voltage drop IZ. This value is added vectorially to the terminal voltage to obtain the induced voltage in the armature. Refer to the vector

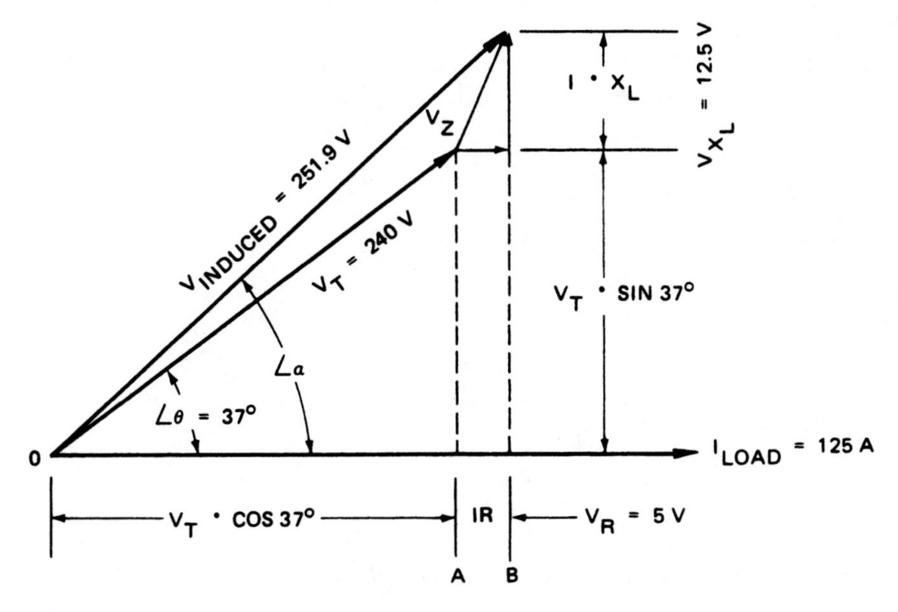

FIGURE 12–19 Alternator with lagging power factor load (*Delmar/Cengage Learning*)

diagram and note that the small armature impedance triangle shifts in a clockwise direction when the current lags the terminal voltage. This shift means that the induced voltage vector has a greater magnitude than it does for the unity power factor load in the previous example.

The induced voltage for a 0.8 lagging power factor (Figure 12–19) can be obtained using the formula

$$\begin{aligned} V_{induced} &= \sqrt{(\cos\angle\theta \cdot V_T + IR)^2 + (\sin\angle\theta \cdot V_T + IX_L)^2} \\ &= \sqrt{(0.8 \times 240 + 5)^2 + (0.6018 \times 240 + 12.5)^2} \\ &= 251.9\ \text{V} \end{aligned}$$

With the lagging power factor, the induced voltage is greater than in the case when the load power factor was unity. The induced voltage increased even though the impedance voltage in the armature is the same. If the induced voltage remains the same, then the terminal voltage decreases with an increase in the angle of lag of the load current behind the terminal voltage. The vector diagram in Figure 12–19 shows that this change in voltage output is due to the angle at which the armature impedance voltage drop subtracts vectorially from the induced voltage.

If the load current leads the terminal voltage, then the induced voltage in the armature is less than the terminal voltage.

Leading Power Factor

The vector diagram in Figure 12–20 for the same 30-kVA single-phase alternator is described in the two previous examples. The load current, however, leads the terminal voltage by a phase angle of 37°. The phase relationship between the impedance voltage drop in the armature and the induced voltage causes the terminal voltage to be greater than the induced voltage. As the angle of lead between the load current and the terminal voltage increases, the armature impedance voltage vector moves counterclockwise. As a result, the

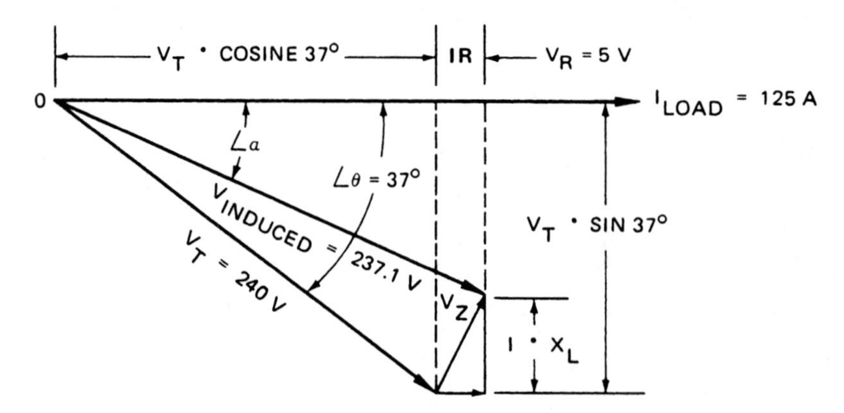

FIGURE 12–20 Alternator with leading power factor load (*Delmar/Cengage Learning*)

magnitude of the induced voltage vector is less for the same terminal voltage. To maintain a terminal voltage of 240 V when the alternator delivers the full-load current of 125 A at a 0.8 leading power factor, the required induced voltage is

$$V_{induced} = \sqrt{(\cos \angle\theta \cdot V_T + IR)^2 + (\sin \angle\theta \cdot V_T - IX_L)^2}$$

$$= \sqrt{(0.8 \times 240 + 5)^2 + (144.43 - 12.5)^2}$$

$$= 237.1 \text{ V}$$

The effects of the IR and IX_L voltage losses in the armature windings on the terminal voltage output were illustrated in Figures 12–18, 12–19, and 12–20. The last example described the case for a leading power factor. However, most alternators are used to supply a load having a lagging power factor.

Effect of Lagging Power Factor

When an alternator operates with a lagging power factor, a magnetomotive force (mmf) is set up by the current in the armature conductors. This magnetizing force opposes the magnetomotive force of the main field and causes a decrease in the main field flux. As a result, there is a decrease in the induced voltage. The lower the value of the lagging power factor, the greater is the armature mmf that opposes and weakens the field. For the few applications where an alternator supplies a load with a leading power factor, the armature current sets up an mmf in the armature. This force aids the mmf of the main field so that the main field flux increases. The voltage of the alternator increases with an increase in the load current. The lower the value of the leading power factor, the greater is the armature mmf. This force aids the mmf of the main field, resulting in an increase in the main field flux.

Effect of Load Power Factor

Figure 12–21 shows how the main field flux is affected by the armature mmf for different load power factors. A two-pole rotating armature-type alternator is used to simplify the illustrations. However, the same conditions occur when a revolving field-type alternator is used.

By studying Figure 12–21, it can be seen that the voltage drop due to the inductive reactance and the armature reaction have the same effect on the terminal voltage. Both of these effects are proportional to the armature current. Generally, these two effects are combined into a single quantity, called the *synchronous reactance* (X_{LS}).

Slot-Type Armature Windings. The stationary armature windings of an alternator may be arranged in the slot formation shown in Figure 12–22. The armature conductors are surrounded by the laminated iron of the stator core. The eddy current losses and the hysteresis losses in the iron core structure mean that there is a loss in power. The current in the armature windings must supply the power expended in the core in overcoming the molecular friction loss (hysteresis loss) and the eddy current losses. The effective resistance will be higher than the pure ohmic resistance. Alternators rated at 25 Hz usually have an effective resistance equal to 1.2 to 1.3 times the pure ohmic resistance. Sixty-hertz generators

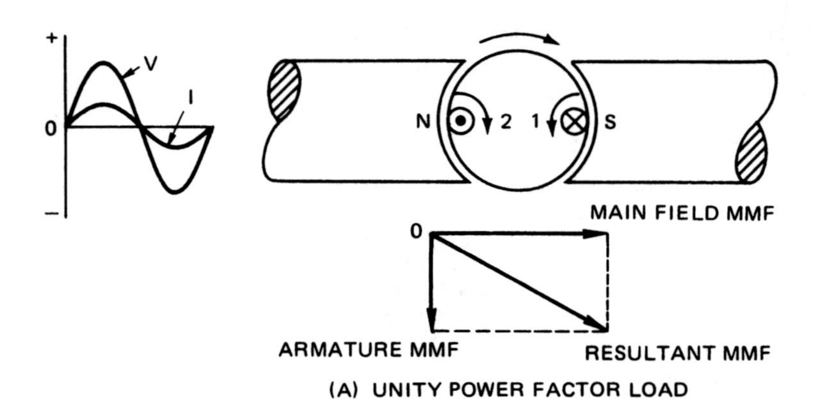

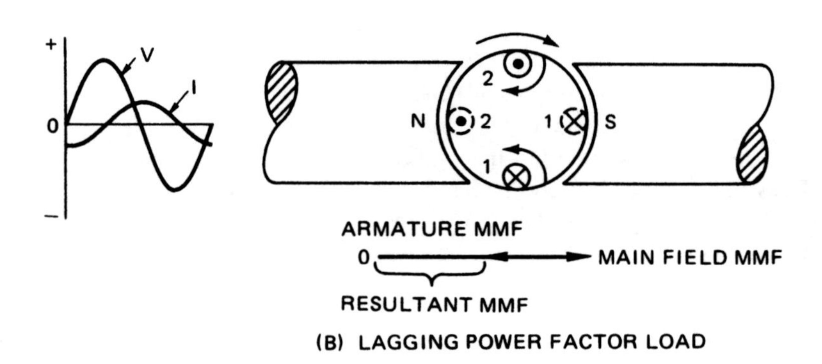

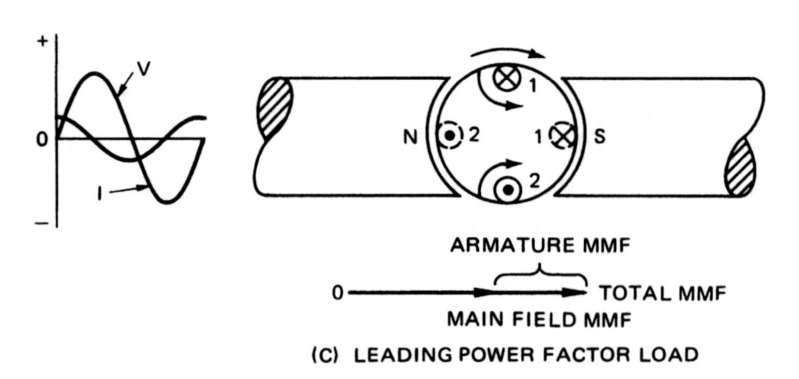

FIGURE 12–21 Effects of armature reaction (*Delmar/Cengage Learning*)

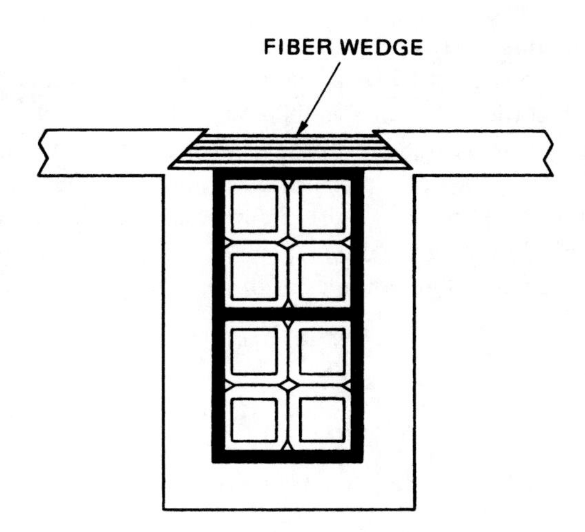

FIGURE 12–22 Typical slot arrangement for stator windings (*Delmar/Cengage Learning*)

have an effective resistance equal to 1.4 to 1.6 times the pure ohmic resistance. For 60-Hz alternators, the dc resistance can be measured and multiplied by 1.5 to obtain the effective resistance.

SYNCHRONOUS IMPEDANCE (WITH ARMATURE REACTION)

Values of R and X_L are usually not given for alternators. A method known as the *synchronous impedance test* can be employed to determine these values. A single-phase alternator is shown in Figure 12–23. To perform this test, the output terminals of the alternator will be shorted by switch S. An ammeter is connected in the alternator circuit to indicate current and a voltmeter is connected across the output terminals to measure voltage.

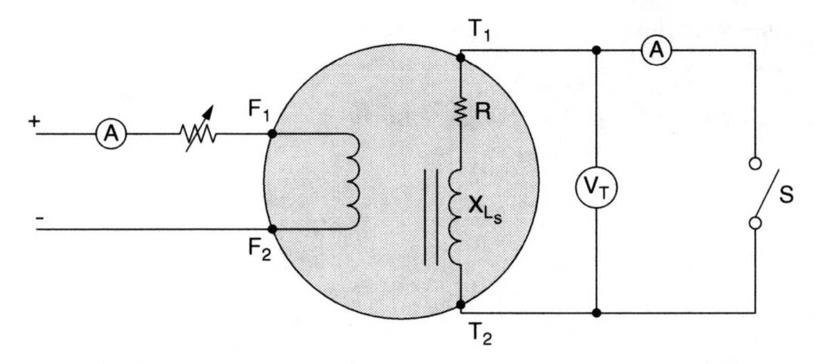

FIGURE 12–23 Connection for a synchronous impedance test for a single-phase alternator (*Delmar/Cengage Learning*)

For discussion purposes, assume that switch S is closed. The field excitation current will be increased until the output current is 150% of the rated full-load current of the alternator. This value of current is recorded. Switch S is then opened and the voltage across the output terminals is recorded. The field excitation current and alternator speed are kept constant. To determine the synchronous impedance of the alternator, the voltage for an open-circuit condition is divided by the current for a short-circuit condition. When the winding is shorted, the voltage is expended in the armature or stator winding. The voltage maintains the current through the impedance of the windings at the operating frequency. The synchronous impedance formula is

$$Z_S = \frac{V_{\text{open circuit}}}{V_{\text{short current}}}$$

The synchronous reactance can be found once the effective resistance is known. The formula for the synchronous reactance is

$$X_{L_S} = \sqrt{Z_S^2 - R_{\text{eff}}^2}$$

PROBLEM 1

Statement of the Problem

The connection in Figure 12–23 is used to determine the synchronous impedance of a 24-kVA, 240-V, single-phase 60-Hz alternator. The field excitation is adjusted until the ammeter indicates 150% of the rated full-load current. At the same values of field excitation current and speed, switch S is opened. The voltmeter indicates 105 V. The ohmic resistance of the armature or stator winding is measured at 0.1 Ω. It will also be necessary to assume an effective resistance value for the alternator. Recall that the effective resistance value in an ac circuit can be greater than the measured ohmic value. This is due to factors such as skin effect, eddy current induction into the iron core of the stator winding, and hysteresis losses. Because the windings are inserted deep in slots formed in the core material, the effective resistance could be high. For this example, it will be assumed that the effective resistance will be 1.5 times greater than the measure resistance value.

At unity power factor, determine

1. the induced voltage when the alternator is delivering the rated current.
2. the percent voltage regulation.

Solution

1. The full load current is

$$I_{\text{full-load}} = \frac{VA}{V} \qquad I_{\text{full-load}} = \frac{2400}{240} \qquad I_{\text{full-load}} = 100\text{A}$$

Current is 150% of the full-load current.

$I_{short\text{-}circuit} = 100 \times 1.50 \qquad I_{short\text{-}circuit} = 150\ A$

Effective resistance is

$$R_{eff} = R_{ohmic} \times 1.5 \qquad R_{eff} = 0.1 \times 1.5 \qquad R_{eff} = 0.15\ \Omega$$

The impedance is

$$Z_S = \frac{V_{open\ circuit}}{I_{short\ current}} \qquad Z_S = \frac{105}{105} \qquad Z_S = 0.7\ \Omega$$

Synchronous reactance is

$$X_{L_S} = \sqrt{Z_S^2 - R_{eff}^2} \qquad X_{L_S} = \sqrt{0.7^2 - 0.15^2} \qquad X_{L_S} = 0.684\ \Omega$$

The voltage drop due to effective resistance is

$$V_R = IR \qquad V_R = 100A \times 0.15\ \Omega \qquad V_R = 15V$$

The voltage drop due to inductive reactance is

$$V_{X_L} = IX_L \qquad V_{X_L} = 100 \times 0.684 \qquad V_{X_L} = 68.4V$$

The induced voltage is

$$V_{induced} = \sqrt{(V_T + V_R)^2 + (V_{XL})^2} \qquad V_{induced} = \sqrt{(240 + 15)^2 + (68.4)^2}$$

$$V_{induced} = 264V$$

2. Percent voltage regulation is

$$\text{Percent Voltage Regulation} = \frac{Voltage_{no\ load} - Voltage_{full-load}}{Voltage_{full-load}} \times 100$$

$$\frac{264 - 240}{240} = \frac{24}{240} \times 100 = 10\%$$

Synchronous Impedance of a Wye-Connected Alternator

The voltage regulation of a three-phase alternator can also be determined using the synchronous impedance method. Figure 12–24 shows the circuit used to determine the synchronous impedance of a wye-connected ac generator. By closing the three-pole shorting switch (S), the field excitation current is increased gradually until the ammeters indicate 150% of the rated output current. The three ammeter readings are recorded. The three-pole switch is then opened and the voltmeter reading is recorded. The field excitation current and speed are kept constant for both readings. Because the alternator is connected in wye, the voltage of each phase winding is equal to the open-circuit voltage across the line terminals divided by the $\sqrt{3}$.

The ammeters connected in the lines indicate the phase current values because line current and phase current are the same in a wye connection. The formula for determining the synchronous impedance of each phase winding is

$$Z_S = \frac{V_S/\sqrt{3}}{I_S} \qquad = \qquad Z_S = \frac{V_S}{\sqrt{3}\ I_S}$$

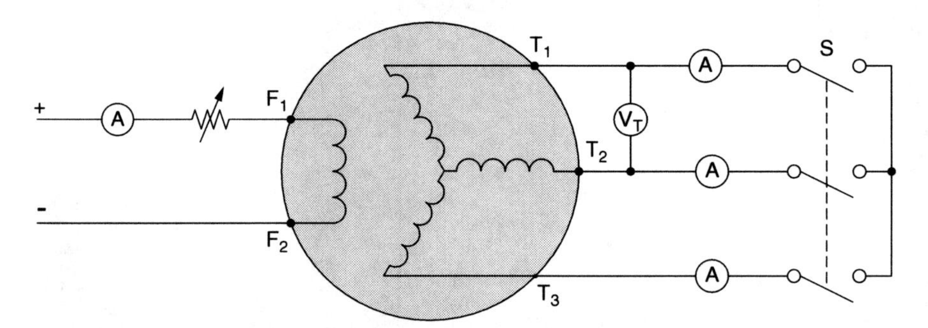

FIGURE 12–24 Connection for a synchronous impedance test for a three-phase alternator *(Delmar/Cengage Learning)*

The following problem shows how to determine the voltage regulation of a wye-connected alternator.

PROBLEM 2

Statement of the Problem

A 2100-kVA, 2400-V, three-phase, wye-connected alternator is short-circuited using three ammeters, as shown in Figure 12–24. A voltmeter is connected across two of the line terminals. The field excitation current is increased until each of the three ammeters indicates 150% of the rated full-load current. The switch is opened and the voltmeter indicates an open circuit voltage of 1125 V. The field excitation current and speed are kept constant for both readings. The ohmic resistance is measured at 0.2 Ω between the line terminals. The ratio of effective resistance to the ohmic resistance is 1.5. Determine

1. the full-load current.
2. the current value for the test.
3. the synchronous impedance.
4. the effective resistance.
5. the synchronous reactance.
6. the voltage regulation at unity power factor.
7. the voltage regulation at 0.866 lagging power factor.
8. the voltage regulation at 0.866 leading power factor.

Solution

1. Full-load current:

$$I_{full\text{-}load} = \frac{VA}{V\sqrt{3}} \qquad I_{full\text{-}load} = \frac{2,100,000}{2400 \times 1.73} \qquad I_{full\text{-}load} = 505.8A$$

2. Test current:

$505.8 \times 1.5 = 758.7A$

3. Synchronous impedance:

$$Z_S = \frac{V_S}{\sqrt{3}\, I_S} \qquad Z_S = \frac{1125}{1.73 \times 758.7} \qquad Z_S = 0.857\ \Omega$$

4. Effective resistance – the measured value of 0.2 Ω is the resistance of two separate phase windings connected in series. Therefore, the resistance of one phase winding is one-half the measured value.

$$R_{eff} = 0.1 \times 1.5 = 0.15\ \Omega$$

5. Synchronous reactance:

$$Z_{X_L} = \sqrt{Z_S^{\,2} - R_{eff}^{\,2}} \qquad Z_{X_L} = \sqrt{0.857^2 - 0.15^2} \qquad Z_{X_L} = 0.844\ \Omega$$

6. Voltage regulation at unity power factor:

Rated phase voltage:

$$V_{phase} = \frac{V_{line}}{\sqrt{3}} \qquad V_{phase} = \frac{2400}{1.73} \qquad V_{phase} = 1387\ V$$

Voltage drop due to effective resistance and synchronous reactance:

$$V_R = I\,R_{eff} \qquad V_R = 505.8 \times 0.15 \qquad V_R = 75.9\ V$$

$$V_{X_L} = I\,X_L \qquad V_{X_L} = 505.8 \times 0.844 \qquad V_{X_L} = 427\ V$$

The vector diagram of Figure 12–25 includes the effects of the armature reactance and the armature reaction for an alternator with a unity power factor load. Note the similarity of this vector diagram with the one given in Figure 12–18. Only the armature impedance, reactance, and the effective resistance were considered.

$$V_{induced} = \sqrt{(V_{phase} + V_R)^2 + (V_{XL})^2}$$

$$V_{induced} = \sqrt{(1387 + 75.9)^2 + (427)^2}$$

$$V_{induced} = 1524\ V$$

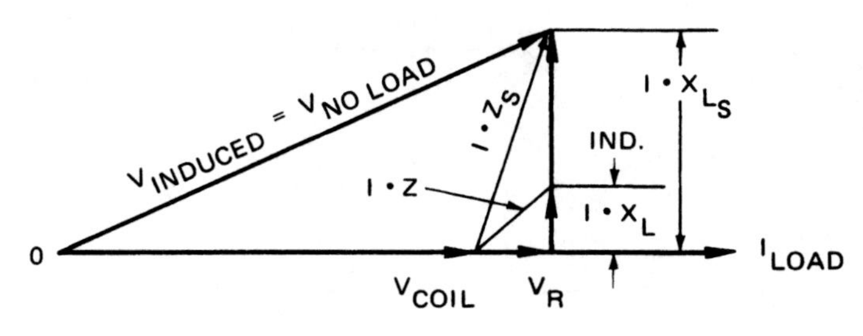

FIGURE 12–25 Synchronous impedance vector diagram (unity power factor load) (*Delmar/Cengage Learning*)

7. Voltage regulation at 0.866 lagging power factor:

$$\text{Percent Voltage Regulation} = \frac{\text{Voltage}_{\text{no load}} - \text{Voltage}_{\text{full-load}}}{\text{Voltage}_{\text{full-load}}} \times 100$$

$$\frac{1524 - 1387}{1387} = \frac{137}{1387} \times 100 = 9.9\%$$

Regardless of the power factor, the full-load current is the same. Therefore, for any power factor load conditions, the voltage loss due to the effective resistance and inductive reactance of the armature or stator winding will be the same. However, the calculated value of the voltage at no load will be different from the value for a unity power factor load. Figure 12–26 and the following calculations show why the voltage at no load is much higher than the voltage at full load with a lagging power factor. This situation means that there is poorer voltage regulation.

To determine the induced voltage at a power factor of 86.6%, or 0.866, it is necessary to determine angle theta. Power factor is the cosine of angle theta. Therefore, angle theta is 30°.

$$V_{\text{induced}} = \sqrt{(\cos \angle \theta \bullet V_{\text{phase}} + V_R)^2 + (\sin \angle \theta \bullet V_{\text{phase}} + V_{XL})^2}$$

$$V_{\text{induced}} = \sqrt{(0.866 \bullet 1387 + 75.9)^2 + (0.5 \bullet 1387 + 427)^2}$$

$$V_{\text{induced}} = 1699 \text{ V}$$

8. Voltage regulation at a 0.866 leading power factor:

$$\text{Percent Voltage Regulation} = \frac{\text{Voltage}_{\text{no load}} - \text{Voltage}_{\text{full-load}}}{\text{Voltage}_{\text{full-load}}} \times 100$$

$$\frac{1699 - 1387}{1387} = \frac{312}{1387} \times 100 = 22.5\%$$

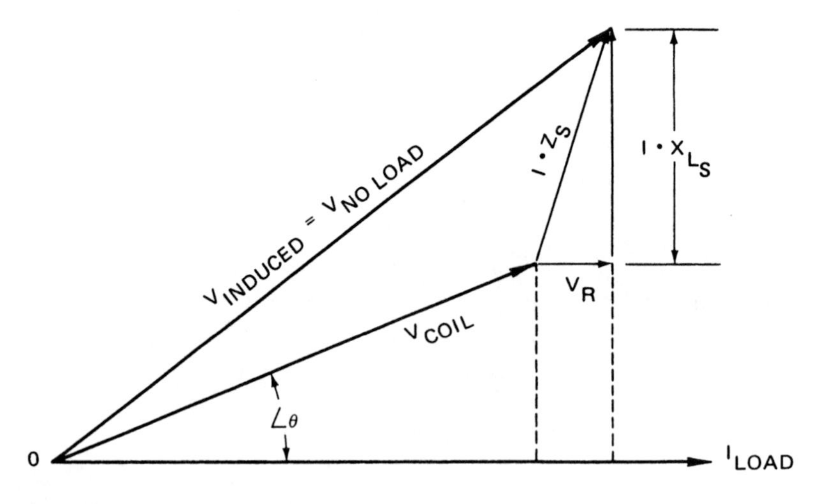

FIGURE 12–26 Synchronous impedance vector diagram (lagging power factor load) (*Delmar/Cengage Learning*)

A vector diagram is given in Figure 12–27 for an alternator supplying a load with a leading power factor of 86.6%, or 0.866. In some cases, the voltage at no load may be less than the voltage at full load, depending on the amount of voltage drop due to effective resistance and the voltage drop due to inductive reactance. As a result, a negative value is obtained for the percent voltage regulation. Whenever the percent voltage regulation is negative, the voltage at no load is less than the voltage at full load. In this case, the voltage at no load ($V_{induced}$) is less than the voltage at full load. Note that in this formula, the voltage drop due to inductive reactance is subtracted from the phase voltage instead of being added to the phase voltage.

$$V_{induced} = \sqrt{(\cos \angle \theta \bullet V_{phase} + V_R)^2 + (\sin \angle \theta \bullet V_{phase} - V_{XL})^2}$$

$$V_{induced} = \sqrt{(0.866 \bullet 1387 + 75.9)^2 + (0.5 \bullet 1387 - 427)^2}$$

$$V_{induced} = 1304.5 \text{ V}$$

$$\text{Percent Voltage Regulation} = \frac{\text{Voltage}_{\text{no load}} - \text{Voltage}_{\text{full-load}}}{\text{Voltage}_{\text{full-load}}} \times 100$$

$$\frac{1304.5 - 1387}{1387} = \frac{-82.5}{1387} \times 100 = -5.9\%$$

Determining Regulation of Three-Phase, Delta-Connected Generator

The synchronous impedance test can also be used to find the voltage regulation of a three-phase, delta-connected alternator (Figure 12–28). For a delta connection, the line current is equal to the phase winding current, multiplied by $\sqrt{3}$. In addition, the phase winding voltage and the line voltage are equal. Therefore, the synchronous impedance of each phase winding for a generator connected in delta is

$$Z_S = \frac{V_S}{\frac{I_{line}S}{\sqrt{3}}}$$

or

$$Z_S = \frac{V_S\sqrt{3}}{I_{line}S}$$

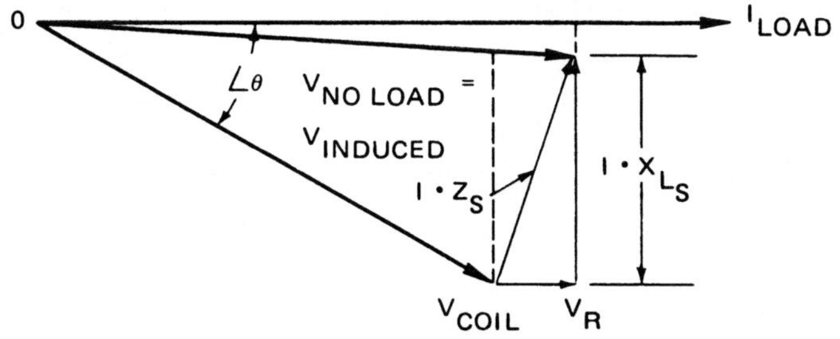

FIGURE 12–27 Synchronous impedance vector diagram (leading power factor load) (*Delmar/Cengage Learning*)

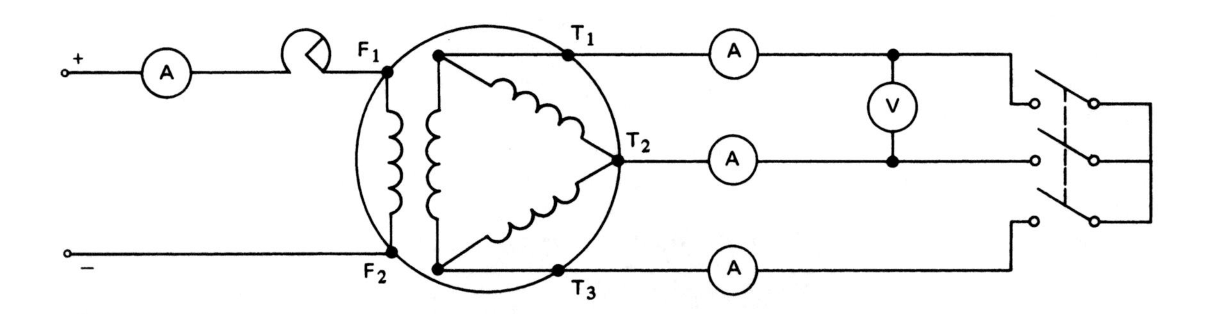

FIGURE 12–28 Connections for the synchronous impedance test for a three-phase alternator (*Delmar/Cengage Learning*)

The voltage regulation is to be found for the alternator in the following example. This is the same alternator described in previous examples, except that the phase windings are connected in delta. It is assumed that each phase winding has the following full-load rating: coil current = 289 A and coil voltage = 1387 V.

If each phase winding is loaded to 150% of its rated value, then the coil current is 289 × 1.5 = 433.5 A.

The three-phase windings are connected in delta. Therefore, the line current for a short-circuit condition is 433.5 × $\sqrt{3}$ = 750 A.

The open-circuit voltage per phase winding is the same for both the wye and delta connections:

$$1125 \div \sqrt{3} = 650 \text{ V}$$

Therefore, the synchronous impedance of each phase winding is

$$Z_S = \frac{V_S}{I}$$

$$= \frac{650}{433.5}$$

$$= 1.5\ \Omega$$

Recall that for an alternator connected in delta, the synchronous impedance was given by the following formula:

$$Z_S = \frac{V_S \sqrt{3}}{I_{line}}$$

This same formula can be used to find the synchronous impedance of an alternator connected in wye:

$$Z_S = \frac{V_S \sqrt{3}}{I_{line}}$$

$$= \frac{1.73 \times 640}{750}$$

$$= 1.5\ \Omega$$

When the ohmic resistance of a delta-connected armature winding is measured across any pair of line terminals, the resulting value includes the resistance of one phase winding in parallel with the other two windings connected in series. For example, when the alternator is connected in wye, the ohmic resistance of each phase winding is 0.2 Ω. If the three windings are connected in delta, then the ohmic resistance is 0.133 Ω measured across any pair of line terminals. For both the delta and wye connections, the ohmic resistance of each phase winding is 0.2 Ω:

$$\frac{1}{0.13} = \frac{1}{R_O} + \frac{1}{2R_O}$$

$$2R_O = 0.4$$

$$R_O = 0.2\ \Omega$$

The effective resistance for each phase winding is

$$R = 1.5 \times 0.2$$

$$= 0.3\ \Omega$$

The synchronous reactance for each phase winding is

$$X_{LS} = \sqrt{Z_S^2 - R^2}$$

$$= \sqrt{1.5^2 - 0.3^2}$$

$$= 1.47\ \Omega$$

The voltage loss, due to the effective resistance and the synchronous reactance, is

$$V_R = IR$$

$$= 289 \times 0.3$$

$$= 87\ V$$

$$V_{XL} = IX_{LS}$$

$$= 289 \times 1.47$$

$$= 425\ V$$

The same procedures are used to determine the voltage at no load and the percent voltage regulation for both delta- and wye-connected alternators. The following calculations give the percent voltage regulation for a delta-connected alternator with a unity power factor load; note that the results are the same as those obtained when the phase windings were connected in wye:

$$V_{\text{no load}} = \sqrt{(V_{\text{coil volts}} + IR)^2 + (IX_{LS})^2}$$
$$= \sqrt{(1387 + 87)^2 + 425^2}$$
$$= 1534 \text{ V}$$

$$\text{Percent voltage regulation} = \frac{\text{voltage at no load} - \text{voltage at full load}}{\text{voltage at full load}} \times 100$$
$$= \frac{1534 - 1387}{1387} \times 100$$
$$= 10.6\%$$

AUTOMATIC VOLTAGE CONTROL

An alternator will experience large changes in its terminal voltage with changes in the load current and the load power factor because of the combined effects of the armature reactance and the armature reaction. However, a relatively constant terminal voltage can be maintained under changing load conditions by the use of an automatic voltage regulator.

Automatic voltage regulators change the alternator field current to adjust for changes in the load current. As the terminal voltage decreases, a relay closes contacts in the regulator that short out a field resistor. There is a resulting increase in the field current. There are also increases in the field flux and the induced voltage. An increase in terminal voltage causes the relay to open the contacts across the field resistor. This action causes a decrease in the field current, the field flux, and the induced voltage.

A simplified schematic diagram of an automatic voltage regulator is given in Figure 12–29. Relay coil A is connected across one phase of the three-phase output of the alternator. During normal operation, relay coil A causes contacts B to open and close several times each second. The exciter generator supplies the alternator field with nearly constant values of dc excitation voltage and current. If the terminal voltage decreases, the voltage across the relay decreases. As a result, the contacts remain closed for longer time intervals. This causes the excitation current supplied to the alternator field to increase. The ac terminal voltage output of the generator rises to its original value. If the load on the alternator changes so that the terminal voltage increases, then the contacts will vibrate at a greater rate. There is a resulting decrease in the time that the resistor in series with the shunt field of the dc exciter generator is short-circuited. The terminal voltage decreases to its normal value. Alternators may be operated with many other types of automatic voltage regulators using vacuum tubes, amplidynes, magnetic amplifiers, ignition rectifiers, and silicon controlled rectifiers.

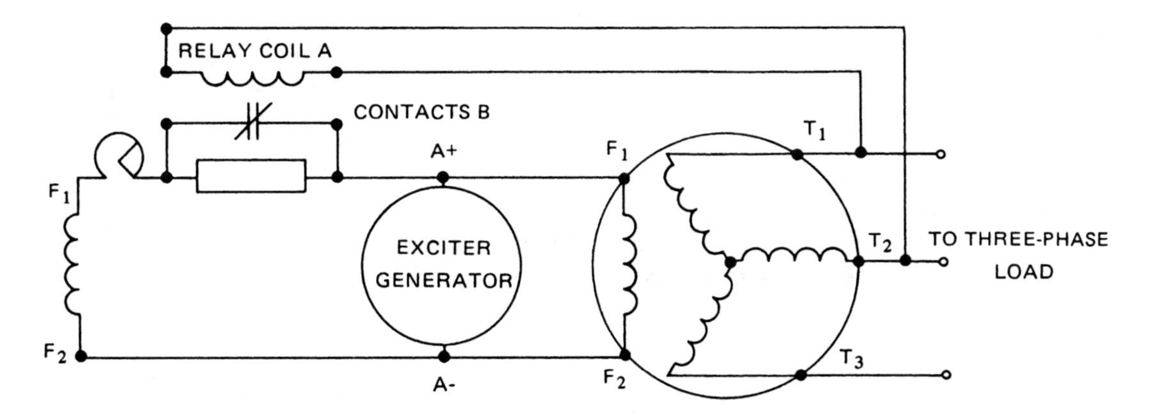

FIGURE 12–29 Automatic voltage regulator for a three-phase alternator (*Delmar/Cengage Learning*)

SATURATION CURVE

For a constant-speed generator, the generated voltage is a direct function of the flux value per pole. At no load, the flux of each pole is determined by the number of ampere-turns of the field pole. Because each field winding consists of a constant number of turns, the flux is proportional to the dc excitation current.

Direct Current Fundamentals explained that when iron is magnetized, the molecules are arranged in a definite pattern. To align more of the iron molecules in this pattern requires a proportional increase in the number of ampere-turns. In other words, as the number of ampere-turns increases, there is an almost proportional increase in the flux. With a decrease in the number of unaligned iron molecules, it is more difficult to increase the flux in the magnetic circuit. A point is reached at which the flux no longer increases in proportion to the increase in the magnetomotive force. This point is called the *saturation point*. As the magnetomotive force is increased beyond the saturation point, the flux increase becomes smaller as there are fewer unaligned molecules in the iron.

Figure 12–30 shows the connections and a typical saturation curve for an ac generator. Note in Figure 12–30A that the field winding is connected to a dc supply. The excitation current is controlled by a rheostat. An ammeter in the field circuit indicates the value of the excitation current. A voltmeter is connected across one pair of line terminals to measure the induced voltage. Because all three voltages should be of the same magnitude, three voltmeters are not required.

Plotting the Saturation Curve

The alternator is operated at the rated speed with the field circuit deenergized. The residual magnetism in the field poles causes a low voltage to appear across the terminals of the alternator. The field rheostat is adjusted to its maximum value, and the dc field circuit is energized. The field current and the induced voltage values are recorded. The field current

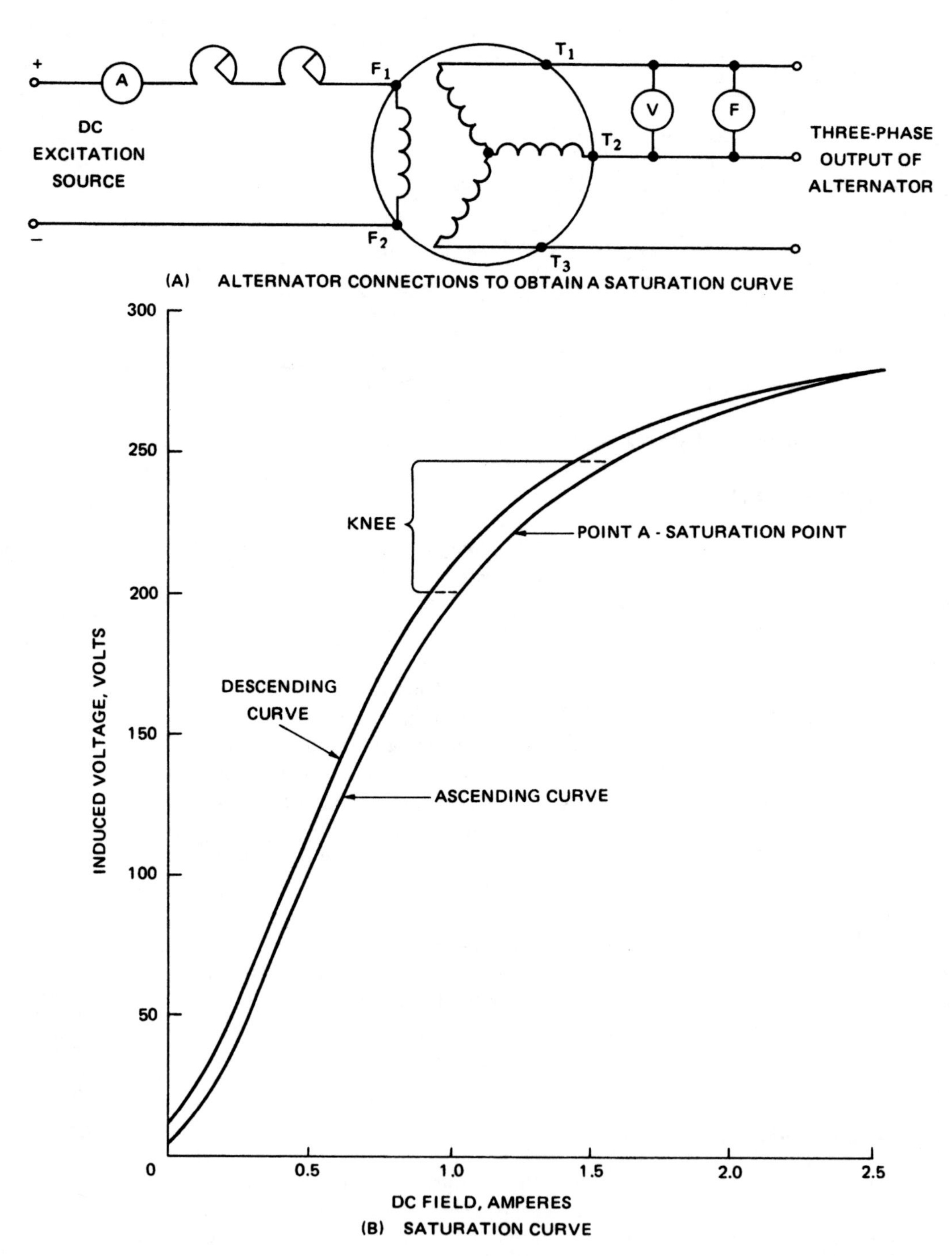

FIGURE 12–30 **Connections and saturation curve for a 7.5-kVA, 220-volt, three-phase alternator (*Delmar/ Cengage Learning*)**

is then increased by fixed increments. The resulting voltage values are recorded for each setting of field current. When the maximum field current is reached, the process can be reversed to obtain a descending curve. The field current is decreased back to zero by the same stepped values.

The ascending section of the saturation curve, as shown in Figure 12–30B, starts out with a small induced voltage at a field current of zero. This voltage is due to residual magnetism. As the field current increases from zero, the induced voltage increases. The voltage increase follows an almost straight-line curve up to point A, the saturation point. As the field current continues to increase, the increase in the flux and the induced voltage becomes smaller. This means that there are few unaligned molecules in the iron of the magnetic circuit. Thus, it is more and more difficult to magnetize the circuit.

Note in Figure 12–30B that the induced voltage values are slightly higher in the descending section of the curve for given values of field current. Compare these values with the voltages in the ascending part of the curve. The slight difference in the curves is due to molecular friction. The molecules of the iron in the magnetic circuit stay aligned even after the magnetomotive force is decreased. A better name for molecular friction is hysteresis effect.

Applications of the Saturation Curve

The saturation curve can be put to a number of practical uses, including the following:

1. The curve shows the approximate value of field excitation current required to obtain the maximum ac voltage output with minimal I^2R losses in the field circuit. This operating point generally occurs near the center of the knee of the curve. In Figure 12–30B, note that this area is in the region of the saturation point A.
2. For small differences between the ascending and descending portions of the curve for given values of the dc field current, the hysteresis effect in the magnetic circuit is small. However, a large spread between the ascending and descending portions of the curve means that the hysteresis effect in the magnetic curve is significant.

ALTERNATOR NAMEPLATE DATA

The capacity of an alternator is given in kilovolt-amperes rather than kilowatts. This practice is followed because the machine may be required to supply a load with a power factor other than unity. Once the kVA output and the power factor are known, the kilowatt output can be determined from the following equation:

$$\mathbf{kW} = \mathbf{kVA} \times \mathbf{cos} \angle \theta$$

The alternator nameplate also contains the following data: the full-load terminal voltage, the rated full-load current per terminal, the number of phases, the frequency, the speed in r/min, the operating power factor, the dc field current and voltage, and the maximum temperature rise. Table 12–1 summarizes the data found on a typical nameplate.

Volt-amperage: 100 kVA	Three phase
Frequency: 60 Hz	1800 r/min
Voltage: 2400 V	Current per terminal: 25 A
Power factor: 80%	Field voltage: 125 V
Field current: 15 A	Temperature rise: 50°C
Manufacturer:	continuous duty
Frame no.:	Model no.:
	Serial no.:

TABLE 12–1 Alternator Nameplate Data

ALTERNATOR EFFICIENCY

DC generators and alternators have nearly the same losses. The fixed or stray power losses include the bearing and brush friction losses, windage loses, and iron losses. (The iron losses include eddy current losses and hysteresis losses.) The copper losses include the I^2R losses in the armature windings and the power expended in the separately excited field circuit.

To determine the efficiency of an alternator, it can be loaded to its rated capacity and the values of the input and output power can then be measured. These values are substituted in the following equation to find the efficiency:

$$\textbf{Efficiency, } \eta = \frac{\textbf{output, watts}}{\textbf{input, watts}} \times \mathbf{100}$$

However, the kVA rating of the alternator can be very high. In this case, it is very difficult to find a suitable loading device having the proper voltage, current, and power factor ratings for a desired load condition. As a result, the efficiency of such alternators is determined using their losses, where

$$\eta = \frac{\textbf{kVA output} \times \angle\,\theta}{\textbf{kVA output} \times \textbf{cos} \angle\,\theta + \textbf{total copper losses} + \textbf{fixed losses}} \times \mathbf{100}$$

In the following problems, alternator losses and percent efficiency are to be determined.

PROBLEM 3

Statement of the Problem

A 480-V, 60-Hz, single-phase alternator delivers 18 kW to a load with a 75% lagging power factor. The generator has an efficiency of 80%. The stray power losses are 2000 W. The separately excited field takes 8 A from a 125-V dc source. Determine

1. the load current.
2. the copper losses in the alternator armature.
3. the effective resistance of the alternator armature.
4. the horsepower delivered by the prime mover.

Solution

1.
$$\text{kW} = \frac{V \times I \times PF}{1000}$$
$$18 = \frac{480 \times I \times 0.75}{1000}$$
$$360\ I = 18{,}000$$
$$I = 50\ \text{A}$$

2.
$$\eta = \frac{\text{output, watts}}{\text{input, watts}}$$
$$0.8 = \frac{18{,}000}{\text{input}}$$
$$\text{Input} = 22{,}500\ \text{W}$$

$$\begin{aligned}\text{Total losses} &= \text{input, watts} - \text{output, watts}\\ &= 22{,}500 - 18{,}000\\ &= 4500\ \text{W}\end{aligned}$$

The loss in the separately excited field is

$$P = VI = 125 \times 8 = 1000\ \text{W}$$

Copper losses in the armature windings

$$\begin{aligned}&= \text{total losses} - (\text{stray power losses} + \text{field losses})\\ &= 4500 - (2000 + 1000)\\ &= 1500\ \text{W}\end{aligned}$$

3.
$$\begin{aligned}P &= I^2R\\ 1500 &= 50^2 \times R\\ 1500 &= 2500\ R\\ R &= 0.6\ \Omega\end{aligned}$$

4. If the prime mover is directly coupled to the alternator, the horsepower output of the prime mover and the input to the alternator are the same. Therefore,

$$\begin{aligned}\text{Horsepower input to alternator} &= \frac{\text{input,watts}}{746}\\ &= \frac{22{,}500}{746}\\ &= 30\ \text{hp}\end{aligned}$$

PROBLEM 4

Statement of the Problem

A 25-kVA, 250-V, 60-Hz, 1800-r/min, single-phase alternator has an effective resistance of 0.1 Ω and an inductive reactance of 0.5 Ω in its armature windings. The generator is delivering the rated load output at a unity power factor to a noninductive heating load. At the rated load and unity power factor, determine

1. the load current.
2. the copper losses in the armature windings.
3. the efficiency of the alternator if the input is 38 hp.

Solution

1. $I = \frac{VA}{V} = \frac{25{,}000}{250} = 100 \text{ A}$

2. $P = I^2R = 100^2 \times 0.1 = 1000 \text{ W}$

3. $P = \text{hp} \times 746$

$$= 38 \times 746$$

$$= 28.348 \text{ W}$$

$$\text{Efficiency} = \frac{\text{output, watts}}{\text{input, watts}} \times 100$$

$$= \frac{25{,}000}{28{,}348} \times 100$$

$$= 88.2\%$$

An alternator should be operated at or near the rated full-load output to obtain the maximum efficiency. At points where the load is light, the fixed losses are a large part of the input. Thus, the efficiency at such points is low. As the load output of an alternator increases, the fixed losses become a much smaller part of the input. This results in a marked increase in the efficiency. Alternators with capacities in the order of 200,000 kVA may have efficiency ratings as high as 96% at the full-load output.

PARALLELING ALTERNATORS

Power-generating stations operate several alternators in parallel. This practice is preferred to the use of a single large generator for the following reasons:

1. The use of several ac generators means that periodic maintenance and repairs can be made to one alternator. Because the other machines are operating in parallel, they can supply the load and prevent a power interruption due to a generator failure.
2. The load requirements for any central generating station change continually. During light-load periods, the load demands can be met by one or two alternators

operating at a high efficiency. This means that these alternators will be operating at a high efficiency. As the load demands increase during certain periods of the day, other alternators can be connected in parallel to meet the peak demands. This procedure is more economical than the use of one huge machine operating, at certain periods of each day, at only a small fraction of its full-load capacity. Such operation results in a low efficiency.

3. Electrical power requirements are continually increasing. To meet these increased demands, utility companies can increase the physical size of their generating plants so that more alternators can be installed. These machines can be operated in parallel with the existing generating equipment. This procedure is a convenient and economical way of increasing the generating capacity of a utility.

Paralleling AC Generators

To parallel dc generators, the voltages and the polarities of the generators must be the same. For ac generators, however, it must be remembered that the output voltages continuously change in both magnitude and polarity at a definite frequency. Therefore, to parallel alternators, the following conditions must be observed:

- The output voltages of the alternators must be equal.
- The frequencies of the alternators must be the same.
- The output voltages of the alternators must be in phase.

When these conditions are met, the alternators are said to be in synchronism. The following steps describe the actual process of synchronizing two three-phase alternators:

1. Assume that alternator 1 is supplying energy to the bus bars of the station at the rated voltage and frequency.
2. An incoming machine, alternator 2, is to be synchronized with alternator 1 for the first time. The speed of alternator 2 is increased until it turns at the value required to give the desired frequency. The voltage of generator 2 is adjusted by means of its field rheostat until it is equal to that of generator 1.
3. The three voltages of the incoming generator must be in phase with the respective voltages of generator 1. To accomplish this, the phase sequence of the two alternators and their frequencies must be the same. The use of synchronizing lamps is a simple way to check these relationships.

Synchronizing Two Alternators

Three-Lamp Method. The circuit shown in Figure 12–31 is used to synchronize two three-phase alternators. Alternator 1 supplies energy to the load. Alternator 2 is to be paralleled with alternator 1. Three lamps are connected across the switches as shown in the figure. Each lamp is rated at the terminal voltage of the alternator. When both machines are operating, either of two actions will occur:

1. The three lamps will go on and off in unison. The rate at which the lamps go on and off depends on the frequency difference between alternator 2 and alternator 1.

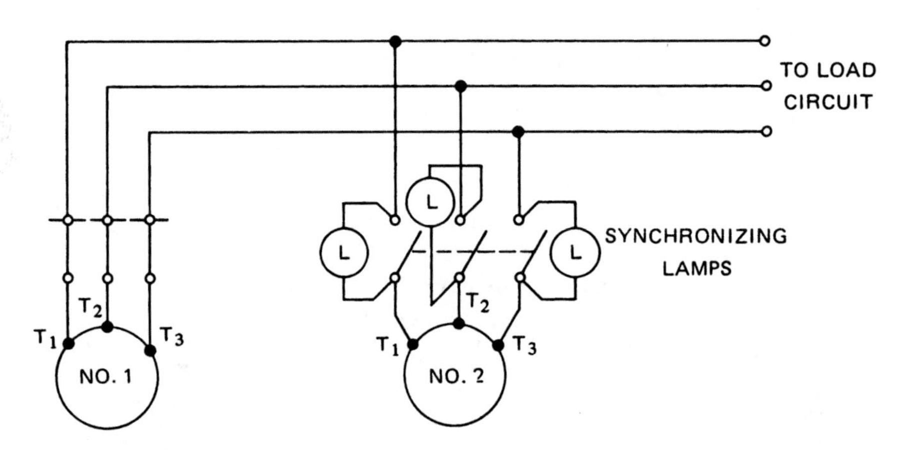

FIGURE 12–31 Synchronization of alternators (*Delmar/Cengage Learning*)

2. The three lamps will light and go off, *but not in unison*. The rate at which this occurs depends on the frequency difference between the two generators. In this case, the phase rotation, or phase sequence, of alternator 2 is not the same as that of alternator 1. The phase sequence of alternator 2 must be corrected so that it will be the same as that of alternator 1. By interchanging the connections of any two leads of alternator 2, the phase sequence can be changed. The three synchronizing lamps should now go on and off in unison, indicating that the phase sequence is correct. A slight adjustment in the speed of the prime mover for alternator 2 makes the frequency of alternator 2 the same as that of alternator 1. As the frequency difference between the alternators decreases, the rate at which the synchronizing lamps increase and decrease in light intensity also decreases. Thus, the rate at which the lamps change in light intensity represents the difference in frequency between the two alternators.

For example, assume that the frequency of alternator 1 is 60 Hz and the frequency of the incoming generator (alternator 2) is 59 Hz. The frequency difference between the alternators is 1 Hz. This means that the synchronizing lamps will come on and go off once each second. When the lamps are off, the instantaneous electrical polarity of alternator 2 is the same as that of alternator 1. The switch for alternator 2 can be closed at this point and the ac generators will be in parallel.

Three Lamps Dark Method. One way of connecting the synchronizing lamps is called the *three lamps dark* method. In this method, the synchronizing lamps are connected directly across the switch from the blade to the jaw. The three lamps dark method can always be used to determine the phase sequence of an alternator. Once the phase sequence is known, permanent connections can be made between the stator windings, the switching equipment, and the station bus bars. From this point on, it is not necessary to repeat the process of determining the phase sequence each time the alternator is paralleled. Figure 12–32A shows the connections for the three lamps dark method to determine the phase sequence. This method is also used to indicate when alternators are in synchronism.

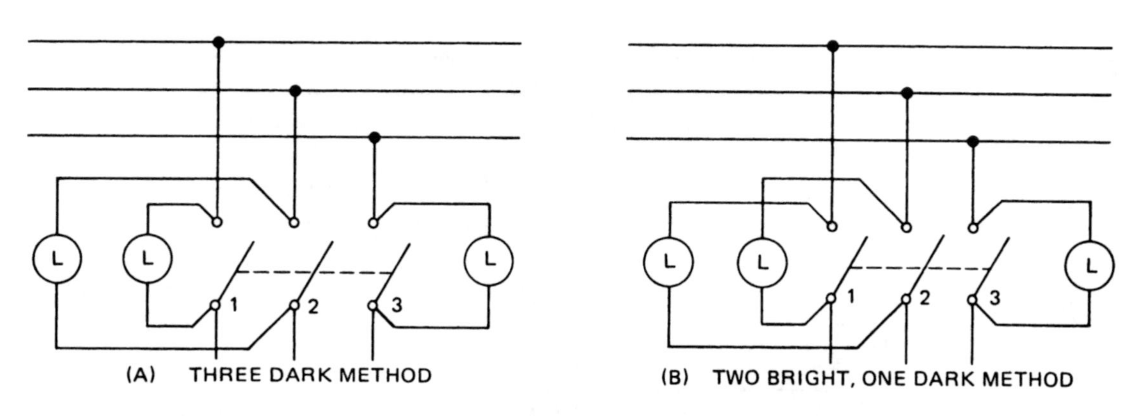

FIGURE 12–32 Connections for synchronizing lamps (*Delmar/Cengage Learning*)

Two Bright, One Dark Method. The connections are shown in Figure 12–32B for another method of synchronizing lamps. This method is called the *two bright, one dark* method. This method is never used to determine the phase sequence. It is used only to indicate the synchronism of two alternators. When the incoming alternator is in synchronism (1) the two lamps in line wires 1 and 2 will have a maximum brightness, and (2) the lamp in line wire 3 will be dark.

There is one disadvantage to both lamp connections shown in Figure 12–32. Using the three dark method, there may be a large voltage difference across the synchronizing lamps (even when they are dark). Thus, a large voltage and phase difference may be present when an attempt is made to bring the incoming alternator into the bus bar circuit system with other machines. A large disturbance in the electrical system may result and damage the alternator windings.

Use of the Synchroscope

Once the phase sequence is known to be correct and permanent connections are made, a *synchroscope* can be used. This single-phase instrument indicates synchronism accurately. Unit 11 gives information on the construction and operation of synchroscopes.

A synchroscope gives an accurate indication of the differences in the frequency–phase (en dash between words, no space between dash and "phase" relationship between two voltages. The voltage from one phase of the three-phase bus bar system is connected to one set of synchroscope coils. The voltage from the same phase of the incoming alternator is connected to another set of synchroscope coils. A pointer is attached to the synchroscope rotor and rotates over a dial face. When the pointer stops, the frequencies of the two alternators are the same. When the pointer stops in a vertical upward position, the frequencies are equal and the voltages are in phase. This means that the alternators are in synchronism and the alternator switch can be closed to parallel both machines.

Figure 12–33 shows the synchroscope connections required to obtain an indication of the synchronism of two alternators. The voltage of the same phase from each alternator is applied to the coils of the synchroscope through special synchroscope switches. Each switch has one position marked “run” and another position marked “incoming.” This flexibility in making connections allows the synchroscope to be used when either alternator is

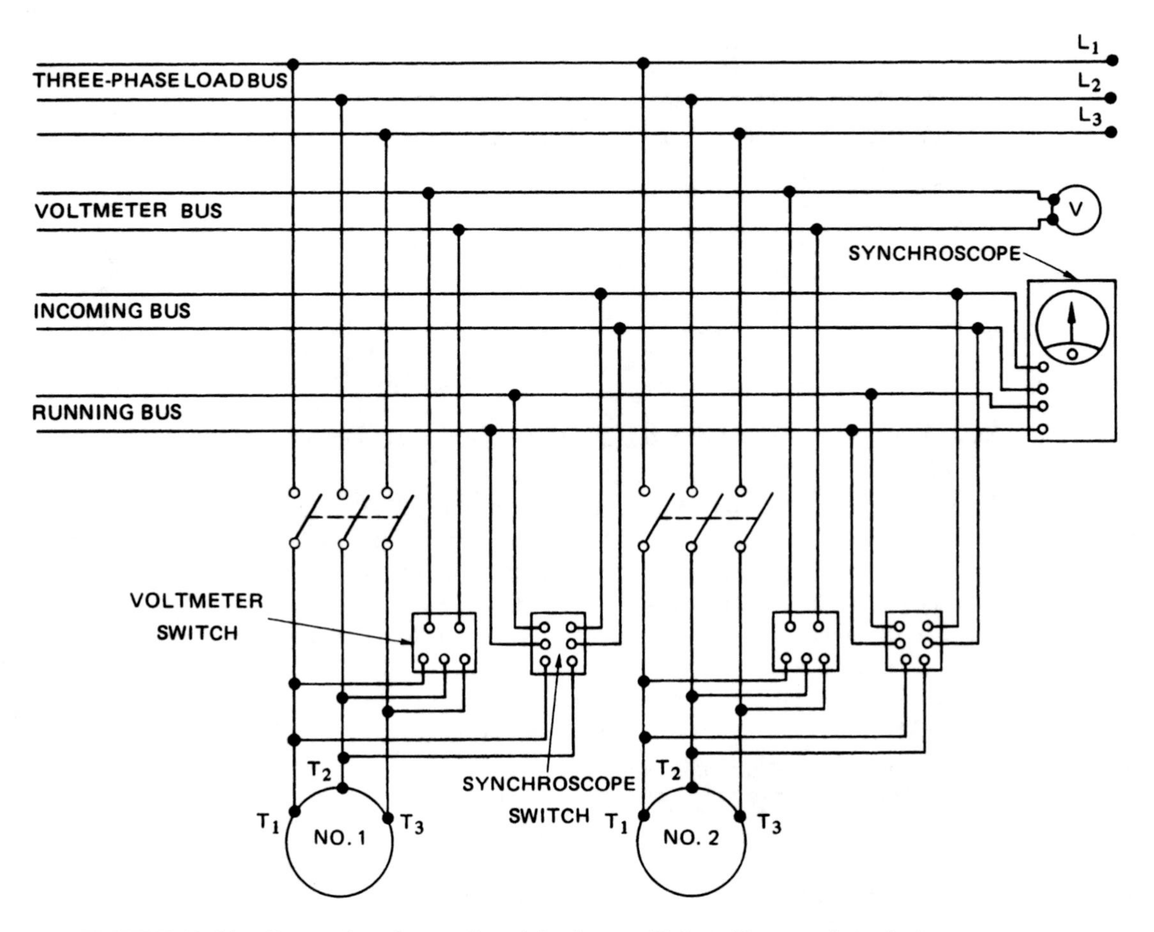

FIGURE 12–33 Connections for synchronizing lamps (*Delmar/Cengage Learning*)

being synchronized with the station bus bars. The use of two voltmeter switches means that one voltmeter can be used to measure the voltages of all three phases of either alternator. Generally, Figure 12–33 shows only the panelboard connections for the synchroscope and the voltmeter. In an actual installation, the alternator panelboard will also contain a three-phase wattmeter, a three-phase power factor meter, ammeters, and control switches.

Load Sharing

Once the two alternators are operating in parallel, the load is shared between them. The load taken by each machine is proportional to the kVA rating of the machine. The division of the load between dc generators is obtained by changing the field excitation of each generator until the load is shared. However, the same method cannot be used to divide the kilowatt load between two alternators in parallel. Keep in mind that alternators in parallel must turn at a fixed speed to maintain a constant frequency. The input to steam turbines, waterwheels, and diesel units is controlled by sensitive governors. Because the governor control holds the input to these prime movers at a fixed value, the input to the alternator will also be a fixed value. Therefore, for machines in parallel, the true power output (in kilowatts) will show very little change even when the field excitation is changed.

A different method must be used to adjust the kilowatt load between alternators in parallel. The prime movers for such alternators should have drooping speed–load characteristic curves. For the alternators shown in Figure 12–33, assume that alternator 1 operates at 60 Hz. Alternator 2 also operates at 60 Hz, once it is paralleled with alternator 1. Alternator 2 should deliver very little load because it cuts the system frequency line (60 Hz) at a point close to zero. On the other hand, alternator 1 is heavily loaded. Its drooping speed–load curve cuts the system frequency line a large distance from zero.

Figure 12–34 illustrates this condition. Point A indicates that alternator 1 supplies most of the kilowatt load. At point B, alternator 2 delivers very little power to the bus bars. To divide the load equally between the two machines, the input to the prime mover must be increased. This is done by slightly opening the governor on the prime mover of alternator 2. As a result, the horsepower input to alternator 2 increases, causing the power output of alternator 2 to increase as well. At the same time, the governor on the prime mover of alternator 1 is closed very slightly. This action decreases the input to the prime mover of generator 1. This decrease in the horsepower input to generator 1 causes a decrease in the power output (in kilowatts) of generator 1. Careful adjustments of the governors of both prime movers can result in both speed–load characteristic curves cutting the system frequency line at the same load point (Figure 12–35).

The governors used on the prime movers usually have electrical controls to ensure that both alternators feed the same amount of power (in kilowatts) into the station bus bars. These controls are operated from the instrument panelboard. Accurate adjustments of the

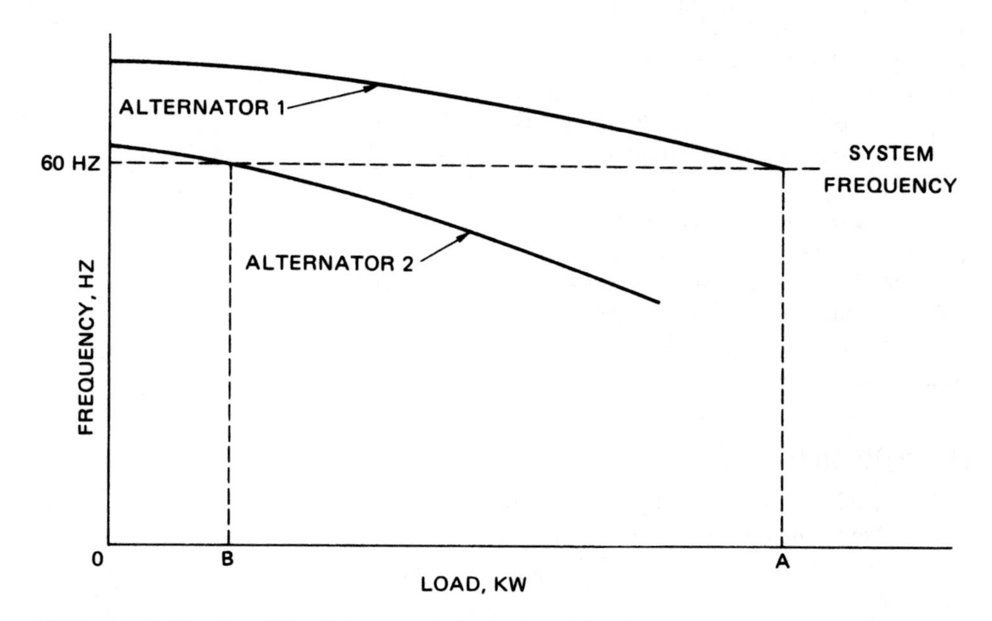

FIGURE 12–34 Speed-load characteristics before changing governor control (*Delmar/ Cengage Learning*)

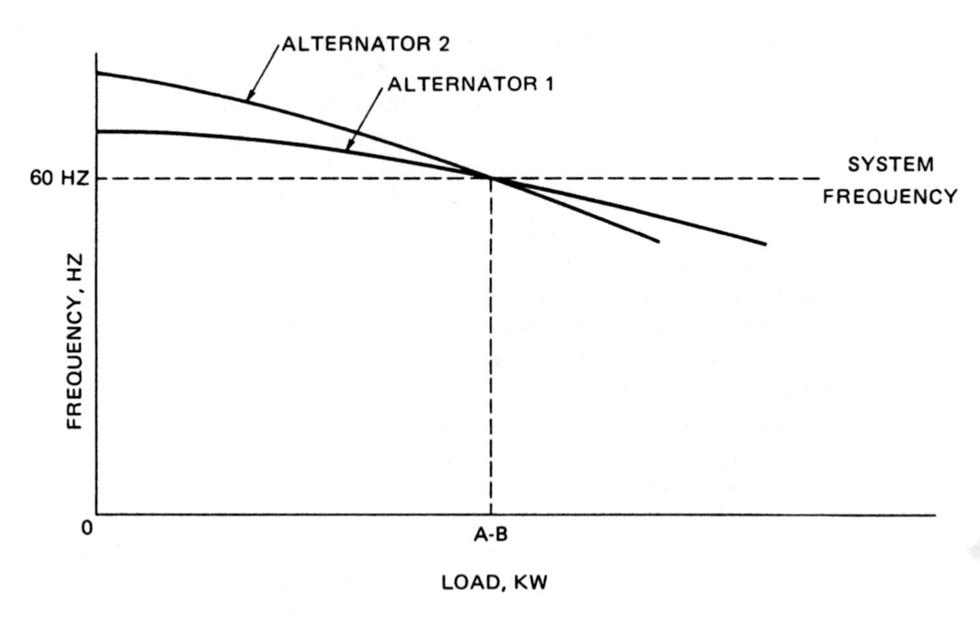

FIGURE 12–35 Speed-load characteristics after changing governor control (*Delmar/ Cengage Learning*)

governors can be made using these controls to obtain a satisfactory load division between machines. For two machines operating in parallel, the load division between the machines must be obtained without causing a change in the frequency. For example, if the governor of alternator 2 is opened slightly, the frequency of the two alternators will increase. To maintain the original frequency, the speed–load curve of alternator 1 must be lowered at the same time that the speed–load curve of alternator 2 is raised. Refer to Figure 12–35 and note that the drooping speed–load curve of alternator 2 is raised and that of alternator 1 is lowered. As a result, both curves cut the original system frequency line (60 Hz) at the same load point.

Alternators operating in parallel in various central stations often will operate in a network power system. A faulty governor will cause the speed of an alternator to increase so that the machine is pulled out of synchronism. However, certain reactions within the generator will prevent this condition. Also, if the governor malfunctions so that it cuts out completely and shuts off the input to the prime mover, the alternator will operate as a synchronous motor until the governor fault is corrected.

Effect of Field Excitation

It was stated earlier in this unit that any change in the division of the kilowatt load between the alternators is due to changing the input to the prime movers of the machines. The field excitation to the alternators is *not* changed. What is the effect of changing the field excitation on alternators operating in parallel?

Following the previous explanation of the paralleling of alternators, assume that even after alternator 2 is connected in parallel with alternator 1, alternator 2 delivers only a small kilowatt input to the bus bars. It is also assumed that both generators operate at unity power factor with a unity power factor load. The vector diagram for this circuit is shown in Figure 12–36. The diagram shows that the currents delivered by the respective alternators (I_1 and I_2) are in phase with the line voltage. The arithmetic sum of the two alternator currents equals the total current supplied to the load (I_{total}). This total current is also in phase with the line voltage (because the load has a unity power factor).

The field excitation of alternator 2 can be increased and the field of alternator 1 can be weakened in an attempt to divide the kilowatt load equally between the two alternators. However, the power output of each generator remains nearly the same. (It is assumed that the input to the prime mover of each alternator is unchanged.) There is an increase in the current output of each alternator. Also, both machines no longer operate at a unity power factor. It is apparent that when the field excitation current of alternator 2 is increased, the internal induced voltage of this machine also increases.

For this case, the power factor decreases from unity to a value in the lagging quadrant. Thus, a greater induced voltage is required to maintain the same terminal voltage. If the field of alternator 1 is weakened, the internal induced voltage of this machine decreases. In addition, the power factor of alternator 1 will decrease from unity to a value in the leading quadrant. A leading power factor means that the same terminal voltage can be maintained with a lower internal induced voltage. (This unit has already explained the effects of armature reactance and armature reaction on the terminal voltage of an alternator for different power factor loads.) The armature reactance and armature reaction give rise to internal operating conditions that adjust themselves. As a result, the terminal voltage and the power output of each alternator in parallel remain nearly constant with changes in the field excitation.

Figure 12–37 shows the circuit conditions after the field excitation of both alternators is changed. When the field excitation of alternator 2 is strengthened, the current value of this machine increases and lags the line voltage. However, the in-phase component of current for this machine is unchanged. When the field of alternator 1 is weakened, the current of this machine increases and leads the line voltage. The in-phase current of alternator 1 also is unchanged by a change in the field excitation. The quadrature current surges back and forth between the alternators. This current causes greater I^2R copper losses in both generators. Because the in-phase current supplied by each alternator remains the same, the power (in kilowatts) supplied by each generator to the bus bars is unchanged. Of course, the current and power will be determined by the load.

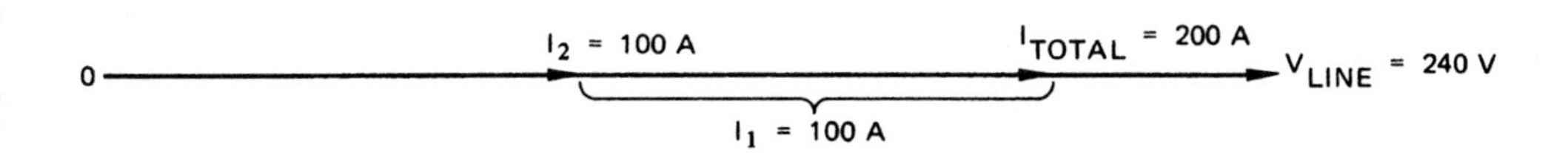

FIGURE 12–36 Alternators in parallel; each alternator operates at a unity power factor. (*Delmar/Cengage Learning*)

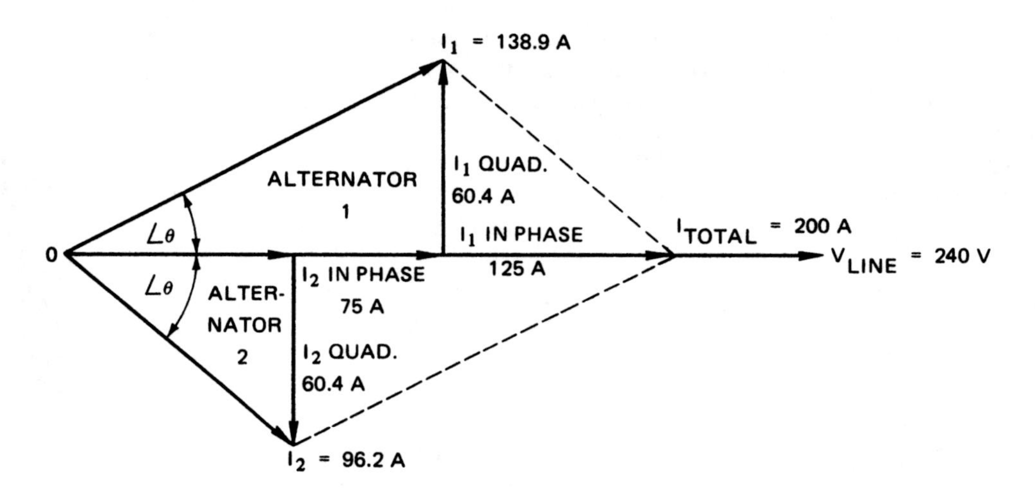

FIGURE 12–37 Alternators in parallel; field excitation for each alternator is changed. (*Delmar/Cengage Learning*)

PROBLEM 5

Statement of the Problem

Two single-phase alternators are each rated at 30 kVA, 240 V, 60 Hz. The alternators are operated in parallel to supply power to a load having a unity power factor. The load requires 48 kW at 240 V. Both alternators operate at a unity power factor. The output of alternator 1 is 30 kW and the output of alternator 2 is 18 kW. Determine

1. the current supplied by each alternator.
2. the total current supplied to the load.

The vector diagram is the same as the one shown in Figure 12–36. Both alternators operate at a unity power factor. Thus, the current of each alternator is in phase with the line voltage. The total line current is also in phase with the line voltage because the load operates at a unity power factor.

Solution

1. $I_1 = \frac{W_1}{V} = \frac{30{,}000}{240} = 125 \text{ A}$

   $I_2 = \frac{W_2}{V} = \frac{18{,}000}{240} = 75 \text{ A}$

2. $I_{total} = \frac{W_{load}}{V} = \frac{48{,}000}{240} = 200 \text{ A}$

   or

   $I_{total} = I_1 + I_2 = 125 + 75 = 200 \text{ A}$

The arithmetic sum of the two alternator currents equals the total current supplied to the load. This is true because each current value is in phase with the line voltage.

PROBLEM 6

Statement of the Problem

An attempt is made to redistribute the load between the alternators in problem 5 by weakening the field excitation of alternator 1 and strengthening the field excitation of alternator 2. The resulting power output of both machines remains nearly the same, as in the previous case. Refer to Figure 12–37. (The input to the prime movers remains the same.) After the field of alternator 1 is weakened and the field of alternator 2 is strengthened, the power factor of alternator 1 is 0.90 leading. Determine

1. the current delivered by alternator 1.
2. the quadrature (quad.) current that circulates between the alternators.
3. the current delivered by alternator 2.
4. the value of lagging power factor for alternator 2.
5. the current required by the load.

Solution

1. $\cos \theta_1 = \frac{I_{in\ phase}}{I_1}$

$$0.9 = \frac{125}{I_1}$$

$$I_1 = 138.9 \text{ A}$$

2. $\cos \theta = 0.9;\ \theta_1 = 25.8°$

$$\sin \theta_1 = \frac{I_{1\ quad}}{I_1}$$

$$0.4352 = \frac{I_{1\ quad}}{138.9}$$

$$I_1 = 60.4 \text{ A}$$

I_1 quadrature current = I_2 quadrature current = 60.4 A

3. $\text{tangent}\ \theta_2 = \frac{I_{1\ quad}}{I_{2\ in\ phase}} = \frac{60.4}{75} = 0.8053$

$$\theta_2 = 38.8° \text{ lag}$$

$$\cos \theta_2 = \frac{I_{2\ in\ phase}}{I_2}$$

$$0.7793 = \frac{75}{I_2}$$

$$I_2 = 96.2 \text{ A}$$

4. $\cos \theta_2 = 0.779$ lagging power factor

5. The power factor of the load is determined by the electrical characteristics of the load alone. Therefore, the power factor remains at unity. If the terminal voltage is kept to 240 V, the load current will remain the same:

$$I_{total} = \frac{\text{watts}}{V_{line}}$$

$$= \frac{48{,}000}{240}$$

$$= 200 \text{ A}$$

When the in-phase components of current for both alternators are added together, the result is the same as I_{total}:

$$125 + 75 \text{ A} = 200 \text{ A}$$

Problem 6 shows that the power output of alternators is not increased by changing the field excitation of the alternators. Only the VARs load on each alternator is increased when the field excitation is changed. To distribute the load properly, the input to the prime mover of alternator 1 must be decreased and the input to the prime mover of alternator 2 must be increased. If the input to the prime mover for alternator 1 is decreased, the drooping speed–load curve of this unit will be lowered. If the input to the prime mover of alternator 2 is increased at the same time, the drooping speed–load curve of this unit will rise. Adjusting the governor controls of both units causes the two drooping speed–load curves to cut the system frequency line (60 Hz) at the same load point. This means that the kilowatt outputs of the generators are equal and the power factor for both units will be unity, if there is no change in the field excitation of either alternator after the machines are in parallel.

The vector diagram for the two alternators is shown in Figure 12–38. An equal distribution of load is obtained by changing the input to the prime movers of the generators. The power factor of each alternator is unity. The currents for both alternators are 100 A. The kilowatt load on each unit is

$$\text{kW} = \frac{V \times I \times PF}{1000}$$

$$= \frac{240 \times 100 \times 1}{1000} = 24 \text{ kW}$$

The power output value of 24 kW at a power factor of 1.00 (unity) is well within the rated capacity of each generator. This means that there is no possibility of an overload.

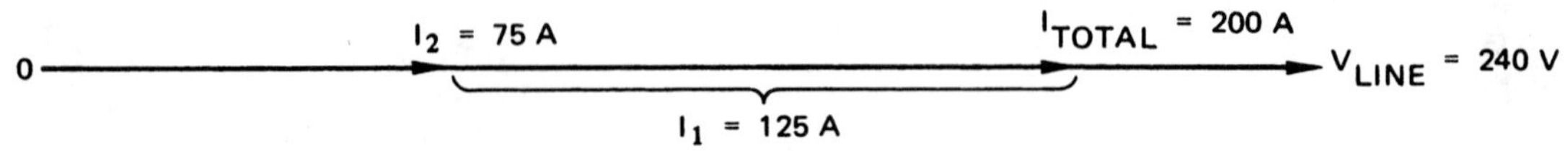

FIGURE 12–38 Alternators in parallel; each alternator is operating at unity power factor, with equal load distribution and supplying equal currents. (*Delmar/Cengage Learning*)

HUNTING

If the torque output of a prime mover pulsates, it may cause the rotor of the alternator to be pulled ahead of, and then behind, its normal running position. A diesel engine is one example of a prime mover that has a pulsating torque output. If an alternator is used with a diesel engine, the alternator rotor periodically will move slightly faster and then slower. This pulsating or oscillating effect is called *hunting*. It causes the current to surge back and forth between alternators operating in parallel. This condition is unsatisfactory and may become cumulative, resulting in such a large increase in the current between the alternators that the overload relays open the circuit.

Hunting can be corrected by the use of a heavy flywheel. A damper winding is often used in the rotating field structure to minimize the pulsating torque. Figure 12–2 showed a damper winding, which is often called an *amortisseur winding*. This winding is embedded in slots in each of the pole faces of the rotating field. The amortisseur winding consists of heavy conductors that are brazed or welded to two end rings. At the instant that hunting develops, the path of the armature flux changes so that it cuts the short-circuited conductors of the amortissuer winding. This change in the flux path produces induced currents in the damper winding. The currents oppose the force producing them (by Lenz's law). The proper design of the damper winding ensures that the effects of hunting are canceled by the induced currents in the short-circuited conductors.

SUMMARY

- An emf is generated
    1. when there is relative motion between the armature conductors and the field.
    2. in armature conductors when rotating in a magnetic field with stationary field poles.
    3. in stationary armature conductors when the field poles rotate past the conductors.
- A dc generator has stationary field poles and rotating armature conductors. The alternating voltage induced in the rotating armature conductors is changed to a direct voltage at the brushes by means of the commutator.
- AC generators (alternators) do not use commutators.
- Alternators are classified into two groups:
    1. Revolving armature machines with stationary fields; the kilovolt-ampere capacity and the low-voltage rating of these machines are usually rather small.
    2. Revolving field machines with stationary armature (stator) conductors.
        a. The field poles rotate inside the stator.
        b. Higher voltages can be generated without insulation failure.
        c. Higher current values can be obtained without arcing or heat production at the brushes and slip rings.

- There are two types of revolving fields:
    1. The salient field rotor (used with slow-speed alternators up to 1800 r/min).
    2. The cylindrical field rotor (used for speeds from 1800 to 3600 r/min; most steam turbine-driven alternators have cylindrical rotors).
- An ac generator cannot supply its own field current:
    1. The field excitation is direct current and is supplied from an external direct-current source.
    2. Slip rings and brushes are used to take the excitation current from the external source to the field windings.
    3. The field voltage is usually in the range between 100 and 250 V.
    4. The amount of power delivered to the field circuit is relatively small.
- The induced field voltage is large enough to damage equipment. To eliminate this danger, a special field discharge switch is used.
    1. In the closed position, the field discharge switch acts like a normal double-pole, single-throw switch.
    2. When the switch is opened, an auxiliary blade closes just before the main switch blades open.
    3. When the main switch blades are fully open, a circuit path still exists through the auxiliary switch blade.
    4. This path goes through the field discharge resistor, bypassing the field rheostat and the ammeter.
    5. The voltage induced in the field coils by the collapsing magnetic field dissipates quickly as a current through the field discharge resistor.
    6. A large machine may use a field contactor or field circuit breaker for the same purpose; each of these devices has two normally open main poles and one overlapping normally closed discharge pole.
- Large three-phase, revolving field-type ac generators can use one of several types of stationary armature (stator) windings; these windings generally consist of an even number of coils spaced around the perimeter of the stator core.
    1. The formed coil is machine wound and insulated before it is installed in the slots of the stator core.
    2. A full-pitch coil spans the distance between poles of opposite polarity.
    3. A fractional-pitch coil is smaller than the distance between poles of opposite polarity. All of the coils of one single-phase winding are known as a phase belt.
        a. Connecting the formed coils of a phase belt (single-phase winding) in series yields a maximum voltage.
        b. Reconnecting the same coils in parallel yields the maximum current possible at a lower voltage value.

4. A three-phase alternator has three separate windings that are placed in the slots of the stator core.
   a. The windings are arranged so that three voltages are produced and are 120 electrical degrees apart.
   b. The three single-phase windings (phase belts) may be connected in either delta or wye.
   c. The wye connection is more commonly used because it gives a higher terminal voltage: 1.73 times the phase winding voltage.

- Eddy currents in the stator core due to the flux of the revolving field can be reduced by the use of a laminated core. Such a core consists of thin strips of steel clamped together.
- Ventilating ducts in the core of the stator and ventilating passages in the steel frame prevent the temperature of the stator windings from becoming too high.
  1. When the alternator has a salient field rotor, it acts as a fan to aid cooling.
  2. High-speed alternators with cylindrical rotors normally have a cooling system that completely encloses the alternator. Either air or hydrogen is used in the system. Hydrogen is more efficient than air because it has almost seven times the heat conductivity of air. However, hydrogen is explosive and is costly to replace.
- When an armature conductor of a generator is cut by a rotating magnetic field, the induced voltage is

$$V_{induced} = \frac{BL_V}{10_8}$$

where

B = flux density, in gauss
L = active length of the armature conductors
v = velocity of the armature conductors

- When a north pole and a south pole move past a coil, one cycle of voltage is generated with a frequency of

$$f = \frac{P}{2} \times \frac{S}{60} = \frac{PS}{120}$$

where

f = frequency, in hertz

P = total number of poles

$\frac{P}{2}$ = number of pairs of poles

S = speed, in r/min

$\frac{S}{60}$ = speed, in r/s

- Controlling the field current:
  1. The voltage induced in any alternator depends on the field strength and the speed of the rotor.
  2. To maintain a fixed frequency, an alternator must operate at a constant speed.
  3. The magnitude of the generated voltage depends on the dc field excitation.
  4. The following method can be used to change or control the terminal voltage:
     a. A rheostat is connected in series with the separately excited field circuit.
     b. If an alternator is operated at a constant speed with a fixed field excitation current, the terminal voltage will change with an increase in the load current.
     c. The actual change in voltage is influenced by the power factor of the load circuit and the impedance of the armature windings.
- Percent voltage regulation is the change in the terminal voltage from a full-load to a no-load condition at a constant speed and a fixed field excitation current:

$$\text{Percent voltage regulation} = \frac{\text{no-load volts} - \text{full-load volts}}{\text{full-load volts}} \times 100$$

- Vector diagrams can be used to determine the induced voltage of an alternator for different load power factors:
  1. Unity power factor:

$$V_{\text{induced}} = \sqrt{(V_T + IR)^2 + (IX_L)^2}$$

  2. Lag ging power factor:

$$V_{\text{induced}} = \sqrt{(V_T \cdot \cos\theta + IR)^2 + (V_T \cdot \sin\theta + IX_L)^2}$$

  3. Leading power factor:

$$V_{\text{induced}} = \sqrt{(V_T \cdot \sin\theta + IR)^2 + (V_T \cdot \sin\theta + IX_L)^2}$$

- Voltage losses in the armature:
  1. The current output of the alternator at rated load is

$$I = \frac{VA}{V}$$

  2. The effective resistance and the inductive reactance cause voltage losses:

$$V_R = I \times R$$
$$V_{X_L} = I \times X_L$$

- At a lagging power factor, the induced voltage is greater than when the load power factor was unity.
- At a leading power factor, the induced voltage is less than when the load power factor was unity. In this case, the induced voltage is less than the terminal voltage.

- Synchronous reactance:
  1. The voltage drop due to the inductive reactance and the armature reaction have the same effect on the terminal voltage.
  2. Both of these effects are proportional to the armature current.
  3. These two effects are known as the synchronous reactance, X_{LS}.
- The synchronous impedance test is used to determine values of R and X_{LS}:
  1. In this test, the field excitation current and the alternator speed are kept at constant values.
  2. The line terminals of the alternator are shorted through an ammeter when the test switch is closed.
  3. The field excitation current increases until the current in the armature is nearly 150% of the rated full-load current. This value of current is recorded.
  4. The test switch is opened and the reading of a voltmeter connected across the generator terminal is recorded.
  5. The synchronous impedance is

     $$Z_S = V_{\text{open circuit}} \div I_{\text{short circuit}}$$

  6. The synchronous reactance can be found once the effective resistance is known:

     $$X_{LS} = \sqrt{Z_S^2 - R^2}$$

  7. The synchronous impedance of each single-phase winding for a wye- or delta-connected alternator is

     $$Z_S = \frac{V_S}{I}$$

  8. For an alternator connected in wye or delta, the synchronous impedance is

     $$Z_S = \frac{V_S\sqrt{3}}{I_{\text{line}}}$$

- A relatively constant alternator terminal voltage can be maintained under changing load conditions by the use of an automatic voltage regulator:
  1. As the terminal voltage decreases, a relay closes contacts in the regulator to short out a field resistor. There is a resulting increase in the field current, field flux, and induced voltage.
  2. An increase in the terminal voltage causes the relay to open the contacts across the field resistor. This action causes a decrease in the field current, field flux, and induced voltage.
  3. Many other types of automatic voltage regulators may be used with alternators. These regulators may use vacuum tubes, amplidynes, magnetic amplifiers, ignition rectifiers, controlled silicon rectifiers, or solid-state control devices.

- Saturation point:
  1. For a constant-speed generator, the generated voltage is a direct function of the flux value per pole.
  2. The flux is determined by the number of ampere-turns of the field pole. Because the number of turns on each field winding is constant, the flux is proportional to the dc excitation.
  3. Increasing the dc excitation increases the flux; therefore, the induced voltage also increases.
  4. A point is reached at which the flux no longer increases in proportion to the increase in dc excitation. This point is called the *saturation point.*
- Reading the saturation curve:
  1. The saturation curve shows an increase in induced voltage with an increase in the field current in an alternator.
  2. The ascending curve shows an increase in induced voltage as the field current is increased; the descending curve represents a decrease in induced voltage as the field current is decreased.
  3. The first part of the ascending curve is nearly vertical because the induced voltage is almost directly proportional to the increase in field current.
  4. When the alternator is being driven and the field current is zero, there is a small induced voltage. This voltage is due to the effects of residual magnetism after the magnetomotive force is removed from the field.
  5. As the ascending curve reaches the saturation point, it flattens out. From this point to the maximum induced voltage on the curve, the increase in induced voltage is not proportional to the increase in field current.
  6. The knee of the ascending curve is located immediately before and after the saturation point.
  7. After reaching the maximum induced voltage, the field current is decreased to zero. The resulting plot of these events is the descending curve.
  8. Note that the descending curve has a slightly higher induced voltage than does the ascending curve. This reaction is due to molecular friction or hysteresis effects in which the molecules of the iron in the magnetic circuit stay aligned even after the magnetomotive force is decreased.
  9. The operating point of a particular generator occurs near the center of the knee of the ascending curve. At this point, there is a maximum ac voltage output with minimal I^2R losses in the field circuit.
- The alternator nameplate contains the following data:
  1. The capacity of the alternator, in kVA
  2. The full-load terminal voltage
  3. The rated full-load current per terminal
  4. The number of phases

5. The frequency
6. The speed in r/min
7. The operating power factor
8. The dc field current and voltage
9. The maximum temperature rise

- Power losses in an alternator consist of fixed or stray losses such as
  1. the bearing and brush friction losses.
  2. windage losses.
  3. iron losses, including eddy current and hysteresis losses.
  4. copper losses, including the I^2R losses in the armature windings and the power expended in the separately excited field circuit.
- Alternator efficiency:
  1. Efficiency, $\eta = \frac{\text{output, watts}}{\text{input, watts}} \times 100$

$$\eta = \frac{\text{kVA output} \times \cos \angle \theta}{\text{kVA output} \times \cos \angle \theta + \text{copper losses} + \text{fixed losses}} \times 100$$

  2. An alternator should be operated at or near the rated full-load output to obtain the maximum efficiency.
  3. Alternators with capacities in the order of 200,000 kVA may have efficiency ratings as high as 96% at the full-load output.
- Advantages of paralleling alternators:
  1. They aid in scheduling of maintenance and emergency repairs on alternators.
  2. They allow the alternators on the line to be operated near their full-load rating (high efficiency range). Another alternator may be paralleled with the first one to meet peak demands.
  3. The generating plant capacity may be expanded to meet increased power demands by installing more alternators. These machines then operate in parallel with the existing generating equipment.
- To parallel ac generators:
  1. Observe the following conditions:
     a. The output voltages of the alternators must be equal.
     b. The frequencies of the alternators must be the same.
     c. The output voltages of the alternators must be in phase.
  2. If these three conditions are met, the alternators are said to be in synchronism.
- Synchronizing two alternators:
  1. Three-lamp method
     a. The three lamps go on and off in unison depending on the frequency difference between the two alternators.

(1) For this case, a slight adjustment in the speed of the prime mover for the alternator coming on line will make the frequency of this machine the same as the alternator presently on the line.

(2) When all three lights go out, the instantaneous electrical polarity of the second machine will equal that of the alternator on the line. The second machine can be brought on line, and the generators will be paralleled.

b. The three lamps will light and go off, but not in unison. In this case, the phase sequence or phase rotation of the second generator is not the same as the alternator on the line. By interchanging the connections of any two leads of the second alternator, the phase sequence can be changed. The steps in part 1 can be followed to adjust the frequency of the second machine so that it can be paralleled.

2. Three lamps dark method
   a. This method is used to determine the phase sequence of an alternator.
   b. Once the phase sequence is known, permanent connections can be made between the stator windings, the switching equipment, and the station bus bars. It is not necessary to determine the phase sequence each time the alternator is paralleled once the equipment is marked correctly.
   c. This method is also used to indicate when alternators are in synchronism.
3. Two bright, one dark method:
   a. This method is never used to determine the phase sequence.
   b. It is used to indicate the synchronism of two alternators.
   c. When the incoming alternator is in synchronism (ready to be paralleled), two lamps in lines 1 and 2 will have a maximum brightness and the lamp in line 3 will be dark.
4. There are disadvantages to both methods of testing:
   a. There may be a large voltage difference across the synchronizing lamps (even when they are dark); thus, a large voltage difference and phase difference may be present.
   b. When an attempt is made to bring the incoming alternator into the bus bar circuit system with the other machines, a large disturbance in the electrical system may result in damage to the alternator windings.

- Use of the synchroscope:
  1. Once the phase sequence is known to be correct and permanent connections are made, a synchroscope can be used.
  2. The single-phase synchroscope indicates synchronism accurately.
  3. A synchroscope gives an accurate indication of the differences in the frequency–phase relationship between two voltages.
     a. A pointer rotates over a dial face. When the pointer stops, the frequencies of the two alternators are the same.

   b. When the pointer stops in a vertical upward position, the frequencies are equal and the voltages are in phase. This means that the alternators are in synchronism and the alternator switch can be closed to parallel both machines.

- Once the two alternators are operating in parallel, the load is shared between them.
  1. The load taken by each machine is proportional to the kVA rating of the machine.
  2. Changing the field excitation of each dc generator will divide the load between the generators connected in parallel. The same method cannot be used to divide the kilowatt load between the two alternators in parallel.
  3. A different method is used to divide the load between two alternators.
     a. The prime movers for such alternators should have drooping speed–load characteristic curves.
     b. To divide the load equally between the two machines, the input to the prime mover must be increased. The governor on the prime mover of the light-load alternator is opened slightly. As a result, there is an increase in both the horsepower of the light-load alternator and the power output of the alternator. At the same time, the governor on the prime mover of the heavy-load alternator is closed very slightly. This action causes decreases in the input to the prime mover, the horsepower input to the heavy-load alternator, and the power output (in kW) of the same alternator.
     c. Careful adjustment of the governors of both prime movers can result in the speed–load characteristic curve of each cutting the system frequency line at the *same* load point. Thus, there is a satisfactory load division between the machines.
     d. This adjustment of the prime movers must be obtained *without* causing a change in the frequency.

- When alternators are operating in parallel, a faulty governor may cause the speed of an alternator to increase; thus, the machine may be pulled out of synchronism.
  1. Certain reactions within the alternator will prevent this condition.
  2. If the governor malfunctions so that it cuts out completely and shuts off the input to the prime mover, the alternator will operate as a synchronous motor until the governor fault is corrected.

- Hunting
  1. is a pulsating or oscillating effect.
  2. of the prime mover causes a torque that may cause the rotor of the alternator to be pulled ahead of, and then behind, its normal running position.
  3. causes the current to surge back and forth between alternators operating in parallel; this condition is unsatisfactory and may become cumulative, resulting in the circuit being opened by the overload relays.
  4. can be corrected by using

a. a heavy flywheel.

b. a damper winding in the rotating field structure. Such a winding is often called an *amortisseur winding*. The proper design of this winding ensures that the effects of hunting are canceled by the induced currents in the short-circuited conductors.

Achievement Review

1. List two advantages of the rotating field alternator, as compared with the rotating armature ac generator.

2. a. Where is the salient pole rotor used?
   b. Where is the cylindrical rotor used?

3. a. Explain why a field discharge resistor is used with the separately excited field circuit.
   b. Draw a schematic wiring diagram of a separately excited field circuit of an alternator. Include a field discharge switch, a field discharge resistor, an ammeter, and a field rheostat. Connect the circuit so that the field rheostat and the ammeter are not in the field discharge circuit path.

4. a. What methods are used to cool the windings of a high-speed turbine-driven alternator having a large kVA capacity?
   b. There are fewer problems in cooling the windings of slow-speed alternators using salient pole rotors. Why?

5. A 25-kVA, 250-V, 60-Hz, 1800-r/min, single-phase alternator has an armature resistance of 0.12 Ω and an armature reactance of 0.5 Ω. The generator delivers the rated load output at a power factor of 1.00 (unity) to a non-inductive load. Determine the induced voltage. (Neglect any armature reaction.)

6. a. Determine the induced voltage of the alternator in question 5 when it delivers the rated output to a load with a lagging power factor of 0.8660.
   b. Determine the induced voltage of the alternator in question 5 when it delivers the rated output to a load with a leading power factor of 0.8660.

7. a. Define voltage regulation as it is used with alternators.
   b. The full-load terminal voltage of an alternator is 240 V. The load is removed. The no-load terminal voltage increases to 265 V at the same speed and field excitation. What is the percent voltage regulation of the alternator?

8. A three-phase, wye-connected alternator is rated at 2000 kVA, 4800 V, 60 Hz. In a short-circuit synchronous impedance test, the field excitation current is increased until the three line ammeters indicate nearly 150% of 360 A, the rated line current. The field excitation current and the speed are kept constant and the three-pole switch is opened. The voltmeter indicates 2250 V. The dc resistance between the line terminals is 0.4 Ω, and the ratio of effective to ohmic resistance is 1.5. Determine

a. the synchronous impedance.
b. the effective resistance.
c. the synchronous reactance.

9. Determine the no-load voltage and the percent voltage regulation for the alternator in question 8. Assume that the alternator is delivering the rated current to a noninductive, unity power factor load.

10. a. Determine the no-load voltage and the percent voltage regulation for the alternator in question 8 assuming that it is delivering the rated current to a load with a lagging power factor of 0.8.
    b. Find the no-load voltage and the percent voltage regulation for the same alternator when it delivers the rated current to a load with a leading power factor of 0.8.

11. Explain what is meant by the terms
    a. synchronous reactance.
    b. synchronous impedance.

12. a. Draw a typical saturation curve for an alternator.
    b. Give two reasons why saturation curves are used.

13. a. What are the fixed losses of an alternator?
    b. What are the copper losses of an alternator?
    c. How is the full-load efficiency of an alternator determined?

14. A three-phase, wye-connected alternator is rated at 500 kVA, 2400 V, 60 Hz. Determine
    a. the full-load kilowatt output of the generator at 80% lagging power factor.
    b. the full-load current per line terminal for the alternator.
    c. the full current rating of each of the phase windings.
    d. the voltage rating of each of the phase windings.

15. A three-phase, delta-connected, diesel-driven alternator is rated at 50 kVA, 240 V, 60 Hz.
    a. Determine
       (1) the full-load current rating per line terminal for the alternator.
       (2) the full-load current rating of each of the phase windings.
       (3) the voltage rating of each of the phase windings.
    b. If the alternator has a rated speed of 240 r/min, how many poles are required for the rotating field?

16. A 5-kVA, 208-V, three-phase alternator is connected in wye.
    a. Determine
       (1) the line current per terminal at full load.
       (2) the coil current at full load.
       (3) the voltage of each phase winding.

b. Assuming that this alternator is reconnected in delta, compute the new terminal voltage and the current at full load.

17. A three-phase, 60-Hz, wye-connected turbine-driven alternator has three single-phase windings. Each winding is at 8000 V and 625 A. The alternator has four poles. Determine
    a. the kilovolt-ampere rating of the alternator.
    b. the kilowatt output of the alternator when it delivers the rated current to a load with an 80% lagging power factor.
    c. the line voltage.
    d. the rated full-load line current.
    e. the speed in r/min of the revolving field of the alternator.
18. For the turbine-driven, high-speed alternator described in question 17, answer the following questions:
    a. Why is the field, rather than the armature, the rotating member?
    b. Why is a cylindrical rotating field used, rather than one with salient poles?
    c. How is the dc excitation current supplied to the rotor of the rotating field alternator?
19. List three reasons why ac generators are operated in parallel.
20. A three-phase, wye-connected alternator is rated at 10,000 kVA, 11,000 V, and 60 Hz. Determine
    a. the full-load kilowatt output of the ac generator at a lagging power factor of 80%.
    b. the full-load line current of the alternator.
    c. the voltage rating of each of the three windings.
    d. the horsepower input to the alternator when it delivers the rated load output at an efficiency of 92% and a lagging power factor of 80%.
21. A 240-V, single-phase, 60-Hz, revolving field alternator delivers 30 kW to a non-inductive load. The generator efficiency is 86%, and the stray power losses are 2000 W. The separately excited field requires 6 A at 240 V, dc. Determine
    a. the full-load current.
    b. the copper losses in the stator winding.
22. a. List the steps, in chronological order, required to place a three-phase alternator in parallel with another ac generator. Assume that this is the first time that the alternator is placed in service.
    b. After paralleling the alternators, what means are used to redistribute the kilowatt load between the two alternators? Assume that the frequency is held constant.
23. a. Diagram the "three lamps dark" method of synchronizing an alternator to the bus bars.

b. How is this method used to determine whether the phase sequence of an alternator is correct with reference to the station bus bars?

24. a. Diagram the "two lamps bright, one lamp dark" method.
    b. How is this method used to determine when an incoming alternator is in phase with the bus bars?

25. a. What does the term *hunting* mean as applied to slow-speed alternators?
    b. How does the amortisseur or damping winding on the field rotor decrease the effects of hunting?
    c. What other way is used to minimize the effects of hunting?

26. Explain how equal load distribution is obtained between alternators in parallel.

27. Explain what happens when an attempt is made to shift the kilowatt load between alternators by changing the field excitation. Assume that the input to the prime movers of the alternators is not changed.

28. List the data commonly found on the nameplate of an alternator.

13

Transformers

Objectives

After studying this unit, the student should be able to

- describe the construction of a simple transformer by naming its parts and showing their relationship to each other.
- define the terms stepdown transformer, stepup transformer, transformer efficiency, exciting current, ampere-turns, and primary winding to secondary winding voltage and current ratios.
- describe leakage flux and compare the primary winding to secondary winding voltage and current ratios.
- measure core losses using the core loss open-circuit test.
- calculate the copper losses in the windings by measurement and by the short-circuit test.
- calculate the efficiency of a transformer using the percent efficiency formula.
- determine the percent voltage regulation of a transformer.
- describe the ASA standard system of marking transformer leads.
- perform standard tests to determine the high-voltage and low-voltage leads of transformers.
- state the standards governing transformer polarity developed by the ASA and NEMA.
- describe the paralleling of transformers and the requirements that must be met before they can be connected.
- explain feedback from one secondary winding to the other primary winding, and state the steps that must be taken to minimize this hazard.
- list the advantages of a single-phase, three-wire service over a single-phase, two-wire service.
- describe the dangers of a three-wire, single-phase circuit with an open neutral.
- explain the most efficient method(s) of cooling and describe the type of insulating oil used with transformers.
- list the information provided on the nameplate of a transformer.
- explain the necessity of meeting the current voltage and frequency operating requirements for transformers.

The transformer may be described as an alternating-current device that transfers energy from one ac circuit to another ac circuit. This transfer takes place with a change in the voltage, but with no change in the frequency. A *stepdown transformer* receives energy at the input winding at one voltage and delivers energy from the output winding at one voltage. A *stepup transformer* delivers energy from the output winding at a higher voltage than at the input. If the rated nameplate voltage is used for the winding selected as the input, then the transformer can be used as either a stepup or a stepdown device. A diagram of a simple transformer is shown in Figure 13–1.

The transformer is a stationary device. In its simplest form, it consists of a laminated iron core. The input and output windings are wound on this common core. The primary winding is the input winding, which is connected to the energy source. The output winding is the secondary winding and is connected to the load. Energy is transferred from the primary winding to the secondary winding by an alternating magnetic flux that is developed by the primary winding.

TRANSFORMER EFFICIENCY

A transformer does not require any moving parts to transfer energy. This means that there are no friction or windage losses, and the other losses are slight. The resulting efficiency of a transformer is high. At full load, the efficiency of a transformer is between 96% and 97%. For a transformer with a very high capacity, the efficiency may be as high as 99%. Transformers can be used for very high voltages because there are no rotating windings and the stationary coils can be submerged in insulating oil. Transformer maintenance and repair costs are relatively low because of the lack of rotating parts.

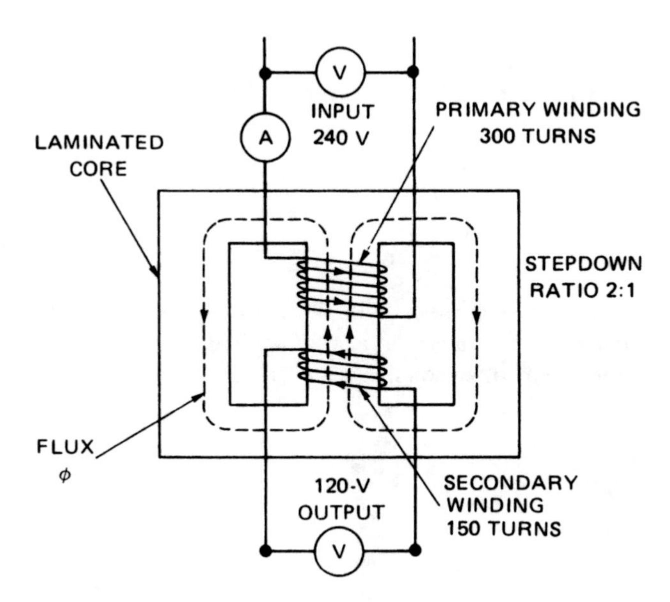

FIGURE 13–1 Diagram of a transformer (*Delmar/Cengage Learning*)

THE EXCITING CURRENT

When the primary winding of a transformer is connected to an alternating voltage, there will be a small current in the input winding. This current is called the *exciting current* and exists even when there is no load connected to the secondary.

The exciting current sets up an alternating flux in the core. This flux links the turns of both windings as it increases and decreases in opposite directions. As the flux links the turns of the secondary winding, an alternating voltage is induced in the secondary. This voltage has the same frequency as, but its direction is opposite that of, the primary winding voltage. The same voltage is induced in each turn of both windings because the same flux links the turns of both windings. As a result, the total induced voltage in each winding is directly proportional to the number of turns in that winding.

PRIMARY AND SECONDARY VOLTAGE RELATIONSHIPS

The relationship between the induced voltage and the number of turns in a winding is given in the following expression

$$\frac{V_P}{V_S} = \frac{N_P}{N_S} = \alpha$$

where

α = transformation ratio, or turns ratio
V_P = voltage in the primary winding
V_S = voltage in the secondary winding
N_P = number of turns in the primary winding
N_S = number of turns in the secondary winding

In this expression, V_P is the voltage induced in the primary according to Lenz's law. This induced voltage is only 1% or 2% less than the applied primary voltage in a typical transformer. Thus, V_P and V_S, respectively, are used to represent the input and output voltages of the transformer.

PROBLEM 1

Statement of the Problem

A transformer has 300 turns on its high-voltage winding and 150 turns on its low-voltage winding. It is used as a stepdown transformer. With 240 V applied to the high-voltage primary winding, determine the induced voltage on the secondary winding.

Solution

$$\frac{V_P}{V_S} = \frac{N_P}{N_S}$$

$$\frac{240}{V_S} = \frac{300}{150}$$

$$36{,}000 = 300\ V_S$$
$$V_S = 120\ V$$

PRIMARY AND SECONDARY CURRENT RELATIONSHIPS

When a load is connected across the terminals of the secondary winding, the instantaneous direction of the current will tend to oppose the effect that is producing the current.

As an example of this effect, consider the simple transformer diagram of Figure 13–2. A noninductive load is connected to the terminals of the secondary winding. The secondary current sets up a magnetomotive force that opposes the flux (ϕ) of the primary winding. As a result, both the primary flux and the counterelectromotive force in the primary winding are reduced. The primary current increases because the impressed primary voltage has less opposition from the counterelectromotive force (induced voltage). The increase in the primary current supplies the energy required by the load connected to the secondary winding.

The ampere-turns of the primary winding increase the magnetizing flux.

It was stated at the beginning of this unit that the exciting current is small when compared to the rated current. Most transformer calculations neglect the exciting current. In addition, it is assumed that the primary and secondary ampere-turns are equal, as determined by the following equation

$$I_P N_P = I_S N_S$$

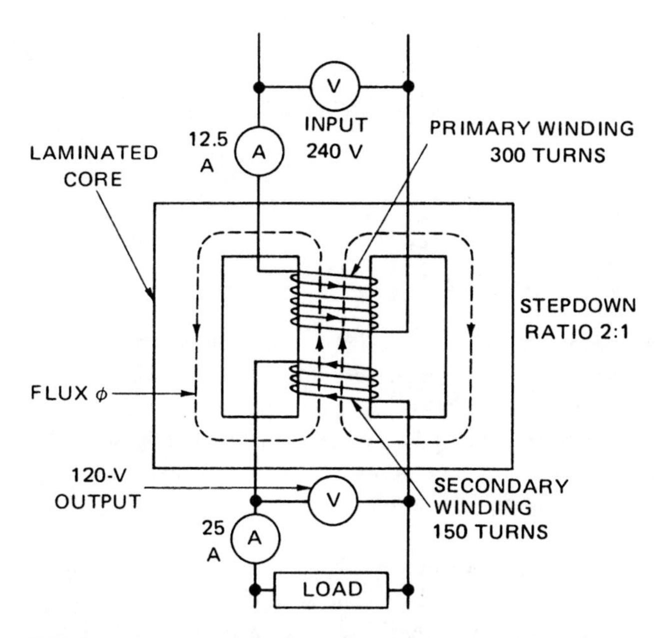

FIGURE 13–2 Transformer diagram (*Delmar/Cengage Learning*)

where

I_P = primary amperes
I_S = secondary amperes
N_P = number of turns in primary winding
N_S = number of turns in secondary winding

PROBLEM 2

Statement of the Problem

The transformer shown in Figure 13–2 delivers 25 A at 120 V to a load with a unity power factor. Neglect the exciting current and determine

1. the primary current.
2. the secondary ampere-turns.
3. the power in watts taken by the load.

Solution

1.
$$NP = 300 \text{ turns}$$
$$N_S = 150 \text{ turns}$$
$$I_S = 25 \text{ A}$$
$$I_P N_P = I_S N_S$$
$$I_P \times 300 = 25 \times 150$$
$$I_P = 12.5 \text{ A}$$

2. Secondary ampere-turns $= I_S \times N_S = 25 \times 150 = 3750$ ampere-turns

3. Output $= V_S \times I_S \cos\angle\theta$

$$= 120 \times 25 \times 1.0 = 3000 \text{ W}$$

LEAKAGE FLUX

Some of the lines of flux produced by the primary winding do not link the turns of the secondary winding in most transformers. The magnetic circuit for this primary leakage flux is in air. In other words, the leakage flux does not follow the circuit path through the core. This flux links the turns of the primary winding, but it does not link the turns of the secondary winding. Because the leakage flux uses part of the impressed primary voltage, there is a reactance voltage drop in the primary winding. The result is that both the secondary flux linkages and the secondary induced voltage are reduced.

A second leakage flux links the secondary turns but not the primary turns. This flux also has its magnetic path in air and not in the core. The secondary leakage flux is proportional to the secondary current. There is a resulting reactance voltage drop in the secondary winding. Both the primary and the secondary leakage fluxes reduce the secondary terminal voltage of the transformer as the load increases. The leakage flux of a transformer can be controlled by the type of core used. The placement of the primary and secondary coils on the legs of the core is also a controlling factor in the amount of leakage flux produced.

EXCITING CURRENT AND CORE LOSSES

When there is no load attached to the secondary winding, the current input to the primary winding usually ranges from 2% to 5% of the full-load current. The primary current at no load is called the *exciting current*. This current supplies the alternating flux and the losses in the transformer core. These losses are known as *core losses* and consist of eddy current losses and hysteresis losses. As the magnitude of the alternating flux increases and decreases, the metal core is cut by the flux, as are the turns of the primary and secondary coils. Voltages are thus induced in the metal core and give rise to eddy currents. These eddy currents circulate through the core and cause I^2R loses, which must by supplied by the exciting current. Eddy current losses can be reduced by laminating the core structure. Each lamination normally is coated with a film of insulating varnish. When a protective film of varnish is not used, the oxide coating on each lamination still reduces the eddy current losses to some extent.

The core structure also experiences a hysteresis loss. In each second, the millions of molecules in the core structure are reversed many times by the alternating flux. Power is required to overcome this molecular friction in the core. This power is supplied by the primary winding. To decrease the amount of power used, a special steel such as silicon steel is used. Eddy current and hysteresis losses are called the *core losses*. In a typical transformer, these losses are relatively small.

To supply the magnetizing flux, the exciting current has a relatively large component of quadrature or magnetizing current. A smaller in-phase component of current supplies the core losses. For an actual transformer, the no-load power factor ranges between 0.05 and 0.10 lagging. This means that the phase angle between the exciting current and the impressed primary voltage is between 84° and 87° lagging. The exciting current can be measured by a method known as the *core loss open-circuit test*.

Measuring Core Losses

The connections for the core loss test are shown in Figure 13–3A. The high-voltage winding of the transformer is rated at 2400 V. The low-voltage winding is rated at 240 V and is used as the primary side of the transformer. This arrangement makes it convenient to use 240 V for the potential circuits of the wattmeter and voltmeter. The test circuit can also use 240 V safely.

> CAUTION: The high-voltage winding leads and terminal connections must be well insulated and barricaded so that no one can contact this high-voltage circuit. This test circuit can be considered hazardous because the transformer operates as a stepup transformer with a 2400-V potential across the leads of the high-voltage secondary winding.

The losses of a transformer at no load are small. This means that instrument errors must be checked. The dashed-line voltmeter connections in Figure 13–3A mean that the

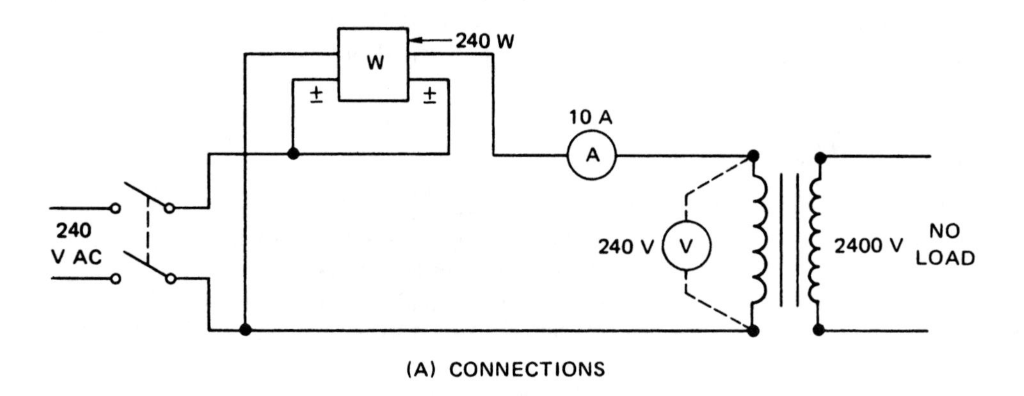

(A) CONNECTIONS

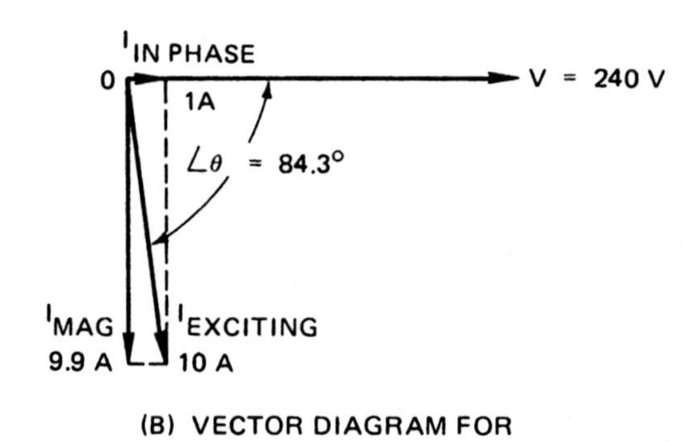

(B) VECTOR DIAGRAM FOR EXCITING CURRENT

FIGURE 13–3 Core loss test (*Delmar/Cengage Learning*)

voltmeter is to be disconnected when a wattmeter reading is to be taken. The voltmeter must be disconnected so that the wattmeter will not indicate the power taken by the voltmeter.

If the rated voltage and frequency are used in this circuit, then the rated alternating flux exists in the core. The resulting core loss is normal. The wattmeter indicates the core loss in watts, and the ammeter indicates the exciting current.

The error introduced by the copper loss in the primary can be neglected. The following example shows why this is so. The primary resistance (R_p) equals 0.007 Ω. The primary copper loss is 102 times 0.007 equals 0.7 W. This value of 0.7 W is small enough to be neglected, when compared to the 140 W indicated by the wattmeter. This assumption still holds for smaller transformers because they have smaller values of exciting current and core losses.

The core loss can also be measured when the high-voltage side of the transformer is used as the primary winding. In this case, a 2400-V source is required. After compensating for instrument losses, it is found that the core loss is the same as for the case when the low-voltage side of the transformer is used as the primary. This result is to be expected because both windings are wound on the same core. Because the same number of ampere-turns will produce the same alternating flux, when either winding is used as the primary, the core loss in watts will be the same for both cases.

Figure 13–3B shows the vector relationship between the exciting current (10 A), its in-phase component (1 A), and its quadrature (magnetizing) lagging component. The phase angle between the line voltage and the exciting current is 84.3°. The core loss is 240 W.

PROBLEM 3

Statement of the Problem

Figure 13–3A shows the connections for a core loss test on a 50-kVA, 60-Hz, single-phase transformer. The high-voltage winding of the transformer is rated at 2400 V and the low-voltage winding is rated at 240 V. The low-voltage winding is used as the primary for the core loss test. With 240 V applied to the primary winding, the wattmeter indicates a core loss of 240 W. The ammeter indicates an exciting current of 10 A. Determine

1. the power factor and the phase angle.
2. the in-phase component of the current.
3. the quadrature lagging or magnetizing component of the current.

Solution

1. $$\text{Power factor} = \frac{W}{V \times I} = \frac{240}{240 \times 10} = 0.10 \text{ lagging}$$

$$= 84.3^\circ \text{ lagging}$$

2. $$\text{Power factor} = \frac{I_{\text{in phase}}}{I}$$

$$0.10 = \frac{I_{\text{in phase}}}{10}$$

$$I_{\text{inphase}} = 1 \text{ ampere}$$

3. $$\sin \angle\theta = 0.9951$$

$$\sin \angle\theta = \frac{I_{\text{quadrature}}}{I}$$

$$0.9951 = \frac{I_{\text{quadrature}}}{10}$$

$$I_{\text{quadrature}} = 9.9 \text{ A}$$

COPPER LOSSES USING DIRECT CURRENT

The copper losses of a transformer consist of the I^2R losses in both the primary and secondary windings. If the effective resistance of each winding is known, the copper losses of a transformer can be found readily. Recall that the approximate ac resistance of an alternator is found by multiplying the dc resistance by 1.4 or 1.5. For transformers, however, the windings are not embedded in the slots of a stator, but consist of coils wound on a core. This means that the difference between the ohmic resistance and the effective resistance is small. Generally, the ac or effective resistance of a transformer is obtained by measuring the dc resistance of the winding and multiplying it by 1.1.

Figure 13–4 shows the connections required to measure the dc, or ohmic resistance, of a winding. A current-limiting resistance is used in this circuit. It is important that small dc currents be used. The voltmeter should be disconnected *before* the circuit is deenergized because the windings are highly self-inductive. The large value of the induced voltage could damage the voltmeter movement and pointer.

PROBLEM 4

Statement of the Problem

The resistance for each winding of a 50-kVA, 2400/240-V, 60-Hz, single-phase, step-down transformer is measured with direct current. The dc resistance of the high-voltage winding is 0.68 Ω. The low-voltage winding has a dc resistance of 0.0065 Ω. Determine

1. the effective resistance of each winding.
2. the total copper losses at full load.

Solution

1. The effective resistance of the primary winding is

   $0.68 \times 1.1 = 0.75\ \Omega$

   The effective resistance of the secondary winding is

   $0.0065 \times 1.1 = 0.0072\ \Omega$

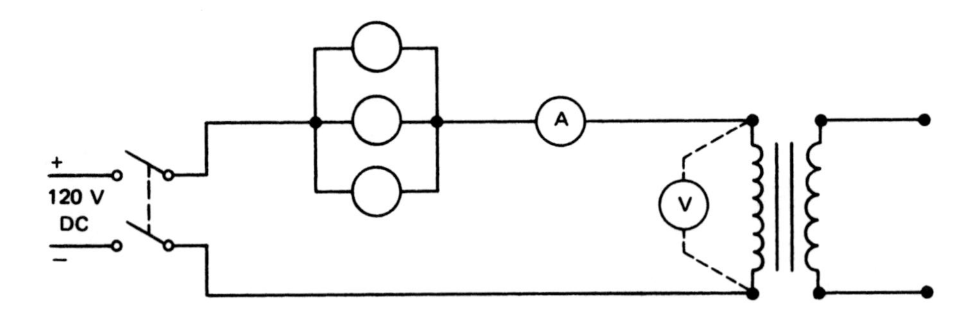

FIGURE 13–4 Measuring dc resistance of transformer windings (*Delmar/Cengage Learning*)

2. The transformer losses are small. Thus, when determining the full-load current rating of each winding, it can be assumed that the volt-ampere input and output are the same:

$$I_P = \frac{VA}{V_P} = \frac{50{,}000}{2400} = 20.83 \text{ A}$$

$$I_S = \frac{VA}{V_S} = \frac{50{,}000}{240} = 208.3 \text{ A}$$

$$\text{Total copper loss} = I_P^2 R_P + I_S^2 R_S$$

$$= (20.83^2 \times 0.75) + (208.3^2 \times 0.0072) = 637.8 \text{ W}$$

TRANSFORMER LOSSES AND EFFICIENCY

Transformer losses consist of copper losses and core losses. Copper losses are the I^2R losses in the primary and secondary windings. These losses increase as the load current in the primary and secondary windings increases. The copper losses can be calculated from the current and the effective resistance for each winding.

Transformer efficiency is the ratio of the output in watts to the input in watts. The load connected to the transformer secondary often has a power factor other than unity. In such cases, the output is the product of the secondary voltage and current plus the power factor of the load. The input equals the output plus the total losses. These losses include the copper losses and the core losses.

The core losses can be measured as shown in Figure 13–3A. These losses consist of eddy current and hysteresis losses. The core losses remain nearly constant at all load points if the frequency and primary voltage remain constant.

If the losses are known or can be calculated, for any given load point, the transformer efficiency can be determined. The basic efficiency formula is as follows:

$$\eta = \frac{\text{output, watts}}{\text{input, watts}} \times 100 = \frac{\text{output, watts}}{\text{output} + \text{losses}} \times 100$$

The Greek letter η (eta) represents the efficiency. A more practical form of this equation is used in transformer calculations

$$\eta = \frac{V_S \bullet I_S \bullet \cos\angle\theta}{V_S \bullet I_S \bullet \cos\angle\theta + \text{core loss, watts} + I_P^2R_P + I_S^2R_S} \times 100$$

where

$V_S \bullet I_S \cos\angle\theta$ = output of the secondary winding, in watts
$I_P^2R_P$ = copper loss in the primary winding
$I_S^2R_S$ = copper loss in the secondary winding

The efficiency of a transformer can be found by loading it at various percentages of the load from no load to full load and measuring the input and output power values. However, the losses are quite small. Unless extremely accurate instruments are used, the results will be of little value. For example, the efficiency of a transformer at full load is usually in the range from 96% to 98%. Typically, an indicating wattmeter can have an error of one percent. This means that the error in the calculations can be as much as 50%. In addition, this method requires various loading devices having the correct current, voltage, and power factor ratings. It is both inconvenient and costly to provide such loading devices. This is particularly true for transformers having extremely large kVA capacities.

The preferred method of determining the efficiency of a transformer is to measure the losses and add this value to the nameplate output to obtain the input. The following example shows how the efficiency is obtained by measuring the losses.

PROBLEM 5

Statement of the Problem

The 50-kVA, 2400/240-V transformer described in problems 3 and 4 delivers the rated load output at a unity power factor. The core loss was found to be 240 W. The primary copper loss was 325.4 W and the secondary copper loss was 312.4 W. Find the efficiency at the rated output and unity power factor.

Solution

The efficiency is determined as follows:

$$\eta = \frac{V_S \bullet I_S \bullet \cos \angle\theta}{V_S \bullet I_S \bullet \cos \angle\theta + \text{core loss, watts} + I_P^2R_P + I_S^2R_S} \times 100$$

$$= \frac{50{,}000 \times 1.0}{50{,}000 \times 1.0 + 240 + 325.4 + 312.4} \times 100 = 98.3\%$$

The Short-Circuit Test

Another method of measuring the copper losses of a transformer is the *short-circuit test*. In this test, the high-voltage winding is used as the primary and the low-voltage winding is short-circuited.

The connections for the short-circuit test are shown in Figure 13–5. A variable resistor is in series with the primary winding. In this way, the current input can be controlled. The series resistor is adjusted until the full-load current circulated in both the primary and secondary windings. When the low-side winding of the transformer is short-circuited, only 3% to 5% of the rated primary voltage is required to obtain the full-load current in both windings. This voltage is called the *impedance voltage*. In other words, the impedance voltage is that voltage required to cause the rated current to flow through the impedances of the primary and secondary windings. The ratio of the impedance voltage to the rated terminal voltage yields the percentage impedance voltage, which is in the range of 3% to 5%.

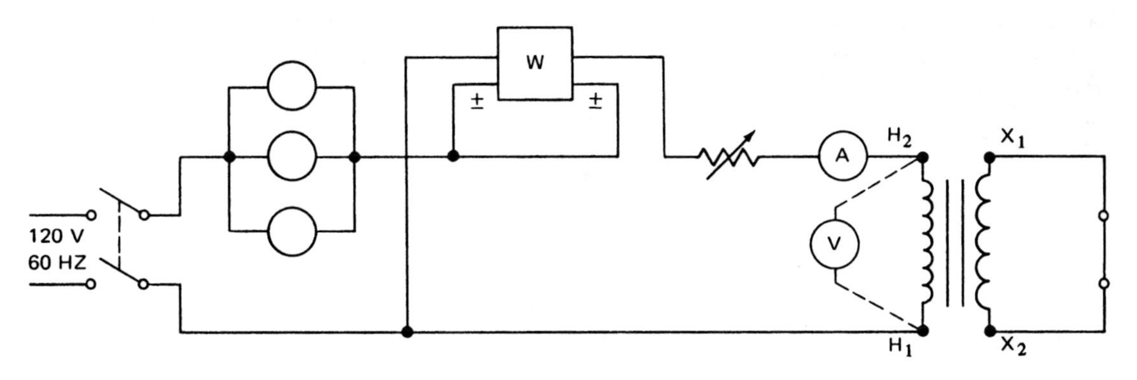

FIGURE 13–5 Short-circuit test (*Delmar/Cengage Learning*)

The wattmeter in Figure 13–5 indicates the total copper losses of the transformer referred to the primary side. The wattmeter reading includes the core losses, which are so small that they can be neglected. The core losses are small because the voltage impressed on the primary winding in this test is very low.

To make the short-circuit test, the core loss of the transformer must be determined using the connections of Figure 13–3A. Using the 50-kVA transformer described in the previous problems, the core loss is 240 W. The readings on the instruments shown in Figure 13–5 will be as follows: a current of 21 A, a power value of 640 W, and a voltage of 80 V. The value of 640 W represents the total copper losses of the transformer. If the dc ohmic values are used to compute the copper losses in the high- and low-voltage windings, then the total losses are

$$I_P^2R_P + I_S^2R_S = 20.83^2 \times 0.68 + 208.3^2 \times 0.0065 = 577 \text{ W}$$

The ratio of 640 W to the total copper loss (577 W) shows how much greater the actual effective resistance is when compared to the dc (ohmic) resistance of the windings:

$$\frac{640}{577} = 1.11$$

This value is nearly the same as the constant of 1.10 used to convert the ohmic resistance to the effective resistance for transformers.

The entire equivalent effective resistance of the transformer referred to the primary side can be found for a wattmeter reading of 640 W and a current of 21 A (at almost full load). When referred to the primary side, this equivalent resistance, R_{OP}, is

$$R_{OP} = \frac{W}{I_P^2}$$

$$= \frac{640}{21^2}$$

$$= \frac{640}{441} = 1.45 \ \Omega$$

The secondary is the output side of the transformer. Therefore, the entire resistance of the transformer must be referred to the secondary side as an equivalent effective resistance, R_{OS}:

$$R_{OS} = R_{OP}\left(\frac{N_S}{N_P}\right)^2$$

$$= 1.45\left(\frac{1}{10}\right)^2 = 0.0145\ \Omega$$

The ratio of turns squared in this equation is very important when matching transformer impedances. The application of the turns squared term is shown in the following equations:

$$\mathbf{V_S = \frac{N_S}{N_P} V_P}$$

$$\mathbf{I_S = \frac{N_P}{N_S} I_P}$$

Because $Z_{OS} = V_S/I_S$, it follows that

$$Z_{OS} = \frac{N_S/N_P\,(V_P)}{N_P/N_S\,(I_P)} = \left(\frac{N_S}{N_P}\right)^2 Z_{OP}$$

Thus, for resistive and inductive loads, we obtain

$$\mathbf{R_{OS} = R_{OP}\left(\frac{N_S}{N_P}\right)^2}$$

$$\mathbf{Z_{OP} = Z_{OP}\left(\frac{N_S}{N_P}\right)^2}$$

The term R_{OS} is defined as the resistance of the primary when referred to the secondary. The equation for R_{OS} is

$$\mathbf{R_{OS} = R_S + \left(\frac{N_S}{N_P}\right)^2 R_P}$$

Values of R_S and R_P were determined previously. Substituting these values in the preceding equation gives

$$R_{OS} = 0.0072 + \left(\frac{1}{10}\right)^2 0.75 = 0.0147$$

This value of R_{OS} checks closely with the previous value of 0.0145.

Similarly

$$R_{OP} = R_P + \left(\frac{N_P}{N_S}\right)^2 R_S$$

At the rated load, the total copper losses of this transformer are determined as follows. The full-load current is

$$I_S = \frac{VA}{240}$$

$$= \frac{50{,}000}{240} = 208.3 \text{ A}$$

$$\text{Total copper losses} = I_S^2 \times R_{OS}$$

$$= 208.3^2 \times 0.0145$$

$$= 629 \text{ W}$$

At the rated load and unity power factor, the efficiency of the transformer is

$$\eta = \frac{V_S \cdot I_S \cdot \cos \angle\theta}{V_S \cdot I_S \cdot \cos \angle\theta + \text{core loss, watts} + \text{total copper losses}} \times 100$$

$$= \frac{240 \times 208.3 \times 1.0}{(240 \times 208.3 \times 1.0) + 240 + 629} \times 100 = 98.3\%$$

This value of efficiency, as determined by the short-circuit test, is the same as the efficiency found in problem 5.

Assume that the same transformer is operated at 50% of the rated current capacity to supply a load with a unity power factor. The core loss is the same because the magnetizing flux is the same. However, the copper loss will decrease to one-fourth of its full-load value. Note that this decrease occurs because of the squaring operation in the I^2R formula. The I^2 multiplier is only one-fourth of the original value. Use the previous formula and calculate the actual efficiency for this load condition:

$$\eta = \frac{240 \times 104.2 \times 1.0}{(240 \times 104.2 \times 1.0) + 240 + 157} \times 100 = 98.4\%$$

If the transformer supplies a load having a power factor other than unity, the output (in watts) will decrease. However, the losses will be the same as those for a unity power factor load. For example, if the 50-kVA transformer supplies the rated output to a load with a power factor of 60% lagging, the efficiency is decreased:

$$\eta = \frac{V_S \cdot I_S \cdot \cos \angle\theta}{V_S \cdot I_S \cdot \cos \angle\theta + \text{core loss, watts} + \text{total copper losses}} \times 100$$

$$= \frac{240 \times 208.3 \times 0.6}{(240 \times 208.3 \times 0.6) + 240 + 629} \times 100 = 97.2\%$$

POLARITY MARKINGS

The American Standards Association (ASA) has developed a standard system of marking transformer leads. The high-voltage winding leads are marked H_1 and H_2. The low-voltage winding leads are marked X_1 and X_2. The H_1 lead is always located on the left-hand side when the transformer is faced from the low-voltage side. When H_1 is instantaneously positive, X_1 is also instantaneously positive.

Transformers with subtractive (buck) and additive (boost) polarities are shown in Figure 13–6. In a transformer with subtractive polarity, the H_1 and X_1 leads are adjacent to or directly across from each other. The H_1 and X_1 leads of a transformer with additive polarity are diagonally across from each other. The arrows in the figure indicate the instantaneous directions of the voltage in the windings.

Standard Test Procedure

Transformer leads normally have identifying tabs or tags marked H_1, H_2 and X_1, X_2. However, because it may be impossible to identify the leads because the tags are missing or disfigured, a standard test procedure can be used.

Figure 13–7A shows a test being made on a transformer with additive (boost) polarity. In this test, a jumper lead is temporarily connected between the high-voltage lead (H_1) and the low-voltage lead directly across from it. A voltmeter is connected between the other high-voltage lead (H_2) and the low-voltage lead directly across from it. If the voltmeter reads the sum of the primary input voltage and the secondary voltage, the transformer has additive (boost) polarity. The sum is 2400 V + 240 V = 2640 V. When H_1 is instantaneously positive, 240 V is induced in the secondary winding. The input voltage (2400 V) is applied to X_2 through the temporary jumper connection. This value adds to the 240 V so that the potential difference is 2640 V, as indicated on a voltmeter connected from X_1 to H_2. Note that the path from X_1 to H_2 has the same direction as the voltage arrows. As a result, the two voltages are added.

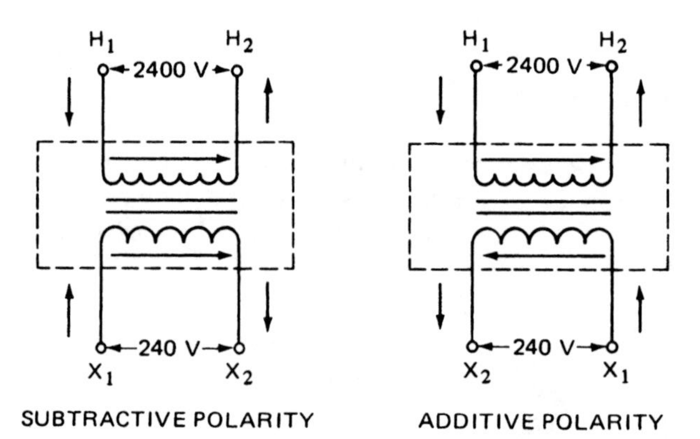

FIGURE 13–6 Marking transformers (*Delmar/Cengage Learning*)

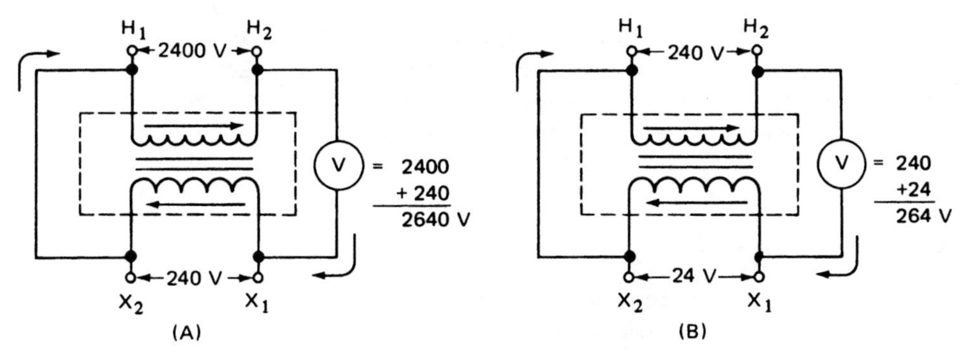

FIGURE 13–7 Test on additive polarity transformer (*Delmar/Cengage Learning*)

Low-Voltage Testing

There is a hazard involved in making the previous test at high voltage values. Thus, a test using a relatively low voltage was developed to determine transformer polarity. For example, in Figure 13–7B, 240 V is used as the test voltage. This potential is usually available in the laboratory or repair shop. By impressing 240 V on the 2400-V winding, a voltage of 24 V is induced in the secondary winding of the transformer. The voltage ratio is 10:1. The voltmeter is connected between H_2 and the low-voltage lead X_1. The reading on the voltmeter is 240 V + 24 V = 264 V. This means that an ac voltmeter, with a range of 0 to 300 V, can be used to determine the polarity of the transformer. The voltmeter is connected between the high-voltage lead (H_2) and the low-voltage lead directly across from it. The meter indicates the sum of the primary and secondary voltages. A transformer with this type of polarity markings is an additive polarity type.

In Figure 13–8A, the same polarity test connections are used for a transformer with subtractive (buck) polarity. The 240 V induced in the secondary winding opposes the 2400 V entering X_1 from the temporary jumper connection. The voltmeter is connected

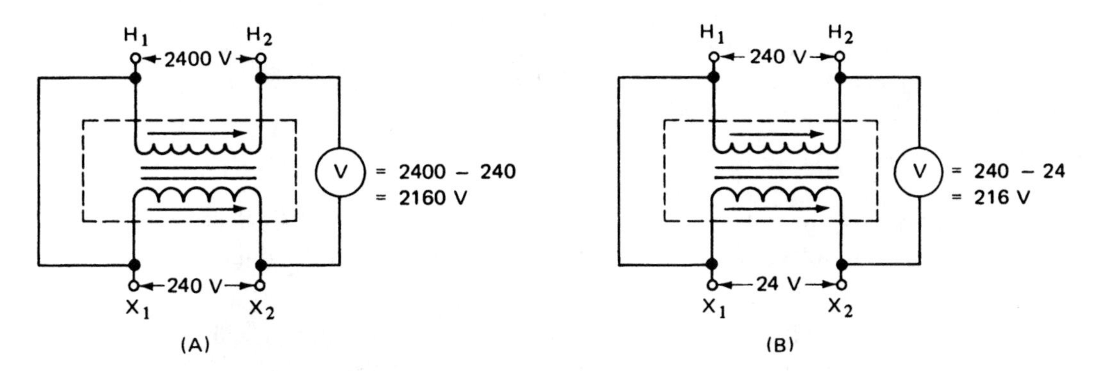

FIGURE 13–8 Test on subtractive polarity transformer (*Delmar/Cengage Learning*)

between H_2 and X_2 and indicates a value of 2400 V − 240 V, or 2160 V. Figure 13–8B shows a polarity test using a low voltage of 240 V. In this case, the voltmeter indicates the difference between the primary and secondary voltages. This difference is 240 V − 24 V = 216 V. The direction from X_2 to H_2 opposes the voltage arrow from X_1 to X_2 and is the same as the voltage arrow from H_1 to H_2. Therefore, the X_1, X_2 voltage is subtracted from the H_1, H_2 voltage.

ASA AND NEMA STANDARDS

The American Standards Association (ASA) and the National Electrical Manufacturers Association (NEMA) developed the following standards that relate to the polarity of transformers:

1. Additive polarity shall be standard for single-phrase transformers up to 200 kVA, and having voltage ratings not in excess of 9000 V.
2. Subtractive polarity shall be standard for all single-phase transformers larger than 200 kVA, regardless of the voltage rating.
3. Subtractive polarity shall be standard for all single-phase transformers in sizes of 200 kVA and below, having high voltage ratings above 9000 V.

The polarity of a single-phase transformer must be known before it can be connected in parallel with other single-phase transformers or in a three-phase bank. This information normally is provided on the transformer lead tags on the nameplate of the machine. However, when such information is not available, the standard polarity test just explained should be used to determine the polarity.

TRANSFORMERS IN PARALLEL

Single-phase transformers often must be operated in parallel. Several conditions must be satisfied to ensure that the current outputs of the transformers will be in proportion to the kVA capacity of the transformers. These conditions are as follows: (1) the transformers must have the same secondary terminal voltages; (2) the transformer polarities must be correct; and (3) each transformer must have the same percent impedance.

Two stepdown transformers are shown in Figure 13–9. If these transformers have the same voltage ratings, percent impedance values, and additive polarity, they can be connected in parallel. The following steps are used to connect the transformers:

1. The high-voltage leads (H_1) of both transformers are connected to one line wire. The other two primary high-voltage leads (H_2) are connected to the other line wire.
2. The low-voltage leads (X_1) of both transformers are connected to one secondary line wire. The other two low-voltage leads are connected to the other low-voltage line wire.

These two transformers satisfy the three conditions listed previously. As a result, they will both deliver secondary currents to the load in proportion to their kVA ratings.

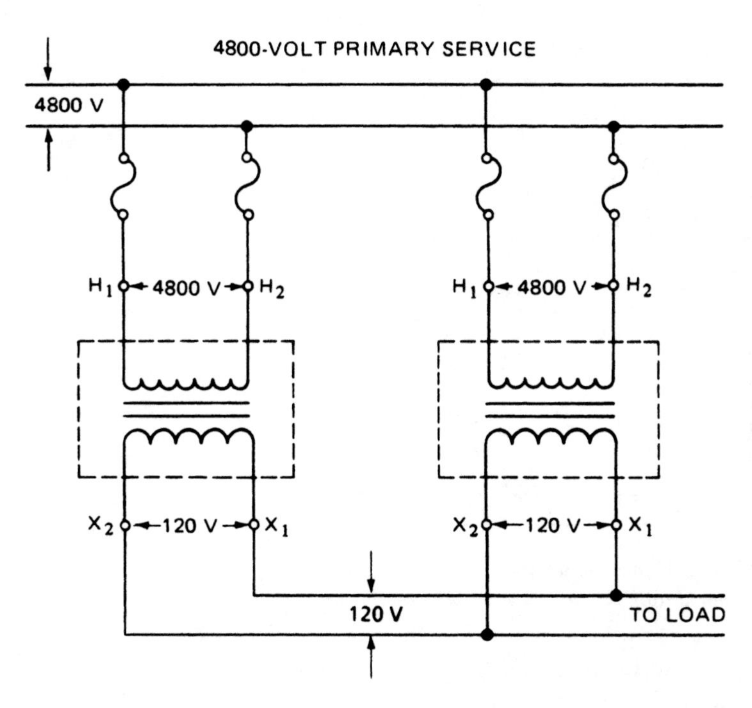

FIGURE 13–9 Single-phase transformers in parallel (*Delmar/Cengage Learning*)

Transformers of Unknown Polarities

When transformers are supplied by different manufacturers and it is not known whether they have additive or subtractive polarity, the following test may be used. It is assumed that one transformer operates as a stepdown transformer to supply energy to the 120-V bus bars. This transformer is called transformer 1. Transformer 2 is to be paralleled with the first transformer. Transformer 2 has the same voltage ratings and percent impedance as transformer 1, but its polarity is not known.

Figure 13–10 shows that transformer 1 has additive polarity. Regardless of the polarity of transformer 2, its H_1 lead is always on the left-hand side when viewed from the low-voltage side of the transformer. This means that the H_1 lead is connected to the same high-voltage line wire as the H_1 lead of transformer 1. Thus, the H_2 lead of transformer 2 is connected to the other side of the high-voltage line. One of the low-voltage leads of transformer 2 is connected to one side of the 120-V secondary. A voltmeter is connected between the other side of the 120-V secondary and that unconnected secondary lead of transformer 2. If transformer 2 has subtractive polarity, the voltmeter reading is twice the secondary coil voltage. In this case, the voltmeter indicates 240 V.

The instantaneous voltage directions are shown in Figure 13–10. The reason why the voltmeter indicates 240 V is evident by reviewing these instantaneous voltage directions. Assume that the X_2 lead of transformer 2 is connected to the secondary line wire

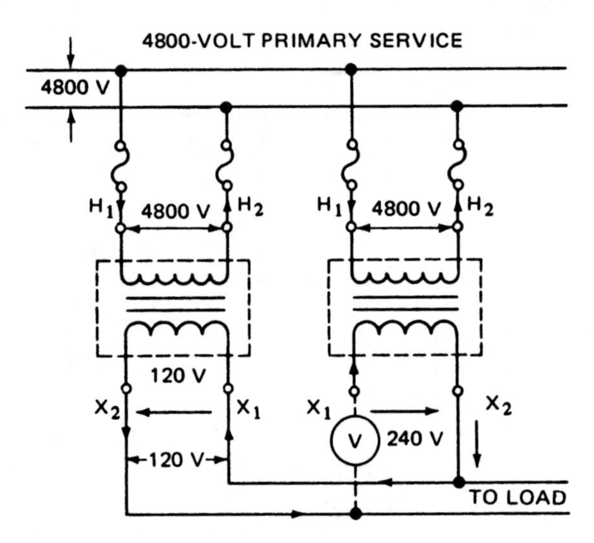

FIGURE 13–10 Checking polarity of transformers previous to paralleling (*Delmar/Cengage Learning*)

where the voltmeter is already connected. There will be a potential difference of 240 V at the connection point, resulting in a short circuit.

In Figure 13–11, the low-voltage lead X_2 of transformer 2 is reconnected to the other secondary line wire. The voltmeter now shows a zero potential because the secondary leads of both transformers have the same instantaneous polarity. The voltmeter can be removed and the final connections made without fear of a short circuit.

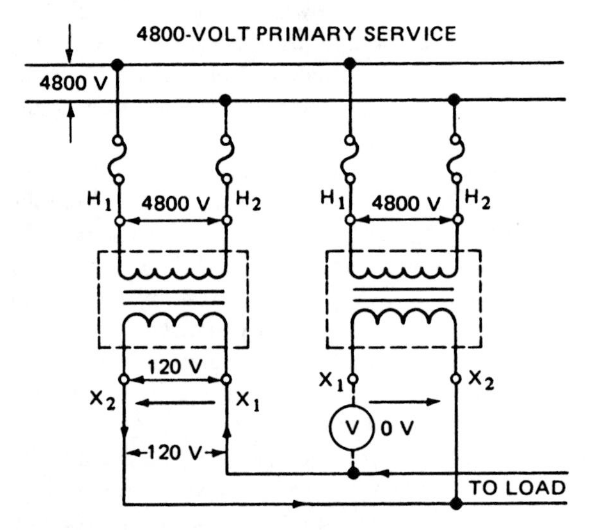

FIGURE 13–11 Checking the polarity of transformers previous to paralleling (*Delmar/Cengage Learning*)

Stepdown Transformers in Parallel

When stepdown transformers are operated in parallel, as in Figure 13–9, it may be necessary to remove one of the transformers from service for repairs. To do this, the low-voltage side of the transformer is always disconnected from the 120-V line wires before the primary fuses are opened. Remember that the 120-V line wires are still energized by the other transformer. If the primary fuses are opened but the low-voltage transformer leads are still connected to the 120-V line, there will be a serious safety hazard. The low-voltage winding will become a high-voltage secondary. A worker may be electrocuted if it is assumed that the high-voltage winding is deenergized because the primary fuses are open. Although the primary fuses are open, there is still 4800 V across the terminals of the high-voltage winding. As a result, a sign reading "DANGER—FEEDBACK" must be placed at each primary fuse to minimize this hazard.

THE DISTRIBUTION TRANSFORMER

A typical distribution transformer is shown in Figure 13–12. The transformer has two high-voltage windings that are rated at 2400 V each. These windings are connected to a terminal block. The block is located slightly below the level of the insulating oil in the transformer case. Small metal links are used to connect the two high-voltage windings either in series, for a 4800-V primary service, or in parallel, for a 2400-V input.

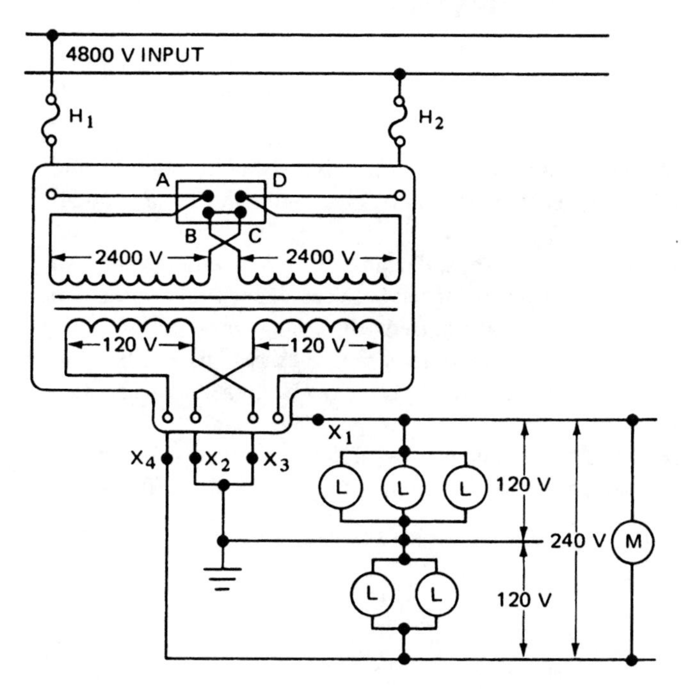

FIGURE 13–12 Connections for a single-phase, three-wire system supplied from a single-phase transformer *(Delmar/Cengage Learning)*

The two 2400-V primary coils shown in Figure 13–12 are connected in series. A metal link connects terminals B and C for 4800-V operation. Operation at 2400 V is obtained by connecting a metal link between terminals A and B. A second link is used to connect terminals C and D to place the two 2400-V coils in parallel. Although the distribution transformer has two high-voltage coils, note that there are only two external high-voltage leads. These leads are marked with the standard designations H_1 and H_2. These leads are permanently connected to the terminal block with lead H_1 attached to terminal A and lead H_2 attached to terminal D.

There are four low-voltage leads. When lead H_1 is instantaneously positive, leads X_1 and X_3 are also instantaneously positive. At the same time, leads X_2 and X_4 are instantaneously negative. If leads X_1 and X_3 are commoned together, and leads X_2 and X_4 are commoned together, then the low-voltage coils are connected in parallel to supply an output of 120 V.

If leads X_2 and X_3 are connected together, the two low-voltage coils are connected in series. The resulting output is 240 V across leads X_1 and X_4.

If there is a requirement for a 120/240-V, single-phase, three-wire service, then the following connections must be made: The two 120-V secondary windings are connected in series and a grounded neutral wire is connected between leads X_2 and X_3. These connections are shown in Figure 13–12. The resulting service provides 120 V for lighting and small-appliance loads and 240 V for heavy-appliance and single-phase, 240-V motor loads.

THE SINGLE-PHASE, THREE-WIRE SYSTEM

Nearly all residential and commercial electrical installations use a single-phase, three-wire service similar to the one shown in Figure 13–12. This type of service has a number of advantages:

1. The system provides two different voltages. The lower voltage supplies lighting and small-appliance loads, and the higher voltage supplies heavy-appliance and single-phase motor loads.
2. There are 240 V across the outside wires. Thus, the current for a given kilowatt load can be reduced by nearly half if the load is balanced between the neutral and the two outside wires. Because of the reduction in the current, the voltage drop in the circuit conductors is reduced, and the voltage at the load is more nearly constant. In addition, the following problems are minimized: dim lights, slow heating, and unsatisfactory appliance performance.
3. A single-phase, three-wire, 120/240-V system uses 37% less copper as compared to a 120-V, two-wire system having the same capacity and transmission efficiency.

Problem 6 shows why less copper is needed in such a single-phase, three-wire circuit.

PROBLEM 6

Statement of the Problem

Figure 13–13 shows a balanced single-phase, three-wire circuit. Two 10-A noninductive heater units are connected to each side of this circuit. The conductors used in this circuit are no. 12 AWG wire. The distance from the source to the load is 100 ft. Determine

1. the voltage drop in the line wires.
2. the percentage voltage drop.
3. the weight of the copper used for the three-wire system.

Solution

1. The total current taken by the two noninductive heater units connected between line 1 and the neutral wire is 10 + 10 = 20 A.
   The two heater units connected between line 2 and the neutral wire also take a total current of 20 A.
   The current in the neutral wire of a single-phase, three-wire noninductive circuit is the difference between the currents in the two line wires. For a balanced circuit, such as that of Figure 13–13, the current in the neutral wire is zero. This means that the actual current path for the 20-A current is the two no. 12 line wires. The actual voltage drop is

$$v = \frac{K \times L \times I}{CM}$$

where

v = voltage drop
K = ohms per mil-foot (see Appendix 7)
L = length in feet
I = current in amperes
CM = circular mil area (see Appendix 8)

$$v = \frac{10.4 \times 200 \times 20}{6530}$$

$$v = 6.4 \text{ V}$$

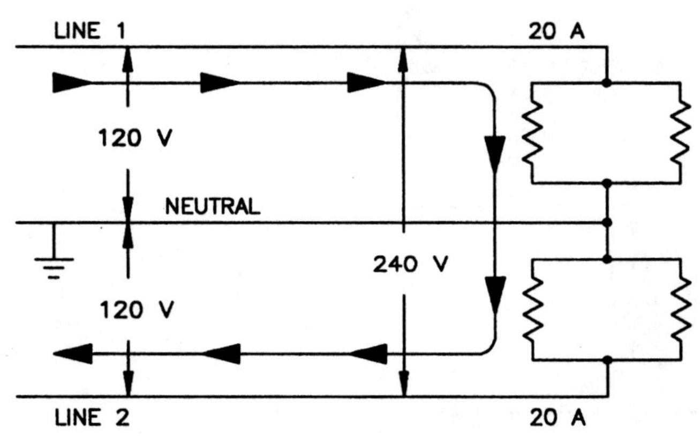

FIGURE 13–13 Balanced single-phase, three-wire circuit (*Delmar/Cengage Learning*)

2. Because the source voltage across the two outside legs of the system is 240 V, the percentage voltage drop is 6.4/240 = 0.02666 = 2.67%.
3. The weight of the copper used in the single-phase, three-wire system is determined as follows:

   Weight of 200 feet of no. 12 AWG wire = 1.98 lb
   Weight of 300 feet of no. 12 AWG wire = 5.94 lb

PROBLEM 7

Statement of the Problem

A 120-V, two-wire circuit is shown in Figure 13–14. Note that the four noninductive heater units are connected in parallel. Each heater unit takes 10 A. The allowable percentage voltage drop is 3%. This value is the same as that used in problem 6 for the single-phase, three-wire system. Determine

1. the wire size in circular mils that will give the same percentage voltage drop and, as a consequence, the same transmission efficiency as specified for the circuit in problem 6.
2. the AWG wire size of the 120-V, two-wire system.
3. the weight of the copper used in the 120-V, two-wire system.
4. the amount of copper saved by using the single-phase, three-wire system.

Solution

1. Desirable voltage drop:

$$v = 120 \times 0.0265 = 3.2 \text{ V}$$

$$CM = \frac{K \times L \times I}{v}$$

$$= \frac{10.4 \times 200 \times 40}{3.2}$$

$$= 26{,}000 \text{ circular mils}$$

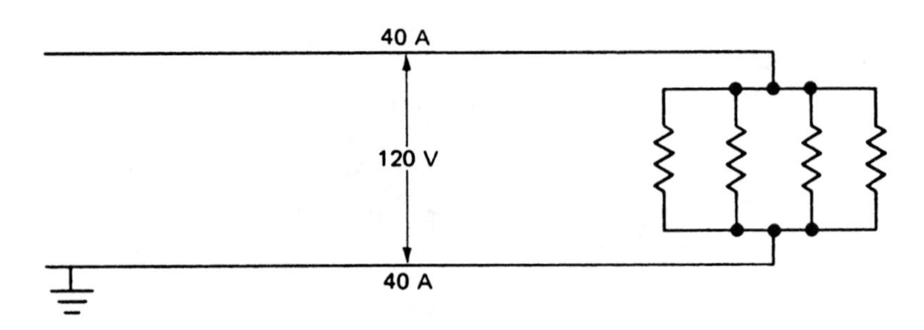

FIGURE 13–14 Single-phase, two-wire circuit (*Delmar/Cengage Learning*)

2. A circular mil area of 26,000 circular mils requires no. 6 AWG wire.
3. The weight of 100 ft of no. 6 AWG wire is 7.95 lb. Therefore, the total weight of the 200 ft of wire used in this circuit is 15.9 lb.
4. The actual percentage of copper used in a single-phase, three-wire system, as compared with an equivalent single-phase, 120-V, two-wire system, is

$$\frac{5.94}{15.9} \times 100 = 37.35\% = 37.4\%$$

Therefore, the maximum amount of copper saved by using the single-phase, three-wire system is 62.65%. The actual amount of copper saved is closer to 50% because the copper used in the three-wire, single-phase system is normally sized larger than the permitted minimum size.

Circuit with Unbalanced Loads

Figure 13–15 shows a single-phase, three-wire system having unbalanced noninductive lighting and heating loads from line 1 to the neutral and from line 2 to the neutral. The function of the neutral now is to maintain nearly constant voltages across the two sides of the system in the presence of different currents. For the ac noninductive circuit in Figure 13–15, the current in the grounded neutral wire is the difference between the currents in the two outside legs. Because this is an ac circuit, the arrows show only the instantaneous directions of current. For the instant shown, X_1 is instantaneously negative and supplies 10 A to line 1. Line 2 returns 15 A to X_4, which is instantaneously positive. The neutral wire conducts the difference between these two line currents. This difference is 5 A.

For the instant shown in Figure 13–15, the current in the neutral wire is in the direction from the transformer to the load. Line 1 supplies a total of 10 A, 5 A to the heater load, and 5 A to the lighting load connected between line 1 and the neutral. However, 10 A is required by the lighting load connected from the neutral to line 2. Therefore, the neutral wire supplies the difference of 5 A to meet the demands of the lighting load connected

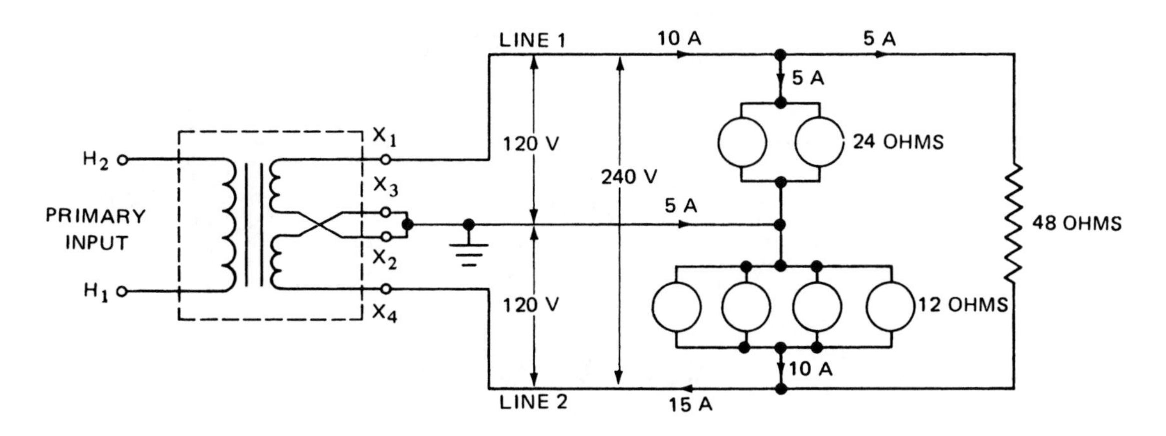

FIGURE 13–15 Single-phase, three-wire circuit with unbalanced load currents (*Delmar/Cengage Learning*)

between line 2 and the neutral wire. Kirchhoff's current law can be applied to this circuit. Recall that the current law states that the sum of the currents at any junction point in a network system is always zero.

In a single-phase, three-wire circuit, the neutral wire is always grounded. Also, it is never fused or broken at any switch control point in the circuit. This direct path to ground by way of the neutral wire means that any instantaneous high voltage, such as that due to lightning, will be instantly discharged to ground. As a result, electrical equipment and the circuit wiring are protected from major lightning damage.

Circuit with an Open Neutral

There is an even more important reason for not breaking the neutral wire. Recall that the neutral wire helps maintain balanced voltages between each line wire and the neutral wire, even with unbalanced loads. Assume that the neutral wire in Figure 13–15 is opened accidentally. The two lighting loads are now in series across 240 V.

Figure 13–16 shows the circuit of Figure 13–15, but with an open neutral. The 48-Ω heater unit still takes 5 A at voltage of 240 V. However, the two lighting loads act like two resistances in series across 240 V. The total resistance of the two lighting loads in series is 24 + 12 = 36 Ω.

The following values can be calculated for this circuit. The current is

$$I = \frac{V}{R} = \frac{240}{36} = 6.67 \text{ A}$$

The voltage across the lighting load connected between line 1 and the neutral wire becomes

$$V = IR = 6.67 \times 24 = 160 \text{ V}$$

The voltage across the lighting load connected between line 2 and the neutral wire becomes

$$V = IR = 6.67 \times 12 = 80 \text{ V}$$

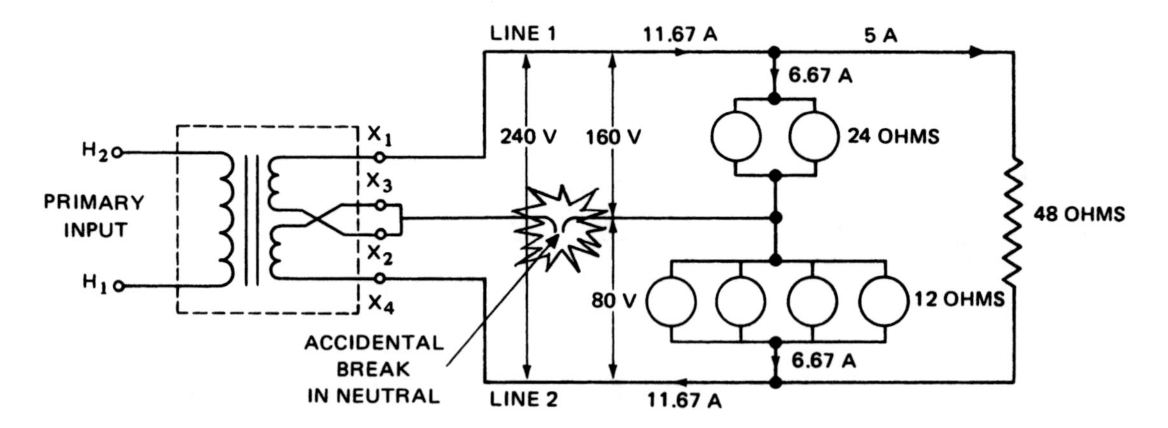

FIGURE 13–16 Single-phase, three-wire circuit with open neutral (*Delmar/Cengage Learning*)

The lighting load connected between line 1 and the neutral wire now has 160 V through it rather than its rated voltage of 120 V. This higher voltage will cause the lamps to burn out. The lighting load connected between line 2 and the neutral wire receives only 80 V, instead of 120 V. This voltage is not sufficient for adequate operation of the load. This type of unbalanced voltage condition would be common if fuses or switch control devices were placed in the neutral wire. In practice, the neutral wire is always carried as a solid conductor through the entire circuit.

TRANSFORMER CHARACTERISTICS

The characteristic curves of a 50-kVA transformer are shown in Figure 13–17. The efficiency of this transformer is very high even when the load is as low as 10% of the rated load. Note that the efficiency curve is nearly flat between 20% of the rated load to about 20% overload. The load–voltage characteristics show that there is little change in

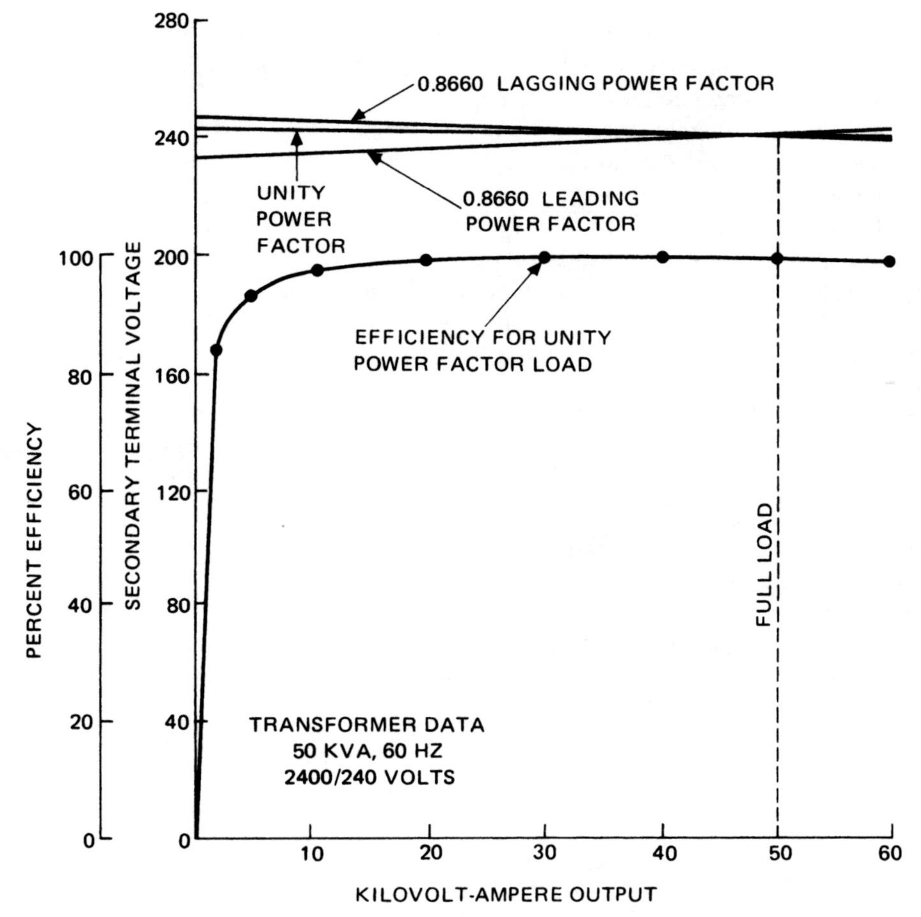

FIGURE 13–17 Transformer efficiency and load-voltage characteristics (*Delmar/ Cengage Learning*)

the terminal voltage from a full-load condition to a no-load condition. This condition was shown in the previous voltage regulation calculations. With a lagging power factor load, the secondary terminal voltage change is slightly greater than it is for a unity power factor load. However, the percentage of voltage regulation is satisfactory. With the leading power factor load, the terminal voltage increases with the load, giving a negative voltage regulation.

TRANSFORMER CONSTRUCTION

The single-phase transformer is a simple alternating-current device. The core consists of thin annealed sheets (laminations) of silicon steel. These laminations may be in the form of rectangular strips or L-shaped stampings. When separate sheets are used to make the core, the sheets are placed with butt joints in each layer. The core can also be made by winding a long strip of silicon steel to the desired size. In the laminated core, the sheets are stacked so that the butt joints are staggered in successive layers to ensure a low reluctance. Because the spiral-wound core has no joints, it has a very low reluctance.

Three basic designs of cores are available for use in transformers. These cores are known as the *core type,* the *shell type,* and the *combined core-and-shell type*. Figure 13–18 shows the core-type structure for a transformer.

In the core-type structure, the windings surround the laminated silicon steel core. The low-voltage winding generally is placed next to the core with the high-voltage winding wound over the low-voltage winding. These windings are carefully insulated from each other. The advantage of placing the low-voltage coils next to the core is that a reduction can be made in the material required to insulate the high-voltage windings. Many simplified transformer drawings show the primary winding on one leg of the core and the secondary winding on the other leg. However, such an arrangement gives rise to excessive leakage flux. In practice, both the primary and secondary windings are placed on the same leg of the core to minimize the leakage.

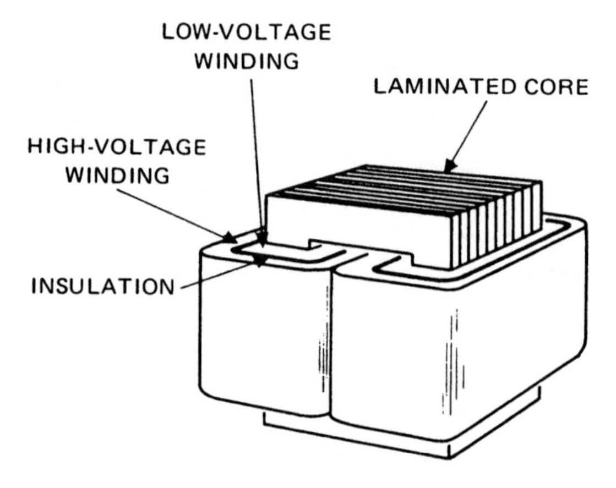

FIGURE 13–18 Coil arrangement for a core-type transformer (*Delmar/Cengage Learning*)

Shell-Type Core

A shell-type core is shown in Figure 13–19. In this structure, the silicon steel core surrounds the windings. The entire flux passes through the center leg of the core and then divides. One-half of the flux passes through each of the two outside legs.

The coil arrangement and the flux path for a shell-type core are shown in Figure 13–20. The low-voltage coil windings are placed next to the laminated core. The high-voltage windings are placed between the low-voltage windings. This coil arrangement provides adequate insulation between the coils. Because the high-voltage coils are not adjacent to the core structure, less insulation is required. *Pancake coils* wound with rectangular copper wire may be used with this type of core. The fact that the coil windings surround the core and the core surrounds the coils means that the leakage flux is minimized.

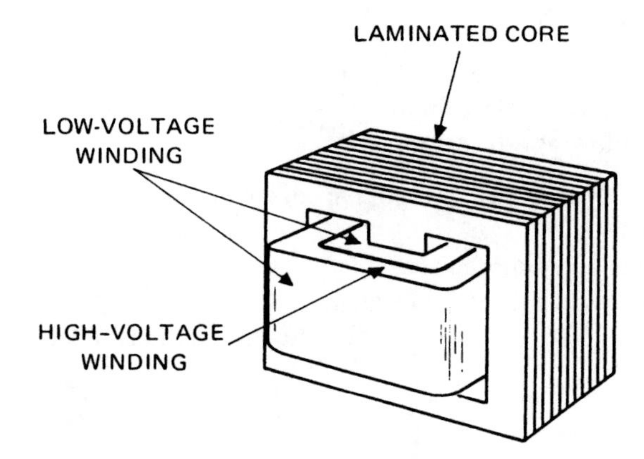

FIGURE 13–19 Coil arrangement for a shell-type transformer (*Delmar/Cengage Learning*)

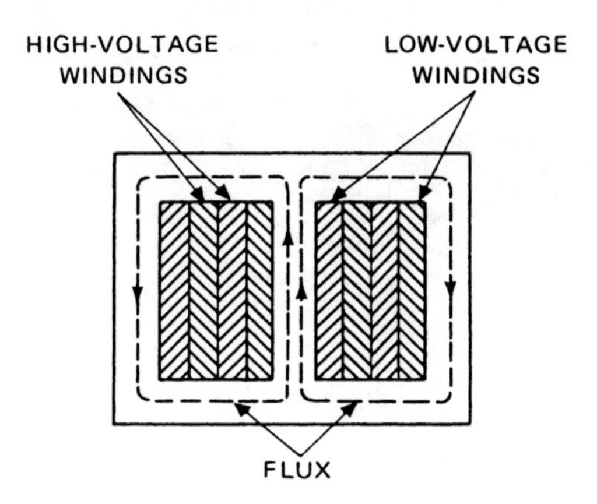

FIGURE 13–20 Flux paths of shell-type core (*Delmar/Cengage Learning*)

Type H Core

A type H core with coil windings is shown in Figure 13–21. The core is shaped like a cross when viewed from above. The coils are constructed so that the high-voltage windings are located between the low-voltage windings. This arrangement of coils minimizes the insulation required. As a result, the only high-voltage insulation required is placed between the high-voltage windings and the low-voltage windings. The structure of this core-and-coil assembly keeps the leakage flux to a minimum because the coil windings are placed on a center leg and are surrounded by the four outside legs of the core structure. The H core is often used for distribution transformers with two high-voltage windings, each rated at 2400 V, and two low-voltage windings, each rated at 120 V. Such a transformer can be used to step down either 2400 V or 4800 V to 120 V, 240 V, or 120/240 V.

The GE 220 Class Transformer

A 220 class transformer developed by the General Electric Company has a wound core. That is, a long strip of silicon steel is wound in a tight spiral around the insulated windings. This type of core has the following advantages:

1. It can be manufactured more easily than the conventional core, consisting of laminations stacked and clamped together.
2. The magnetic circuit path is relatively short and has a large cross section.
3. The construction of the core helps reduce the flux leakage.
4. The flux path direction is always along the grain of the silicon steel, thus reducing the iron losses.

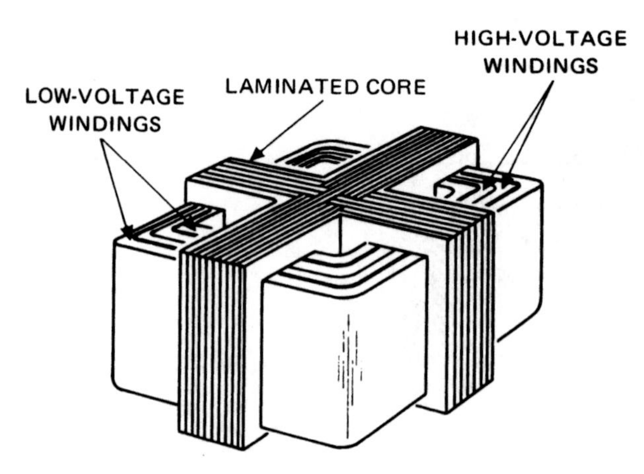

FIGURE 13–21 Core and coils for type H transformer
(Delmar/Cengage Learning)

TRANSFORMER INRUSH CURRENT

Although transformers and reactors are both inductive devices, there is a great difference in their operating characteristics. Reactors are often used to prevent inrush current from becoming excessive when a circuit is first turned on. Transformers, however, can produce extremely high inrush currents when power is first applied to the primary winding. The type of core used is primarily responsible for this difference in characteristics.

MAGNETIC DOMAINS

Magnetic materials contain tiny magnetic structures in their molecular material, known as *domains*. These domains can be affected by outside sources of magnetism. Figure 13–22 illustrates a magnetic domain that has not been polarized by an outside magnetic source.

Now assume that the north pole of a magnet is placed toward the top of the material that contains the magnetic domains (Figure 13–23). Notice the structure of the domain has changed to realign the molecules in the direction of the outside magnetic field.

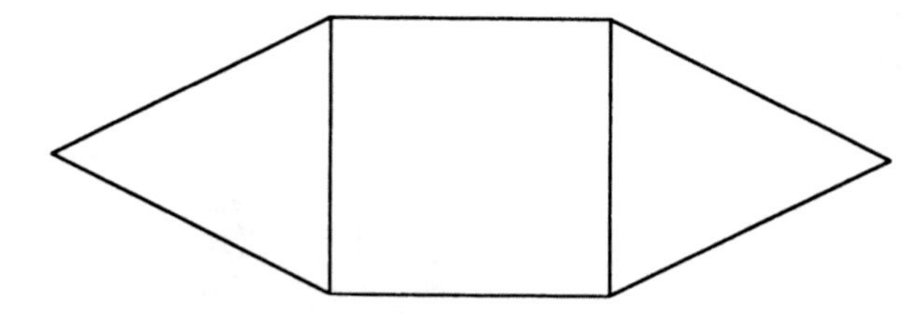

FIGURE 13–22 Magnetic domain in neutral position ***(Delmar/Cengage Learning)***

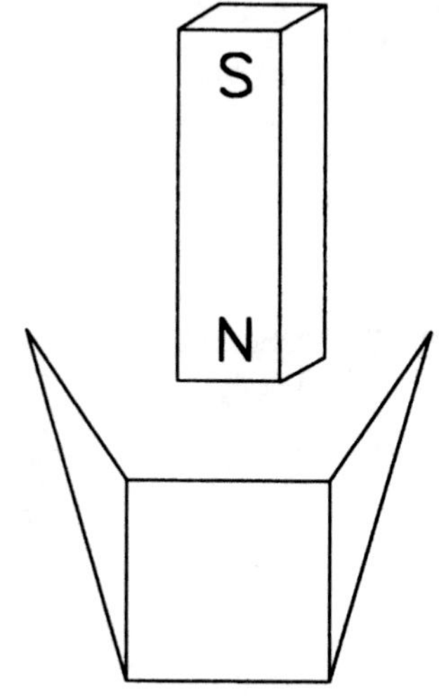

FIGURE 13–23 Magnetic domain influenced by a north magnetic field ***(Delmar/Cengage Learning)***

If the polarity of the magnetic pole is changed (Figure 13–24), the molecular structure of the domain will change to realign itself with the new magnetic lines of flux. This external influence can be produced by an electromagnet as well as a permanent magnet.

In certain types of cores, the molecular structure of the domain will snap back to its neutral position when the magnetizing force is removed. This type of core is used in the construction of reactors or chokes (Figure 13–25). A core of this type is constructed by

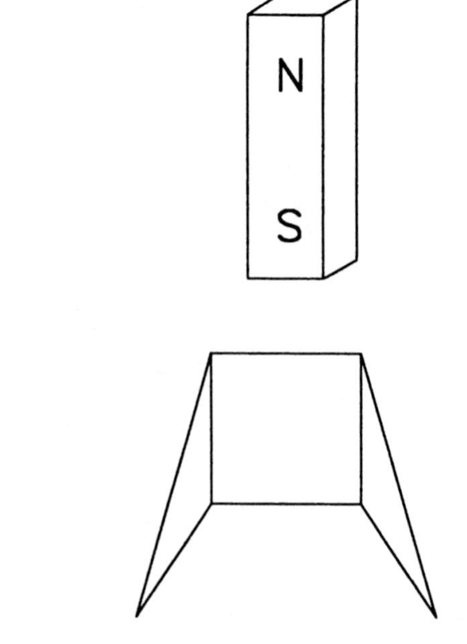

FIGURE 13–24 Magnetic domain influenced by a south magnetic field (*Delmar/Cengage Learning*)

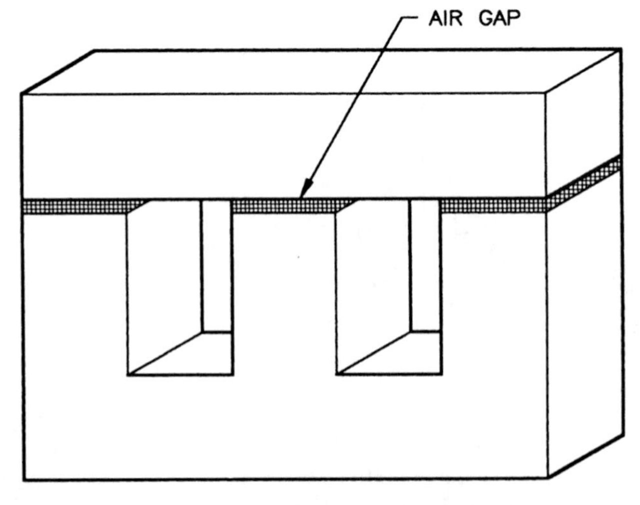

FIGURE 13–25 Core of an inductor (*Delmar/Cengage Learning*)

separated sections of the steel laminations with an airgap. This airgap breaks the path of the magnet through the core material and is responsible for the domains returning to their neutral position once the magnetizing force is removed.

The core construction of a transformer, however, does not contain an airgap. The steel laminations are connected together in such a manner as to produce a very low reluctance path for the magnetic lines of flux. In this type of core, the domains remain in their set position once the magnetizing force has been removed. This type of core "remembers" where it was last set. This was the principle of operation of the core memory of early computers. It is also the reason why transformers can have extremely high inrush currents when they are first connected to the power line.

The amount of inrush current in the primary of a transformer is limited by the following three factors:

1. The amount of applied voltage.
2. The resistance of the wire in the primary winding.
3. The flux change of the magnetic field in the core. The amount of flux change determines the amount of inductive reactance produced in the primary winding when power is applied.

Figure 13–26 illustrates a simple isolation-type transformer. The alternating current applied to the primary winding produces a magnetic field around the winding. As the current changes in magnitude and direction, the magnetic lines of flux change also. Because the lines of flux in the core are continually changing polarity, the magnetic domains in the core material are changing also. As stated previously, the magnetic domains in the core of a transformer remember their last set position. For this reason, the point on the waveform where current is disconnected from the primary winding can have a great bearing on the amount of inrush current when the transformer is reconnected to power. For example, assume that the power supplying the primary winding is disconnected at the zero crossing point (Figure 13–27). In this instance, the magnetic domains would be set at the neutral point. When power is restored to the primary winding, the core material can be magnetized by either magnetic polarity. This permits a change of

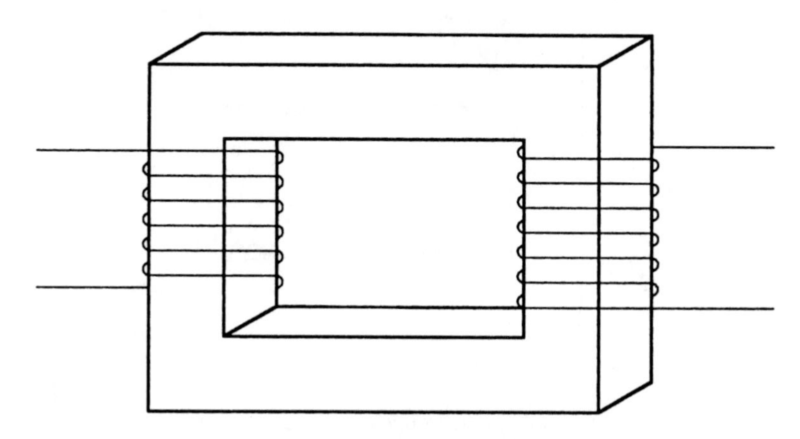

FIGURE 13–26 Isolation transformer (*Delmar/Cengage Learning*)

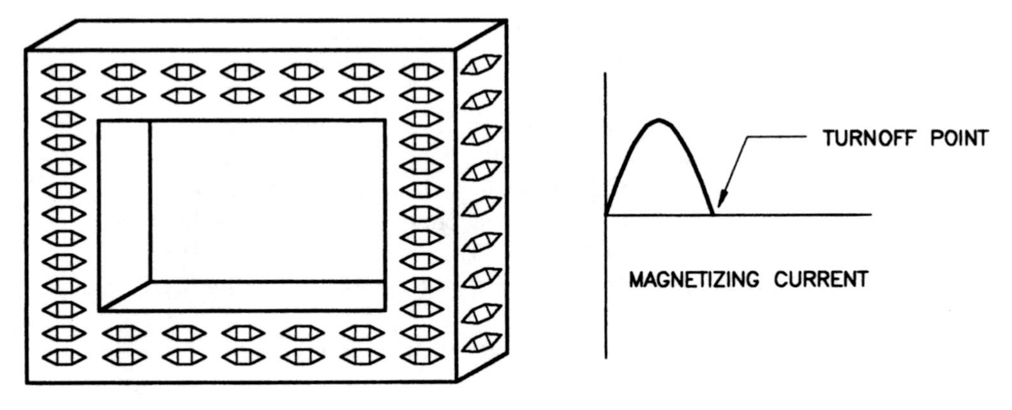

FIGURE 13–27 Magnetic domains set in the neutral position ***(Delmar/Cengage Learning)***

flux, which is the dominant current-limiting factor. In this instance, the amount of inrush current would be relatively low.

If the power supplying current to the primary winding is interrupted at the peak point of the positive or negative half-cycle, however, the domains in the core material will be set at that position. Figure 13–28 illustrates this condition. It is assumed that the current was stopped as it reached its peak positive point. If the power should be reconnected to the primary winding during the positive half-cycle, only a very small amount of flux change can take place. Because the core material is saturated in the positive direction, the primary winding of the transformer is essentially an air core inductor, which greatly decreases the inductive characteristics of the winding. The inrush current in this situation would be limited by the resistance of the winding and a very small amount of inductive reactance.

This characteristic of transformers can be demonstrated with a clamp-on ammeter that has a "peak hold" capability. If the ammeter is connected to one of the primary leads, and power is switched on and off several times, it can be seen that the amount of inrush current will vary over a wide range.

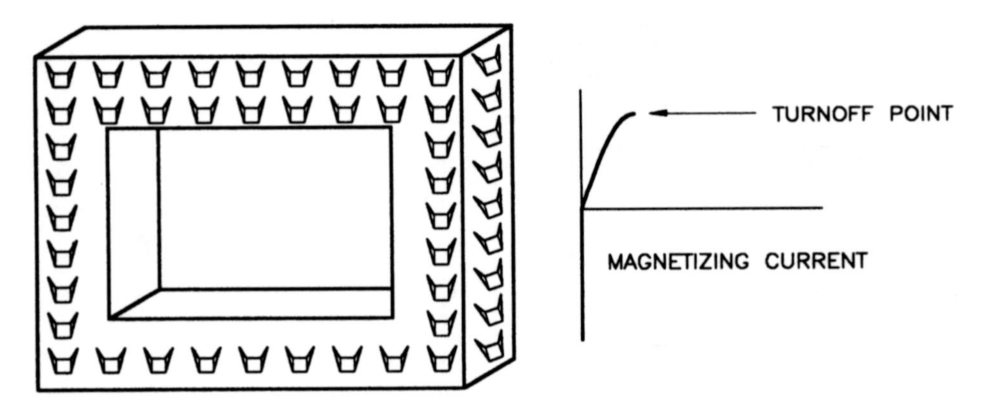

FIGURE 13–28 Domains set at one end of magnetic polarity ***(Delmar/Cengage Learning)***

MULTIPLE TAPPED WINDINGS

It is not uncommon for transformers to be designed with windings that have more than one set of lead wires connected to the primary or secondary. The transformer shown in Figure 13–29 contains a secondary winding rated at 24 V. The primary winding contains several taps, however. One of the primary lead wires is labeled C and is the common for the other leads. The other leads are labeled 120, 208, and 240, respectively. This transformer is designed in such a manner that it can be connected to different primary voltages without changing the value of the secondary voltage. In this example, it is assumed that the secondary winding has a total of 120 turns of wire. To maintain the proper turns ratio, the primary would have 600 turns of wire between C and 120, 1040 turns between C and 208, and 1200 turns between C and 240.

The transformer shown in Figure 13–30 contains a single primary winding. The secondary winding, however, has been tapped at several points. One of the secondary lead wires is labeled C and is common to the other lead wires. When rated voltage is applied to

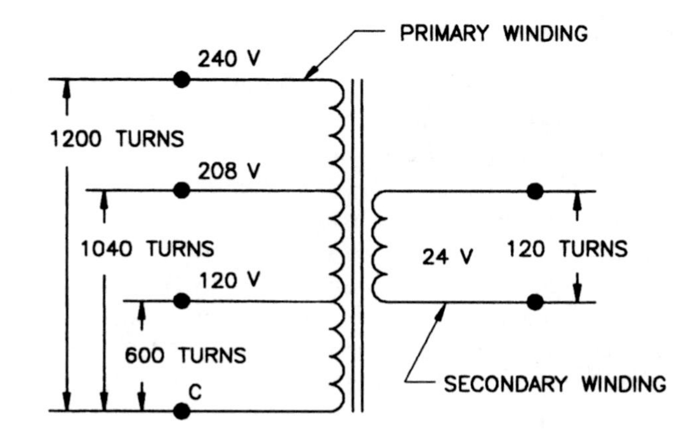

FIGURE 13–29 Transformer with multiple taps on the primary winding (*Delmar/Cengage Learning*)

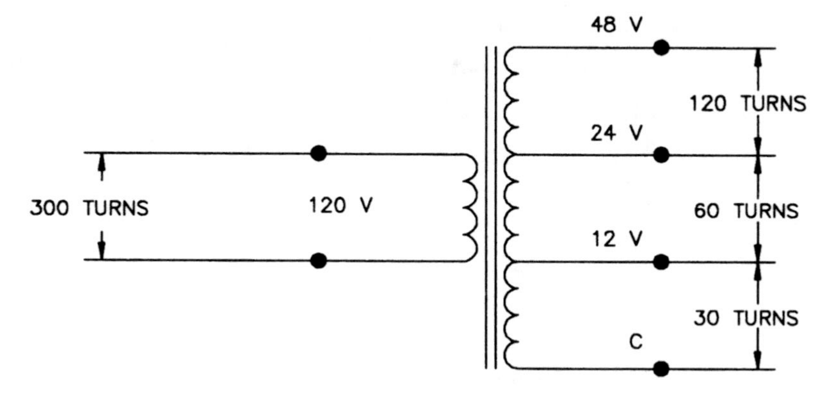

FIGURE 13–30 Transformer with multiple taps on the secondary winding (*Delmar/Cengage Learning*)

the primary, voltages of 12, 24, and 48 V can be obtained at the secondary. It should also be noted that this arrangement of taps permits the transformer to be used as a center-tapped transformer for two of the voltages. If a load is placed across the lead wires labeled C and 24, the lead wire labeled 12 becomes a center tap. If a load is placed across the C and 48 lead wires, the 24 lead wire becomes a center tap.

In this example, it is assumed the primary winding has 300 turns of wire. To produce the proper turns ratio, it would require 30 turns of wire between C and 12, 60 turns of wire between C and 24, and 120 turns of wire between C and 48.

The transformer shown in Figure 13–31 is similar to the transformer in Figure 13–30. The transformer in Figure 13–31, however, has multiple secondary windings instead of a single secondary winding with multiple taps. The advantage of the transformer in Figure 13–31 is that the secondary windings are electrically isolated from each other. These secondary windings can be either stepup or stepdown depending on the application of the transformer.

COOLING TRANSFORMERS

The core-and-coil assembly is placed in a pressed steel tank and is completely covered with an insulating oil. This insulating oil removes heat from the core and the coil windings and insulates the windings from the core and the transformer case. The core-and-coil assembly for a typical transformer contains channels or ducts. These ducts permit the oil to circulate and remove the heat. As the oil gains heat, its density decreases and it rises. As it rises and circulates in the transformer, it contacts the tank walls and the heat is transferred from the oil to the tank walls. The walls are cooled by air circulation. The specific gravity of the oil increases as it loses heat. As a result, the oil flows down to the bottom of the tank and again circulates up through the coil ducts to repeat the cooling process. Transformers having a very large

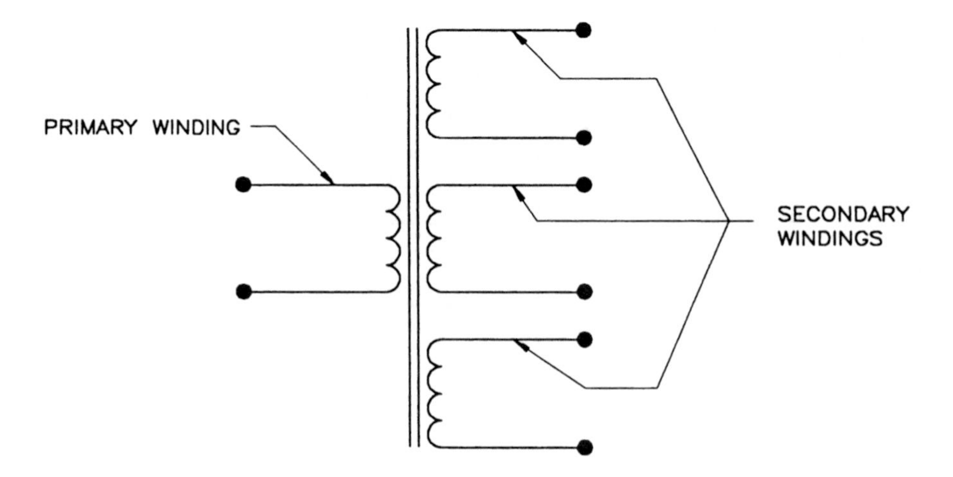

FIGURE 13–31 Transformer with multiple secondary windings (*Delmar/Cengage Learning*)

kVA capacity require more than the available surface area of the transformer case for cooling.

The surface area of a transformer can be increased by the use of tubes or fins added to the steel tank assembly (Figure 13–32). This increased surface area will take more heat from the oil and will radiate it faster to the surrounding air in a given time.

Transformer Oil

The insulating oil used in transformers is a high-grade oil that must be kept clean and moisture-free. This type of oil should be checked periodically to determine whether there is any change in its insulating ability. If traces of moisture or foreign materials are found, the oil must be filtered or replaced. The insulating fluid in some transformers is a nonflammable, nonexplosive liquid. One fluid of this type is Pyranol, manufactured by the General Electric Company. This liquid is a synthetic dielectric that is an effective cooling and insulating agent.

Oil-cooled transformers are considered a fire hazard in some locations. Air-cooled transformers are used under these conditions. Such transformers permit the natural circulation of air to remove the heat from the coils and the core. A perforated metal enclosure provides mechanical protection to the coil and windings and permits air to circulate through the windings.

In addition to the method described, other ways of cooling transformers include forced air circulation, natural air circulation, natural circulation of oil with water cooling, and forced oil circulation (for large transformers).

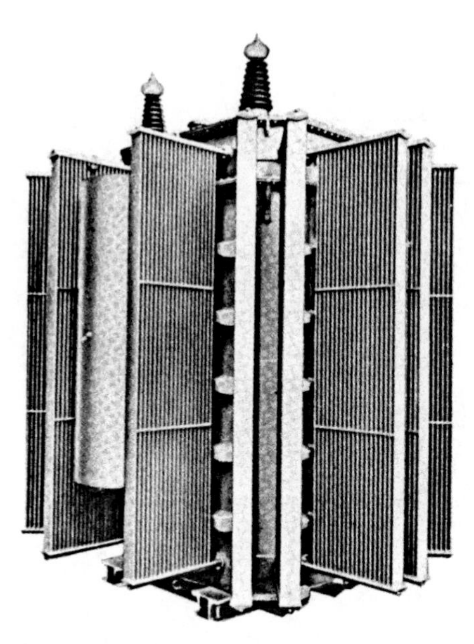

FIGURE 13–32 Oil-filled transformer (*Delmar/Cengage Learning*)

TAP-CHANGING TRANSFORMERS

There is some voltage loss in high-voltage distribution circuits. This means that the impressed voltage on the primary winding of a stepdown transformer may be lower than the primary voltage indicated on the nameplate. As a result, the secondary terminal voltage is also lower than the nameplate value.

Because this condition is undesirable, provisions can be made for changing taps to obtain the voltage expected. In this way, minor adjustments can be made in the transformer ratio to compensate for the voltage loss in the high-voltage primary line. A simple method of doing this is to bring out the tap points of the high-voltage winding to a terminal block inside the transformer. The tap connections can be changed on the terminal board to vary the number of turns of the primary winding.

Tap-Changing Switch

Figure 13–33 shows the connections for a typical tap-changing switch used with the high-voltage windings of a stepdown transformer. Each full coil winding on the high-voltage side is rated at 2400 V. The voltage rating of the two coil windings in series is 4800 V. The transformer steps the voltage down to 120/240 V. The tap-changing switch contacts are shown in position 1. This means that both full coil windings are in series. If the switch is moved to position 2, section A is removed from the primary coil winding on the left, resulting in a decrease in the turns ratio between the primary and secondary windings. Thus, the secondary voltage increases. When the tap-changing switch is moved in sequence to

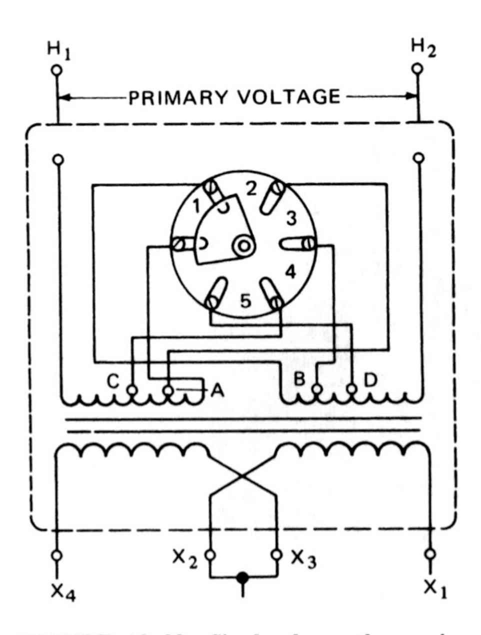

FIGURE 13–33 Single-phase, three wire output connections for tap-changing transformer (*Delmar/Cengage Learning*)

positions 3, 4, and 5, sections B, C, and D of the primary windings are removed in the same order. As each section of the primary winding is removed from the circuit, the turns ratio decreases and the secondary voltage increases.

Special tap switches, immersed in oil, are used on some large transformers. This type of switch permits the tap connections to be changed under load. In general, the tap-changing mechanism is placed in a compartment separated from the transformer tank. As a result, oil is confined near the arc-interrupting area of the tap-changing mechanism and there is no reduction in the high dielectric strength of the oil in the transformer case. A small motor drives the tap-changing mechanism. The motor is operated from a control panel.

FREQUENCY AND VOLTAGE

A transformer must be operated on an ac circuit at the frequency for which it is designed. If a lower frequency is used, the reactance of the primary winding will decrease. As a result, there will be a marked increase in the exciting current. This increase causes the flux density in the core to increase, with the result that the saturation of the core is above normal. This greater saturation causes a greater heat loss and the efficiency of the transformer is lowered. If the frequency is greater than the nameplate frequency value, the reactance will increase and the exciting current will decrease. There will be a lower flux density, but the core loss will remain nearly constant.

If the voltage is increased to a high enough value above the nameplate rating, excessive heating will take place at the windings. The flux density of the core will increase and the saturation of the core will be above normal. Transformers are designed so that they can be operated, without overheating, at a voltage that is 5% above the nameplate rating. If a transformer is operated at a voltage lower than the nameplate rating, the power output will be reduced in proportion to the reduction in voltage.

TRANSFORMER IMPEDANCE

Transformer impedance is determined by the physical construction of the transformer. Factors such as the amount and type of core material, wire size used to construct the windings, the number of turns, and the degree of magnetic coupling between the windings greatly affect the transformer's impedance. Impedance is expressed as a percent and is measured by connecting a short circuit across the low-voltage winding of the transformer and then connecting a variable voltage source to the high-voltage winding (Figure 13–34). The variable voltage is then increased until rated current flows in the low-voltage winding. The transformer impedance is determined by calculating the percentage of variable voltage as compared to the rated voltage of the high-voltage winding.

Example

Assume that the transformer shown in Figure 13–34 is a 2400/480-V, 15-kVa transformer. To determine the impedance of the transformer, first compute the full-load current rating of the secondary winding:

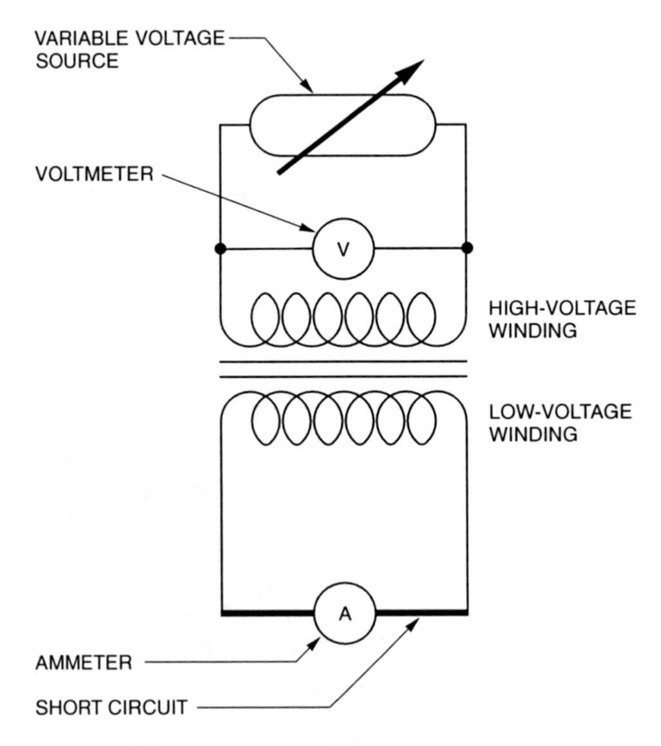

FIGURE 13–34 Determining transformer impedance (*Delmar/Cengage Learning*)

$$I = \frac{VA}{V}$$

$$= \frac{15{,}000}{480} = 31.25 \text{ A}$$

Next, increase the source voltage connected to the high-voltage winding until a current of 31.25 A flows in the low voltage winding. For the purpose of this example, assume that voltage value to be 138 V. Finally, determine the percentage of applied voltage as compared to the rated voltage:

$$\%Z = \frac{\text{source voltage}}{\text{rated voltage}} \times 100$$

$$= \frac{138}{2400} \times 100$$

$$= 0.0575 \times 100 = 5.75$$

The impedance of this transformer is 5.75%.

Discussion

Transformer impedance is a major factor in determining the amount of voltage drop that a transformer will exhibit between no-load and full-load conditions and in determining the amount of current flow in a short-circuit condition. Short-circuit current can be computed using the following formula:

$$\text{(Single phase)}I_{SC} = \frac{VA}{V \times \%Z}$$

$$\text{(Three phase) } I_{SC} = \frac{VA}{V \times \sqrt{3} \times \%Z}$$

Because one of the formulas for determining current in a single-phase circuit is

$$I = \frac{VA}{V}$$

and one of the formulas for determining current in a three-phase circuit is

$$I = \frac{VA}{V \times \sqrt{3}}$$

these formulas can be modified to show that the short-circuit current can be computed by dividing the rated secondary current by the %Z:

$$I_{SC} = \frac{I_{rated}}{\%Z}$$

TRANSFORMER NAMEPLATE DATA

The information listed on the nameplate for a typical transformer is shown in Table 13–1. The transformer capacity is rated in kilovolt-amperes, not in kilowatts. The kilovolt-ampere capacity is used because the load connected to the secondary determines the power factor of the output circuit of the transformer. The kVA rating represents the full-load output of the transformer, rather than the input. The nameplate also lists the voltage ratings of the high- and low-voltage windings, the frequency, the phase, the percent impedance, the polarity, the maximum temperature rise, and the gallons of transformer oil required.

Manufacturer's Name	
Volt-amperage: 50 kVA	Single-phase
HV winding: 2400 V	LV winding: 240 V
Frequency: 60 Hz	Percent impedance: 3.5%
Additive polarity:	Temperature rise: 55°C
Serial no.:	continuous duty
Type no.:	Gallons of insulating oil:
	Model no.:

TABLE 13–1 Typical transformer nameplate data

SUMMARY

- A transformer transfers energy from one ac circuit to another. This transfer is made with a change in the voltage, but with no change in the frequency.
- A transformer is a stationary device.
    1. In its simplest form, it consists of a laminated iron core. The input and output windings are wound on this common core.
    2. A transformer does not require any moving parts to transfer energy. There are no friction or windage losses and other losses are slight. Maintenance and repair costs are relatively low because of the lack of rotating parts.
- At full load, the efficiency of a transformer is between 96% and 97%. For a transformer with a very high capacity, the efficiency may be as high as 99%.
- Exciting current:
    1. The exciting current is present when the primary winding of a transformer is connected to an alternating voltage source.
    2. This current exists even when there is no load connected to the secondary.
    3. It sets up an alternating flux in the core. This flux links the turns of both windings and induces a voltage in both windings. The total induced voltage in each winding is directly proportional to the number of turns in that winding. The alternating voltage induced in the secondary has the same frequency, but is opposite in direction to the primary winding voltage. The induced voltage in the primary winding opposes the source voltage and limits the excitation current of the primary.
- The induced voltages in the primary and secondary windings are directly proportional to the number of turns:

$$\frac{V_P}{V_S} = \frac{N_P}{N_S} = \alpha$$

where α = transformation ratio or turns ratio.

- The ampere-turns of the primary and secondary windings are equal:

$$I_P N_P = I_S N_S$$

- Leakage flux:
    1. This is a flux that does not link all of the turns of all of the coil windings.
    2. This flux has its magnetic path in air and not in the core.
    3. Leakage flux from the primary voltage does not link the turns of the secondary winding.
    4. Leakage flux from the secondary links the secondary windings, but not the primary winding.
    5. The leakage flux uses part of the impressed primary voltage to cause a reactance voltage drop; thus, both the secondary flux linkages and the secondary induced voltage are reduced.

6. Leakage flux from the secondary voltage is proportional to the secondary current; thus, there is a reactance voltage drop in the secondary winding. (Both the primary and the secondary leakage fluxes reduce the secondary terminal voltage of the transformer as the load increases.)
7. The leakage flux of a transformer can be controlled by the type of core used. The placement of the primary and secondary windings on the legs of the core is also a controlling factor.

- The primary current at no load is called the *exciting current.*
    1. This current is usually 2% to 5% of the full-load current.
    2. It supplies the alternating flux and the losses in the transformer core. These losses are known as *core losses* and consist of eddy current losses and hysteresis losses. The losses cause I^2R losses.
    3. Exciting current consists of a large quadrature (magnetizing) component and a small in-phase component to supply the core losses.
    4. The power factor for the exciting current is from 5% to 10% lagging.
- Voltages are induced in the metal core by the large quadrature (magnetizing) component of the exciting current.
    1. Eddy currents result and circulate through the core.
        a. Eddy currents can be reduced by assembling the core structure from lamination coated with insulating varnish.
        b. If the laminations are not varnished, the oxide coating on each lamination still reduces the eddy current losses.
    2. The core structure also has a hysteresis loss.
        a. This loss is due to molecular friction or the opposition offered by the molecules to changes in their direction.
        b. To reduce the amount of power needed to overcome these losses, a special silicon steel is used in the core construction.
- Core losses can be measured by the core loss open-circuit test.
    1. The losses of a transformer at no load are small.
    2. In the open-circuit test, the voltmeter measures the primary voltage.
    3. The voltmeter must be disconnected before a reading is taken with the wattmeter. This is done so that the wattmeter will not indicate the power taken by the voltmeter.
    4. The wattmeter indicates the core loss in watts, and the ammeter indicates the exciting current.
    5. The effective resistance of the windings of a transformer is approximately 10% more than the dc or ohmic resistance.
    6. Generally, the ac or effective resistance of a transformer is obtained by measuring the dc resistance of the winding and multiplying it by 1.1.
- The efficiency of a transformer is a ratio of the output in watts to the input in watts.

$$(\text{Percent efficiency})\ \eta = \frac{\text{output, watts}}{\text{input, watts}} \times 100 = \frac{\text{output, watts}}{\text{output + losses}} \times 100$$

or

$$\eta = \frac{V_S \bullet I_S \bullet \cos \angle\theta}{V_S \bullet I_S \bullet \cos \angle\theta + \text{core loss, watts} + I_P^2R_P + I_S^2R_S} \times 100$$

- The preferred method of determining the efficiency of a transformer is to measure the losses and add this value to the nameplate output to obtain the input.
- The impedance voltage is that voltage required to cause the rated current to flow through the impedance of the primary and secondary windings.
- The ratio of the impedance voltage to the rated terminal voltage yields the percentage impedance voltage, which is in the range of 3% to 5%.
- The voltage regulation of a transformer is the percentage change in the secondary terminal voltage from the full-load condition to the no-load condition, whereas the impressed primary voltage is held constant.

$$\text{Percent voltage regulation} = \frac{\text{no-load } V_S - \text{full-load } V_S}{\text{full-load } V_S} \times 100$$

- The polarity of the transformer lead-in wires is marked according to a standard developed by the American Standards Association (ASA).
    1. The high-voltage winding leads are marked H_1 and H_2. H_1 is on the left-hand side when the transformer is faced from the low-voltage side.
    2. The low-voltage winding leads are marked X_1 and X_2.
    3. When H_1 is instantaneously positive, X_1 is also instantaneously positive.
    4. In a transformer with subtractive polarity, the H_1 and X_1 leads are adjacent or directly across from each other.
    5. In a transformer with additive polarity, the H_1 and X_1 leads are diagonally across from each other.
- If paralleled transformers are to deliver an output that is proportional to the kVA ratings
    1. the percentage impedance values of each transformer must be the same.
    2. the secondary terminal voltages must be the same.
    3. the instantaneous polarity of the secondary leads of each transformer must be correct.
- If it is impossible to identify the transformer leads because the identifying tags are missing or disfigured, a standard test procedure can be used to determine the polarity markings.
    1. In this test, a jumper lead is temporarily connected between the high-voltage lead and the low-voltage lead directly across from it.
    2. A voltmeter is connected from the other high-voltage lead to the second low-voltage lead.

3. If the voltmeter reads the sum of the primary input voltage and the secondary voltage, the transformer has additive polarity.
4. If the voltmeter reads the primary input voltage minus the secondary voltage, the transformer has subtractive polarity.
5. There is a hazard involved in making the test at high-voltage values. Thus, a test using a relatively low voltage was developed to determine transformer polarity.

- NEMA and ASA standards for transformers:
  1. Additive polarity shall be standard for single-phase transformers up to 200 kVA, and having voltage ratings not in excess of 9000 V.
  2. Subtractive polarity shall be standard for all single-phase transformers larger than 200 kVA, regardless of the voltage rating.
  3. Subtractive polarity shall be standard for all single-phase transformers in sizes of 200 kVA and below, having high-voltage ratings above 9000 V.
- When stepdown transformers are operated in parallel, it may be necessary to remove one from service for repairs.
  1. The low-voltage side of the transformer is always disconnected from the line wires before the primary fuses are opened.
  2. A sign reading "DANGER—FEEDBACK" must be placed at each primary fuse to minimize the hazard of feedback voltage from one secondary winding to the other primary winding.
- Distribution transformer:
  1. A distribution transformer has two high-voltage windings, rated at 2400 V each.
  2. Small metal links are used to connect the two high-voltage windings either in series, for 4800-V primary service, or in parallel, for a 2400-V input. (Note that there are only two external high-voltage leads, which are marked with the standard designations H_1 and H_2.)
  3. The connections are made from the two windings at a terminal block slightly below the level of the insulating oil in the transformer case.
  4. There are two secondary windings. These windings may be connected in parallel for 120-V service, or in series for 240-V service.
  5. To obtain 120/240-V service, the two 120-V secondary windings are connected in series and a grounded neutral wire is connected between leads X_2 and X_3.
- Advantages of a single-phase, three-wire service:
  1. This type of service provides two different voltages. The lower voltage supplies lighting and small appliance loads, and the higher voltage supplies heavy appliance and single-phase motor loads.
  2. There are 240 V across the outside wires. Thus, the current for a given kilowatt load can be reduced by nearly half if the load is balanced between the neutral and the two outside wires.

   a. Because of the reduction in the current, the voltage drop in the circuit conductors is reduced. Thus, the size of the conductor may be reduced and the voltage at the load is more nearly constant.
   b. In addition, the following problems are minimized: dim lights, slow heating, and unsatisfactory appliance performance.
3. A single-phase, three-wire, 120/240-V system uses 37% less copper as compared to a 120-V, two-wire system having the same capacity and transmission efficiency.

- For the ac noninductive circuit, the current in the grounded neutral wire is the difference between the currents in the two outside legs.
  1. For a balanced load on a three-wire, single-load service, the grounded neutral wire carries no current.
  2. In a single-phase, three-wire circuit, the neutral wire is always grounded. The neutral wire is never fused or broken at any switch control point in the circuit.
  3. With a direct path to ground by way of the neutral wire, any instantaneous high-voltage surge, such as that due to lightning, is instantly discharged to ground. In this way, the electrical equipment and the circuit wiring are protected from major lightning damage.
- Danger of an open neutral:
  1. On a three-wire, single-phase system, the neutral wire helps maintain balanced voltages between each line wire and the neutral, even with unbalanced loads.
  2. If the neutral wire is opened accidentally, there is an open circuit in the return path for the current from each device connected to the two outside wires back to the transformer.
  3. Because of the open circuit, the devices are connected in series across the outside wires, or 240 V.
  4. Some devices, such as lamps and motors, may have too great a voltage drop and will burn out. Other devices may receive a reduced voltage that does not permit proper operation of the load.
- A characteristic curve can be plotted showing the percent efficiency of a transformer.
  1. The load–voltage characteristic curve usually shows that there is little change in the terminal voltage from a full-load condition to no-load condition on a transformer whose efficiency is very high.
  2. With a lagging power factor load, the secondary terminal voltage change is slightly greater than it is for a unity power factor load.
  3. With a leading power factor load, the terminal voltage increases with the load, giving a negative voltage regulation.
- Cooling of transformers
  1. The core-and-coil assembly is placed in a pressed steel tank and is completely covered with an insulating oil.

2. The insulating oil removes heat from the core and the coil windings. It also insulates the windings from the core and the transformer case.
3. The core-and-coil assembly contains channels or ducts to permit the circulation of the oil to remove the heat.
4. As the oil gains heat, its density decreases and it rises. As it rises, it contacts the tank walls and transfers the heat to the tank. The walls are cooled by air circulation. As the oil loses heat, its specific gravity increases and it returns to the bottom of the tank. This action is repeated over and over.
5. To obtain better cooling, the surface area of a transformer can be increased by adding tubes or fins to the steel tank assembly.
6. The insulating oil used in transformers is a high-grade oil that must be kept clean and moisture free.
7. Pyranol (manufactured by the General Electric Company) is a synthetic dielectric that is an effective, nonflammable, nonexplosive liquid cooling and insulating agent.
8. Other methods of cooling transformers include forced air circulation, natural air circulation, natural circulation of oil with water cooling, and forced oil circulation (for large transformers).

- Some voltage is lost in high-voltage distribution circuits.
  1. The impressed voltage on the primary winding may be lower than indicated on the nameplate of the transformer.
  2. As a result, the secondary terminal voltage is also lower than the nameplate value. This condition is undesirable, and provisions can be made for changing taps to obtain the desired voltage.
  3. The turns ratio of the transformer is adjusted by changing the taps.
     a. The change in the tap connections can be made at the terminal board inside the transformer.
     b. A tap-changing switch can be used to vary the number of turns of the primary winding, resulting in a change in the turns ratio.
     c. A special tap switch, immersed in oil, may be used on some large transformers. This type of switch permits the tap connections to be changed under load. A small motor drives the tap-changing mechanism and is operated from a control panel.
- A transformer must be operated on an ac circuit at the frequency for which it is designed.
  1. If a lower frequency is used, the reactance of the primary will decrease. The exciting current will increase, causing an increase in the flux density. The saturation of the core is above normal and there is a greater heat loss. Thus, the efficiency of the transformer is decreased.
  2. If the frequency is greater than the nameplate value, the reactance will increase and the exciting current will decrease. There will be a lower flux density, but the core loss will remain nearly constant.

- A transformer must be operated at its rated primary voltage, as indicated on the nameplate.
    1. If the voltage is increased to a high enough value above the nameplate rating, excessive heating will take place at the windings. The flux density of the core will increase, and saturation of the core will be above normal. Transformers are designed to operate without overheating at a voltage that is 5% above the rated voltage.
    2. If the transformer is operated at a voltage lower than the nameplate rating, the power output will be reduced in proportion to the voltage decrease.

- Inrush current for a transformer can be very high compared to a choke.
    1. The magnetic domains in the core material of a transformer "remember" where they were set when the power is turned off.
    2. Three factors that determine the amount of inrush current for a transformer are the amount of applied voltage, the resistance of the wire in the primary winding, and the flux change of the magnetic field in the core.
    3. Magnetic domains or molecules can be made to reset to a neutral position by inserting an airgap in the core material.

- Transformer impedance is determined by the physical construction of the transformer.
    1. Transformer impedance is expressed as a percent.
    2. Transformer impedance is used to calculate the short-circuit current of a transformer.

Achievement Review

1. Define the term *static transformer*.
2. Explain what is meant by the terms *primary* and *secondary,* as applied to transformers.
3. a. What is a stepdown transformer?
   b. What is a stepup transformer?
4. a. What is the relationship between the induced voltages on the primary and secondary windings and the number of turns of the two windings?
   b. What is the difference between the impressed voltage and the induced voltage in the primary winding?
5. The primary winding of a transformer is rated at 115 V and the secondary winding is rated at 300 V. The primary winding has 500 turns. How many turns does the secondary winding have?

6. The transformer in question 5 has a full-load secondary output of 300 VA at 300 V.
   a. What is the full-load secondary current?
   b. Determine the full-load primary current (neglect all losses).
   c. What is the relationship between the primary current and the secondary current and the number of turns on the two windings?

7. a. What is an exciting current?
   b. What two functions does it perform?

8. a. Explain what is meant by core losses.
   b. How can the core losses be kept to a minimum?
   c. Draw a circuit diagram to show how the core loss, in watts, is measured for a 5-kVA transformer rated at 2400 to 240 V. The low-voltage side is used as the primary winding.
   d. What safety precautions should be observed in measuring the core loss on the low-voltage side?
   e. Using accurate instruments, should the core loss be the same for the high-voltage winding as for the primary?

9. The core loss of the transformer in question 8 is measured using the low-voltage side as the primary. The following data are obtained:

   Voltmeter reading = 240 V
   Exciting current = 1.0 A
   Wattmeter reading = 24 W

   Determine
   a. the core loss, in watts.
   b. the power factor and the phase angle.
   c. the in-phase component of the exciting current required for the core losses.
   d. the lagging quadrature (magnetizing) component of the current.

10. Explain why the secondary ampere-turns oppose the magnetizing flux of the primary winding.

11. List, in sequence, the reactions that take place in a transformer that result in an increase in the primary current input with an increase in the secondary current output to the load.

12. a. What is the primary leakage flux?
    b. What is the secondary leakage flux?
    c. What effect does leakage flux have on the operation of a transformer?
    d. How can the leakage flux of a transformer be kept to a minimum?

13. Using diagrams, describe the three types of core construction used in transformers.

14. a. List the losses that occur in a transformer.
    b. Which losses remain nearly constant at all load points? Why?
    c. Which losses change when there is a change in the load? Why?

15. A 5-kVA, 240/120-V, 60-Hz transformer has a core loss of 32 W. The transformer has an effective resistance of 0.05 Ω in the high-voltage winding and 0.0125 Ω in the low-voltage winding. Determine the efficiency of the transformer at the rated load and unity power factor.

16. A 25-kVA, 2400/240-V, 60-Hz stepdown transformer has a core loss of 120 W. The effective resistance of the primary winding is 1.9 Ω. The effective resistance of the secondary winding is 0.02 Ω. Determine
    a. the full-load current rating of the high-voltage and low-voltage windings. (Neglect any losses and assume that the input and the output are the same.)
    b. the total copper losses of the transformer at the rated load.
    c. the efficiency of the transformer at the rated load and a unity power factor.

17. Determine the efficiency of the transformer in question 15 at 50% of the rated load output and 0.175 lagging power factor.

18. A 50-kVA, 4600/230-V, 60-Hz, single-phase, stepdown transformer is designed so that at the condition of full load and a unity power factor, the core losses equal the copper losses. The copper losses are equally divided between the two windings. At full load the efficiency is 96.5%. Determine
    a. the core losses.
    b. the total copper losses.
    c. the rated current of the primary and secondary windings. (Neglect the losses and assume that the input and the output are the same.)
    d. the effective resistance of the primary and secondary windings.

19. Explain why it is hard to arrive at an accurate value of the efficiency of a transformer using measurements of the power input and output.

20. a. Draw a diagram of the connections for the impedance short-circuit test for a single-phase transformer, rated at 20 kVA, 60 Hz, 4800/240 V. The high-voltage winding is the input side and is supplied from a 240-V ac source with a variable resistor in series.
    b. Why should the measurements for the short-circuit test (part a) be made on the high-voltage side of the transformer?

21. The following data were obtained during the short-circuit test for the 20-kVA, 4800/240-V transformer in question 20:

    Impedance voltage = 160 V
    Ammeter reading = 4.2 A
    Wattmeter reading = 280 W

    Another test shows the core loss to be 120 W. Determine
    a. the efficiency at the rated load and a unity power factor.
    b. the efficiency at 50% of the rated load and 0.80 lagging power factor.

22. a How are transformer leads marked?
    b. Explain what is meant by *additive polarity* and *subtractive polarity*.

23. A transformer is used to step down a primary voltage of 4800 V to 240 V to supply a single-phase load. Additional equipment is installed, making the kVA load greater than the full-load rating of the transformer. A second transformer, having the same kVA capacity, is to be operated in parallel with the first transformer.
    a. List three requirements that are to be observed if the transformers are to share the load properly.
    b. Explain the procedure used in connecting the second transformer in parallel.

24. a. There is a question about the accuracy of the turns ratio of a transformer rated at 2400/240 V. A 240-V test source is available in the laboratory. Assuming that voltmeters are available with the proper scales, explain how the voltage ratio can be determined. Draw a schematic diagram to clarify the answer.
    b. Explain how the polarity of this transformer may be obtained using the 240-V test source.

25. A 30-kVA, 2400/240-V, 60-Hz transformer is used as a stepdown transformer. The core loss is 150 W, the rated primary current is 12.5 A, the resistance of the primary winding is 1.5 Ω, the rated secondary current is 125 A, and the resistance of the secondary winding is 0.015 Ω. At the rated kVA capacity and a 0.75 lagging power factor, determine
    a. the total copper losses.
    b. the efficiency of the transformer.

26. A 20-kVA standard distribution transformer is rated at 2400/4800 V on the high-voltage side, and 120/240 V on the low-voltage side. The transformer has an efficiency of 97% when it is operating at the rated load as a stepdown transformer supplying a noninductive lighting load.
    a. Draw a diagram to show the internal and external transformer connections required to step down the voltage from 2400 V to a 120/240-V, single-phase, three-wire service. Show the polarity markings for the transformer terminals for additive polarity.
    b. What is the full-load secondary current if the load is balanced?
    c. What is the power input to the primary side at the rated load?
    d. If the no-load voltage of the 240-V secondary is 4 V more than the full-load voltage, what is the percentage voltage regulation?

27. Explain why the single-phase, 120/240-V, three-wire system is preferred to the single-phase, 120-V, two-wire system.

28. a. Why is oil used in transformers?
    b. What advantage is there in using Pyranol, instead of oil, in a transformer?

29. a. Explain the process of cooling a transformer by the natural circulation of oil and air.
    b. How can the heat-dissipating surface be increased for transformers with very large kVA ratings?

30. a. What is the purpose of a tap-changing switch as used on a transformer?
    b. What are some of the applications of a transformer with tap-changing facilities?

31. Explain why a transformer has a high efficiency from 10% of the rated load to full load.

32. List the data commonly found on a transformer nameplate.

33. A 50-kVA single-phase transformer has a secondary voltage of 240 V. The nameplate indicates that the transformer has an impedance of 2.35%. What is the short-circuit current for this transformer?

34. A three-phase transformer is rated at 75 kVA and has a secondary voltage of 208 V. The nameplate indicates a transformer impedance of 3.5%. What is the short-circuit current for this transformer?

PRACTICE PROBLEMS FOR UNIT 13

Find the missing values in the following problems. Refer to Figure 13–35 and the formulas under the Transformers section of Appendix 15.

1. V_P = 13,800 V | V_S = 480 V
   I_P = ________ | I_S = ________
   N_P = 5750 | N_S = ________
   R = 12 Ω

2. V_P = 120 V | V_S = ________
   I_P = ________ | I_S = 0.25 A
   N_P = 162 | N_S = 11,475
   R = ________

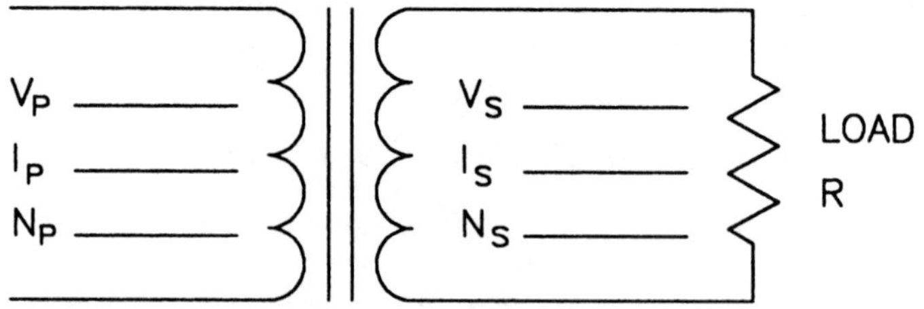

FIGURE 13–35 Isolation transformer with one secondary winding (*Delmar/Cengage Learning*)

3. V_P = ________ V_S = 24 V

I_P = 0.8 A I_S = 4 A

N_P = ________ N_S = 60

R = ________

4. V_P = 480 V V_S = 277 V

I_P = ________ I_S = ________

N_P = 338 N_S = ________

R = 4 Ω

Find the missing values in the following problems. Refer to Figure 13–36 and the formulas listed under the Transformers section of Appendix 15.

[*Note:* Total primary current is equal to the sum of the primary currents needed to supply power to each secondary.]

5.

V_P = 240 V	V_{S_1} = 360 V	V_{S_2} = 48 V	V_{S_3} = 24 V
I_P = ________	I_{S_1} = ________	I_{S_2} = ________	I_{S_3} = ________
N_P = 400	N_{S_1} = ________	N_{S_2} = ________	N_{S_3} = ________
	R_1 = 480 Ω	R_2 = 16 Ω	R_3 = 3 Ω

6.

V_P = 208 V	V_{S_1} = ________	V_{S_2} = ________	V_{S_3} = ________
I_P = ________	I_{S_1} = ________	I_{S_2} = ________	I_{S_3} = ________
N_P = 520	N_{S_1} = 300	N_{S_2} = 150	N_{S_3} = 30
	R_1 = 200 Ω	R_2 = 24 Ω	R_3 = 4 Ω

7.

V_P = 208 V	V_{S_1} = ________	V_{S_2} = ________	V_{S_3} = ________
I_P = ________	I_{S_1} = 0.5 A	I_{S_2} = 4 A	I_{S_3} = 4 A
N_P = 400	N_{S_1} = 200	N_{S_2} = 100	N_{S_3} = 20
	R_1 = ________	R_2 = ________	R_3 = 6 Ω

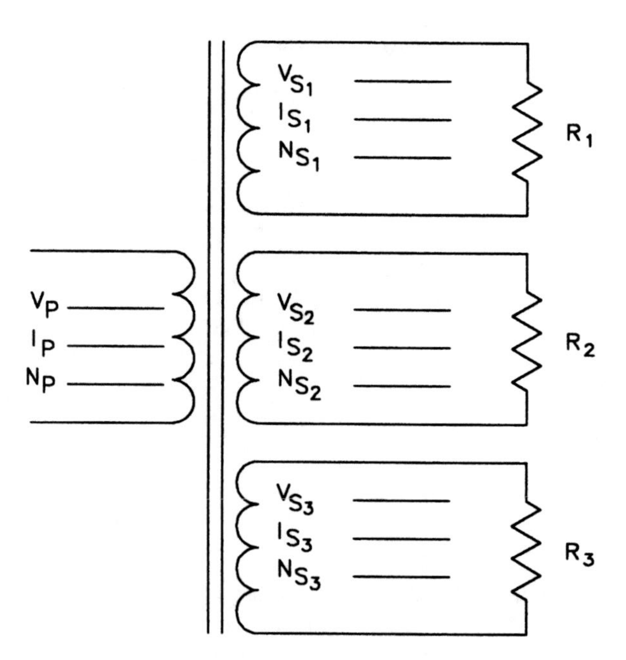

FIGURE 13–36 Isolation transformer with multiple secondary windings (*Delmar/Cengage Learning*)

14

Transformer Connections for Three-Phase Circuits

Objectives

After studying this unit, the student should be able to

- explain the differences between a large three-phase transformer and a bank of single-phase transformers, including the advantages and disadvantages of two units.
- select the type of transformer connection to be used on particular three-phase power system.
- calculate the following quantities using the appropriate formulas:
  1. The total capacity of a bank of three single-phase transformers connected in closed delta; the total capacity of two single-phase transformers connected in open delta
  2. The primary coil voltage, the line voltage, the coil current, and the line current for the following cases:
     a. Three single-phase transformers connected in delta–delta
     b. Three single-phase transformers connected in wye–delta
- describe a dual-load, three-phase, four-wire service connected in delta–delta.
- diagram and explain the standard procedure for making a closed-delta connection and a wye–wye connection for a three-phase service using three single-phase transformers.
- list the advantages of a wye connection over a delta connection with regard to the insulation of the transformers.
- explain why a delta connection has an advantage of $\sqrt{3}$ or 1.73 times the line current over a wye connection.
- describe what a harmonic is.
- discuss the problems concerning harmonics.
- identify the characteristics of different harmonics
- perform a test to determine whether harmonic problems exist.
- discuss methods of dealing with harmonic problems.

Most electrical energy is generated by three-phase alternating-current generators. Three-phase systems are also used to transmit and distribute this electrical energy. Many applications require that the voltage on such three-phase systems be transformed to either a higher value or a lower value.

Three single-phase transformers can be used to transform the voltages of three-phase systems. These transformers may be connected in any one of four standard ways: delta–delta, wye–wye, delta–wye, and wye–delta. Another possible connection is the open delta (or V). This connection uses only two single-phase transformers to transform voltages on a three-phase system.

THE DELTA CONNECTION

A delta–delta connection is used when three single-phase transformers are used to step down as three-phase voltage of 2400 V to 240 V, three phase.

The connections for the three single-phase transformers are shown in Figure 14–1. The primary and secondary windings of the transformers are connected in delta. The high-voltage leads of the primary winding of each transformer are marked H_1 and H_2. The leads of the low-voltage secondary winding of each transformer are marked X_1 and X_2.

Primary Winding Connections

The high-voltage primary windings are connected in a closed-delta arrangement. H_1 is assumed to be the beginning of each high-voltage winding and H_2 is assumed to be the end. The end (H_2) of each primary winding is connected to the beginning (H_1) of the next primary winding to form a series arrangement. One three-phase line wire is connected to each junction of two windings. In other words, the primary winding of each transformer is connected directly across the line voltage. Each of the three line voltages is 2400 V and the primary winding of each transformer is correctly rated at 2400 V. Once the high-voltage primary connections are made, the three-phase 2400-V input may be energized and tested for the correct phase rotation.

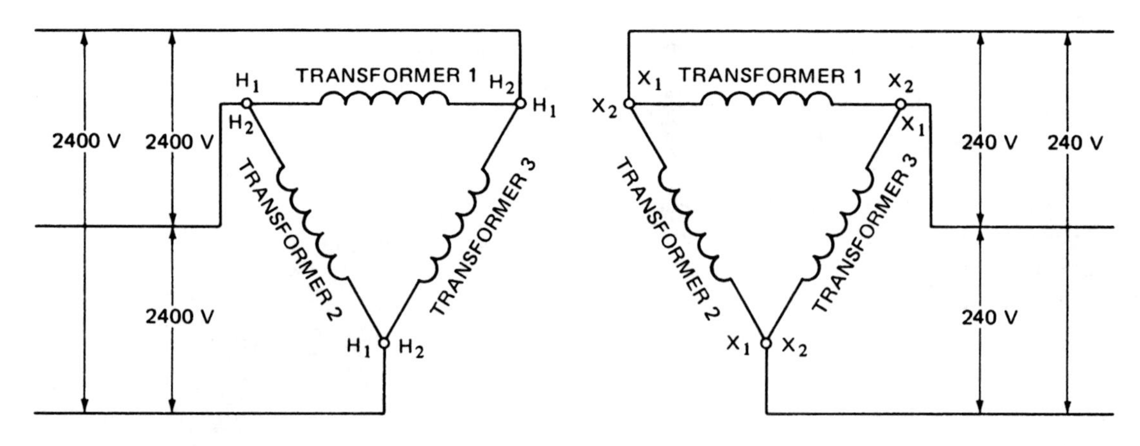

FIGURE 14–1 Delta–delta transformer connections (*Delmar/Cengage Learning*)

Unit 13 described the high-voltage lead (H_1) of any transformer as being on the left when the transformer is viewed from the low-voltage side. This means that the positions of H_1 and H_2 are fixed. Unit 13 also explained that transformers are standardized with regard to polarity. They have additive or subtractive polarity, according to their voltage ratings and capacities in kilovolt-amperes. However, there are many exceptions to these standards. To determine whether a transformer has additive or subtractive polarity, check the terminal markings or the nameplate data.

Phase Inversion

The primary side of the delta-connected transformer may be connected so as to reverse the polarity of one of its legs. Such a phase inversion of the primary side must be corrected in the secondary side. The following method can be used to correct the polarity:

1. Determine whether the voltage output of each of the three transformers is the same as the voltage rating on the nameplate.
2. Connect the end of one secondary winding to the beginning of another secondary winding, as shown in Figure 14–2A. If the connection is correct, the voltage across

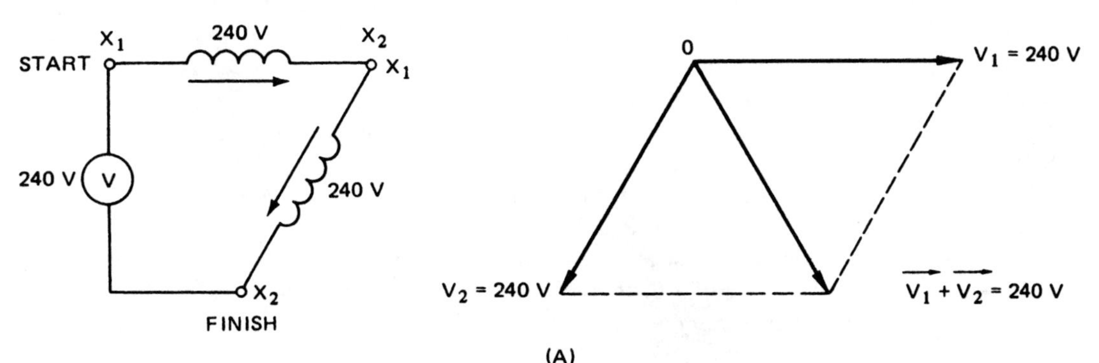

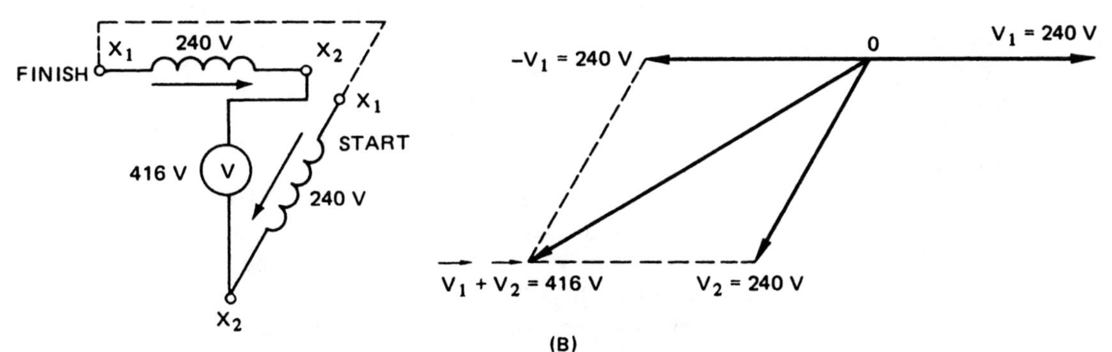

FIGURE 14–2 Delta connection (*Delmar/Cengage Learning*)

the open ends of the two transformers should be the same as the output of each transformer. In this case, the voltage is 240 V. The resultant voltage in the vector diagram is the same as the two secondary winding voltages and is equal to 240 V. Note that the path from start to finish is in the direction of both voltage arrows and does not cause a reversal of the vectors. The connections of the transformer windings are shown reversed in Figure 14–2B. The path from start to finish is in the direction of the voltage arrow and opposes the other voltage arrow. The V_1 vector is reversed, and the resultant voltage is equal to $\sqrt{3}$ times the secondary coil voltage. The resultant voltage is 416 V. This condition is corrected by changing the connections of the secondary leads of one of the transformers. Thus, the new connections will be the same as in Figure 14–2A.

3. The proper connections are shown in Figure 14–3A for one end of the secondary coil of the third transformer. The vector diagram shows that there is a resultant voltage opposite to each secondary voltage and having a magnitude equal to the secondary voltage. Therefore, the voltage at any corner of the delta is zero. If the voltage is zero across the last pair of open leads, they can be connected together. A line wire is attached at each of the connection points. These three line wires form the 240-V, three-phase output. The three line voltages and each of the three transformer secondary voltages have the same value of 240 V. In Figure 14–3B the connections of the third secondary winding are reversed. As shown in the vector diagram, reversing the secondary voltage of the third transformer (V_3) causes the resultant voltage to be twice the secondary voltage. In this case, the resultant voltage is 480 V. This condition is corrected by reversing the connections of the secondary leads of the third transformer. As a result, the voltage across the last pair of open leads is zero and the delta connection may be closed. (CAUTION: Never connect the last pair of open leads if there is a voltage difference across them. The potential difference must be zero, indicating that the winding connections are correct.)

Transformer Banks

When the primary and secondary windings of three transformers are all connected in delta, the resulting arrangement is called a *delta–delta* (D–D) *connection*. The delta connection of the primary windings is indicated by the first delta symbol. The second delta symbol indicates the connection of the secondary windings. When two or three single-phase transformers are used to transform voltages on a three-phase system, they are known as a *transformer bank*.

The connections shown in Figure 14–4 are for a bank of transformers connected in delta–delta. Note that it is not necessary to represent delta–delta connections as triangles.

The delta–delta connection is used to supply an industrial load by stepping down a 2400-V, three-phase, three-wire service to a 240-V, three-phase, three-wire service. Assume that the industrial load consists of three-phase motors. The current in each wire is nearly the same and is considered to be balanced. For most applications, the load is balanced and the three single-phase transformers have the same kilovolt-ampere capacity. To determine the total capacity in kilovolt-amperes of the delta–delta-connected transformer bank, the three ratings are added. For example, if each transformer is rated at 50 kVA, the total capacity is 150 kVA.

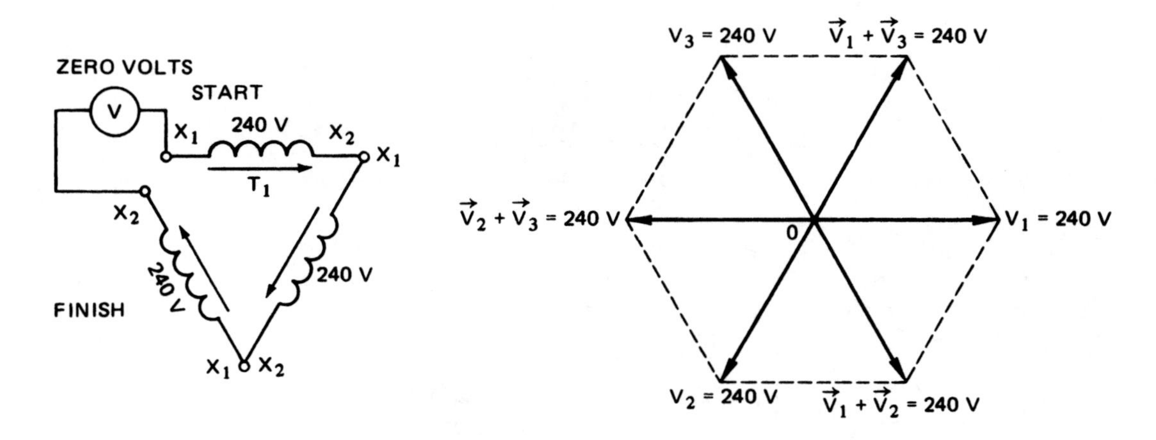

(A)

CORRECT DELTA CONNECTION—VOLTAGE ACROSS LAST PAIR OF OPEN TERMINALS EQUALS ZERO.

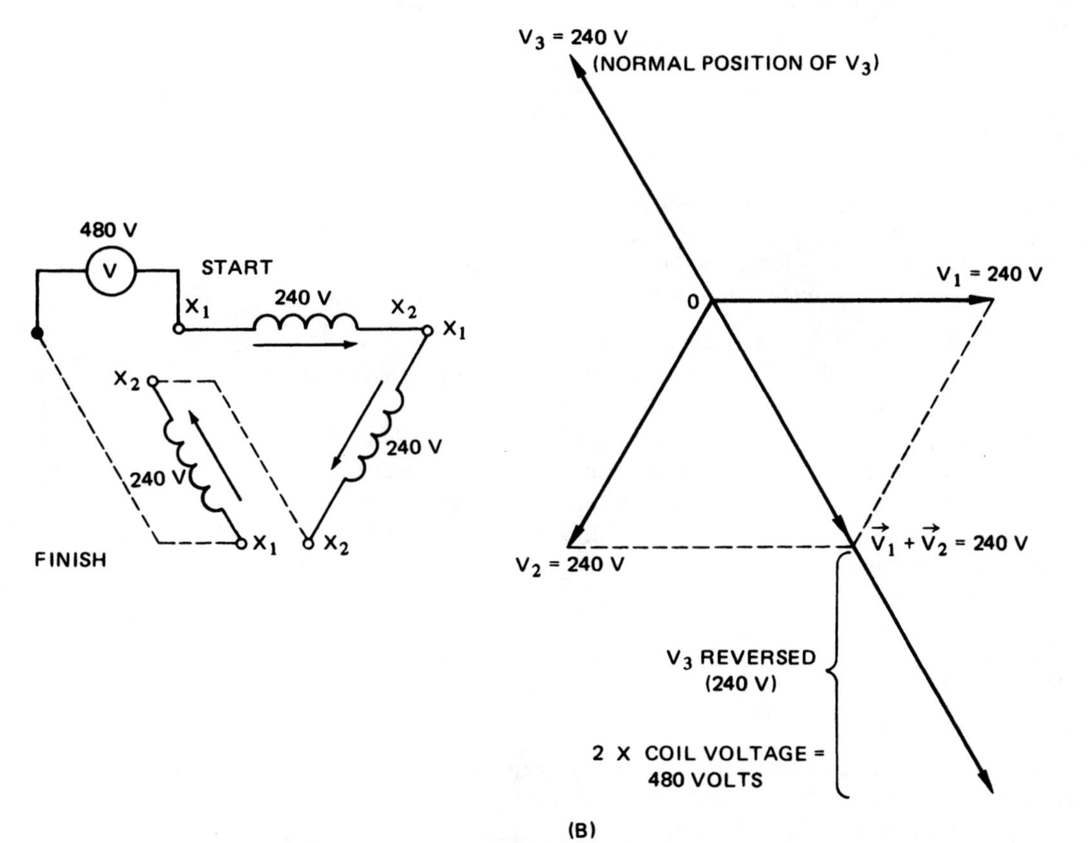

(B)

INCORRECT DELTA CONNECTION—SECONDARY LEADS OF THIRD TRANSFORMER ARE REVERSED.

FIGURE 14–3 Delta connection (*Delmar/Cengage Learning*)

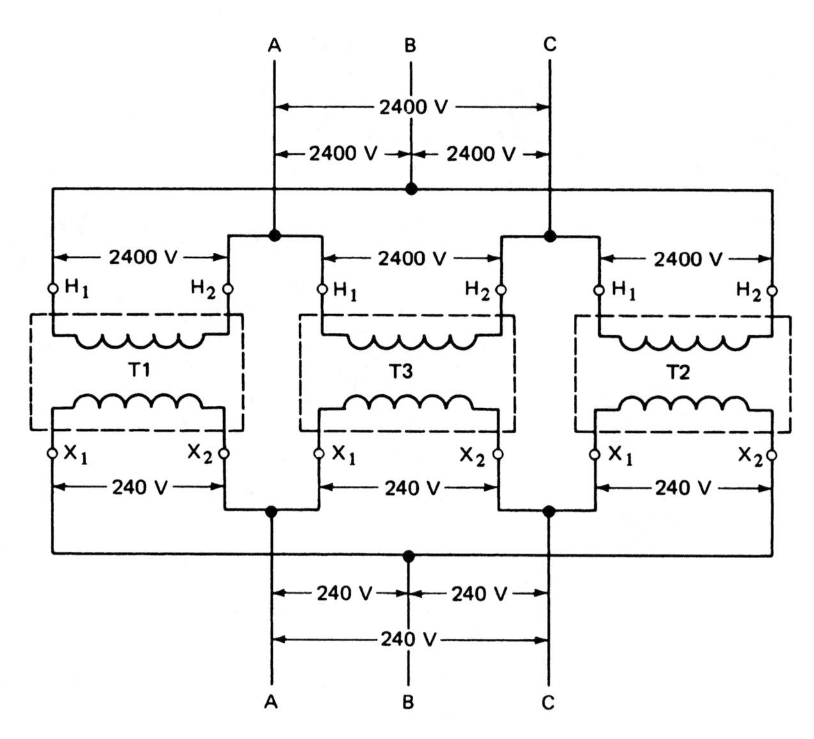

FIGURE 14–4 Delta–delta transformer connection (*Delmar/Cengage Learning*)

The current–voltage relationships for a delta-connected circuit were described in Unit 10. The following information applies to both delta-connected circuits and systems.

1. In any delta connection, each transformer winding is connected across two line leads. This means that the line voltage and the transformer coil winding voltage are the same.
2. The line current in a delta-connected transformer bank is equal to 1.73 times the coil winding current. In a closed-delta transformer connection, each line wire is fed by two transformer coil currents that are out of phase. Because the two coil currents are out of phase, they do not add together directly, but must be added vectorially to obtain the line current.

PROBLEM 1

Statement of the Problem

Three single-phase transformers are connected in delta–delta, as shown in Figure 14–4. Each transformer is rated at 50 kVA, 2400/240 V. The current in each of the three secondary line leads is 300 A. The secondary line voltage is 240 V across all three phases. The three-phase load connected to the transformer has a lagging power factor of 0.75. Determine

1. the kilovolt-ampere load on the transformer bank.
2. the kilowatt load on the transformer bank.
3. the percentage load on the transformer bank, in terms of its rated kVA capacity.
4. the current in each secondary winding.
5. the voltage across each secondary winding.
6. the kVA load on each transformer.
7. the line current on the primary side (assuming that the losses are negligible).
8. the kVA input to the transformer bank, if the losses are neglected.

Solution

1. $$\text{kVA} = \frac{\sqrt{3} \times V_{line} \times I_{line}}{1000}$$

$$= \frac{1.73 \times 240 \times 300}{1000} = 124.6 \text{ kVA}$$

2. $$\text{kW} = \frac{\sqrt{3} \times V_{line} \times I_{line} \times \cos\theta}{1000} = 93.4 \text{ kW}$$

$$= \frac{1.73 \times 240 \times 300 \times 0.75}{1000} = 93.4 \text{ kW}$$

3. The total capacity of the transformer bank, in kVA, is obtained by adding the kVA ratings of the three single-phase transformers:

$$50 + 50 + 50 = 150 \text{ kVA}$$

The percentage load in the transformer bank is

$$\frac{\text{Actual load on bank, kVA}}{\text{Capacity of bank, kVA}} \times 100 = \frac{124.6}{150} \times 100 = 83\%$$

4. The current in each secondary winding is

$$I_{coil} = \frac{I_{line}}{\sqrt{3}} = \frac{300}{1.73} = 173 \text{ A}$$

5. The voltage across each secondary winding is 240 V.

6. The kVA load on each of the three single-phase transformers is the product of the secondary winding current and the secondary voltage divided by 1000:

$$\text{Transformer kVA} = \frac{V_S \times I_S}{1000} = \frac{240 \times 173}{1000} = 41.5 \text{ kVA}$$

7. The secondary coil current of each transformer is 173 A. If the losses are neglected, the primary coil current is

$$173 \div 10 = 17.3 \text{ A}$$

The primary windings are connected in delta. Therefore, the primary line current is

$$\sqrt{3}\ I_{coil} = 1.73 \times 17.3 = 30\ A$$

8. Input, kVA $= \dfrac{\sqrt{3} \times V_{line} \times I_{line}}{1000} = \dfrac{1.73 \times 2400 \times 30}{1000} = 124.6\ kVA$

FEEDING A DUAL LOAD

Some power companies use a delta–delta-connected transformer bank to feed two types of loads. These loads consist of a 240-V, three-phase industrial load and a 120/240-V, single-phase, three-wire lighting load.

One single-phase transformer supplies the single-phase, three-wire lighting load. This transformer usually is larger than the other two transformers. The 120/240-V, single-phase, three-wire service is obtained from this transformer by bringing out a tap from the midpoint of the 240-V, low-voltage secondary winding. Many transformers have two 120-V windings. As explained in Unit 13, these windings can be connected in series, with a tap brought out at the midpoint to give 120/240-V service.

Three single-phase transformers are connected as a delta–delta bank in Figure 14–5. Each transformer has two 120-V windings. When these windings are connected in series, each transformer has a total output of 240 V. The high-voltage primary windings are connected in closed delta. The low-voltage output windings or secondary windings are also connected in closed delta to give three-phase, 240-V service to the industrial power load. Because the middle transformer also feeds the single-phase, three-wire, 120/240-V lighting load, a tap is made at the midpoint on the secondary output side of the transformer to give 120/240-V service. This tap feeds to the neutral wire of the single-phase, three-wire system and is grounded.

A check of the connections in Figure 14–5 shows that there is 120 V to ground on both lines A and C of the three-phase, 240-V secondary system. However, line B will have 208 V to ground. The condition can be a serious hazard and cannot be used for lighting service.

THE WYE CONNECTION

The wye connection is another standard method of connecting single-phase transformers to obtain three-phase voltage transformation. The wye connections must be made systematically to avoid errors. The student must be familiar with and able to use the basic voltage and current relationships for this type of connection. In Unit 10, the following information was given for the three-phase wye connection:

- The line current and the coil winding current are equal.
- The line voltage is equal to $\sqrt{3}$ times the coil winding voltage.

Figure 14–6 shows three single-phase transformers connected in wye–wye. The H_2 leads of the high-voltage windings are considered to be the ends of each of the high-voltage windings and are connected together. The H_1 (beginning) lead of each transformer is

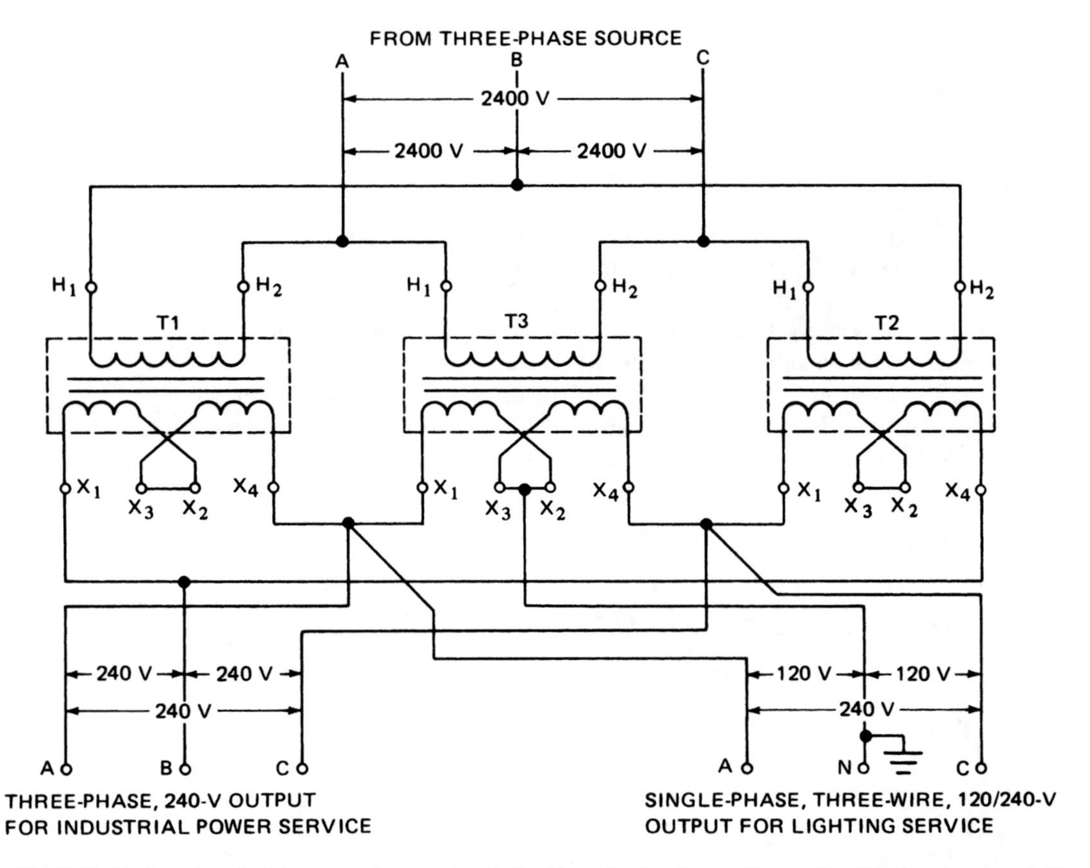

FIGURE 14–5 **Closed-delta transformer bank feeding single-phase, three-wire lighting load and three-phase, three-wire power load** ***(Delmar/Cengage Learning)***

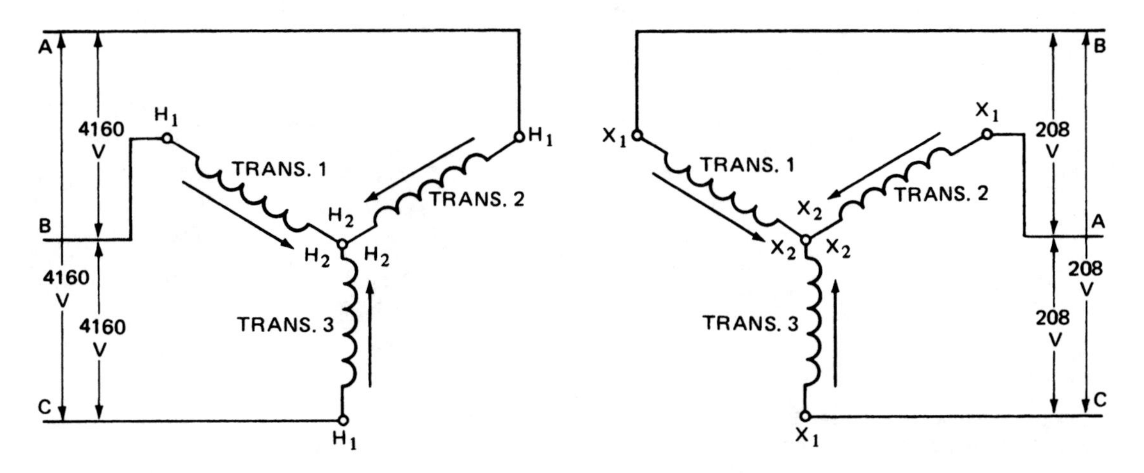

FIGURE 14–6 **Wye–wye transformer connections** ***(Delmar/Cengage Learning)***

connected to one of the three line leads. When this connection is shown in a schematic diagram, it looks like the letter Y (which is written as wye). The connection may also be called a *star* connection.

In general, the low-voltage winding leads are marked and the polarity is shown on the transformer nameplate. However, the following procedure should be used to make the low-voltage secondary connections:

1. Energize the three-phase, wye-connected, high-voltage side of the transformer bank. The voltage output of each of the three transformers must be the same as the nameplate rating.
2. Deenergize the primary. Connect the X_2 ends of two low-voltage secondary windings, as shown in Figure 14–7A. With all three X_1s open and clear, energize the primary. If the connections are made correctly, the voltage across the open ends is $\sqrt{3}$ times the secondary winding voltage. In this case, the voltage across the open ends is 208 V, as indicated in the vector diagram.

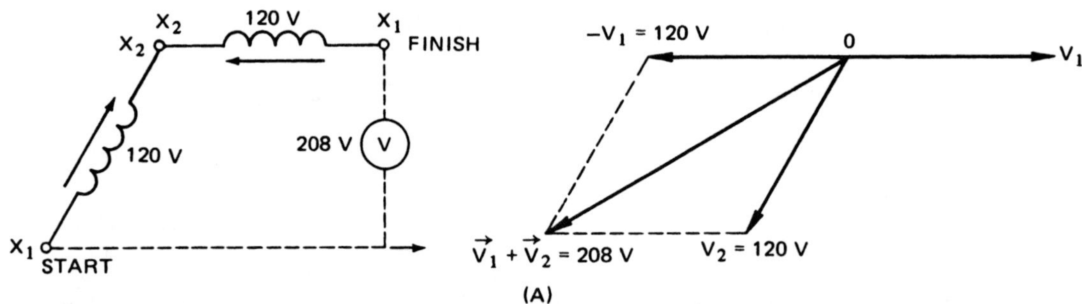

(A)

CORRECT WYE CONNECTION—VOLTAGE ACROSS THE OPEN ENDS EQUALS $\sqrt{3}$ X THE COIL VOLTAGE

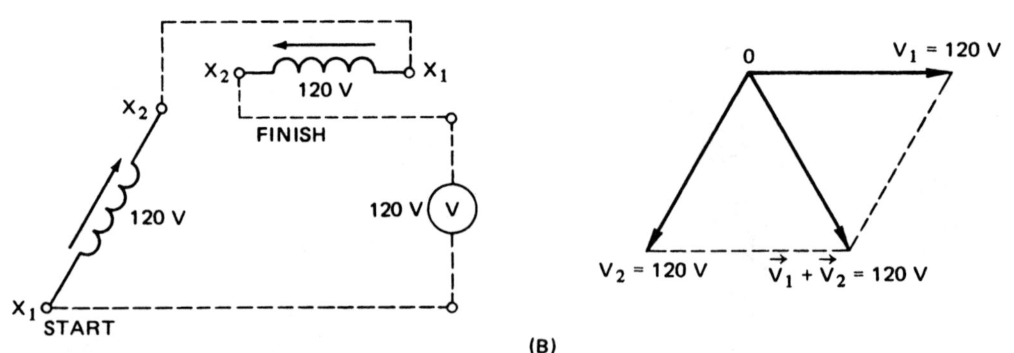

(B)

INCORRECT WYE CONNECTION—SECONDARY LEADS OF ONE TRANSFORMER ARE REVERSED AND THE VOLTAGE ACROSS THE OPEN ENDS EQUALS THE COIL VOLTAGE

FIGURE 14–7 Correct and incorrect wye connections (*Delmar/Cengage Learning*)

Secondaries Incorrectly Connected

The secondaries of the two transformers are shown connected incorrectly in Figure 14–7B. In this case, the voltage across the open ends is the same as the secondary voltage of each transformer. According to the vector diagram, the resultant voltage is only 120 V. The connections can be corrected by reversing the leads of transformer 2. As a result, the voltage across the open ends will be 208 V.

Secondaries Properly Connected

The correct wye connections are shown in Figure 14–8A for the secondary windings of the three transformers. The voltage across each pair of line leads is equal to $\sqrt{3}$ times the secondary coil voltage or 208 V. The vector diagram in Figure 14–8B shows the relationship between the coil voltages and the line voltages in a three-phase, wye-connected system.

The wye–wye connection can be used in those applications where the load on the secondary side is balanced. If the load consists of three-phase motor loads only, and the load currents are balanced, then the wye–wye connection can be used. This connection cannot be used if the secondary load becomes unbalanced. An unbalanced load causes a serious imbalance in the three voltages of the transformer bank.

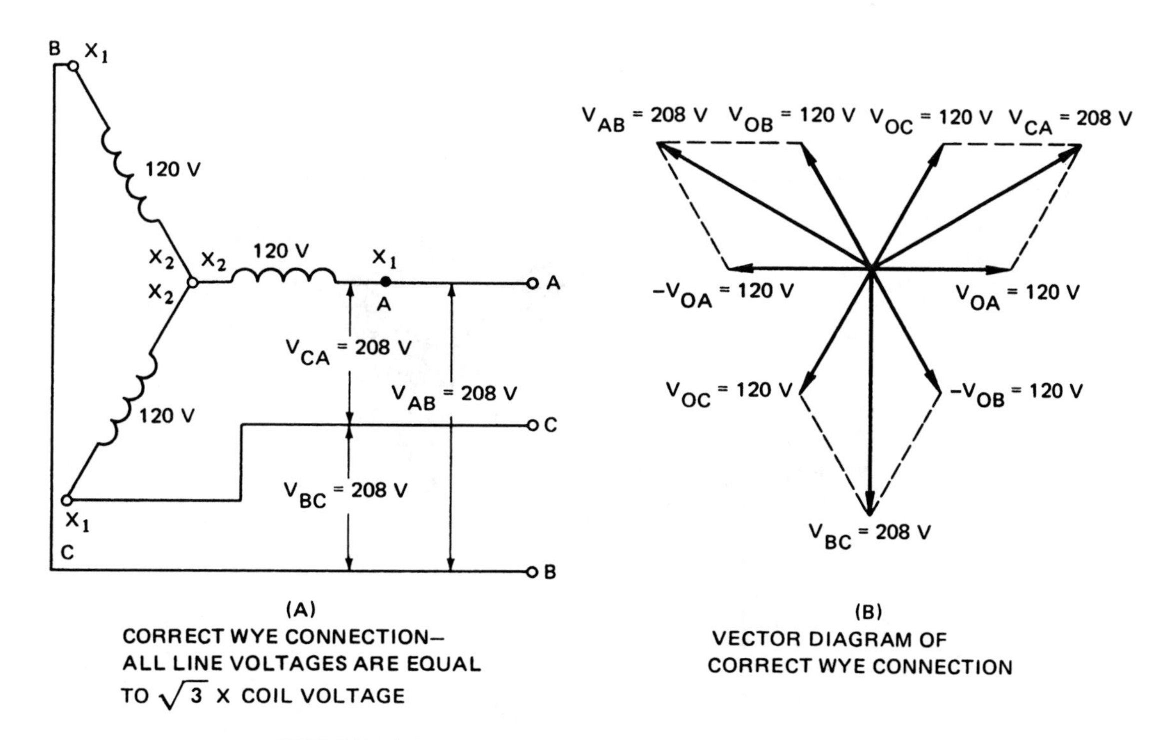

FIGURE 14–8 Wye connection (*Delmar/Cengage Learning*)

The Neutral Wire

A fourth wire known as the *neutral wire* is added to eliminate unbalanced voltages. The neutral wire is connected between the source and the common point on the primary side of the transformer bank.

The diagram in Figure 14–9 shows a wye–wye-connected transformer bank having a three-phase, four-wire, 2400/4160-V input and a three-phase, four-wire, 120/208-V output. On the high-voltage input side, the neutral wire is connected to the common point. This is the point where all three high-voltage primary winding (H_2) leads terminate. The voltage from the neutral to any one of the three line leads is 2400 V. Each high-voltage winding is connected between the neutral and one of the three line leads. This means that each high-voltage winding is connected across 2400 V. The voltage across the three line leads is $\sqrt{3} \times$ 2400 V, or 4160 V. The neutral wire maintains a nearly constant voltage across each of the high-voltage windings, even when the load is unbalanced. The neutral conducts any unbalance of current between the source and the neutral point on the input side of the transformer bank. The neutral wire is grounded and helps protect the three high-voltage windings from lightning surges.

For the transformer bank shown in Figure 14–9, the three-phase, four-wire system feeds from the low-voltage side of the bank to the load. Each low-voltage winding is connected between the secondary neutral and one of the three line leads. The voltage output of each secondary winding is 120 V. Thus, there is 120 V between the neutral and any one of the three secondary line leads. The voltage across the line wires is $\sqrt{3} \times 120$, or 208 V. The use of a three-phase, four-wire secondary provides two voltages that can be used for different load types:

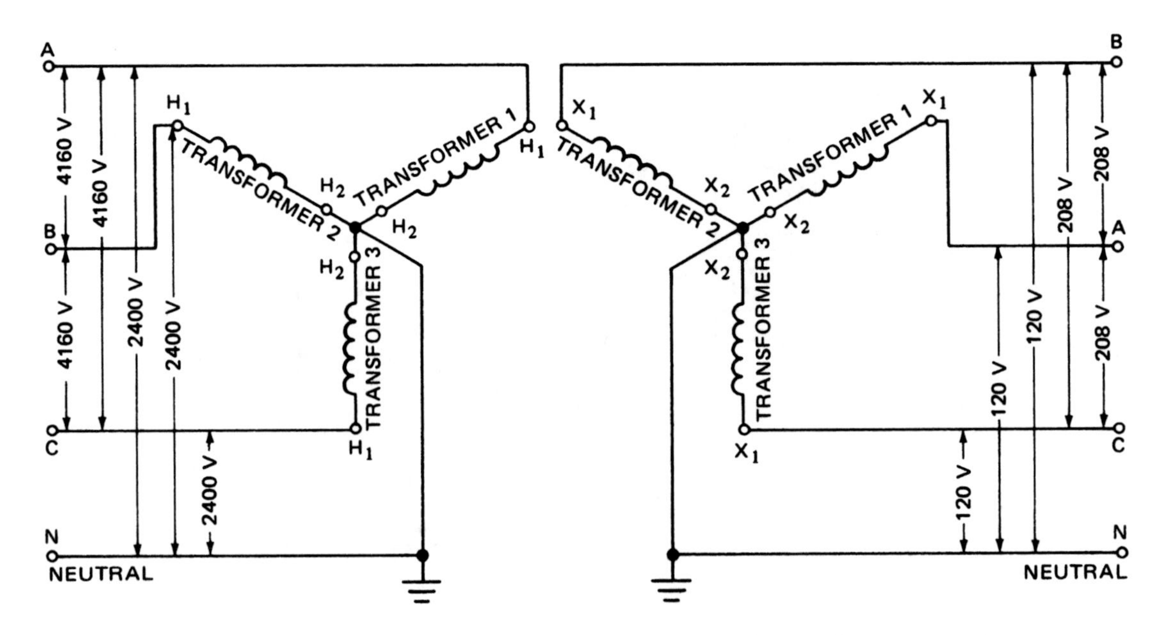

FIGURE 14–9 Wye–wye transformer bank with neutral connections (*Delmar/Cengage Learning*)

1. A 208-V, three-phase service is available for industrial power loads such as three-phase motors.
2. The use of a neutral wire means that 120 V is available for lighting loads.

Neutral Wire Used on Primary and Secondary

Figure 14–10 shows the connections for a wye–wye-connected transformer bank. Note that both the primary and secondary sides contain a neutral wire. Each transformer has two 120-V, low-voltage windings connected in parallel. The voltage output for each single-phase transformer is 120 V. Because this is a three-phase, four-wire system, the

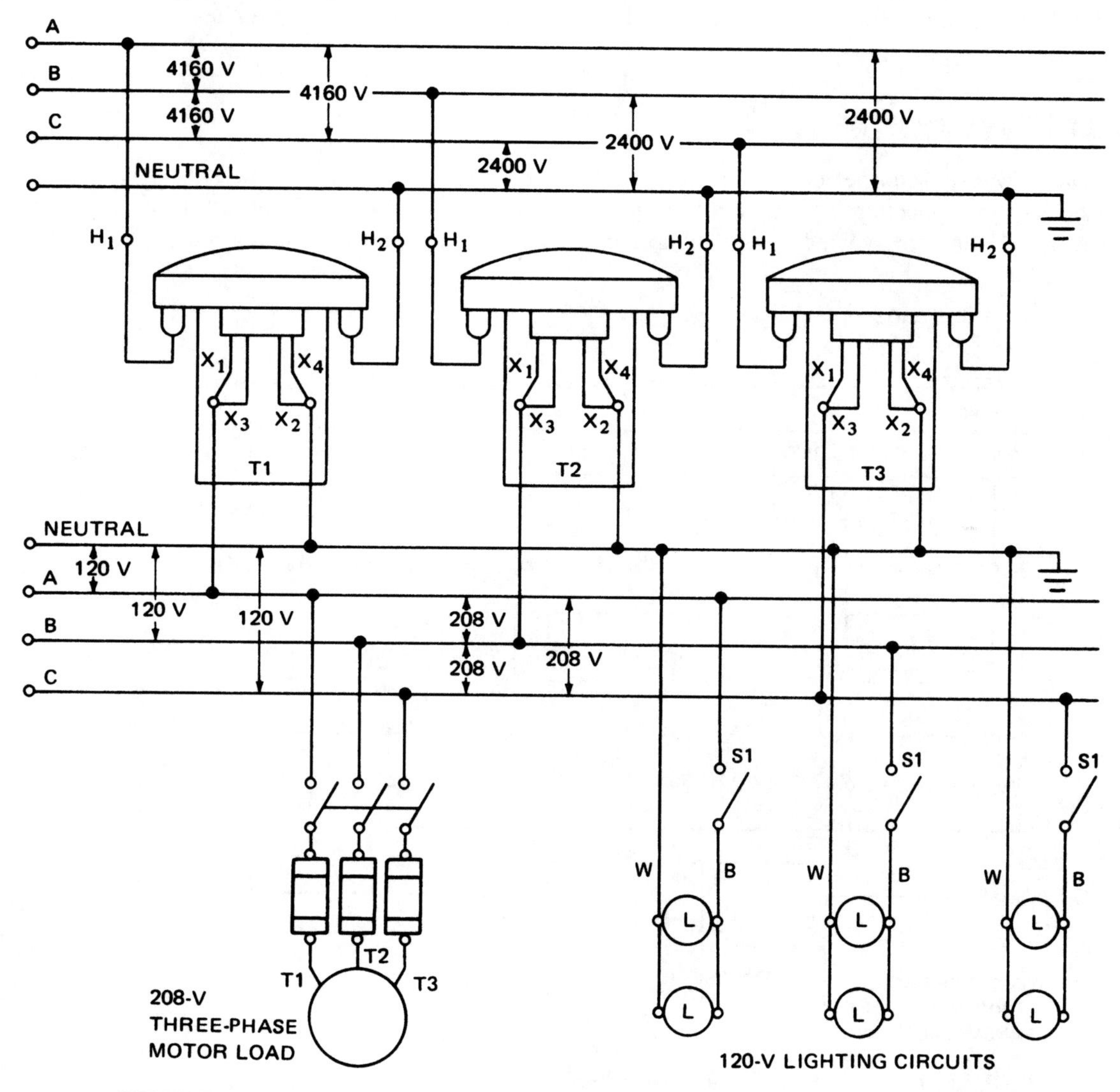

FIGURE 14–10 Wye–wye transformer bank connections (*Delmar/Cengage Learning*)

following voltages are available: (1) a three-phase, 208-V service for motor loads and (2) a 120-V, single-phase service for the lighting loads. The lighting load should be distributed evenly between the three transformers. Thus, an attempt is always made to balance the lighting circuits from the line wire to the neutral.

Nearly all wye–wye-connected transformer banks use three single-phase transformers having the same kilovolt-ampere capacity. The individual kVA ratings are added to find the capacity of the transformer bank. For example, if a bank consists of three transformers, each rated at 25 kVA, the total rating is 25 + 25 + 25 = 75 kVA.

In a wye–wye connection, a defective transformer must be replaced before the wye–wye transformer bank can be reenergized. Unlike a delta–delta bank, a wye–wye transformer bank *cannot* be temporarily reconnected in an emergency using two single-phase transformers only.

DELTA–WYE CONNECTION

Delta–wye transformer connections may also be used for voltage transformation. The delta–wye connection is used both to step up and step down voltage. As shown in Figure 14–11, the primary windings are connected in delta and the secondary windings are connected in wye.

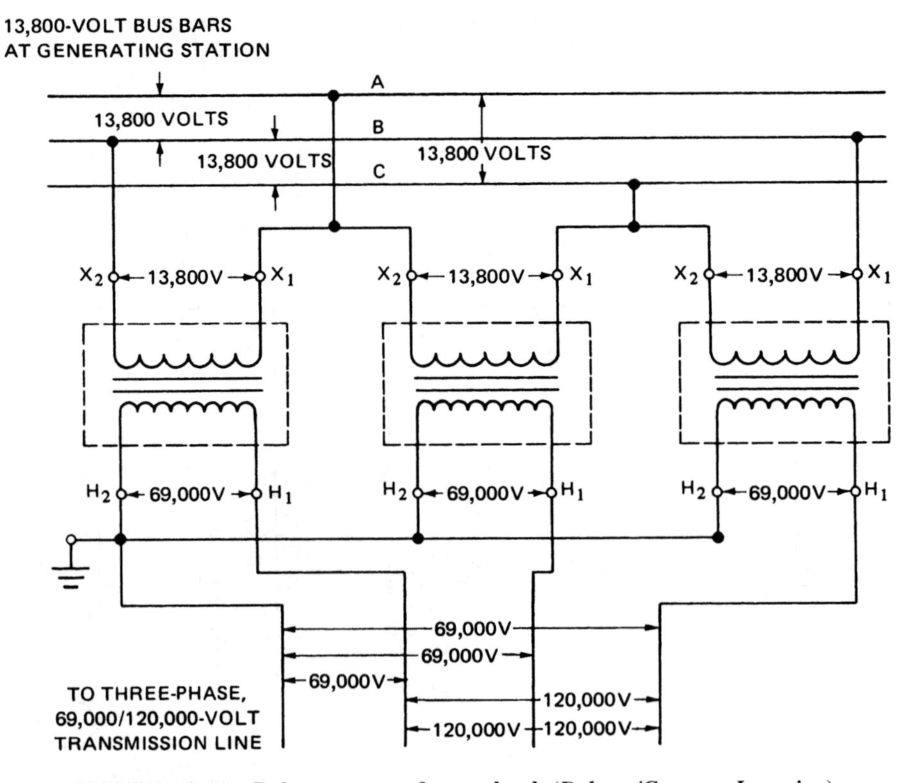

FIGURE 14–11 Delta–wye transformer bank (*Delmar/Cengage Learning*)

Delta–Wye Transformer Bank

The delta–wye transformer bank shown in Figure 14–11 is used to step up the voltage at a generating station. The input voltage is stepped up by the transformer ratio. This voltage is further increased by the factor of 1.73. The high-voltage output is then connected directly to three-phase transmission lines. These transmission lines deliver the energy to users who may be miles away from the generating station. The use of the delta–wye connection means that the insulation requirements are reduced for the secondary windings. This fact is very important when the secondary side has a very high voltage. (Recall that the coil voltage is only 58% of the line voltage, where $1 \div \sqrt{3} = 1 \div 1.73 = 0.58 = 58\%$.)

In Figure 14–11, the ac generators deliver energy to the generating station bus bars at a three-phase potential of 13,800 V. For the three single-phase transformers, the primary windings are each rated at 13,800 V. The windings are connected in delta to the bus bars of the generating station. The coil and line voltages are the same in a delta connection. Therefore, each primary winding has 13,800 V applied to it. The stepup ratio of the transformers shown is 1:5. This means that the voltage output of the secondary of each single-phase transformer is $5 \times 13{,}800 = 69{,}000$ V. The three secondary windings are connected in wye. Each high-voltage secondary winding is connected between the secondary neutral and one of the three line wires. The voltage between the neutral and any one of the three line wires is the same as the secondary coil voltage, or 69,000 V. The voltage across the three line wires is $\sqrt{3} \times 69{,}000 = 119{,}370$ V, or 120,000 V.

Three-phase output voltages can be balanced because of the neutral wire on the high-voltage secondary. This is true even when the load current is unbalanced. The neutral wire is grounded at the transformer bank. It is also grounded at intervals on the transmission line. The neutral wire helps protect the high-voltage secondary windings of the three single-phase transformers from damage due to lightning surges.

Stepping Down Voltages Using the Delta–Wye Connection

The delta–wye connection can also be used for applications where the voltage is stepped down. For example, assume that energy is to be transferred from a 13,800-V, three-phase, three-wire distribution system to a 277/480-V, three-phase, four-wire system. This voltage is then used to supply the power and lighting needs of a large office building.

Figure 14–12 shows the delta–wye-connected transformer bank used in this case. The primary side of the transformer bank is connected in delta to a three-phase, three-wire, 13,800-V distribution circuit. The line voltage and the voltage across each primary coil winding are all equal to 13,800 V. The transformer ratio is 50:1. As a result, the secondary coil winding voltage of each transformer is $13{,}800 \div 50 = 276$ V.

The secondary side of the transformer bank is connected in wye. The voltage from the grounded neutral to each of the three line wires is 277 V. The voltage across the line wires is $\sqrt{3} \times 277$, or 480 V.

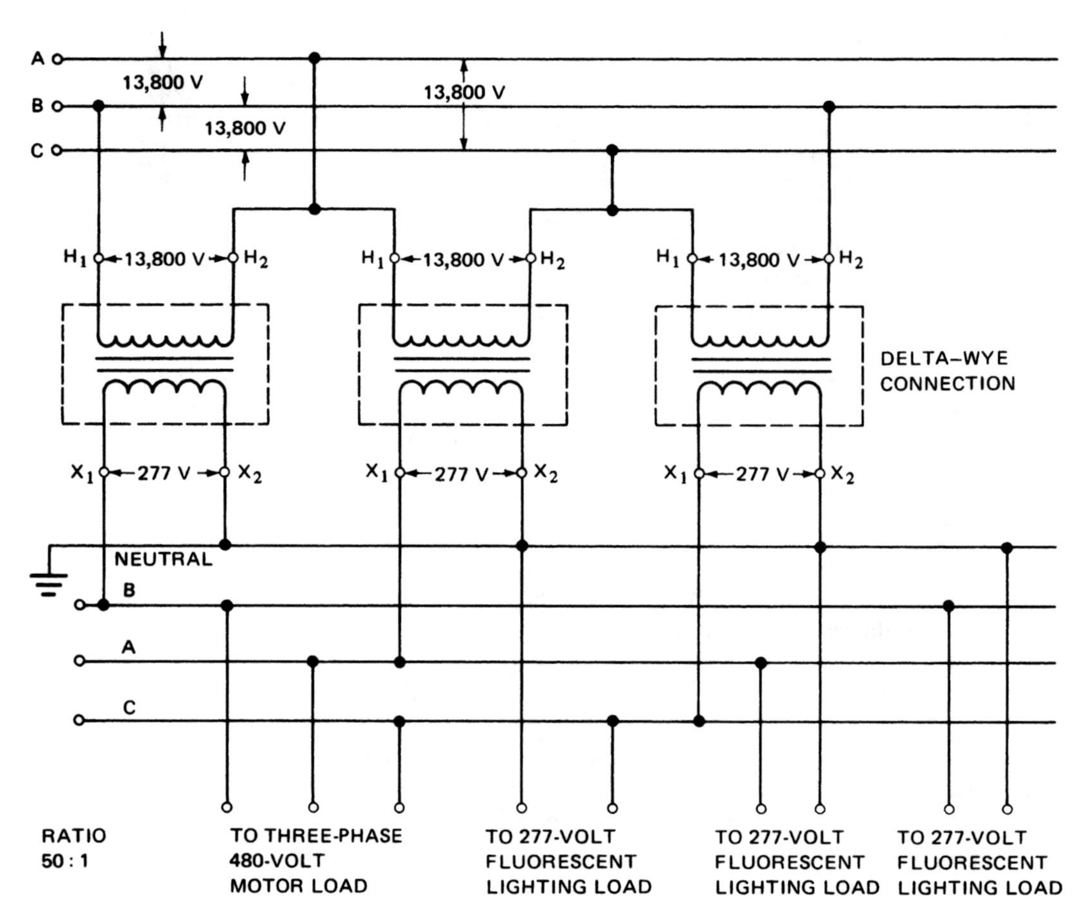

FIGURE 14–12 Delta–wye transformer bank supplying 277/480-V lighting and power loads (*Delmar/Cengage Learning*)

Three-Phase, Four-Wire System of Power and Lighting

The standard voltage for lighting circuits is 120 V. Industrial power applications normally use either 208 or 240 V. However, power and lighting applications also can be served by a 277/480-V, three-phase, four-wire system.

Modern lighting applications generally require a high level of light intensity. Fluorescent lighting units furnish light having the required intensity. Standard 120-V fluorescent lighting fixtures are used with special ballasts for operation on 277-V circuits. Motors wound for 480 V, rather than 208 or 220 V, can be used for air-conditioning units, fans, pumps, and elevators. In office buildings, the lighting demand can be as much as 7 to 10 volt-amperes per square foot (VA/ft^2). The motor load may average as much as 4 VA/ft^2. A 277/480-V, three-phase, four-wire system has the following advantages:

- The voltage drop is reduced in feeders and branch circuits, resulting in an increase in the operating efficiency.

- Smaller sizes of copper conductors, conduits, and equipment can be used to save up to 25% of the installation costs.
- The load demands on the 277/480-V system can be increased with a minimum of changes and expense.

All office buildings have miscellaneous loads requiring 120 V. These loads include desk lamps, office machines, and communications equipment. Under normal conditions, such loads are only a fraction of the total load and can be supplied from small air-cooled transformers. These transformers are located on each floor of the building. They are connected to the 277/480-V system and step down the voltage from 480 V to 120 V.

WYE–DELTA CONNECTIONS

The wye–delta transformer bank is used to step down relatively high transmission line voltages at the load center. A transformer bank of this type is commonly used to step down three-phase voltages of 60,000 V or more. There are two advantages in using the wye–delta connection. The first is that the three-phase voltage is reduced by the transformer ratio times 1.73. The second advantage is that there is a reduction in the insulation requirements for the high-voltage windings. Less insulation is required because the actual primary coil voltage is only 58% of the primary line voltage.

The diagram of a wye–delta transformer bank is shown in Figure 14–13. This bank is located at the end of a three-phase, four-wire transmission line. The primary three-phase voltage of 60,900 V is stepped down to 4400 V, three phase. Three single-phase transformers are used. The high-voltage side of each transformer is rated at 1000 kVA, 35,200 V. Each low-voltage side is rated at 4400 V. The voltage ratio of each transformer is 8:1. Assume that the three-phase primary line voltage is 60,900 V. As shown in Figure 14–13, the three single-phase transformers are connected in wye on the high-voltage side. The primary line voltage is 60,900 V. Therefore, the voltage impressed across the primary winding of each transformer is

$$V_{\text{primary winding}} = \frac{V_{\text{primary line}}}{\sqrt{3}}$$

$$= \frac{60{,}900}{1.73} = 35{,}200 \text{ V}$$

The voltage ratio of each single-phase transformer is 8:1. This means that the voltage across the secondary winding of each transformer is

$$35{,}000 \div 8 = 4400 \text{ V}$$

Balanced three-phase voltages are obtained even when there are unbalanced load currents because of the neutral wire on the high-voltage primary input. The neutral wire is grounded and gives lightning surge protection.

For both delta–wye and wye–delta connections, the three single-phase transformers generally have the same kVA capacity. The total capacity of the transformer bank in kVA is obtained by adding the kVA ratings of the three transformers. In Figure 14–13, for example, each single-phase transformer is rated at 1000 kVA and the total capacity of the transformer bank is 3000 kVA.

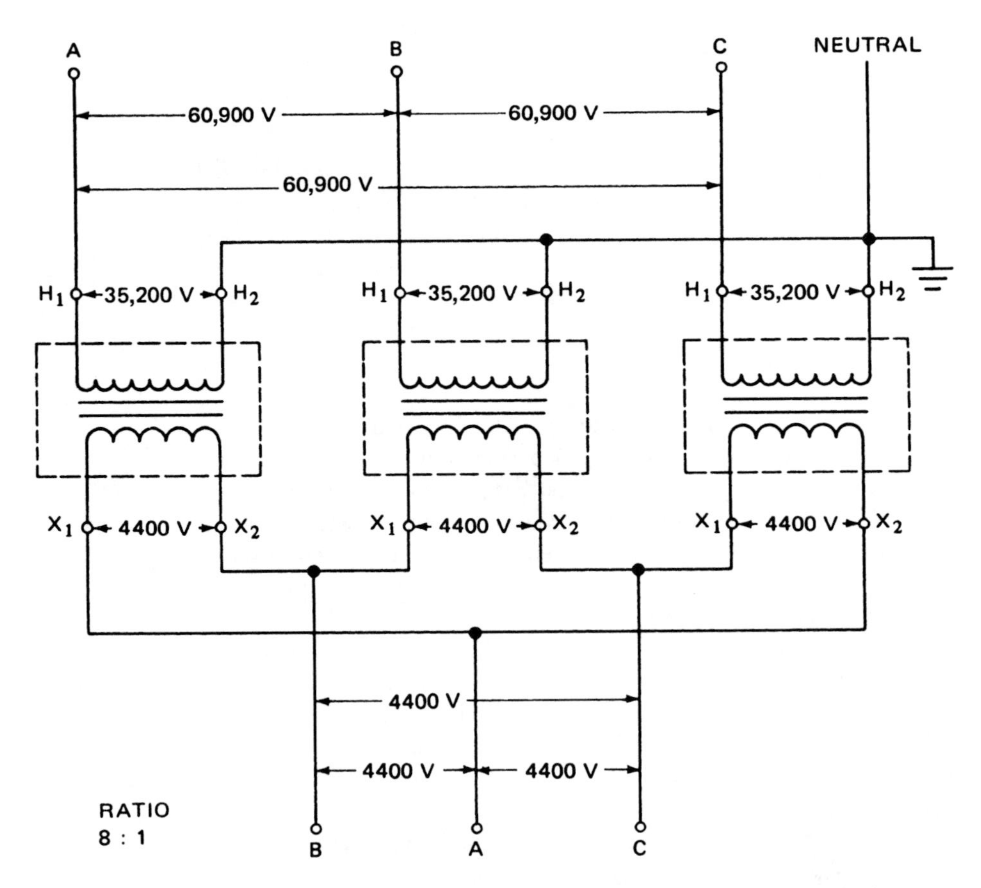

FIGURE 14–13 Wye–delta transformer bank (*Delmar/Cengage Learning*)

OPEN-DELTA CONNECTION

It is possible to achieve three-phase transformation of energy by using two transformers only. One connection that will do this is called the *open-delta connection* (V connection). On occasion, one of the three transformers in a delta–delta bank will become defective. To restore three-phase service to consumers as soon as possible, the defective transformer is cut out of the system and the configuration of the open-delta connection is used.

The following example describes a typical use of the open-delta connection. A delta–delta connection is made using three 50-kVA transformers. Each one is rated at 2400 V on the high-voltage side and 240 V on the low-voltage side. This closed-delta transformer bank steps down 2400 V, three phase to 240 V, three phase to supply an industrial consumer. One of the transformers is damaged by lightning, resulting in a power failure. The three-phase service must be restored at once.

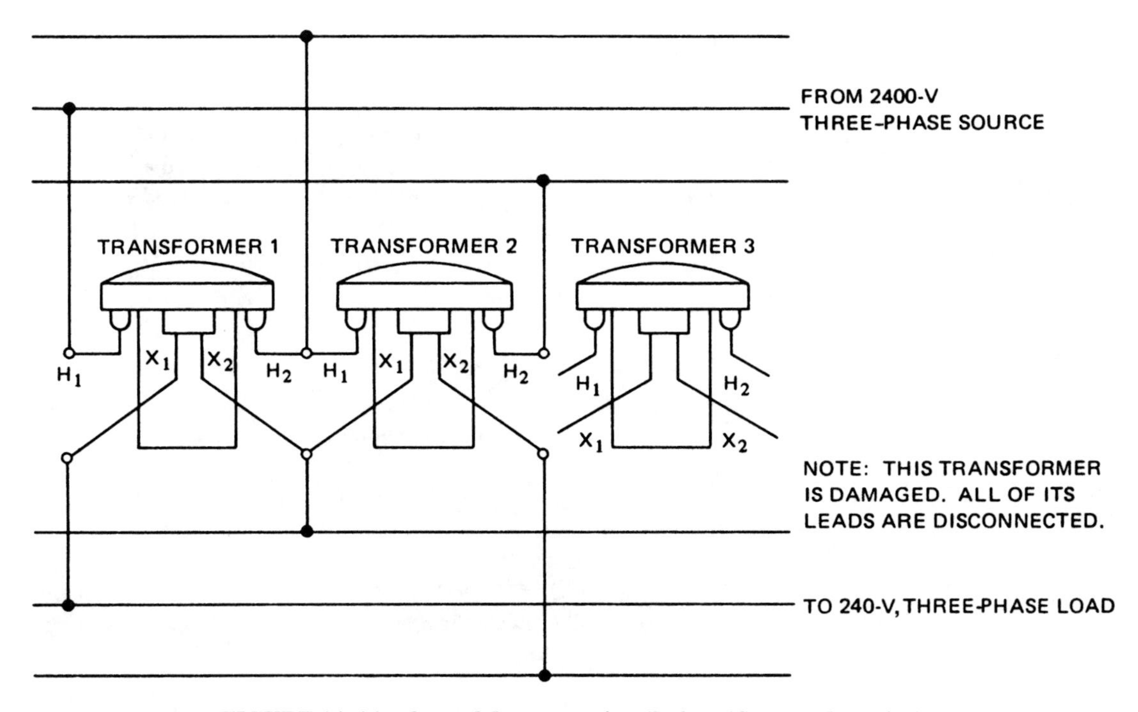

FIGURE 14–14 Open-delta connection (*Delmar/Cengage Learning*)

By disconnecting all of the leads of the damaged transformer, the closed-delta bank becomes an open-delta bank, as shown in Figure 14–14.

The student may expect that the total kVA capacity of the open-delta bank will be two-thirds of the capacity of the closed-delta bank. Actually, the capacity of an open-delta bank is only 58% that of a closed-delta bank. For this example, the total capacity of the delta–delta bank is equal to the sum of the kVA capacities of the three transformers: 50 + 50 + 50 = 150 kVA. When one transformer is disconnected, an open-delta connection is formed. The total kVA capacity is now only 58% of the capacity of the closed-delta connection: 150 × 0.58 = 87 kVA (or 86.6% of the total capacity of the two remaining transformers).

Capacity of the Open-Delta Connection

In Figure 14–15, three 1-kVA transformers are connected to form a closed-delta connection. The secondary voltage of each transformer is 100 V. The maximum current for each winding is 10 A (1000 kVA/100 V = 10 A). The total power for this connection can be found by using the formula

$$3 \times V \times I \times \cos\theta$$

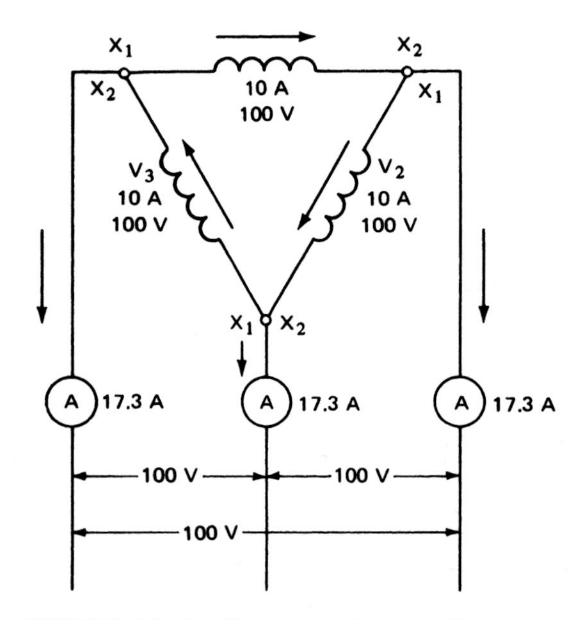

FIGURE 14–15 Three secondary windings, connected in closed-delta (*Delmar/Cengage Learning*)

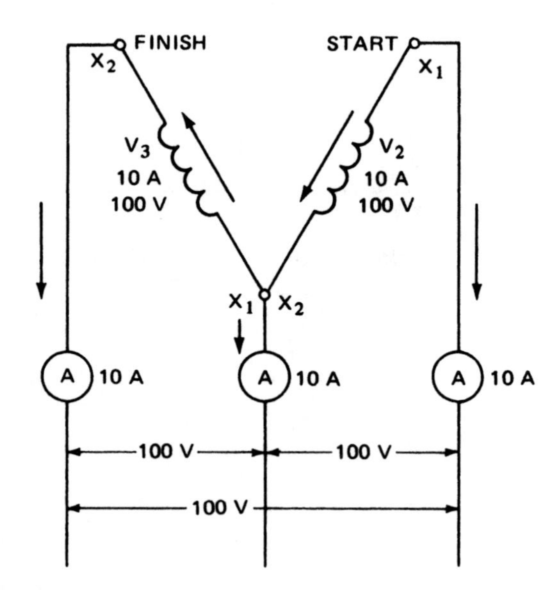

FIGURE 14–16 Two secondary windings, connected in open-delta (*Delmar/Cengage Learning*)

If it is assumed that the transformers are being operated at unity power factor, the total output for this connection is

$$3 \times 100 \times 10 \times 1 = 3000 \text{ VA}, \quad \text{or} \quad 3 \text{ kVA}$$

If one of the transformers is removed from the connection, as shown in Figure 14–16, an open-delta connection is formed. The total power for this connection can be found using the formula

$$\sqrt{3} \times \text{V} \times \text{I} \times \cos \theta$$

If a unity power factor is again assumed for this connection, the total output for this transformer bank is

$$\sqrt{3} \times 100 \times 10 \times 1 = 1730 \text{ VA}, \quad \text{or} \quad 1.73 \text{ kVA}$$

This is 58% of the total output of the closed-delta connection, or 86.6% of the capacity of the two 1-kVA transformers used to make the open-delta connection:

$$\frac{3000}{1730} = 0.58 \times 100, \quad \text{or} \quad 58\%$$

$$\frac{2000}{1730} = 0.866 \times 100, \quad \text{or} \quad 86.6\ \%$$

THREE-PHASE TRANSFORMERS

Electrical energy may be transferred from one three-phase circuit to another three-phase circuit. A three-phase transformer is used to accomplish this transfer, resulting in a change in the voltage. The core structure of this transformer consists of three legs. For each phase, the low-voltage and high-voltage windings are wound on one of the three legs.

Figure 14–17 shows the assembled core of a three-phase transformer, including the low-voltage coil windings. The flux in each coil leg is 120° out of phase with the flux values. This means that each flux will reach its maximum value at a different instant. At any point in time, at least one of the core legs will act as the return path for the fluxes in each of the other phases.

FIGURE 14–17 Three-phase Pyranol-filled transformer *(Delmar/Cengage Learning)*

The core structure and the coil windings of the three-phase transformer are placed in a single case, or tank. They are then covered with transformer oil or a nonflammable liquid such as Pyranol.

The connections between the coil windings are made inside the transformer case. Delta–delta, wye–wye, delta–wye, or wye–delta connections can be made.

Delta–delta-connected, three-phase transformers have three high-voltage leads and three low-voltage leads that are brought out through insulated bushings on the transformer case.

Four leads are brought out when the individual coil windings are connected in wye. The fourth lead is necessary for the neutral wire connection.

Figure 14–18 shows a three-phase transformer with both the high-voltage windings and the low-voltage windings connected in wye. The three-phase line voltage on the input side is 4160 V. The voltage across each high-voltage winding is $4160 \div \sqrt{3} = 2400$ V.

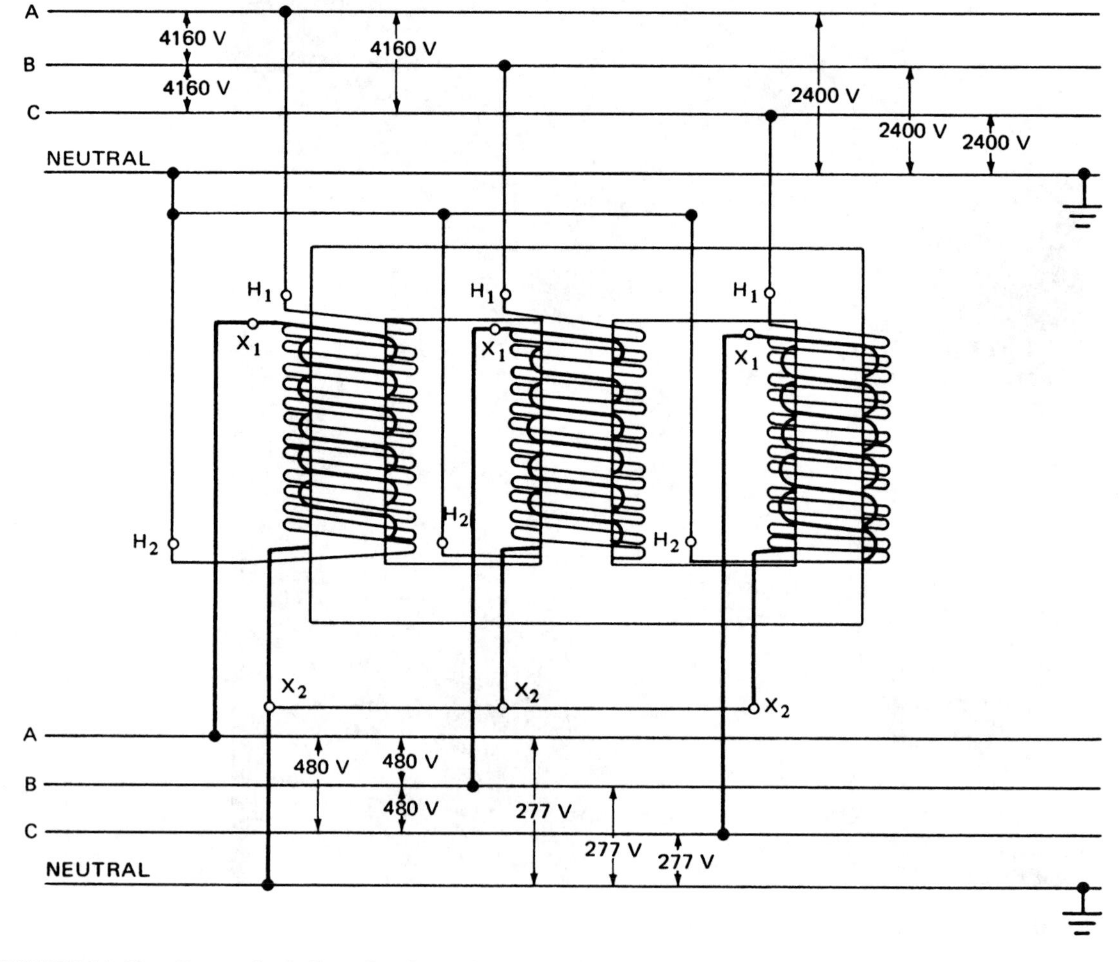

FIGURE 14–18 Core and windings for three-phase transformer, connected in wye–wye; primary is 2400/4160 V and secondary is 277/480 V. (*Delmar/Cengage Learning*)

The voltage induced in each low-voltage secondary winding is 277 V. Because the three secondary windings are connected in wye, the line voltage on the output side is $\sqrt{3} \times 277 = 480$ V. The rating for this three-phase transformer is 2400/4160 V to 277/480 V.

Advantages of Three-Phase Transformers

Three-phase transformers are commonly used for both stepdown and stepup applications for the following reasons:

- The operating efficiency of a three-phase transformer is slightly higher than the overall efficiency of three separate single-phase transformers.
- The three-phase transformer weighs less and requires less space than do three separate single-phase transformers.
- One three-phase transformer supplying the same kVA output costs less than three single-phase transformers.
- The necessary bus bar structure, switchgear, and wiring is installed in either an outdoor or an indoor substation. For a three-phase transformer, this equipment is easier to install and is less complex than that required by a transformer bank consisting of three single-phase transformers.

Disadvantage of the Three-Phase Transformer

The three-phase transformer has one disadvantage. If one of the phase windings becomes defective, then the entire three-phase unit must be taken out of service. A defective single-phase transformer in a three-phase bank can be disconnected. Partial service can be restored using the remaining transformers until a replacement unit is obtained. However, because transformers have a high reliability, most applications requiring large transformers use three-phase transformers.

The three problems that follow show how single-phase transformers are used in three-phase transformer banks.

PROBLEM 2

Statement of the Problem

A three-phase transformer bank is used to step down a 2400-V, three-phase, three-wire primary service to a 240-V, three-phase, three-wire secondary service. The transformer bank consists of three 20-kVA transformers. Each transformer has additive polarity. The high-voltage side of each transformer is rated at 2400 V. The low-voltage side of each transformer has two 120-V windings.

1. Draw a schematic diagram of the connections for this circuit. The leads of each transformer are to be marked for additive polarity.
2. At the rated load and a lagging power factor of 0.80, determine
   a. the rating of the transformer bank, in kVA.
   b. the output at the rated load and lagging power factor of 0.80, in kW.
   c. the secondary line current.

d. the secondary coil current and the coil voltage.
e. the primary coil current.
f. the primary line current.

Solution

1. In a delta connection, the coil and line voltages are equal. The primary line voltage is 2400 V. The high-voltage winding of each transformer is also rated at 2400 V. The line voltage on the secondary is to be 240 V. The two 120-V windings on the low side of each transformer can be connected in series to give 240 V. As shown in Figure 14–19, the transformer bank is connected in delta–delta.

2. a. The kVA capacity of the transformer bank is

$$20 + 20 + 20 = 60 \text{ kVA}$$

b. The output of the transformer bank at the rated load and a lagging power factor of 0.80 is

$$\text{Output, in kW} = \text{kVA} \times \cos\theta = 60 \times 0.80 = 48 \text{ kW}$$

c. The secondary line current at the rated load output is determined as follows:

$$\text{kVA} = \frac{\sqrt{3} \times V_{line} \times I_{line}}{1000}$$

$$60 = \frac{1.73 \times 240 \times I_{line}}{1000}$$

$$415.2\, I_{line} = 60{,}000$$

$$I_{line} = 144.5 \text{ A}$$

d. The coil voltage and the line voltage are the same in a delta connection. Thus, if the secondary line voltage is 240 V, the secondary coil voltage is also 240 V. The line current is equal to $\sqrt{3}$ times the coil winding current. The line current was found to be 144.5 A. The secondary coil current is

$$I_{coil} = \frac{I_{line}}{\sqrt{3}} = \frac{144.5}{1.73} = 83.5 \text{ A}$$

e. The ratio of the primary and secondary coil voltages is

$$2400/240 = 10/1$$

This ratio is also the ratio between the turns on the high-voltage and low-voltage sides. Unit 13 states that the turns on the windings are inversely proportional to the current. This means that

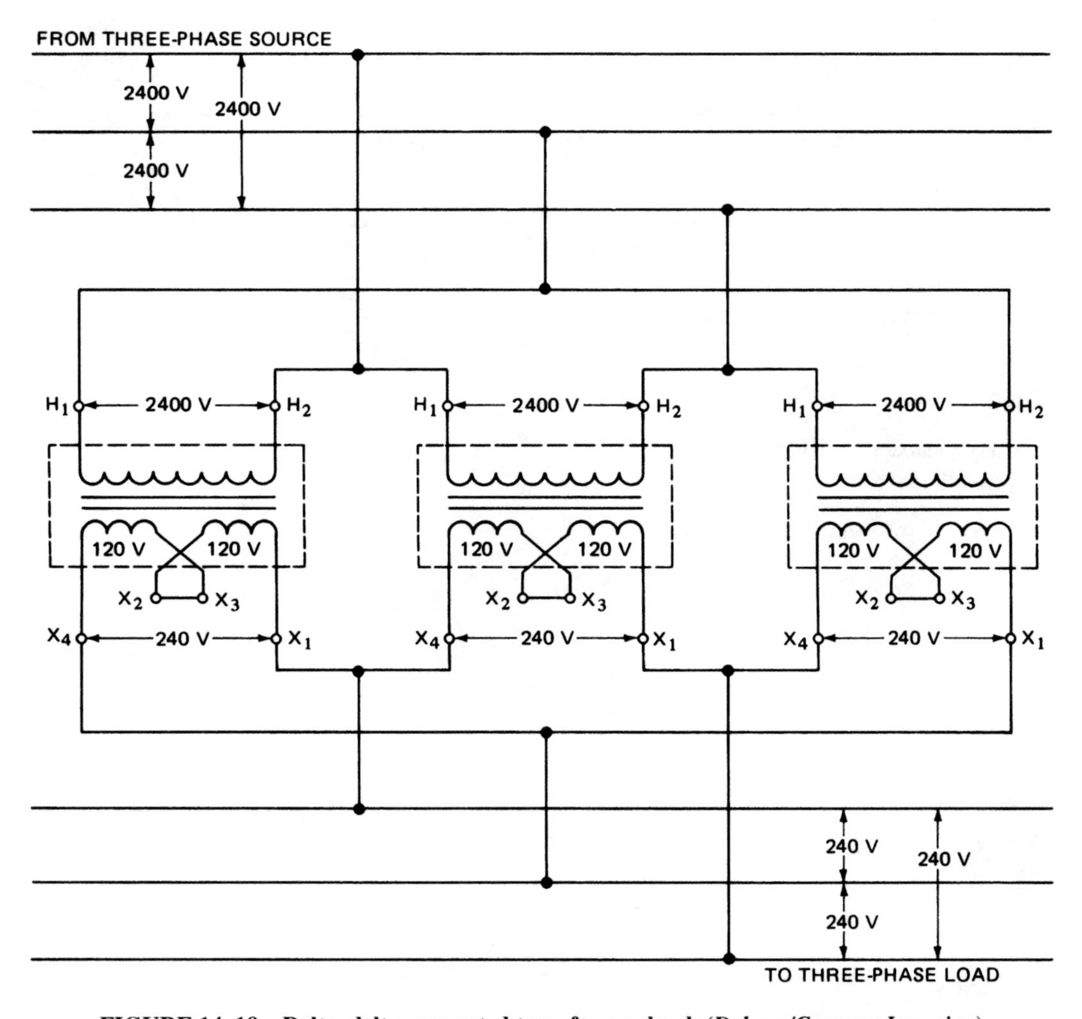

FIGURE 14–19 Delta–delta-connected transformer bank (*Delmar/Cengage Learning*)

$$\frac{I_P}{I_S} = \frac{N_S}{N_P}$$

$$\frac{I_P}{83.5} = \frac{1}{10}$$

Therefore,

$$IP = 8.35 \text{ A}$$

f. The line current for the delta-connected primary side of the transformer bank is

$$I_{line} = \sqrt{3}\, I_{coil}$$

$$= 1.73 \times 8.35 = 14.45 \text{ A}$$

PROBLEM 3

Statement of the Problem

For the delta–delta transformer bank described in problem 2, one transformer is damaged. The remaining two transformers are reconnected in open delta:

1. What is the capacity of the open-delta bank in kVA?
2. Assuming that the transformer bank is loaded to the rated kVA capacity with a balanced load having a lagging power factor of 0.80, determine
   a. the kW output.
   b. the line current on the secondary side.

Solution

1. The kVA capacity of the open-delta bank is 58% of the capacity of the original closed-delta bank:

$$\text{k VA} = 60 \times 0.58 = 34.8 \text{ kVA}$$

2. a. $\text{kW} = \text{kVA} \cos \theta = 34.8 \times 0.8 = 27.8 \text{ kW}$

   b. $$\text{kVA} = \frac{\sqrt{3} \times V_{\text{line}} \times I_{\text{line}}}{1000}$$

$$34.8 = \frac{1.73 \times 240 \times I_{\text{line}}}{1000}$$

Thus, the full-load secondary line current is

$$I_{\text{line}} = 83.8 \text{ A}$$

PROBLEM 4

Statement of the Problem

A 4800-V, three-phase, three-wire primary voltage is stepped down to a 120/208-V, three-phase, four-wire, secondary service. The transformer bank used consists of three single-phase transformers. Each transformer is rated at 15 kVA, 4800/120 V. The load is a noninductive heater unit consisting of three 1-Ω sections connected in wye to the three-phase, four-wire, secondary system.

1. Draw a schematic diagram of the connections for the transformer bank. Assume each transformer has additive polarity.
2. Determine
   a. the kVA capacity of the transformer bank.
   b. the kVA load on the transformer bank.
   c. the secondary line current.
   d. the primary line current.

Solution

1. The primary line voltage is 4800 V. The high-voltage windings are also rated at 4800 V. Thus, the primary windings are connected in delta. The low-voltage windings of the transformer are rated at 120 V. These windings are connected in wye to give a three-phase, four-wire, 120/208-V service. For wye connections, the line voltage is $\sqrt{3}$ times the coil voltage. In this case, the line voltage is $\sqrt{3} \times 120 = 208$ V. See Figure 14–20.

2. a. The capacity of the transformer bank is $15 + 15 + 15 = 45$ k VA
   b. The current taken by each heater element is

$$I_{\text{heater section}} = \frac{V}{R} = \frac{120}{1} = 120 \text{ A}$$

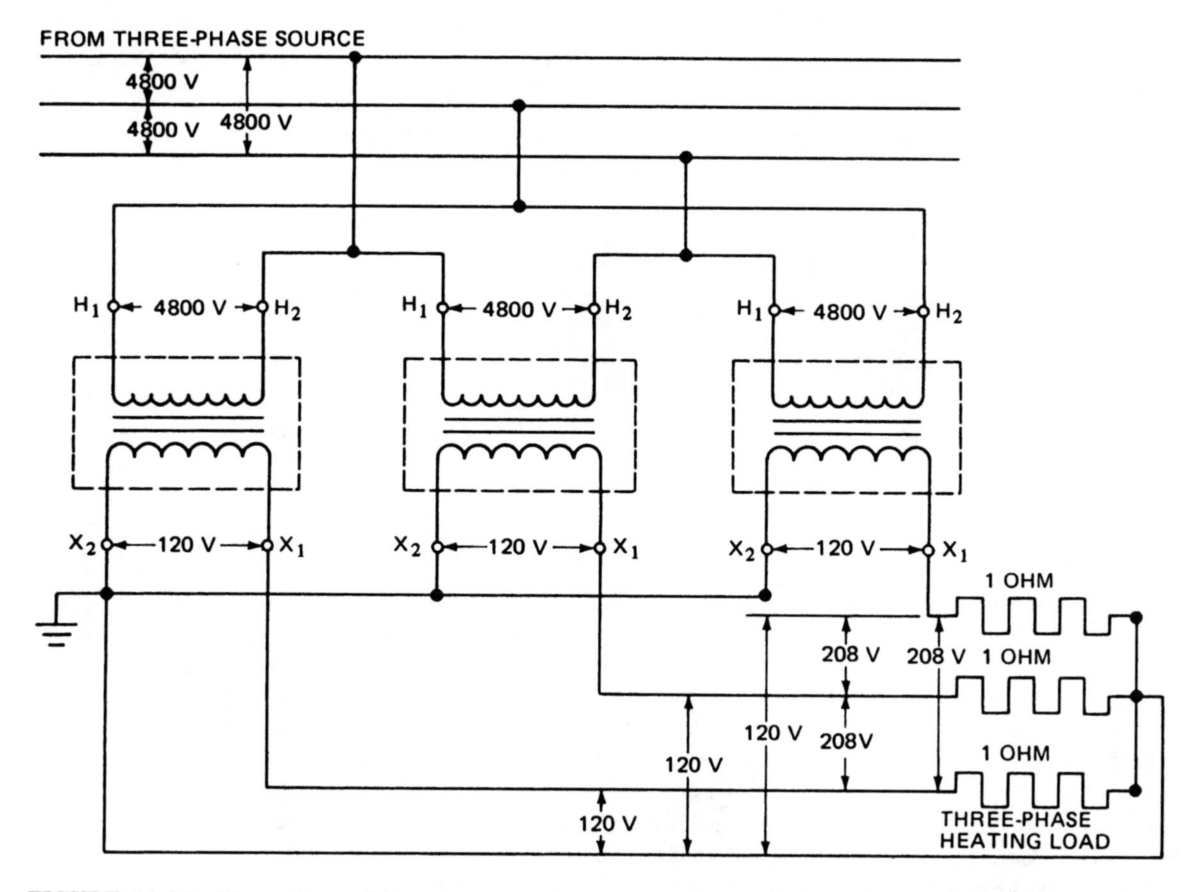

FIGURE 14–20 **Three-phase, delta–wye connections to step down 4800 V to 120/208 V** *(Delmar/Cengage Learning)*

The line current in each of the three line wires of the wye system is 120 A. The total kVA load is

$$kVA = \frac{\sqrt{3} \times V_{line} \times I_{line}}{1000}$$
$$= \frac{1.73 \times 208 \times 120}{1000} = 43.2\ kVA$$

The total load can also be obtained by multiplying the power used by each single-phase heater element by 3:

$$kVA = V \times I \div 1000$$
$$= 120 \times 120 \div 1000 = 14.4\ \text{kVA per heater element}$$

The total load in kVA is $3 \times 14.4 = 43.2$ kVA. Because the load has a power factor of unity, this value is also equal to the load in kW on the transformer bank.

c. The secondary line current is

$$I = V/R = 120/1 = 120\ A$$

d. The secondary windings of the transformer are connected in wye. This means that both the secondary line and the coil currents are 120 A. The primary coil current and the turns ratios of each transformer are

$$\frac{I_P}{I_S} = \frac{N_S}{N_P}$$
$$\frac{I_P}{120} = \frac{1}{40}$$
$$I_P = \frac{120}{40} = 3\ A$$

The primary windings are connected in delta. Therefore, the primary line current is

$$I_{line} = \sqrt{3} \times I_{coil} = 1.73 \times 3 = 5.2\ A$$

T-CONNECTED TRANSFORMERS

Another connection involving the use of two transformers to supply three-phase power is the *T connection* (Figure 14–21). In this connection, one transformer is generally referred to as the *main transformer* and the other is called the *teaser transformer*. The main transformer must contain a center or 50% tap for both the primary and secondary winding, and it is preferred that the teaser transformer contain an 86.6% voltage tap for both the primary and secondary winding. Although the 86.6% tap is preferred, the connection can be made with a teaser transformer that has the same voltage rating as the main transformer. In this instance, the teaser transformer is operated at reduced flux (Figure 14–22). This connection permits two transformers to be connected T instead of open delta in the event that one transformer of a delta–delta bank should fail.

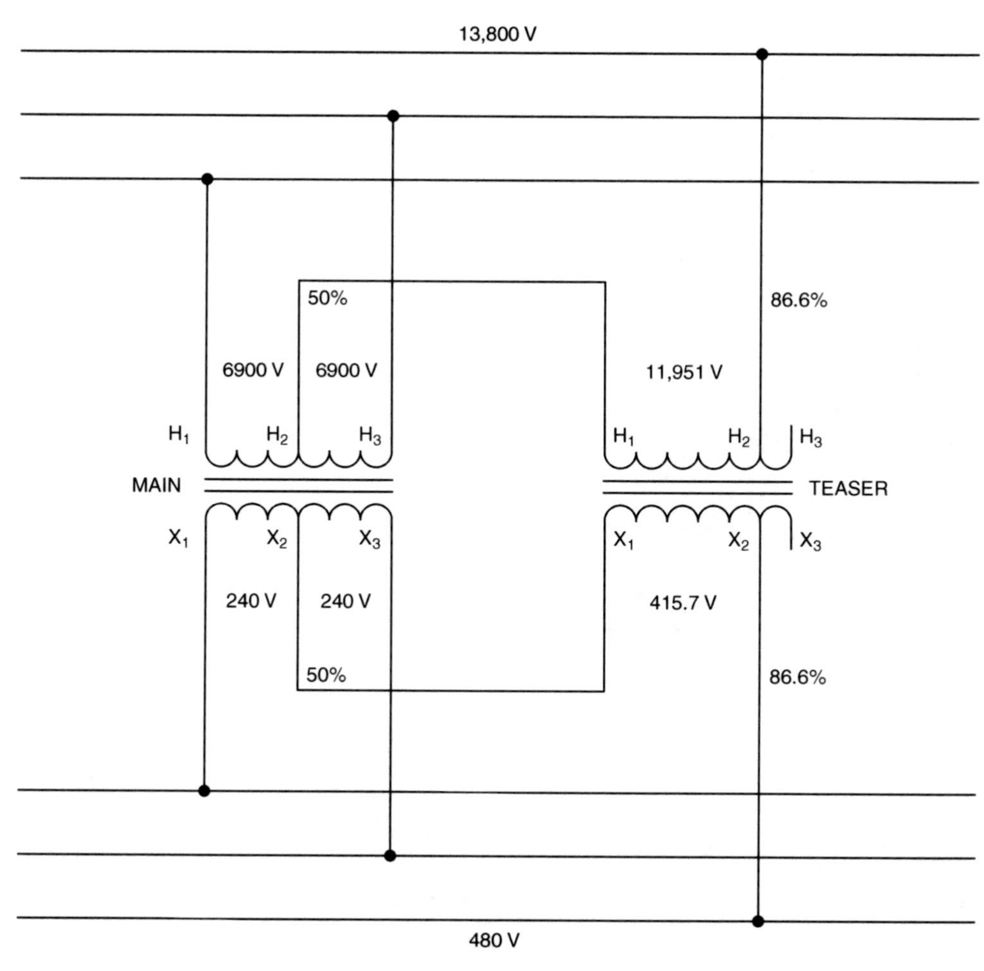

FIGURE 14–21 T-connected transformers (*Delmar/Cengage Learning*)

Transformers intended for use as T-connected transformers are often specially wound for the purpose, and both transformers are often contained in the same case. When making the T connection, the main transformer is connected directly across the power line. One primary lead of the teaser transformer is connected to the center tap of the main transformer, and the 86.6% tap is connected to the power line. The same basic connection is made for the secondary. A vector diagram illustrating the voltage relationships of the T connection is shown in Figure 14–23. The greatest advantage of the T connection over the open-delta connection is that it maintains a better phase balance, and the T-connected transformer can be connected to provide a three-phase, four-wire output similar to that of a four-wire wye connection. T-connected transformers used to provide three-phase, four-wire power generally have voltages of 480/277 or 208/120 V. The greatest disadvantage of the T connection is that one transformer must contain a center tap of both its primary and secondary windings.

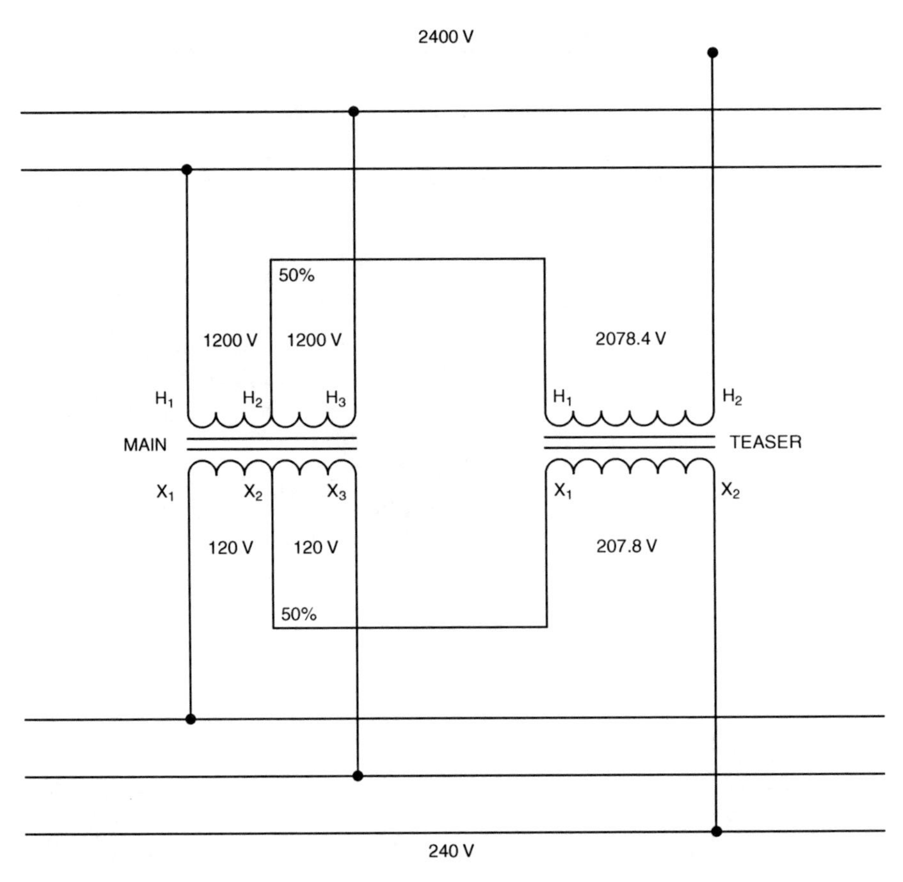

FIGURE 14–22 T-connected transformers with same voltage rating (*Delmar/Cengage Learning*)

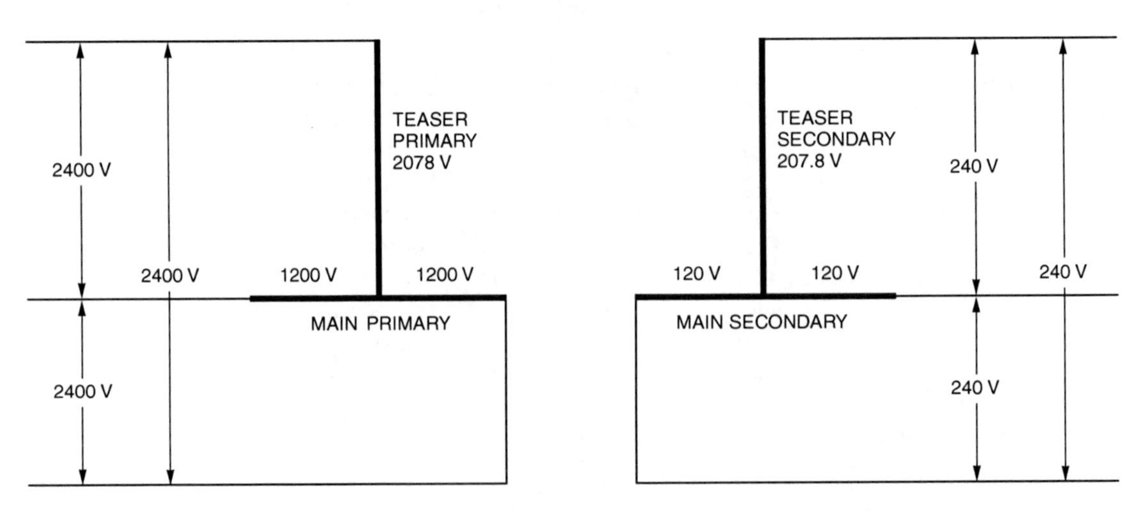

FIGURE 14–23 Voltage vector relationships of a T connection (*Delmar/Cengage Learning*)

SCOTT CONNECTION

The *Scott connection* is used to convert three-phase power into two-phase power using two single-phase transformers. The Scott connection is very similar to the T connection in that one transformer, called the *main transformer,* must have a center or 50% tap, and the second or teaser transformer must have an 86.6% tap on the primary side. The difference between the Scott and T connections lies in the connection of the secondary windings (Figure 14–24). In the Scott connection, the secondary windings of each transformer provide the phases of a two-phase system. The voltages of the secondary windings are 90° out of phase with each other. The Scott connection is generally used to provide two-phase power for the operation of two-phase motors.

ZIGZAG CONNECTION

The *zigzag* or *interconnected wye* transformer is primarily used for grounding purposes. It is used mainly to establish a neutral point for the grounding of fault currents. The zigzag connection is basically a three-phase autotransformer whose windings are divided into six equal parts (Figure 14–25). In the event of a fault current, the zigzag connection forces the current to flow equally in the three legs of the autotransformer, offering minimum impedance to the flow of fault current. A schematic diagram of the zigzag connection is shown in Figure 14–26.

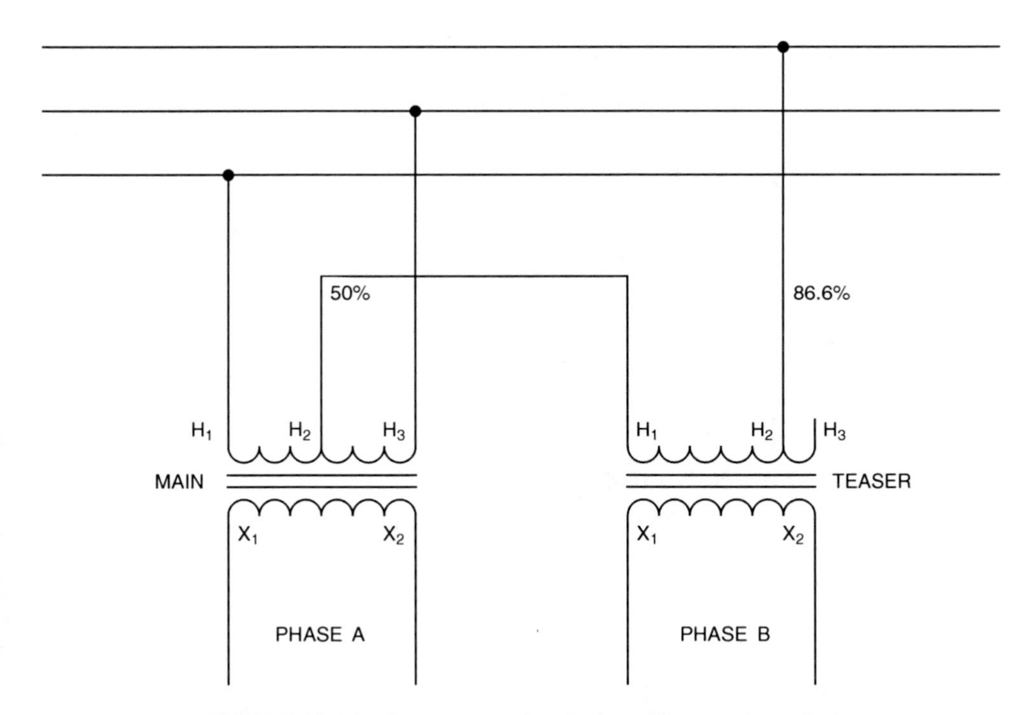

FIGURE 14–24 Scott connection (*Delmar/Cengage Learning*)

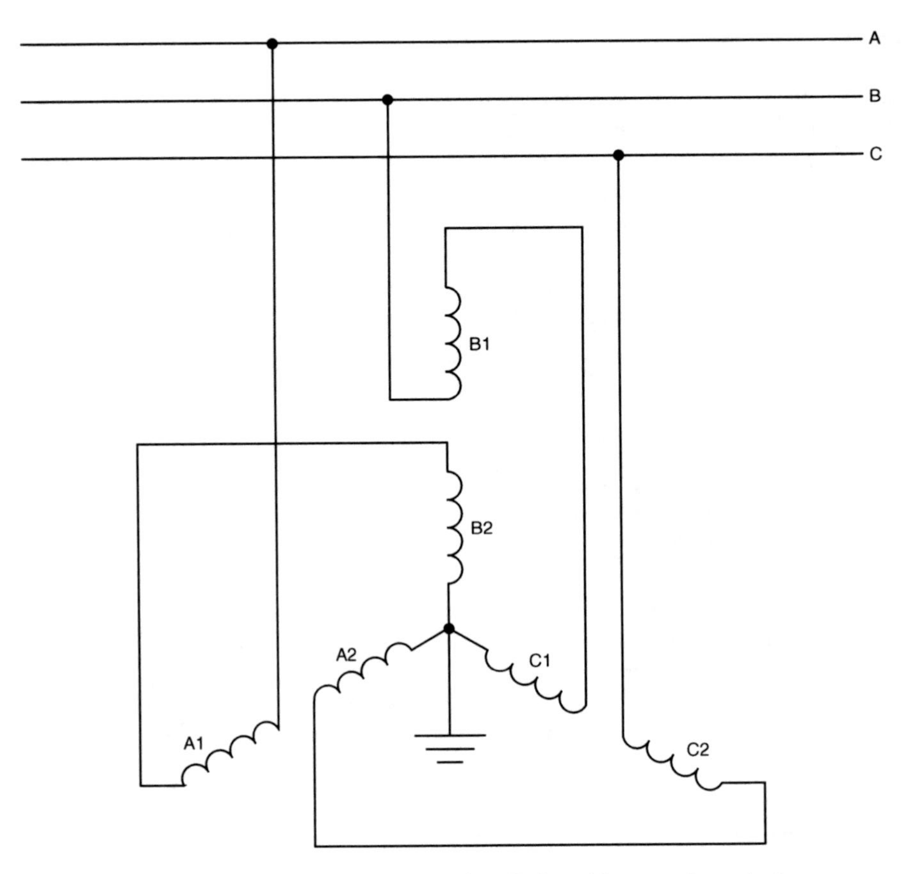

FIGURE 14–25 Zigzag connection (*Delmar/Cengage Learning*)

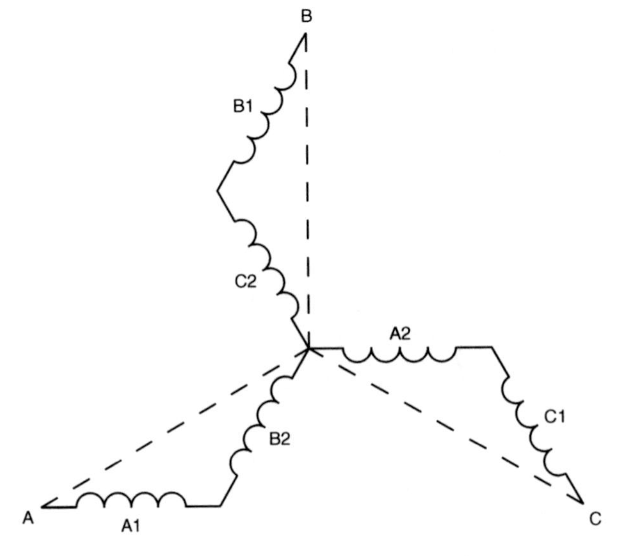

FIGURE 14–26 Schematic diagram of a zigzag connection (*Delmar/Cengage Learning*)

THREE-PHASE-TO-SIX-PHASE CONNECTIONS

There are some instances when it is desirable to have a power system with more than three phases. A good example of this is when it is necessary to convert or rectify alternating current into direct current with a minimum amount of ripple (pulsations of voltage). Power supplies that produce a low amount of ripple require less filtering. One of the most common three-phase-to-six-phase connections is the *diametrical connection* (Figure 14–27). The diametrical connection is preferred because it requires only one low-voltage winding on each transformer. If these windings are center-tapped, a neutral conductor can be provided for the six-phase output, permitting half-wave rectification to be used. The high-voltage windings can be connected in wye or delta, but the delta is preferred because it helps reduce harmonics in the secondary winding. A schematic diagram of a diametrical connection with a delta-connected primary and three-phase half-wave rectifier is shown in Figure 14–28.

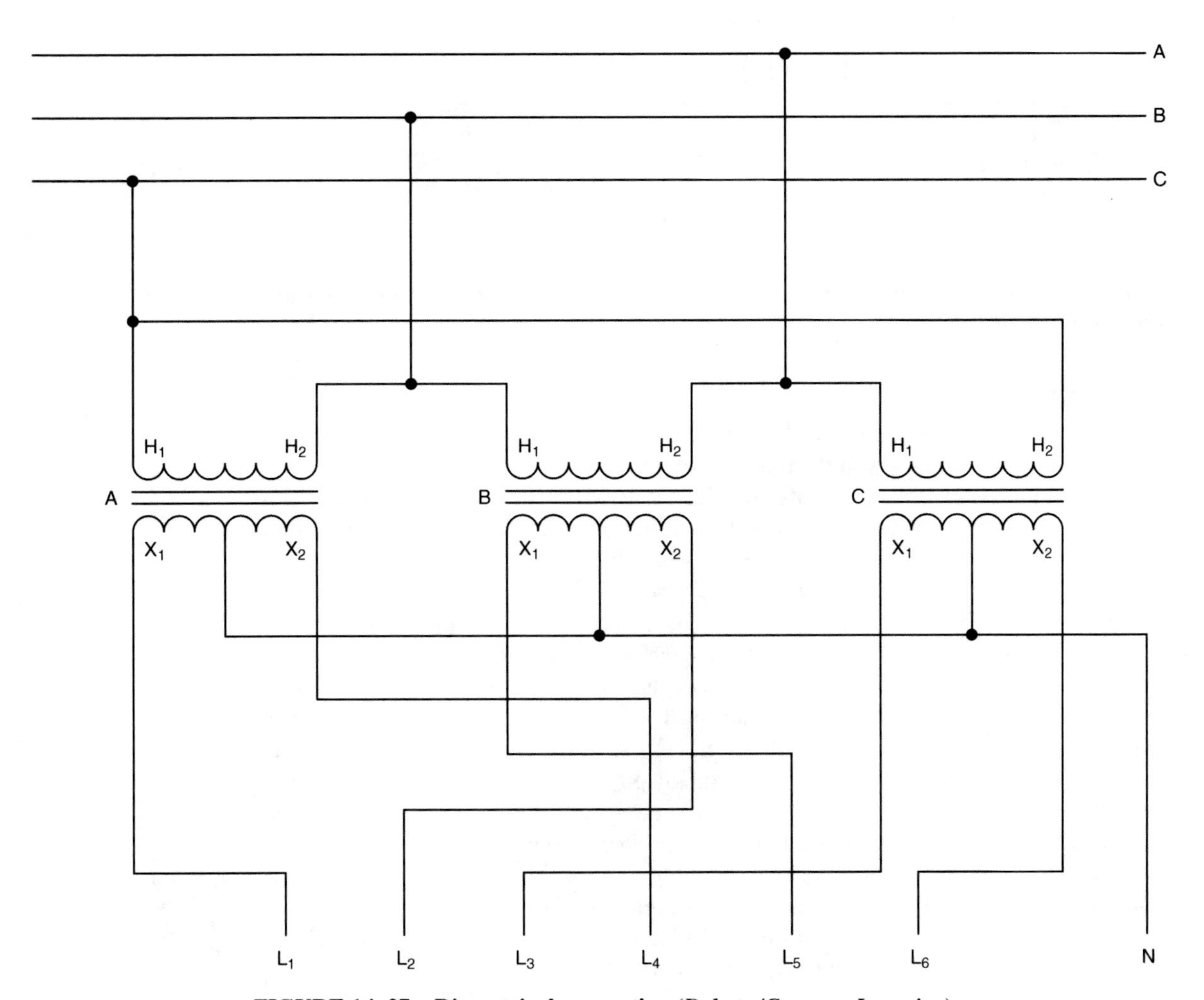

FIGURE 14–27 Diametrical connection (*Delmar/Cengage Learning*)

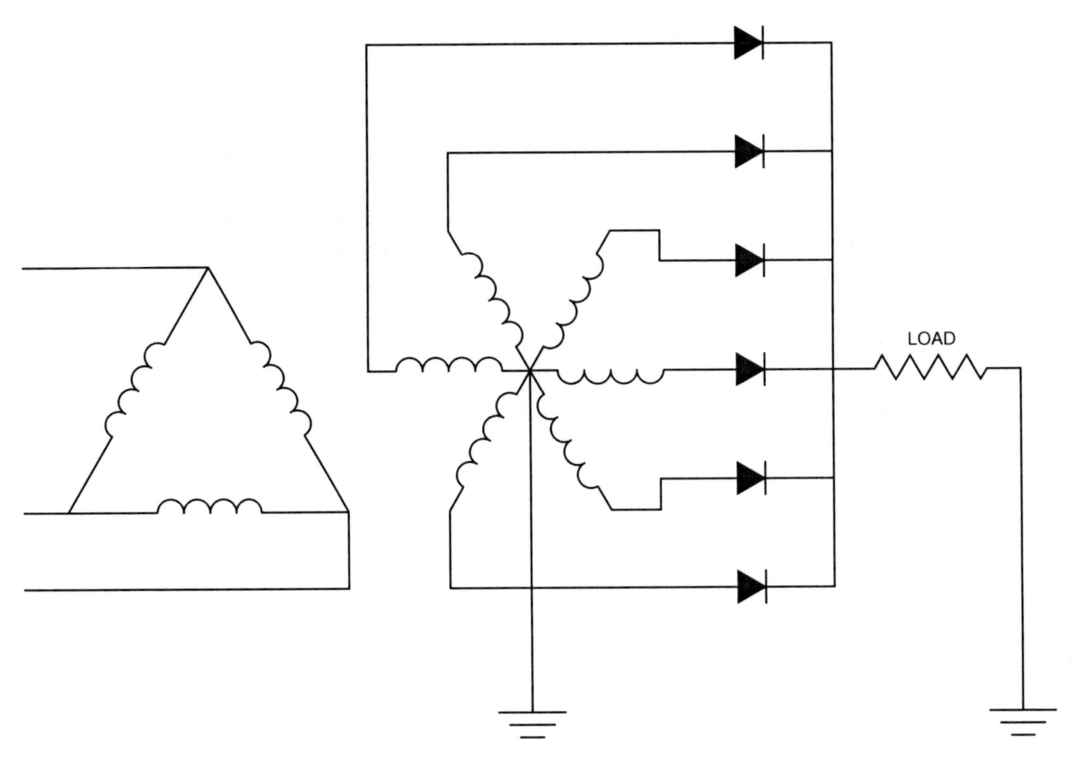

FIGURE 14–28 Schematic diagram of a diametrically connected transformer with six-phase, half-wave rectifier (*Delmar/Cengage Learning*)

HARMONICS

Harmonics are voltages or currents that operate at a frequency that is a multiple of the fundamental power frequency. If the fundamental power frequency is 60 Hz, for example, the second harmonic would be 120 Hz, the third harmonic would be 180 Hz, and so on. Harmonics are produced by nonlinear loads that draw current in pulses rather than in a continuous manner. Harmonics on single-phase power lines are generally caused by devices such as computer power supplies, electronic ballasts in fluorescent lights, triac light dimmers, and so on. Three-phase harmonics are generally produced by variable-frequency drives for ac motors and electronic drives for dc motors. A good example of a pulsating load is one that converts ac current into dc and then regulates the dc voltage by pulsewidth modulation (Figure 14–29). Many regulated power supplies operate in this manner. The bridge rectifier in Figure 14–29 changes the alternating current into pulsating direct current. A filter capacitor is used to smooth the pulsations. The transistor turns on and off to supply power to the load. The amount of time the transistor is turned on compared with the time it is turned off determines the output dc voltage. Each time the transistor turns on, it causes the capacitor to begin discharging. When the transistor turns off, the capacitor will begin to charge again. Current is drawn from the ac line each time the capacitor charges. These pulsations of current produced by the charging capacitor can cause the ac sine wave to become distorted. These distorted current and voltage waveforms flow back into the other parts of the power system (Figure 14–30).

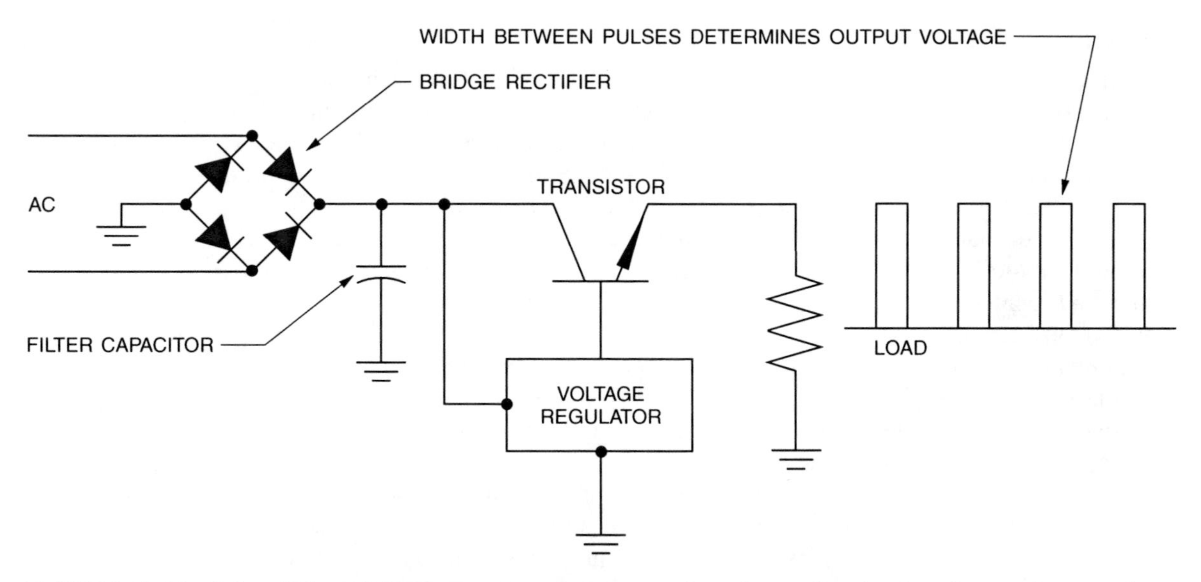

FIGURE 14–29 Pulsewidth modulation regulates the output voltage by varying the time the transistor conducts as compared to the time it is turned off. (*Delmar/Cengage Learning*)

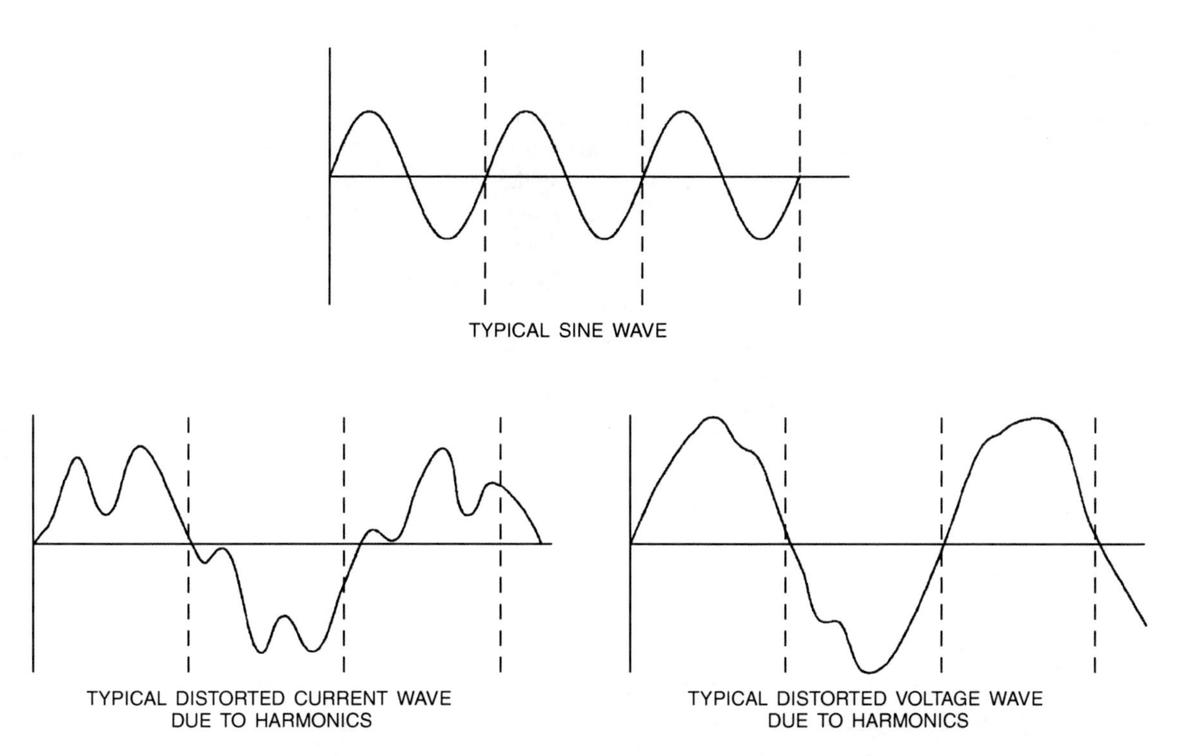

FIGURE 14–30 Harmonics cause an ac sine wave to become distorted. (*Delmar/Cengage Learning*)

Harmonic Effects

Harmonics can have very detrimental effects on electrical equipment. Some common symptoms of harmonics are overheated conductors and transformers and circuit breakers that seem to trip when they should not. *Harmonics* are classified by name, frequency, and sequence. The name refers to whether the harmonic is the second, third, fourth, and so on of the fundamental frequency. The *frequency* refers to the operating frequency of the harmonic. The second harmonic operates at 120 Hz, the third at 180 Hz, the fourth at 240 Hz, and so on. The sequence refers to the phasor rotation with respect to the fundamental waveform. In an induction motor, a positive sequence harmonic would rotate in the same direction as the fundamental frequency. A negative sequence harmonic would rotate in the opposite direction of the fundamental frequency. A particular set of harmonics called "triplens" has a zero sequence. Triplens are the odd multiples of the third harmonic (third, ninth, fifteenth, twenty-first, and so on). A chart showing the sequence of the first nine harmonics is shown in Figure 14–31.

Harmonics with a positive sequence generally cause overheating of conductors and transformers, and circuit breakers. Negative-sequence harmonics can cause the same heating problems as positive harmonics plus additional problems with motors. Because the phasor rotation of a negative harmonic is opposite that of the fundamental frequency, it will tend to weaken the rotating magnetic field of an induction motor, causing it to produce less torque. The reduction of torque causes the motor to operate below normal speed. The reduction in speed results in excessive motor current and overheating.

Although triplens do not have a phasor rotation, they can cause a great deal of trouble in a three-phase, four-wire system, such as a 208/120-V or 480/277-V system. In a common 208/120-V, wye-connected system, the primary is generally connected in delta and the secondary is connected in wye (Figure 14–32).

Single-phase loads that operate on 120 V are connected between any phase conductor and the neutral conductor. The neutral current will be the vector sum of the phase currents. In a balanced three-phase circuit (where all phases have equal current), the neutral current will be zero. Although single-phase loads tend to cause an unbalanced condition, the vector sum of the currents will generally cause the neutral conductor to carry less current than any of the phase conductors. This is true for loads that are linear and draw a continuous sine-wave current. When pulsating (nonlinear) currents are connected to a three-phase, four-wire system, triplens harmonic frequencies disrupt the normal phasor relationship of the phase currents and can cause the phase currents to add in the neutral conductor instead of cancel. Because the neutral conductor is not protected by a fuse or circuit breaker, there is real danger of excessive heating in the neutral conductor.

Name	Fund.	2nd	3rd	4th	5th	6th	7th	8th	9th
Frequency(Hz)	60	120	180	240	300	360	420	480	540
Sequence	+	–	0	+	–	0	+	–	0

FIGURE 14–31 Name, frequency, and sequence of the first nine harmonics (*Delmar/Cengage Learning*)

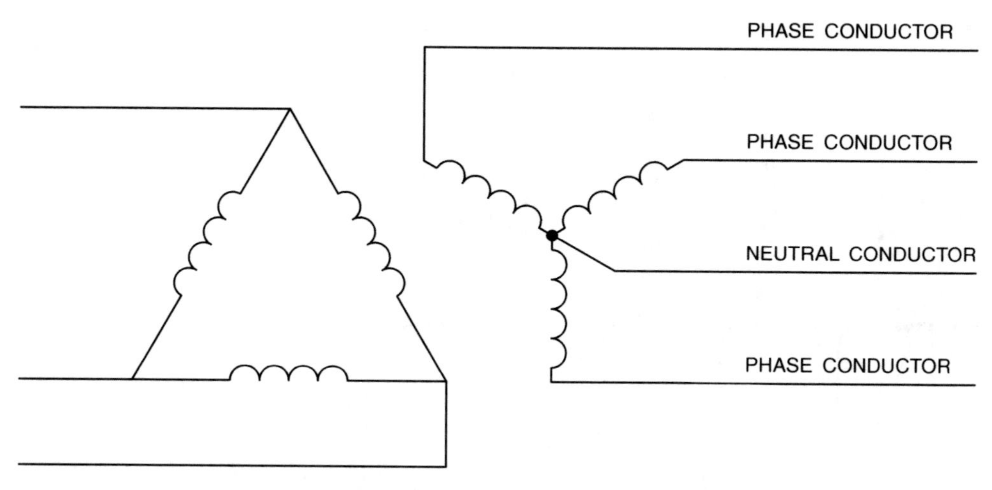

FIGURE 14–32 In a three-phase, four-wire, wye-connected system, the center of the wye-connected secondary is tapped to form a neutral conductor. (*Delmar/Cengage Learning*)

Harmonic currents are also reflected in the delta primary winding, where they circulate and cause overheating. Other heating problems are caused by eddy current and hysteresis losses. Transformers are typically designed for 60-Hz operation. Higher harmonic frequencies produce greater core losses than the transformer is designed to handle. Transformers that are connected to circuits that produce harmonics must sometimes be derated or replaced with transformers that are specially designed to operate with harmonic frequencies.

Transformers are not the only electrical component to be affected by harmonic currents. Emergency and standby generators can be affected in the same way as transformers. This is especially true for standby generators used to power data-processing equipment in the event of a power failure. Some harmonic frequencies can even distort the zero crossing of the waveform produced by the generator.

CIRCUIT BREAKER PROBLEMS

Thermomagnetic circuit breakers use a bimetallic trip mechanism that is sensitive to the heat produced by the circuit current. These circuit breakers are designed to respond to the heating effect of the true RMS current value. If the current becomes too great, the bimetallic mechanism trips the breaker open. Harmonic currents cause a distortion of the RMS value, which can cause the breaker to trip when it should not, or not to trip when it should. Thermomagnetic circuit breakers, however, are generally better protection against harmonic currents than electronic circuit breakers. Electronic breakers sense the peak value of current. The peaks of harmonic currents are generally higher than the fundamental sine wave (Figure 14–33). Although the peaks of harmonic currents are generally higher than the fundamental frequency, they can be lower. In some cases, electronic breakers may trip at low currents and in other cases they may not trip at all.

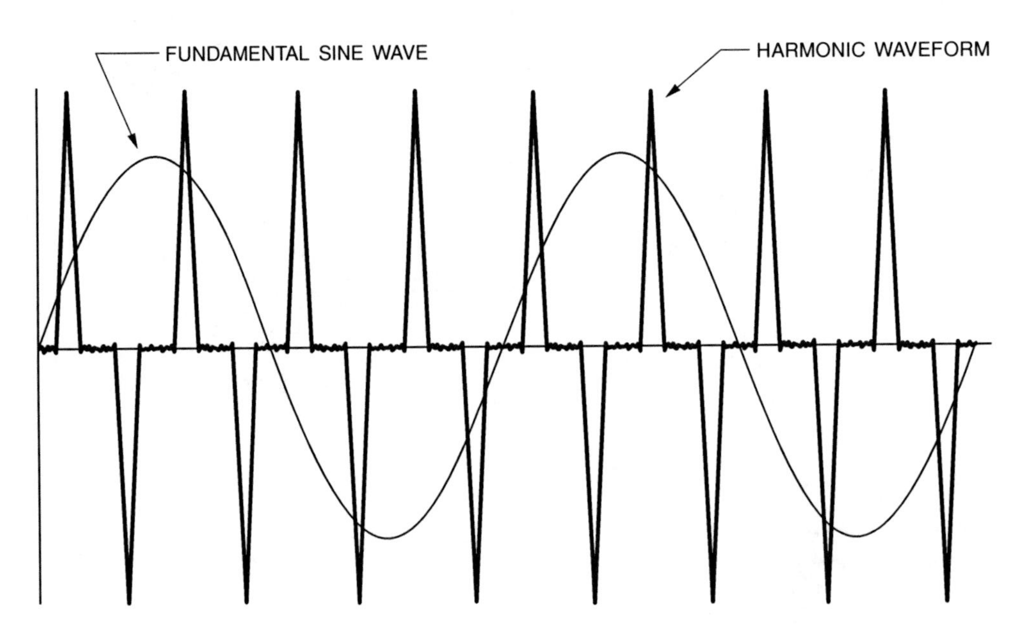

FIGURE 14–33 Harmonic waveforms generally have higher peak values than the fundamental waveform. (*Delmar/Cengage Learning*)

BUS DUCTS AND PANEL PROBLEMS

Triplen harmonic currents can also cause problems with neutral bus ducts and connecting lugs. A neutral bus is sized to carry the rated phase current. Because triplen harmonics can cause the neutral current to be higher than the phase current, it is possible for the neutral bus to become overloaded.

Electrical panels and bus ducts are designed to carry currents that operate at 60 Hz. Harmonic currents produce magnetic fields that operate at higher frequencies. If these fields should become mechanically resonant with the panel or bus duct enclosures, the panels and bus ducts can vibrate and produce buzzing sounds at the harmonic frequency.

Telecommunications equipment is often affected by harmonic currents. Telecommunication cable is often run close to power lines. To minimize interference, communication cables are run as far from phase conductors as possible and as close to the neutral conductor as possible. Harmonic currents in the neutral conductor induce high-frequency currents into the communication cable. These high-frequency currents can be heard as a high-pitched buzzing sound on telephone lines.

Determining Harmonic Problems on Single-Phase Systems

There are several steps that can be followed in determining whether there is a problem with harmonics. One step is to do a survey of the equipment. This is especially important in determining whether there is a problem with harmonics in a single-phase system.

1. Make an equipment check. Equipment such as personal computers, printers, and fluorescent lights with electronic ballast are known to produce harmonics. Any piece of equipment that draws current in pulses can produce harmonics.
2. Review maintenance records to see whether there have been problems with circuit breakers tripping for no apparent reason.
3. Check transformers for overheating. If the cooling vents are unobstructed and the transformer is operating excessively hot, harmonics could be the problem. Check transformer currents with an ammeter capable of indicating a true RMS current value. Make sure that the voltage and current ratings of the transformer have not been exceeded.

It is necessary to use an ammeter that responds to true RMS current when making this check. Some ammeters respond to the average value, not the RMS value. Meters that respond to the true RMS value generally state this on the meter. Meters that respond to the average value are generally less expensive and do not state that they are RMS meters.

Meters that respond to the average value use a rectifier to convert the alternating current into direct current. This value must be increased by a factor of 1.111 to change the average reading into the RMS value for a sine-wave current. True RMS responding meters calculate the heating effect of the current. The chart in Figure 14–34 shows some of the differences between average indicating meters and true RMS meters. In a distorted waveform, the true RMS value of current will no longer be average × 1.111 (Figure 14–35). The distorted waveform generally causes the average value to be as much as 50% less than the RMS value.

Another method of determining whether a harmonic problem exists in a single-phase system is to make two separate current checks. One check is made using an ammeter that indicates the true RMS value and the other is made using a meter that indicates the average value (Figure 14–36). In this example, it is assumed that the true RMS ammeter indicates a

AMMETER TYPE	SINE WAVE RESPONSE	SQUARE WAVE RESPONSE	DISTORTED WAVE RESPONSE
AVERAGE RESPONDING	CORRECT	APPROXIMATELY 10% HIGH	AS MUCH AS 50% LOW
TRUE RMS RESPONDING	CORRECT	CORRECT	CORRECT

FIGURE 14–34 Comparison of average responding and true RMS responding ammeters (*Delmar/Cengage Learning*)

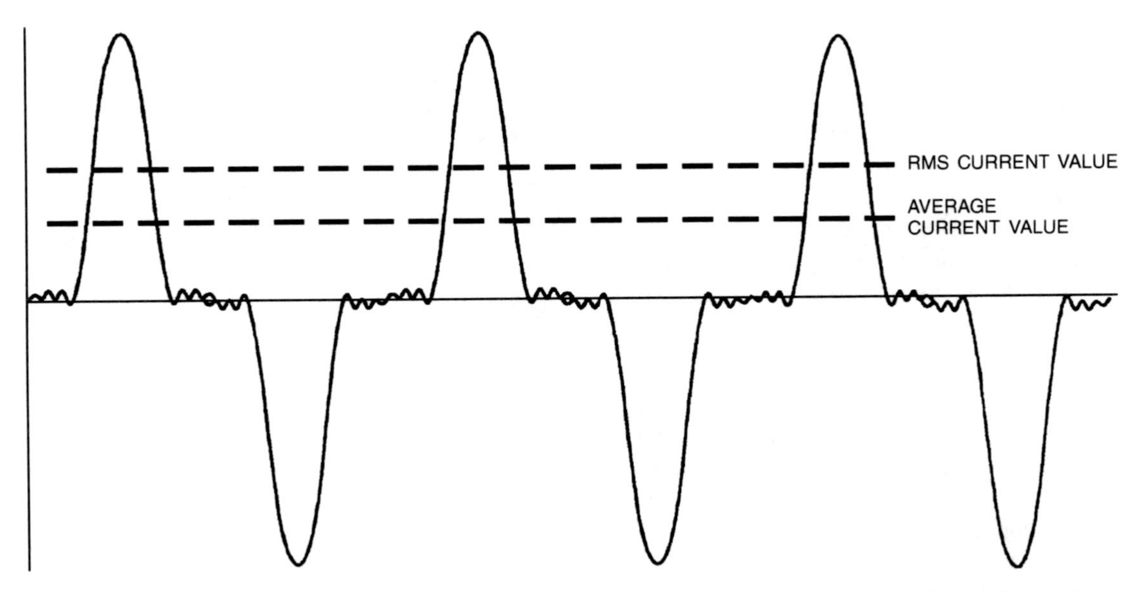

FIGURE 14–35 Average current values are generally less than the true RMS value in a distorted waveform. *(Delmar/Cengage Learning)*

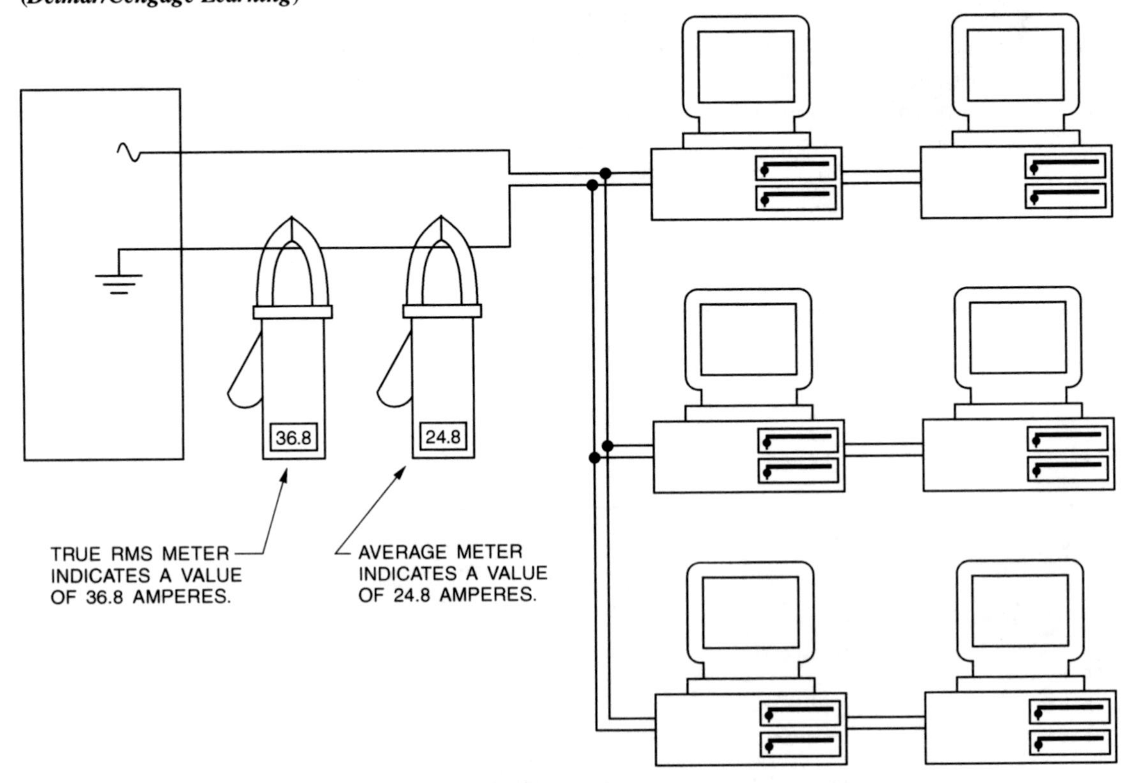

FIGURE 14–36 Determining harmonic problems using two ammeters *(Delmar/Cengage Learning)*

value of 36.8 A, and the average ammeter indicates a value of 24.8 A. Determine the ratio of the two measurements by dividing the average value by the true RMS value:

$$\text{Ratio} = \frac{\text{average}}{\text{RMS}}$$
$$= \frac{24.8}{36.8}$$
$$= 0.674$$

A ratio of 1 would indicate no harmonic distortion. A ratio of 0.5 would indicate extreme harmonic distortion. This method does not reveal the name or sequence of the harmonic distortion, but it does indicate whether there is a problem with harmonics.

The most accurate method for determining whether there is a harmonics problem is to use a harmonic analyzer. The harmonic analyzer will determine the name, sequence, and amount of harmonic distortion present in the system.

Determining Harmonic Problems on Three-Phase Systems

Determining whether a problem with harmonics exists in a three-phase system is similar to determining the problem in a single-phase system. Because harmonic problems in a three-phase system generally occur in a wye-connected, four-wire system, this example will assume a delta-connected primary and wye-connected secondary with a center-tapped neutral as shown in Figure 14–32. To test for harmonic distortion in a three-phase, four-wire system, measure all phase currents and the neutral current with both a true RMS indicating ammeter and an average indicating ammeter. It will be assumed that the three-phase system being tested is supplied by a 200-kVA transformer, and the current values shown in Figure 14–37 were recorded. The current values indicate that a problem with harmonics does exist in the system. Note the higher current measurements made with the true RMS indicating ammeter, and also the fact that the neutral current is higher than any phase current.

Conductor	True RMS Responding Ammeter	Average Responding Ammeter
Phase 1	365	292
Phase 2	396	308
Phase 3	387	316
Neutral	488	478

FIGURE 14–37 Measuring phase and neutral currents in a three-phase, four-wire, wye-connected system (*Delmar/Cengage Learning*)

Dealing with Harmonic Problems

After it has been determined that harmonic problems exist, something must be done to deal with them. It is generally not practical to remove the equipment causing the harmonic distortion, so other methods must be employed. It is a good idea to consult a power quality expert to determine the exact nature and amount of harmonic distortion present. Some general procedures for dealing with harmonics follow:

1. In a three-phase, four-wire system, the 60-Hz part of the neutral current can be reduced by balancing the current on the phase conductors. If all phases have equal current flow, the neutral current would be zero.
2. If triplen harmonics are present on the neutral conductor, harmonic filters can be added at the load. These filters can help reduce the amount of harmonics on the line.
3. Pull extra neutral conductors. The ideal situation would be to use a separate neutral for each phase, instead of using a shared neutral.
4. Install a larger neutral conductor. If it is impractical to supply a separate neutral conductor for each phase, increase the size of the common neutral.
5. Derate or reduce the amount of load on the transformer. Harmonic problems generally involve overheating of the transformer. In many instances it is necessary to derate the transformer to a point that it can handle the extra current caused by the harmonic distortion. When this is done, it is generally necessary to add a second transformer and divide the load between the two.

Determining Transformer Harmonic Derating Factor

Probably the most practical and straightforward method for determining the derating factor for a transformer is recommended by the Computer & Business Equipment Manufacturers Association. To use this method, two ampere measurements must be made. One is the true RMS current of the phases and the second is the instantaneous peak phase current. The instantaneous peak current can be determined with an oscilloscope connected to a current probe or with an ammeter capable of measuring the peak value. Many of the digital clamp-on ammeters are capable of measuring the average, true RMS, and peak values of current. For this example, it will be assumed that peak current values are measured for the 200-kVA transformer discussed previously. These values are added to the previous data obtained with the true RMS and average indicating ammeters (Figure 14–38). The formula for determining the transformer harmonic derating factor (THDF) is

$$\text{THDF} = \frac{(1.414)(\text{RMS phase current})}{\text{instantaneous peak phase current}}$$

This formula will produce a derating factor somewhere between 0 and 1.0. Because the instantaneous peak value of current is equal to the RMS value × 1.414, if the current waveforms are sinusoidal (no harmonic distortion), the formula will produce a derating factor of 1.0. Once the derating factor is determined, multiply the derating factor by the kVA capacity of the transformer. The product will be the maximum load that should be placed on the transformer.

Conductor	True RMS Responding Ammeter	Average Responding Ammeter	Instantaneous Peak Current
Phase 1	365	292	716
Phase 2	396	308	794
Phase 3	387	316	737
Neutral	488	478	957

FIGURE 14–38 Peak currents are added to the chart. (*Delmar/Cengage Learning*)

Assuming that the phase currents are unequal, find an average value by adding the currents together and dividing by 3:

$$\text{Phase (RMS)} = \frac{365 + 396 + 387}{3} = 382.7$$

$$\text{Phase (peak)} = \frac{716 + 794 + 737}{3} = 749$$

$$\text{THDF} = \frac{(1.414)(382.7)}{749} = 0.722$$

The 200-kVA transformer in this example should be derated to 144.4 kVA (200 kVA $\times$ 0.722).

SUMMARY

- Most electrical energy is generated by three-phase alternating-current generators.
    1. Three-phase systems are used to transmit and distribute this electrical energy.
    2. For economy, the generated three-phase voltage is stepped up to an extremely high voltage by three-phase transformers for long-distance transmission over the three-phase system.
    3. Three-phase transformers are used at distribution points to step down the high voltage to a safe and usable level.
- A large three-phase transformer is completely enclosed in one container. Three single-phase transformers, connected in a bank, can be used to transform the voltages of three-phase systems at distribution points.

1. These transformers may be connected in any one of four standard ways:
   a. Delta–delta
   b. Wye–wye
   c. Delta–wye
   d. Wye–delta
2. Another method is the open-delta (or V) connection. This connection uses only two single-phase transformers to transform the voltages on a three-phase system.

- The delta–delta connection:
  1. The delta–delta connection is used to supply an industrial load by stepping down a 2400-V, three-phase, three-wire service to a 240-V, three-phase, three-wire service.
  2. For most applications, the load is balanced and the three single-phase transformers have the same kilovolt-ampere rating.
  3. The primary and the secondary windings are connected in delta.
  4. To determine the total capacity in kilovolt-amperes of a delta–delta-connected transformer bank, the three kVA ratings are added.
- For any delta-connected circuits and systems:
  1. Each transformer winding is connected across the two line leads; thus, the line voltage and the transformer coil winding voltage are the same.
  2. The line current is equal to $\sqrt{3}$ or 1.73 times the coil winding current.
  3. In a closed-delta transformer connection, each line wire is fed by two transformer coil currents that are out of phase. These coil currents do not add directly, but must be added vectorially to obtain the line current.
- Feeding a dual load using a delta–delta connection—some power companies use a delta–delta-connected transformer bank to feed two types of loads:
  1. A 240-V, three-phase industrial load.
  2. A 120/240-V, single-phase, three-wire lighting load.
     a. The transformer supplying the lighting load is larger than the other two transformers of the three-phase system.
     b. The neutral wire is tied to the midpoint of the 240-V, low-voltage secondary winding.
     c. Many transformers have two 120-V windings. These windings are connected in series and the neutral is brought out at the midpoint to give a 120/240-V service.
  3. Both lines A and C have 120 V to the grounded neutral. Line B has approximately 208 V to ground. This connection can be a serious hazard and cannot be used for lighting service.
- Procedure for making a closed-delta connection:

1. Two primary winding leads are brought out of the transformer and are marked H_1 and H_2. It is assumed that H_1 is the beginning and H_2 is the ending of each high-voltage winding. The H_2 end of each primary winding is connected to the beginning (H_1) of the next primary winding to form a series arrangement.
2. One three-phase primary source line is connected to each junction of two windings. With a 2400-V source, each of the primary windings has a line voltage of 2400 V impressed across it.
3. Once the high-voltage primary connections are made, the three-phase, 2400-V input may be energized and tested for the correct phase rotation. If the phase rotation is incorrect, the circuit is deenergized and any two line wires are interchanged.
4. To determine whether a transformer has additive or subtractive polarity, the terminal markings or the nameplate data can be checked.
5. The secondary closed-delta connection is the same as the primary connection, with X_1 substituted for H_1 and X_2 substituted for H_2. The primary side of the delta-connected transformer may be connected so as to cause a polarity reversal in one of its legs. A phase inversion on the primary side must be corrected on the secondary side.
   a. Determine whether the voltage output of each of the three transformers is the same as the voltage rating on the nameplate.
   b. Connect the end of one secondary winding to the beginning of another secondary winding. If the connection is correct, the voltage across the open ends of the two transformers should be the same as the output of each transformer. If the connections are not correct, the resultant voltage is $\sqrt{3}$ times the secondary rated voltage. The phase reversal is corrected by interchanging the connections to one of the coils.
   c. With the correct voltage on the two coils, add the end of the third secondary winding coil to the beginning of the second coil. The resultant voltage is opposite to each secondary voltage and has a magnitude equal to the secondary voltage. If the voltage is zero across the last pair of open leads, they can be connected together. A line wire is attached at each of the connection points. These three line wires form the three-phase output. If the voltage reading is not zero, reverse the connections of the last transformer coil before the delta connection is closed and the output lines are added.

- The kVA load on a transformer bank connected in closed delta is

$$\text{kVA} = \frac{\sqrt{3} \times V_{\text{line}} \times I_{\text{line}}}{1000}$$

- The kilowatt load on a transformer bank connected in closed delta is

$$\text{kW} = \frac{\sqrt{3} \times V_{\text{line}} \times I_{\text{line}} \times \cos\theta}{1000}$$

- The percentage load on a transformer bank connected in closed delta, in terms of its rated capacity, is

$$\text{Percentage load} = \frac{\text{actual load on bank, kVA}}{\text{capacity of bank, kVA}} \times 100$$

- The total output of a closed-delta connection is

$$\text{Total output, in watts} = 3 \times V_{\text{coil}} \times I_{\text{coil}} \times \cos\theta$$

- The wye–wye connection:
  1. This connection can be used when the load on the secondary side is balanced. If the secondary load consists of three-phase motors only and the load currents are balanced, the wye–wye connection can be used.
  2. This connection cannot be used if the secondary load becomes unbalanced. An unbalanced load causes a serious imbalance in the three voltages of the transformer bank.
  3. A fourth wire, known as the *neutral wire,* is added to eliminate unbalanced voltages. It is connected between the source and the common point on the primary side of the transformer bank.
     a. Each high-voltage winding is connected between the neutral and one of the three line leads. If the source is 2400/4160 V, the voltage across each high-voltage winding is 2400 V. The voltage across the three line leads is $\sqrt{3} \times 2400$, or 4160 V.
     b. The neutral wire maintains a constant voltage across each of the high-voltage windings, even when the load is unbalanced.
     c. The neutral wire is grounded and helps protect the three high-voltage windings from lightning surges.
  4. With the secondary connected in wye, each low-voltage winding is connected between the secondary neutral and one of the three line leads.
     a. For a four-wire, three-phase system rated at 2400/4160 V and 120/280 V, the voltage output of each secondary winding is 120 V. There are 120 V between the neutral wire and any one of the three secondary line leads.
     b. The voltage across the line wires is $\sqrt{3} \times 120$, or 208 V.
        (1) 208-V, three-phase service is available for industrial power loads such as three-phase motors.
        (2) 120 V is available for lighting loads.
  5. Nearly all wye–wye-connected transformers use three single-phase transformers having the same kVA capacity:

$$\text{Total kVA rating} = \text{kVA}_1 + \text{kVA}_2 + \text{kVA}_3$$

  6. In a wye–wye connection, a defective transformer must be replaced before the bank can be reenergized.

- The delta–wye connection:
  1. This connection is also used for transformation to step up or step down a voltage.
  2. The primary windings are connected in delta and the secondary windings are connected in wye.
  3. This connection is used at a generator station to step up the voltage. The input voltage is stepped up by the transformer ratio. This voltage is further increased by $\sqrt{3}$, or 1.73 times the secondary coil voltage. The high voltage is then connected directly to three-phase transmission lines.
  4. The insulation requirements are reduced for the secondary windings because the coil voltage is only 58% of the line voltage. (This fact is very important when the secondary side has a very high voltage.)
  5. This connection can also be used at the distribution point to step down the voltage.
  6. The delta–wye connection can be used for a three-phase, four-wire system for power and lighting.
     a. In a 208/120-V system, 208 V supplies the three-phase power load and 120 V supplies the lighting load.
     b. In a 480/277-V system, 480 V supplies the three-phase power load and 277 V supplies the lighting load.
        (1) Standard 120-V fluorescent lighting fixtures are used with special ballasts for operation on 277-V circuits.
        (2) A 277/480-V, three-phase, four-wire system has the following advantages:
            (a) The I^2R drop is reduced in feeders and branch circuits, resulting in an increase in the operating efficiency.
            (b) Smaller sizes of copper conductors, conduits, and equipment can be used to save up to 25% of the installation cost.
            (c) The load demands on the 277/480-V system can be increased with a minimum of changes and expense.
     c. In a building using a 277/480-V, three-phase system, the additional requirement of a 120-V service uses little power. A transformer is added to step down the voltage from 480 V to 120/240 V. A small air-cooled transformer can be used to supply the 120-V service and is centrally located.
- The wye–delta connection:
  1. This connection is used to step down relatively high transmission line voltages at the load center. A transformer bank of this type is commonly used to step down voltages of 60,000 V or more.
  2. Advantages of the delta–wye connection are
     a. the three-phase voltage is reduced by the transformer ratio times 1.73.

   b. the insulation requirements are reduced for the high-voltage windings because the wye primary coil voltage is only 58% of the primary line voltage.

- For the delta–wye and wye–delta connections, the three single-phase transformers generally have the same kVA capacity, and the total capacity of the transformer bank, in kVA, is obtained by adding the kVA ratings of the three transformers.
- The open-delta connection (V connection):
  1. This connection uses two transformers only.
  2. It can be used in emergency situations when one of three transformers is damaged in a three-phase service. The damaged transformer is cut out of the system, and the configuration of the open-delta connection is used.
  3. With two transformers supplying the service, the total capacity is only 58% of the capacity of the closed-delta connection.
  4. On the secondary side, the line voltage and the line current are the same as the secondary coil current and voltage.
  5. The original transformer installation may consist of an open-delta bank. As the industrial power load requirements increase, a third transformer may be added. When the third transformer is added, a delta–delta bank (closed-delta bank) is formed.
  6. The total kVA capacity for an open-delta bank is

$$\text{Total kVA} = \text{kVA}_1 + \text{kVA}_2 \times 0.866$$

- Advantages of an enclosed three-phase transformer:
  1. The operating efficiency of a three-phase transformer is slightly higher than the overall efficiency of three separate single-phase transformers.
  2. The three-phase transformer weighs less and requires less space than do three separate single-phase transformers.
  3. The three-phase transformer supplies the same kVA output and costs less than three single-phase transformers.
  4. The installation equipment required (such as bus bars, switchgear, and wiring) is easier to install and is less complex than that required by a transformer bank consisting of three single-phase transformers.
- If the three-phase transformer develops a problem, such as a defective phase winding, then the entire three-phase unit must be taken out of service.
- T-connected transformers are similar to an open-delta connection in that only two transformers are required to make the connection.
  1. Before two transformers can be connected T, one transformer must have a 50% tap on both the primary and secondary windings. It is preferable for the second transformer to have an 86.6% tap.
  2. T-connected transformers are generally wound specially for the purpose and are contained in the same case.
  3. The phase balance of the T-connected transformer is better than that of an open-delta bank.

4. The T-connected transformer can be connected to supply three-phase, four-wire service in a manner similar to that for a four-wire wye connection.

- Scott-connected transformers are similar to T-connected ones in that the main transformer must have a 50% tap on both the primary and secondary, and the teaser transformer must have an 86.6% tap.
  1. The Scott connection is used to produce two-phase power from a three-phase connection.
  2. A two-phase system has voltages 90° out of phase with each other.
- The zigzag connection is generally used for grounding purposes.
- Harmonics can cause heating problems in ac circuits.
- Harmonics are generally caused by devices that cause pulsations on the line such as switching power supplies and variable-frequency drives.
- The third harmonic generally causes overload conditions on neutral conductors.
- It is sometimes necessary to derate transformers because of harmonic problems.

Achievement Review

1. An industrial plant has a 2400-V, three-phase, three-wire service. It uses three 100-kVA, single-phase transformers. Each transformer is rated at 2400/240 V. The transformers supply a 240-V, three-phase, three-wire system. Each transformer has additive polarity.
   a. Draw a connection diagram showing how the transformer bank is connected. Mark the polarity of all transformer leads.
   b. A balanced load of 200 kW at a lagging power factor of 0.70 is supplied by the transformer bank. Determine
      1. the secondary line current.
      2. the secondary coil current.
      3. the primary coil current.
      4. the primary line current.
2. What is the percentage of load in kVA on the transformer bank described in question 1?
3. Explain a procedure that may be used to connect the secondary windings of single-phase transformers in closed delta.
4. a. Assume that one of the transformers in the delta–delta transformer bank in question 1 is cut out of service because it is damaged by lightning. Assuming that plant operations are cut to a minimum so that the peak load does not exceed the capacity of the two remaining transformers, show how they could be connected during this emergency.

b. What is the maximum balanced load, in kilowatts, at a lagging power factor of 0.70, that may be connected to the transformer bank during the emergency described in part a of this question?

5. Explain a procedure that may be used to connect the secondary windings of three single-phase transformers in wye.

6. Three single-phase, 20-kVA, 2400/120-V transformers are connected in a three-phase transformer bank. The transformers have additive polarity. The bank steps down the 2400/4160-V, three-phase, four-wire primary service to a 120/208-V, three-phase, four-wire secondary service. This service supplies both a 120-V, single-phase lighting load and a 208-V, three-phase motor load. Draw a schematic diagram of the connections for this transformer bank.

7. a. What is the maximum balanced three-phase load, in kVA, that can be connected to the transformer bank in question 6?

   b. If the load has a lagging power factor of 0.80, what is the maximum output in kW that can be obtained from the transformer bank?

8. Give two practical applications for a delta–wye transformer bank. Give one practical application for a wye–delta transformer bank.

9. A three-phase, four-wire, wye-connected transmission system has a voltage of approximately 60,550 V across the three line wires. In addition, there is 35,000 V from each line wire to the neutral. This voltage is to be stepped down at a substation to supply energy to a 5000-V, three-phase, three-wire distribution system. Each transformer is rated at 35,000/5000 V, 2000 kVA. Each transformer has subtractive polarity. Draw a schematic wiring diagram for this circuit and determine

   a. the full-load capacity of the bank, in kVA.

   b. the line current on the secondary side when the transformer bank is loaded to its rated capacity.

   c. the line current on the primary side when the transformer bank is loaded to its rated capacity, neglecting any losses.

10. What are typical applications of a 277/480-V, three-phase, four-wire, wye-connected system? What are the advantages of such a system?

11. a. List several advantages of a three-phase transformer, as compared with a three-phase bank consisting of three single-phase transformers.

    b. Give one disadvantage of the use of a three-phase transformer.

12. The high-voltage windings of a three-phase transformer are connected in wye. The low-voltage windings are connected in delta. The actual ratio between the high-voltage windings and their respective low-voltage windings is 10:1. The primary side of the transformer is supplied from a three-phase, four-wire, 2400/4160-V circuit. The secondary output is a three-phase, three-wire service for a three-phase industrial motor load. Determine

a. the secondary voltage.
b. the kVA output of the bank, when the current supplied to the motor load on the secondary is 60 A in each line wire.
c. the line current in each line wire on the high-voltage primary side of the transformer, neglecting any losses.

13. What is the frequency of the second harmonic?
14. Of the following, identify those that are considered triplen harmonics: third, sixth, ninth, twelfth, fifteenth, and eighteenth.
15. Would a positive rotating harmonic or a negative rotating harmonic be more harmful to an induction motor? Explain your answer.
16. What instrument should be used to determine what harmonics are present in a power system?
17. A 22.5-kVA single-phase transformer is tested with a true RMS ammeter and an ammeter that indicates the peak value. The true RMS reading is 94 A. The peak reading is 204 A. Should this transformer be derated, and if so, by how much?

PRACTICE PROBLEMS FOR UNIT 14

In the following problems, three single-phase control transformers have been connected to operate with a primary voltage of 240 V and a secondary voltage of 120 V. This gives each transformer a turns ratio of 2:1. It will be assumed that a line-to-line voltage of 208 V is connected to the primary winding, and that the load connected to the secondary has an impedance of 4 Ω in each phase. Find the unknown values in each of the following problems. It will be necessary to use the formulas in the Transformers and Three-Phase Connections sections of Appendix 15.

[*Note:* When computing values of voltage and current between the primary and secondary windings, use *phase* values and not *line* values.] In the following problems:

V_{P_P} = phase voltage of the primary
I_{P_P} = phase current of the primary
V_{L_P} = line-to-line voltage of the primary
I_{L_P} = line current of the primary
V_{P_S} = phase voltage of the secondary
I_{P_S} = phase current of the secondary
V_{L_S} = line voltage of the secondary
I_{L_S} = line current of the secondary
V_{P_L} = phase voltage of the load
I_{P_L} = phase current of the load

1. Refer to Figure 14–39 to find the following unknown values:

V_{P_P} = ______	V_{P_S} = ______	V_{P_L} = ______
I_{P_P} = ______	I_{P_S} = ______	I_{P_L} = ______
V_{L_P} = 208 V	V_{L_S} = ______	Ratio = 2:1
I_{L_P} = ______	I_{L_S} = ______	Z = 4 Ω per phase

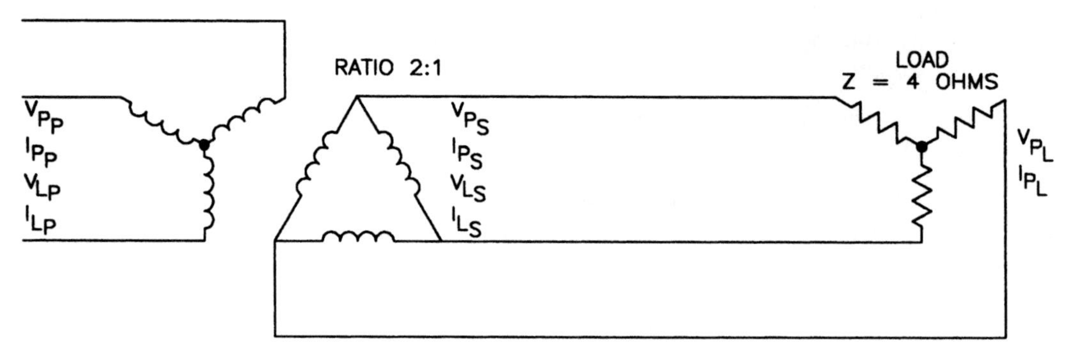

FIGURE 14–39 Wye–delta transformer with wye load (*Delmar/Cengage Learning*)

2. Refer to Figure 14–40 to find the following unknown values:

V_{P_P} = ______	V_{P_S} = ______	V_{P_L} = ______
I_{P_P} = ______	I_{P_S} = ______	I_{P_L} = ______
V_{L_P} = 208 V	V_{L_S} = ______	Ratio = 2:1
I_{L_P} = ______	I_{L_S} = ______	Z = 4 Ω per phase

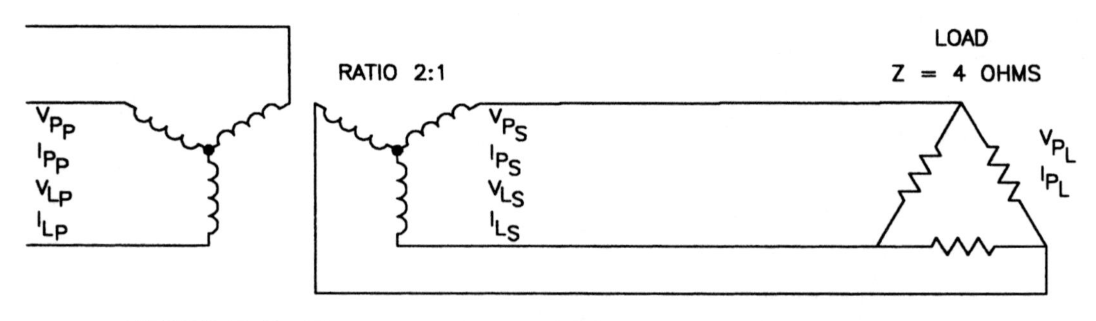

FIGURE 14–40 Wye–wye transformer with delta load (*Delmar/Cengage Learning*)

3. Refer to Figure 14–41 to find the following unknown values:

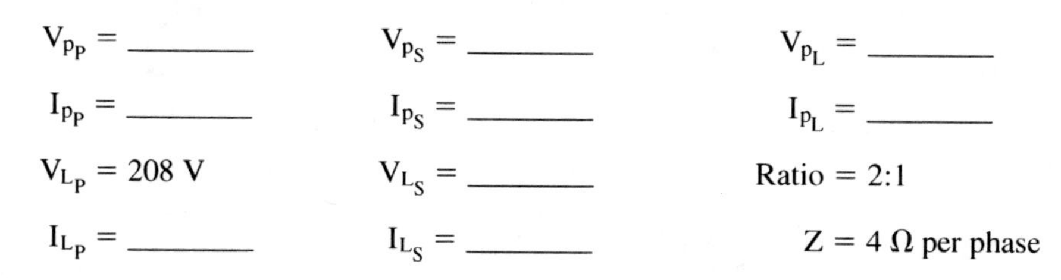

V_{P_P} = ______	V_{P_S} = ______	V_{P_L} = ______
I_{P_P} = ______	I_{P_S} = ______	I_{P_L} = ______
V_{L_P} = 208 V	V_{L_S} = ______	Ratio = 2:1
I_{L_P} = ______	I_{L_S} = ______	Z = 4 Ω per phase

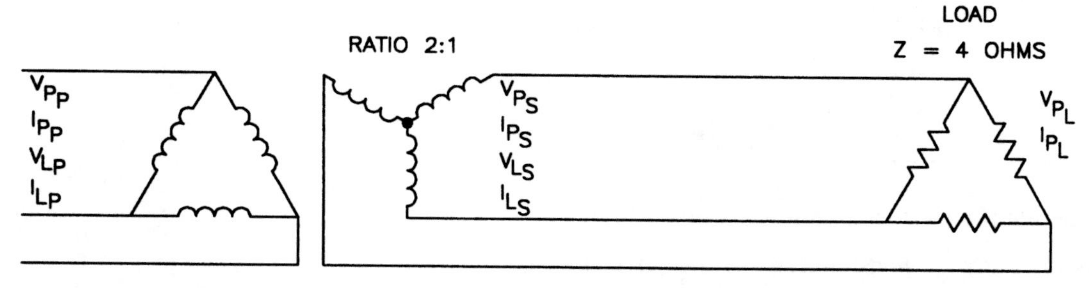

FIGURE 14–41 Delta–wye transformer with delta load (*Delmar/Cengage Learning*)

4. Refer to Figure 14–42 to find the following unknown values:

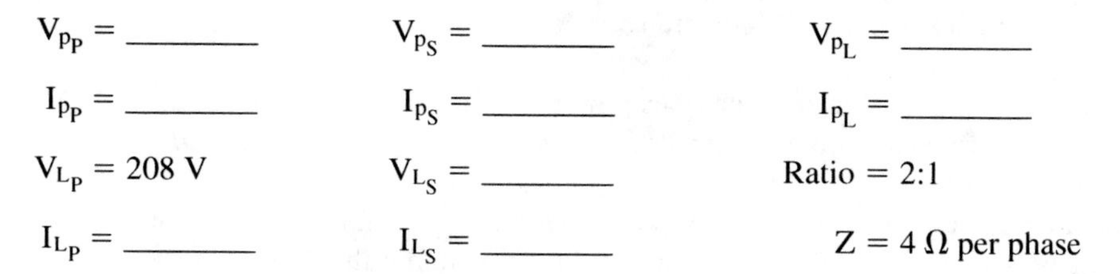

V_{P_P} = ______	V_{P_S} = ______	V_{P_L} = ______
I_{P_P} = ______	I_{P_S} = ______	I_{P_L} = ______
V_{L_P} = 208 V	V_{L_S} = ______	Ratio = 2:1
I_{L_P} = ______	I_{L_S} = ______	Z = 4 Ω per phase

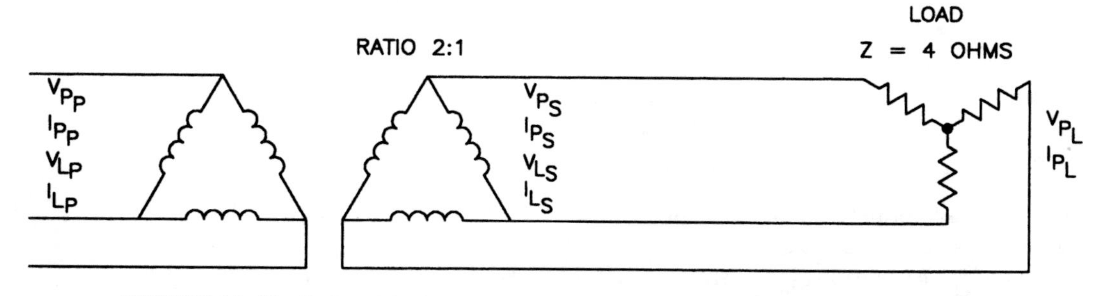

FIGURE 14–42 Delta–delta transformer with wye load (*Delmar/Cengage Learning*)

15

Special Transformer Applications

Objectives

After studying this unit, the student should be able to

- classify instrument transformers according to their use and application.
- describe the construction of intrument transformers, comparing their construction and capacity to that of power or distribution transformers.
- demonstrate how potential and current transformers are connected to meter single-phase and three-phase services.
- calculate the primary-to-secondary ratios of instrument transformers when used as potential transformers and as current transformers.
- list the advantages and disadvantages of using autotransformers.
- describe the hazards involved in the use of autotransformers.
- define the terms conductive power and transformed power as they relate to autotransformers.
- describe how a constant-current transformer takes power into the primary at a constant voltage and a variable current and delivers power from the secondary at a constant current and a variable voltage.
- explain the operation of a single-phase induction voltage regulator and a three-phase induction voltage regulator, including how these regulators maintain the delivery of a constant line voltage to a distribution point.

INSTRUMENT TRANSFORMERS

Instrument transformers are used with instruments and relays to measure and control ac circuits. Large and expensive devices are required to measure high voltages and currents directly. However, instrument transformers used with small standard devices provide a way of measuring high voltage and current values safely. The use of instrument transformers also means increased safety for the operator, less chance of control equipment damage due to high voltages, more accurate measurements, and greater convenience.

There are two types of instrument transformers: the instrument potential transformer and the instrument current transformer.

The Potential Transformer

The potential transformer is similar to a power transformer or a distribution transformer. However, its capacity is relatively small when compared to that of power transformers. Typical ratings of potential transformers range from 100 to 500 VA. The low-voltage side of the transformer generally is wound for 120 V. The load on the secondary (low-voltage) side of the potential transformer consists of the potential coils of various instruments. In some cases, the potential coils of relays and other control equipment are also connected to the secondary. In almost all cases, the load is light. Thus, potential transformers require a capacity no greater than 100 to 500 VA.

The primary circuit voltage and the voltage rating of the primary winding of a potential transformer are the same. As an example of the use of a potential transformer, assume that the voltage of a 4800-V, single-phase line must be measured. In this case, the primary of the potential transformer is rated at 4800 V. The low-voltage secondary is rated at 120 V. The ratio between the primary and secondary voltages is $4800/120 = 40/1$.

A voltmeter can be connected across the secondary of the potential transformer to measure the primary voltage, or 120 V. To determine the actual voltage of the high-voltage circuit, the voltmeter reading is multiplied by the transformer ratio. Thus, $120 \times 40 = 4800$ V. In some cases, the voltmeter is calibrated to indicate the actual value of the voltage on the primary side. As a result, it is not necessary to apply the multiplier to the instrument reading. This means that errors are minimized.

Figure 15–1 shows the connections for a potential transformer with a 4800-V primary input and a 120-V output to a voltmeter. Note that the transformer has subtractive polarity. All instrument potential transformers now manufactured have subtractive polarity. One of the secondary leads is grounded to minimize high-voltage hazards.

The ratio between the primary and secondary voltages of a potential transformer is very accurate. Normally, the percentage error is less than 0.5%.

The Current Transformer

The second type of instrument transformer is the current transformer. Current transformers are used to avoid the need to connect ammeters, instrument current coils, and relays directly to high-voltage lines. Current transformers step down the current by a known ratio. Thus, small and accurate instruments and control devices can be used because they are insulated from the high-voltage line.

The primary winding of a current transformer is connected in series with one of the line wires. The primary winding consists of a few turns of heavy wire wound on a laminated iron core. The secondary coil has more turns of smaller wire and is wound on the same core as the primary coil. The current rating of the primary winding is the maximum current that the winding will be required to conduct. For higher currents, the line in question may be fitted through a toroidal core with no turns at all. The core than acts as the primary winding and has no connection to the line. The secondary winding is always rated at 5 A.

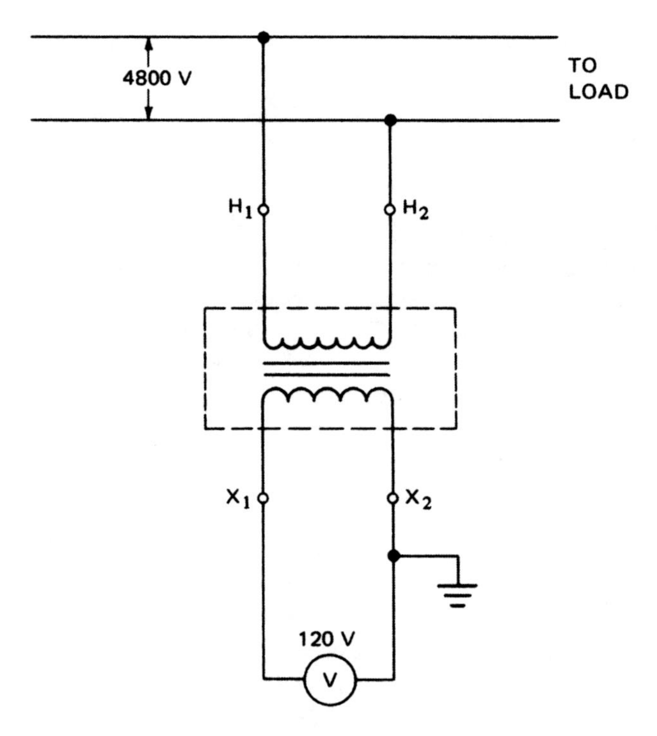

FIGURE 15–1 Connections for a potential transformer (*Delmar/Cengage Learning*)

To illustrate the operation of a current transformer, assume that the current rating of the primary winding is 100 A. The secondary winding has the standard rating of 5 A. The primary winding consists of three turns of wire, and the secondary winding has 60 turns. The ratio between the primary and the secondary currents is 100 A/5 A, or 20:1. In other words, the primary current is 20 times greater than the secondary current. Note that the number of turns and the current in the primary and secondary windings are related by an inverse proportion.

Stepping Down Current. The current transformer shown in Figure 15–2 is used to step down the current in a 4800-V, single-phase circuit. The primary is rated at 100 A. Because the secondary has the standard 5-A rating, the transformer ratio is 20:1. In other words, there is 20 A in the primary winding for each ampere in the secondary winding. If the ammeter indicates 4 A, the actual current in the primary is 20 times this secondary current, or 80 A.

Polarity Markings. The polarity markings of the current transformer are also shown in Figure 15–2. The high-voltage primary terminals are marked H_1 and H_2. The secondary terminals are marked X_1 and X_2. Electrons enter H_1 and leave X_1 at the same instant. Some manufacturers mark only the H_1 and X_1 leads. The H_1 lead connects to the line wire feeding from the source. The H_2 lead connects to the line wire feeding to the load. The secondary

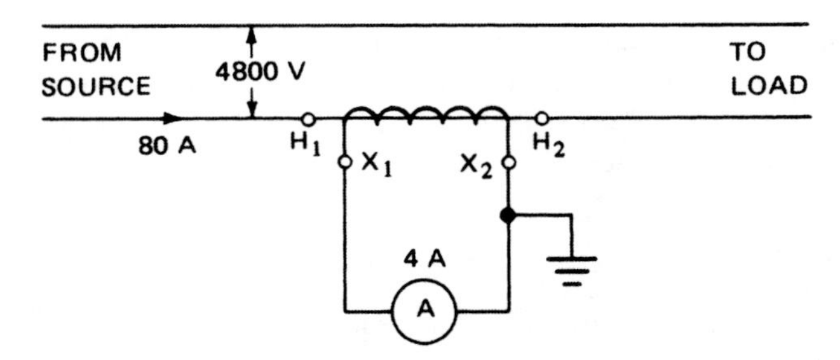

FIGURE 15–2 Current transformer used with an ammeter (*Delmar/Cengage Learning*)

leads connect directly to the ammeter. One of the secondary leads is grounded to reduce the high-voltage hazard. A common current transformer with a ratio of 200:5 is shown in Figure 15–3.

Current Transformer Precaution. The secondary circuit of a current transformer must not be opened when there is current in the primary winding. An absence of current in the secondary winding means that there is no secondary magnetomotive force opposing the primary mmf. The primary current becomes an exciting current. The flux of this current is unopposed by a secondary mmf. The primary flux causes a high voltage to be induced in the secondary

FIGURE 15–3 Current transformer (*Delmar/ Cengage Learning*)

winding. This voltage is great enough to be a hazard. If the instrument circuit must be opened when there is current in the primary winding, a short-circuiting switch is installed at the secondary terminals of the current transformer. This switch is closed before the instrument circuit is opened to make repairs or rewire the metering circuit.

Primary-to-Secondary Ratios. Current transformers have very accurate ratios between the primary and secondary currents. The percent error generally is less than 0.5.

Construction. A bar-type construction is often used when the primary winding of a current transformer has a large rating. The primary winding consists of a straight copper bus bar that passes through the center of a hollow metal core. The secondary winding is wound on the core. All standard current transformers rated at 1000 A or larger have this type of structure. In some cases, current transformers with ratings smaller than 1000 A also have the bar-type structure.

Measuring Current, Voltage, and Power

A potential transformer and a current transformer are shown in Figure 15–4. These transformers are being used with standard instrument movements to measure the voltage,

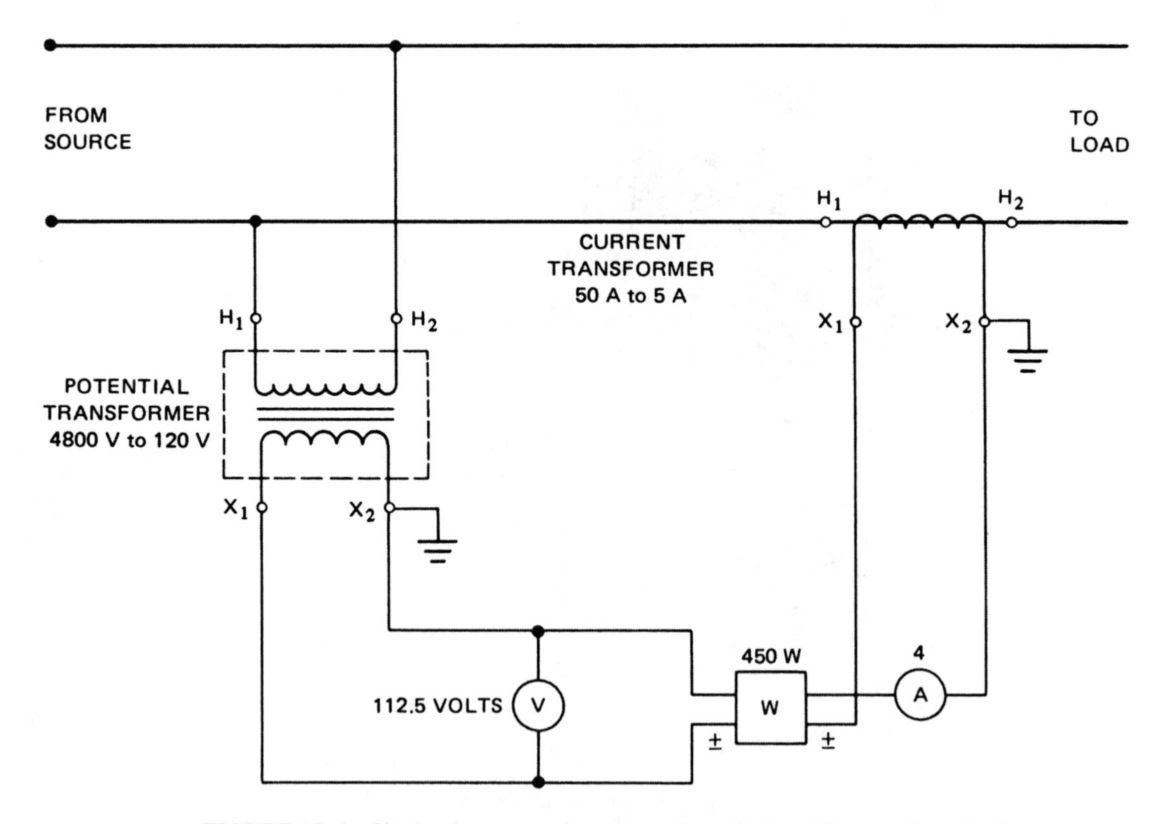

FIGURE 15–4 Single-phase metering connections (*Delmar/Cengage Learning*)

current, and power for a 4600-V single-phase circuit. The potential transformer is rated at 4800/120 V, and the current transformer is rated at 50/5 A.

The voltmeter and the potential coil of the wattmeter are connected in parallel across the low-voltage output of the potential transformer. The voltage across the potential coils of both instruments is the same.

The ammeter and the current coil of the wattmeter are connected in series across the secondary output of the current transformer. This means that the same current appears in the current coils of both instruments.

When lead H_1 of the potential transformer and lead H_1 of the current transformer are instantaneously positive, the X_1 leads of both transformer secondaries are also instantaneously positive. The current and voltage terminals of the wattmeter (marked ±) have the same instantaneous polarity. Thus, the torque on the wattmeter movement causes the pointer to move upscale. The secondary side of each instrument transformer is grounded to minimize high-voltage hazards.

Calculating the Primary Voltage and Current. In Figure 15–4, the voltmeter reading is 112.5 V, the ammeter reading is 4 A, and the wattmeter reading is 450 W. The primary voltage can be found as follows:

$$\text{Voltmeter mutiplier} = \frac{4800}{120} = 40$$

$$\text{Primary voltage} = 112.5 \times 40 = 4500 \text{ V}$$

The primary current is as follows:

$$\text{Current mutiplier} = \frac{50}{5} = 10$$

$$\text{Primary current} = 4 \times 10 = 40 \text{ A}$$

Calculating Power and Power Factor. The primary power must be determined using a power multiplier. Recall that power is the product of the voltage. the current, and the power factor. For this example, the power multiplier is the product of the current multiplier and the voltage multiplier:

$$\begin{aligned}\text{Wattmeter mutiplier} &= \text{voltmeter mutiplier} \times \text{current mutiplier}\\ &= 40 \times 10 = 400\\ \text{Primary power, in watts} &= 450 \times 400 = 180{,}000 \text{ W} = 180 \text{ kW}\end{aligned}$$

The apparent power of the primary circuit is obtained by multiplying the primary voltage and the current.

$$\begin{aligned}\text{Apparent power, in volt-amperes} &= \text{V} \times \text{I}\\ &= 4500 \times 40 = 180{,}000 \text{ VA} = 180 \text{ kVA}\end{aligned}$$

The power factor is determined as follows:

$$\text{Power factor} = \frac{\text{kW}}{\text{kVA}} = \frac{180}{180} = 1.0, \text{ or unity}$$

Metering Connections

Three-phase circuits are used for most high-voltage and current transmission and distribution lines. This means that the three-phase, three-wire system requires two potential transformers having the same rating, and two current transformers having the same rating.

The connections shown in Figure 15–5 are for a three-phase, three-wire system using instrument transformers and measuring devices. The two potential transformers are connected in open delta to the 4800-V, three-phase line. There are three secondary voltages of 120 V each. Two current transformers are used. The primary winding of one current transformer is connected in series with line A. The primary winding of the other current transformer is connected in series with line C.

It can be seen by checking the respective primary and secondary circuit paths for the ammeters that each ammeter is connected correctly. Other instruments that may be added to the circuit include a three-phase wattmeter, a three-phase watt-hour meter, or a three-phase power factor meter. The proper phase relationships must be maintained when three-phase instruments are connected into a secondary circuit. If the instruments are not connected properly, their readings will be incorrect. For the three-phase, three-wire metering system

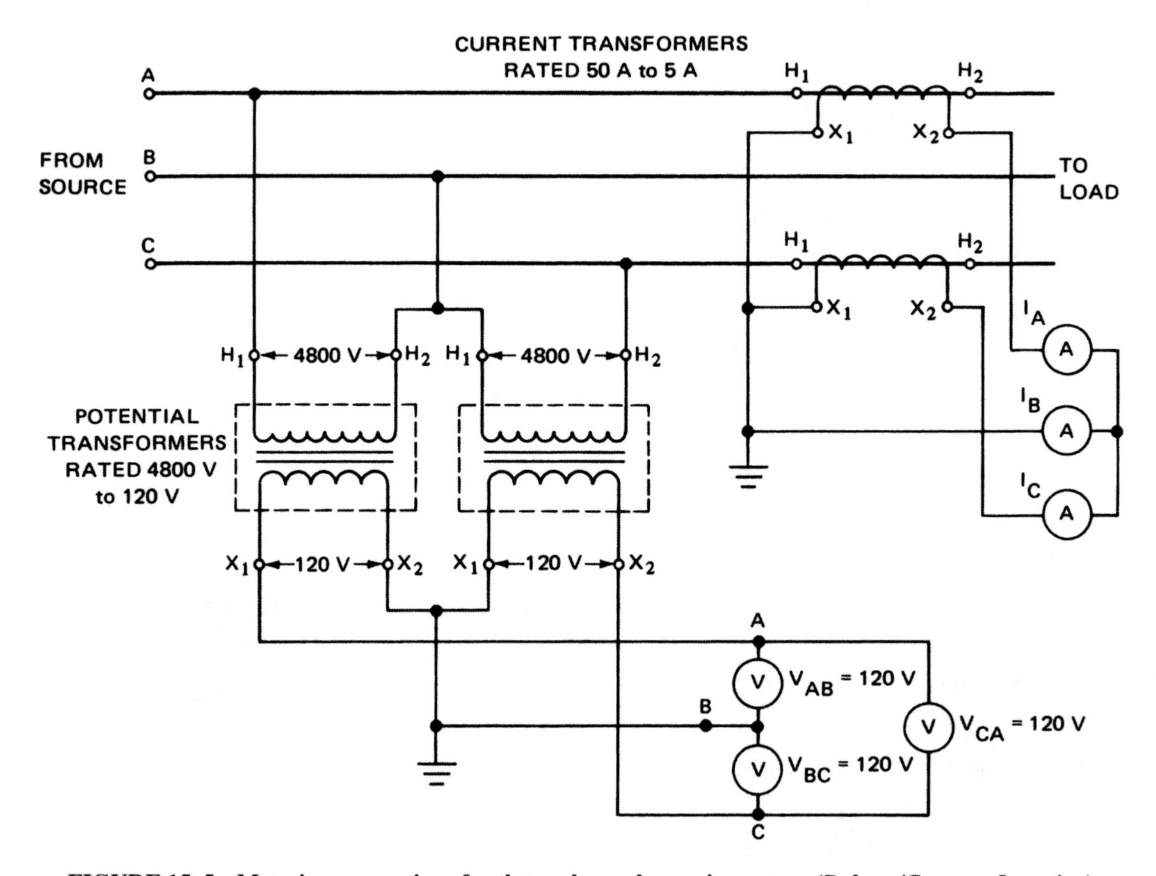

FIGURE 15–5 Metering connections for three-phase, three-wire system (*Delmar/Cengage Learning*)

of Figure 15–5, both the interconnected secondary potential circuit and the interconnected secondary current circuit are grounded. Grounding serves as protection from high-voltage hazards.

Testing Metering Connections

When a test is to be made on industrial equipment such as a three-phase motor, temporary metering connections are commonly made using portable instruments and instrument transformers. Figure 15–6 shows the connections required to obtain measurements of the currents, the voltage, and the power in watts for a 20-hp, 480-V, three-phase motor. Two potential transformers, rated at 480/120 V, are connected in open delta. Two current transformers, each rated at 50/5 A, are connected in line wires A and C.

The circuit in Figure 15–6 shows one method of measuring the three-phase power using the two-wattmeter method. The current coil of wattmeter 2 is in the secondary circuit of the current transformer and is in series with line wire A. The potential coil of wattmeter 2 is connected across the secondary of the potential transformer whose primary winding is

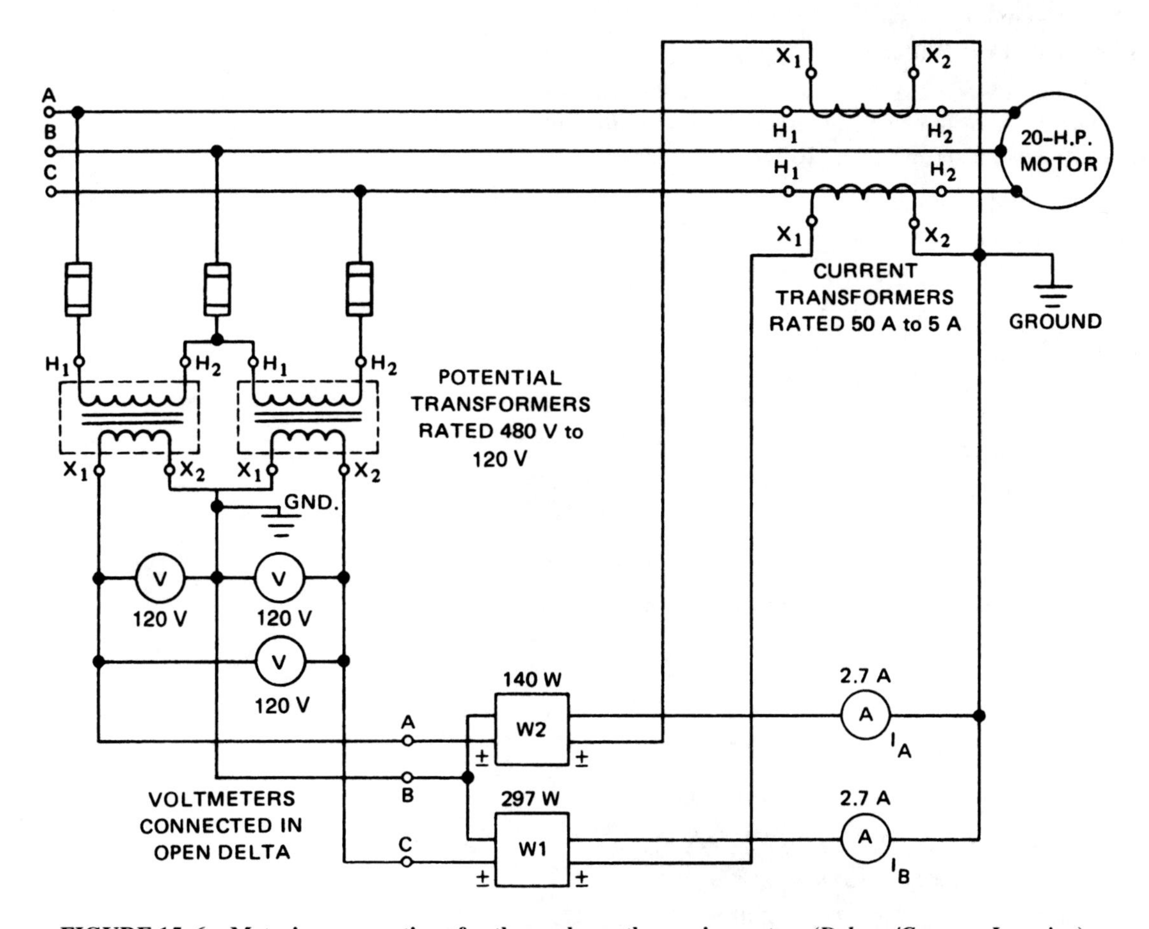

FIGURE 15–6 Metering connections for three-phase, three-wire system (*Delmar/Cengage Learning*)

connected to line wire A and B. The current coil of wattmeter 1 in the secondary circuit of the current transformer is in series with line wire C. The potential coil of wattmeter 1 is connected across the secondary of the potential transformer whose primary winding is connected to line wires B and C. The polarity terminals of both wattmeters are marked and are connected into the circuit correctly.

To check the connections, note that when primary line wires A and C are instantaneously positive, line wire B is instantaneously negative. The H_1 terminals on the current transformers are positive. The X_1 terminals on both current transformers are positive at the same instant. This means that the marked polarity current terminal on each wattmeter is instantaneously positive. Leads A and C from the low-voltage side of the potential transformers feed to the marked polarity potential terminals of the two wattmeters. These leads are also positive at this same instant. Therefore, both wattmeters indicate upscale if the power factor of the load is greater than 0.50 lag.

PROBLEM 1

Statement of the Problem

Figure 15–6 shows that the three secondary voltages are 120 V each. Both ammeters indicate 2.7 A. Wattmeter 1 reads 317 W, and wattmeter 2 indicates 121 W. The motor is delivering its rated capacity of 20 hp. Determine

1. the input, in watts.
2. the power factor.
3. the motor efficiency.

Solution

1. The total true power input is determined as follows:

$$\begin{aligned}
\text{Voltage mutiplier} &= 480/120 = 4 \\
\text{Current mutiplier} &= 50/5 = 10 \\
\text{Wattmeter mutiplier} &= 4 \times 10 = 40 \\
\text{Primary } W_1 &= 317 \times 40 = 12{,}680 \text{ W} \\
\text{Primary } W_2 &= 121 \times 40 = 4840 \text{ W} \\
\text{Primary watts total} &= W_1 + W_2 = 12{,}680 + 4840 \\
&= 17{,}520 \text{ W}
\end{aligned}$$

2. The power factor is

$$\begin{aligned}
V_{primary} &= 120 \times 4 = 480 \text{ V} \\
I_{primary} &= 2.7 \times 10 = 27 \text{ A} \\
\text{Volt-amperes} &= \sqrt{3} \times V_{line} \times I_{line} \\
&= 1.732 \times 480 \times 27 = 22{,}447
\end{aligned}$$

$$\cos\theta = \frac{W_1 + W_2}{\sqrt{3} \times V_{line} \times I_{line}} = \frac{17{,}520}{22{,}447} = 0.78 \text{ lagging power factor}$$

The ratio of the two power values may be called the *power factor ratio* for the two-wattmeter method given in Unit 10. This ratio can be determined using the two-wattmeter readings or the primary power values:

$$\text{Ratio, using secondary values} = \frac{W_2}{W_1} = \frac{121}{317} = 0.38$$

$$\text{Ratio, using primary values} = \frac{W_2}{W_1} = \frac{4840}{12{,}680} = 0.38$$

If the ratio of 0.38 is projected on the power factor ratio curve, it will be 0.78. This value agrees with the calculated power factor of 0.78 lag.

3. When the motor is delivering its rated horsepower output, its efficiency is

$$\eta = \frac{\text{output, watts}}{\text{input, watts}} \times 100 = \frac{\text{hp output} \times 746}{\text{input, watts}} \times 100$$

$$= \frac{20 \times 746}{17{,}520} \times 100 = \frac{14{,}920}{17{,}520} \times 100 = 85.2\%$$

Secondary Metering Connections

Figure 15–7 shows the secondary metering connections for a 2400/4160-V, three-phase, four-wire system. The three potential transformers are connected in wye. The three-phase output of these transformers consists of three secondary voltages of 120 V to neutral. A 50-to-5-A current transformer is connected in series with each of the three line wires.

When wattmeters, watt-hour meters, and other three-phase instruments are used in a circuit, their readings will be correct only if they are connected with the proper phase relationship. Three single-phase wattmeters are shown in the circuit in Figure 15–7. These instruments make it easier to check the instantaneous directions of current and voltage. In an actual installation, a three-element indicating wattmeter and possibly a three-element watt-hour meter would be used. However, the connections are the same. Note that both the low-voltage potential network and the secondary current network are grounded to minimize high-voltage hazards.

AUTOTRANSFORMERS

The conventional transformer has separate primary and secondary windings that are electrically insulated from each other. The autotransformer, on the other hand, has a primary winding and a secondary winding in the form of one continuous winding. This winding is tapped at certain points to obtain the desired voltages. The winding is wound on a laminated silicon steel core. Therefore, the primary and secondary sections of the winding are in the same magnetic circuit.

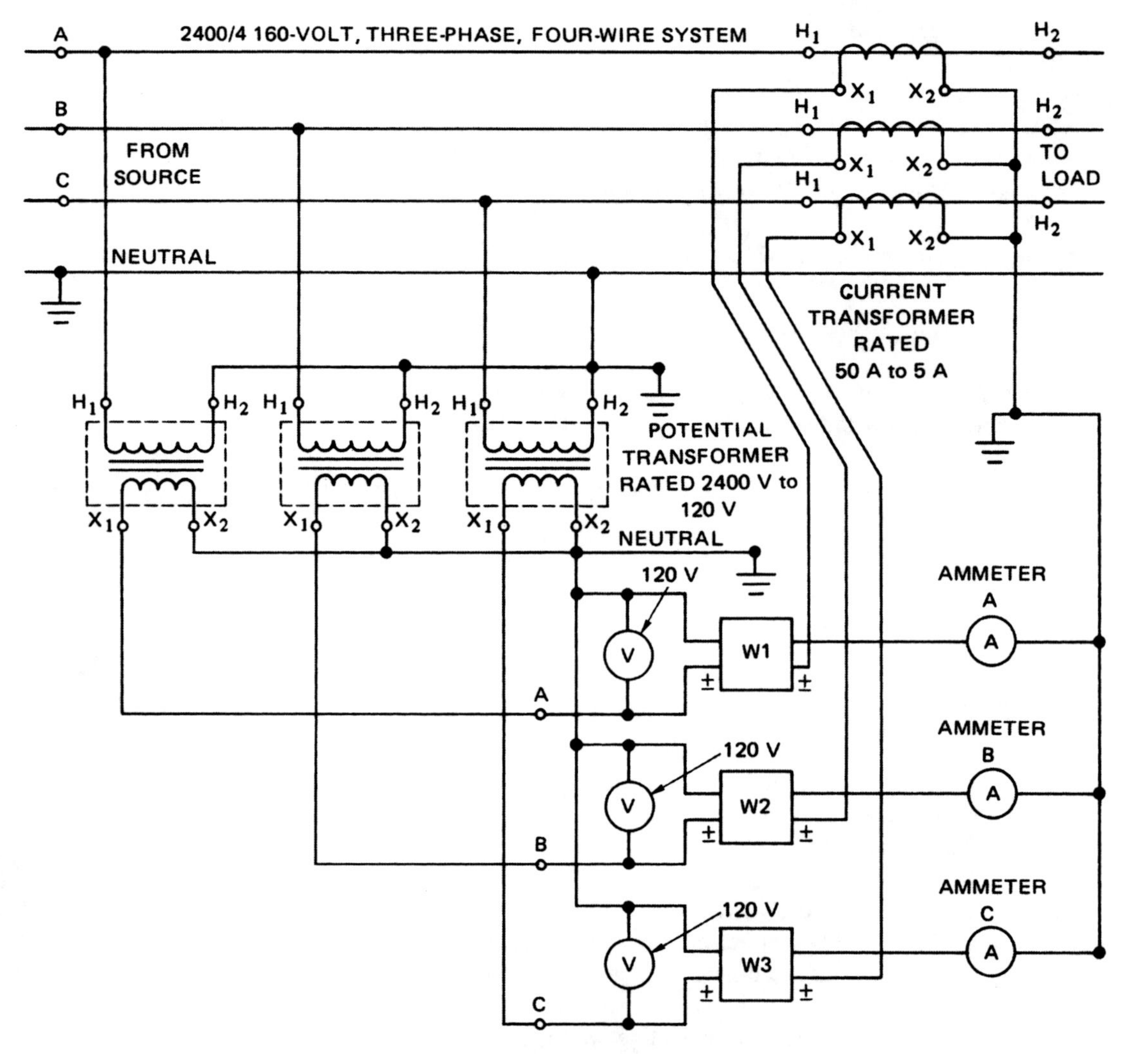

FIGURE 15–7 Metering connections for three-phase, four-wire system (*Delmar/Cengage Learning*)

Figure 15–8 shows how the autotransformer is connected into a typical circuit. The autotransformer can be viewed as having a primary side of terminals 1–3 and a secondary side of terminals 2–3. The calculations necessary for autotransformer circuits involve two new quantities: conductive power and transformed power.

Conductive power is the product of the current conducted *to* the secondary side *from* the primary side and the voltage on the secondary side.

Transformed power is the product of the difference between the primary and secondary voltages and the secondary current.

As an example of the calculation of these power values, consider the circuit shown in Figure 15–8. The 240-V input is connected across the entire winding at terminals 1 and 3.

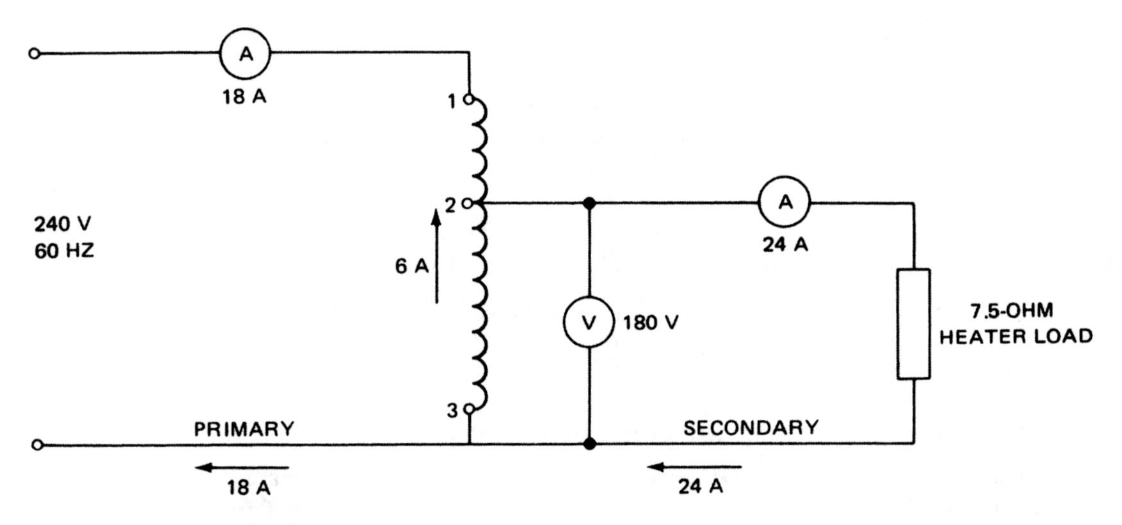

FIGURE 15–8 Autotransformer with a stepdown ratio of 4 to 3 (*Delmar/Cengage Learning*)

The 180-V output is obtained between terminals 2 and 3. This voltage is applied to the noninductive 7.5-Ω heating load. The current taken by the noninductive heating load is determined as follows:

$$I = V/R = 180/7.5 = 24\ A$$

The output of the autotransformer, in watts, is

$$W = V \times I \times \cos\theta$$
$$= 180 \times 24 \times 1.0 = 4320\ W$$

If losses are neglected and it is assumed that the input in watts equals the output, then the input is 4320 W as well. The current input from the source to the entire winding is

$$I_{input} = \frac{W_{input}}{V_{input}} = \frac{4320\ W}{240\ V} = 18\ A$$

Figure 15–8 indicates that 18 A is supplied from the source to terminal 1. This current passes from terminal 1 down through the winding to terminal 2. Note that the load requires 24 A. As a result, there must be a current of 6 A up from terminal 3 to terminal 2 at this same instant: 6 A + 18 A = 24 A, as required by the load. The current entering terminal 1 is 18 A at a potential of 240 V. The current at terminal 2 is at a potential of only 180 V. There is a potential drop of 60 V between terminals 1 and 2. This change represents a power value of 60 × 18 = 1080 W. This power is the transformed power. The product of 180 V and 6 A also equals 1080 W. The section of the winding between terminals 1 and 2 represents the primary winding with an input of 1080 W. The section of the winding between terminals 3 and 2 represents the secondary winding with an output of 1080 W.

The 1080-W output provides only a part of the power required by the load. The load requires a total of 4320 W. The 18-A current from terminal 1 leaves the winding at terminal 2 at a potential of 180 volts. The product of 180 V and 18 A is a part of the power, in watts, that is conducted directly through a section of the windings to supply the load:

$$180 \times 18 = 3240 \text{ W}$$

This value of power is the conductive power. The sum of the transformed power and the conductive power equals the total power supplied to the load, or 4320 W:

$$\text{Output, watts} = \text{transformed power} + \text{conductive power}$$
$$= 3240 \text{ W} + 1080 \text{ W} = 4320 \text{ W}$$

Autotransformer Used to Step Up Voltage

The autotransformer can also be used to step up a voltage. If, in Figure 15–8, a voltage of 180 V is impressed across the section of the winding from terminals 2 to 3, the output voltage across terminals 1 and 3 will be 240 V.

An autotransformer used to step up a voltage is shown in Figure 15–9. A heating load is connected across the full winding, which has an output of 250 V. The current supplied to the load is

$$I = V \div R = 250 \div 12.5 = 20 \text{ A}$$

If the load is noninductive and the 20-A current is supplied to it at an output voltage of 250 V, the power output to the load is

$$W = V \times I \times \cos\theta$$
$$= 250 \times 20 \times 1.0 = 5000 \text{ W}$$

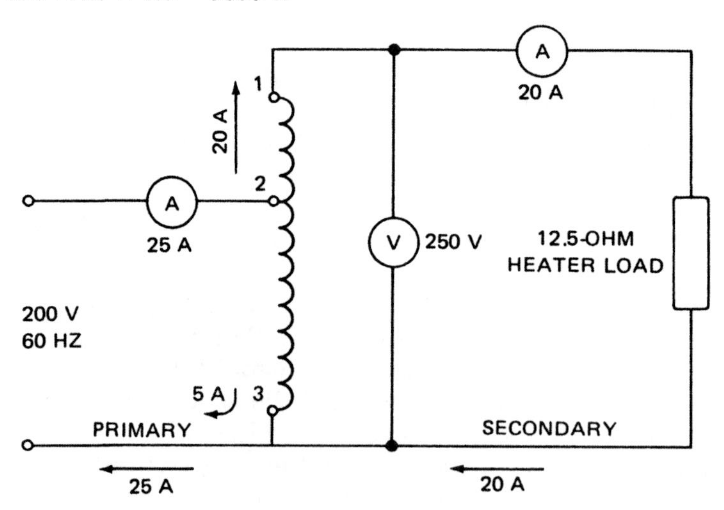

FIGURE 15–9 Autotransformer with a stepup ratio of 4 to 5 (*Delmar/Cengage Learning*)

This means that if losses are neglected, the input to the transformer will be 5000 W. The current input to the autotransformers at an input voltage of 200 V can be found knowing the power input:

$$I = W \div V = 5000 \div 200 = 25 \text{ A}$$

It is assumed that the line wire connected to terminal 2 is negative. A current of 25 A enters the winding at terminal 2. The load connected between terminals 1 and 3 requires only 20 A. Therefore, 20 A appears in the winding from terminal 2 to terminal 1. The remaining 5 A at terminal 2 appears in the winding between terminal 2 and terminal 3. Between terminal 2 and terminal 3, there is a loss of 200 V at 5 A. This change represents a transfer of 1000 W of power from the winding between terminals 2 and 3 to the flux of the core. Therefore, this section of the coil is the primary winding with a power input of 1000 W. The 20-A current between terminals 2 and 1 increases in potential by 50 V. The product of 50 V and 20 A is 1000 W. The winding between terminals 2 and 1 is the secondary winding and has a power output of 1000 W. This value is the part of the total power output that is transferred by transformer action from one section of the winding to a second section of the winding. The conductive power is considered to be the actual 20-A current at 200 V between terminals 2 and 1:

$$P = V \times I = 200 \times 20 = 4000 \text{ W}$$

If the two power values are added, the sum is equal to the transformer output supplied to the load, or 5000 W:

$$\begin{aligned}\text{Total power, watts} &= \text{conductive power} + \text{transformed power} \\ &= 4000 + 1000 = 5000 \text{ W}\end{aligned}$$

As compared to a constant-voltage transformer, the core and the copper losses in an autotransformer are smaller. The autotransformer has a slightly better efficiency than does the conventional transformer. The exciting current and the leakage reactance in the autotransformer are small. Therefore, it is an accepted practice to assume that the input and output are the same.

Applications of Autotransformers

Autotransformers are used in applications requiring small increases or decreases in voltage. Several typical uses of autotransformers are as follows:

- On long single-phase and three-phase distribution lines where the line voltage must be increased to compensate for voltage drops in the line wires
- In electronic circuits where several voltages are required
- In three-phase manual and automatic motor starters to reduce the startup voltage applied to three-phase motors; the decreased voltage reduces the starting current surge
- To provide a multivoltage source, as required by electronics laboratories

Multivoltages from an Autotransformer

Figure 15–10 shows one method of providing a number of voltages from one source. Taps are connected to the single winding of an autotransformer to provide a range of voltages, from very low values to values above the input voltage. For example, voltages ranging from 15 to 260 V may be obtained from an autotransformer with a 240-V input. This type of multivoltage supply can be used on some types of single-phase variable-speed motors, such as the repulsion motor. When the tap points are connected to a switching mechanism, a range of above- and below-normal speeds is available from the motor.

Another common autotransformer is often referred to as a variable transformer. This type of transformer has a sliding brush that can provide any voltage from 0 to full output voltage. The sliding brush permits the turns ratio to be changed to any setting. Many of these transformers can be connected to provide a higher output voltage than the input voltage. Autotransformers that are connected to 120 Vac can often provide output voltages of 140 Vac. The connection diagram of a variable autotransformer is shown in Figure 15–11. In the circuit shown in Figure 15–12, 120 volts is connected to terminals 2 and 4. The load is connected between the sliding brush connected to terminal 3 and terminal 4. This connection will permit the transformer to supply any voltage between 0 and 120 Vac to the load. If the 120-volt supply is changed to terminals 1 and 4, Figure 15–13, the autotransformer becomes a stepup transformer.

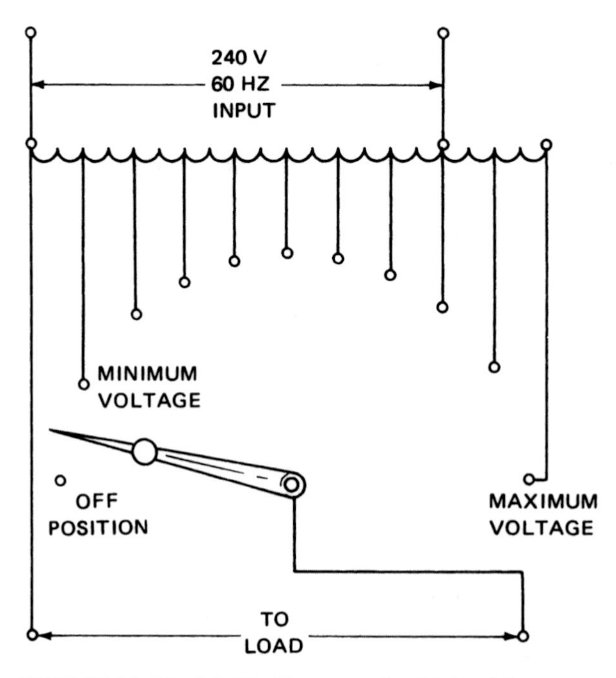

FIGURE 15–10 Multivoltage supply obtained from an autotransformer (*Delmar/Cengage Learning*)

FIGURE 15–11 Connection diagram of a typical variable autotransformer ***(Delmar/Cengage Learning)***

The autotransformer can now supply 0 to 140 Vac to the load. The part of the winding connected to power is the primary, and the part of the winding connected to the load is the secondary. In Figure 15–13, the primary winding is connected across terminals 1 and 4. Therefore, the winding between these two terminals is the primary. The sliding brush, however, can access all the windings between terminals 4 and 2. The number of turns between terminals 2 and 4 is greater than between terminals 1 and 4, permitting an increase in voltage between terminals 3 and 4. A variable transformer is shown in Figure 15–14.

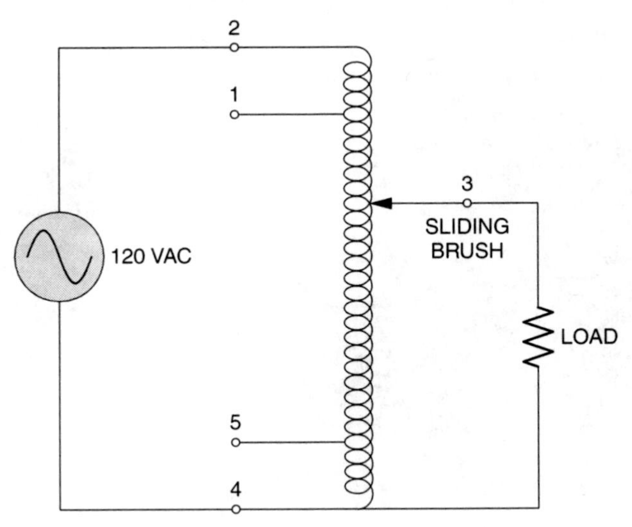

FIGURE 15–12 The autotransformer provides output voltage of 0 to 120 Vac. ***(Delmar/Cengage Learning)***

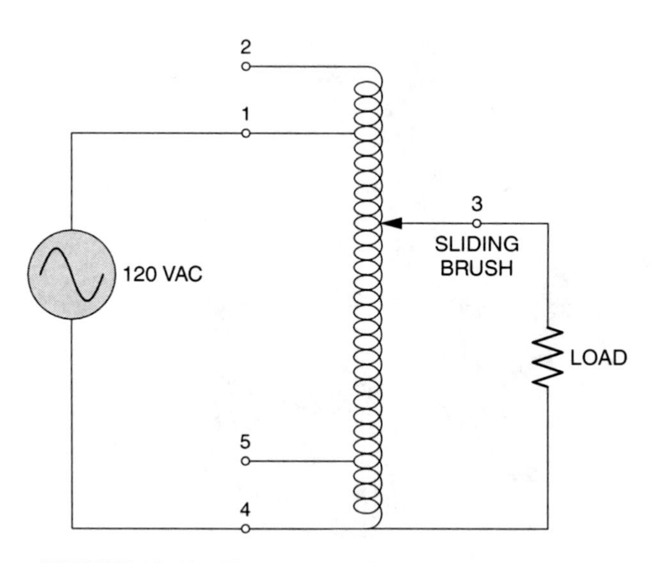

FIGURE 15–13 The autotransformer provides output voltage of 0 to 140 Vac. (***Delmar/Cengage Learning***)

FIGURE 15–14 A variable autotransformer (***Delmar/ Cengage Learning***)

Autotransformer Limitations

There are certain applications for which autotransformers should not be used. For example, an autotransformer should not be used to step down 2400 to 120 V for a lighting load. Recall that the primary and secondary windings of the autotransformer are connected electrically through the same winding. Thus, if the low-voltage section is open circuited, the entire high-voltage input will be applied across the low-voltage output.

Possible Hazards with Autotransformers

The normal operation of an autotransformer is shown in Figure 15–15A. Figure 15–15B shows a break in the low-voltage section of the winding. The entire 2400-V input is applied across the break in the winding and across the lighting load. Another voltage hazard will exist if one of the high-voltage leads becomes grounded. In this case, there will be a high voltage between ground and one of the low-voltage wires. Because of these hazards, an autotransformer is not used in applications requiring a high voltage to be transformed to 120 or 240 V.

CONSTANT-CURRENT TRANSFORMERS

Street lamp circuits may be connected in series or in parallel. The method used depends on the policies of the local utility company. If the lamps are connected in series, some method must be provided to short out any defective lamps. Remember that if one lamp in a series circuit burns out, all of the lamps in the circuit will go out. Shunting in a series circuit is accomplished by placing a small cutout device in the lamp socket. The device consists of two metal contacts that are separated by a film of insulating material. If one lamp filament burns out,

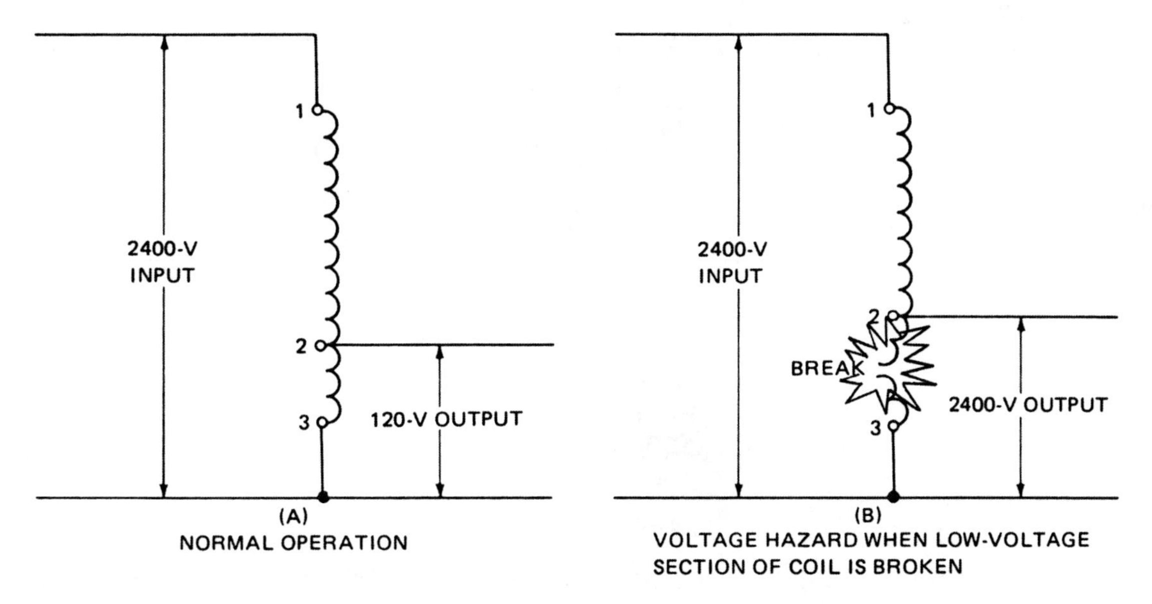

FIGURE 15–15 Voltage hazard with an autotransformer (*Delmar/Cengage Learning*)

the entire voltage of the series circuit is applied across the cutout device. This momentary voltage punctures the insulation between the metal contacts and shorts out the lamp.

Figure 15–16 is a drawing of a typical lamp socket used on a series street lighting circuit. As the lamps burn out, the total impedance of the series circuit decreases. If a constant voltage were maintained, the current would increase and cause other lamps to burn out.

Most series street lighting circuits require a current of 6.6 A. The voltage of the circuit will depend on the number of lamps in series. A constant 6.6 A is maintained in the circuit by a *constant-current transformer*. This type of transformer takes power into its primary winding at a constant voltage and a variable current. It delivers power from the secondary winding at a constant current and a variable voltage.

Functioning of the Constant-Current Transformer

The primary and secondary coils of the constant-current transformer are both placed on a silicon steel laminated core. The primary winding is stationary. Its input comes from a constant-voltage source. The secondary winding moves on the center leg of the core. Its movement is balanced by a counterweight–dashpot arrangement. The output of the secondary winding is connected to the series street lighting circuit.

A constant-current transformer is shown in Figure 15–17. The primary winding is energized from a constant-voltage source. The position of the movable secondary winding is adjusted by a counterweight until the desired value of secondary current is obtained. If there are a number of lamp failures in the series circuit, the total impedance of the series circuit decreases. At the same time, the current in the series circuit, the current in the movable secondary coil, and the current in the stationary primary coil will increase.

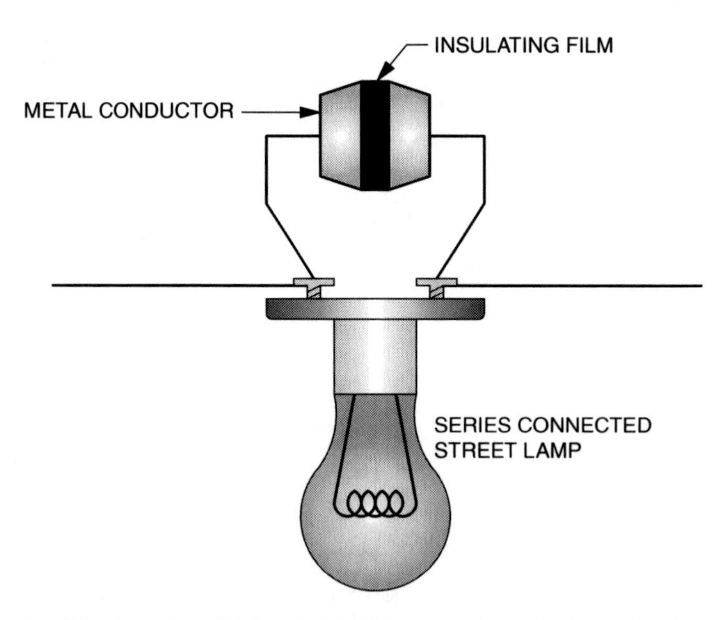

FIGURE 15–16 Cutout device is connected to the lamp. In the event the lamp filament opens, the cutout device will short and permit the current path to continue. (*Delmar/Cengage Learning*)

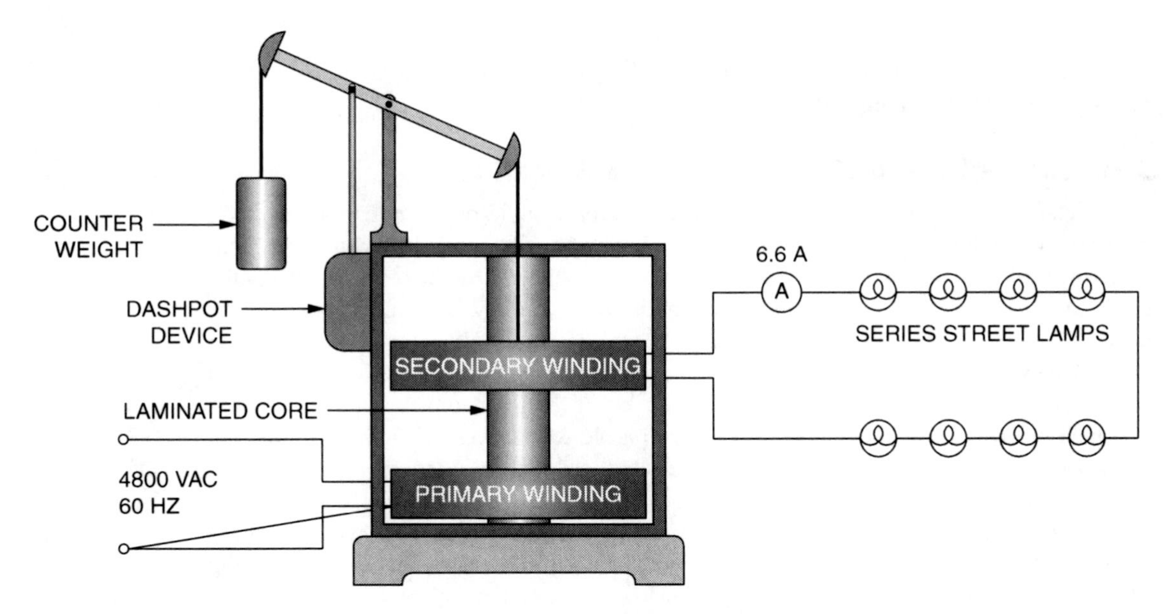

FIGURE 15–17 Constant-current transformer (*Delmar/Cengage Learning*)

The force of repulsion between the coils will increase (as described by Lenz's law). This force causes the secondary coil to move away from the primary coil. As a result, there is less primary flux linking the turns of the secondary winding, and the induced voltage in the secondary will decrease. The secondary coil continues to move away from the primary coil until the secondary current returns to its normal value of 6.6 A. When the current reaches this value, the repulsion force between the coils is balanced by the counterweight system.

Energizing the Primary

The primary of the constant-current transformer may be energized from an oil switch. The transformer is operated by a phototube and an amplifier circuit combined with a relay system. For street lighting applications, this type of control operates as follows. When the normal light intensity decreases below a set level, the phototube causes the amplifier circuit to operate the relay system and energize the transformer primary. The secondary coil moves so that the required 6.6 A is maintained in the circuit. When the light intensity increases above the set level, the phototube and the amplifier circuit operate the relays to open the coil switch. Thus, both the transformer and the street lighting circuit are deenergized.

INDUCTION VOLTAGE REGULATOR

A nearly constant voltage can be maintained on long distribution circuits by compensating for the voltage losses due to varying load conditions. An almost constant voltage can be maintained at the load center of the distribution circuit. In this case, an induction voltage regulator is used to compensate for the resistive and reactive voltage drops in the line wires.

The primary winding of a single-phase induction voltage regulator is wound in the slots of a laminated rotor core. The rotor can be turned through an angle of 180°. A small motor and a worm gear mechanism drive the rotor. The stationary secondary winding is placed in the slots of a laminated stator core.

Operation of the Induction Voltage Regulator

A schematic diagram is shown in Figure 15–18 for a single-phase induction regulator. The primary winding is connected across the line voltage. The secondary winding is connected in series with one of the line wires. In one position, the flux of the primary will cut the secondary turns and induce a maximum voltage in the secondary winding. This voltage is added to the original line voltage to compensate for the line drop. When the primary winding is turned, there is less primary flux linking the secondary turns. Thus, less voltage is induced in the secondary winding. Assume that the rotor is moved 90° from the point where the maximum voltage was induced. The rotor is now placed so that none of its lines of flux link the secondary turns. This means that no voltage is induced in the secondary winding. Under these conditions, the secondary winding must not be allowed to act like a choke coil. Therefore, a short-circuited winding is placed in the rotor slots. This winding is known as a tertiary winding and acts like the short-circuited secondary of a transformer. The tertiary winding reduces the inductance of the secondary winding to a very small value that can be neglected.

The rotor is now turned until it is 180° from its original position. A maximum voltage is induced in the secondary winding again. However, at 180°, the voltage induced in the secondary is in a direction that causes this voltage to oppose the line voltage.

When the maximum flux of the primary links the secondary turns in one position, a maximum voltage is induced in the secondary. In this instance, the voltage regulator acts like a voltage booster. When the primary is moved 180° from this position, the induced voltage in the secondary is also a maximum value. However, this voltage now opposes the line voltage. The amount of voltage increase or opposition can be controlled by turning the rotor to various positions.

The motor that turns the primary rotor is controlled by relays. These relays are energized by a contact-making voltmeter. When the voltage decreases, one set of contactors closes. The motor then turns in a given direction to boost the voltage back to its normal value. If the voltage becomes too high, another set of contactors is activated. The motor then turns in the opposite direction. As the primary moves to a new position, the voltage is reduced to its normal value.

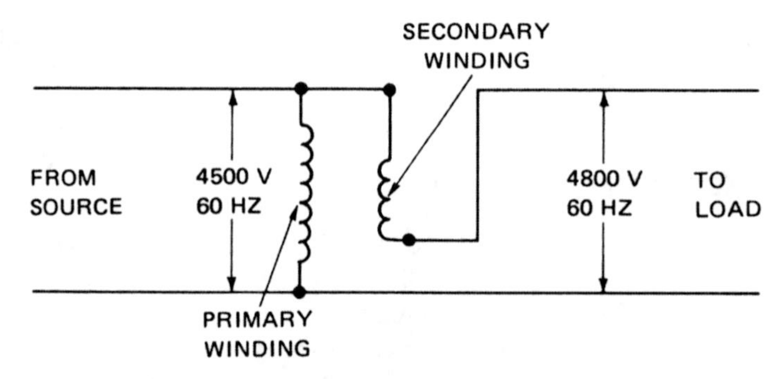

FIGURE 15–18 Single-phase induction regulator boosting voltage of a distribution circuit (*Delmar/Cengage Learning*)

The construction of the three-phase induction voltage regulator differs from that of the single-phase regulator. Three stationary primary windings are placed in the slots of a laminated stator core. The windings are connected in wye or delta to the three line wires of the distribution circuit. The three secondary windings are placed in the slots of the laminated rotor core. The secondary windings are insulated from each other. Each winding is connected in series with one of the three line wires of the three-phase circuit.

When the three-phase stationary primary windings are energized, a uniform magnetic field is set up inside the stator core. This means that the voltage induced in the three secondary windings is nearly independent of the rotor position. However, by moving the rotor with respect to the stator, there will be a change in the phase relationship between each induced voltage and its respective line voltage. This regulation gives the desired increase or decrease in the three-phase voltage of the distribution circuit.

Another common type of three-phase induction voltage regulator consists of three single-phase induction voltage regulators. These standard regulators are mounted in one enclosure. They are geared together so that they may be driven by one motor.

SUMMARY

- Instrument transformers are used with instruments and relays
    1. to measure and control ac circuits.
    2. to measure high voltage and current values directly.
    3. to provide safety for the operator and to protect control equipment from damage due to high voltages.
    4. to provide more accurate measurements and greater convenience.
- The two types of instrument transformers are the potential transformer and the current transformer.
    1. The potential transformer
        a. is similar in construction to a power or distribution transformer.
        b. has relatively small capacity in the range of 100 to 500 VA.
        c. has a primary winding designed and rated to be operated at a designated line voltage, such as 4800 V; the primary circuit voltage and the voltage rating of the primary winding of a potential transformer are the same.
        d. has a low-voltage secondary rated at 120 V; thus, the ratio between the primary and secondary voltage is 4800/120 = 40:1 (transformer ratio).
        e. normally has a subtractive polarity.
        f. has one of its secondary leads grounded to minimize high-voltage hazards.
        g. has a percentage error of less than 0.5%.

        The actual voltage of the high-voltage circuit is determined as follows:
            a. A voltmeter is used to measure the 120-V secondary.
            b. The voltmeter reading is multiplied by the transformer ratio of 40:1; thus, $120 \times 40 = 4800$ V.
            c. In some cases, a panel voltmeter is calibrated to read the actual primary value (this value includes the transformer ratio multiplier).

2. The current transformer
   a. is used so that ammeters, relays, and instrument current coils need not be connected to high-voltage lines.
   b. steps down the current by a known ratio.
   c. permits the use of small and accurate instruments because they are insulated from the high-voltage line.
   d. has a primary winding that is sized to carry the line current. This winding is connected in series with one of the lines. For higher currents, the line in question may be fed through a toroidal core (having no turns).
   e. is always rated at 5 A at the secondary. Any standard current ratio can then be applied to the transformer.
   f. generally has standard polarity markings. The H_1 lead is connected to the source end of the line. The H_2 lead is connected to the load end of the line. The X_1 and X_2 leads are not always marked by manufacturers.
   g. has a grounded secondary lead to reduce the high-voltage hazard.

   Also,
   a. the X_1 and X_2 leads are connected directly to the ammeter.
   b. the secondary circuit transformer must not be open when there is current in the primary. A voltage great enough to be a hazard can be induced in the secondary winding.
   c. a short-circuit switch installed at the secondary terminals of the transformer is closed when the instrument circuit is to be opened for repairs or rewiring of the metering circuit.

- When using potential transformers and current transformers with wattmeters,
  1. the potential coil of the wattmeter is connected across the low-voltage output of the potential transformer.
  2. the current coil of the wattmeter is connected in series with the output of the current transformer.
  3. the current and voltage terminals of the wattmeter (marked $\pm$) have the same instantaneous polarity.
  4. the torque on the wattmeter movement causes the pointer to move upscale.
- Calculating power and power factor:
  1. The power multiplier is the product of the current multiplier and the voltage multiplier.

     $$\text{Wattmeter mutiplier} = PT_{ratio} \times CT_{ratio}$$

     where PT is the potential transformer and CT is the current transformer.
     a. Primary power in watts is equal to the wattmeter reading times the wattmeter multiplier.
     b. Apparent power, in volt-amperes, is equal to the primary voltage multiplied by the line current.

2. The power factor is determined as follows:

$$\text{Power factor} = \frac{\text{kW}}{\text{kVA}}$$

- A three-phase, three-wire system requires two potential transformers having the same rating and two current transformers having the same rating.
- Other instruments may be added to the three-phase, three-wire circuit, including a three-phase wattmeter, a three-phase watt-hour meter, and a three-phase power factor meter.
- Portable instruments and instrument transformers are used to measure the current, the voltages, and the power in watts when testing industrial equipment.
- The secondary metering connections for a 2400/4160-V, three-phase, four-wire system are as follows:
  1. The three potential transformers are connected in wye.
  2. The three-phase output of these transformers consists of three secondary voltages of 120 V to neutral.
  3. A 50-to-5-A current transformer is connected in series with each of the three line wires.
  4. The proper phase relationship must be established when wattmeters, watt-hour meters, and other three-phase instruments are used in the circuit.
- The primary and secondary windings of an autotransformer are in the form of one continuous winding.
  1. This winding is tapped at certain points to obtain the desired voltages.
  2. The winding is wound on a laminated silicon steel core. As a result, the primary and secondary sections of the winding are in the same magnetic circuit.
  3. The calculations necessary for autotransformer circuits involve two new quantities.
     a. Conductive power is the product of the current conducted to the secondary side from the primary side and the voltage on the secondary side:

     $$P_{\text{conductive}} = I_P \times V_S$$

     b. Transformed power is the product of the difference between the primary and secondary voltages and the secondary current:

     $$P_{\text{transformed}} = (V_P - V_S) \times I_S$$

  4. Total power, watts = conductive power + transformer power.
  5. The autotransformer can be used to step up a voltage.
     a. The load is connected across the full winding.
     b. The source voltage is impressed across only part of the winding.
     c. With a noninductive load, $I = V \div R$ and $W = V \times I \times \cos\theta$.
     d. The current input to the autotransformer at a known input voltage can be found knowing the power input:

     $I = P \div V$

6. The core and the copper losses in an autotransformer are smaller than those in a constant-voltage transformer.
7. The efficiency of the autotransformer is slightly better than that of the conventional transformer.
8. The exciting current and the leakage reactance in the autotransformer are small. As a result, it is an accepted practice to assume that the input and the output are the same.

- Autotransformers are used when a small increase or decrease in voltage is required
  1. on long-distance distribution lines, both single-phase and three-phase, to compensate for the line voltage drop by increasing the line voltage.
  2. in electronic circuits where several voltages are required.
  3. in manual and automatic motor starters for three-phase systems to reduce the startup voltage applied to three-phase motors; the decreased voltage reduces the starting current surge.
  4. to provide a multivoltage source as needed in electronics laboratories and for some types of single-phase variable speed motors, such as the repulsion motor.
- Autotransformers are not used between a high-voltage source and a low-voltage circuit such as a lighting circuit. If the low-voltage section is open circuited, the entire high-voltage input is applied across the low-voltage output.
- A constant-current transformer takes power into its primary winding at a constant voltage and a variable current. It delivers power from the secondary winding at a constant current and a variable voltage.
- A series street lamp circuit is one application of the constant-current transformer. If a street lamp burns out, the total impedance of the series circuit decreases. If a constant voltage were maintained, the current would increase and cause other lamps to burn out. To prevent this, a small cutout device is placed in the socket of each lamp. When a lamp filament does burn out, the entire voltage of the circuit is applied across the cutout device, which short-circuits the socket and closes the series circuit.
- When used in a series street lighting circuit, a typical constant-current transformer functions as follows:
  1. The primary is stationary. Its input comes from a constant voltage source.
  2. The secondary winding moves on the center leg of the core. Its movement is balanced by a counterweight–dashpot arrangement. The output of the secondary is connected to the series street lighting circuit.
  3. If the total impedance of the series circuit decreases because lamps burn out, there is an increase of current in the series circuit, the movable secondary coil, and the primary coil.
  4. The force of repulsion between the primary and secondary coils will increase and cause the secondary coil to move away from the primary coil. As a result, there is less primary flux linking the secondary, and the induced voltage in the secondary will decrease. Thus, the secondary current also decreases.
  5. When the counterweight system and the repulsion effect between the coils are equal and opposite, the secondary coil maintains a constant current.

- The primary of the street lighting system transformer (constant-current transformer) is generally energized from an oil switch. The oil switch is operated by a relay that is controlled by a phototube and an amplifier circuit.
- An induction voltage regulator is used to compensate for the resistive and reactive voltage drops in long-distance distribution circuits.
    1. The primary winding of the regulator is wound in the slots of a laminated rotor core. The winding is connected across the line voltage.
    2. The secondary winding is stationary and is placed in the slots of a laminated stator core. The secondary winding is connected in series with one of the line wires.
    3. The rotor can be turned through an angle of 180°. A small motor and worm gear mechanism drive the rotor.
    4. When the maximum flux of the primary links the secondary turns in one position, a maximum voltage is induced in the secondary. This voltage is added to the incoming line voltage.
    5. At the center position of the rotor (90°), none of the flux from the primary coil cuts the secondary winding. Thus, no voltage is induced in the secondary.
        a. A short-circuited winding known as a *tertiary winding* is placed in the rotor slot.
        b. This winding acts like the short-circuited secondary of a transformer. It prevents the secondary from acting as a *choke coil* when there is no induced voltage in it. The tertiary winding reduces the inductance of the secondary winding to a very small value that can be neglected.
    6. As the rotor is moved from the 90° position to the 180° position, the voltage increases until it reaches a maximum at 180°. This induced voltage is in a direction that causes the secondary voltage to oppose the line voltage. Thus, this voltage is subtracted from the line voltage.
    7. The amount of voltage increase or opposition can be controlled by turning the rotor to various positions.
        a. The motor that turns the primary rotor is controlled by relays. The relays are energized by a contact-making voltmeter.
        b. One set of contacts closes when the line voltage decreases.
        c. The motor then turns in a given direction to boost the voltage back to its normal value.
        d. If the voltage becomes too high, another set of contactors closes and the motor turns in the opposite direction.
        e. As the primary rotor moves to a new position, the voltage is reduced to its normal value.
- Construction of the three-phase induction voltage regulator:
    1. The regulator has three primary windings connected in wye or delta to the three line wires of the distribution circuit.
    2. Each of the three secondary windings is connected in series with one of the three line wires of the three-phase circuit.

3. The primary windings are stationary and the secondary windings are movable.
4. The motor control system is the same as for a single-phase voltage regulator.

Achievement Review

1. a. What are the two functions of an instrument potential transformer?
   b. What are the two functions of an instrument current transformer?
2. In what ways does a current transformer differ from a constant-potential transformer?
3. Explain why the secondary circuit of an instrument current transformer must be closed when there is current in the primary circuit.
4. A 2300/115-V potential transformer and a 100/5-A current transformer are installed and connected on a single-phase line. A voltmeter, an ammeter, and a wattmeter are connected to the secondaries of the instrument transformers. The voltmeter indicates 110 V, the ammeter reads 4 A, and the wattmeter indicates 352 W. Draw a schematic wiring diagram for this circuit. Indicate the polarities of the terminals of the instrument transformers. Also indicate which current and voltage terminals of the wattmeter should be marked with the polarity.
5. Using the circuit in question 4, determine
   a. the primary voltage.
   b. the primary current, in amperes.
   c. the primary power, in watts.
   d. the circuit power factor.
6. Instrument potential and current transformers are installed to meter the voltage, current, and power of a 2400-V, three-phase, three-wire distribution circuit. Two 2400/120-V instrument potential transformers and two 100/5-A instrument current transformers are used. All three secondary voltages are 116 V. Two ammeters are connected to the secondaries of the current transformers. Each meter indicates 3.5 A. Two single-phase wattmeters are used. Wattmeter 1 indicates 400 W and wattmeter 2 reads 160 W. Draw a schematic wiring diagram of this circuit. Indicate the polarities on the instrument transformers. Also show which current and potential terminals on each of the wattmeters are to have polarity markings.
7. Using the circuit in question 6, determine
   a. the primary voltage.
   b. the primary current.
   c. the total three-phase primary power, in kVA.

   d. the total primary apparent power, in kVA.
   e. the circuit power factor.

8. Determine the power factor of the circuit in question 6. Use the power factor curve for the two-wattmeter method given in Unit 10. Compare this value with the calculated power factor obtained in question 7.

9. a. What is an autotransformer?
   b. What is one advantage to the use of an autotransformer?
   c. What is one limitation of an autotransformer?

10. Explain what is meant by the following terms as they relate to autotransformers:
    a. Transformed power
    b. Conductive power

11. An autotransformer is shown in Figure 15–19. It is used to step down 600 V to 480 V. The 480-V output is to supply a 24-kW noninductive load. Neglect the losses of the transformer. Determine
    a. the load current.
    b. the input current.
    c. the current between terminals 1 and 2.
    d. the current between terminals 3 and 2.
    e. what part of the power, in watts, supplied to the load is the transformed power.
    f. what part of the power supplied to the load is the conductive power.

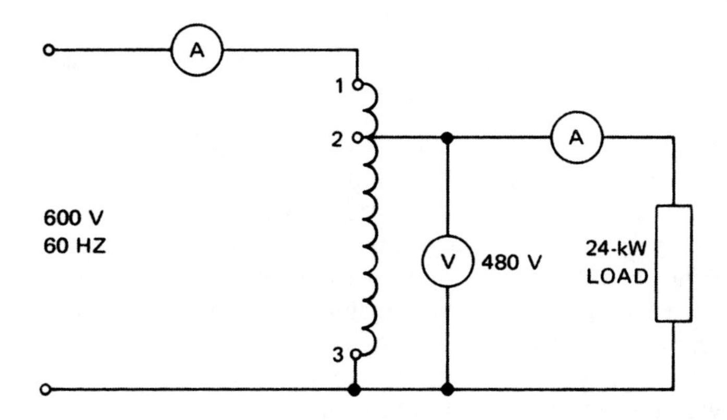

FIGURE 15–19 ***(Delmar/Cengage Learning)***

12. The autotransformer shown in Figure 15–20 is used to boost the voltage on a long, single-phase line. The voltage is to be changed from 2000 V to 2400 V. The 2400-V output supplies a load of 72 kW. The load has a power factor of unity. Neglect the losses of the transformer and determine
    a. the current output to the load.
    b. the current input to the transformer.

c. the current in coil section 2–1.
d. the current in coil section 2–3.
e. the part of the power, in watts, supplied to the load due to
   (1) transformed power.
   (2) conductive power.

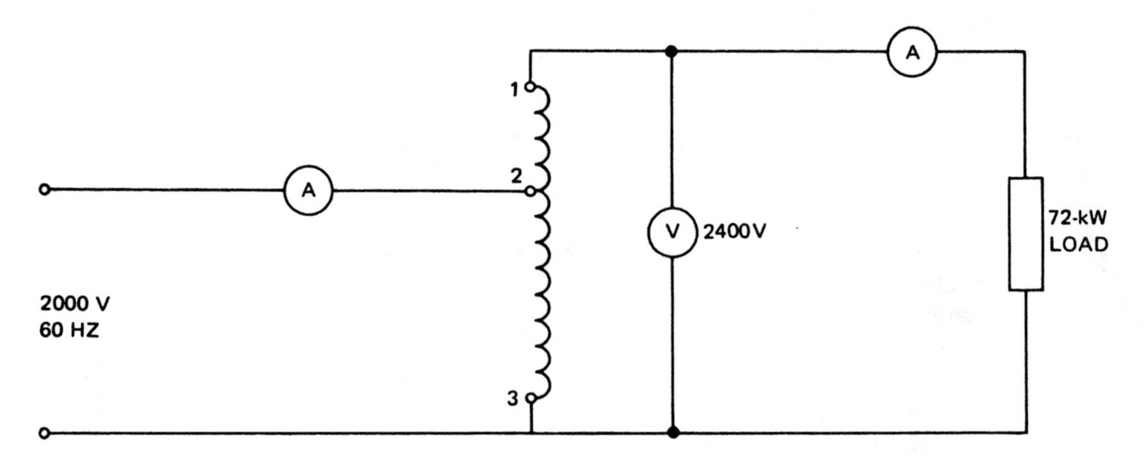

FIGURE 15–20 ***(Delmar/Cengage Learning)***

13. a. What is a constant-current transformer?
    b. Explain the reactions that take place in a constant-current transformer so that a constant current is maintained when there are changes in the load.

14. A constant-current transformer is located in an outdoor substation. The transformer delivers an output of 29.7 kW at a potential of 4500 V to a street lighting series circuit. The circuit consists of 50 lamps having the same rating. The power factor of this load is unity.
    a. What current flows in this street lighting series circuit?
    b. If 20 lamps are cut out of this series circuit, what output voltage is required from the constant-current transformer so that the current is the same as that for the 50-lamp circuit?
    c. Explain why a series street lighting circuit does not go out when the filament of one lamp burns out.

15. Explain the operations of a single-phase induction voltage regulator and a three-phase induction voltage regulator.

16

Three-Phase Induction Motors

Objectives

After studying this unit, the student should be able to

- describe the construction of a three-phase, squirrel-cage induction motor, including the function(s) of each component.
- define the terms synchronous speed and percent slip.
- explain how the stator winding and the rotor winding react with each other to produce a torque for the three-phase, squirrel-cage induction motor.
- use formulas to calculate the rotor frequency of the three-phase, squirrel-cage induction motor.
- describe the effects of a heavy overload on a three-phase, squirrel-cage motor.
- compare the starting current, the developed torque, and the efficiency of a three-phase, squirrel-cage induction motor to that of a single-phase, squirrel-cage induction motor.
- construct a diagram showing the connections used for the two-wattmeter test to find the horsepower rating and the power rating of a three-phase, squirrel-cage induction motor.
- describe the procedure for measuring motor losses.
- show how a speed controller with resistance can be used on a wound-rotor, three-phase, squirrel-cage induction motor.
- list the information that is given on the motor nameplate.
- show how the loss of one phase of the three-phase supply to the three-phase, squirrel-cage induction motor affects the starting operation, and efficiency of the motor.
- using electrical schematic diagrams as an aid, explain the operation of an across-the-line magnetic motor starter.
- describe how the direction of rotation is reversed for a three-phase, squirrel-cage induction motor and a three-phase, wound-rotor induction motor.
- compare the performance and application of a squirrel-cage induction motor and a wound-rotor induction motor.

THREE-PHASE, SQUIRREL-CAGE INDUCTION MOTOR

A three-phase, squirrel-cage induction motor is shown in Figure 16–1. This motor is simple in construction and is easy to maintain. For a given horsepower rating, the physical size of this motor is small, when compared with other types of motors. It has very good speed regulation under varying load conditions. This motor is used for many industrial applications because of its low purchase price, rugged construction, and operating characteristics.

Construction

The basic structure of a three-phase, squirrel-cage induction motor consists of a stator, a rotor, and two end shields that house the bearings supporting the rotor shaft.

The stator is a three-phase winding that is placed in the slots of a laminated steel core. The winding itself is made of formed coils that are connected to give three single-phase windings spaced 120 electrical degrees apart. The three separate single-phase windings are connected in wye or delta. Three line leads from the three-phase stator windings are brought out to a terminal box mounted on the frame of the motor.

The rotor has a cylindrical core consisting of steel laminations (Figure 16–2). Aluminum bars are mounted near the surface of the rotor. These bars are brazed or welded to two aluminum end rings. Some types of squirrel-cage induction motors are smaller than others

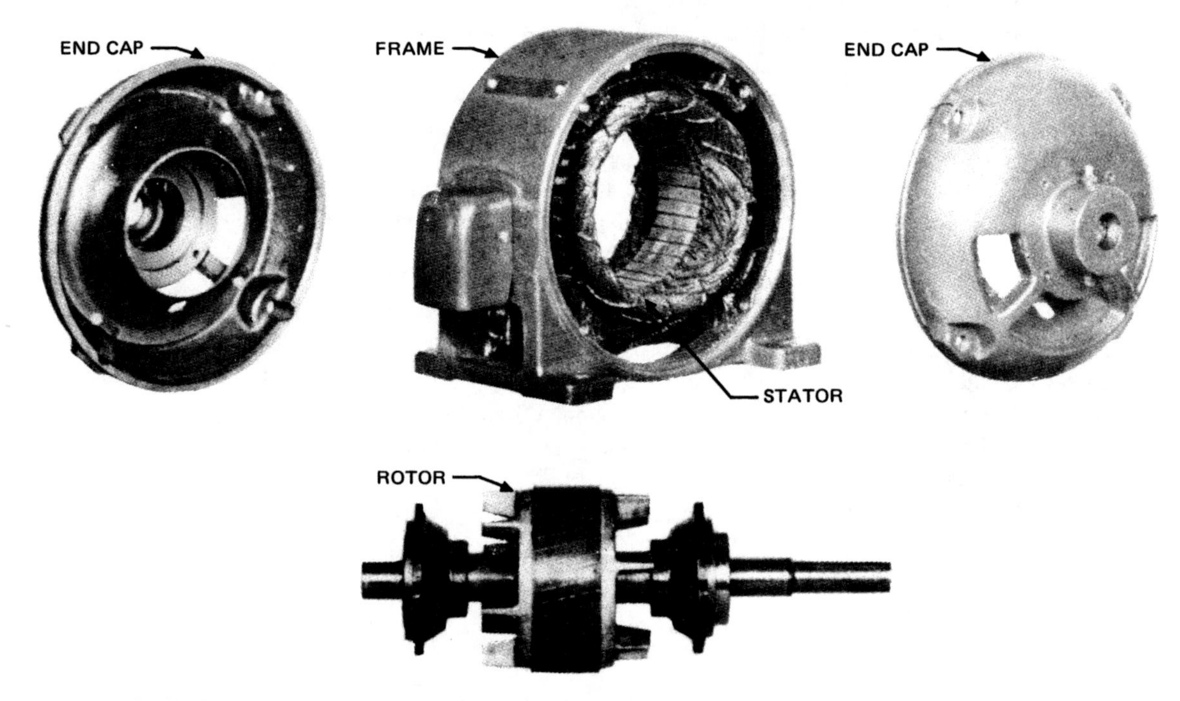

FIGURE 16–1 Essential parts of a squirrel-cage induction motor (*Delmar/Cengage Learning*)

FIGURE 16–2 Cutaway view of a squirrel-cage rotor (*Delmar/ Cengage Learning*)

and have aluminum end rings that are cast in one piece. The rotor shaft is supported by bearings housed in the end shields.

THE ROTATING MAGNETIC FIELD

The principle of operation for all three-phase motors is the rotating magnetic field. There are three factors that cause the magnetic field to rotate:

1. The fact that the voltages of a three-phase system are 120° out of phase with each other
2. The fact that the three voltages change polarity at regular intervals
3. The arrangement of the stator windings around the inside of the motor

Figure 16–3A shows three ac voltages 120° out of phase with each other, and the stator winding of a three-phase motor. The stator illustrates a two-pole, three-phase motor.

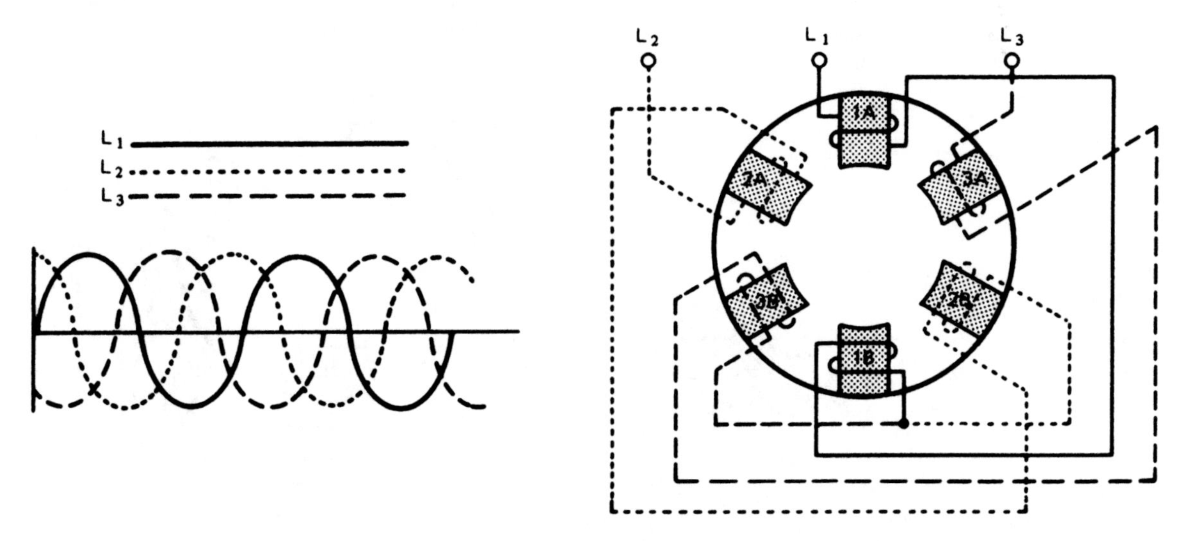

FIGURE 16–3A (*Delmar/Cengage Learning*)

Two-pole means that there are two poles per phase. AC motors seldom have actual pole pieces as shown in Figure 16–3A, but they will be used here to aid in understanding how the rotating magnetic field is created in a three-phase motor. Notice that pole pieces 1A and 1B are located opposite each other. The same is true for poles 2A and 2B, and 3A and 3B. Pole pieces 1A and 1B are wound with wire that is connected to phase 1 of the three-phase system. Notice also that the pole pieces are wound in such a manner that they will always have opposite magnetic polarities. If pole piece 1A has a north magnetic polarity, pole piece 1B will have a south magnetic polarity at the same time.

The windings of pole pieces 2A and 2B are connected to line 2 of the three-phase system. The windings of pole pieces 3A and 3B are connected to line 3 of the three-phase system. These pole pieces are also wound in such a manner as to have the opposite polarity of magnetism.

To understand how the magnetic field rotates around the inside of the motor, refer to Figure 16–3B. Notice that a line labeled A has been drawn through the three voltages of the system. This line is used to illustrate the condition of the three voltages at this point in time. The arrow drawn inside the motor indicates the greatest strength of the magnetic field at the same point in time. It is to be assumed that the arrow is pointing in the direction of the north magnetic field. Notice in Figure 16–3B that phase 1 is at its maximum positive peak and that phases 2 and 3 are less than maximum. The magnetic field is, therefore, strongest between pole pieces 1A and 1B.

In Figure 16–3C, line B indicates that the voltage of line 3 is zero. The voltage of line 1 is less than maximum positive, and line 2 is less than maximum negative. The magnetic field at this point is concentrated between the pole pieces of phases 1 and 2.

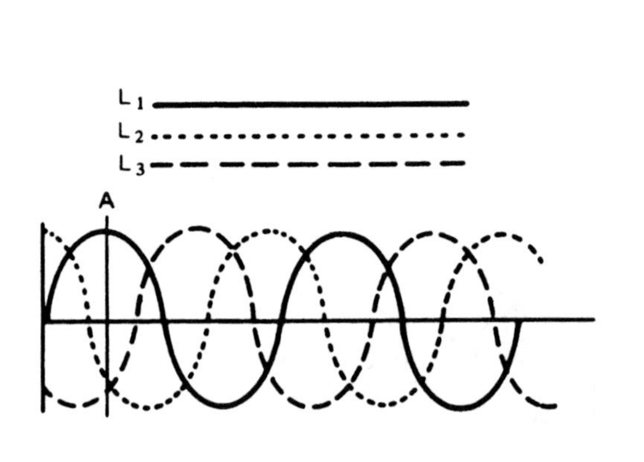

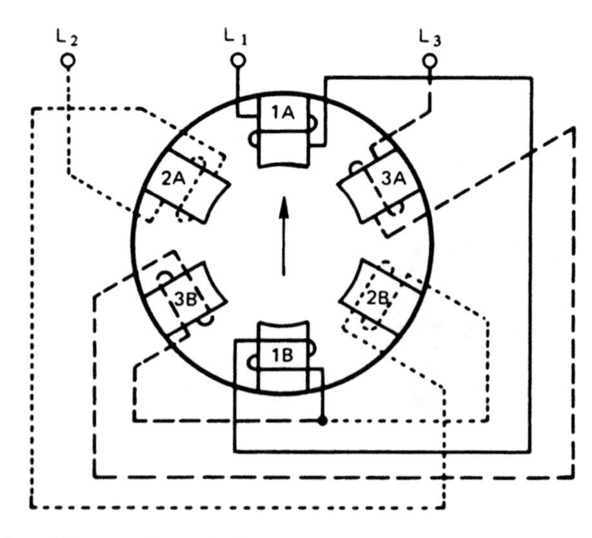

FIGURE 16–3B (*Delmar/Cengage Learning*)

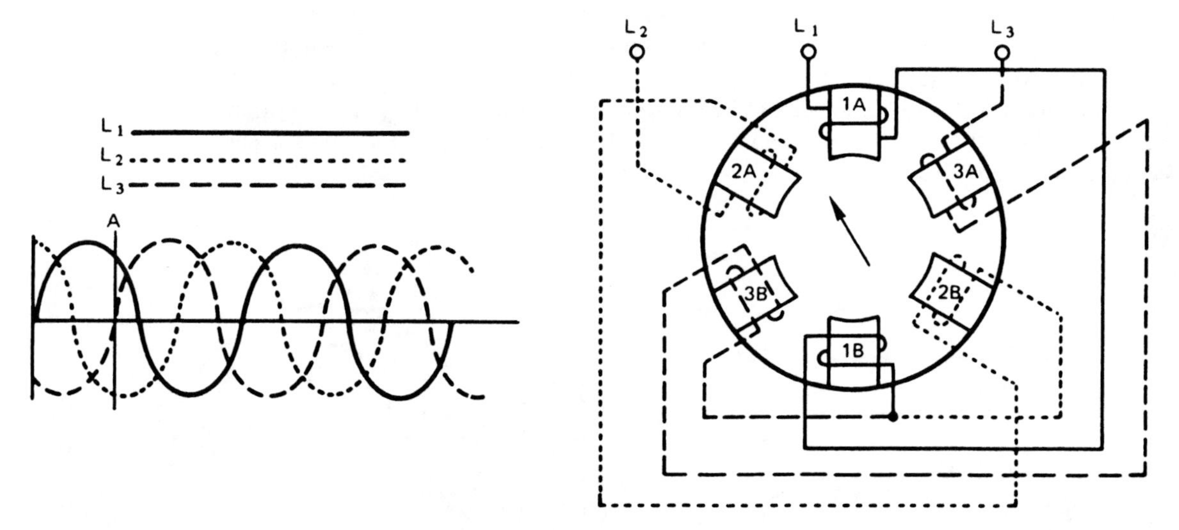

FIGURE 16–3C (*Delmar/Cengage Learning*)

In Figure 16–3D, line C indicates that line 2 is at its maximum negative peak and that lines 1 and 3 are less than maximum positive. The magnetic field at this point is concentrated between pole pieces 2A and 2B.

In Figure 16–3E, line D indicates that line 1 is zero. Lines 2 and 3 are less than maximum and in opposite directions. At this point in time, the magnetic field is concentrated between the pole pieces of phase 2 and phase 3.

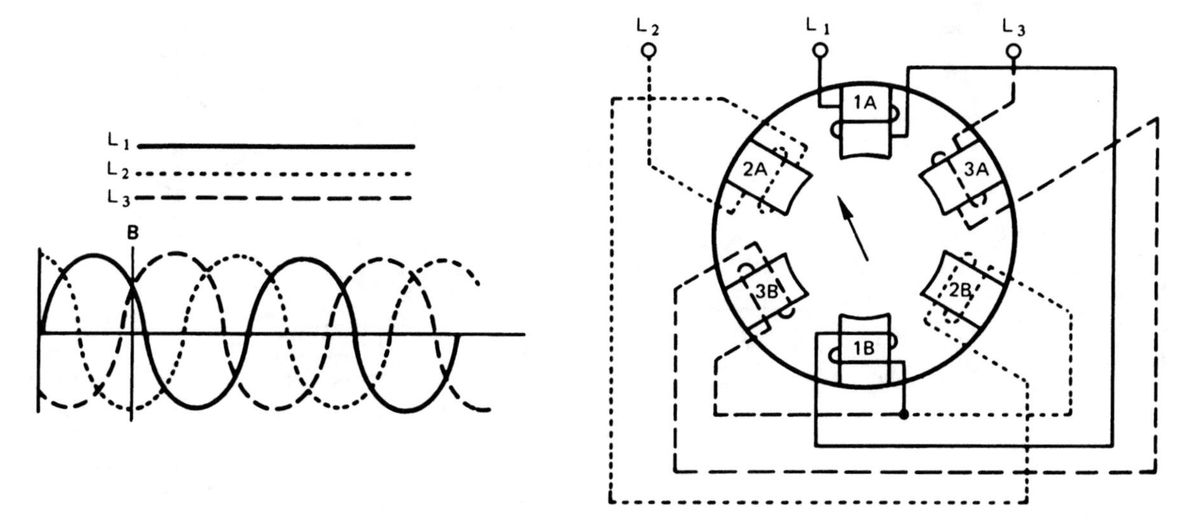

FIGURE 16–3D (*Delmar/Cengage Learning*)

In Figure 16–3F, line E indicates that phase 3 is at its maximum positive peak, and lines 1 and 2 are less than maximum and in the opposite direction. The magnetic field at this point is concentrated between pole pieces 3A and 3B.

In Figure 16–3G, line F indicates that phase 2 is zero. Line 3 is less than maximum positive, and line 1 is less than maximum negative. The magnetic field at this time is concentrated between the pole pieces of phase 1 and phase 3.

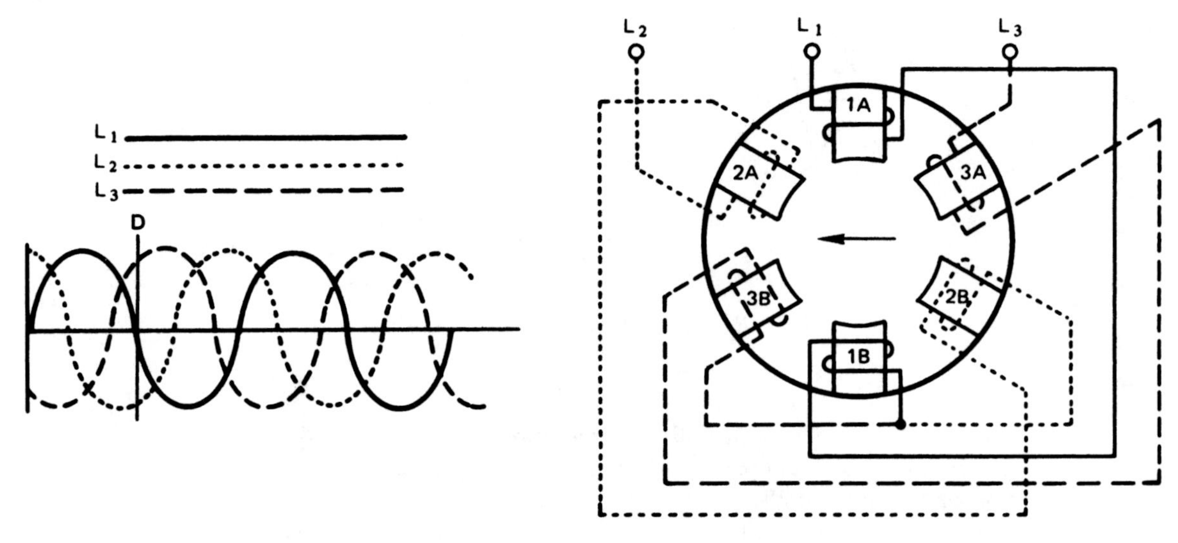

FIGURE 16–3E *(Delmar/Cengage Learning)*

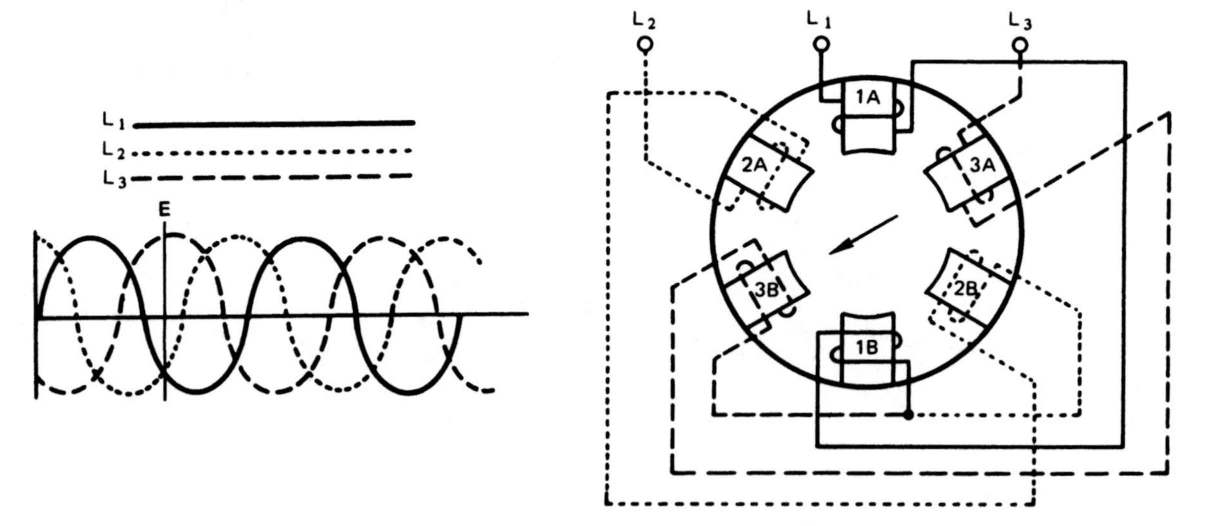

FIGURE 16–3F *(Delmar/Cengage Learning)*

In Figure 16–3H, line G indicates that phase 1 is at its maximum negative peak, and phases 2 and 3 are less than maximum and in the opposite direction. Notice that the magnetic field is again concentrated between pole pieces 1A and 1B. This time, however, the magnetic polarity is reversed because the current has reversed in the stator winding.

In Figure 16–3I, line H indicates that phase 2 is at its maximum positive peak and phases 1 and 3 are less than maximum and in the negative direction. The magnetic field is concentrated between pole pieces 2A and 2B.

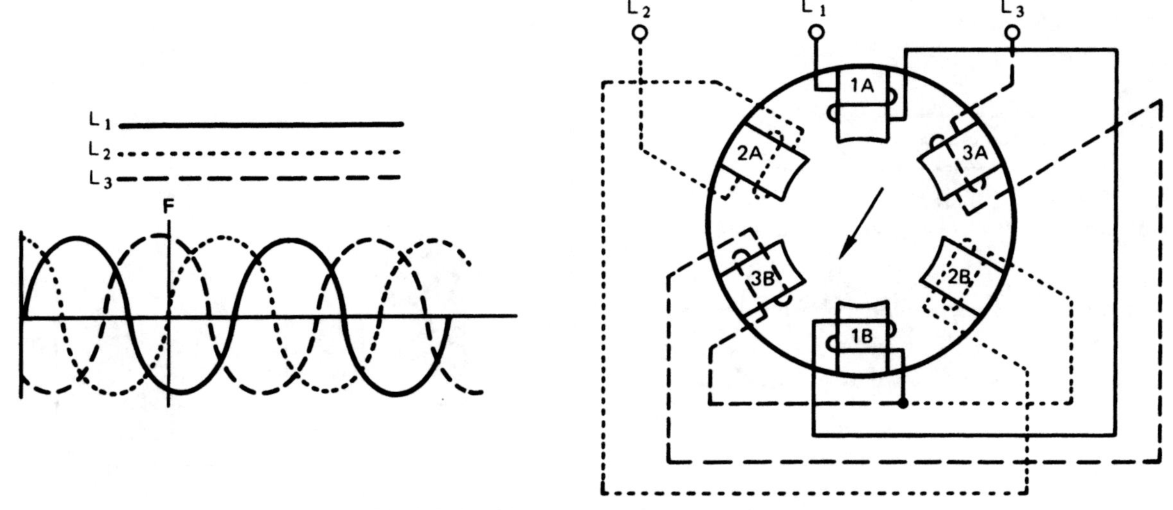

FIGURE 16–3G (*Delmar/Cengage Learning*)

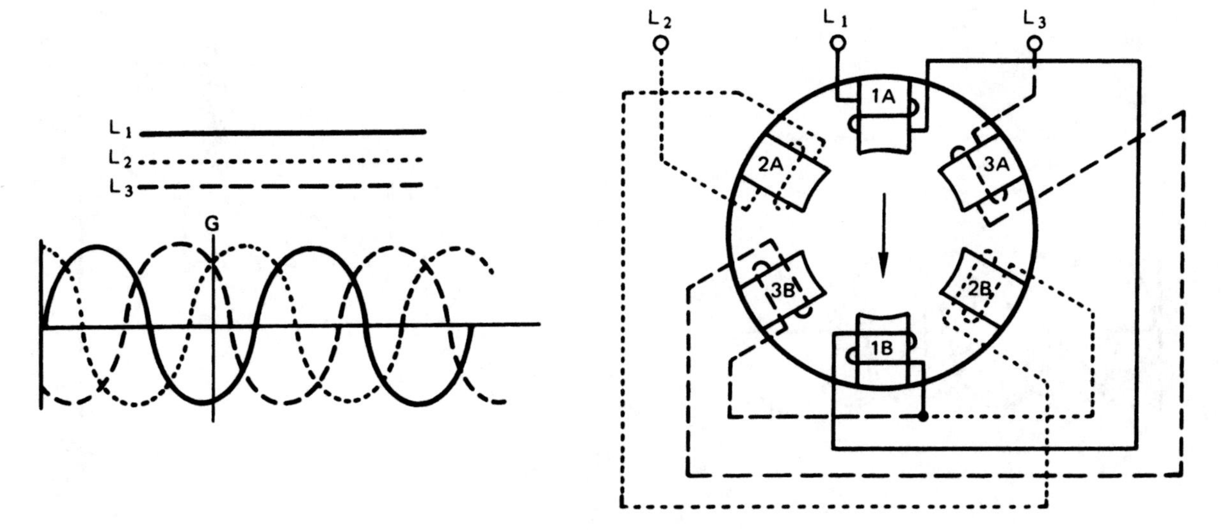

FIGURE 16–3H (*Delmar/Cengage Learning*)

In Figure 16–3J, line I indicates that phase 3 is maximum negative, and phases 1 and 2 are less than maximum in the positive direction. The magnetic field at this point is concentrated between pole pieces 3A and 3B.

In Figure 16–3K, line J indicates that phase 1 is at its positive peak, and phases 2 and 3 are less than maximum and in the opposite direction. The magnetic field is again concentrated between pole pieces 1A and 1B. Notice that in one complete cycle of the three-phase voltage, the magnetic field has rotated 360° around the inside of the stator winding.

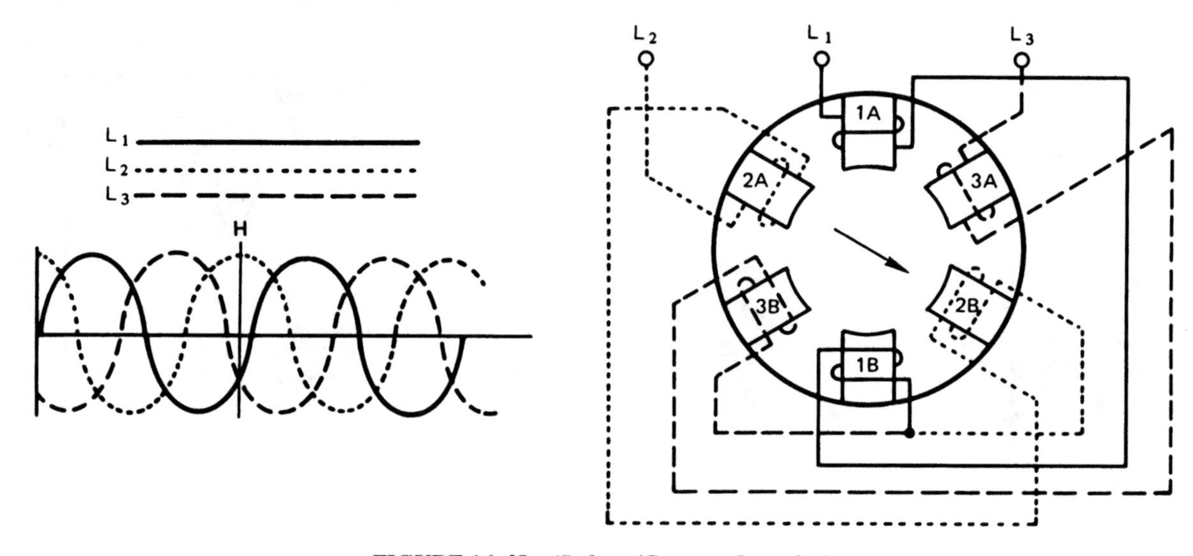

FIGURE 16–3I (*Delmar/Cengage Learning*)

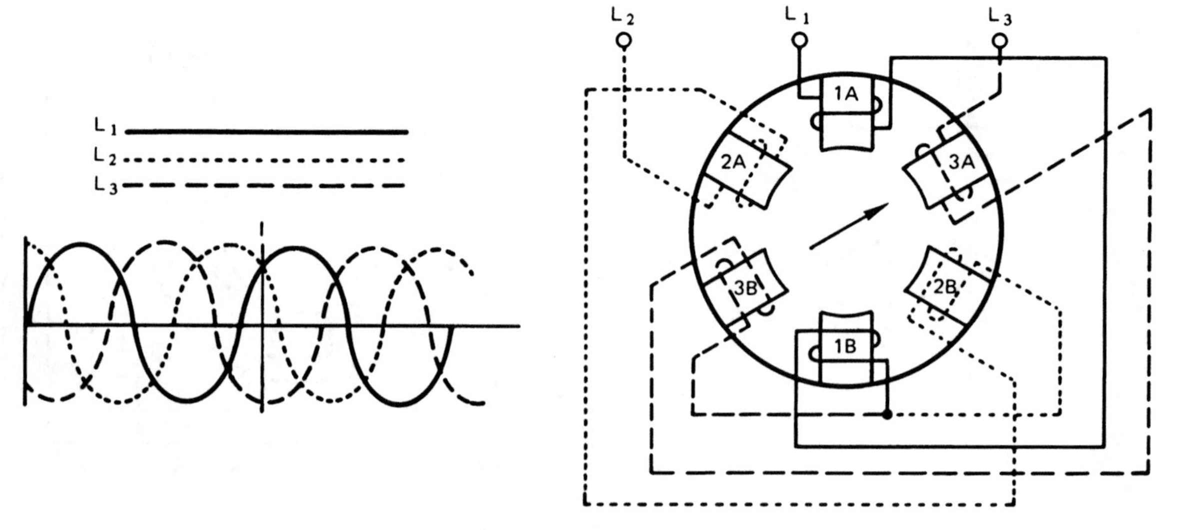

FIGURE 16–3J (*Delmar/Cengage Learning*)

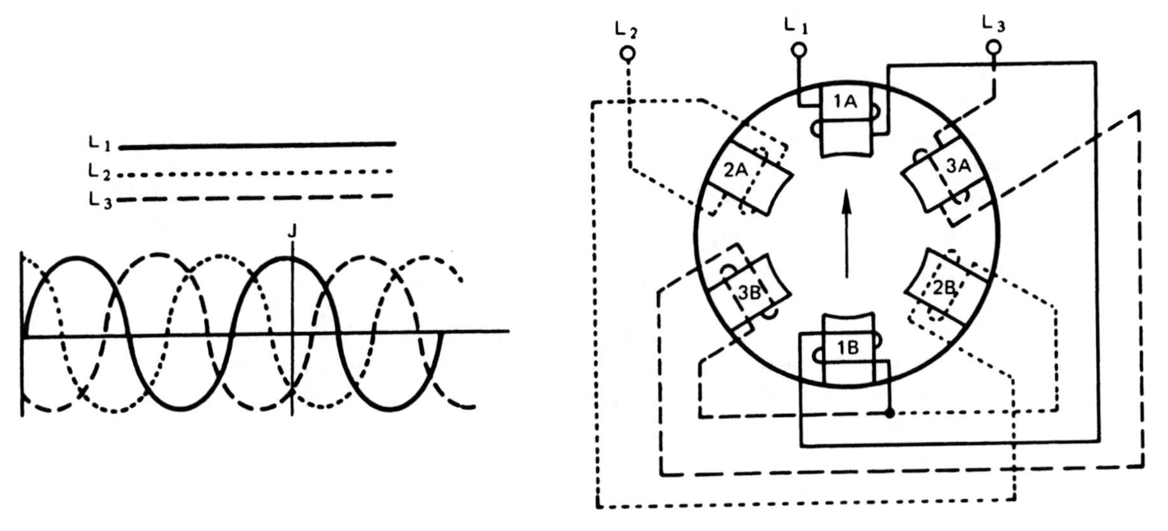

FIGURE 16–3K (*Delmar/Cengage Learning*)

If any two of the stator leads is connected to a different line, the relationship of the voltages will change and the magnetic field will rotate in the opposite direction. The direction of rotation of a three-phase motor can be reversed by changing any two stator leads.

SYNCHRONOUS SPEED

The speed at which the magnetic field rotates is known as the *synchronous speed*. The synchronous speed of a three-phase motor is determined by two factors:

1. The number of stator poles
2. The frequency of the ac line

Because 60 Hz is a standard frequency throughout the United States and Canada, the following gives the synchronous speeds for motors with different numbers of poles:

2 poles	3600 r/min
4 poles	1800 r/min
6 poles	1200 r/min
8 poles	900 r/min

The synchronous speed can be determined using the formulas

$$\mathbf{f} = \frac{\mathbf{P \times S}}{\mathbf{120}}$$

or

$$S = \frac{120\,f}{P}$$

where

S = synchronous speed in r/min
f = frequency in Hz
P = number or poles

Speed Performance

The field set up by the stator windings cuts the copper bars of the rotor. Voltages induced in the squirrel-cage winding set up currents in the rotor bars. As a result, a field is created on the rotor core. The attraction between the stator field and the rotor field causes the rotor to follow the stator field. The rotor always turns at a speed that is slightly less than that of the stator field (less than the synchronous speed). In this way, the stator field cuts the rotor bars and induces the necessary rotor voltages and currents to create the rotor field.

The torque produced by an induction motor results from the interaction between the *stator flux* and the *rotor flux*. If the rotor is turned at the same speed as the stator field, there will be no relative motion between the rotor bars and the stator field. This means that no torque can be produced. A torque is produced *only* when the rotor turns at a speed that is less than synchronous speed. At no load, the mechanical losses of the motor can be overcome by a small torque. The rotor speed will be slightly less than the synchronous speed of the stator field.

As a mechanical load is applied to the motor shaft, the rotor speed will decrease. The stator field turns at a constant synchronous speed and cuts the rotor bars at a faster rate per second. The voltages and currents induced in the rotor bars increase accordingly, causing a greater induced rotor voltage. The resulting increase in the rotor current causes a large torque at a slightly lower speed.

The squirrel-cage winding was described as consisting of heavy copper bars welded to two end rings. The impedance of this winding is relatively low. Therefore, a slight decrease in the speed causes a large increase in the currents in the rotor bars. Because the rotor circuit of a squirrel-cage induction motor has a low impedance, the speed regulation of this motor is very good.

PERCENT SLIP

The speed performance of squirrel-cage induction motors is measured in terms of *percent slip*. In determining percent slip, the synchronous speed of the stator field is used as a reference point. The synchronous speed for a particular motor is constant, because the number of poles and the frequency remain the same. *Slip* is the number of revolutions per minute by which the rotor falls behind the speed of the rotating field of the stator. Slip is determined by subtracting the speed of the rotor from the synchronous speed of the stator field. For example, a three-phase, two-pole induction motor has a full-load speed of 3480 r/min. The synchronous speed of the stator field is

$$S = \frac{120\,f}{P} = \frac{120 \times 60}{2} = 3600 \text{ r/min}$$

Slip = 3600 − 3480 = 120 r/min (number of revolutions per minute by which the rotor speed falls behind the speed of the stator field)

The slip is usually expressed as a percentage, using the synchronous speed as a reference point. For the motor in this example, the percent slip is

$$\text{Percent slip} = \frac{\text{synchronous speed} - \text{rotor speed}}{\text{synchronous speed}} \times 100$$

$$= \frac{120 \times 100}{3600} = 3.33\%$$

Smaller values of percent slip mean that the motor has better speed regulation. When determined at the rated load, the percent slip of most squirrel-cage induction motors varies from 2% to 5%. This type of motor is considered to be a constant-speed motor because there is a small decrease in the speed between the no-load and full-load points.

ROTOR FREQUENCY

In the previous example, the rotor slips behind the speed of the stator field by 120 revolutions per minute. The flux of the two stator poles passes a given rotor bar of the squirrel-cage winding only 120 times every minute. Thus, the voltages and currents induced in the rotor will have a very low frequency. The rotor frequency is

Method 1: $$f = \frac{P \times S_R}{120}$$

$$= \frac{2 \times 120}{120} = 2 \text{ Hz}$$

where

f = rotor frequency, in hertz
P = poles
S_R = percent slip

Unit 1 showed that when a conductor passes a pair of unlike poles, one cycle (Hz) of voltage is induced in the conductor. In this example, a pair of stator poles passes a given bar in the squirrel-cage rotor 120 times per minute or twice per second. Thus, the frequency must be 2 Hz. If the slip is increased, the rotor frequency will increase, because the flux of the revolving field will cut a given bar in the squirrel-cage winding more times per second. This relationship can be expressed as a formula

Method 2: $$f_R = f_S \times S_R$$

$$= 2 \text{ Hz}$$

$$= 60 \times 0.0333$$

where

f_R = rotor frequency, in hertz

f_S = stator frequency, in hertz

S_R = percent slip, changed to a decimal value

Note that methods 1 and 2 both give the same frequency for the rotor. This frequency is an important factor in the operation of the motor. A change in the rotor frequency causes a change in the inductive reactance component ($X_L = 2\pi fL$) of the rotor impedance. Thus, a change in the frequency will affect the starting and running characteristics of the motor.

PROBLEM 1

Statement of the Problem

A 5-hp, 220-V, three-phase, 60-Hz, squirrel-cage induction motor has eight poles. At the rated load, it has a speed of 870 r/min. Determine

1. the synchronous speed.
2. the percent slip, at the rated load.
3. the rotor frequency, at the instant of start-up.
4. ssthe rotor frequency, at the rated load.

Solution

1. The synchronous speed of the stator field is found using the frequency formula transposed to solve for S:

$$S = \frac{120 \times f}{P} = \frac{120 \times 60}{8} = 900 \text{ r/min}$$

2. At the rated load, the percent slip is

$$\text{Percent slip} = \frac{\text{synchronous speed} - \text{rotor speed}}{\text{synchronous speed}} \times 100$$

$$= \frac{900 - 870}{900} \times 100 = 3.33\%$$

3. At start-up, the rotor is not turning. The slip at this instant is unity, or 100%. Therefore, the rotor frequency and the stator frequency are both 60 Hz.

4. The rotor turns at 870 r/min at the rated load. The rotor frequency at this speed can be determined as follows:

Method 1:

$$\text{Slip, in r/min} = \text{synchronous speed} - \text{rotor speed}$$

$$= 900 - 870 = 30 \text{ r/min}$$

$$\text{Rotor frequency} = \frac{P \times S}{120} = \frac{8 \times 30}{120} = 2 \text{ Hz}$$

Method 2:

$$\text{Rotor frequency} = \text{stator frequency} \times \text{percentage slip}$$

$$= 6 \times 0.033 = 2 \text{ Hz}$$

TORQUE AND SPEED CHARACTERISTICS

The torque produced by an induction motor depends on the strengths of the stator and rotor fields and the phase relationship between the fields:

$$\mathbf{T = K_T \times \phi_S \times I_R \times \cos \theta_R}$$

where

T = torque, in pound • feet
K_T = torque constant
ϕ_S = stator flux (constant at all speeds)
I_R = rotor current (a function of slip)
$\cos \theta_R$ = rotor power factor (a function of slip)

This torque formula is similar to the formula for the torque of a dc motor: $T = k \times \theta_f \times I_A$. The difference between the formulas is in the $\cos \theta_R$ function. The equivalent diagram of the rotor is an inductor and resistor in series. Because rotor frequency is a function of slip, it follows that $\cos \theta_R$ varies with slip.

STARTING CHARACTERISTICS

At the instant the motor is started, the rotor is not turning and there is 100% slip. The rotor frequency at this moment is equal to the stator frequency. The inductive reactance of the rotor is very large compared to the effective resistance component. Also, the rotor has a very low lagging power factor. This means that the rotor flux lags the stator flux by a large phase angle. As a result, the interaction between the two fields is small and the starting torque is low.

As the speed of the motor increases, the percent slip and the frequency of the rotor decrease. The decrease in the rotor frequency causes the inductive reactance and the impedance of the rotor to decrease. Thus, the phase angle between the stator and rotor fluxes is reduced. The torque then increases to its maximum value at about 20% slip. As the rotor continues to accelerate, the torque decreases until it reaches the value required to turn the mechanical load applied to the motor shaft. The slip at this point is between 2% and 5%.

Starting Current

At start-up the stator field cuts the rotor bars at a faster rate than when the rotor is turning. The large voltage induced in the rotor causes a large rotor current. As a result, the stator current will also be high at start-up. The squirrel-cage induction motor resembles a static transformer during this brief instant. That is, the stator may be viewed as the primary or input winding, and the squirrel-cage rotor winding as the secondary winding.

Most three-phase, squirrel-cage induction motors are started with the rated line voltage applied directly to the motor terminals. This means that the starting surge of current reaches a value as high as three to five times the full-load current rating of the motor. This high starting current requires induction motors to have starting protection. This protection may be rated as high as three times the full-load current rating of the motor. In some instances, very large induction motors are started with auxiliary starters. These devices reduce the motor voltage at start-up to limit the starting surge of current. As a result, there is less voltage disturbance on the feeder circuit supplying the motor load.

Starting with Reduced Voltage

There are problems in starting a large induction motor with a reduced voltage. For example, assume that the voltage applied to the motor terminals at start-up is reduced to 50% of the rated nameplate voltage. The magnetizing flux of the stator is also reduced to half of the normal value. The voltages and currents induced in the rotor are similarly reduced by half. The resulting torque output of the motor is reduced to one-fourth of its original value. Figure 16–4 shows that a 50% reduction in voltage causes the torque to decrease to 25% of its normal value.

The torque formula given previously shows why the large reduction in the torque output occurs. Both the stator flux (ϕ_S) and the rotor current (I_R) are reduced to half of their original values. This means that the product (torque) of these terms is only one-fourth of its original value. For a given value of slip, the torque varies as the square of the impressed voltage.

As explained previously, an increase in slip increases the rotor frequency and the inductive reactance of the rotor. In the normal operating range of the motor from no load to full load, the rotor frequency seldom is greater than 2 to 3 Hz. Therefore, a change in the frequency has negligible effects on the impedance of the rotor at full load, and even at 125% of the rated load.

Motor with Overload

When a motor has a heavy overload, the percent slip will increase, causing an increase in the rotor frequency. The increased frequency causes an increase in the inductive reactance and the impedance of the rotor circuit. Two effects result from the increase in the inductive reactance of the rotor circuit. First, the power factor of the rotor decreases, causing the rotor current to lag the induced rotor voltage. The rotor field flux will not reach its maximum value until the peak value of the stator flux wave has passed it. Although the currents in the stator and rotor circuits increase because of the overload, the fluxes of the stator and the rotor fields are out of phase with each other. Therefore, there is less interaction between the fields and the torque decreases. The second effect is that the increase in the inductive reactance and the impedance of the rotor decrease the rate at which the rotor current increases with an increase in slip. Because of these two effects, the torque increase will be less rapid. The torque reaches its maximum value at approximately 20% slip in the typical squirrel-cage induction motor.

Breakdown Torque

In Figure 16–4, note that the torque curve increases as a straight line well beyond the rated load. As the percent slip increases between 10% and 20%, there is a reduction in the rate at which the torque increases. Finally, at approximately 20% slip, the torque reaches its maximum value. The point of the maximum torque output is called the *breakdown point.* An increase in the load beyond this point results in less torque being developed by the motor and the rotor stops. As shown in the figure, this breakdown point is reached between 200% and 300% of the rated torque.

The following example shows that for a given value of slip, the torque varies as the square of the impressed voltage. Assume that a 240-V, squirrel-cage motor is operated on a 208-V, three-phase circuit. The value of 208 V is 87% of the rated voltage of the motor. The torque output is $0.87^2 = 0.75$. This means that the breakdown torque is reduced to 75% of its rated value.

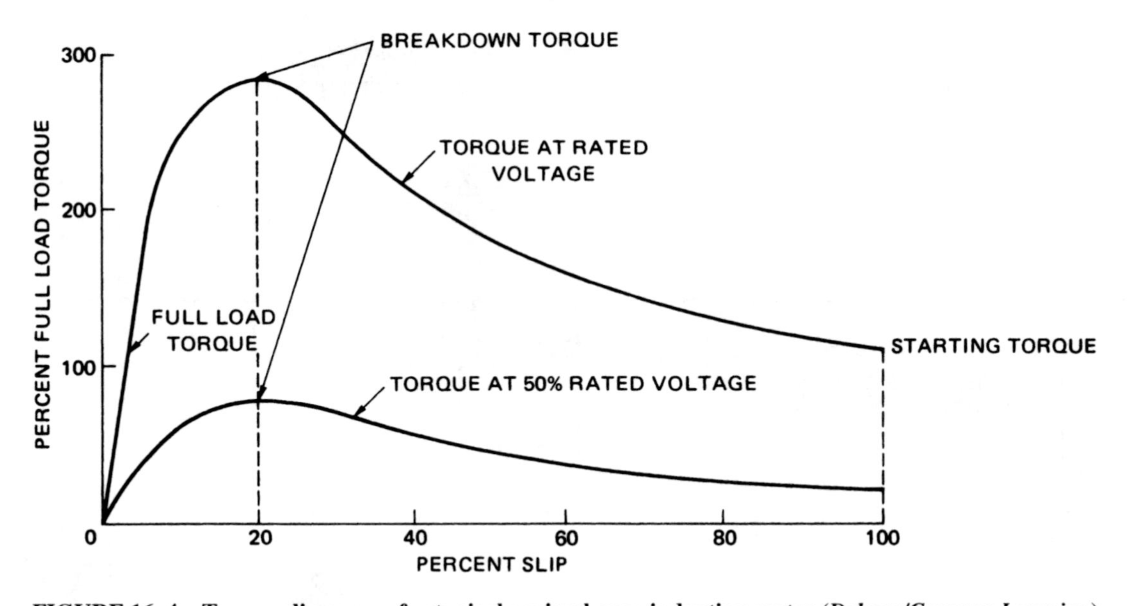

FIGURE 16–4 Torque-slip curves for typical squirrel-cage induction motor (*Delmar/Cengage Learning*)

POWER FACTOR

A squirrel-cage induction motor operating at no load has a low power factor in the range of 10% to 15% lag. The current input to an induction motor at no load consists of a large component of quadrature magnetizing current and a very small component of in-phase current to supply the losses. The resulting power factor angle between the coil voltage and the coil current for each phase winding is large (Figure 16–5A).

As the load on the motor increases, the in-phase current supplied to the motor increases. At the same time, there is little change in the magnetizing component. The resultant current is more nearly in phase with the voltage, as shown in Figure 16–5B. A smaller angle of lag results in a higher power factor. At the rated load, the power factor may be as high as 85% to 90% lagging.

SUMMARY OF OPERATING CHARACTERISTICS

The three-phase, squirrel-cage induction motor operates at a relatively constant speed from no load to full load. The rotor has a very low impedance. As a result, only a slight decrease in speed will cause a large increase in the rotor current. This current develops the necessary torque to turn the increased load. The percent slip at no load is less than 1%. At full load, the percent slip is usually between 3% and 5%. A squirrel-cage induction motor is considered to be a constant-speed motor because of this small change in percent slip from no load to full load. The slip increases as a straight-line characteristic, as shown in Figure 16–6. The rotor current will likewise increase in practically a direct proportion. Thus, the torque increases as a straight-line characteristic.

Figure 16–6 shows the characteristic curves of a 5-hp, three-phase, 220-V, 60-Hz, four-pole, squirrel-cage induction motor. The speed, percent slip, torque, efficiency, and power factor curves are included in this figure.

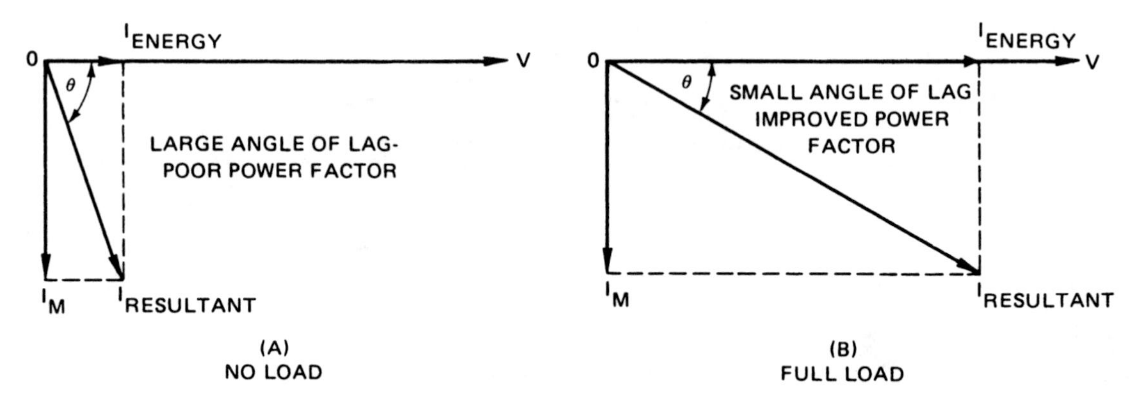

FIGURE 16–5 Power factor at no load and full load (*Delmar/Cengage Learning*)

Induction Motor Losses

The losses in an induction motor consist of the stray power losses and the copper losses. The stray power losses include mechanical friction losses, windage losses, and iron losses. These losses are nearly constant at all load points and are often called *fixed losses*.

The copper losses consist of the I^2R losses in the windings of the motor. An increase in the load causes the current to increase in the motor windings. As a result, the I^2R losses increase. At light loads, the percent efficiency is low because the fixed losses are a large part of the input. As the load on the motor increases, the losses become a smaller part of the input. Thus, the efficiency increases to its maximum value. However, when the rated capacity of the motor is exceeded, the copper losses become excessive and the efficiency decreases.

The efficiency of an ac induction motor is given by the following equation:

$$\eta = \frac{\text{output, watts}}{\text{input, watts}} \times 100$$

$$= \frac{\text{input, watts} - (\text{copper losses} + \text{stray power losses})}{\text{input, watts}} \times 100$$

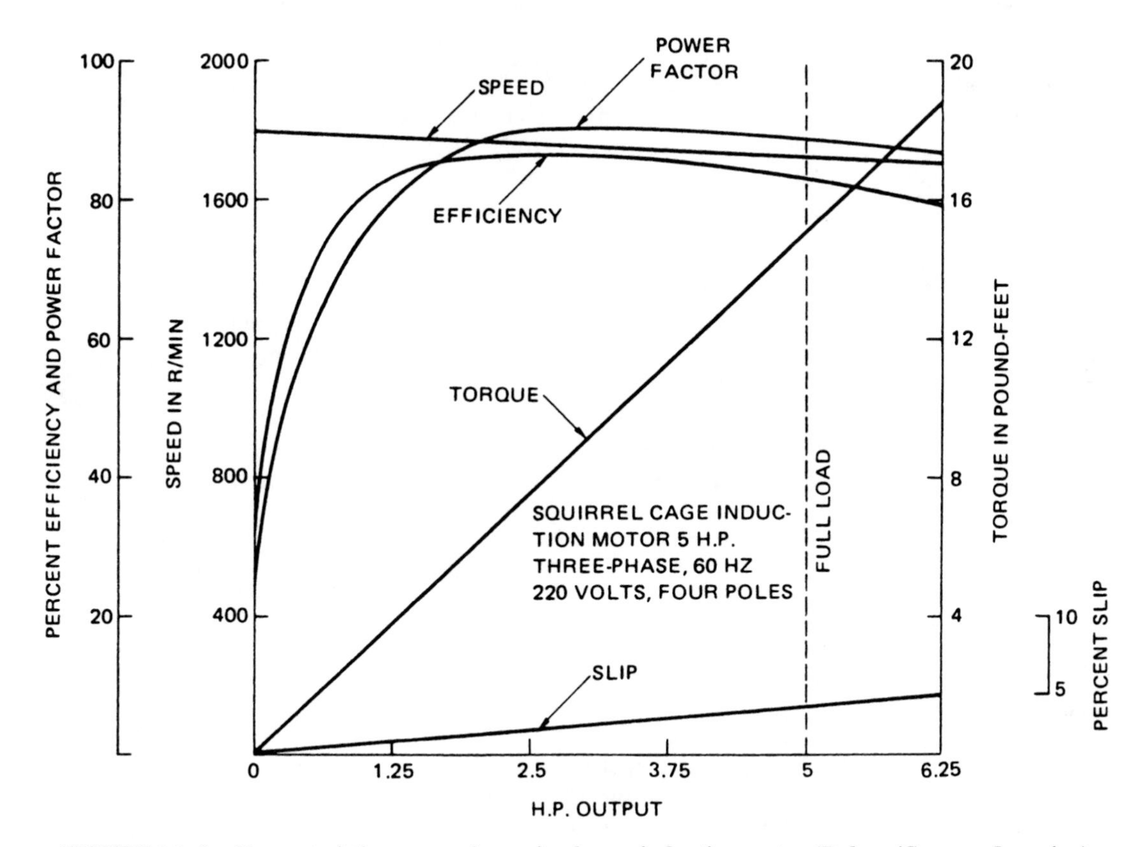

FIGURE 16–6 Characteristic curves of a squirrel-cage induction motor (*Delmar/Cengage Learning*)

Power Factor of Induction Motor

The power factor curve in Figure 16–6 shows a value at no load of approximately 0.15 lag. The no-load current consists mainly of magnetizing current. This current produces the mmf required to send the stator flux across the airgap and through the magnetic circuit. The in-phase component of the no-load current is low because the losses are small. Therefore, the no-load current lags the voltage by a large phase angle and the power factor is low. As the load on the motor increases, the in-phase current component supplied to the motor increases, the phase angle decreases, and the power factor increases. In practice, the power factor of the inductive motor at the rated load is between 0.85 and 0.90 lag.

Advantages of Induction Motor

The squirrel-cage induction motor has several advantages:

- It has excellent speed regulation with a small percent slip. This means that it is ideal for constant-speed applications.
- The motor is simple in construction and requires little maintenance or repairs.
- Brushes and slip rings are not required. Thus, it can be used in locations such as chemical plants and flour and lumber mills where there is the possibility of explosions due to arcing.

The major disadvantagse of the squirrel-cage induction motor is that there is no practical method of providing stepless speed control. Thus, this motor cannot be used in applications where variable speeds are required.

PROBLEM 2

Statement of the Problem

A test is made on a squirrel-cage induction motor when it delivers the rated load output. The motor is delta-connected and is a 10-hp, three-phase machine. The two-wattmeter method is used to obtain the following values:

Line voltage = 220 V
Line current = 28 A
Wattmeter 1 = +6500 W
Wattmeter 2 = +2500 W

At the rated load, determine

1. the percent efficiency.
2. the power factor.

Solution

1. The percent efficiency is the ratio of the output in watts to the input in watts. The rated nameplate output is 10 hp. This value must be expressed in watts. The total input to the motor is the sum of the two wattmeter readings:

$$\eta = \frac{\text{output, watts}}{\text{input, watts}} \times 100 = \frac{\text{hp} \times 746}{W_1 + W_2} = 100$$

$$= \frac{10 \times 746}{6500 + 2500} = 100 = 82.9\%$$

2. The power factor is the ratio of the true power to the apparent power. This ratio is expressed by the following formula. This formula can be used for both delta- and wye-connected three-phase loads:

$$PF = \frac{W1 + W2}{\sqrt{3} \times V_{line} \times I_{line}}$$

$$= \frac{6500 + 2500}{1.73 \times 220 \times 28} = 0.8445 \text{ lag}$$

PROBLEM 3

Statement of the Problem

A three-phase, 220-V, 60-Hz, six-pole, squirrel-cage induction motor takes 13.4 A per terminal at the rated load. The power factor is 0.88 lag. The efficiency is 83%. The percent slip at the rated load is 4%.

1. What is the full-load speed?
2. What is the rotor frequency?
3. What is the horsepower output at the rated load?
4. Determine the torque in pound • feet at the rated load and the rated speed.

Solution

1. The synchronous speed of this motor is

$$S = \frac{120\ f}{P} = \frac{120 \times 60}{6} = 1200 \text{ r/min}$$

$$\text{Percent slip} = \frac{\text{synchronous speed} - \text{rotor speed}}{\text{synchronous speed}} \times 100$$

$$0.04 = \frac{1200 - X}{1200}$$

$$48 = 1200 - X$$

$$X = 1152 \text{ r/min}$$

2. The rotor frequency is

$$f = \frac{P \times F}{120} = \frac{6 \times 48}{120} = 2.4 \text{ Hz}$$

3. The horsepower output at the rated load can be determined once the true power input to the motor is known. The percent efficiency given in the problem is used to determine the output in watts. This value is then converted to horsepower:

$$\text{Watts} = \sqrt{3} \times V_{line} \times I_{line} \times \cos \angle\theta$$

$$= 1.73 \times 220 \times 13.5 \times 0.88 = 4522 \text{ W, input to motor}$$

$$\eta = \frac{\text{output, watts}}{\text{input, watts}} = 0.83 = \frac{X}{4522}$$

$$X = 3753 \text{ watts output}$$

$$\text{hp output} = \frac{\text{output, watts}}{746} = \frac{3753}{746} = 5.0 \text{ hp}$$

4. In the text *Direct Current Fundamentals*, a formula was given for determining the horsepower output by the prony brake method:

$$\text{hp output} = \frac{2\pi \times \text{torque} \times \text{r/min}}{33{,}000}$$

The prony brake method is not commonly used to determine the horsepower output of electric motors. However, the formula can be simplified and transposed to give the torque output of a motor at a given load point if the speed in r/min is known. The transposed formula used to find torque in pound • feet is

$$\text{Torque, in pound} \bullet \text{feet} = \frac{\text{hp output}}{0.00019 \times \text{r/min}}$$

$$= \frac{5}{0.00019 \times 1152} = 22.8 \text{ lb} \bullet \text{ft}$$

where the constant $2\pi/33{,}000$ is given as 0.00019.

PROBLEM 4

Statement of the Problem

A three-phase induction motor is connected to a 480-V, 60-Hz power source. An ammeter indicates a current of 54 amperes supplying the motor. A three-phase wattmeter indicates a true power of 29 kW for the motor.

1. Determine the power factor of the motor.
2. Determine the amount of capacitance necessary to correct the power factor to 95%. It is to be assumed that the capacitors will be connected in parallel with the motor and the capacitors will be wye-connected.
3. Determine the amount of current the circuit should draw after the power factor has been corrected to 95%.

Solution

1. Determine the power factor of the motor.
   Determine the apparent power:

$$VA = E_{line} \times I_{line} \times \sqrt{3}$$

$$VA = 480 \times 54 \times 1.73$$

$$VA = 44{,}841.6$$

$$PF = \frac{P}{VA}$$

$$PF = \frac{29{,}000}{44{,}841.6}$$

$$PF = 0.647 \quad \text{or} \quad 64.7\%$$

2. Determine the capacitance necessary to correct the power factor to 95%.

To determine the amount of capacitance needed, it is first necessary to determine the apparent power at a power factor of 95%.

$$VA = \frac{P}{PF}$$

$$VA = \frac{29{,}000}{0.95}$$

$$VA = 30{,}526.3$$

To determine the amount of capacitance necessary to correct the power factor, subtract the present apparent power from the desired apparent power. This will indicate the capacitive VARs necessary to correct the power factor to 95%.

$$VARs_C = 44{,}841.6 - 30{,}526.3$$
$$VARs_C = 14{,}315.3$$

Determine the amount of capacitive current necessary to produce 14,315.3 capacitive VARs.

$$I_C = \frac{VARs_C}{E_L \times \sqrt{3}}$$

$$I_C = \frac{14{,}315.3}{480 \times 1.73}$$

$$I_C = 17.2 \text{ A}$$

Determine the amount of capacitive reactance necessary to produce a current of 36.8 amperes. Because the capacitors form the phases of a wye connection, the voltage across the capacitors will be the phase value of the wye connection, which is less than the line voltage by a factor of 3, or 277 volts.

$$X_C = \frac{E_{phase}}{I_C}$$

$$X_C = \frac{277}{17.2}$$

$$X_C = 16.1\ \Omega$$

Determine the amount of capacitance necessary to produce 16.1 Ω of capacitive reactance.

$$C = \frac{1}{2\pi f X_C}$$

$$C = \frac{1}{377 \times 16.1}$$

$$C = 0.000165 \text{ farads} \quad \text{or} \quad 165\ \mu F$$

3. Determine the amount of current at a power factor of 95%.

$$I_{line} = \frac{VA}{E_{line} \times 1.73}$$

$$I_{line} = \frac{30{,}526.3}{830.4}$$

$$I_{line} = 36.8 \text{ amperes}$$

PROCEDURE FOR MEASURING MOTOR LOSSES

On many occasions, it is necessary to measure the losses of a squirrel-cage induction motor to determine the motor output and efficiency. The two types of losses for this motor are the copper losses and the fixed losses. The output of the motor is equal to the input of the motor minus the losses.

For example, assume that a test is made of a three-phase, squirrel-cage induction motor. The motor is rated at 5 hp, 220 V, 13.3 A, 1735 r/min, and 60 Hz. The motor has four poles and is wye-connected.

Equivalent Resistance

The first step is to obtain the equivalent resistance of the motor windings. The three-phase power and current input to the motor are measured with the rotor stationary. A very low value of three-phase voltage, at the rated frequency, is applied to the test circuit. This voltage is increased until the rated current is indicated by the three ammeters. This test is called the *blocked rotor test*. The data from this test are given in Table 16–1.

Line Voltage Volts			Line Current Amperes			Watts	
AB	BC	CA	A	B	C	Wattmeter 1	Wattmeter 2
48	48	48	13.3	13.3	13.3	555	−5

TABLE 16–1 Data for blocked rotor test

The total power taken by the motor is

$$\text{Total power} = W_1 - W_2 = 555 - 5 = 550 \text{ watts}$$

Note that the rated current of 13.3 A is achieved at only 48 V, three phase. The core losses are negligible at this low voltage. Therefore, it is assumed that the total power input of 550 W is used to supply the copper losses.

It is now possible to find the "equivalent ac resistance" per phase for the three wye-connected windings. The power loss in one phase winding is I^2R. For the three windings, the power loss is $P = 3\ I^2R$, where W is the power taken by the blocked rotor motor and R is the equivalent resistance per phase:

$$P = 3\ (I^2R)$$

$$550 = 3 \times 13.3^2 \times R$$

$$R = 1.036\ \Omega$$

It is simple to calculate the copper losses for any load current knowing the effective resistance of each single-phase winding.

Stray Power Losses

The stray power losses include mechanical friction losses, windage losses, and iron losses. To measure these losses, the motor is operated at no load with the rated voltage applied to the motor terminals. The data shown in Table 16–2 are the result of this part of the test.

Line Voltage Volts			Line Current Amperes			Watts		Speed
AB	BC	CA	A	B	C	Wattmeter 1	Wattmeter 2	r/min
220	220	220	4.2	4.2	4.2	515	−255	1795

TABLE 16–2 Motor operated at no load

The following procedure is used to determine the fixed losses or the stray power losses.

Total power input at no load:

$$P_T = W_1 + W_2 = 515 - 255 = 260 \text{ W}$$

Copper losses at no load:

$$P = 3 \times I^2 \times R = 3 \times 4.2^2 \times 1.036 = 54.8, \quad \text{or} \quad 55 \text{ W}$$

The fixed losses at no load are obtained by subtracting the no load copper losses from the power input at no load:

$$\text{Stray power losses} = 260 - 55 = 205 \text{ W}$$

The stray power losses can be measured at no load only. It is assumed that these losses remain nearly constant from no load to full load. This assumption can be made because the speed of a squirrel-cage motor remains almost constant from no load to full load. Also, the magnetizing current is nearly constant throughout the load range of the motor.

Total Losses, Output, and Efficiency

The motor can now be loaded to any desired load point to determine the total losses, the output, and the efficiency. The data obtained with the motor operating at full load are given in Table 16–3.

Line Voltage Volts			Line Current Amperes			Watts		Speed
AB	BC	CA	A	B	C	Wattmeter 1	Wattmeter 2	r/min
220	220	220	13.3	13.3	13.3	2880	1620	1795

TABLE 16–3 Motor operating at full load

The total power input to the motor, at the rated load, is

$$W_1 + W_2 = 2880 + 1620 = 4500 \text{ W}$$

The stray power losses are assumed to be the same as those found at no load, or 205 W.

The copper losses increase greatly at the rated load because the current in each winding of the wye-connected stator is 13.3 A. As determined previously, the actual copper losses are 550 W.

The output is equal to the input minus the total losses. Therefore, if the copper losses and the fixed losses are subtracted from the power input, the result is the power output of the motor:

$$\begin{aligned}\text{Output, watts} &= \text{input, watts} - (\text{copper losses} + \text{stray power losses})\\ &= 4500 - (550 + 205) = 3745 \text{ W}\end{aligned}$$

Total power input at no load:

$$P_T = W_1 + W_2 = 515 - 255 = 260 \text{ W}$$

Copper losses at no load:

$$P = 3 \times I^2 \times R = 3 \times 4.2^2 \times 1.036 = 54.8, \quad \text{or} \quad 55 \text{ W}$$

The fixed losses at no load are obtained by subtracting the no-load copper losses from the power input at no load.

The actual efficiency of this motor at the rated load is

$$\eta = \frac{\text{input, watts} - (\text{copper losses} + \text{stray power losses})}{\text{input, watts}} \times 100$$

$$= \frac{4500 - (550 + 205)}{4500} \times 100$$

$$= \frac{3745}{4500} \times 100 = 83.2\%$$

The full-load speed of this motor is 1735 r/min. At this load point, the torque output in pound • feet can be obtained as follows:

$$\text{hp output} = 0.00019 \text{ T} \times \text{S}$$

$$5 = 0.00019 \times \text{T} \times 1735$$

$$\text{T} = 15.2 \text{ lb} \cdot \text{ft}$$

The discussion on motor losses does not include a detailed study of the variations of certain factors that were assumed to be constants. A more complete study of motor losses can be obtained by consulting electrical engineering texts.

SPEED CONTROL

The synchronous speed of the magnetic field of the stator is determined by the number of stator poles and the frequency of the ac source. Generally, it is not possible to vary the speed of an induction motor by changing its frequency. (Nearly all motors are operated from an ac source having a fixed frequency.) In a few applications, a single alternator may supply one or two motors. In these cases, the frequency can be changed by varying the speed of the prime mover. As a result, the motor speed changes.

Squirrel-cage motors may be provided with special stator windings. When these windings are reconnected by special switch controls, different numbers of stator poles are formed. In this way, different speeds are obtained from a squirrel-cage induction motor connected to an ac source having a fixed frequency.

One type of multispeed squirrel-cage motor is designed for two synchronous speeds, where one speed is twice the other. This two-speed motor has one stator winding. A switch control device is used with this winding to provide two synchronous speeds. For example, speeds of 900 r/min and 1800 r/min may be obtained, or 1800 r/min and 3600 r/min.

CODE LETTER IDENTIFICATION

Squirrel-cage rotors are not all the same. Rotors are made with different types of bars. The type of rotor bars used in the construction of the rotor determines the operating characteristics of the motor. AC squirrel-cage motors are given a code letter on their nameplate. These code letters should not be confused with the NEMA code letter found on motors manufactured since 1999. The code letter indicates the type of bars used in the rotor. Figure 16–7A shows a rotor with type A bars. A type A rotor has the highest resistance of any squirrel-cage rotor. This means that the starting torque will be high per ampere of starting current because the rotor current is closer to being in phase with the induced voltage than any other type of rotor. Also, the high resistance of the rotor bars limits the amount of current flow in the rotor when starting. This produces a low starting current for the motor. A rotor with type A bars has very poor running characteristics, however. Because the bars are resistive, a large amount of voltage will have to be induced into the rotor to produce an increase in rotor current and, therefore, an increase in the rotor magnetic field. This means that when load is added to the motor, the rotor must slow down a great amount to produce enough current in the rotor to increase the torque. Motors with type A rotors have the highest percent slip of any squirrel-cage motor. Motors with type A rotors are generally used in applications where starting is a problem, such as a motor that must accelerate a large flywheel from 0 r/min to its full speed. Flywheels can have a very large amount of inertia, which may require several minutes to accelerate them to their running speed when they are stated.

Figure 16–7B shows a rotor with bars similar to those found in rotors with code letters B through E. These rotor bars have lower resistance than the type A rotor. Rotors of this type have fair starting torque, low starting current, and fair speed regulation.

Figure 16–7C shows a rotor with bars similar to those found in rotors with code letters F through V. This rotor has low starting torque per ampere of starting current. The starting current is high, and these motors exhibit good running torque. Motors containing rotors of this type generally have very good speed regulation and low percent slip. It should be noted that although motors with rotors that fall into this range exhibit poor starting torque per ampere of starting current, the starting torque is still greater than the amount of running torque. These motors do not generally exhibit difficulty starting unless the load requires some time to reach normal speed. The extended time will cause the high starting current to overheat the windings.

The Double-Squirrel-Cage Rotor

Some motors use a rotor that contains two sets of squirrel-cage windings (Figure 16–8). The outer winding consists of bars with a relatively high resistance located close to the top of the iron core. Because these bars are located close to the surface, they have a relatively low reactance. The inner winding consists of bars with a large cross-sectional area that gives them a low resistance. The inner winding is placed deeper in the core material, which causes it to have a much higher reactance.

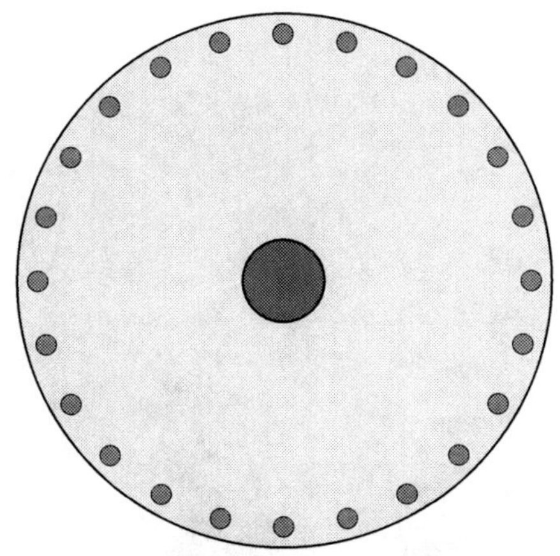

This type of rotor has a high starting torque per ampere of starting current, and low starting current. It is constructed with small rotor bars located near the surface of the rotor. It is used in motors that power metal shears, punch presses, and metal drawing equipment.

INDUCTION MOTORS WITH CODE LETTER A

(a)

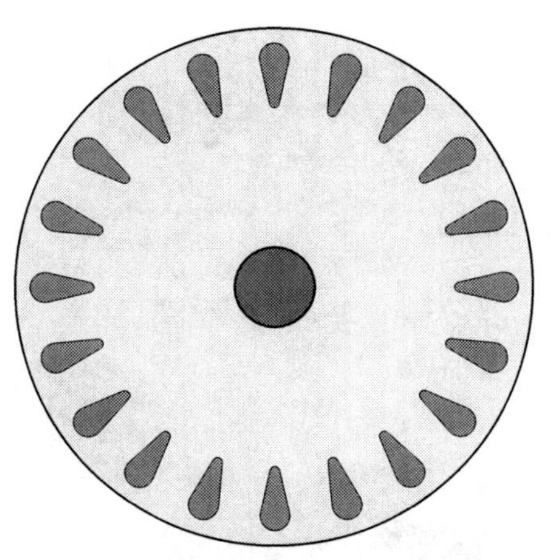

This type of rotor exhibits high reactance and low resistance. Motors with rotors of this type have relatively low starting current and fair starting torque per ampere of starting current. They are generally used for motor generator sets, blowers, centrifugal pumps, and other applications that do not require high starting torque.

INDUCTION MOTORS WITH CODE LETTERS B–E

(b)

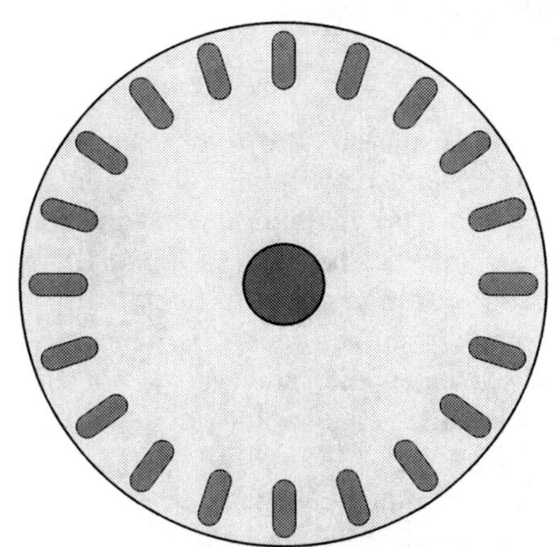

This type of rotor has relatively low resistance and low inductive reactance. It exhibits high starting current and relatively poor starting torque per ampere of starting current. Motors using rotors that fall into this range are used in motor generator sets, pumps, blowers, and other applications that do not require starting for extended periods of time.

INDUCTION MOTORS WITH CODE LETTERS F–V

(c)

FIGURE 16–7 Various types of squirrel-cage rotors (*Delmar/Cengage Learning*)

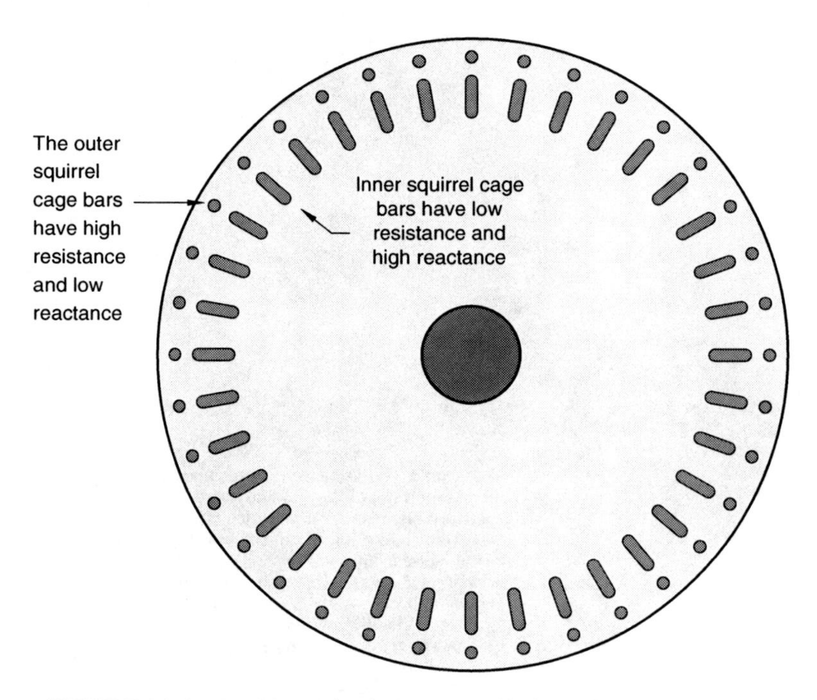

FIGURE 16–8 Double-squirrel-cage rotor (*Delmar/Cengage Learning*)

When the double-squirrel-cage motor is started, the rotor frequency is high. Because the inner winding is inductive, its impedance will be high as compared to the resistance of the outer winding. During this period of time, most of the rotor current flows through the outer winding. The resistance of the outer winding limits the current flow through the rotor, which limits the starting current to a relatively low value. Because the current is close to being in phase with the induced voltage, the rotor flux and stator flux are close to being in phase with each other, and a strong starting torque is developed. The starting torque of a double-squirrel-cage motor can be as high as 250% of rated full-load torque.

When the rotor reaches its full-load speed, rotor frequency decreases to 2 or 3 Hz. The inductive reactance of the inner winding has now decreased to a low value. Most of the rotor current now flows through the low-resistance inner winding. This type motor has good running torque and excellent speed regulation.

MOTOR NAMEPLATE DATA

The data contained on the typical nameplate for a squirrel-cage induction motor include the following items: horsepower rating, full-load speed, full-load amperes, voltage, number of phases, frequency, frame number, permissible temperature rise, model number, manufacturer name, locked-rotor ampere code letter, service factor, and NEMA code letter. The NEMA code letter should not be confused with the code letter for locked-rotor amperes. The NEMA code letter is used to determine the fuse or circuit breaker

Mod. No. XXXXXXX		Manufacturer Name
PH 3	Hp1/3	Volts 230/460
Frame 56	Amps 1.3/.9	RPM 1725
NEMA Design B	SF 1.35	Max AMB 40°C
LR kVA Code L	Duty Cont.	Hz 60

FIGURE 16–9 Typical motor nameplate (*Delmar/Cengage Learning*)

size for the motor circuit. The locked-rotor amperes code letter describes the type of bars in the rotor and is used in conjunction with *NEC® Section 430* to determine the starting current of the motor. A typical nameplate for a squirrel-cage induction motor is shown in Figure 16–9.

STATOR WINDING CONNECTIONS

Many of the three-phase motors used in industry are designed to be operated on two voltages, such as 240 or 480 V. Motors of this type contain two sets of windings per phase. Most dual voltage motors bring out nine T leads at the terminal box. There

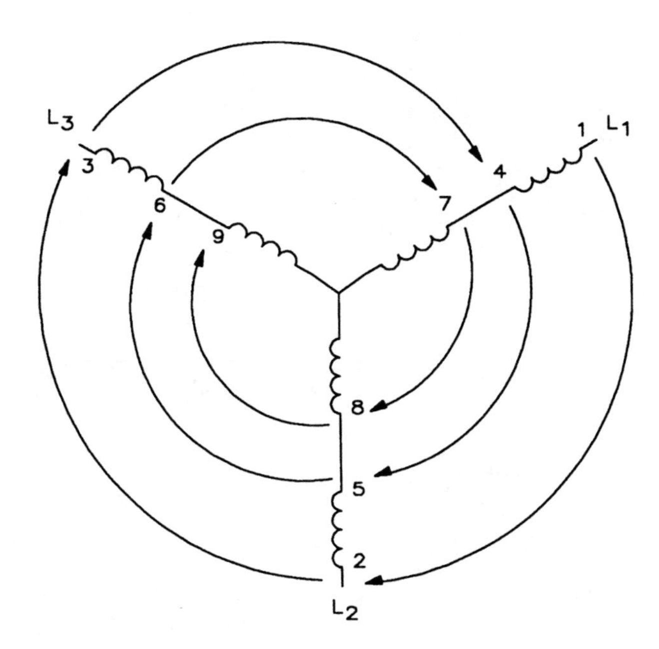

FIGURE 16–10 Leads are numbered in a spiral. (*Delmar/ Cengage Learning*)

is a standard method used to number these leads, as shown in Figure 16–10. Starting with terminal 1, the leads are numbered in a decreasing spiral. Another method of determining the proper lead numbers is to add three to each terminal. For example, starting with lead 1, add three to one. Three plus one equals four. The phase winding that begins with 1 ends with 4. Now add three to four. Three plus four equals seven. The beginning of the second winding for phase one is seven. This method will work for the windings of all phases. If in doubt, draw a diagram of the phase windings and number them in a spiral.

Three-phase motors can be constructed to operate in either wye or delta. If a motor is to be connected to high voltage, the phase windings will be connected in series. In Figure 16–11, a schematic diagram and terminal connection chart for high voltage are

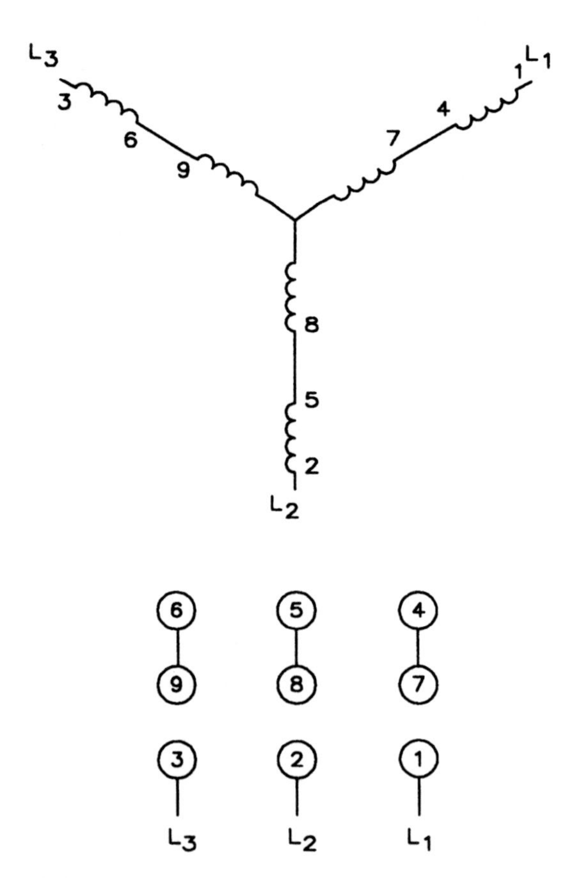

FIGURE 16–11 High-voltage wye connection (*Delmar/Cengage Learning*)

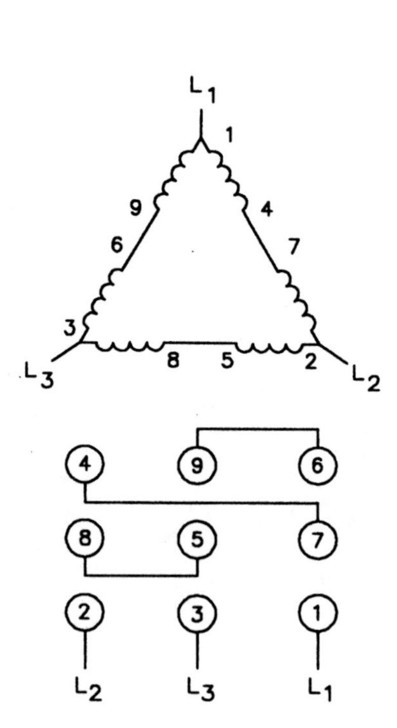

FIGURE 16–12 High-voltage delta connection (*Delmar/Cengage Learning*)

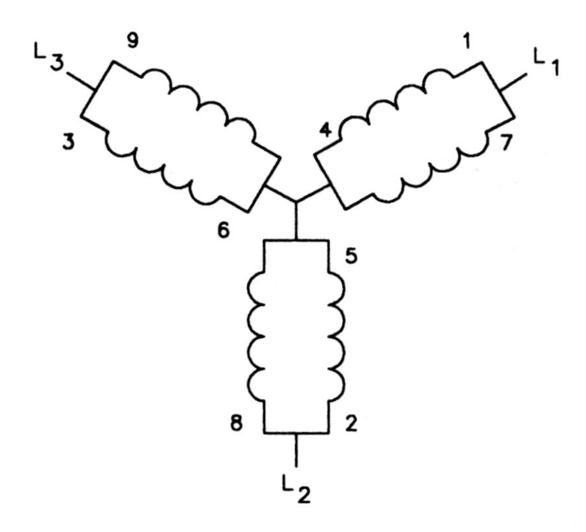

FIGURE 16–13 Stator windings connected in parallel (*Delmar/Cengage Learning*)

shown for a wye-connected motor. In Figure 16–12, a schematic diagram and terminal connection chart for high voltage are shown for a delta-connected motor.

When a motor is to be connected for low-voltage operation, the phase windings must be connected in parallel. Figure 16–13 shows the basic schematic diagram for a wye-connected motor with parallel phase windings. In actual practice, however, it is not possible to make this exact connection with a nine-lead motor. The schematic shows that terminal 4 connects to the other end of the phase windings that starts with terminal 7. Terminal 5 connects to the other end of winding 8, and terminal 6 connects to the other end of winding 9. In actual motor construction, the opposite ends of windings 7, 8, and 9 are connected together inside the motor and are not brought outside the motor case. The problem is solved, however, by forming a second wye connection by connecting terminals 4, 5, and 6 together, as shown in Figure 16–14.

The phase winding of a delta-connected motor must also be connected in parallel for use on low voltage. A schematic for this connection is shown in Figure 16–15. A connection diagram and terminal connection for this hookup is shown in Figure 16–16.

Some dual-voltage motors will contain twelve T leads instead of nine. In this instance, the opposite ends of terminals 7, 8, and 9 are brought out for connection. Figure 16–17 shows the standard numbering for both delta- and wye-connected motors. Twelve leads are brought out if the motor is intended to be used for wye–delta starting. When this is the case, the motor must be designed for normal operation with its windings connected in delta. If the windings are connected in wye during starting, the starting current of the motor is greatly reduced.

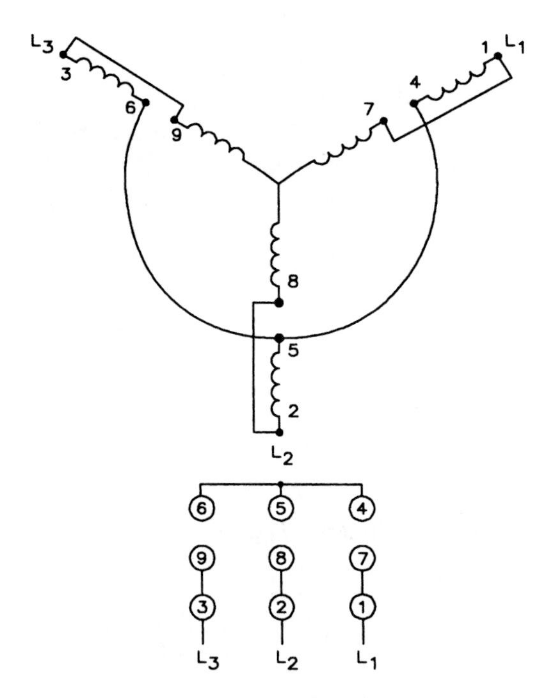

FIGURE 16–14 Low-voltage wye connection (*Delmar/Cengage Learning*)

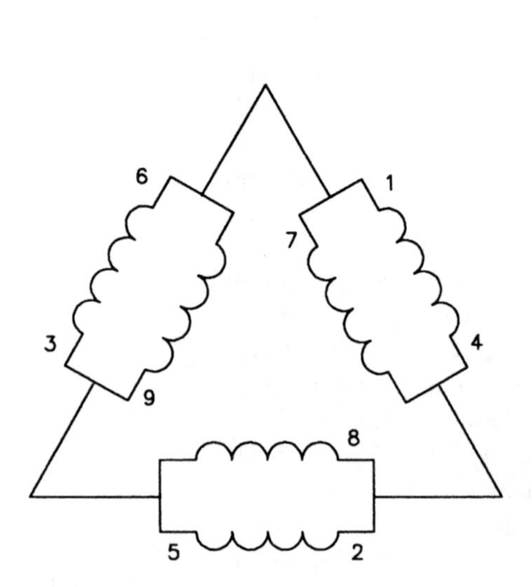

FIGURE 16–15 Parallel delta connection (*Delmar/Cengage Learning*)

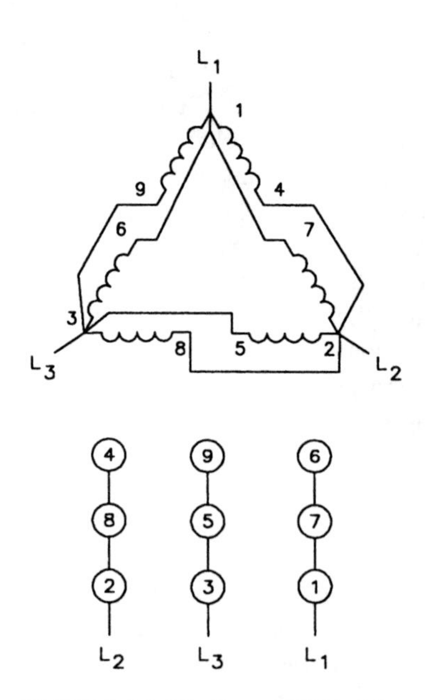

FIGURE 16–16 Low-voltage delta connection (*Delmar/Cengage Learning*)

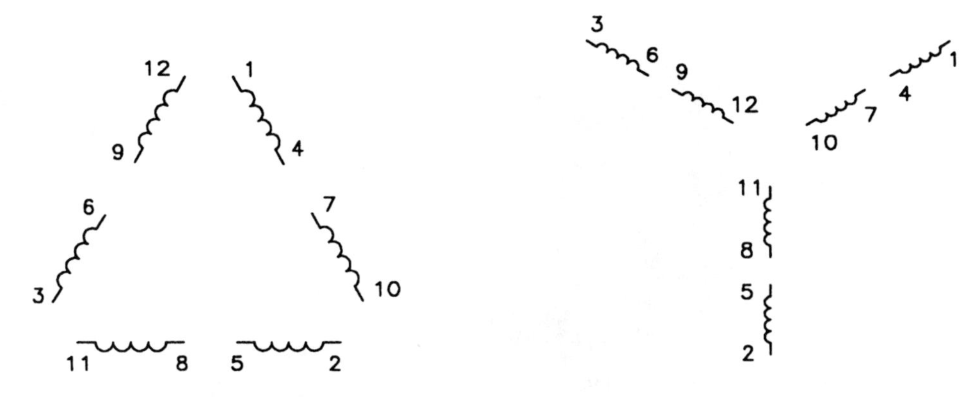

FIGURE 16–17 Twelve-lead motor (*Delmar/Cengage Learning*)

SINGLE-PHASE OPERATION OF THREE-PHASE INDUCTION MOTORS

A line wire feeds the stator windings of a three-phase induction motor. If this wire is opened, the motor will operate as a single-phase induction motor. It will not have enough torque to start when energized from a single-phase source. However, if the three-phase motor is running when the break in the line wire occurs, it will continue to operate with a greatly reduced capacity. If the rated load is applied to the motor when it is operating as a single-phase motor, it will overheat. The insulation of the windings may be damaged as a result.

The three-phase motor will not start on a single phase because the induced voltage and currents in the rotor set up a magnetic field in the rotor. This field opposes the stator field. (This situation is an application of Lenz's law.) The rotor current produces a rotor field in which the rotor poles are centered with the stator field poles, as shown in Figure 16–18. As a result, there is no torque in either a clockwise or counterclockwise direction.

If the three-phase motor is operating at the rated speed when the single-phase condition develops, the rotor continues to turn. The moving rotor cuts the stator field flux and causes induced voltages and currents in the rotor bars. The rotor currents create a rotor field with poles midway between the stator poles. The rotor has high-reactance and low-resistance components. Therefore, the rotor current will lag behind the induced voltage in the rotor by nearly 90%. As a result, the rotor and stator fields are practically 90% out of phase with each other. The rotor current produces magnetic polarities that are 90° out-of-phase with those produced in the stator. The motor continues to operate due to attraction and repulsion of magnetism in the same manner as a single-phase induction motor. The three-phase motor will continue to run, but at reduced capacity. Once the motor has been stopped, it cannot restart because a rotating magnetic field cannot be produced with a single phase.

FIGURE 16–18 The rotor develops magnetic poles 90° out of phase with the stator magnetic poles. The rotor continues to turn due to attraction and repulsion of magnetism. (*Delmar/Cengage Learning*)

THE WOUND-ROTOR INDUCTION MOTOR

Many industrial applications require three-phase motors with variable-speed control. The basic squirrel-cage induction motor is a constant speed motor. Thus, another type of induction motor is required for variable-speed applications. The wound-rotor induction motor meets these needs.

The wound-rotor induction motor has nearly the same stator construction and winding arrangement as the squirrel-cage induction motor. Figure 16–19 shows a typical stator for a wound-rotor induction motor.

FIGURE 16–19 Wound stator (*Delmar/Cengage Learning*)

A wound rotor is shown in Figure 16–20. The cylindrical core of the rotor is made up of steel laminations. Slots are cut into the cylindrical core to hold the formed coils for three single-phase windings. These windings are placed 120 electrical degrees apart. The insulated coils of the rotor winding are grouped to form the same number of poles as in the stator windings. The three single-phase rotor windings are connected in wye.

Three leads from these windings terminate at three slip rings mounted on the rotor shaft. Carbon brushes ride on these slip rings and are held securely by adjustable springs mounted in the brush holders. The brush holders are fixed rigidly, because it is not necessary to vary their position. Leads from the carbon brushes are connected to an external speed controller.

Principle of Operation

When the stator windings of a wound-rotor motor are energized from a three-phase source, the rotating magnetic field formed travels around the inside of the stator core, just as in a squirrel-cage induction motor. The speed of the rotating magnetic field depends on the number of stator poles and the frequency of the source. The formula used to find the synchronous speed for squirrel-cage induction motors can also be used for this type of motor:

$$S = 120\,f \div P$$

As the rotating field travels at the synchronous speed, it cuts the wound-rotor windings and induces voltages in these windings. The induced voltages set up currents that form a closed-circuit path from the rotor windings through the slip rings and brushes to a wye-connected speed controller. Figure 16–21 shows a wound-rotor induction motor connected to a wye-connected speed controller.

FIGURE 16–20 Wound rotor (*Delmar/Cengage Learning*)

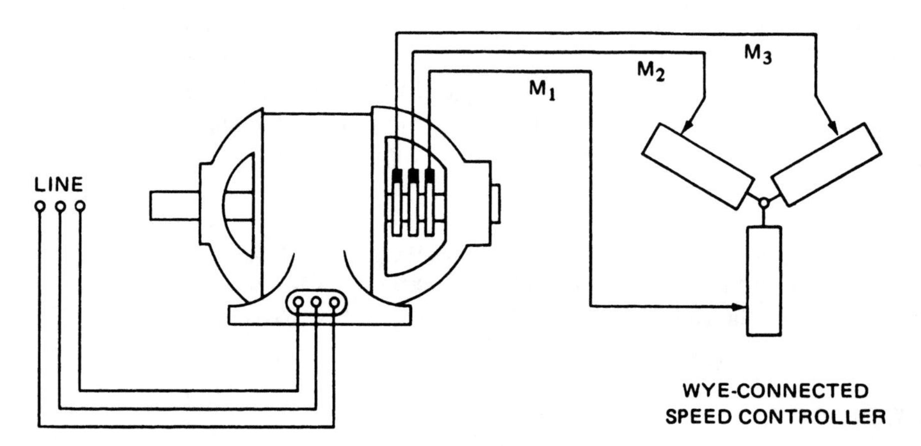

FIGURE 16–21 Connection of a faceplate speed controller to a wound-rotor motor ***(Delmar/Cengage Learning)***

Speed Control

At start-up, all of the resistance of the wye-connected speed controller is inserted in the rotor circuit. This additional resistance causes an excellent starting torque and a large percent slip. The added resistance in the rotor circuit increases the impedance. Because the rotor circuit has a large resistance component and a small reactive component, the rotor current is nearly in phase with the stator field flux. Thus, there is a maximum interaction between the two fields, resulting in a strong starting torque.

As the motor accelerates, steps of resistance are cut out of the wye-connected speed controller. When all of the resistance is cut out, the rotor slip rings are short-circuited. The motor then operates at the rated speed like a squirrel-cage induction motor. The speed of the wound-rotor motor can be changed by inserting or removing resistance in the rotor circuit using a wye-connected speed controller.

This motor can be operated at heavy loads by cutting in resistance to the rotor circuit to obtain a below-normal speed. However, the I^2R losses in the rotor circuit are high and cause a large reduction in the motor efficiency. Additional resistance inserted in the rotor circuit leads to poor speed regulation. This effect is due to the large increase in slip that is necessary to obtain the required torque increase with an increase in the load.

Torque Performance

The curves in Figure 16–22 show the torque performance of a wound-rotor induction motor. When the proper value of resistance is inserted in the rotor circuit, the starting torque has its maximum value at 100% slip (at start-up). If all of the resistance is cut out of the speed controller, and the motor is started, the starting torque is poor.

In this case, the rotor circuit has a large reactive component and a small resistance component. This means that the motor will have the same starting torque characteristic as a squirrel-cage induction motor.

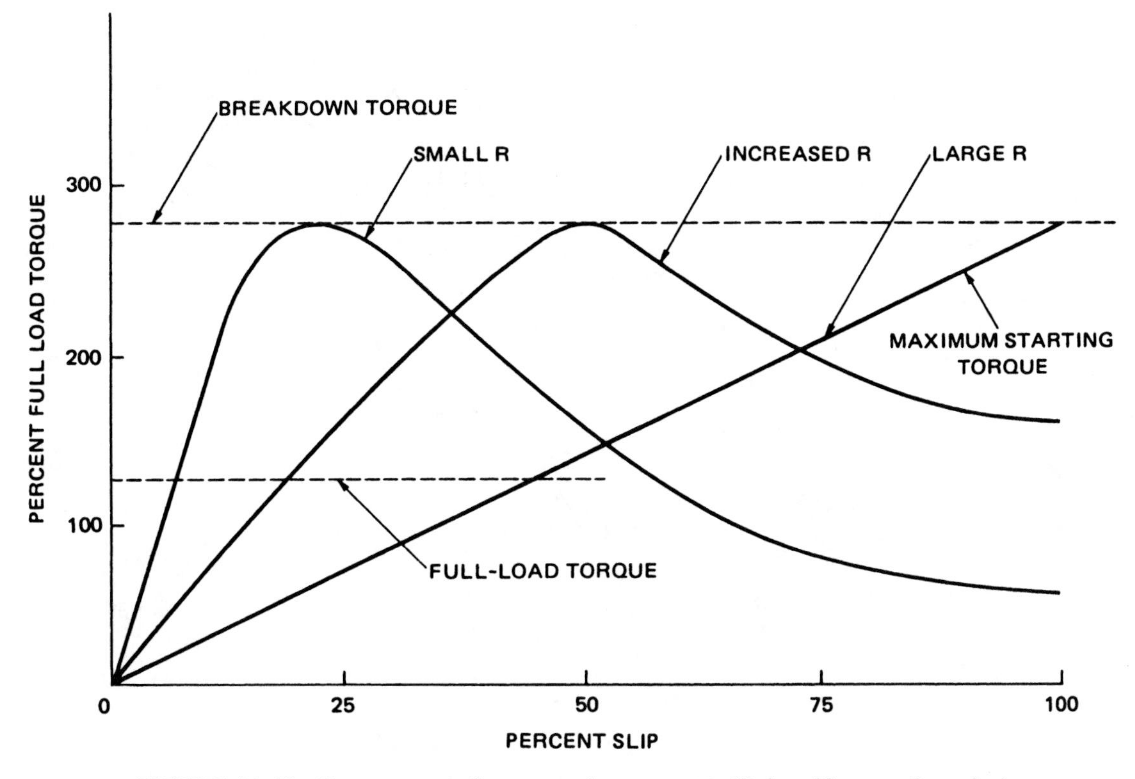

FIGURE 16–22 Torque curves for a wound-rotor motor (*Delmar/Cengage Learning*)

REVERSE ROTATION

The direction of rotation of a wound-rotor induction motor can be reversed. To do this, interchange the connections of any two of the three line leads feeding to the stator windings. Note that both the wound-rotor and the squirrel-cage induction motors are reversed using the same procedure. Reversing the phase sequence of the three-phase input to the stator windings changes the direction of rotation of the magnetic field produced by these windings. Therefore, the direction of rotation of the rotor is reversed. Figure 16–23 shows the connection changes that are required to reverse the direction of rotation. There is *no* reversal in the direction of rotation of the motor when any of the leads feeding from the slip rings of the speed controller are interchanged.

TERMINAL MARKINGS

The stator leads of a three-phase, wound-rotor induction motor are marked T_1, T_2, and T_3. This is the same marking system used with three-phase, squirrel-cage induction motors. The rotor leads are marked M_1, M_2, and M_3. The M_1 lead connects to the rotor ring nearest the bearing housing. The M_2 lead connects to the middle slip ring, and the M_3 lead connects to the slip ring nearest the rotor windings.

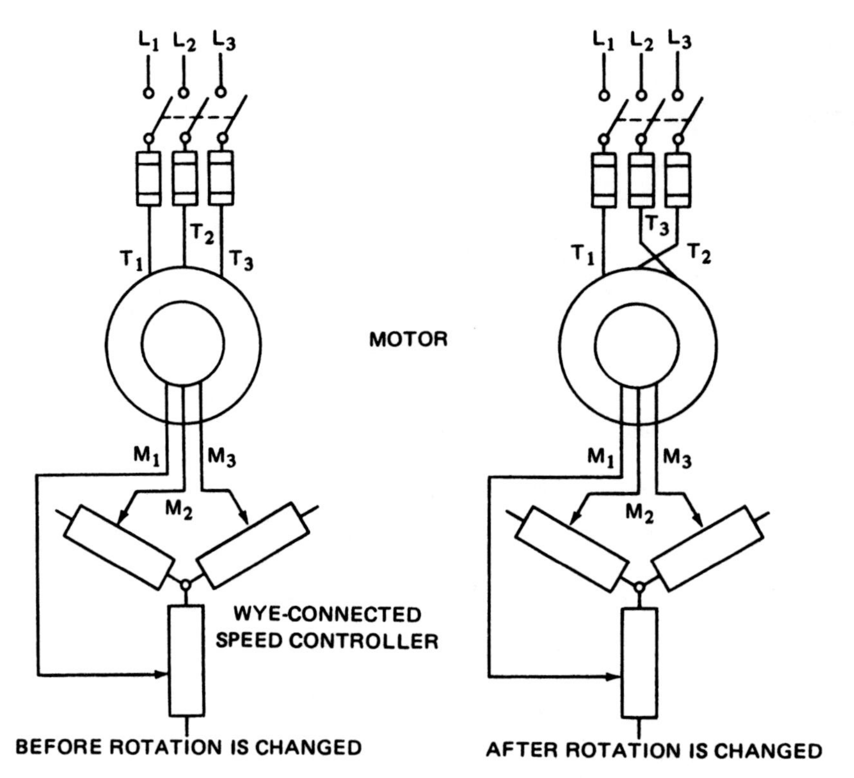

FIGURE 16–23 Changes necessary to reverse the direction of rotation of a wound-rotor motor (*Delmar/Cengage Learning*)

OPERATING CHARACTERISTICS

The wound-rotor motor and the squirrel-cage motor have the same percent efficiency, power factor, percent slip, speed, and torque characteristics. This statement is true if the wound-rotor motor is operated from no load to full load with all of the resistance cut out of the rotor circuit. If the motor is operated at or near the rated load while there is resistance in the rotor circuit, there will be I^2R losses in the resistance components of the speed controller. These losses cause a decrease in the motor efficiency. When the motor is operated with resistance in the rotor circuit, there is also a sharp increase in percent slip with an increase in the load. The percent slip must increase to obtain the required increase in the rotor current so that there is a larger value of torque to meet the increased load demands.

If the motor is started with all of the resistance of the speed controller inserted in the rotor circuit, the starting torque will be at a maximum with 100% slip. The starting surge of current to the stator is limited to a relatively low value. The current is low because of the high resistance inserted in the rotor.

The wound-rotor induction motor is used when a strong starting torque and a range of speed control are required. Typical applications for this motor include cranes, large compressors, elevators, and pumps. This type of motor is also used in the heavy-steel industry for applications requiring adjustable speed.

COMPARISON BETWEEN SQUIRREL-CAGE AND WOUND-ROTOR INDUCTION MOTORS

The wound-rotor induction motor has the following advantages:

- High starting torque and low starting current if the motor is started with the maximum resistance inserted in the rotor circuit
- Variable speed
- Smooth acceleration under heavy loads

There are several disadvantages to this motor, including the following:

- Greater initial cost
- Higher maintenance and repair costs
- Low efficiency and poor speed regulation (when the motor is operated with resistance inserted in the rotor circuit)

SUMMARY

- A three-phase, squirrel-cage induction motor
  1. is simple in construction and is easy to maintain.
  2. is smaller in physical size than other motors for a given horsepower rating.
  3. has very good speed regulation under varying load conditions.
  4. is used for many industrial applications because of its low cost, rugged construction, and operating characteristics.

  Other characteristics are as follows:

  1. The basic components of a squirrel-cage motor are a stator, a rotor, and two end shields housing the bearings supporting the rotor shaft.
  2. The stator is a three-phase winding that is placed in the slots of a laminated steel core. Three single-phase windings are spaced 120 electrical degrees apart. These windings may be connected in wye or delta.
  3. The rotor consists of a cylindrical core made from steel laminations. Aluminum bars are mounted near the surface and are connected at the ends to two aluminum end rings.

- The squirrel-cage induction motor is also known as a *synchronous speed motor*.
  1. The synchronous speed is determined by the number of poles in the stator winding and the frequency of the ac source

  $$S = \frac{120\ f}{P}$$

where

S = synchronous speed, in r/min
f = frequency, in hertz
P = number of poles (*not* pairs of poles)

2. The following table lists the speeds of a motor at two different frequencies for different numbers of poles:

Poles	Speed in r/min	
	25 Hz	60 Hz
2	1500	3600
4	750	1800
6	500	1200
8	375	900

- Slip of an induction motor:
  1. *Slip* is the difference between the synchronous speed of the motor and the actual speed of the rotor. The rotor always turns at a speed that is slightly less than synchronous speed.
  2. The speed performance of an induction motor is usually measured in terms of percent slip. The percent slip for an induction motor is generally in the range of 2% to 5%:

$$\text{Percent slip} = \frac{\text{synchronous speed} - \text{rotor speed}}{\text{synchronous speed}} \times 100$$

  3. The standard squirrel-cage induction motor has excellent speed regulation, from no load to full load. Because of its good speed regulation, an induction motor is considered to be a constant-speed motor:

$$\text{Percent speed regulation} = \frac{\text{no-load speed} - \text{full-load speed}}{\text{full-load speed}} \times 100$$

  4. The torque output of an induction motor varies as the square of the terminal voltage for a given percent slip.
  5. The power factor of the induction motor is low at no load and high at full load. At full-load conditions, the in-phase component of current is large and the magnetizing component is small; the opposite is true at no-load conditions.
  6. The losses of an induction motor consist of the stray power losses and the copper losses. The stray power losses are nearly constant at all loads. The copper losses increase as the current of the motor increases.

7. The direction of rotation of a three-phase, squirrel-cage induction motor can be changed by reversing two of the three incoming supply leads.

- The torque of an induction motor
  1. results from the interaction between the stator flux and the rotor flux.
  2. is produced only when the rotor turns at a speed that is less than the synchronous speed of the stator flux pattern.
  3. is small at no load because there is only a slight difference between the rotor speed and the synchronous speed of the stator flux.
  4. will increase to a larger value as the load on the shaft of the rotor increases. The resulting increase in the rotor current causes a larger torque at a slightly lower rotor speed.

- Rotor frequency:
  1. This frequency is an important factor in the operation of the motor. A change in the rotor frequency causes the inductive reactance and the impedance of the rotor to decrease. A change in the frequency will affect the starting and running characteristics of the motor.
  2. Rotor frequency is expressed by the following formulas:

$$f = \frac{P \times S_R}{120}$$

or

$$f_R = f_S \times S_R$$

where

P = number of poles
S_R = slip of rotor
f_R = rotor frequency, in hertz
f_S = stator frequency, in hertz
S_R = percent slip, changed to a decimal value

- At the instant of start-up
  1. the rotor is not turning and there is 100% slip. The rotor frequency is equal to the stator frequency.
  2. the inductive reactance of the rotor is very large compared to the effective resistance component. The rotor has a very low lagging power factor and the starting torque is low.
  3. the stator field cuts the rotor bars at a faster rate; thus, the induced voltage in the rotor causes a large rotor current and the stator current will also be high at start-up.

In addition:

1. Because of the high starting current, induction motors must have starting protection. This protection may be rated as high as three times the full-load current rating of the motor.

2. Some very large induction motors are started with auxiliary starters. These devices reduce the motor voltage at start-up to limit the starting surge of current. As a result, there is less voltage disturbance on the feeder circuit supplying the motor load.

- The point of the maximum torque output is called the *breakdown point*.
  1. An increase in the load beyond this point means that less torque is developed by the motor and the rotor stops.
  2. The breakdown point occurs at 200% to 300% of the rated torque.
- The efficiency of an ac induction motor is given by the following equation:

$$\eta = \frac{\text{output, watts}}{\text{input, watts}} \times 100$$

or

$$\eta = \frac{\text{input, watts} - (\text{copper losses} + \text{stray losses})}{\text{input, watts}} \times 100$$

- To find the percent efficiency, it may be necessary to find the true power input to the motor in watts first:

$$\text{Watts} = \sqrt{3} \times V_{\text{line}} \times I_{\text{line}} \times \cos\theta$$

$$\text{hp output} = \frac{2\pi \times \text{torque} \times \text{r/min}}{33{,}000}$$

or

$$\text{hp output} = \frac{\text{output, watts}}{746}$$

$$(1000 = 1.34 \text{ hp})$$

or

$$\text{hp output} = 0.00019 \times \text{torque} \times \text{motor speed}$$

- Brushes and slip rings are not required on a squirrel-cage induction motor. Thus, this type of motor can be used in locations where there is a possibility of explosion due to arcing, such as in chemical plants and flour or lumber mills.
- There is no practical method of providing stepless speed control for the induction motor.
- The blocked rotor test is used to determine the equivalent resistance of the motor winding, per phase.
  1. Two wattmeters are used in this test:

$$\text{Total power} = W_1 - W_2 \text{ (used to supply the copper losses)}$$

$$P = I^2R$$

$$\text{Equivalent resistance} = \frac{P}{3 \times I^2R}$$

- Stray power losses:
  1. Stray power losses include mechanical friction losses, windage losses, and iron losses.
  2. To measure these losses, the motor is operated at no load with the rated voltage applied to the motor terminals:

     Total power input at no load: $P_T = W_1 + W_2$

     Copper losses at no load: $P = 3 \times I^2R$

  3. The fixed losses at no load are obtained by subtracting the no-load copper losses from the power input at no load:

     Fixed losses $= P_T - P$

- Speed control of an induction motor:
  1. The synchronous speed of the magnetic field of the stator is determined by the number of stator poles and the frequency of the ac source.
  2. Generally, the speed of an induction motor cannot be varied by changing the frequency.
  3. The speed of a squirrel-cage motor may be changed if the motor is provided with special stator windings. When these windings are reconnected using special switch controls, different numbers of stator poles are formed. Thus, different speeds can be obtained for a squirrel-cage induction motor connected to an ac source at a fixed frequency.
- Code letter identification for motors:
  1. A system of code letters is used to identify certain induction motors. These motors are grouped according to their torque and starting current characteristics.

Code Letter	Torque	Starting Current	Application
A	High	Low	Used for metal shears, punch presses, and metal-drawing equipment
B–E	Fair	Relatively low	Used for motor-generator sets, fans, blowers, centrifugal pumps (where high torque is not required)
F–V	Fair	High	Same as B–E usages

2. The code letter designates the ratio between the starting and full-load currents. This letter appears on the nameplate of all squirrel-cage induction motors.
3. By referring to the *National Electrical Code,* this code letter can be used to determine the current ratings of circuit breakers, fuses, and other overload protective devices.

- A motor nameplate contains the following information:
  Full-load horsepower output (hp)
  Full-load speed (r/min)
  Full-load current amperage (FLA)
  Locked-rotor current amperage (LRA)
  Manufacturer name
  Terminal voltage (volts)
  Number of phases (phase)
  Frequency (cycles) (Hz)
  Temperature rise (rise °C) (or insulation system class and rated ambient temperature)
  Time rating (5, 15, 30, 60 min, or continuous)
  Code letter

  [*Note:* Other general information may be placed on the motor nameplate.]

- A double-squirrel-cage motor has a low starting current, a strong starting torque, and excellent speed regulation. The starting torque for this type of motor can be as high as 250% of the rated torque.

- Single-phase operation of three-phase induction motors:
  1. If a three-phase induction motor is running when it is subjected to single-phase conditions, it will continue to operate but at a greatly reduced capacity.
  2. The motor will not have enough torque to start when it is energized from a single-phase source.
  3. If the rated load is applied to the motor when it is operating as a single-phase motor, it will overheat and the insulation of the windings may be damaged.

- The wound-rotor induction motor:
  1. This motor can be used for variable-speed applications. The squirrel-cage induction motor is a constant-speed motor.
  2. The formula used to find the synchronous speed for a squirrel-cage induction motor can also be used for the wound-rotor motor:

     $$S = 120\,f \div P$$

  3. A wye-connected speed controller provides control current through slip rings to the windings of the rotor.
     a. At start-up, all of the resistance of the wye-connected speed controller is inserted in the rotor circuit. This additional resistance causes an excellent starting torque and a large percent slip.

   b. As the motor accelerates, steps of resistance are cut out of the wye-connected speed controller.
   c. When all of the resistance is cut out, the rotor slip rings are short-circuited. The motor then operates at the rated speed like a squirrel-cage induction motor.
   d. This motor can be operated at heavy loads by cutting in resistance to the rotor circuit to obtain a below-normal speed at a lower motor efficiency.
4. The direction of rotation can be changed by reversing any two leads of the three line leads feeding to the stator windings.
5. A wound-rotor induction motor is used when a strong starting torque and a range of speed control are required. Typical applications include cranes, large compressors, elevators, and pumps.
6. Compared to the squirrel-cage induction motor, the wound-rotor motor has the following advantages:
   a. High starting torque and low starting current if it is started with the maximum resistance inserted in the motor circuit
   b. Variable speed
   c. Smooth acceleration under heavy loads
7. Compared to the squirrel-cage induction motor, the wound-rotor motor has the following disadvantages:
   a. Greater initial cost
   b. Higher maintenance and repair costs
   c. Low efficiency and poor speed regulation (when it is operated with resistance inserted in the rotor circuit)

Achievement Review

1. Explain what is meant by the synchronous speed of a three-phase induction motor.
2. What two conditions determine the synchronous speed of a three-phase induction motor?
3. Explain what is meant by the following terms:
   a. Revolutions slip
   b. Percent slip
   c. Rotor frequency
4. a. What is the rotor frequency of a three-phase, 60-Hz, squirrel-cage induction motor at the instant of start?
   b. What is the approximate rotor frequency of a three-phase, 60-Hz, squirrel-cage induction motor when it is operating at the rated load?

5. What is the reason for the poor starting torque of a squirrel-cage induction motor?
6. A six-pole, three-phase, 60-Hz induction motor has a full-load speed of 1140 r/min. Determine
   a. the synchronous speed.
   b. the revolutions slip.
   c. the percent slip.
   d. the rotor frequency.
7. a. Draw characteristic curves for a three-phase, squirrel-cage induction motor for the speed, percent slip, percent efficiency, power factor, and torque.
   b. Discuss each of the five characteristic curves developed in part a of this question.
8. For a given value of slip, the torque output of an induction motor varies as the square of the impressed terminal voltage. Explain what is meant by this statement.
9. A 10-hp, 220-V, three-phase, 60-Hz, squirrel-cage induction motor is rated at 28 A per terminal. The full-load speed is 855 r/min, and the full-load power factor is 0.90 lag. The motor has eight poles. At the rated load, determine
   a. the synchronous speed.
   b. the slip in r/min.
   c. the percent slip.
   d. the rotor frequency at the rated speed.
10. A three-phase, 60-Hz, four-pole, 220-V, squirrel-cage induction motor takes 52 A per terminal at full load. The power factor is 0.85 lag, and the efficiency is 88%. The slip is 3.0%. At the rated load, determine
    a. the speed in r/min.
    b. the horsepower output of the motor.
    c. the total losses.
11. A three-phase, 60-Hz, six-pole, 220-V, squirrel-cage induction motor has a full-load output of 15 hp. The full-load efficiency is 87%, and the power factor is 0.88 lag. The windings of the motor are connected in delta. At the rated load, determine
    a. the line current.
    b. the phase winding voltage.
12. Assuming that the three-phase motor in question 11 is reconnected in wye with the same load and power factor, determine
    a. the new rated line voltage.
    b. the line current per motor terminal.
13. Explain how the direction of rotation is reversed for
    a. a three-phase squirrel-cage induction motor.
    b. a three-phase wound-rotor induction motor.

14. Explain why a three-phase squirrel-cage induction motor will not start when energized from a single-phase source.

15. Neither of the following alternating-current, three-phase induction motors is operating properly. Give a possible reason for the motor failure described, and state what should be done to correct each condition.
    a. A 15-hp, 220 V, three-phase squirrel-cage induction motor overheats while operating at a normal load. The motor circuit is deenergized. When an attempt is made to restart the motor, it will not turn.
    b. A 5-hp, 220-V, three-phase squirrel-cage induction motor stops as soon as the start push button for the across-the-line motor starter is released.
    c. A newly installed 10-hp, 220-V, three-phase squirrel-cage induction motor has dual voltage ratings of 220 V and 440 V. The motor is supplied from a 220-V, three-phase source. At no load, the motor operates at a speed that is slightly below synchronous speed. When the rated load is applied, the motor stalls.

16. Show the connections for the nine terminal leads of a wye-connected, three-phase motor rated at 220/440 V for operation at
    a. a line voltage of 440 V.
    b. a line voltage of 220 V.

17. Compare a three-phase squirrel-cage induction motor and a three-phase wound-rotor induction motor, with regard to
    a. construction.
    b. starting torque.
    c. speed control.
    d. initial cost and maintenance.
    e. efficiency.

18. a. Explain why the power factor of an induction motor is low at no load.
    b. Explain what happens to the power factor as the load on an induction motor is increased.

19. Explain how both a good starting torque and small percent slip can be obtained using a squirrel-cage induction motor with a double-squirrel-cage rotor.

20. How is speed control provided for a squirrel-cage induction motor?

21. The following data are obtained in a prony brake test of a three-phase squirrel-cage induction motor:

Line Voltage	Line Current	Wattmeters		Rotor Speed	Brake Arm Length	Spring Scale	Tare
		1	2				
220 V	13.2 A	2885 W	1615 W	1740 r/min	2 ft	9¼ lb	1¾ lb

Determine
a. the power factor.
b. the horsepower output.
c. the efficiency.
d. the torque output in pound • feet.

22. a. What are the losses in a squirrel-cage induction motor?
    b. Which of these losses are constant and which of the losses vary with a change in load?

23. List the nameplate data of a typical squirrel-cage motor.

24. Code letters are included in the nameplate data for three-phase squirrel-cage induction motors. What is the purpose of these code letters?

25. What is the purpose of the protective starting device used on faceplate speed controllers and drum-type speed controllers?

26. The following test data are obtained on a 7.5-hp, three-phase, 220-V, wye-connected, four-pole squirrel-cage induction motor. The data are for a no-load condition and a full-load condition. The effective resistance of each single-phase winding of the wye-connected stator is 0.65 Ω. Determine the stray power losses of the motor.

Line Voltage Volts			**Line Current Amperes**			**Watts**		**Speed**	**Load Condition**
AB	BC	CA	A	B	C	Wattmeter 1	Wattmeter 2	r/min	
220	220	220	5.0	5.0	5.0	680	−360	1795	No load
220	220	220	20.0	20.0	20.0	4400	2300	1740	Full load

27. Use the data given in question 26 and determine
    a. the copper losses of the motor at full load.
    b. the efficiency of the motor at rated load.
    c. the horsepower output at full load.
    d. the torque output in pound • feet at full load.

28. A three-phase, squirrel-cage induction motor must carry an additional load. List the sequence of steps, in chronological order, showing how the motor will adjust itself to carry this additional load.

17

The Synchronous Motor

Objectives

After studying this unit, the student should be able to

- describe the construction of the synchronous motor, listing all of its components, and explain the function(s) of each component.
- describe the steps in the operation of the synchronous motor from the moment it is energized until the rotor locks in at the synchronous speed.
- calculate the synchronous speed of the three-phase synchronous motor.
- explain the function of the field discharge resistor.
- state how the synchronous motor compensates for changes in the mechanical load on the motor.
- define torque angle, pullout torque, and normal field excitation.
- list several industrial applications in which synchronous motors are used.
- use formulas to calculate the apparent power, the input in watts, and reactive VARs for a synchronous motor.
- explain how a synchronous motor may be used to correct the power factor.
- use schematic wiring diagrams to show the connections for selsyn transmitter, differential, and receiver units when used in a remote indicating system.

The synchronous motor is a three-phase ac motor that operates at a constant speed from no load to full load. It is similar to a three-phase ac generator. That is, it has a revolving field that must be separately excited from a direct-current source. The dc field excitation current can be varied to obtain a wide range of lagging and leading power factor values.

The synchronous motor is used in many industrial applications for the following reasons: it has a fixed speed from no load to full load, a high efficiency, and a low initial cost. It is also used to improve the power factor of three-phase ac industrial circuits.

CONSTRUCTION

The three-phase synchronous motor consists of the following components:

- A laminated stator core with a three-phase armature winding
- A revolving field with an amortisseur winding and slip rings
- Brushes and brush holders
- Two end shields housing the bearings that support the rotor shaft

The stator core and windings of a synchronous motor are similar in construction to those of a three-phase squirrel-cage induction motor or a wound-rotor induction motor. The leads for the stator windings are marked T_1, T_2, and T_3. These leads end in a terminal box, which normally is mounted on the side of the motor frame.

The rotor has salient field poles. The poles are connected to give alternate polarity. There must be the same number of rotor field poles as stator field poles. The field circuit leads are brought out to two slip rings mounted on the rotor shaft. A squirrel-cage (amortisseur) winding is provided as a means of starting the motor. The synchronous motor is not self-starting without this auxiliary winding.

Figure 17–1 shows a rotor having salient poles and an amortisseur winding. The amortisseur winding consists of copper bars embedded in the laminated metal structure of each pole face. The copper bars of this special squirrel-cage winding are brazed to rings mounted on each end of the rotor.

Carbon brushes, mounted in brush holders, contact the two slip rings. The terminals of the field circuit are brought out from the brush holders to a second terminal box, which is mounted on the motor frame. The two leads for the field circuit are marked F_1 and F_2.

FIGURE 17–1 Synchronous motor rotor with amortisseur winding (*Delmar/Cengage Learning*)

OPERATING PRINCIPLES

The rated three-phase voltage is applied to the stator windings, resulting in a rotating magnetic field. This field travels at synchronous speed. The speed is determined by the same factors that govern the synchronous speed of induction motors. The synchronous speed is found using the following equation:

$$\textbf{Synchronous speed, in r/min} = \frac{\textbf{120} \times \textbf{frequency}}{\textbf{number of poles}}$$

$$\mathbf{S} = \frac{\mathbf{120\,f}}{\mathbf{P}}$$

The rotating magnetic field set up by the stator windings cuts across the amortisseur (squirrel-cage) winding of the rotor. The amortisseur winding is a squirrel-cage winding very similar to the type A winding in a squirrel-cage induction motor. Due to its relatively high resistance, it provides low starting current and high starting torque per ampere of starting current. Voltages and currents are induced in the bars of this winding. A magnetic field is set up in the squirrel-cage windings. This field reacts with the stator field and causes rotation of the rotor. The speed of the rotor increases until it is just below the synchronous speed of the stator field. In other words, there is a slight slip of the rotor behind the magnetic field set up by the stator windings. When the synchronous motor is started as an induction motor with the amortisseur windings, the rotor accelerates to about 85% to 97% of the synchronous speed.

The field circuit is excited from an outside direct-current source. Magnetic poles of fixed polarity are set up in the rotor field cores. The fixed magnetic poles of the rotor are attracted to unlike poles of the rotating magnetic field. (This field was set up by the stator windings.)

Figure 17–2 shows how the rotor field poles lock with unlike poles of the stator field. As a result, the rotor speed becomes the same as the speed of the stator field. This speed is the synchronous speed. When the rotor begins turning at synchronous speed, there is no longer any cutting action across the bars of the amortisseur winding. At this point there is no induced voltage in the amortisseur winding, and it therefore has no effect on the operation of the motor.

DC FIELD EXCITATION

The direct current for synchronous motors is obtained from a dc exciter circuit. Such a circuit may supply field excitation to several ac machines. A dc generator may be coupled directly to the synchronous motor shaft. Other installations may use electronic rectifiers to supply the dc excitation current.

The dc connections to a synchronous motor are shown in Figure 17–3. A field rheostat controls the current in the separately excited field circuit. When the field switch is open, the field discharge resistor is connected directly across the field winding.

The Brushless Exciter

Most large synchronous motors use an exciter that does not depend on brushes and slip rings. This is accomplished by adding a separate small alternator of the

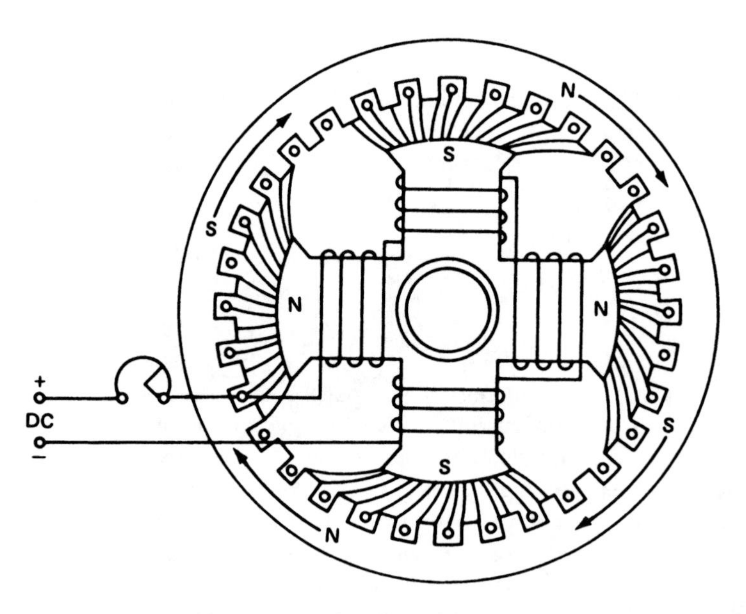

FIGURE 17–2 Principle of operation of a synchronous motor ***(Delmar/Cengage Learning)***

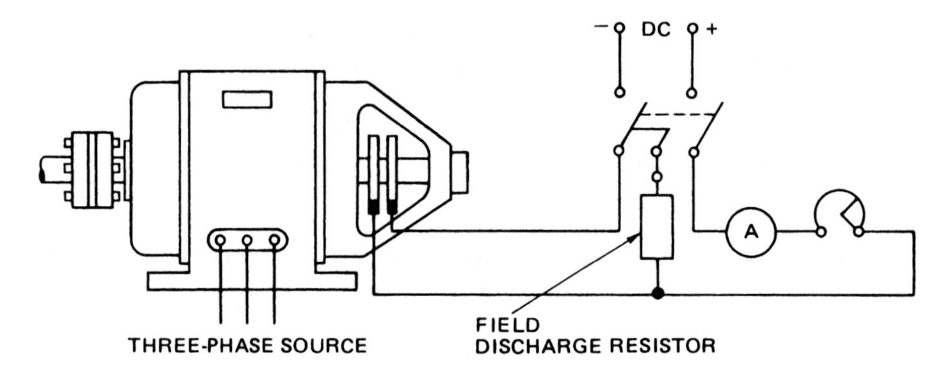

FIGURE 17–3 External connections for a synchronous motor ***(Delmar/Cengage Learning)***

armature type on the shaft of the synchronous motor. The armature rotates between wound electromagnets. The dc excitation current is connected to the wound stationary magnets (Figure 17-4). The amount of voltage induced in the armature can be controlled by varying the amount of dc current supplied to the electromagnets. The output voltage of the armature is connected to a three-phase bridge rectifier mounted on the rotor shaft (Figure 17-5). The three-phase bridge rectifier converts the three-phase alternating current produced in the armature to direct current before it is applied to the rotor of the synchronous motor. Because the three-phase armature and rectifier are contained on the rotor shaft, they all turn together and no brushes or slip rings are needed to provide excitation for the rotor of the large synchronous motor. The rotor of a synchronous motor with a brushless exciter is shown in Figure 17–6.

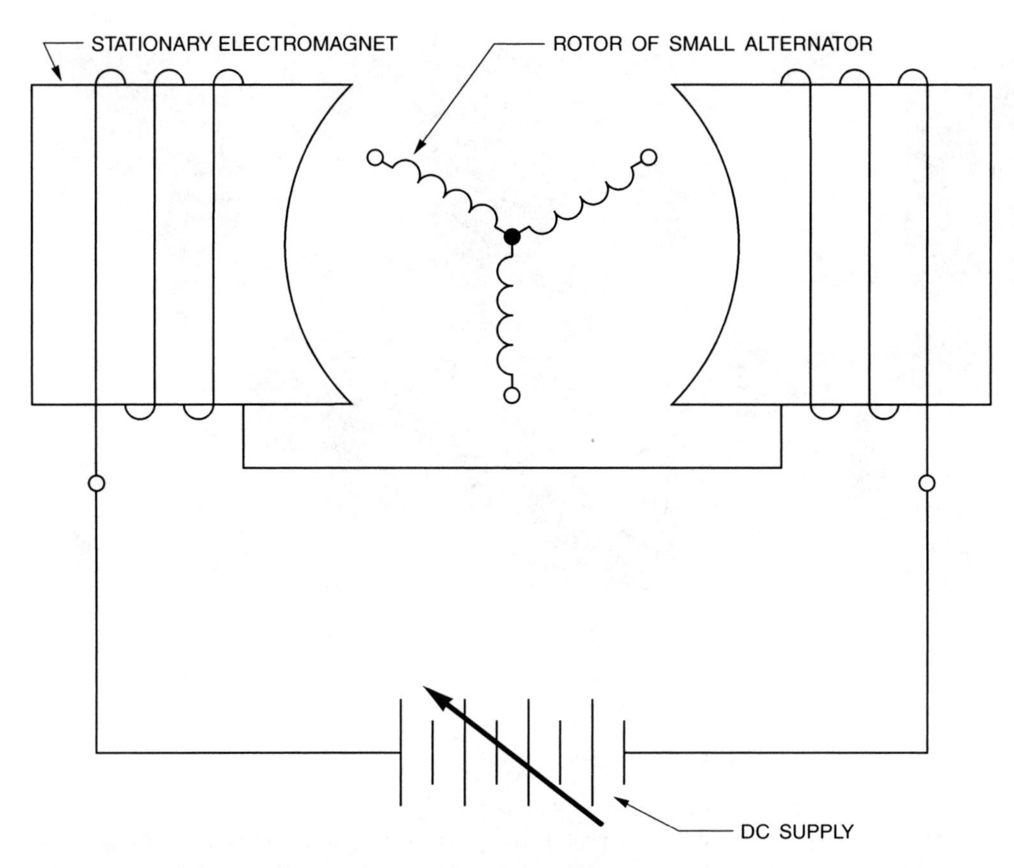

FIGURE 17–4 The brushless exciter uses stationary electromagnetics (*Delmar/Cengage Learning*)

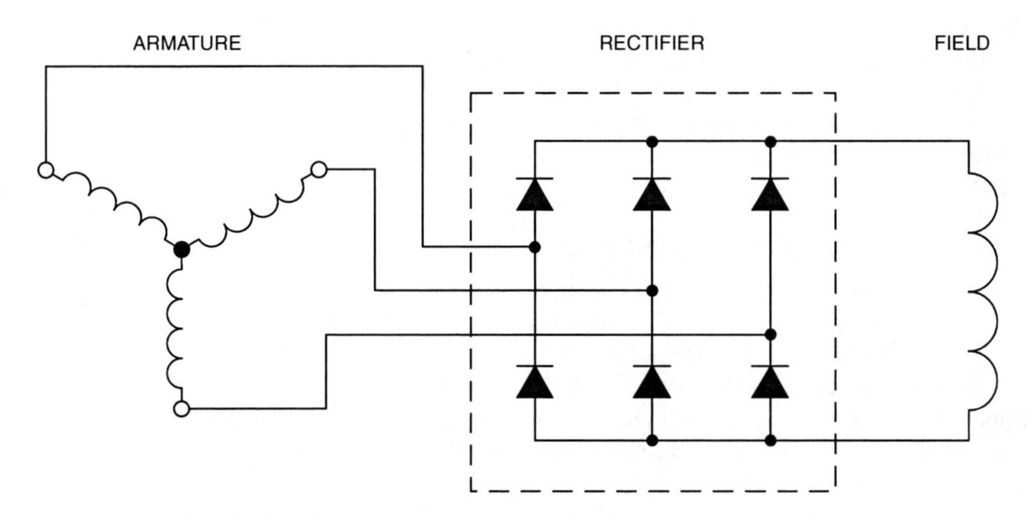

FIGURE 17–5 Basic brushless exciter circuit (*Delmar/Cengage Learning*)

FIGURE 17–6 Rotor of a synchronous motor with brushless exciter ***(Courtesy of GE Industrial Systems, Fort Wayne, Indiana)***

Starting

The dc field circuit is never energized when the synchronous motor is started because the rapidly rotating field produces an alternating torque on the stationary rotor poles. In general, a synchronous motor is started by connecting its armature winding (stator) to the ac line and its field winding (rotor) to a field discharge resistor. The motor is started as an induction motor.

At the instant of startup, the rotating stator field cuts the turns of the dc field coils many times per second. The stator field turns at the synchronous speed and induces a high voltage in the field windings. This voltage may reach 1500 V. This means that the field circuits must be well insulated and enclosed to protect personnel. The field discharge resistor is connected across the field windings so that the energy in the field circuit is spent in the resistor. This arrangement also reduces the voltage at the field terminals, although it is still high enough to be a shock hazard.

Achieving Synchronous Speed

Once the motor accelerates to nearly 95% of synchronous speed, the field circuit is energized from the dc source. The field discharge resistor is then disconnected. The rotor will pull into synchronism with the revolving armature (stator) flux. Thus, the motor will operate at a constant speed. If the load has a high inertia and is hard to start, special automatic equipment is required to apply the field. To ensure that there is a successful transition from induction motor operation to synchronous operation, the field must be applied at the best position of the rotor slip cycle.

Field Discharge Resistor

To shut down the motor, the field circuit is deenergized by opening the field discharge switch. As the field flux collapses, a voltage is induced in the field windings. This voltage may be large enough to damage the insulation of the windings. To prevent such a high voltage, the field discharge resistor is connected across the field circuit. As a result, the energy stored in the magnetic field is spent in the resistor and a lower voltage is induced in the field circuit.

LOAD ON A SYNCHRONOUS MOTOR

When the mechanical load is increased on a dc motor or an ac induction motor, there is a decrease in the speed. This decrease results in a reduction of the counterelectromotive force. Thus, the source can supply more current to meet the increased load demands. This method of compensating for an increased load cannot be used with a synchronous motor because the rotor must turn at synchronous speed at all loads. The relative positions of a stator pole and a rotor pole are shown at no load in Figure 17–7A for a synchronous motor. Note that the centers of both poles are in line with each other.

Once the rated load is applied, the relative positions of the stator pole and the rotor pole are as shown in Figure 17–7B. There is now an angular displacement of the rotor pole with respect to the stator pole. The speed is unchanged because the rotor will continue to rotate at synchronous speed. The angular displacement between the centers of the stator and rotor field poles is called the *torque angle, α*.

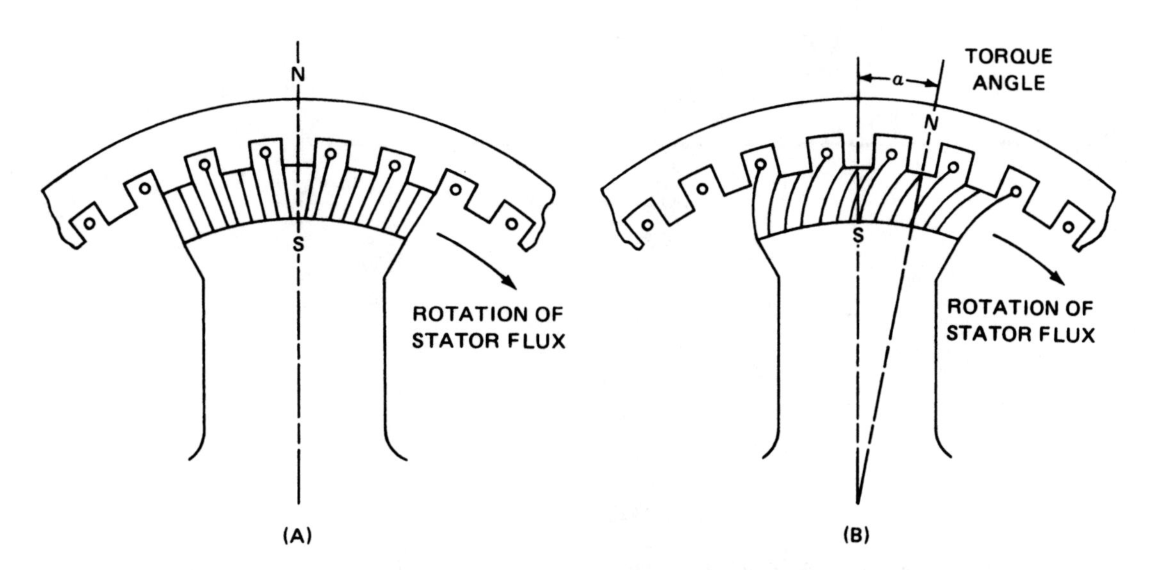

FIGURE 17–7 Relative positions of stator and rotor poles (*Delmar/Cengage Learning*)

Operation with Different Loads

A synchronous motor operating at no load has a torque angle of nearly 0°. The counterelectromotive force in this case is almost equal to the impressed voltage (neglecting no–load motor losses). As the mechanical load increases, the torque angle increases. The phase angle between the impressed voltage and the counter-emf also increases. This increase allows the impressed voltage to cause more current in the stator windings to meet the additional load demands.

In Figure 17–8A, the counter emf is equal and opposite in direction to the impressed voltage at no load. The torque angle here is zero. The counter-emf shown in Figure 17–8B is shifted by the angle *a* from its no-load position in Figure 17–8A. The shift is due to the addition of the load, which causes the rotor pole centers to shift behind the stator pole centers by the angle *a*. The line voltage and the counter-emf are not opposite each other at this point. The resultant voltage V_R gives rise to the current I_S in the stator windings. The stator windings have a high reactance, which causes current I_S to lag the resultant voltage V_R by nearly 90°. The power input for one phase of the three-phase motor is equal to

$$V_{stator} \times I_{stator} \times \cos\theta$$

The synchronous motor can carry an increased mechanical load by shifting the relative positions of the stator and rotor poles. There is no decrease in the speed of the synchronous motor.

A serious overload will cause the angle between the centers of the stator and rotor poles to become too great. In this case, the rotor will pull out of synchronism. With the aid of the amortisseur winding, the motor will operate as an induction motor. The *pull-out torque* is the maximum torque value that can be developed by a synchronous motor

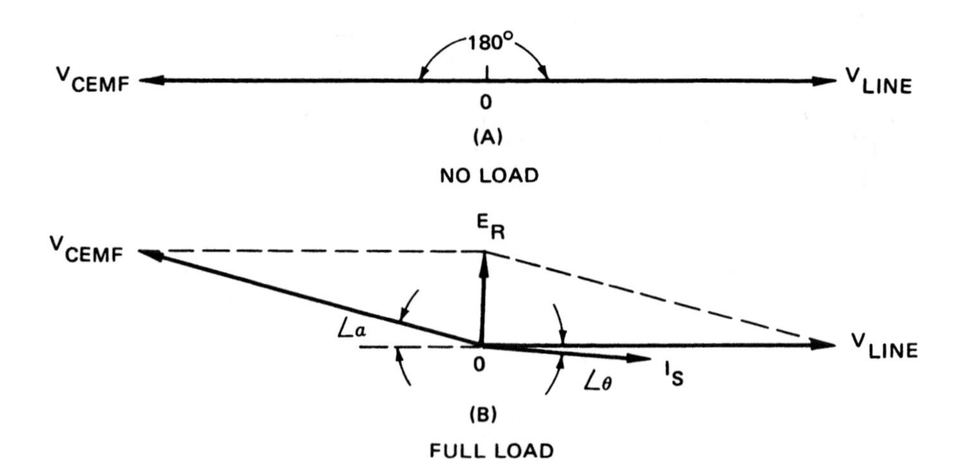

FIGURE 17–8 Vector diagram for a synchronous motor, showing no-load and full-load current and voltage conditions (*Delmar/Cengage Learning*)

without dropping out of synchronism. For most synchronous motors, the pullout torque is 150% to 200% of the rated torque output.

POWER FACTOR

Changes in the dc field excitation do not affect the motor speed. However, such changes do alter the power factor of a synchronous motor. If all of the resistance of the rheostat is inserted in the field circuit, the field current drops below its normal value. A poor lagging power factor results. If the dc field is weak, the three-phase ac circuit to the stator supplies a magnetizing current to strengthen the field. This magnetizing component lags the voltage by 90 electrical degrees. The magnetizing current becomes a large part of the total current input. This gives rise to a low lagging power factor.

If a weak dc field is strengthened, the three-phase ac circuit to the stator supplies less magnetizing current. Because this current component becomes a smaller part of the total current input to the stator winding, the power factor increases. The field strength can be increased until the power factor is unity, or 100%. When the power factor reaches unity, the three-phase ac circuit supplies energy current only. The dc field circuit supplies all of the current required to magnetize the motor. The amount of dc field excitation required to obtain a unity power factor is called *normal field excitation*.

The magnetic field of the rotor can be strengthened still more by increasing the dc field current above the normal excitation value. The power factor in this case decreases. The circuit feeding the stator winding delivers a demagnetizing component of current. This current opposes the rotor field and weakens it until it returns to the normal magnetic strength.

Interaction between DC and AC Fields

Figure 17–9 shows how the magnetic field set up by the ac windings aids or opposes the dc field. The dc field is assumed to be stationary. The revolving armature is connected to the ac source. (In practice, most synchronous motors have stationary ac windings and a revolving dc field. However, the principle involved is the same.)

The dc excitation current is below its normal value in Figure 17–9A. The ac stator windings supply a magnetizing component of current. This current lags the impressed voltage by 90°. This means that the current reaches its maximum value as shown. For this position, the flux created by the magnetizing component of current aids the weakened dc field. An in-phase component of current creates the torque required by the load. (This in-phase component is not shown in the figure.) The synchronous motor will have a lagging power factor because of the magnetizing component of current.

In Figure 17–9B, the dc excitation current is increased to its normal value. The ac input now supplies an in-phase current only. This current meets the torque requirements of the motor, but there is no magnetizing current component. Therefore, the power factor is unity.

The dc field is overexcited in Figure 17–9C. The ac input supplies a leading quadrature current that is really a demagnetizing component of current. This current sets up an mmf that opposes the mmf of the overexcited dc field. In this way, the dc field flux is limited to its normal full-strength value. Although not shown in Figure 17–8C, there is also an in-phase component of current. This in-phase component supplies the torque requirements

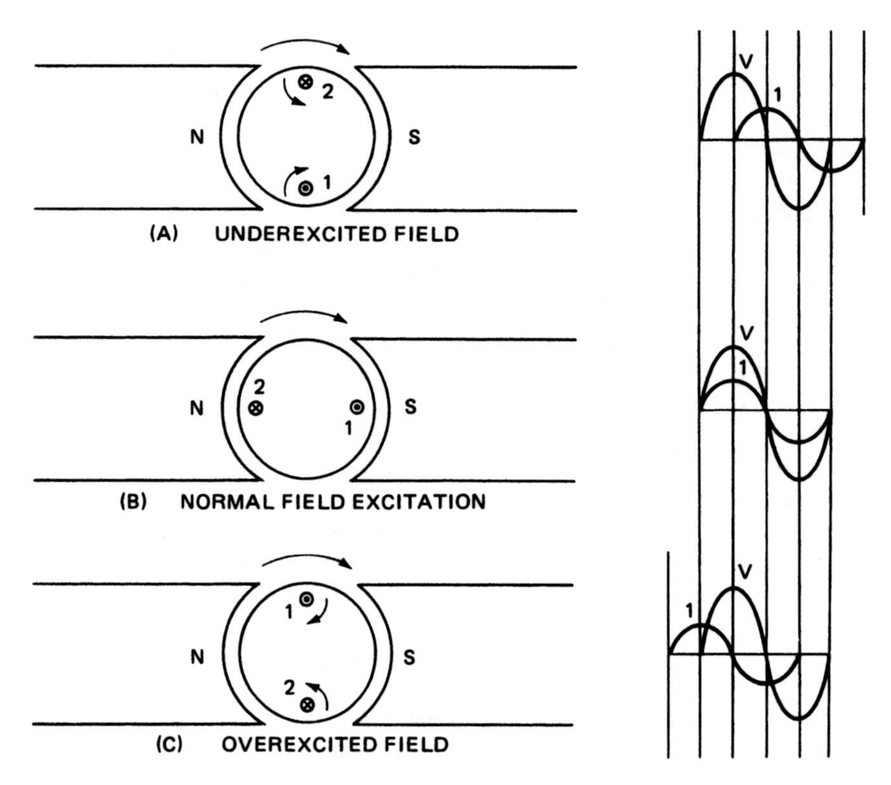

FIGURE 17–9 Field excitation in synchronous motor (*Delmar/Cengage Learning*)

of the motor. When a synchronous motor with an overexcited dc field is supplied with a demagnetizing component of current (quadrature lead current), the motor has a leading power factor.

Typical Characteristic Curves

Typical characteristic curves for a synchronous motor with a constant mechanical load are shown in Figure 17–10. These curves show the changes that occur in the stator current and power factor as the dc field excitation current is varied.

At the normal dc field excitation current, the power factor has a peak value of unity. At the same point, the ac stator current is at its lowest value. As the field current is decreased, the power factor decreases in the lag quadrant. There is a resultant rise in the ac stator current. If the dc field current is increased above the normal field excitation value, the power factor decreases in the lead quadrant and the ac stator current increases.

INDUSTRIAL APPLICATIONS

Synchronous motors rated at 20 horsepower (hp) or more are used for constant speed applications. They are used to drive large air and gas compressors that must be operated at a

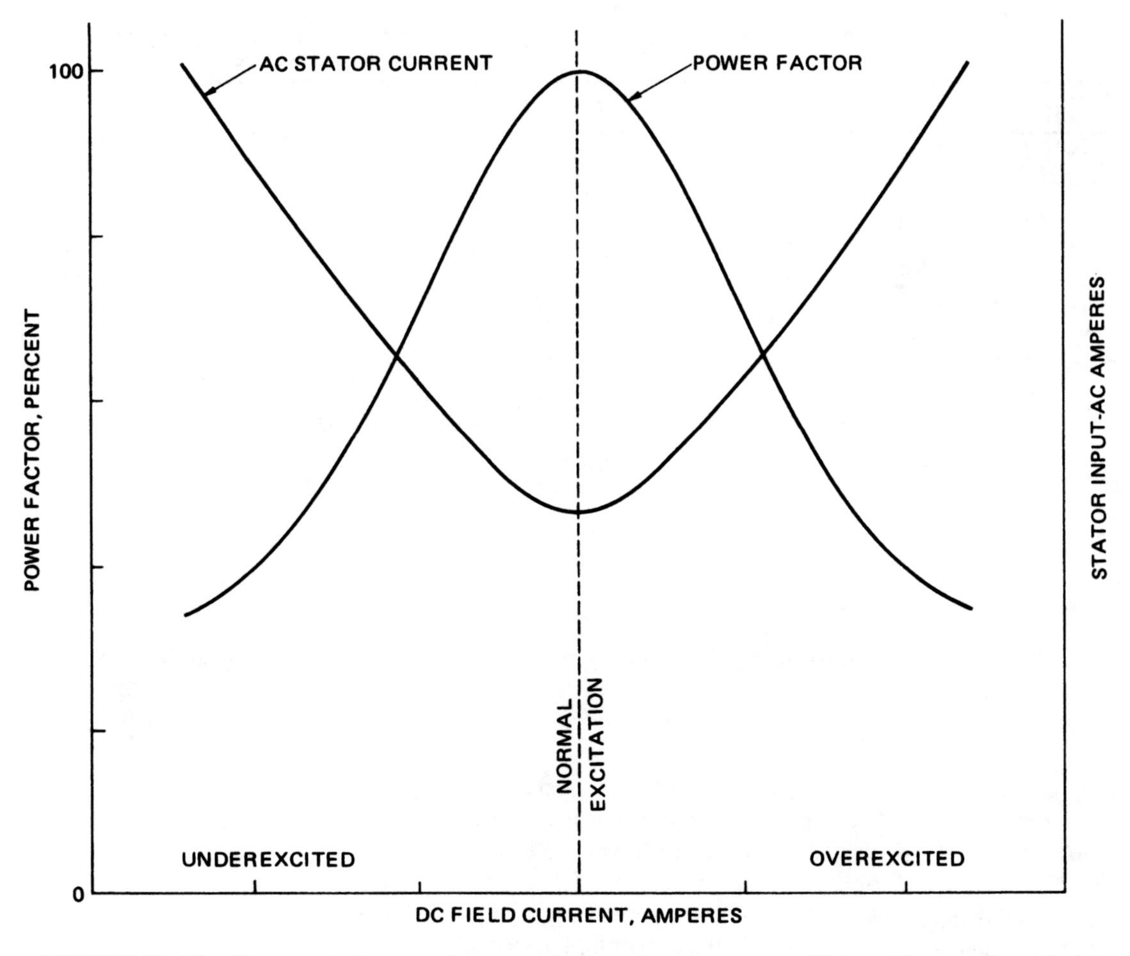

FIGURE 17–10 Characteristic operating curves for synchronous motors (*Delmar/Cengage Learning*)

fixed speed to maintain a constant output at the maximum efficiency. Synchronous motors are used to drive dc generators, fans, blowers, and large pumps in water-pumping stations.

Some industrial applications use three-phase synchronous motors to drive mechanical loads and correct power factor values. A typical industrial feeder is shown in Figure 17–11. The feeder has a lagging power factor condition due to two induction motors. A synchronous motor is connected to this same feeder and is operated with an overexcited field. The synchronous motor supplies leading reactive kilovars to compensate for the lagging kilovars due to the induction motors or other inductive load on the same three-phase distribution system. The dc field of the synchronous motor can be overexcited enough to supply a value of leading kilovars equal to the lagging kilovars. As a result, the power factor of the distribution system is corrected to unity.

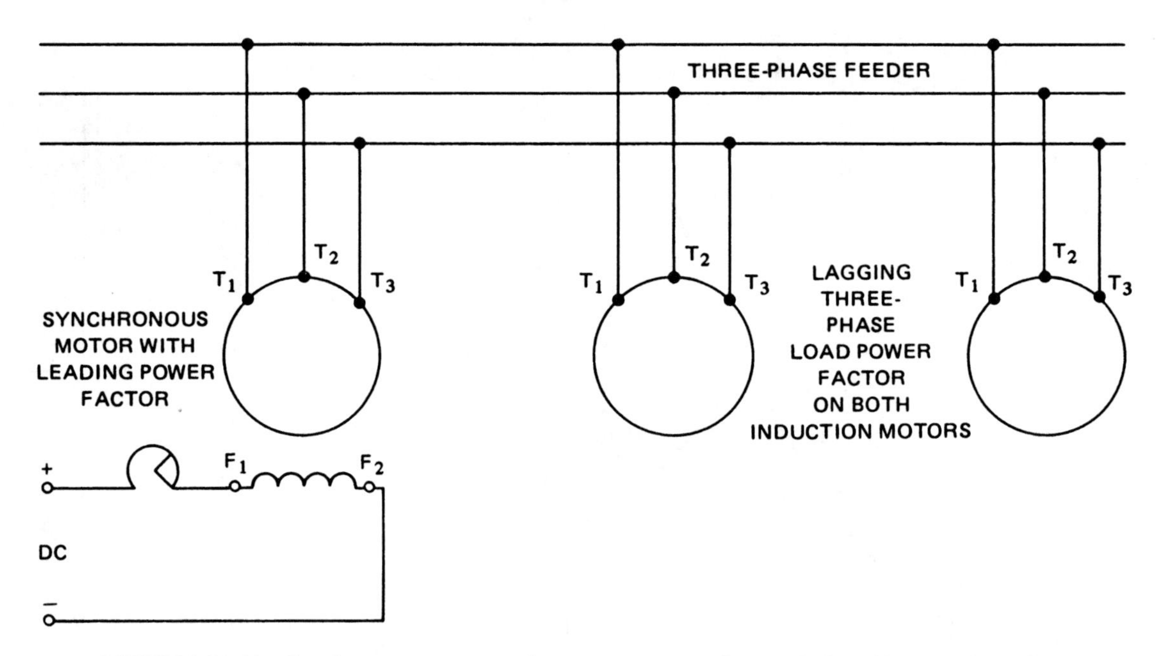

FIGURE 17–11 Synchronous motor used to correct power factor (*Delmar/Cengage Learning*)

The synchronous motor may be used only to correct the power factor and not for driving any mechanical load. It then has the same function as a bank of capacitors. When used to correct the power factor only, the synchronous motor is called a *synchronous capacitor* or a *synchronous condenser*.

The following problems show how a synchronous motor is used to overcome the lagging kilovars due to induction motors on a three-phase distribution system.

PROBLEM 1

Statement of the Problem

A three-phase, 240-V feeder supplies two motors. One motor is a three-phase, wound-rotor induction motor. It takes 40 A at 81% lag power factor. The other motor is a three-phase, synchronous motor that takes 30 A at 65% lead power factor. Determine

1. the watts, volt-amperes, and lagging VARs of the wound-rotor induction motor.
2. the watts, volt-amperes, and leading VARs of the synchronous motor.
3. the total load, in kilowatts, supplied to the two motors.
4. the power factor of the three-phase feeder circuit.
5. the line current of the three-phase feeder circuit.

Solution

1. The apparent power in volt-amperes taken by the wound-rotor induction motor is

$$\begin{aligned} \text{VA} &= \sqrt{3} \times \text{V}_{\text{line}} \times \text{I}_{\text{line}} \\ &= 1.73 \times 240 \times 40 = 16{,}608 \text{ volt-amperes (VA)} \end{aligned}$$

The input in watts to the wound-rotor induction motor is

$$\begin{aligned} \text{P} &= \sqrt{3} \times \text{V}_{\text{line}} \times \text{I}_{\text{line}} \times \cos\theta \\ &= 1.73 \times 240 \times 40 \times 0.81 = 13{,}452 \text{ W} \end{aligned}$$

The VARs taken by the motor is

$$\begin{aligned} \cos\theta &= 0.81 \text{ lag} \\ \theta &= 35.9^\circ \text{ lag} \\ \sin\theta &= 0.5864 = \frac{\text{VARs}}{16{,}608} = 9739 \text{ VARs} \end{aligned}$$

2. The apparent power, in volt-amperes, taken by the synchronous motor is

$$\begin{aligned} \text{VA} &= \sqrt{3} \times \text{V}_{\text{line}} \times \text{I}_{\text{line}} \\ &= 1.73 \times 240 \times 30 = 12{,}456 \text{ VA} \end{aligned}$$

The input in watts to the synchronous motor is

$$\begin{aligned} \text{P} &= \sqrt{3} \times \text{V}_{\text{line}} \times \text{I}_{\text{line}} \times \cos\theta \\ &= 1.73 \times 240 \times 30 \times 0.65 = 8096 \text{ W} \end{aligned}$$

The VARs for the synchronous motor is

$$\begin{aligned} \cos\theta &= 0.65 \\ \theta &= 49.5^\circ \text{ lead} \\ \sin\theta &= 0.7604 = \frac{\text{VARs}}{\text{VA}} \end{aligned}$$

$$0.7604 = \frac{\text{VARs}}{12{,}456} = 9516 \text{ VARs}$$

3. The total true power, in kilowatts, taken by the two motors is the arithmetic sum of the individual power values for the two motors:

$$\text{Total watts} = 13{,}452 + 8096 = 21{,}548 \text{ W} = 21{,}548 \text{ kW}$$

4. The reactive power for the entire three-phase feeder is the difference between the lagging and leading VARs:

$$9739 - 9516 = 223 \text{ VARs}$$

Thus, there are 245 VARs of lagging quadrature power that are not overcome by the synchronous motor. This value of lagging VARs combined with the total power in watts gives the apparent power in volt-amperes:

$$\begin{aligned} \text{VA} &= \sqrt{\text{watts}^2 + \text{VARs}^2} \\ &= \sqrt{21{,}548^2 + 223^2} \\ &= 21{,}549 \text{ volt-amperes} = 21{,}549 \text{ kVA} \end{aligned}$$

The power factor of the entire system is the ratio of the total true power, in kilowatts, to the total apparent power, in kilovolt-amperes:

$$\begin{aligned} \cos\theta &= \frac{\text{kW}}{\text{kVA}} \\ &= \frac{21{,}548}{21{,}549} \\ &= 0.9999 \text{ (unity power factor)} \end{aligned}$$

5. If the total apparent power is known for the three-phase system, the line current is determined as follows:

$$\begin{aligned} \text{kVA} &= \frac{\sqrt{3} \times V_{\text{line}} \times I_{\text{line}}}{1000} \\ 21{,}549 &= \frac{1.73 \times 240 \times I_{\text{line}}}{1000} \\ I_{\text{line}} &= 51.9 \text{ A} \end{aligned}$$

PROBLEM 2

Statement of the Problem

The load of an industrial plant consists of 600 kVA at a power factor of 0.75 lag. One large motor is to be added to the load and will require 150 kW. Determine the new load, in kVA, and the power factor, given that

1. the motor installed is an induction motor operating at a power factor of 0.85 lag.
2. the motor installed is a synchronous motor operating at a power factor of 0.80 lead.

Solution

1. **For the induction motor:**

$$\text{Original kW load} = 600 \times 0.75 = 450 \text{ kW}$$

$$\cos\theta = 0.75 \text{ lag} = 41.4^\circ \text{ lag}$$

$$\sin \theta = 0.6613 = \frac{\text{kVAR}}{699}$$

Kilovars = 396.6 kVAR, lagging

Added kVA load:

$$\cos \theta = \frac{\text{kW}}{\text{kVA}}$$

$$0.85 = \frac{150}{\text{kVA}}$$

$$\text{kVA} = 176.5$$

$$\cos \theta = 0.85$$

$$\theta = 31.8^\circ \text{ lag}$$

$$\sin \theta = 0.5270 = \frac{\text{kVAR}}{176.5}$$

$$\text{kVAR} = 0.5270 \times 176.5 = 93.0 \text{ kVAR, lagging}$$

Total load including induction motor:

$$\text{Total load, kW} = 450 + 150 = 600 \text{ kW}$$

$$\text{Total kilovar lag} = 396.8 + 93 = 489.8 \text{ kVAR lag}$$

$$\text{Total load, kVA} = \sqrt{\text{kW}^2 + \text{kVAR}^2}$$

$$= \sqrt{600^2 + 489.8^2} = 774.5 \text{ kVA}$$

Power factor of entire load including induction motor:

$$\cos \theta = \frac{\text{kW}}{\text{kVA}}$$

$$= \frac{600}{774.5} = 0.7747 \text{ lag}$$

2. **For the synchronous motor:**

   Original load, kW = 450 kW

   Original lagging kVAR load = 396.8 kVAR

kVA load:

$$\cos \theta = \frac{\text{kW}}{\text{kVA}}$$

$$0.80 = \frac{150}{\text{kVA}}$$

$$\text{kVA} = 187.5 \text{ kVA}$$
$$\cos \theta = 0.80 \text{ lead}$$
$$\theta = 36.9^\circ \text{ lead}$$
$$\sin \theta = 0.6004$$
$$\frac{\text{kVAR}}{\text{kVA}} = 0.6004$$
$$\frac{\text{kVAR}}{187.5} = 112.6 \text{ lead kVAR}$$

Total load including synchronous motor:

$$\text{Total load, kW} = 450 + 150 = 600 \text{ kW}$$

The original load has a quadrature power component of 396.8 lag kilovars. The synchronous motor has a quadrature power component of 112.6 lead kilovars. Therefore, the net load in kilovars is $396.8 - 112.6 = 284.2$ lag kVAR.

The total load in kVA, including the synchronous motor, is

$$\text{kVA} = \sqrt{\text{kW}^2 + \text{kVAR}^2} = \sqrt{600^2 + 284.2^2} = 664 \text{ kVA}$$

Compare the kVA load for a synchronous motor, 664 kVA, with that for the induction motor, 774.5 kVA. The actual true power is the same for both motors. The lower kVA value for the synchronous motor indicates that the power factor has improved greatly:

$$\cos \theta = \frac{\text{kW}}{\text{kVA}} = \frac{600}{664} = 0.9036 \text{ lag power factor for the entire system}$$

The synchronous motor can supply a leading reactive component of 112.6 kVAR. This value compensates in part for the lagging reactive component of the original load, 396.8 kVAR. A lagging component of 284.2 kVAR remains. It must be supplied from the source, as well as the true power component of 600 kW. Therefore, the power factor is still a lagging power factor, although it is improved. Compare this power factor with the system power factor determined for the first part of this problem.

STARTING LARGE SYNCHRONOUS MOTORS

The earlier part of this unit described how a synchronous motor is started. In general, the starting current for synchronous motors is less than that for squirrel-cage induction motors having the same horsepower and speed ratings. In most cases, it is preferred to start a synchronous motor at full voltage. Motors as powerful as 10,000 or 15,000 hp can be started at full voltage. This type of starting does not harm a motor that is properly designed. Full-voltage starting means that a simpler, less expensive starter is required as compared to reduced-voltage starting.

When the motor starting inrush causes voltage disturbances, reduced-voltage starting current can be used. Approximately 50% to 65% of the rated line voltage is applied to the motor through a reactor or an autotransformer starter.

At a reduced voltage, the motor starts as an induction motor. It then accelerates to a speed close to the synchronous speed when the autotransformer, or reactor starter, is shifted into the running position. At this point, the dc field circuit is energized.

Many automatic motor starters use special relays to energize the field circuit. The circuit must be energized at the correct instant to ensure that the maximum pullin torque is obtained with a minimum amount of disturbance to the system. After the motor is operating properly as a synchronous motor, the dc field excitation current is adjusted to give the desired power factor.

SYNCHRONOUS MOTOR RATINGS

The nameplate data of a synchronous motor contains the same information as the nameplate of an ac generator. The only exception is the output rating because the alternator is rated in kVA and the synchronous motor is rated in horsepower.

The nameplate of a synchronous motor also gives a power factor rating. This type of motor is usually rated at unity power factor, 90% lead power factor, or 80% lead power factor. The motor may be rated at other power factor values for special applications. If a motor is rated at unity power factor, it can be operated with a leading power factor. In this case, the mechanical load must be decreased so that only the rated ac stator current will flow at the reduced power factor. In other words, the mechanical load, in horsepower output, and the electrical load, in leading kilovars, must not exceed the rating of the motor. When a synchronous motor is rated at a leading power factor of 80% or 90%, it will have a larger current capacity for a given horsepower output. This is necessary if the rated horsepower output is to be supplied for the larger current at the reduced power factor.

Synchronous motor losses are the same as those for an ac generator. The losses include the power spent in the separately excited field, the mechanical friction losses consisting of the copper losses in the three-phase stator winding and the stray power losses, the windage losses, and the iron losses. The efficiency of the three-phase synchronous motor is slightly higher than that of an induction motor having the same speed and horsepower ratings.

SMALL SINGLE-PHASE SYNCHRONOUS MOTORS

The Warren Motor

Small single-phase synchronous motors are used in many timing applications, including electric clocks, time switches, graphical recording instruments, and stroboscope devices. These motors do not require dc field excitation. A widely used motor of this type is the Warren (or General Electric) clock motor. It consists of a laminated stator core with an exciting coil. Normally, this coil is wound for 120-V operation. The coil has two poles, and each pole is divided into two sections. One turn of heavy copper wire is placed over half of each pole to produce a rotating field effect.

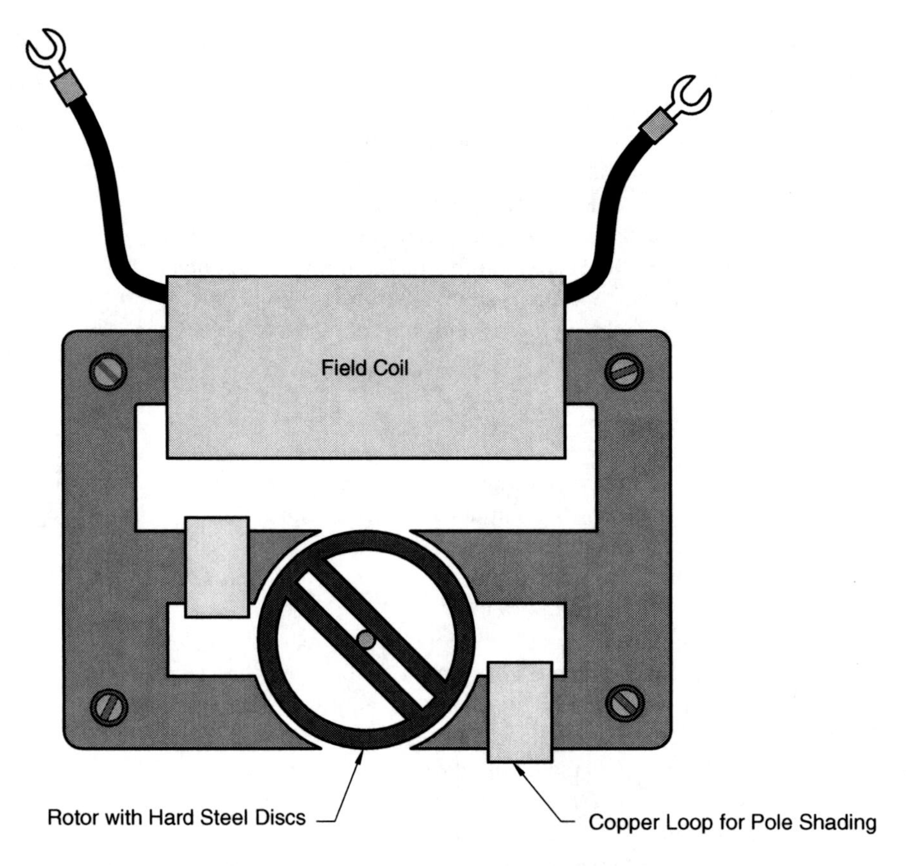

FIGURE 17–12 G.E. (Warren) clock motor (*Delmar/Cengage Learning*)

A General Electric (Warren) clock motor is shown in Figure 17–12. The rotor consists of several hardened steel discs that are pressed on the small rotor shaft. The discs have a high hysteresis loss. The rotating field acts on the rotor to produce strong torque. The rotor accelerates to a speed near synchronous speed. It then locks into synchronism with the rotating field. The flux of the rotating field seeks the path of minimum reluctance (resistance) through the two small rotor crossbars.

Operation with Two-Pole Stator

How is a rotating field effect obtained with the two-pole stator? As the flux increases in value, part of the flux attempts to pass through the section of the pole having the copper shading loop. This flux induces a voltage and current in the copper loop. These induced values oppose the flux that produces them. Thus, most of the flux passes through the nonshaded section of each pole. After the flux reaches its maximum value, there is no change in the flux instantaneously. Then, the induced voltage in the loop decreases to zero. The current in the copper loop also drops to zero, as well as the opposing mmf of the short-circuited coil. A large part of the main field flux now passes through this section of the pole piece. When the main field flux begins to decrease, voltage and current are induced again in the copper loop. This time, the mmf developed attempts to prevent the flux from decreasing in the shaded portion of each pole piece. In this way, a type of rotating field is created. The flux first reaches its maximum value in the nonshaded section of each pole and later reaches its maximum value in the shaded section.

The rotor speed for the single-phase, two-pole synchronous clock motor is 3600 r/min when operated from a 60-Hz source. A gear train is housed in a sealed case containing a light lubricating oil. This train reduces the speed of the motor to the value required for use in various timing devices.

The Holtz Motor

The Holtz motor (Figure 17–13) is another type of single-phase synchronous motor. This motor has the same shaded-pole arrangement as the General Electric (Warren) motor. For the Holtz motor, however, the rotor has six slots that hold a small squirrel-cage

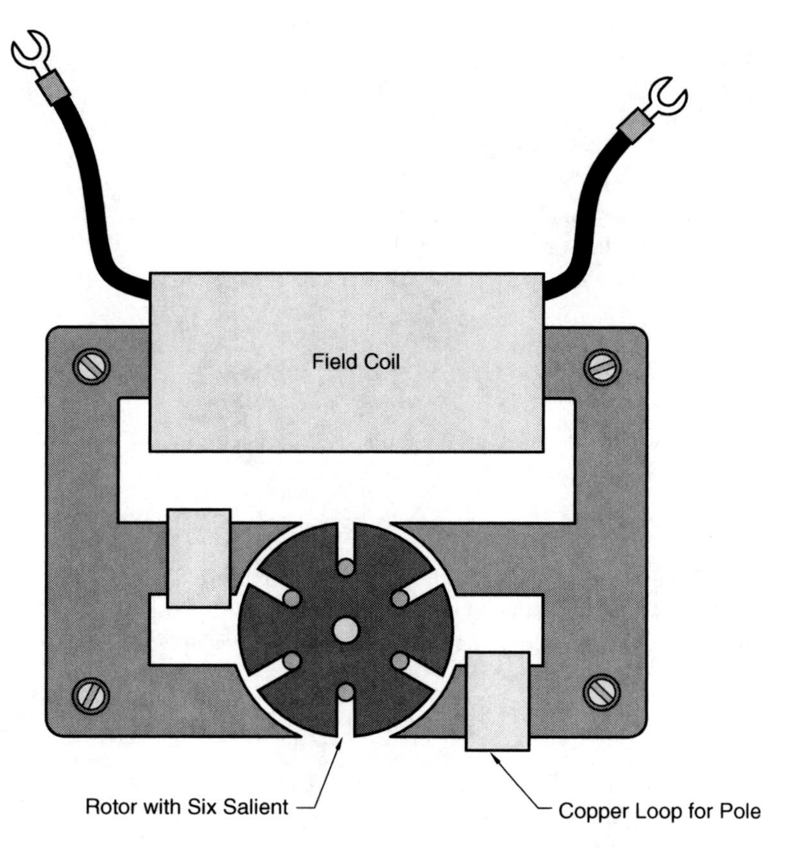

FIGURE 17–13 Holtz motor (*Delmar/Cengage Learning*)

winding. The rotor slots are arranged so that six salient field poles are formed. The Holtz motor starts as a squirrel-cage motor. When operated from a 60-Hz service, the salient poles of the rotor lock with the sections of the field poles each half-cycle. The resulting speed is one-third of the synchronous speed, or 1200 r/min.

SELSYN MOTORS

Selsyn is a contraction of the term *self-synchronous*. Selsyn motors provide a means of electrically interconnecting two or more remote points of a system. This means that they give automatic synchronized control or indication at one point, with respect to the other point. Selsyn units are extremely flexible and are used widely in industry.

For example, selsyn motors are used to give indications of the positions of remote devices such as generator rheostats, steam turbine governors, waterwheel governors, transformer tap connections, swing bridges, gates or valves, elevators, and the roll height in steel rolling mills. They are also used in many automatic control systems, signaling systems, and remote control systems.

Figure 17–14 shows the internal wiring of a selsyn motor. The rotor terminals are marked R_1 and R_2. The terminals marked S_1, S_2, and S_3 are the stator terminals. The rotor field circuit is excited from an external single-phase source. The stator windings consist of a conventional three-phase winding.

Two selsyn motors are shown in Figure 17–15. One motor is called the *transmitter*, and the other is called the *receiver*. Stators S_1, S_2, and S_3 of the transmitter are connected to S_1, S_2, and S_3, respectively, of the receiver. The rotors (R_1 and R_2) are tied together and are connected in common to an alternating-current source.

The transmitter can be turned by either manual or mechanical means. As it turns, the rotor follows at the same speed and in the same direction. The terms *transmitter* and *receiver* can be used interchangeably. Either selsyn can be the active transmitter or the passive receiver.

OPERATION

Figure 17–16 shows the vector relationships between the induced voltages of the stator for different positions of the rotor.

Case I

The rotor is lined up with coil S_3, resulting in the maximum magnetic coupling. There is a maximum induced voltage in S_3 and a partial induced voltage in both S_1 and S_2. Refer to the vector diagram and note that the resultant field of stators S_1, S_2, and S_3 lines up with the rotor field R_1.

Case II

The rotor is now in a position 30° from S_3. Therefore, no voltage is induced in S_1. A partial voltage is induced in S_2 and S_3. Again, the resultant field of stators S_1, S_2, and S_3 lines up with rotor field R_1.

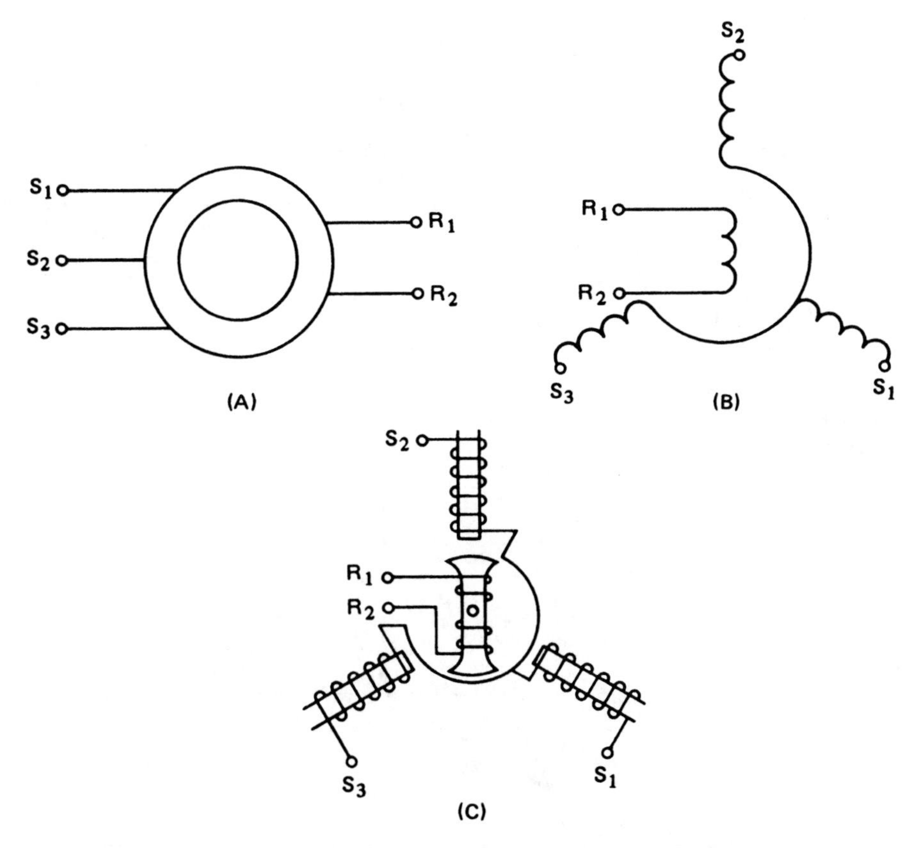

FIGURE 17–14 Internal wiring diagram of selsyn motor (*Delmar/Cengage Learning*)

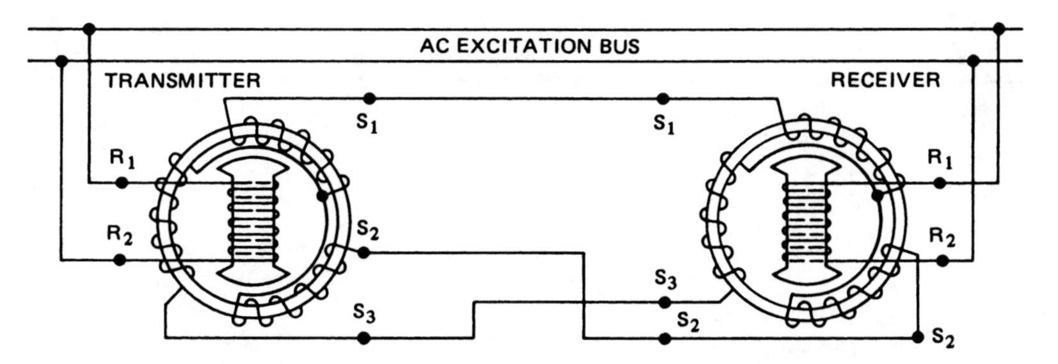

FIGURE 17–15 Two selsyn motors connected to an excitation source (*Delmar/Cengage Learning*)

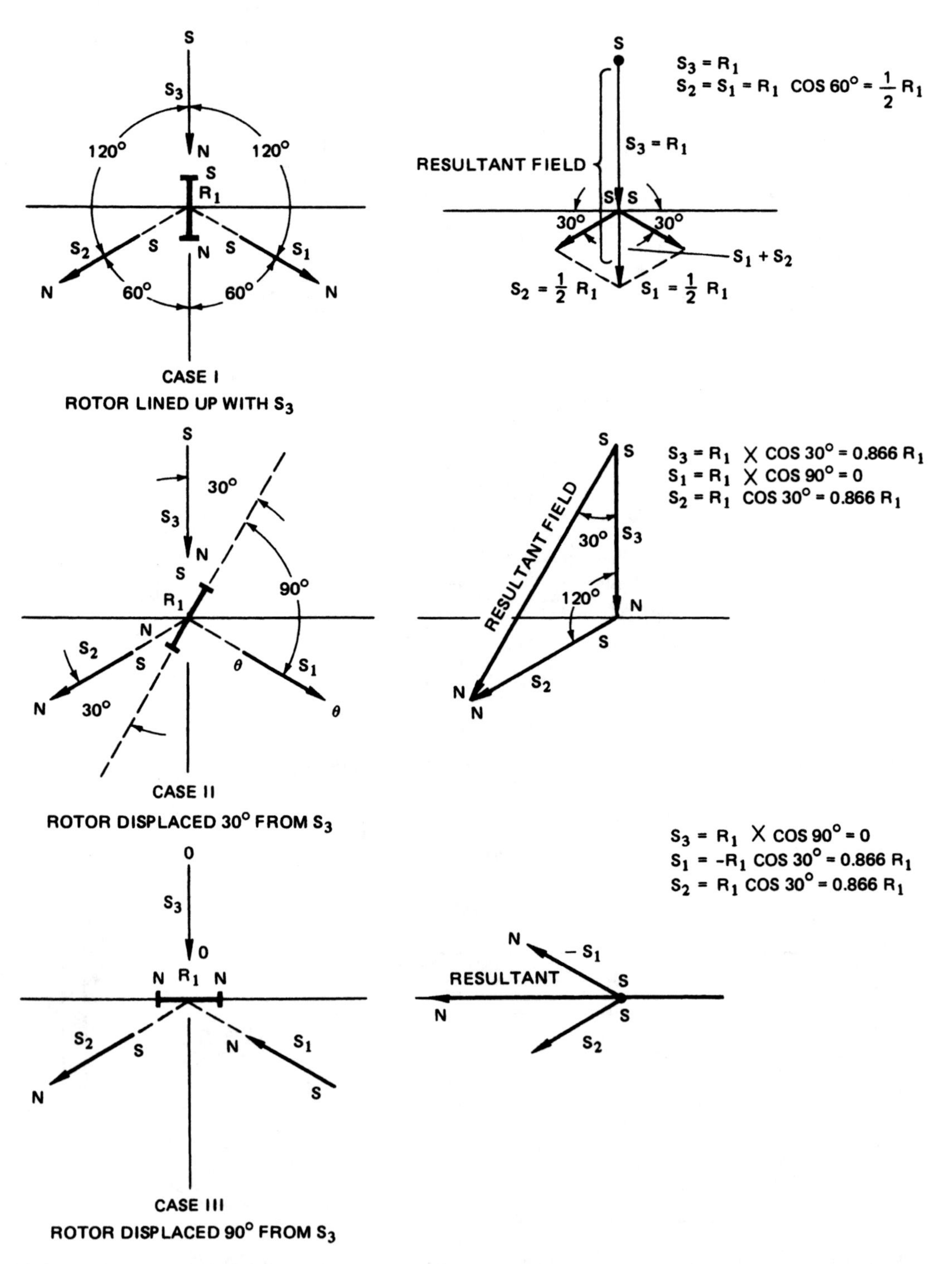

FIGURE 17–16 **Vector relationships of the stator electromotive forces as a function of the rotor position (*Delmar/Cengage Learning*)**

Case III

The rotor is now lined up on the X axis. The polarity of S_1 changes because the south pole of the rotor induces a north pole in S_1. In cases I and II, the north pole of the rotor induces a south pole in S_1. The change in polarity of S_1 means that its vector must be reversed. Thus, the resultant field vector lines up with the rotor field R_1.

Figure 17–14 shows that the three legs of each stator winding are connected in wye. The three voltages induced in the legs of the secondary windings are not equal in value. These voltages vary with the position of the rotor.

When the two rotors are in exactly corresponding positions, the voltages induced in the transmitter secondary winding are equal and opposite to those induced in the receiver secondary windings. Therefore, there is no current in the secondary windings.

If the transmitter rotor is moved from its original position, the induced voltages are not equal and opposite. As a result, current flows in the secondary or stator windings. This current sets up a torque. Because of this torque, the receiver rotor moves to the same position as the transmitter rotor.

SELSYN DIFFERENTIAL SYSTEM

A selsyn differential system consists of a transmitter, a differential, and a receiver, Figure 17–17. The differential selsyn resembles a miniature wound-rotor induction motor. There are three rotor windings, usually connected in wye. The windings are brought out through three slip rings to external terminals R_1, R_2, and R_3. The stator windings are also connected in wye. They are brought out to terminals S_1, S_2, and S_3. The differential selsyn is really a single-phase transformer. Three-phase voltages and currents are *not* present.

The stator windings serve as the primary side. The rotor windings serve as the secondary side. The differential selsyn is used to modify the electrical angle transmitted by the transmitter selsyn. This means that the selsyn receiver will take up a position that is either the sum or the difference of the angles applied to the selsyn transmitter and the differential selsyn. Thus, two selsyn generators (transmitters) can be connected through a differential selsyn and turned through any angle. The differential selsyn will indicate the difference between the two angles.

For the differential system shown in Figure 17–17, the voltage distribution in the primary winding is the same as that in the secondary winding of the excited selsyn. If any one of the three selsyns is fixed in position, and a second selsyn is displaced by a certain angle, the third selsyn will turn through the same angle. The direction of rotation can be reversed by reversing any pair of leads to the rotor or stator windings of the differential selsyn. If any two of the selsyns are rotated at the same time, the third selsyn will rotate through an angle equal to the algebraic sum of the movement of the first two selsyns. The algebraic sign of the angle depends on the physical direction of rotation of the rotors and the phase rotation of the windings.

The differential selsyn is not connected directly to the excitation source. The excitation current for the differential must be furnished through one, or both, of the standard selsyn units. Generally the excitation current is supplied to the primary (stator) windings of the differential selsyn from the exciter selsyn. For this reason, the exciter selsyn is always designed to be larger in capacity than the selsyn transmitter or receiver.

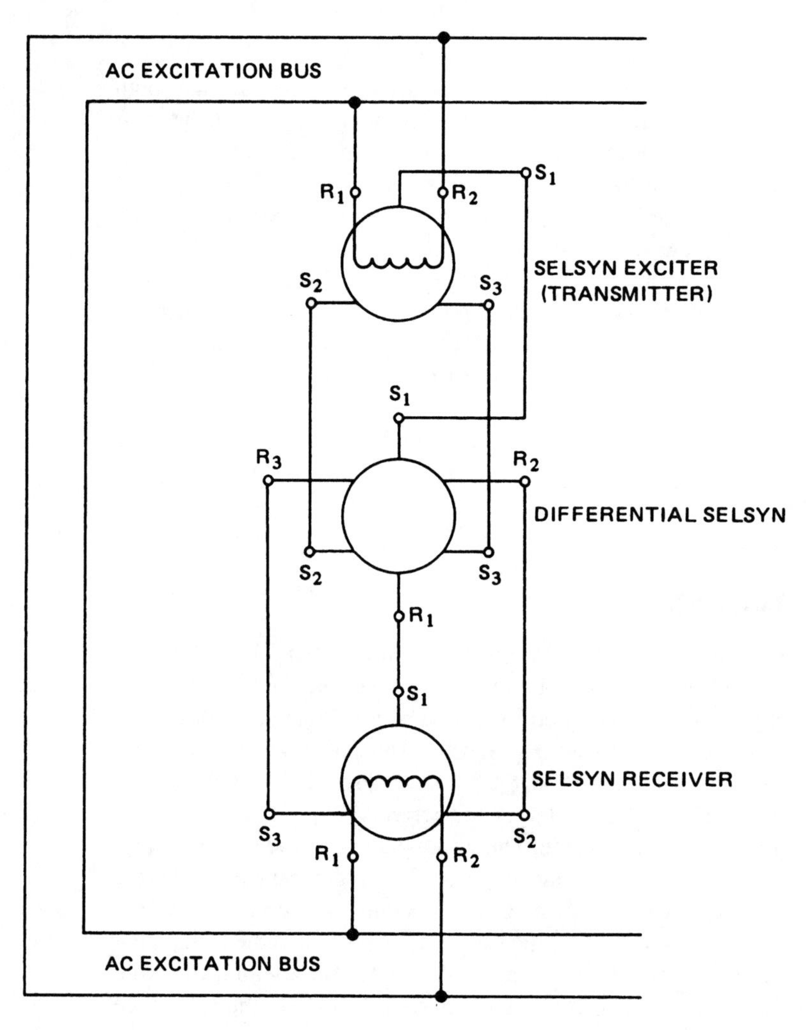

FIGURE 17–17 **Schematic diagram of a differential selsyn system (*Delmar/Cengage Learning*)**

SUMMARY

- The synchronous motor
  1. It is a three-phase ac motor that operates at a constant speed from no load to full load.
  2. It is similar to a three-phase generator.
  3. It has a revolving field. This field is separately excited from a direct-current source, which can be varied to obtain a wide range of lagging and leading power factor values.

4. The synchronous motor is used in many industrial applications because
   a. it has a fixed speed from no load to full load.
   b. it has a high efficiency.
   c. the initial cost is low.
   d. it can improve the power factor of three-phase ac industrial circuits.
5. Construction
   a. The three-phase synchronous motor consists of
      (1) a laminated stator core with a three-phase armature winding.
      (2) a revolving field with an amortisseur winding and slip rings.
      (3) brushes and brush holders.
      (4) two end shields that house the bearings supporting the rotor shaft.
   b. The leads for the stator winding are marked T_1, T_2, and T_3.
   c. The field is marked F_1 and F_2.
6. Operating principles
   a. $\text{Synchronous speed} = \dfrac{120 \times \text{frequency}}{\text{number of poles}}$
   b. A three-phase voltage applied to the stator windings causes a rotating magnetic field.
      (1) This field travels at the synchronous speed.
      (2) The field cuts across the amortisseur (squirrel-cage) winding of the rotor.
      (3) A magnetic field is set up in the rotor. This field reacts with the stator field and causes the rotation of the rotor.
      (4) The rotor speed reached 95% to 97% of synchronous speed.
   c. A synchronous motor is started by connecting its armature winding (stator) to the ac line and its field windings (rotor) to a field discharge resistor. The motor is started as an induction motor.
      (1) The voltage induced in the field windings may reach 1500 V. Therefore, the field circuits must be well insulated and enclosed to protect personnel.
      (2) The field discharge resistor is connected across the field windings so that the energy in the field circuit is spent in the resistor.
   d. Once the motor accelerates to nearly 95% of synchronous speed
      (1) the field circuit is energized from the dc source.
      (2) the field discharge resistor is disconnected.
      (3) the rotor pulls into synchronism with the revolving armature (stator) flux. Thus, the motor will operate at a constant speed.
   e. To shut down the motor, the field circuit is deenergized by opening the field discharge switch. The field discharge resistor is connected across the field circuit to reduce the induced voltage in the field as the field flux collapses. The energy stored in the magnetic field is spent in the resistor, and a lower voltage is induced in the field circuit.

- Load on a synchronous motor
  1. Once the rated load is applied, there is an angular displacement of the rotor pole with respect to the stator pole.
  2. The speed is unchanged because the rotor continues to rotate at the synchronous speed.
  3. The angular displacement between the centers of the stator and rotor field poles is called the *torque angle* and is represented by the Greek letter alpha, α.
     a. At no load, the torque angle is nearly 0°.
     b. As the mechanical load increases, the torque angle increases. The phase angle between the impressed voltage and the counter emf also increases. This increase means that the impressed voltage will cause more current in the stator windings to meet the additional load demands.
     c. A serious overload causes too large a value of the phase angle. In this case, the rotor pulls out of synchronism. With the aid of the amortisseur winding, the motor will operate as an induction motor.
     d. The pullout torque is the maximum torque value that can be developed by a synchronous motor without dropping out of synchronism. The pullout torque is usually 150% to 200% of the rated torque output.

- Power factor of a synchronous motor
  1. Changes in dc field excitation alter the power factor of a synchronous motor.
     a. Increasing the resistance of the field rheostat lowers the field current. A poor lagging power factor results. As the field current decreases, more magnetizing current is supplied by the ac stator windings. This ac current lags the voltage by 90° and causes a low lagging power factor.
     b. As the field current increases, the stator supplies a smaller magnetizing component of current. Thus, the power factor increases.
  2. The field strength can be increased until the power factor is unity or 100%.
  3. When the power factor reaches unity, the three-phase ac circuit supplies energy current only.
  4. The amount of dc field excitation required to obtain a unity power factor is called the *normal field excitation.*
  5. Overexciting the dc field produces a demagnetizing component of current called the *quadrature lead current.* The motor has a leading power factor.

- Industrial applications
  1. Synchronous motors rated at 20 hp or more are used for constant-speed applications.
  2. They are used to drive large air and gas compressors that are operated at fixed speeds to maintain a constant output at the maximum efficiency.
  3. They are also used to drive dc generators, fans, blowers, and large pumps in water-pumping stations.

4. Three-phase synchronous motors can be used to drive mechanical loads and correct power factor values.
5. When the synchronous motor is used to correct the power factor only, it is called a *synchronous capacitor* or a *synchronous condenser.* It then has the same function as a bank of capacitors.

- Formulas for synchronous motors
  1. Apparent power: $VA = \sqrt{3} \times V_{line} \times I_{line}$
  2. Input, in watts: $P = \sqrt{3} \times V_{line} \times I_{line} \times \cos\theta$
  3. VARs for the synchronous motor: $VARs = V \times A \times \sin\theta$
- Synchronous motor used as a synchronous capacitor (syn. cap.)
  1. Total watts taken by the circuit:

     $$\text{Total watts} = P_{motor} + P_{syn.\,cap.}$$

  2. The reactive power for the entire three-phase feeder is the difference between the lagging and leading VARs:

     $$VARs_T = VARs_{\text{Taken by motor}} - VARs_{\text{Supplied by syn. cap.}}$$

  3. $\text{Total VA} = \sqrt{\text{watts}^2 + \text{VARs}^2}$
  4. Power factor of the entire system:

     $$PF = \cos\theta = \frac{kW}{kVA}$$

  5. If the total apparent power is known, the line current is

     $$I_{line} = \frac{VA}{\sqrt{3} \times V_{line}}$$

- Generally, a synchronous motor is started at full voltage. Motors rated as high as 10,000 or 15,000 hp may be started at full voltage.
  1. When the starting inrush current is high enough to cause voltage disturbances, reduced voltage starting current can be used.
  2. Fifty percent to 65% of the rated line voltage is applied to the motor through a reactor or an autotransformer starter.
  3. At a reduced voltage, the motor starts as an induction motor. It accelerates to a speed close to the synchronous speed. Then the autotransformer, or reactor starter, is shifted into the running position. At this point, the dc field circuit is energized.
- Nameplate data
  1. The nameplate data for the synchronous motor is the same as that for an ac generator. A horsepower rating is given for the synchronous motor instead of a kVA rating.
  2. The mechanical load, in horsepower output, and the electrical load, in leading kilovars, must not exceed the rating of the motor.

- Synchronous motor losses are the same as those for an ac generator.
- The efficiency of the three-phase synchronous motor is slightly higher than that of an induction motor having the same speed and horsepower ratings.
- Small single-phase synchronous motor
  1. This type of motor is used in many applications, including electric clocks, time switches, graphical recording instruments, and stroboscope devices.
  2. These motors do not require dc field excitation.
  3. The Warren (or General Electric) clock motor is a commonly used motor of this type.
     a. It consists of a laminated stator core with an exciting coil. Normally, this coil is wound for 120-V operation.
     b. The coil has two poles, and each pole is divided into two sections. One turn of heavy copper wire is placed over half of each pole to produce a rotating field effect.
     c. The rotor consists of several hardened steel discs that are pressed on the small rotor shaft. The rotating field acts on the rotor to produce a strong torque. The rotor accelerates to a speed near synchronous speed. It then locks into synchronism with the rotating field. The flux of the rotating field seeks the path of minimum reluctance (resistance) through the two small rotor crossbars.
  4. The Holtz motor
     a. This type of motor has the same shaded-pole arrangement as the Warren motor.
     b. The rotor has six slots that hold a small squirrel-cage winding.
     c. The motor starts as a squirrel-cage motor. The salient poles of the rotor then lock with the sections of the field poles twice each cycle. The resulting speed at 60 Hz is one-third of the synchronous speed, or 1200 r/min.
  5. Selsyn motors
     a. Selsyn is a contraction of the term *self-synchronous.*
     b. Selsyn motors are used to interconnect two or more remote points of a system by electrical means.
     c. A selsyn system indicates the positions of remote devices such as generator rheostats, steam turbine governors, waterwheel governors, transformer tap connections, swing gates or valves, elevators, and the roll height in steel rolling mills. Selsyn motors are also used in automatic control systems, signaling systems, and remote control systems.
     d. The rotor terminals are marked R_1 and R_2. The stator terminals are marked S_1, S_2, and S_3.
     e. The rotor field circuit is excited from an external single-phase source.
     f. The stator windings consist of a conventional three-phase winding.
     g. One motor of a selsyn system is called the *transmitter* and the other motor is called the *receiver.*

(1) The stator terminals of one motor are wired to the stator terminals of the other motor in sequence.

(2) The rotors (R_1 and R_2) are tied together and are connected in common to an alternating-current source. Turning the rotor of one motor causes the rotor of the other motor to be displaced an equal amount.

6. Selsyn differential system
   a. A differential unit is added between the transmitter and the receiver of the common selsyn system.
   b. The differential resembles a miniature wound-rotor induction motor. There are three rotor windings. Leads R_1, R_2, and R_3 are brought out through three slip rings. The stator windings are marked S_1, S_2, and S_3.
   c. The differential selsyn is used to modify the electrical angle transmitted by the transmitter selsyn. The differential selsyn indicates the sum or difference between the two selsyns.

Achievement Review

1. What are the factors that determine the speed of a synchronous motor?
2. Explain how the operation of a synchronous motor is different from that of a squirrel-cage induction motor.
3. What is the purpose of the amortisseur winding in a synchronous motor?
4. Explain how a synchronous motor adjusts its electrical input with an increase in the mechanical output.
5. What is the purpose of the field discharge resistor that is used with the separately excited dc field circuit of a synchronous motor?
6. A motor generator set is used for frequency conversion from 60 Hz to another frequency. A three-phase, 60-Hz synchronous motor with 24 poles is used as the prime mover. The ac generator has 20 poles.
   a. Find the speed of the synchronous motor.
   b. Find the frequency of the output voltage of the ac generator.
7. When is a synchronous motor underexcited? When is it overexcited?
8. Describe a method for starting a large three-phase synchronous motor.
9. A three-phase, three-wire, 208-V feeder supplies two motors. One motor is a three-phase, squirrel-cage induction motor and requires 50 A at 0.80 lagging power factor. The other motor is a three-phase synchronous motor that requires 40 A at a leading power factor of 0.75.
   a. For the squirrel-cage induction motor, determine
      (1) the apparent power, in kilovolt-amperes.

   (2) the true power, in kilowatts.
   (3) the reactive power, in kVARs.
   b. For the three-phase synchronous motor, determine
      (1) the apparent power, in kilovolt-amperes.
      (2) the true power, in kilowatts.
      (3) the reactive power, in kVARs.

10. For the circuit given in question 9, determine
    a. the total true power, in kilowatts, supplied by the feeder circuit.
    b. the total apparent power, in kVA, supplied by the feeder circuit.
    c. the power factor of the feeder circuit. Indicate whether it is a leading or lagging power factor.
    d. the quadrature power, in kVARs, supplied by the feeder circuit.
    e. the total current, in amperes, supplied by the three-phase, three-wire feeder to operate both motors.

11. A three-phase, three-wire, 2400-V feeder supplies an industrial plant that has a 500-kVA load operating at a lagging power factor of 0.75. A 2400-V, three-phase synchronous motor is to be installed. This motor will require 200 kVA at a power factor of 0.80 lead. With the synchronous motor in operation, determine
    a. the total load, in kW, on the 2400-V feeder.
    b. the total load, in kVA, on the 2400-V feeder.
    c. the power factor on the 2400-V feeder. Indicate whether it is a leading or lagging power factor.

12. What are some practical applications for three-phase synchronous motors?

13. Explain the operation of a General Electric (Warren) clock.

14. Explain the operation of a single-phase Holtz motor.

15. Draw a circuit diagram for a selsyn circuit consisting of a selsyn transmitter and a selsyn receiver.

16. Explain the operation of the selsyn circuit in question 15.

17. Draw a circuit diagram for a selsyn circuit, including a selsyn transmitter, a differential selsyn, and a selsyn receiver.

18. Explain the operation of the differential selsyn system in question 17.

19. List some practical uses for selsyn units.

18

Single-Phase Motors

Objectives

After studying this unit, the student should be able to

- discuss the single-phase induction motor and its general characteristics, principles of operation, and applications.
- describe the construction and connections of a resistance-start, induction-run motor.
- describe the capacitor-start, induction-run motor, including the principle of operation.
- compare the capacitor-start, induction-run motor to a capacitor-start, capacitor-run motor.
- compare the operating characteristics of a repulsion-start, induction-run motor with those of a resistance-start, induction-run motor.
- describe how the brush holder of a repulsion-start, induction-run motor is shifted to change the direction of rotation of the motor.
- list the common applications for repulsion-type motors.
- explain why a dc series or shunt motor cannot be operated on an ac supply.
- use sketches to show the location of the shading poles for a shaded-pole motor.
- explain why the shaded-pole motor does not require a centrifugal starting device or other starting means.

INTRODUCTION

There are a number of types of single-phase motors. The types covered in this unit are the single-phase induction motor, the repulsion motor, the repulsion-induction motor, the series motor, and the shaded-pole motor. Three-phase motors generally perform better than single-phase motors. However, in many instances, only single-phase service is available. Most single-phase motors have fractional horsepower ratings. In general, their use is limited to commercial and residential applications.

SINGLE-PHASE INDUCTION MOTORS

There are two main types of single-phase induction motors: the resistance-start, induction-run motor and the capacitor-start, induction-run motor. These motors have fractional horsepower ratings. The resistance-start induction motor is used in appliances and with other small loads where a strong starting torque is not required. The capacitor-start induction motor is used on refrigerators, compressors, and similar loads. Both types of motors are low in cost, are rugged, and have good performance. These induction motors are also called *split-phase motors*. That is, the capacitor or resistance is used to obtain a phase change for one winding, resulting in a rotating field.

RESISTANCE-START, INDUCTION-RUN MOTOR

Construction

The basic parts of a resistance-start, induction-run motor are

- the stator (stationary part).
- the rotor (revolving part).
- a centrifugal switch, located inside the motor.
- two end shields bolted to the steel frame; these shields house the rotor shaft bearings.
- a cast steel frame; the stator core is pressed into this frame.

Stator. The stator consists of two windings, held in place in the slots of a laminated steel core. The windings are made up of insulated coils. The coils are placed so that they are 90 electrical degrees apart. One winding is the main (running) winding. The other winding is the auxiliary (starting) winding.

The running winding is made of a heavy insulated copper wire. It is located at the bottom of the stator slots. A small wire size is used in the starting winding. This winding is placed near the top of the slots above the running winding.

At start-up, both windings are connected in parallel to the single-phase line. Once the motor accelerates to two-thirds or three-quarters of the rated speed, a centrifugal switch disconnects the starting winding automatically.

Rotor. The rotor of the resistance-start, induction-run motor (Figure 18–1) is the same as that of a three-phase, squirrel-cage induction motor. The cylindrical core of the rotor

FIGURE 18–1 Cast aluminum rotor, shaft, and bearing of a single-phase induction motor (*Delmar/Cengage Learning*)

consists of steel laminations. Copper bars are mounted near the surface of the rotor and are brazed or welded to two copper end rings. The rotor may also be a one-piece case aluminum unit. This type of rotor requires very little maintenance. It contains no windings, brushes, slip rings, or commutator where faults may develop. Fans are provided with the rotor to keep the temperature of the windings at a reasonable level.

Centrifugal Switch. The centrifugal switch is mounted inside the motor. It disconnects the starting winding after the rotor reaches a predetermined speed. The switch has a stationary part that is mounted on one of the end shields. This part has two contacts that act like a single-pole, single-throw switch. The centrifugal switch also has a rotating part that is mounted on the rotor. A typical centrifugal switch is shown in Figure 18–2. This switch is commonly used on split-phase induction motors.

The operation of a centrifugal switch is shown in Figure 18–3. When the rotor is not moving, the pressure of a spring on the fiber ring of the rotating part of the switch keeps the contacts closed. At three-quarters of the rated motor speed, the centrifugal action of the

FIGURE 18–2 Centrifugal switch (*Delmar/Cengage Learning*)

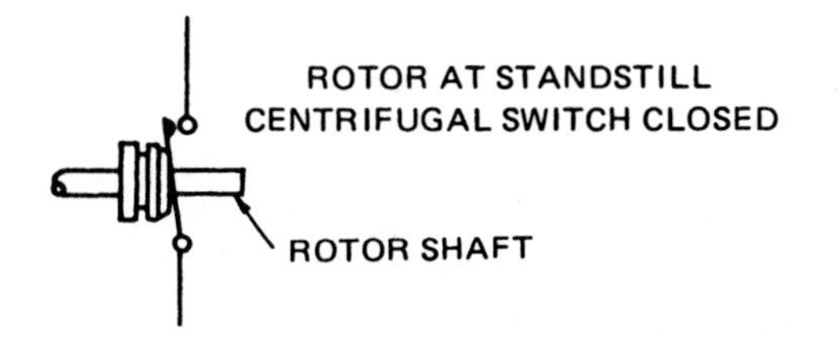

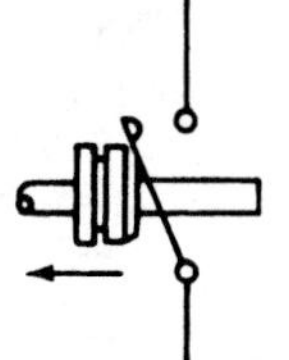

FIGURE 18–3 Centrifugal switch operation (*Delmar/Cengage Learning*)

rotor causes the spring and fiber ring to release its pressure and opens the switch contacts. As a result, the starting winding circuit is disconnected from the line.

Frame and End Shields. The stator core is pressed directly into the cast steel frame. The two end shields are bolted to the frame. The shields contain bearings that support the rotor and center it in the stator. Thus, the shaft rotates with little friction and does not strike or rub against the stator core.

Principle of Operation

At the instant the motor circuit is closed, both the starting and running windings are energized. Because the running winding uses large-size wire, it has a low resistance. But the running winding is placed at the bottom of the stator core slots. Thus, its inductive reactance is high. Because of its low resistance and high inductive reactance, the current of the running winding lags behind the voltage.

The current in the starting winding is more nearly in phase with the voltage. Small wire is used in this winding, resulting in a high resistance. Because the winding is near the top of the stator slots, the mass of iron surrounding it is small. This means that its inductive reactance is small. Because the starting winding has a high resistance and a low inductive reactance, its current is more nearly in phase with the voltage.

Figure 18–4 shows the relationship between the currents in the windings and the voltage. The current of the main winding (I_M) lags the current of the starting winding (I_S) by nearly 90 electrical degrees. When a current passes through each of these windings, the resulting pulsating field effect gives rise to a rotating field around the inside of the stator core. The speed of this rotating magnetic field is determined in the same way it was found for a three-phase induction motor.

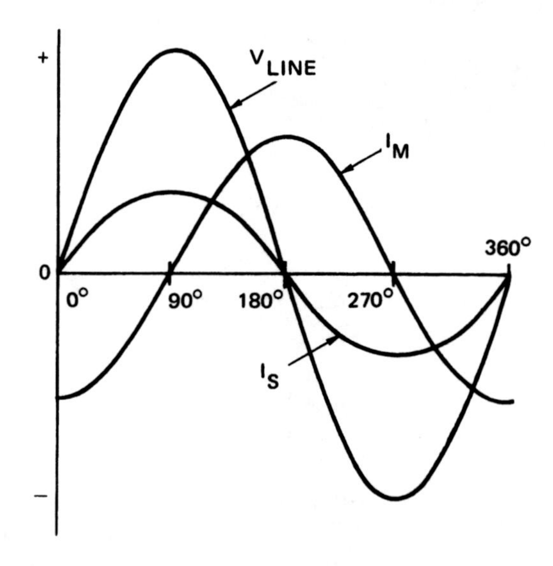

FIGURE 18–4 Phase relationship between the currents in the running and starting windings (*Delmar/Cengage Learning*)

Synchronous Speed. Consider a single-phase, resistance-start, induction-run motor with four poles wound into both the main windings and the starting winding. If this motor is energized from a 60-Hz source, the synchronous speed of the revolving field is

$$S = \frac{120\,f}{4} = \frac{120 \times 60}{4} = 1800 \text{ r/min}$$

where

S = synchronous speed, in r/min
f = frequency, in hertz

While traveling at the synchronous speed, the rotating field cuts the copper bars of the squirrel-cage winding. Voltages are induced in the windings and cause currents in the rotor bars. A rotor field is created. This field reacts with the stator field to develop the torque that causes the rotor to turn.

As the rotor accelerates to approximately 80% of the rated speed, the centrifugal switch disconnects the starting winding from the line. The connections for the centrifugal switch are shown in Figure 18–5. At start-up, the switch is closed. As the motor accelerates to its normal running speed, the centrifugal switch opens. The motor then continues to operate, using only the running winding.

Once the motor is running, current is induced in the rotor for two reasons: (1) the alternating stator flux induces "transformer voltage" in the rotor, and (2) "speed voltage" is induced in the rotor bars as they cut across the magnetic field of the stator. The combined effect of these alternating voltages produces a torque that keeps the rotor turning.

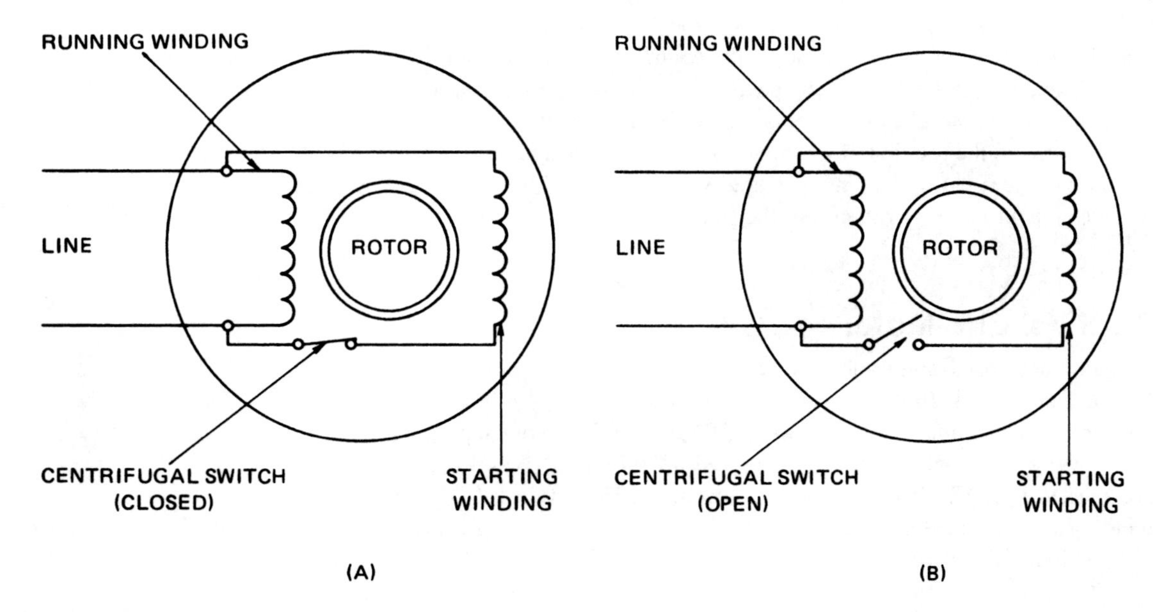

FIGURE 18–5 Connections of a centrifugal switch at (A) start-up and (B) when the motor is running (*Delmar/ Cengage Learning*)

Energized Windings. A resistance-start, induction-run motor must have the starting winding and the main winding energized at start-up. The motor resembles a two-phase induction motor because the currents of the windings are approximately 90 electrical degrees out of phase with each other. However, a single-phase source supplies the motor. Therefore, the motor is called a *split-phase motor* because it starts like a two-phase motor from a single-phase line. Once the motor accelerates to a value near the rated speed, it operates on the running winding as a single-phase induction motor.

If the centrifugal switch mechanism fails to close the switch contacts when the motor stops, the starting winding circuit will be open. This means that when the motor circuit is reenergized, the motor will not start. Both the starting and running windings must be energized at the instant the motor circuit closes if the necessary starting torque is to be formed. If the motor does not start, but a low humming sound is present, then one winding is open. The centrifugal switch should be checked to determine whether its mechanism is faulty or the switch contacts are pitted.

If only one stator winding is energized, an alternating field, rather than a rotating field, is formed. If the rotor is at rest, this field induces an alternating current in the rotor winding. This current acts as the secondary of a transformer. Rotor poles are developed by this induced current. These poles are exactly aligned with the stator poles. Thus, no starting torque is developed in either direction of rotation.

If a split-phase motor is started with too great a mechanical load, it may not accelerate enough to open the centrifugal switch. Also, if a low terminal voltage is applied to the motor, the motor may fail to reach the speed required to operate the centrifugal switch.

Starting Winding. The starting winding uses a small size of wire, resulting in a large resistance. The starting winding is designed to operate across the line voltage for just three or four seconds as the motor accelerates to the rated speed. Therefore, the starting winding must be disconnected from the line by the centrifugal switch as soon as the motor accelerates to three-quarters of the rated speed. If the motor operates on its starting winding for more than sixty seconds, the winding may char or burn out.

To reverse the rotation of the motor, simply interchange the leads of either the starting winding or the running winding, preferably the starting winding. Interchanging these leads reverses the direction of rotation of the stator field (and the rotor).

Dual-Voltage Motor

In many cases, single-phase motors have dual voltage ratings such as 120 V and 240 V. The running winding consists of two sections, each of which is rated at 120 V. One section of the running winding is marked T_1 and T_2. The other section is marked T_3 and T_4. The starting winding consists of only one 120-V winding. The leads of the starting winding are marked T_5 and T_6. To operate the motor on 240 V, the two 120-V sections of the running winding are connected in series across the 240-V line. The starting winding is paralleled with one section of the running winding.

To operate the motor on 120 V, the two 120-V sections of the running winding are connected in parallel across the 120-V line. The starting winding is connected in parallel with both sections of the running winding.

Figure 18–6 shows the circuit connections for a dual-voltage motor connected for 120-V operation. The connections for 240-V operation are shown in Figure 18–7. In this case, the jumpers are changed in the terminal box so that the two 120-V running windings are connected in series. These sections are then connected across the 240-V line. Note that the 120-V starting winding is connected in parallel with one section of running winding. If the voltage drop across this section of the running winding is 120 V, then the voltage across the starting winding is 120 V. The starting winding may also consist of two sections for some dual-voltage, resistance-start, induction-run motors. As before, the two running winding sections are marked T_1 and T_2 and T_3 and T_4. The two starting winding sections are marked "T_5 and T_6" and "T_7 and T_8" (Figure 18–8).

The table in Figure 18–8 shows the correct connections for 120-V operation and for 240-V operation.

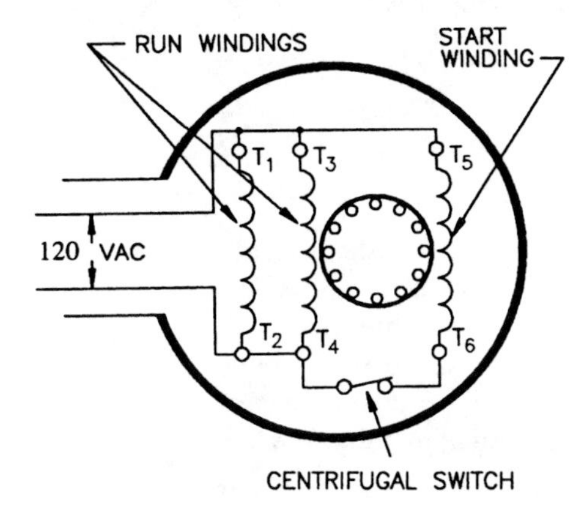

FIGURE 18–6 Dual-voltage motor connection for 120 V (*Delmar/Cengage Learning*)

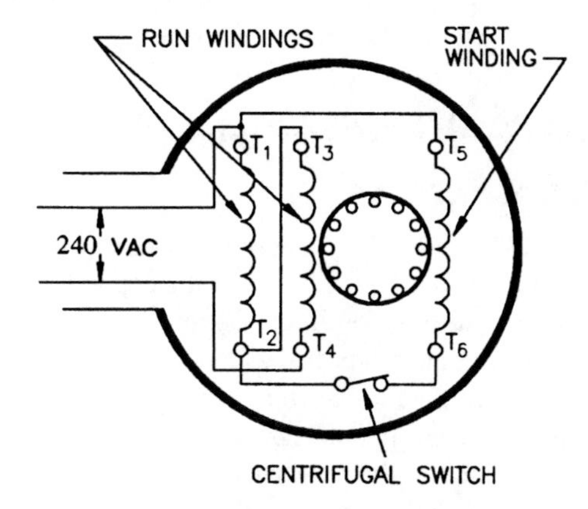

FIGURE 18–7 Dual-voltage motor connection for 240 V (*Delmar/Cengage Learning*)

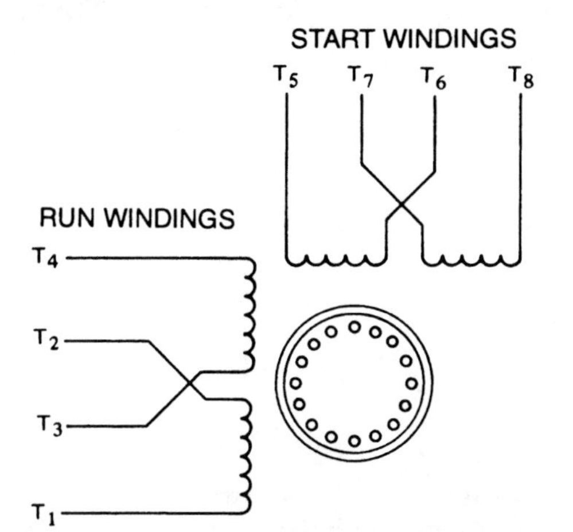

Voltage Rating	L_1	L_2	Tie Together
120 Volts	T_1, T_3, T_5, T_7,	T_2, T_4, T_6, T_8,	—
240 Volts	T_1, T_5,	T_4, T_8,	T_2 and T_3, T_6 and T_7

FIGURE 18–8 Winding arrangement for dual-voltage motor with two starting and two running windings (*Delmar/Cengage Learning*)

Speed Regulation. A resistance-start, induction-run motor has very good speed regulation. From no load to full load, the speed performance of this type of motor is about the same as that of a three-phase, squirrel-cage induction motor. The percent slip on most fractional horsepower split-phase motors ranges from 4% to 6%.

Starting Torque. The starting torque of the motor is poor. Although the main winding consists of large wire, it has some resistance. Also, the starting winding has an inductive reactive component, even though it is placed near the top of the stator slots. As a result, the current in the main winding does not lag the line voltage by a full 90° because of the resistance. The current in the starting winding is not in phase with the line voltage because of the inductive reactance.

Figure 18–9 shows the phase angle between the current in the main winding and the current in the starting winding for a typical resistance-start, induction-run motor. A phase angle in the order of 30° to 50° is large enough to provide a weak rotating magnetic field. As a result, the starting torque is small.

CAPACITOR-START, INDUCTION-RUN MOTOR

Construction

Physically, the capacitor-start, induction-run motor is similar to the resistance-start, induction-run motor. However, the capacitor-start motor has a capacitor connected in series with the starting windings. Generally, the capacitor is mounted in any convenient external location on the motor frame. In some cases, the capacitor is mounted inside the motor housing. With regard to the starting torque, the capacitor-start motor has a higher torque than the resistance-start motor. The capacitor also limits the starting surge of current to a lower value as compared to the resistance-start motor.

The capacitor-start, induction-run motor is used in refrigeration units, compressors, oil burners, small-machine equipment, and in any application where split-phase induction motors are used. Because of its improved starting torque characteristics, the capacitor-start motor is replacing the split-phase motor in many applications.

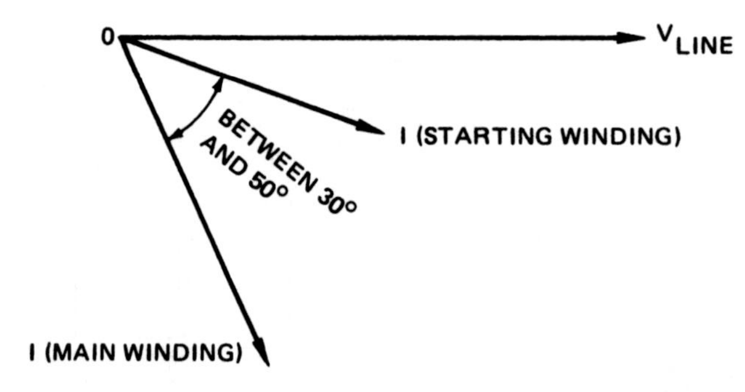

FIGURE 18–9 Phase angle between the starting and main winding currents in a resistance-start, induction-run motor (*Delmar/Cengage Learning*)

Principle of Operation

A typical capacitor-start, induction-run motor is shown in Figure 18–10. At start-up, both the running and starting windings are connected in parallel across the line voltage while the centrifugal switch is closed. The starting winding is also connected in series with the capacitor. When the motor reaches 75% of the rated speed, the centrifugal switch opens. The starting winding and the capacitor are disconnected from the line. The motor then operates on the running winding as a single-phase induction motor.

Starting Torque. As stated previously, the starting torque is due to to a revolving magnetic field that is set up by the stator windings. By adding a capacitor of the correct value in series with the starting winding, the current in this winding will lead the running winding current by 90 electrical degrees.

The angle between the starting winding current and the running winding current is almost 90° (Figure 18–11). The magnetic field set up by the stator windings is almost identical to that of a two-phase induction motor. Therefore, the starting torque for the capacitor-start motor is much better than the torque for a resistance-start, induction-run motor.

Capacitor Starting. This motor starts with the capacitor in the starting winding circuit. For normal running, only the running winding is energized and the motor operates as a single-phase induction motor. The capacitor is used to improve the starting torque. Because it is energized for just two or three seconds at start-up, the capacitor cannot make any improvement in the power factor.

FIGURE 18–10 Cutaway view of a fractional horsepower, capacitor-start motor (*Delmar/Cengage Learning*)

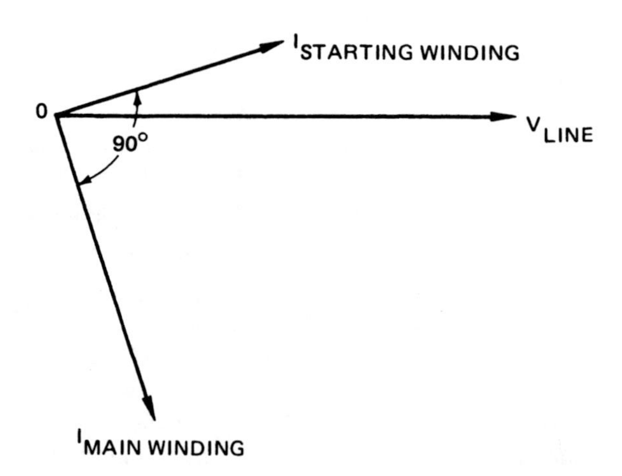

FIGURE 18–11 Phase angle of 90° between the starting winding and main winding currents in a capacitor start, induction-run motor (*Delmar/Cengage Learning*)

Defective capacitors often cause problems in the motors using them. For example, the capacitor may short-circuit and blow the fuse on the branch motor circuit. If the fuse rating is high so that the fuse does not interrupt the power supply to the motor, the starting winding may burn out. Starting capacitors may short-circuit if the motor is turned on and off many times in a short period of time. To prevent failures, many manufacturers recommend that this type of motor be started no more than twenty times per hour.

Paper or oil filled capacitors are used in these motors. If the motor is started too often in a short period of time, the current surge at start-up gradually damages the dielectric of the capacitor. Eventually, the dielectric breaks down and shorts (short-circuits) the capacitor plates. The capacitor-start motor is recommended for use only in those applications where there are few starts in a short period of time.

Speed Regulation. The speed regulation for a capacitor-start, induction-run motor is very good. The percent slip at full load is in the range from 4% to 6%. This means that the speed performance is the same as that of a resistance-start, induction-run motor.

Reversing the Motor. To reverse the direction of rotation of this motor, the leads of the starting winding circuit are reversed. The magnetic field developed by the start windings in the stator then rotates around the stator core in the opposite direction. Thus, the rotation of the rotor is reversed. The direction of rotation can also be reversed by interchanging the leads of the running windings.

The circuit connections for the capacitor-start motor are shown in Figure 18–12. In Figure 18–13, the starting winding leads are interchanged to reverse the direction of rotation of the motor.

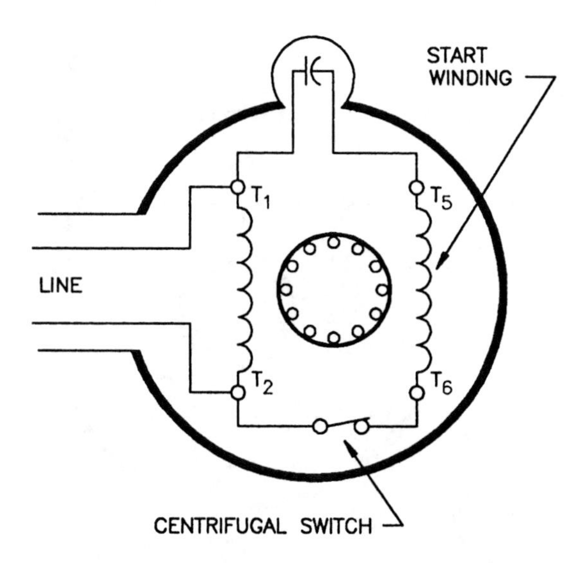

FIGURE 18–12 Connections for a capacitor-start, induction-run motor *(Delmar/Cengage Learning)*

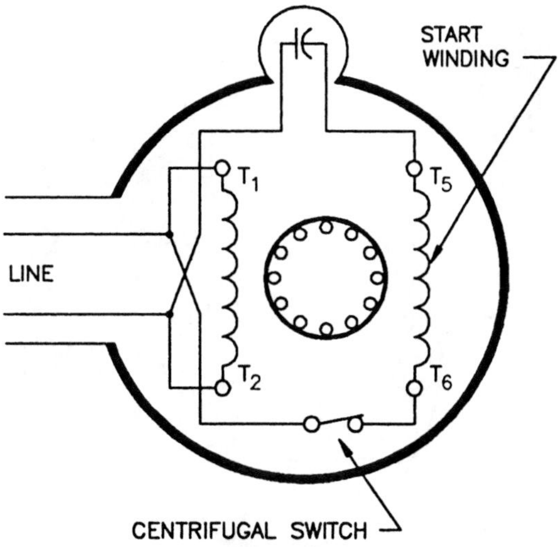

FIGURE 18–13 Connections for reversing a capacitor-start, induction-run motor *(Delmar/Cengage Learning)*

Dual-Voltage Ratings. Capacitor-start, induction-run motors can also have dual-voltage ratings of 120 V and 240 V. The connections in this case are the same as those for split-phase induction motors.

CAPACITOR-START, CAPACITOR-RUN MOTOR

In the capacitor-start, capacitor-run (CSCR) motor, the starting winding and the capacitor are connected into the circuit at all times. This motor has a very good starting torque. Because the capacitor is used at all times, the power factor at the rated load is 100%, or unity.

Operation

Several different versions of the CSCR motor are available. In one type of motor, two stator windings are placed 90 electrical degrees apart. The running winding is connected directly across the rated line voltage. A capacitor is connected in series with the starting winding. This winding is also connected across the rated line voltage. Because the starting winding is energized as long as the motor is operating, a centrifugal switch is not required. Figure 18–14 gives the internal connections for a CSCR motor.

Using One Capacitor. The motor shown in Figure 18–14 is quiet in operation. It is used on oil burners, fans, and small woodworking and metalworking machines. The direction of rotation of this motor is reversed by interchanging the lead connections of either winding.

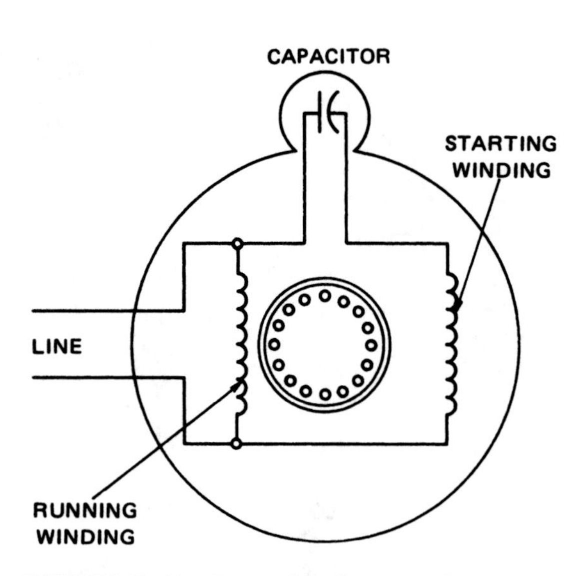

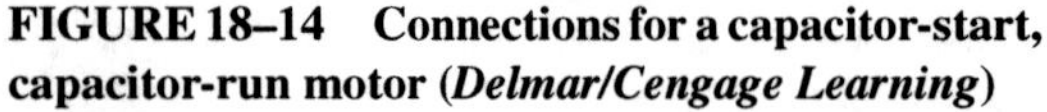

FIGURE 18–14 Connections for a capacitor-start, capacitor-run motor (*Delmar/Cengage Learning*)

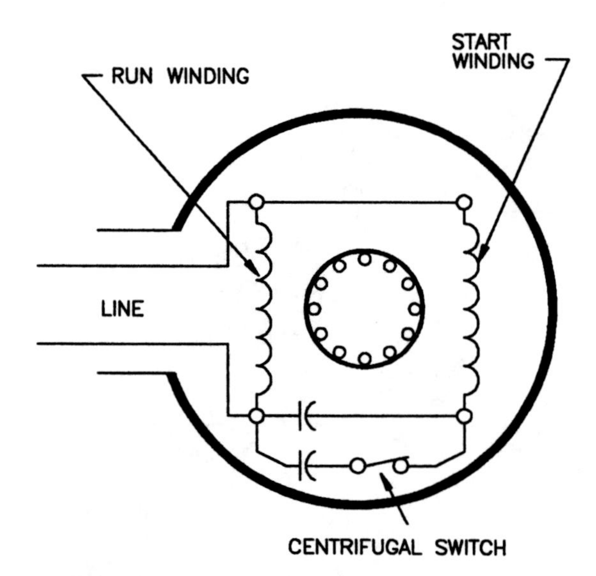

FIGURE 18–15 Connections for a capacitor-start, capacitor-run motor using two capacitors (*Delmar/Cengage Learning*)

Using Two Capacitors. A second type of motor uses two capacitors. At start-up, a large value of capacitance is connected in series with the starting winding, as shown in Figure 18–15.

At this moment, the two capacitors shown are in parallel. When the motor accelerates to nearly 75% of the rated speed, the centrifugal switch disconnects the larger capacitor. The motor then operates with the small-value capacitor connected in series with the starting winding.

This type of motor has a very good starting torque, good speed regulation, and a power factor of almost 100% at the rated load. It is used on furnace stokers, refrigerators, and compressors, and in other applications where its strong starting torque and good speed regulation are required.

Autotransformer and One Capacitor. Another version of the CSCR motor uses an autotransformer. One capacitor is used to obtain a high starting torque and a large power factor.

The internal connections for this motor are shown in Figure 18–16. As the motor starts, the centrifugal switch connects winding 2 to point A on the tapped autotransformer. The capacitor is connected across nearly 500 V. A large leading current is formed in winding 2 and a strong starting torque results.

When the motor reaches nearly 75% of the rated speed, the centrifugal switch disconnects the starting winding from point A and reconnects it to point B on the autotransformer. In this way, less voltage is applied to the capacitor. The motor operates with both windings energized. The capacitor maintains the power factors at nearly unity at the rated load.

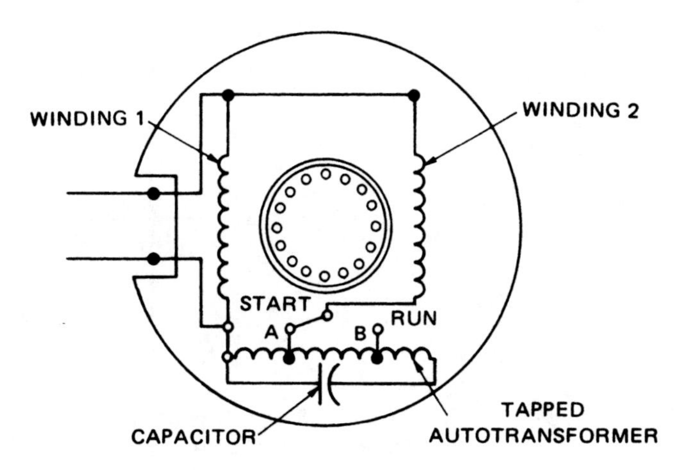

FIGURE 18–16 Connections for a capacitor-start, capacitor-run motor with an autotransformer (*Delmar/ Cengage Learning*)

The starting torque of this motor is very good, and its speed regulation is satisfactory. This motor is used in refrigerators, in compressors, and in other applications in which strong starting torque and good speed regulation are required.

MULTISPEED MOTOR

There are two basic types of multispeed motors. One type is known as the *consequent pole motor*. The other type is generally a capacitor-start, capacitor-run motor.

The Consequent Pole Motor

The speed of the rotating magnetic field of an ac induction motor can be changed in either of two ways:

1. Change the frequency of the ac voltage.
2. Change the number of stator poles.

The consequent pole motor changes motor speed by changing the number of its stator poles. The run winding in Figure 18–17 has been tapped in the center. If the ac line is connected to each end of the winding as shown, current flows through the winding in only one direction. Therefore, only one magnetic polarity is produced in the winding. If the winding is connected as shown in Figure 18–18, current flows in opposite directions in each half of the winding. Because current flows through each half of the winding in opposite directions, the polarity of the magnetic field is different in each half of the winding. The run winding now has two polarities instead of one. If the windings of a two-pole motor were to be tapped in this manner, the motor could become a four-pole motor. The synchronous speed of a two-pole motor is 3600 r/min, and the synchronous speed of a four-pole motor is 1800 r/min.

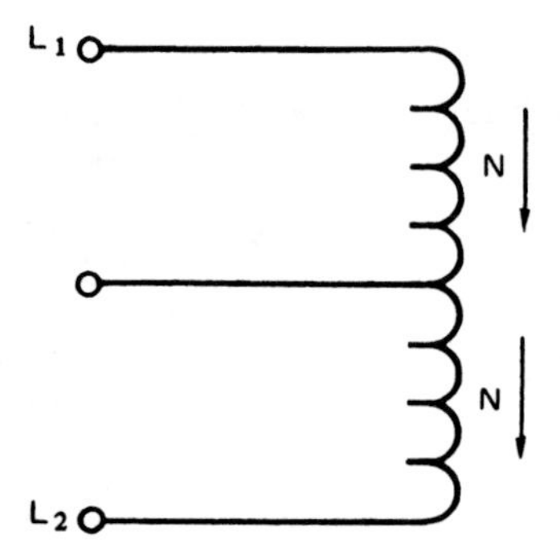

FIGURE 18–17 Center-tapped run winding (*Delmar/Cengage Learning*)

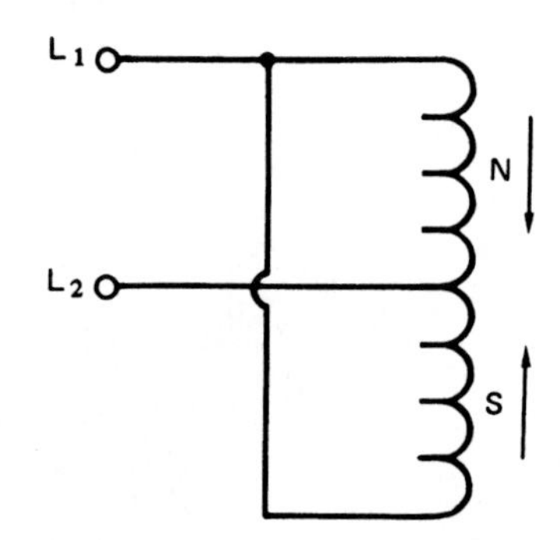

FIGURE 18–18 Two magnetic poles are produced (*Delmar/Cengage Learning*)

The consequent pole motor has the disadvantage of having a wide variation in speed. When the speed is changed, it changes from a synchronous speed of 3600 r/min to 1800 r/min. The speed cannot be changed by a small amount. This wide variation in speed makes the consequent pole motor unsuitable for some loads such as fans and blowers.

The consequent pole motor, however, does have some advantages over the other type of multispeed motor. When the speed on the consequent pole motor is reduced, its torque increases. For this reason, the consequent pole motor can be used to operate heavy loads.

Multispeed Fan Motors

Multispeed fan motors have been used in industry for many years. These motors are generally wound for two to five steps of speed and are used to operate fans and squirrel-cage blowers. A schematic drawing of a three-speed motor is shown in Figure 18–19. Notice that the run winding has been tapped to produce low, medium, and high speed. The start

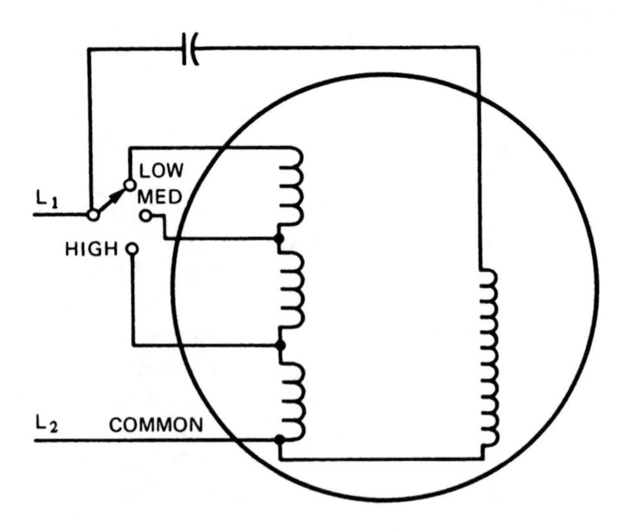

FIGURE 18–19 Three-speed fan motor (*Delmar/Cengage Learning*)

winding is connected in parallel with the run winding section. The other end of the start lead is connected to an external oil-filled capacitor. This motor obtains a change in speed by inserting inductance in series with the run winding. The actual run winding for this motor is between the terminals marked "High" and "Common." The windings shown between "High" and "Medium" are connected in series with the main run winding. When the rotary switch is connected to the medium-speed position, the inductive reactance of this coil limits the amount of current flow through the run winding. When the current of the run winding is reduced, the strength of the magnetic field of the run winding is reduced and the motor produces less torque. This causes the motor speed to decrease.

If the rotary switch is changed to the low position, more inductance is connected in series with the run winding. This causes less current to flow through the winding and another reduction in torque. When the torque is reduced, the motor speed decreases again.

Common speeds for a four-pole motor of this type are 1625, 1500, and 1350 r/min. Notice that this motor does not have the wide range between speeds that the consequent pole motor does. Most induction motors would overheat and damage the motor windings if the speed were to be reduced to this extent. This motor, however, has much higher impedance windings than do most motors. The run windings of most split-phase motors have a wire resistance of 1 to 4 Ω. This motor will generally have a resistance of 10 to 15 Ω in its run winding. It is the high impedance of the windings that permits the motor to be operated in this manner without damage.

Because this motor is designed to slow down when load is added, it is not used to operate high-torque loads. This type of motor is generally used to operate only low-torque loads such as fans and blowers.

REPULSION-TYPE MOTORS

There are three general types of repulsion motors: the repulsion motor; the repulsion-start, induction-run motor; and the repulsion-induction motor. These three types of motors differ in their construction, operating characteristics, and industrial applications.

REPULSION MOTOR

Construction

A repulsion motor has the following basic components:

1. A laminated stator core; one winding of the stator is similar to the main (running) winding of a split-phase motor. The stator is usually wound with four, six, or eight poles.
2. A rotor consisting of a slotted core containing a winding. This rotor is called an *armature* because it is similar to the armature of a dc motor. The coils of this armature winding are connected to the commutator. The axial-type commutator has segments, or bars, parallel to the armature shaft.

3. Two cast steel end shields that act as protection for the motor bearings. These bearings are mounted on the motor frame.
4. Carbon brushes that contact with the commutator surface. A brush holder assembly, mounted on one of the end shields, holds the brushes in place. The brushes are connected together by heavy copper jumpers. The brush holder assembly can be moved. In this way, the brushes can contact the commutator at different points to obtain the correct rotation and maximum torque output.
5. Two bearings that support and center the armature shaft in the stator core. Either sleeve bearings or ball bearings may be used.
6. A cast steel frame; the stator core is pressed into this frame.

Operation of a Repulsion Motor

When the stator winding is connected to a single-phase line, a field is set up by the current in the windings. A voltage and a resultant current are induced in the rotor windings by this field. By placing the brushes in the proper position on the commutator segments, the current induced in the armature windings causes magnetic poles in the armature. These poles have a set relationship to the stator field poles.

The armature field poles are displaced by 15 electrical degrees from the field poles of the stator windings. Because the instantaneous polarity of the rotor poles is the same as that of the adjacent stator poles, a repulsion torque is formed. This torque causes the motor armature to rotate. This motor is known as a *repulsion motor* because it is based on the principle that like poles repel.

Positioning the Brushes. Figure 18–20, parts A–D, shows how the positioning of the brushes affects the torque. In Figure 18–20A, there is no torque when the brushes are placed at right angles to the stator poles. In this position, there are equal induced voltages in the two halves of the armature winding. These voltages oppose each other at the connection between the two sets of brushes. Therefore, there is no current in the windings, and no flux is set up by the armature windings.

In Figure 18–20B, the brushes are now under the center of the stator poles. In this position, there is a heavy current in the armature windings. However, a torque is not created. The armature poles due to the current are in line with the stator poles. Thus, there is no twisting force (torque) in either the clockwise or counterclockwise direction.

Figure 18–20C shows the brushes shifted in a counterclockwise direction from the stator pole centers by 15 electrical degrees. The magnetic poles set up in the armature are 15 electrical degrees from the stator pole centers and have the same polarity. Thus, a repulsion torque is formed between the stator and rotor field poles. This torque causes the armature to rotate counterclockwise.

To reverse the motor, the brushes are shifted 15 electrical degrees to the other side of the stator field pole centers (Figure 18–20D). The armature poles formed again have the same polarity but are now 15° in a clockwise direction from the stator pole centers. This torque causes the armature to rotate clockwise.

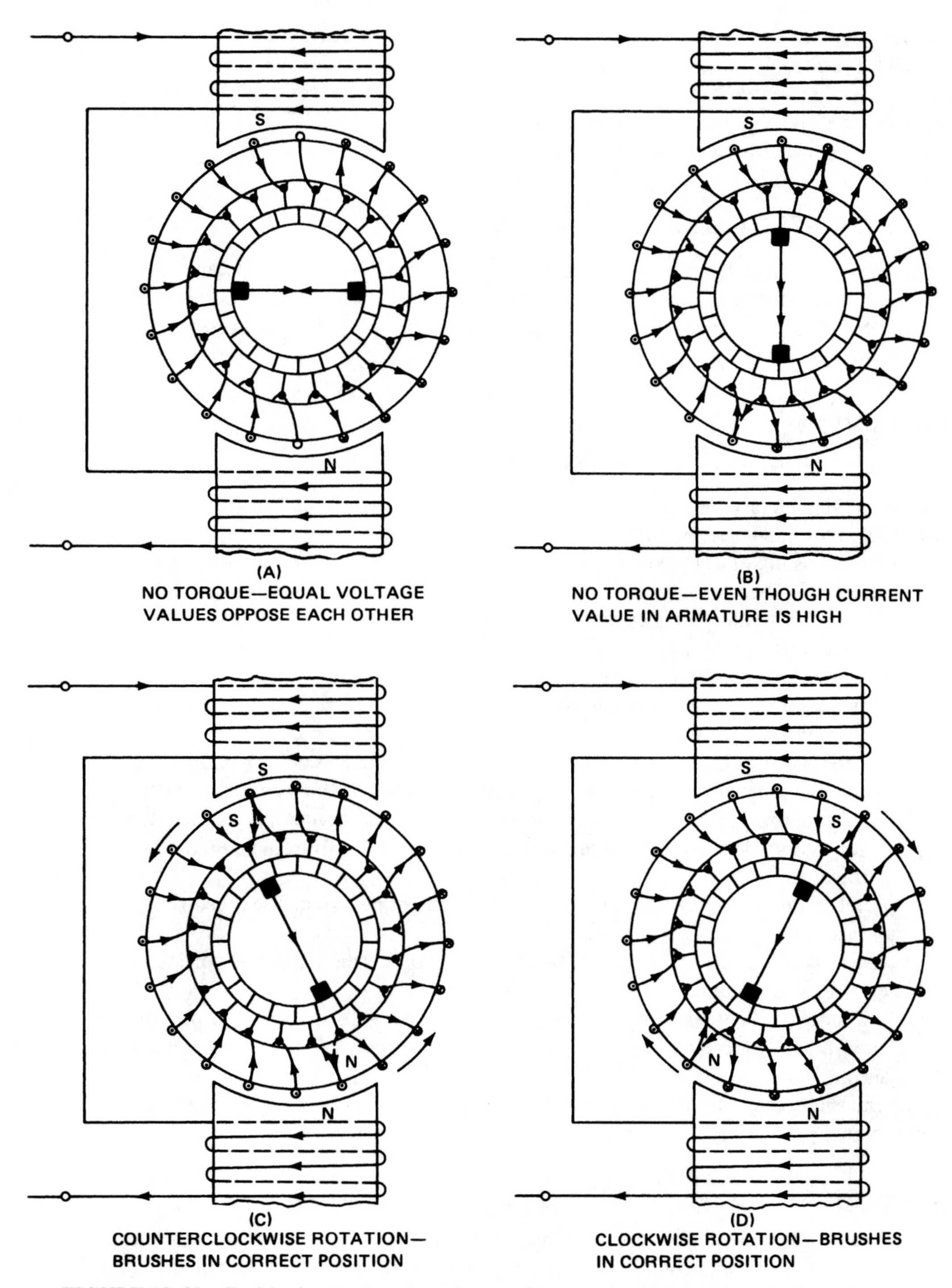

FIGURE 18–20 Positioning the brushes of a repulsion motor (*Delmar/Cengage Learning*)

Torque and Speed. The repulsion motor has excellent starting torque. The speed of this motor can be varied by changing the value of the supply voltage. However, the repulsion motor has unstable speed characteristics. It may race to a very high speed if there is no mechanical load. The repulsion motor can be used for applications requiring a strong starting torque and a range of speed control.

REPULSION-START, INDUCTION-RUN MOTOR

There are two basic types of repulsion-start, induction-run motors. For the *brush-lifting* type of motor, the brushes lift off the commutator surface once the motor accelerates to nearly 75% of the rated speed. The second type of motor is known as the *brush-riding* type. In this motor, the brushes ride on the commutator surface at all times.

Construction

A repulsion-start, induction-run motor has the following components:

1. A laminated stator core. This core has one winding that is similar to the running winding of a split-phase motor.
2. A rotor consisting of a slotted core. The core contains a winding that is connected to a commutator. The rotor core and winding are similar to the armature of a dc motor. Thus, the rotor is called an *armature*.
3. a. For the brush-lifting motor, a centrifugal device lifts the brushes from the commutator surface at 75% of the rated speed. The centrifugal mechanism consists of governor weights, a short-circuiting necklace, spring barrel, spring, push rods, brush holders, and brushes.
   b. For the brush-riding motor, a centrifugal device also operates at 75% of the rated speed. This mechanism consists of governor weights, a short-circuiting necklace, and a spring barrel. The centrifugal device short-circuits the commutator segments but does not lift the brushes and brush holders from the commutator surface.
4. A special radial-type commutator is used in the brush-lifting motor. The brush-riding motor has an axial-type commutator.
5. a. The brush holder assembly for the brush-riding motor is the same as the one used on a repulsion motor.
   b. The brush holder assembly for the brush-lifting motor is designed so that the centrifugal device can lift the brush holders and the brushes from the commutator surface.
6. The end shields, bearings, and motor frame are the same as those used in the repulsion motor.

Operation

A repulsion-start, induction-run motor and a repulsion motor are started in the same way. The repulsion-start, induction-run motor, however, operates as an induction motor once it accelerates to 75% of the rated speed.

Figure 18–21 is an exploded view of an armature, radial commutator, and centrifugal device used on the brush-lifting motor.

Operation of the Centrifugal Mechanism. As the push rods of the centrifugal mechanism move forward, they push the spring barrel forward. The short-circuiting necklace then contacts the bars of the radial commutator and all of the bars are short-circuited. At the same time, the brush holders and brushes are lifted from the commutator surface. In this way, there is no unnecessary wear on the brushes and the commutator surface. Lifting the brushes from the commutator also eliminates a great deal of noise.

By short-circuiting the commutator bars, the armature is converted to a form of squirrel-cage rotor. As a result, the motor operates as a single-phase induction motor. This type of motor starts as a repulsion motor and runs as an induction motor.

Brush-Riding Motor. The brush-riding type of motor uses an axial commutator. The centrifugal mechanism consists of a number of copper segments held in place by a spring (Figure 18–22). This device is mounted next to the commutator. At nearly 75% of the rated speed, centrifugal force causes the short-circuiting necklace to short-circuit the commutator segments. When this happens, the motor operates as an induction motor.

Torque and Speed. The starting torque and speed performances are very good for both types of repulsion-start, induction-run motors. Therefore, these motors are suitable for applications requiring a strong starting torque and a fairly constant speed from no load to full load. Typical uses include commercial refrigerators, compressors, and pumps.

To reverse the direction of rotation of the repulsion-start, induction-run motor, follow the steps outlined for a repulsion motor.

Schematic Diagrams. A schematic diagram of a repulsion-start, induction-run motor is shown in Figure 18–23. The same diagram also represents a repulsion motor.

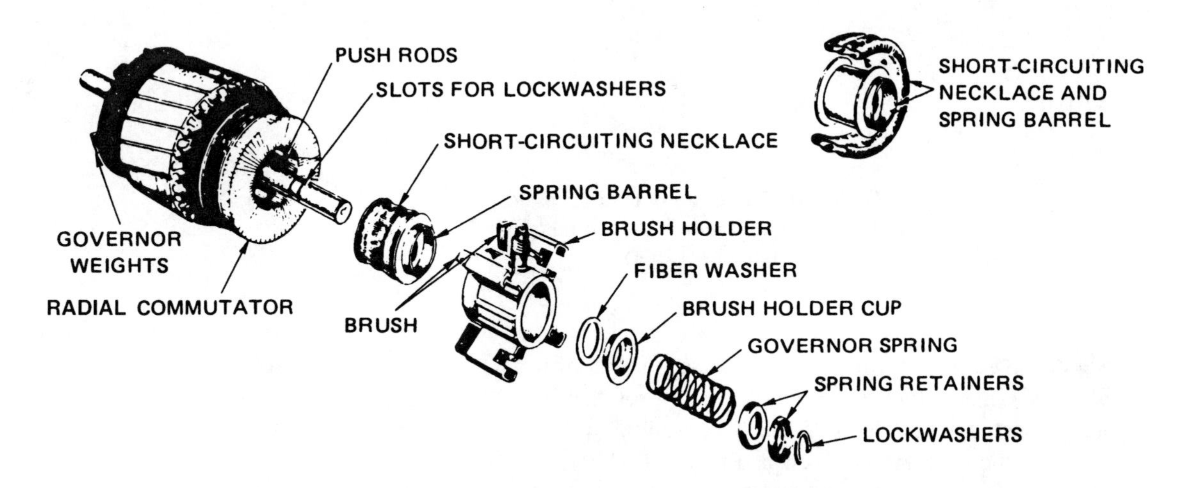

FIGURE 18–21 Exploded view of radial commutator and centrifugal brush-lifting device for a repulsion-start, induction-run motor *(Delmar/Cengage Learning)*

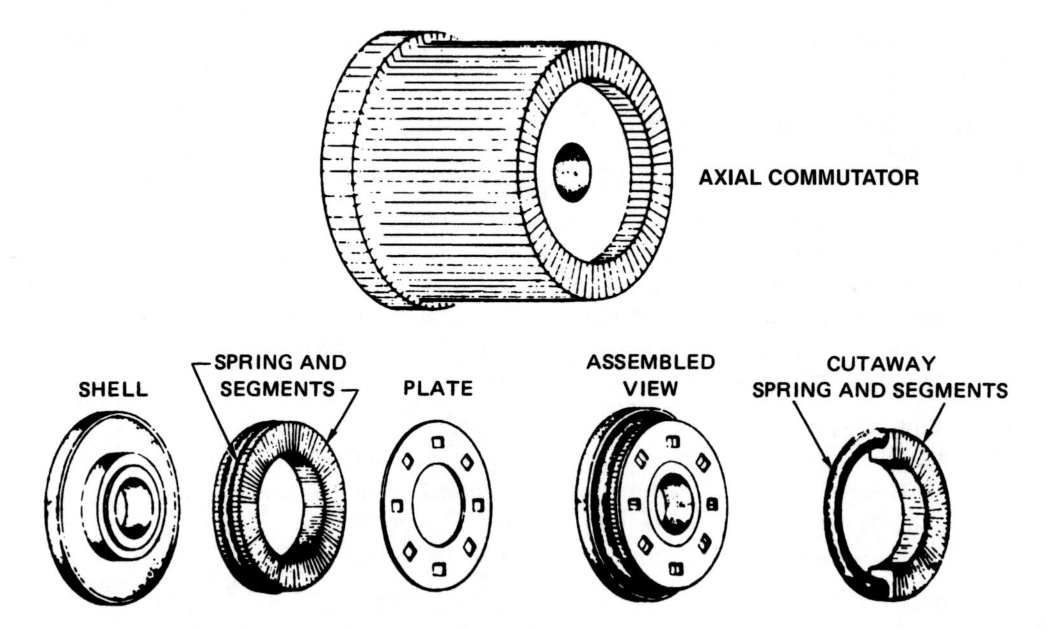

FIGURE 18–22 Exploded view of a short-circuiting device used with a brush-riding, repulsion-start, induction-run motor *(Delmar/Cengage Learning)*

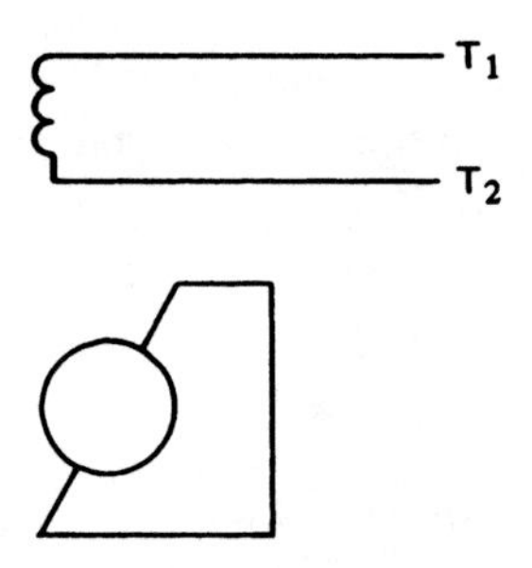

FIGURE 18–23 Schematic diagram of a repulsion-start, induction run motor *(Delmar/Cengage Learning)*

Repulsion-start, induction-run motors are often constructed for operation on either 120 or 240 V. That is, two stator windings are provided. For 120-V operation, the windings are connected in parallel. For 240-V operation, the windings are connected in series.

A schematic diagram is given in Figure 18–24 for a dual-voltage, repulsion-start, induction-run motor. The connection table provided shows how the leads of this motor are connected for 120- or 240-V operation. The same connections also apply to dual-voltage repulsion motors.

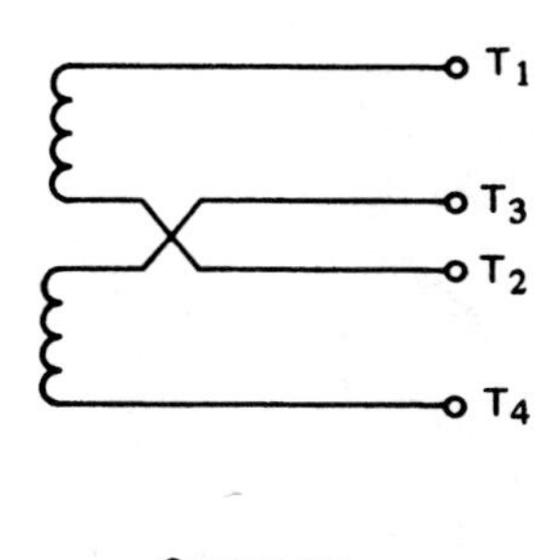

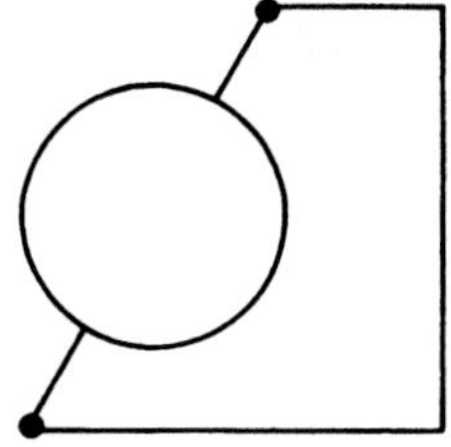

Voltage	L_1	L_2	Tie Together
120 volts	T_1 T_3	T_2 T_4	——
240 volts	T_1	T_4	T_2 T_3

FIGURE 18–24 Schematic diagram of a dual-voltage, repulsion-start, induction-run motor (*Delmar/Cengage Learning*)

REPULSION–INDUCTION MOTOR

A repulsion-induction motor has nearly the same operating characteristics as a repulsion-start, induction-run motor. However, this type of motor has no centrifugal mechanism. The repulsion-induction motor has the same type of armature and commutator as the repulsion motor, but it does not have a centrifugal mechanism. There is a squirrel-cage winding beneath the slots of the armature of the repulsion-induction motor, as shown in Figure 18–25.

Both the repulsion-induction, motor and the repulsion-start, induction-run motor have very good starting torque because they start as repulsion motors. The squirrel-case winding of the repulsion-induction motor means that it has fairly constant speed regulation from no load to full load. The start-up of this motor and that of the repulsion-start, induction-run motor are similar. This motor is called a *repulsion-induction motor* to distinguish it from the repulsion-start, induction-run motor having a centrifugal system.

The torque–speed performance of a repulsion-induction motor is similar to that of a dc compound motor.

The repulsion-induction motor is designed for use on 120 or 240 V. The stator windings have two sections that are connected in parallel for 120-V operation and in series for 240-V operation. The markings of the motor terminals and the connection arrangement

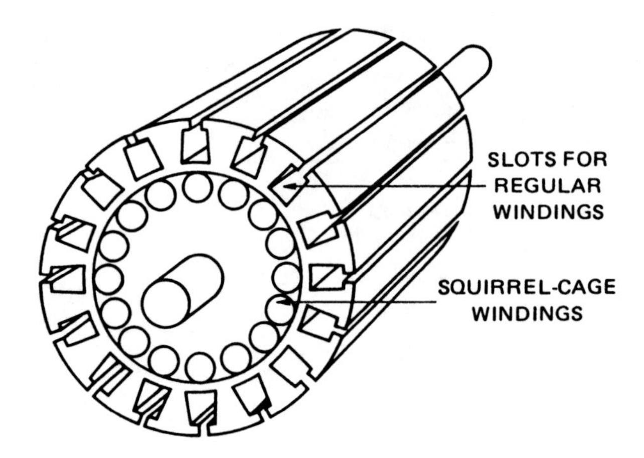

FIGURE 18–25 Rotor of a repulsion-induction motor (*Delmar/Cengage Learning*)

for these leads are the same as for a repulsion-start, induction-run motor. The repulsion-induction motor, the repulsion motor, and the repulsion-start, induction-run motor all have the same schematic diagram symbol.

ALTERNATING-CURRENT SERIES MOTOR

Without careful examination, it may appear that the typical dc series or shunt motor can be operated from an ac supply. If the line terminals to the dc motor are reversed, the current and the magnetic flux are reversed in both the field and armature circuits. For a motor operating from an ac source, the net torque will be in the same direction. As a result, it is impractical to operate a dc shunt motor in this manner because the high inductance of the shunt field causes the field current and the field flux to lag the line voltage by almost 90°. The motor thus develops little torque.

Universal Motors

Another reason why the series dc motor should not be operated on ac is the large amount of heat due to eddy currents in the field poles. In addition, there is an excessive voltage drop across the series field windings due to the high reactance. The field poles may be laminated to reduce the eddy currents. The voltage loss across the field poles can be reduced by using a small number of field turns on a low reluctance core. The core is then operated at a low flux density. This type of motor is known as a *universal motor* and can operate on ac or dc. Universal motors are available with fractional horsepower ratings and are commonly used in household appliances.

A concentrated field universal motor has two salient poles and a winding of few turns connected to give the opposite magnetic polarity. Figure 18–26 shows the laminated field structure of a typical concentrated field universal motor.

The field windings and armature of a universal motor are connected in series, as shown in Figure 18–27.

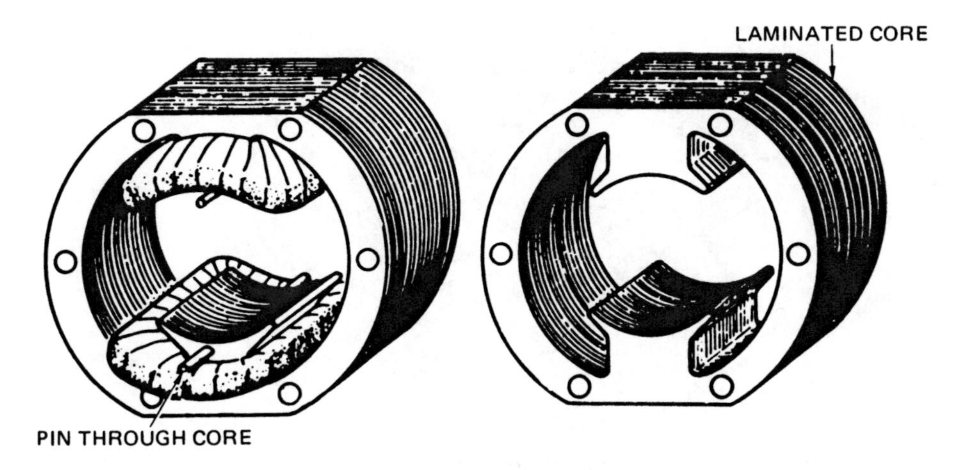

FIGURE 18–26 **Field core of two-pole universal motor** (*Delmar/Cengage Learning*)

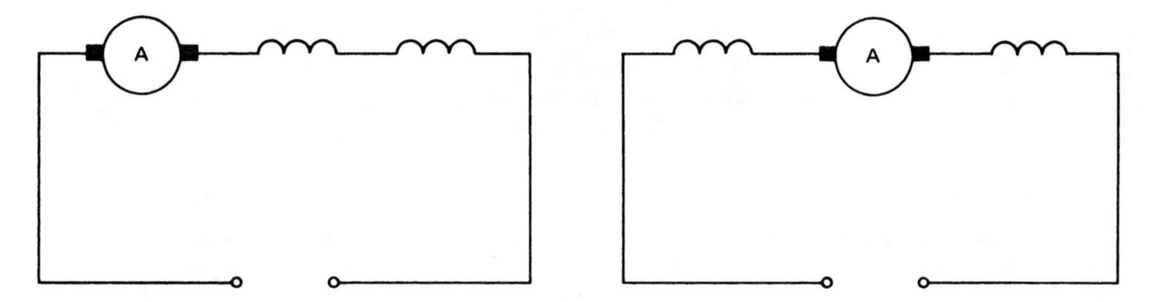

FIGURE 18–27 **Field and armature connections for a universal motor** (*Delmar/Cengage Learning*)

Universal Motor Type

Another type of universal motor is the distributed field motor. One version of this motor is a single-field compensated motor. A second version is a two-field compensated motor. The field windings of a two-pole, single-field compensated motor resemble the stator winding of a two-pole, split-phase ac motor. A two-field compensated motor has a stator with a main winding and a compensating winding. These windings are spaced 90 electrical degrees apart. When the motor is operated from an ac source, the compensating winding reduces the reactance voltage developed in the armature by the alternating flux.

Universal motors have aluminum, cast iron, and rolled steel frames (Figure 18–28). The field poles are bolted to the frame. The field cores consist of laminations bolted together. The armature core is also laminated. A conventional commutator and brushes are used.

Speed Control

Universal motors operate at nearly the same speed on dc and single-phase ac. These series-wound motors operate at excessive speeds with no load. To overcome this, they are

FIGURE 18–28 Disassembled view of concentrated field universal motor (*Delmar/Cengage Learning*)

permanently connected to the devices being driven. The speed regulation for universal motors consists of resistance inserted in series with the motor. Tapped resistors, rheostats and tapped nichrome wire coils, wound over a single-field pole, are used. The speed can also be controlled by varying the inductance through taps on one of the field poles.

Motor Reversal

Any series-wound motor can be reversed by changing the direction of the current in either the field or armature circuit. Universal motors are sensitive to brush position. Severe arcing at the brushes results when the direction of rotation is changed and the brushes are not shifted to the neutral (sparkless) plane.

CONDUCTIVE AND INDUCTIVE COMPENSATION

AC motors larger than ½ horsepower (hp) are used to drive loads when a high starting torque is required. For these motors, there is excessive armature reaction under load. One method of overcoming the armature reaction is known as *conductive compensation*. In this method, an additional compensating winding is placed in slots cut in the pole faces. The strength of this field increases with an increase in the load current. Thus, there is a reduction in the distortion of the main field flux by the armature flux.

The compensating winding is connected in series with the series field winding and the armature (Figure 18–29). A motor with conductive compensation has a high starting torque and poor speed regulation. However, resistor-type speed controllers can be used to obtain a wide range of speed control.

A second method of overcoming armature reaction in ac series motors uses an inductively coupled winding (Figure 18–30). This winding acts like a short-circuited secondary winding in a transformer. This winding links the cross magnetizing flux of the armature. (The armature can be compared to the primary winding of a transformer.) Because the magnetomotive force of the secondary is equal in magnitude to the primary magnetomotive

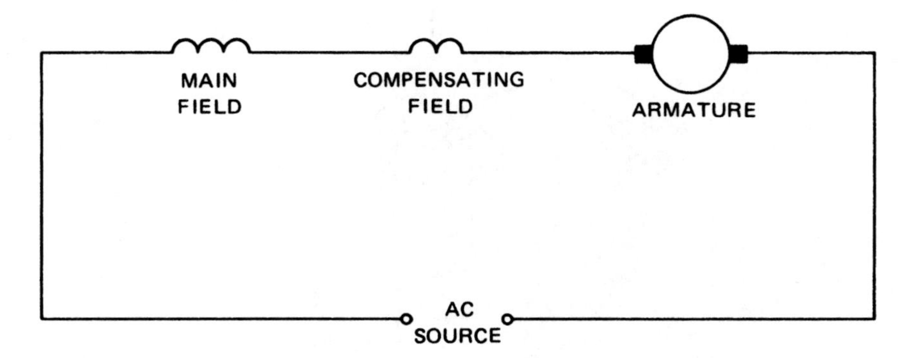

FIGURE 18–29 Schematic diagram of a universal motor with conductive compensation (*Delmar/Cengage Learning*)

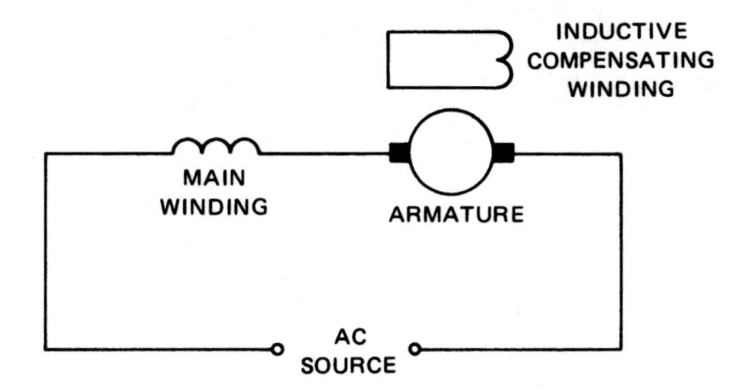

FIGURE 18–30 Connections for a universal motor with inductive compensation (*Delmar/Cengage Learning*)

force but is nearly opposite in phase, the compensating winding flux almost neutralizes the armature cross-flux. An ac series motor of this type cannot be used on dc current. Its operating characteristics are similar to those of the universal motor with conductive compensation.

SHADED-POLE INDUCTION MOTOR

If a single-phase induction motor has a small fractional horsepower rating, it may be started by shading coils mounted on one side of each of the stator poles. A standard squirrel-cage rotor is used in this motor, which is known as a *shaded-pole induction motor*. The motor does not require a starter mechanism such as a centrifugal switch. Thus, there is no possibility of motor failure due to a faulty centrifugal switch mechanism.

Figure 18–31 shows a typical shaded-pole induction motor with four stator poles. The four poles are wound in alternate directions. Note that a *shading coil* is wound on one section of each of the stator field poles. The shading coil is a low-resistance copper loop.

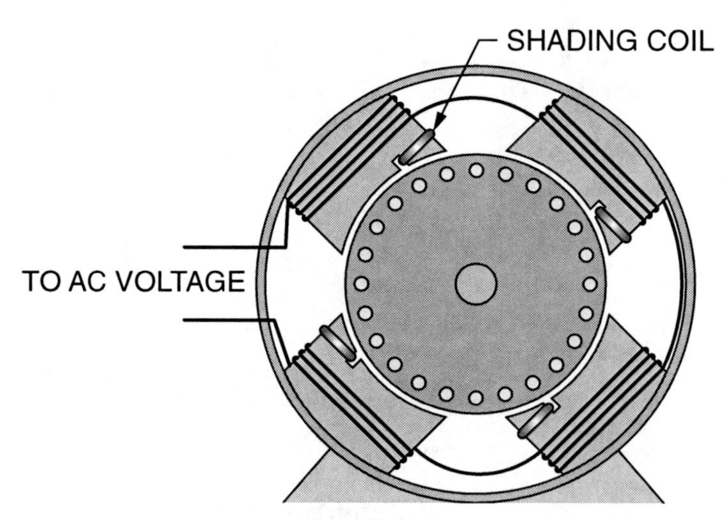

FIGURE 18–31 Basic construction of a shaded pole induction motor (*Delmar/Cengage Learning*)

Action of Flux

As the current in the stator circuit increases, the stator flux also increases. A voltage is induced in each shading coil. The induced current in the shading loop opposes any increase in the main field flux through the loop. As a result, the flux increases in the other section of each pole face (Figure 18–32A).

When the stator current and flux both reach maximum values, there is an instant when no other change occurs in the current or the flux. At this instant, there is no voltage or current in the shading coil. Thus, the shading coil does not set up a magnetomotive force to oppose the stator flux. The resulting stator field is uniform across the pole face (Figure 18–32B). When the stator current and flux decrease, the induced voltage and current in the shading coil set up a magnetomotive force. This force aids the stator field. As a result, the flux decreases less rapidly in the section of the pole face where the shading coil is mounted than it does in the other part of the pole face (Figure 18–32C).

Figure 18–32 shows that the shading coil causes the field flux to shift across the pole face. This shifting flux can be likened to a rotating magnetic field. The torque produced is small. Thus, the shaded-pole type of starting is used only on small motors rated at no more than 1⁄10 hp. Such motors are used in applications where a strong starting torque is not essential. Typical applications include driving small devices such as fans and blowers.

VARIABLE-SPEED MOTORS

The use of small variable-speed motors has increased greatly because of an increase in demand for products that employ their use. These motors are commonly used to operate light loads such as ceiling fans and blower motors. Two types of motors are used for these applications: the shaded-pole and the capacitor-start, capacitor-run motors. These motors

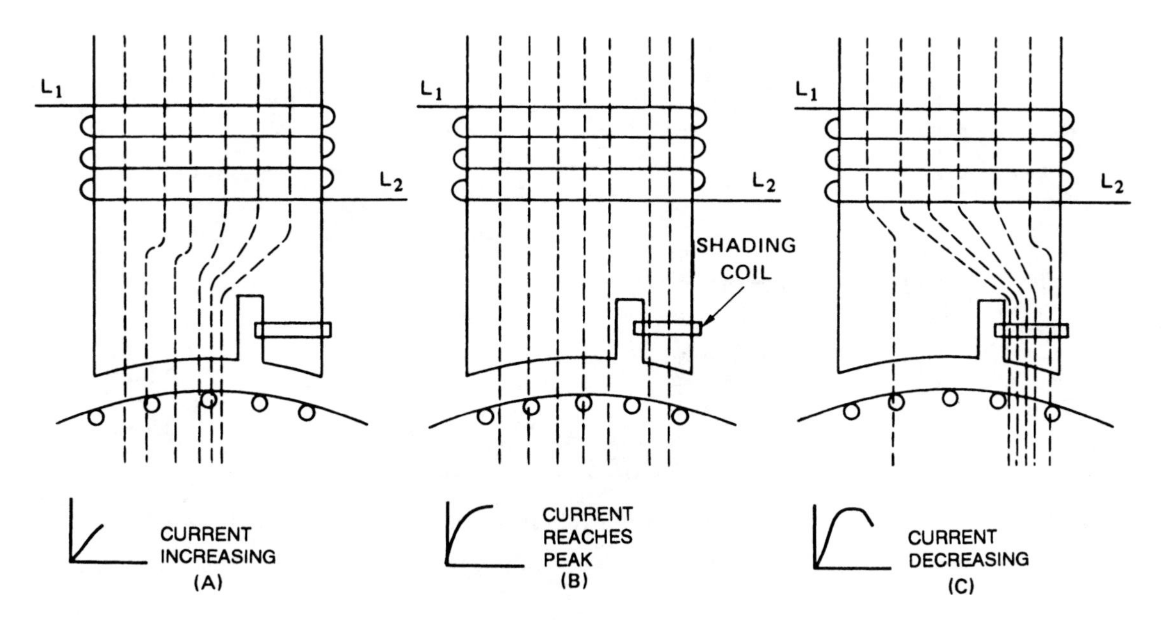

FIGURE 18–32 Shifting of air gap flux in shaded-pole motor (*Delmar/Cengage Learning*)

are used because they operate without having to disconnect a set of start windings with a centrifugal switch. Motors intended to be used in this manner are wound with high-impedance stator windings. The high impedance of the stator limits the current flow through the motor when the speed of the rotor is decreased. Speed control for these motors is accomplished by controlling the amount of voltage applied to the motor or by inserting impedance in series with the stator winding.

Variable Voltage Control

The amount of voltage applied to the motor can be controlled by several methods. One method is to use an autotransformer with several taps (Figure 18–33). This type of controller has several steps of speed control. Notice that the applied voltage, 120 V in this illustration, is connected across the entire transformer winding. When the rotary switch is moved to the first tap, 30 V is applied to the motor. This produces the lowest motor speed for this controller. When the rotary switch is moved to the second tap, 60 V is applied to the motor. This provides an increase in motor speed. When the switch has been moved to the last position, the full 120 V is applied to the motor and it operates at its highest speed.

Another type of variable voltage control uses a triac to control the amount of voltage applied to the motor (Figure 18–34). This type of speed control provides a more linear control because the voltage can be adjusted from zero to the full applied voltage. At first appearance, many people assume this controller to be a variable resistor connected in series with the motor. A variable resistor large enough to control even a small motor would produce several hundred watts of heat and could never be mounted in a switch box. The

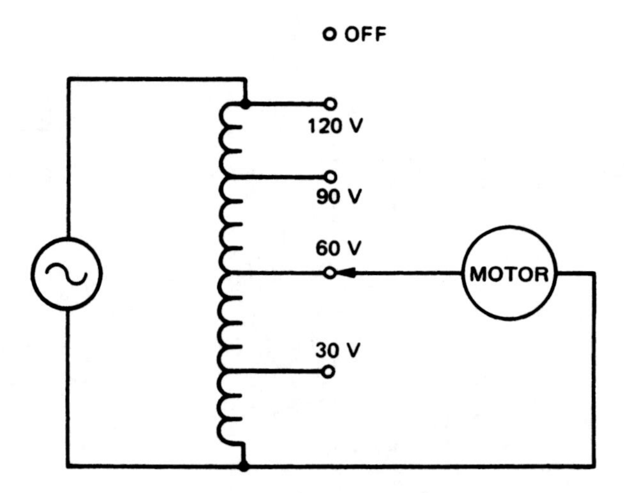

FIGURE 18–33 Autotransformer controls motor voltage (*Delmar/Cengage Learning*)

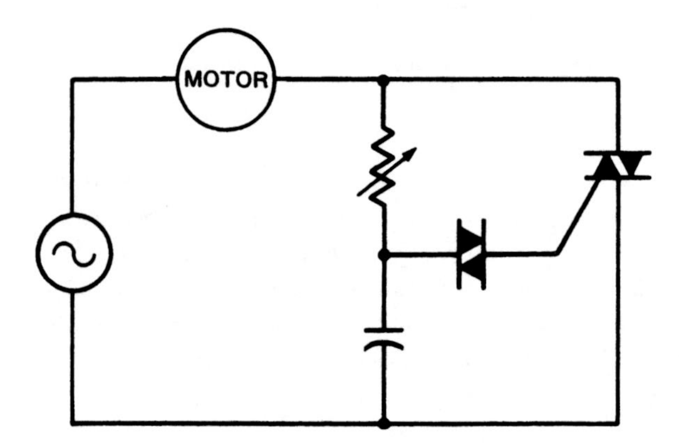

FIGURE 18–34 Triac used to control motor speed (*Delmar/Cengage Learning*)

variable resistor in this circuit is used to control the amount of phase shift for the triac. The triac controls the amount of voltage applied to the motor by turning on at different times during the ac cycle.

A triac speed control is very similar to a triac light dimmer used in many homes. A light dimmer, however, should never be used as a motor speed controller. Triac light dimmers are intended to be used with resistive loads such as incandescent lamps. Light-dimmer circuits will sometimes permit one-half of the triac to start conducting before the other half. The waveform shown in Figure 18–35 illustrates this condition. Notice that only the positive half of the waveform is being conducted to the load. Because only positive voltage is being applied to the load, it is dc. Operating a resistive load such as an incandescent lamp with dc will do no damage. Operating an inductive load such as the winding of a motor can do a great deal of damage, however. When direct current is applied to a motor winding, there is no inductive reactance to limit the current. The actual wire resistance of

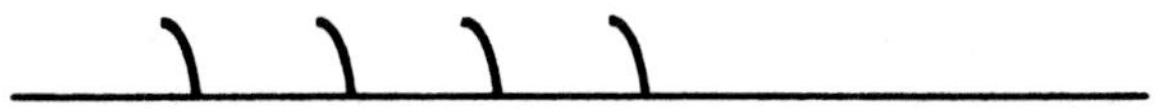

FIGURE 18–35 Triac conducts only the positive half of the waveform. (*Delmar/Cengage Learning*)

the stator is the only current-limiting factor. The motor winding or the controller can easily be destroyed if direct current is applied to the motor. For this reason, only triac controllers designed for use with inductive loads should be used for motor control. A photograph of a triac speed controller is shown in Figure 18–36.

Series Impedance Control

Another common method of controlling the speed of small ac motors is to connect impedance in series with the stator winding. This is the same basic method of control used with multispeed fan motors. The circuit in Figure 18–37 shows a tapped inductor connected in series with the motor. When the motor is first started, it is connected directly to the full voltage of the circuit. As the rotary switch is moved from one position

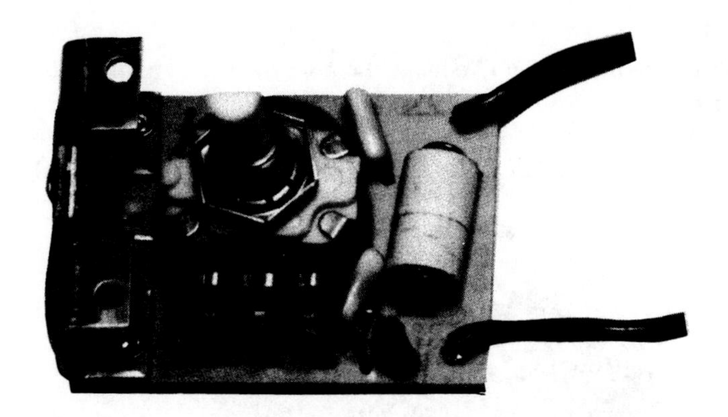

FIGURE 18–36 A triac speed controller (*Delmar/Cengage Learning*)

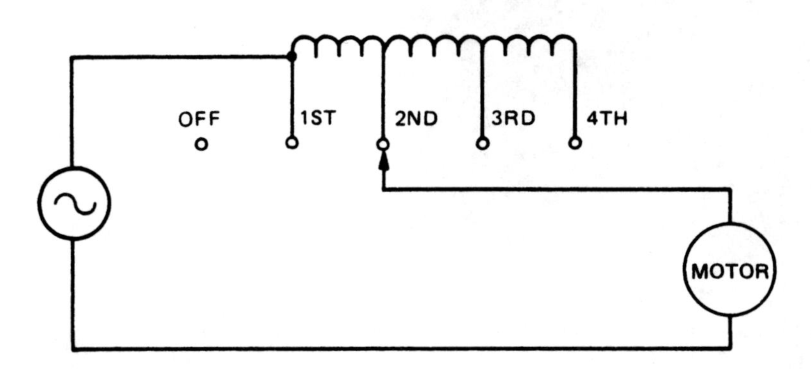

FIGURE 18–37 Series inductor changes impedance of circuit (*Delmar/Cengage Learning*)

to another, steps of inductance are connected in series with the motor. As more inductance is connected in series with the stator, the amount of current flow decreases. This produces a weaker magnetic field in the stator. Rotor slip increases because of the weaker magnetic field, and the motor speed decreases. A photograph of this type of controller is shown in Figure 18–38.

STEPPING MOTORS

Stepping motors are devices that convert electrical impulses into mechanical movement. Stepping motors differ from other types of dc or ac motors in that their output shaft moves through a specific angular rotation each time the motor receives a pulse. Each time a pulse is received, the motor shaft moves by a precise amount. The stepping motor allows a load to be controlled with regard to speed, distance, or position. These motors are very accurate in their control performance. Generally, less than 5% error per angle of rotation exists, and this error is not cumulative regardless of the number of rotations. Stepping motors are operated on dc power but can be used as a two-phase synchronous motor when connected to ac power.

Theory of Operation

Stepping motors operate on the theory that like magnetic poles repel and unlike magnetic poles attract. Consider the circuit shown in Figure 18–39. In this illustration, the rotor is a permanent magnet and the stator winding consists of two electromagnets. If current flows through the winding of stator pole A in such a direction that it creates a north

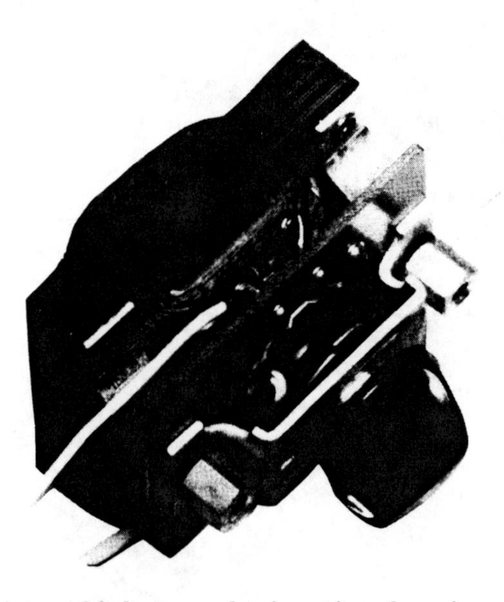

FIGURE 18–38 A tapped inductor used to insert impedance in series with the stator winding (*Delmar/Cengage Learning*)

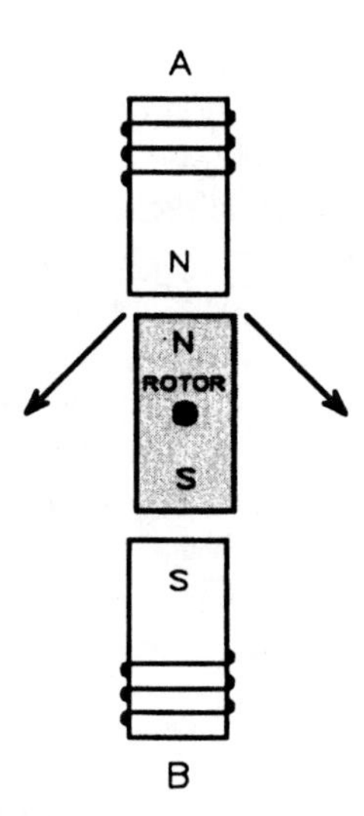

FIGURE 18–39 The rotor could turn in either direction. ***(Delmar/Cengage Learning)***

magnetic pole, and through B in such a direction that it creates a south magnetic pole, it would be impossible to determine the direction of rotation. In this condition, the rotor could turn in either direction.

Now consider the circuit shown in Figure 18–40. In this circuit, the motor contains four stator poles instead of two. The direction of current flow through stator pole A is still in such a direction as to produce a north magnetic field; the current flow through pole B

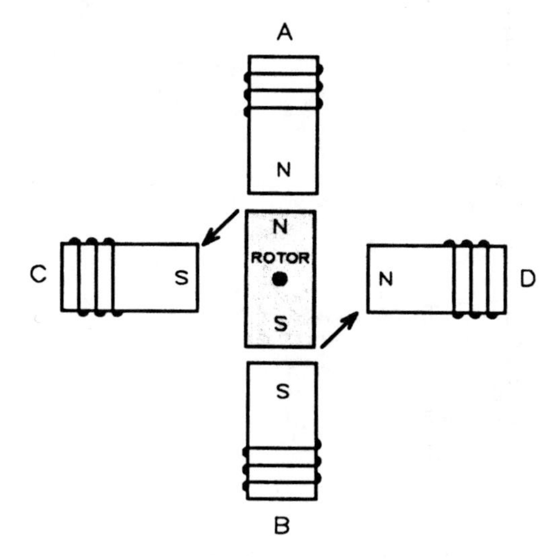

FIGURE 18–40 Direction of rotation is known. ***(Delmar/Cengage Learning)***

produces a south magnetic field. The current flow through stator pole C, however, produces a south magnetic field, and the current flow through pole D produces a north magnetic field. As illustrated, there is no doubt regarding the direction or angle of rotation. In this example, the rotor shaft will turn 90° in a counterclockwise direction.

Figure 18–41 shows yet another condition. In this example, the current flow through poles A and C is in such a direction as to form a north magnetic pole, and the direction of current flow through poles B and D forms south magnetic poles. In this illustration, the permanent-magnet rotor has rotated to a position between the actual pole pieces.

To allow for better stepping resolution, most stepping motors have eight stator poles, and the pole pieces and rotor have teeth machined into them as shown in Figure 18–42. In practice, the number of teeth machined in the stator and rotor determines the angular rotation achieved each time the motor is stepped. The stator–rotor tooth configuration shown in Figure 18–42 produces an angular rotation of 1.8° per step.

Winding

There are different methods for winding stepper motors. A standard three-lead motor is shown in Figure 18–43. The common terminal of the two windings is connected to ground of an aboveground–belowground power supply. Terminal 1 is connected to the common of a single-pole, double-throw switch (switch 1), and terminal 3 is connected to the common of another single-pole, double-throw switch (switch 2). One of the stationary contacts of each switch is connected to the positive or aboveground voltage,

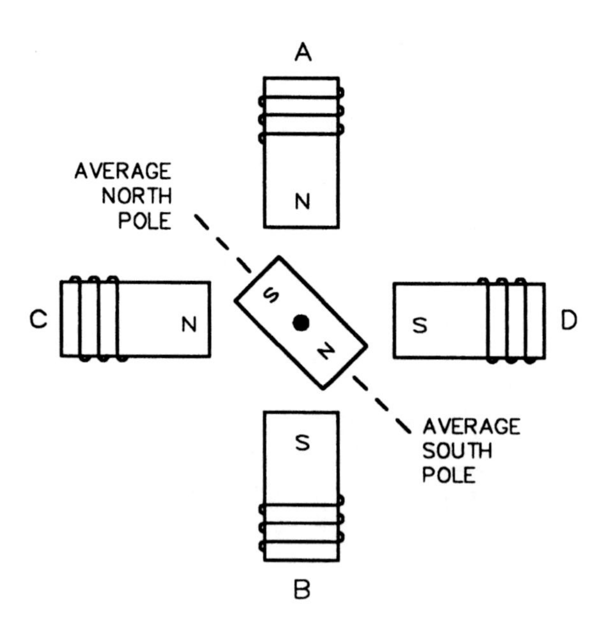

FIGURE 18–41 The rotor is between poles A and C. (*Delmar/Cengage Learning*)

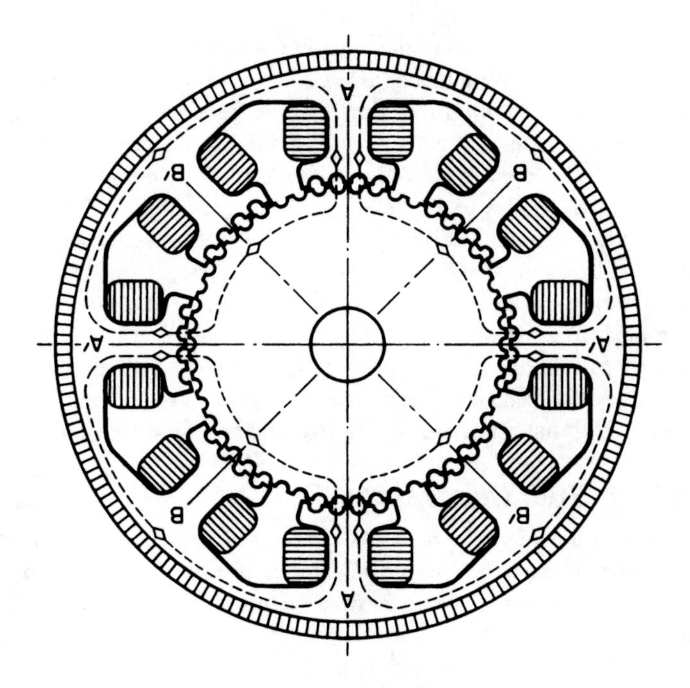

FIGURE 18–42 Construction of a dc stepping motor (*Courtesy of Superior Electric Company*)

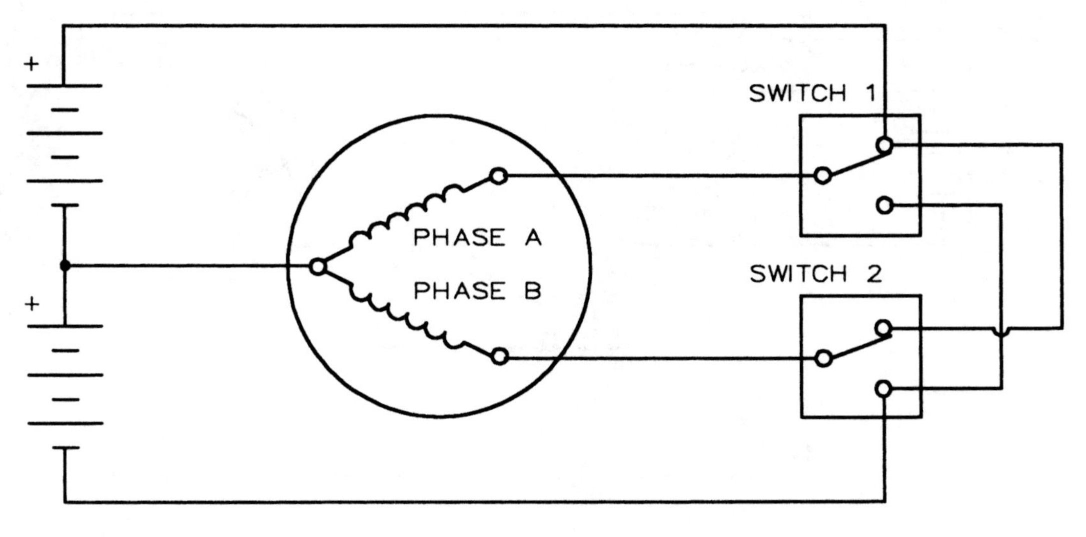

FIGURE 18–43 Standard three-lead motor (*Delmar/Cengage Learning*)

and the other stationary contact is connected to the negative or belowground voltage. The polarity of each winding is determined by the position setting of its control switch.

Stepping motors can also be wound *bifilar*, as shown in Figure 18–44. The term *bifilar* means that there are two windings wound together. This is similar to a transformer winding with a center-tap lead. A bifilar stepping motor has twice as many windings as does the three-lead type, which makes it necessary to use smaller wire in the windings. This results in higher wire resistance in the winding, producing a better inductive–resistive (L/R) time constant for the bifilar wound motor. The increased L/R time constant results in better motor performance. The use of a bifilar stepper motor also simplifies the drive circuitry requirements. Notice that the bifilar motor does not require an aboveground–belowground power supply. As a general rule, the power supply voltage should be about five times greater than the motor voltage. A current-limiting resistance is used in the common lead of the motor. This current-limiting resistor also helps improve the L/R time constant.

Four-Step Switching (Full Stepping)

The switching arrangement shown in Figure 18–44 can be used for a four-step sequence. Each time one of the switches changes position, the rotor will advance one-fourth of a tooth. After four steps, the rotor has turned the angular rotation of one "full" tooth. If the rotor and stator have 50 teeth, it will require 200 steps for the motor to rotate one full revolution. This corresponds to an angular rotation of 1.8° per step (360°/200 steps = 1.8° per step). Figure 18–45 illustrates the switch positions for each step.

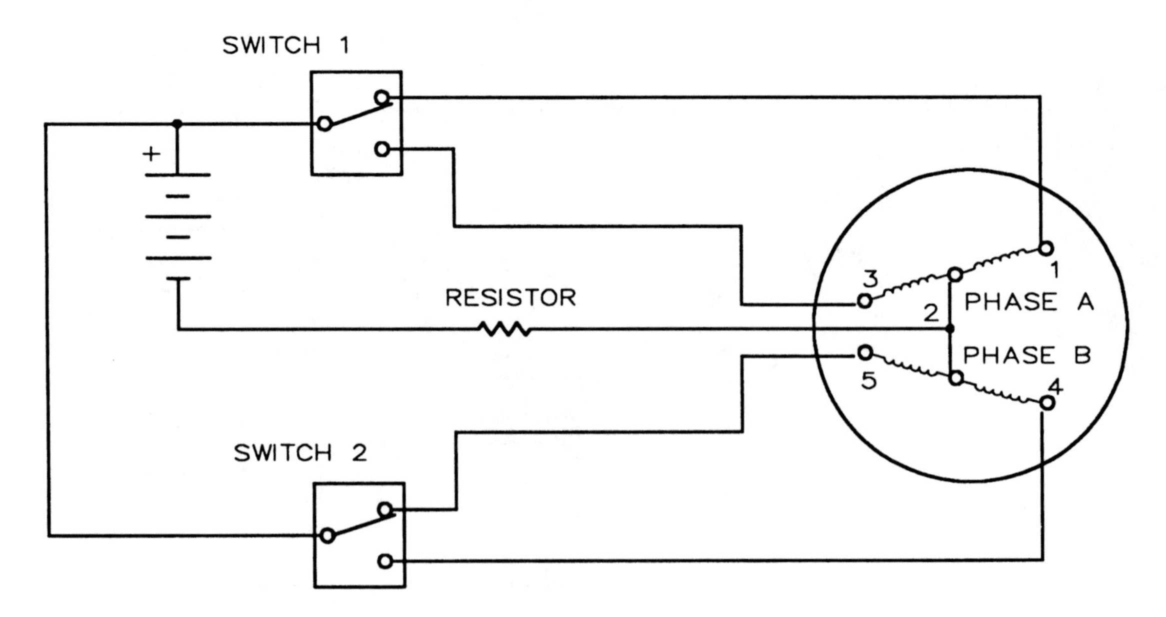

FIGURE 18–44 Bifilar wound stepping motor (*Delmar/Cengage Learning*)

STEP	SWITCH 1	SWITCH 2
1	1	5
2	1	4
3	3	4
4	3	5
1	1	5

FIGURE 18–45 Four-step switching sequence (*Delmar/Cengage Learning*)

Eight-Step Switching (Half-Stepping)

Figure 18–46 illustrates the connections for an eight-stepping sequence. In this arrangement, the center-tap lead for phases A and B are connected through their own separate current-limiting resistors back to the negative of the power supply. This circuit contains four separate single-pole switches instead of two switches. The advantage of this arrangement is that each step causes the motor to rotate one-eighth of a tooth instead of one-fourth of a tooth. The motor now requires 400 steps to produce one revolution, which

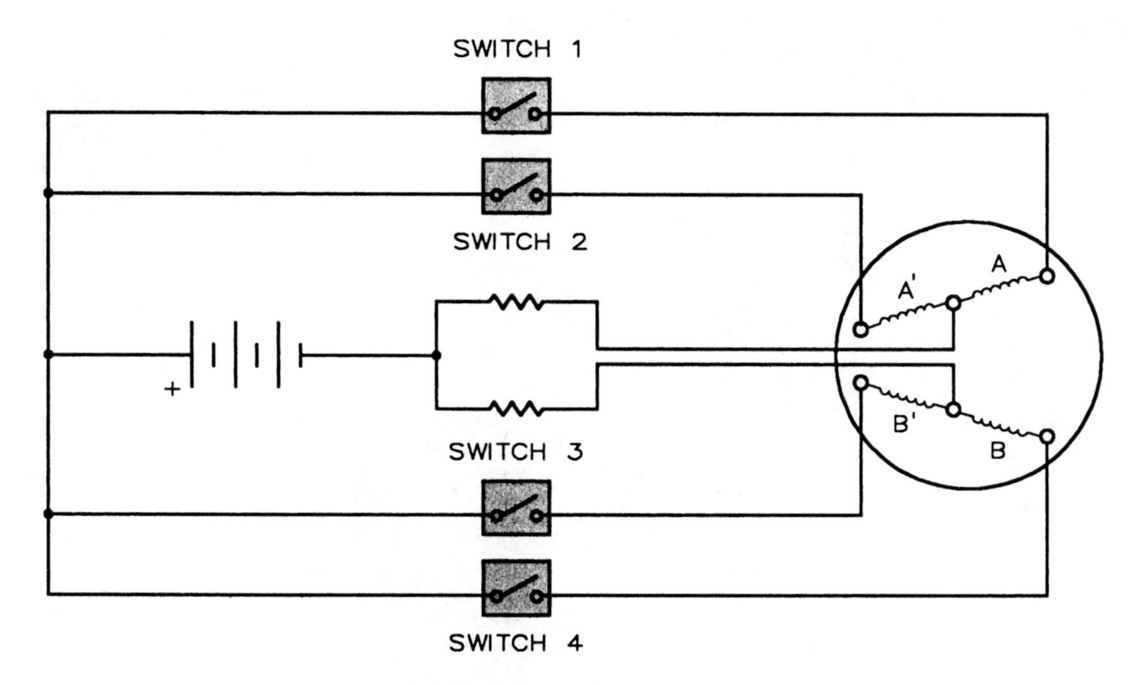

FIGURE 18–46 Eight-step switching (*Delmar/Cengage Learning*)

produces an angular rotation of 0.9° per step. This results in better stepping resolution and greater speed capability. The chart in Figure 18–47 illustrates the switch position for each step. Figure 18–48 depicts a solid-state switching circuit for an eight-step switching arrangement. A stepping motor is shown in Figure 18–49.

STEP	SW #1	SW #2	SW #3	SW #4
1	ON	OFF	ON	OFF
2	ON	OFF	OFF	OFF
3	ON	OFF	OFF	ON
4	OFF	OFF	OFF	ON
5	OFF	ON	OFF	ON
6	OFF	ON	OFF	OFF
7	OFF	ON	ON	OFF
8	OFF	OFF	ON	OFF
1	ON	OFF	ON	OFF

FIGURE 18–47 Eight-step switching sequence (*Delmar/Cengage Learning*)

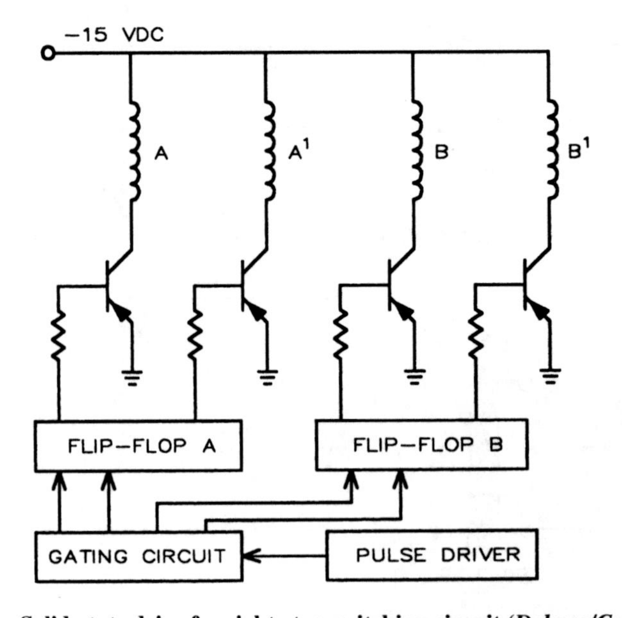

FIGURE 18–48 Solid-state drive for eight-step switching circuit (*Delmar/Cengage Learning*)

FIGURE 18–49 DC stepping motor (*Courtesy of Superior Electric Company*)

AC Operation

Stepping motors can be operated on ac voltage. In this mode of operation, they become two-phase ac synchronous constant-speed motors and are classified as *permanent-magnet induction motors*. Refer to the exploded diagram of a stepping motor in Figure 18–50. Notice that this motor has no brushes, slip rings, commutator, gears, or belts. Bearings maintain a constant airgap between the permanent-magnet rotor and the stator windings. A typical eight-stator pole-stepping motor will have a synchronous speed of 72 r/min when connected to a 60-Hz two-phase ac power line.

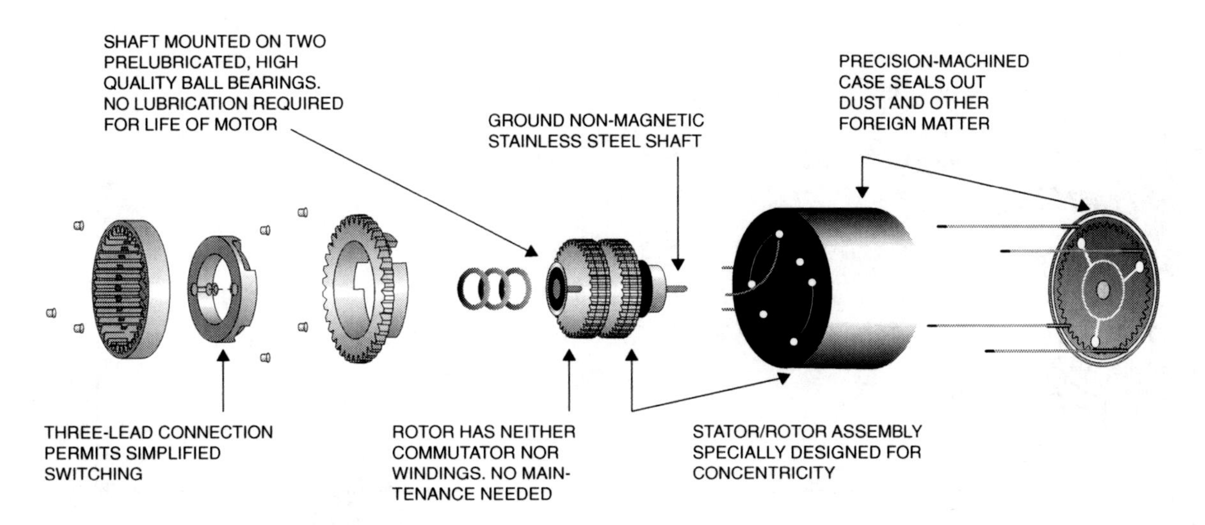

FIGURE 18–50 Exploded diagram of a stepping motor (*Courtesy of Superior Electric Company*)

A resistive–capacitive network can be used to provide the 90° phase shift needed to change single-phase ac into two-phase ac. A simple forward–off–reverse switch can be added to provide directional control. A sample circuit of this type is shown in Figure 18–51. The correct values of resistance and capacitance are necessary for proper operation. Incorrect values can result in random direction of rotation when the motor is started, change of direction when the load is varied, and erratic and unstable operation, as well as failure of the motor to start. The correct values of resistance and capacitance will be different with different stepping motors. The manufacturer's recommendations should be followed for the particular type of stepping motor used.

Motor Characteristics

When stepping motors are used as two-phase synchronous motors, they have the ability to virtually start, stop, or reverse direction of rotation instantly. The motor will start within about 1½ cycles of the applied voltage and stop within 5 to 25 ms. The motor can maintain a stalled condition without harm to it. Because the rotor is a permanent magnet, no induced current is in the rotor, and no high inrush of current occurs when the motor is started. The starting and running currents are the same. This simplifies the power requirements of the circuit used to supply the motor. Because of the permanent-magnet structure of the rotor, the motor does provide holding torque when turned off. If more holding torque is needed, dc voltage can be applied to one or both windings when the motor is turned off. An example circuit of this type is shown in Figure 18–52. If dc voltage is applied to one winding, the holding torque will be approximately 20% greater than the *rated* torque of the motor. If dc voltage is applied to both windings, the holding torque will be about 1½ times greater than the rated torque.

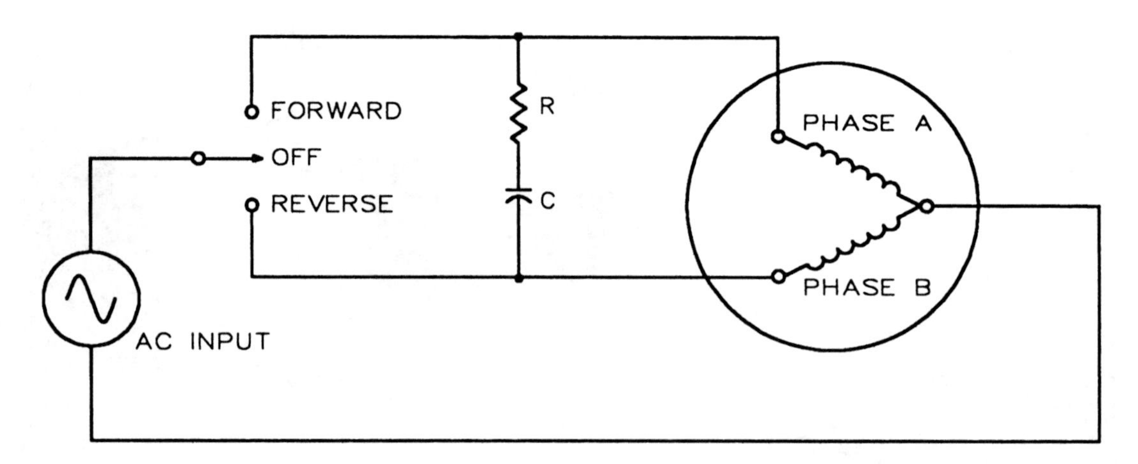

FIGURE 18–51 Phase shift circuit converts single-phase ac into two-phase ac (*Delmar/Cengage Learning*)

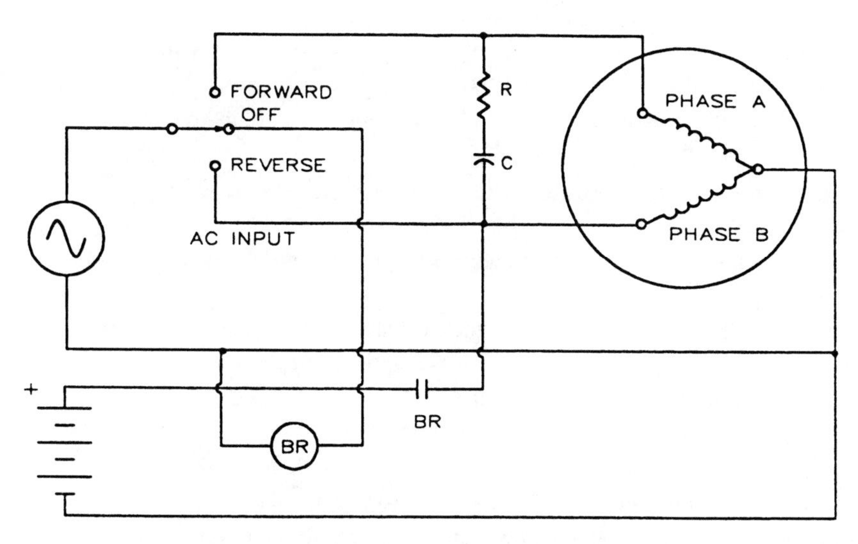

FIGURE 18–52 Applying dc voltage to increase holding torque (*Delmar/Cengage Learning*)

SUMMARY

- Types of single-phase motors (split-phase)
  1. Single-phase induction motor
  2. Repulsion motor
  3. Repulsion-induction motor
  4. Series motor (universal motor)
  5. Shaded-pole motor
- Three-phase motors generally perform better than do single-phase motors. In many instances, however, only single-phase service is available.
- Most single-phase motors have fractional horsepower ratings. In general, their use is limited to commercial and residential applications.
- Resistance-start, induction-run motor
  1. A resistance-start, induction-run motor has the following basic parts:
     a. A stator (stationary part).
     b. A rotor (revolving part).
     c. A centrifugal switch, located inside the motor.
     d. Two end shields bolted to the steel frame; these shields house the rotor shaft bearings.
     e. A cast steel frame; the stator core is pressed into this frame.
  2. The stator contains two windings, which are placed 90 electrical degrees apart.
     a. The main or running winding
     b. The starting or auxiliary winding

3. At start-up, both windings are connected in parallel to the single-phase line. Once the motor accelerates to two-thirds or three-quarters of the rated speed, a centrifugal switch disconnects the starting winding automatically.
4. The rotor of the resistance-start, induction-run motor is the same as that of a three-phase, squirrel-cage induction motor.
   a. The rotor contains no windings, brushes, slip rings, or commutator where faults may develop.
   b. Fans are provided with the rotor to keep the temperature of the windings at a reasonable level.
5. The centrifugal switch is mounted inside the motor. It disconnects the starting winding after the rotor reaches approximately 66% to 75% of its rated speed.
6. The running winding has a low resistance and a high inductive reactance. Thus, the current of the running winding lags behind the voltage.
7. The current in the starting winding is more nearly in phase with the voltage because the winding has a high resistance and a low inductive reactance.
8. When a current passes through each of these windings, the resulting pulsating field effect gives rise to a rotating field around the inside of the stator core.
9. While traveling at the synchronous speed, $S = 120\ f/P$, the rotating field cuts the copper bars of the squirrel-cage winding (rotor). Voltages are induced in the winding and cause currents in the rotor bars.
10. A rotor field is created. This field reacts with the stator field to develop the torque that causes the rotor to turn.
11. As the rotor nears the rated speed, the centrifugal switch disconnects the starting winding from the line. The motor then continues to operate, using only the running winding and the current induced in the rotor. The alternating voltages produced generate a torque that keeps the rotor turning.
12. The motor resembles a two-phase induction motor because the currents of the windings are approximately 90 electrical degrees out of phase with each other. However, a single-phase source supplies the motor. Therefore, the motor is called a *split-phase motor*.
13. If the motor does not start, but a low humming sound is present, then one winding is open. The centrifugal switch should be checked to determine whether its mechanism is faulty or whether the switch contacts are pitted.
14. If the motor operates on its starting winding for more than 60 seconds, the winding may char or burn out.
15. To reverse the rotation of the motor, interchange the leads of either the starting winding or the running winding (preferably the starting winding). Interchanging these leads reverses the direction of rotation of the field.
16. Dual-voltage motors
   a. Some single-phase motors have dual-voltage ratings such as 115/230 V. The running winding consists of two sections, each of which is rated at 115 V.

b. The windings are marked "T_1 and T_2" and "T_3 and T_4."
c. For the higher voltage, the two windings are connected in series across 230 V.
d. For the lower voltage, the windings are connected in parallel across 115 V.
e. The starting winding consists of one winding rated at 115 V.
f. It is connected in parallel across 115 V with the two running windings.
g. For the higher voltage, the starting winding is connected in parallel across one of the series running windings.

17. Speed regulation
    a. A resistance-start, induction-run motor has very good speed regulation from no load to full load.
    b. The regulation compares to that of a three-phase, squirrel-cage induction motor.
    c. The percent slip ranges from 4% to 6%.
18. Starting torque
    a. The starting torque of the motor is poor.
    b. A phase angle of 30° to 50° is large enough to provide a weak rotating magnetic field and the starting torque is small.

- Capacitor-start, induction-run motor
  1. This type of motor is similar to the resistance-start, induction-run motor. A capacitor is connected in series with the starting winding. The capacitor is mounted in or on the motor.
  2. The starting torque is higher than that of a resistance-start motor.
  3. The capacitor limits the starting surge of current to a lower value as compared to the resistance-start motor.
  4. This motor is used in refrigeration units, compressors, oil burners, small-machine equipment, and any application where split-phase induction motors are used.
  5. The basic operation of this motor is nearly the same as that of the resistance-start, induction-run motor. The capacitor is energized for two or three seconds only at start-up. Thus, it cannot improve the power factor.
  6. Defective capacitors often cause problems in the motors using them.
     a. The capacitor may short-circuit and blow the fuse on the branch motor circuit.
     b. If the fuse rating is high, and the fuse does not interrupt the power supply to the motor, the starting winding may burn out.

- Multispeed motors
  1. Two basic types of motors are used for multispeed applications:
     a. The consequent pole motor.
     b. The capacitor-start, capacitor-run motor.

2. The consequent pole motor changes speed by changing the number of stator poles. This causes the synchronous speed of the rotating magnetic field to change.
3. The capacitor-start, capacitor-run (CSCR) type of multispeed motor changes speed by inserting impedance in series with the run winding.
4. Multispeed motors commonly have from two to five ranges of speed.
5. Capacitor-start, capacitor-run motors designed for use as multispeed motors have higher impedance windings than other CSCR motors.

- Repulsion-type motors
  1. There are three types of repulsion motors:
     a. The repulsion motor
     b. The repulsion-start, induction-run motor
     c. The repulsion-induction motor
  2. These three types of motors differ in their construction, operating characteristics, and industrial applications.
     a. Repulsion motor
        (1) The stator is usually wound with four, six, or eight poles.
        (2) The rotor is a slotted core containing a winding. The rotor is similar to the armature of a dc motor. The coils are connected to the commutator through segments, or bars, parallel to the armature shaft.
        (3) Carbon brushes contact the commutator surface. The brush holders are movable. They are connected together by heavy copper jumpers. The brushes are movable so that they can make contact at different points to obtain the correct rotation and maximum torque output.
        (4) To reverse the motor, the brushes are shifted 15 electrical degrees to the other side of the stator field pole centers.
        (5) This motor has excellent starting torque.
        (6) The speed of the motor can be changed by changing the value of the supply voltage.
        (7) The motor has very unstable speed characteristics. It may race to a very high speed if there is no mechanical load.
     b. Repulsion-start, induction-run motor
        (1) This motor is available in two basic types: the brush-lifting motor and the brush-riding motor.
        (2) For the brush-lifting motor, a centrifugal device lifts the brushes from the commutator surface at 75% of the rated speed. A short-circuiting necklace shorts out the commutator. The brushes are lifted from the commutator, thus saving wear and tear on all parts.
        (3) The brush-riding type of motor has the same short-circuiting necklace, but the brushes ride on the commutator at all times.

(4) This motor starts like a repulsion motor. After it reaches 75% of the rated speed, it runs as an induction motor.

(5) To reverse the direction of rotation, follow the steps outlined for a repulsion motor.

(6) These motors may be designed for dual-voltage connections.

c. Repulsion-induction motor

(1) This type of motor has the same characteristics as a repulsion-start, induction-run motor. It does not have a centrifugal mechanism.

(2) It does have a squirrel-cage winding beneath the slots of the armature. The squirrel-cage winding improves the speed regulation from no load to full load.

(3) The speed for this type of motor is similar to that of the dc compound motor.

(4) The motor may be designed for dual-voltage windings.

- The same schematic diagram symbol is used for the repulsion-induction motor, the repulsion motor, and the repulsion-start, induction-run motor.
- The dc series or shunt motor cannot be operated from an ac supply.
- Universal motor
  1. The field poles of a universal motor are laminated to reduce eddy currents.
  2. The voltage loss across the field poles is reduced by using a small number of field turns on a low-reluctance core.
  3. A universal motor can be operated on an ac or dc supply.
  4. Universal motors are available with fractional horsepower ratings and are commonly used in household appliances.
  5. The field windings and the armature of a universal motor are connected in series.
  6. There are two versions of the universal distributed field motor: the two-field motor and the single-field compensated motor. The two-field motor has a main winding and a compensating winding to reduce the reactance voltage developed in the armature by the alternating flux.
  7. The speed of a universal motor can be controlled by varying the inductance through taps on one of the field poles. The motor may be operated at an excessive speed with no load. Universal motors are permanently connected to the devices being driven.
  8. To reverse the direction of rotation, the current flow is reversed through the field circuit or the armature circuit.
- Conductive and inductive compensation
  1. Conductive compensation is achieved by placing special windings in slots cut in the pole faces to overcome the armature reaction under load.
     a. The strength of this field increases with an increase in the load current.

   b. The compensating winding is connected in series with the series field winding and the armature.
   c. A motor with conductive compensation has high starting torque and poor speed regulation.
2. An inductively coupled winding is also used to overcome armature reaction in ac series motors.
   a. This winding acts like a short-circuited secondary winding in a transformer.
   b. The operating characteristics of an inductively compensated motor are similar to those of the conductively compensated motor.

- Shaded-pole induction motor
  1. Small fractional horsepower induction motors may be started by shading coils mounted on each of the stator poles.
  2. A standard squirrel-cage rotor is used.
  3. The motor does not require a starter mechanism such as a centrifugal switch.
  4. The shading coil is a low-resistance copper loop.
  5. The shading poles distort the main flux pattern, and therefore set up a rotating magnetic field.
  6. The torque produced is small.
  7. Shaded poles are used only on motors rated at 1/10 hp or smaller.
  8. A shaded-pole motor is used in applications where a strong torque is not needed, such as for fans and blowers.

- Variable-speed motors
  1. Two types of motors are used for variable-speed control:
     a. The capacitor-start, capacitor-run motor
     b. The shaded-pole induction motor
  2. These motors are generally used because they do not require the use of a centrifugal switch.
  3. Motors used for variable-speed control have higher impedance windings than do other types of motors not designed for use as variable-speed motors.
  4. There are two general methods used to control the speed of variable-speed motors:
     a. Variable-voltage control
     b. Inserting impedance in series with the motor winding
  5. Two common methods of obtaining variable voltage are
     a. A solid-state controller using a triac
     b. An autotransformer
  6. Only triac controllers designed for use as variable-speed motor controllers should be used.

- Stepping motors
    1. Stepping motors generally operate on direct current and are used to produce angular movements in steps.
    2. Stepping motors are generally used for position control.
    3. Stepping motors can be used as single-phase synchronous motors when connected to two-phase ac.
    4. Most stepping motors operate at a synchronous speed of 72 r/min when connected to 60-Hz ac.
    5. Stepping motors can produce a holding torque by applying a continuous direct current to their stator windings.
    6. Full stepping produces an angular rotation of 1.8° per step for a total of 200 steps per revolution.
    7. Half stepping produces an angular rotation of 0.9° per step for a total of 400 steps per revolution.

Achievement Review

1. Explain how a motor is started using the split-phase method.
2. Explain why the starting torque of a capacitor-start, induction-run motor is better than that of a resistance-start, induction-run motor.
3. How is the direction of rotation reversed for each of the following?
    a. Resistance-start, induction run motor
    b. Capacitor-start, induction-run motor
4. The centrifugal switch contacts fail to close when a resistance-start, induction-run motor is deenergized. Explain what happens when the motor is energized.
5. A resistance-start, induction-run motor is started with a heavy overload. The motor fails to accelerate to a speed high enough to open the centrifugal switch contacts. What will happen?
6. A resistance-start, induction-run motor has a dual-voltage rating of 120/240 V. This motor has two running windings, each rated at 120 V, and one starting winding rated at 120 V. Draw a schematic connection diagram for this motor connected for 240 V.
7. Show the connections for a 120-V, capacitor-start, induction-run motor.
8. a. Compare the operating characteristics of a resistance-start, induction-run motor with those of a capacitor-start, induction-run motor.
    b. List three applications for
        (1) a resistance-start, induction-run motor.
        (2) a capacitor-start, induction run motor.

9. Explain the differences in construction and operation between a capacitor-start, induction-run motor and a capacitor-start, capacitor-run motor.
10. Explain how a repulsion motor operates.
11. a. Explain why torque is not developed by a repulsion motor when the brushes are placed directly in line with the stator pole centers.
    b. The brushes are shifted 90 electrical degrees from the position described in part a of this question. Is a torque developed by the motor in this case?
    c. At what position should the brushes be placed relative to the stator pole centers to obtain the maximum torque?
12. Explain how a repulsion-start, induction-run motor operates.
13. Explain the difference between a repulsion-start, induction-run motor (brush-lifting type) and a repulsion-start, induction-run motor (brush-riding type).
14. How are the following motors reversed?
    a. A repulsion motor
    b. A repulsion-start, induction-run motor
15. Compare the operating characteristics of a repulsion motor with those of a repulsion-start, induction-run motor.
16. List three practical applications for
    a. a repulsion motor.
    b. a repulsion-start, induction-run motor.
17. Explain the difference between a repulsion-start, induction-run motor and a repulsion-induction motor.
18. Explain why shunt motors cannot be operated satisfactorily on ac service.
19. What is the difference between an ac series motor and a dc series motor?
20. Why are ac series motors with small fractional horsepower ratings called *universal motors?*
21. What is the difference between a series motor with conductive compensation and a series motor with inductive compensation?
22. List four practical applications for ac series motors.
23. Describe the construction and operation of a shaded-pole motor.
24. Where is the shaded-pole motor used?
25. A 120-V, two-pole, 60-Hz, capacitor-start, induction-run motor has a full-load speed of 3450 r/min. Determine
    a. the synchronous speed.
    b. the percent slip.
26. Name two methods of changing the speed of the rotating magnetic field.

27. What type of multispeed motor changes speed by changing the synchronous speed of the rotating magnetic field?
28. Name a major difference between capacitor-start, capacitor-run motors designed for multispeed use and CSCR motors that are not.
29. What two types of motors are used as variable-speed motors?
30. Why are these types of motors used?
31. Name two methods of controlling the speed of variable-speed motors.
32. What solid-state component is used to control the voltage applied to a variable-speed motor?
33. Explain the difference in operation between a stepping motor and a common dc motor.
34. What is the principle of operation of a stepping motor?
35. Why do stepping motors have teeth machined into the stator poles and rotor?
36. When a stepping motor is connected to ac power, how many phases must be applied to the motor?
37. How many degrees out of phase are the voltages of a two-phase system?
38. What is the synchronous speed of an eight-pole stepping motor when connected to a two-phase, 60-Hz ac line?
39. How can the holding torque of a stepping motor be increased?

19

Control Circuits

Objectives

After studying this unit, the student should be able to

- define the difference between a schematic diagram and a wiring diagram.
- determine the operation of a control circuit using a schematic diagram.
- discuss the operation of an on-delay timer.
- discuss the operation of an off-delay timer.
- discuss interlocking.
- discuss plugging and inching controls.

Schematics and wiring diagrams are the written language of control circuits. It will be impossible for you to become proficient in troubleshooting electrical faults if you cannot read and interpret electrical diagrams. Learning to read electrical diagrams is not as difficult as many people first believe it to be. Once a few basic principles are understood, reading schematics and wiring diagrams will become no more difficult than reading a newspaper.

TWO-WIRE CIRCUITS

Control circuits are divided into two basic types: the two-wire and the three-wire. Figure 19–1 shows a simple two-wire control circuit. In this circuit, a simple switch is used to control the power applied to the coil of a motor starter. If the switch is open, there is no complete path for current flow, and the motor will not operate. If the switch is closed, power is supplied to the motor starter and it closes the M contacts, connecting the three-phase motor to the power line.

THREE-WIRE CIRCUITS

Three-wire control circuits are used because they are more flexible than two-wire circuits. Three-wire circuits are characterized by the fact that they are operated by a magnetic relay or motor starter. These circuits are generally controlled by one or more pilot devices.

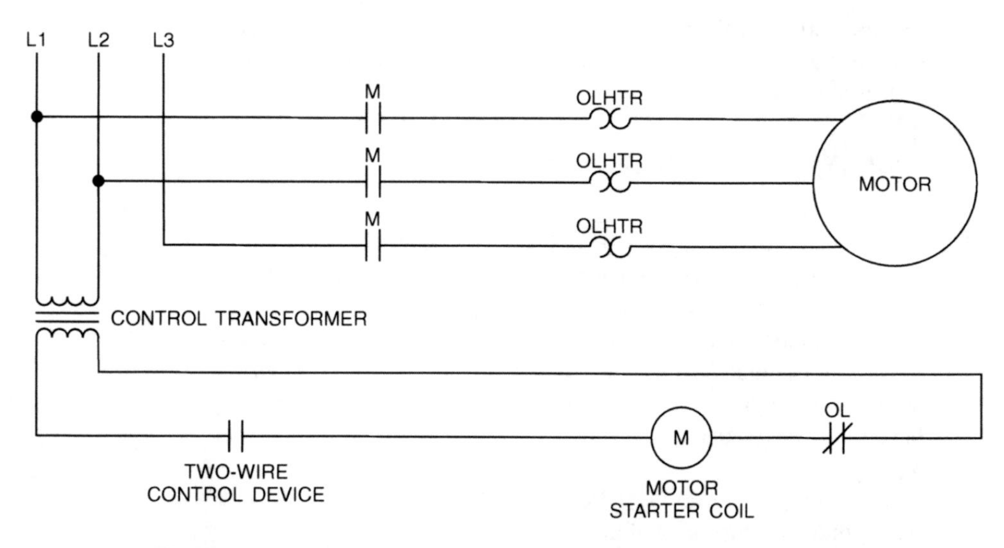

FIGURE 19–1 Two-wire control circuit (*Delmar/Cengage Learning*)

ELECTRICAL SYMBOLS FOR PILOT/CONTROL DEVICES

When people first begin to learn to read, they learn a set of symbols that are used to represent different sounds. This set of symbols is generally referred to as the *alphabet*. When learning to read electrical diagrams, it is necessary to learn the symbols used to represent different devices and components. The symbols that follow are commonly used on control schematics and wiring diagrams. These are not all the symbols used. Unfortunately, there is no real set standard for the use of electrical symbols. Most of these symbols are approved by the National Electrical Manufacturers Association (NEMA). These symbols are as follows:

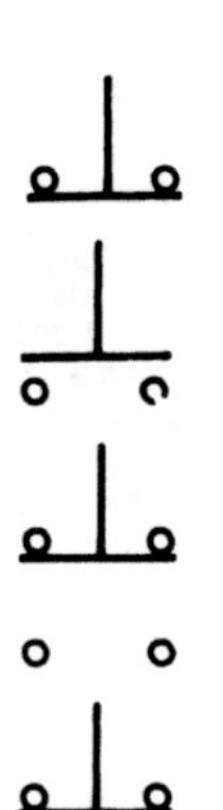

1. Normally closed push button. Generally used to represent a stop button.
2. Normally open push button. Generally used to represent a start button.
3. Double-acting push button. Contains both normally closed and normally open contacts on one push button.
4. Double-acting push button, drawn differently but meaning the same thing.

5. Double-acting push button. The dashed line indicates mechanical intertie. This means that when one button is pushed, the other moves at the same time.

6. Single-pole, single-throw (SPST) switch.

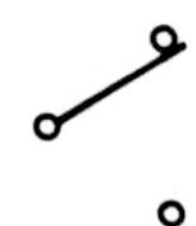

7. Single-pole, double-throw (SPDT) switch. Notice that this switch has only one pole, the switch arm, but it has two stationary contacts. In the diagram, the switch arm makes contact with the upper stationary contact. When the switch is thrown, contact will be made between the switch arm and the lower stationary contact. The switch arm or pole of the switch is generally referred to as the "common" because it can make contact to either of the two stationary contacts.

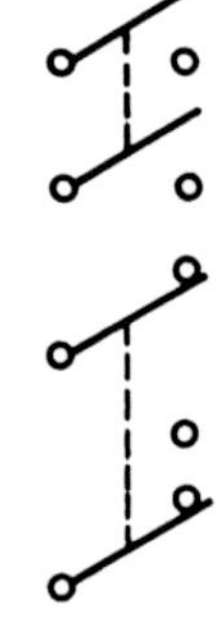

8. Double-pole, single-throw (DPST) switch. Notice the dashed line, which indicates mechanical intertie between the two switch arms.

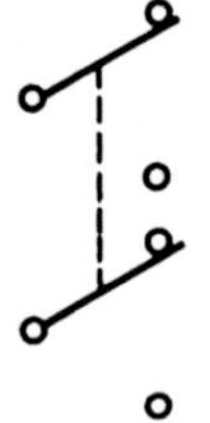

9. Double-pole, double-throw (DPDT) switch.

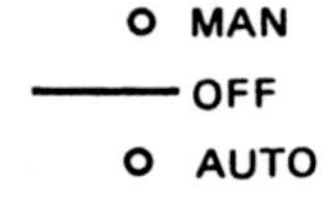

10. Off–automatic–manual control switch. This switch is basically a single-pole, double-throw switch that has a center-off position.

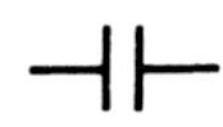

11. Normally open relay contact.

12. Normally closed relay contact.

13. Fuse.

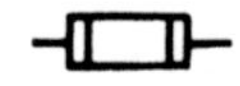

14. Fuse.

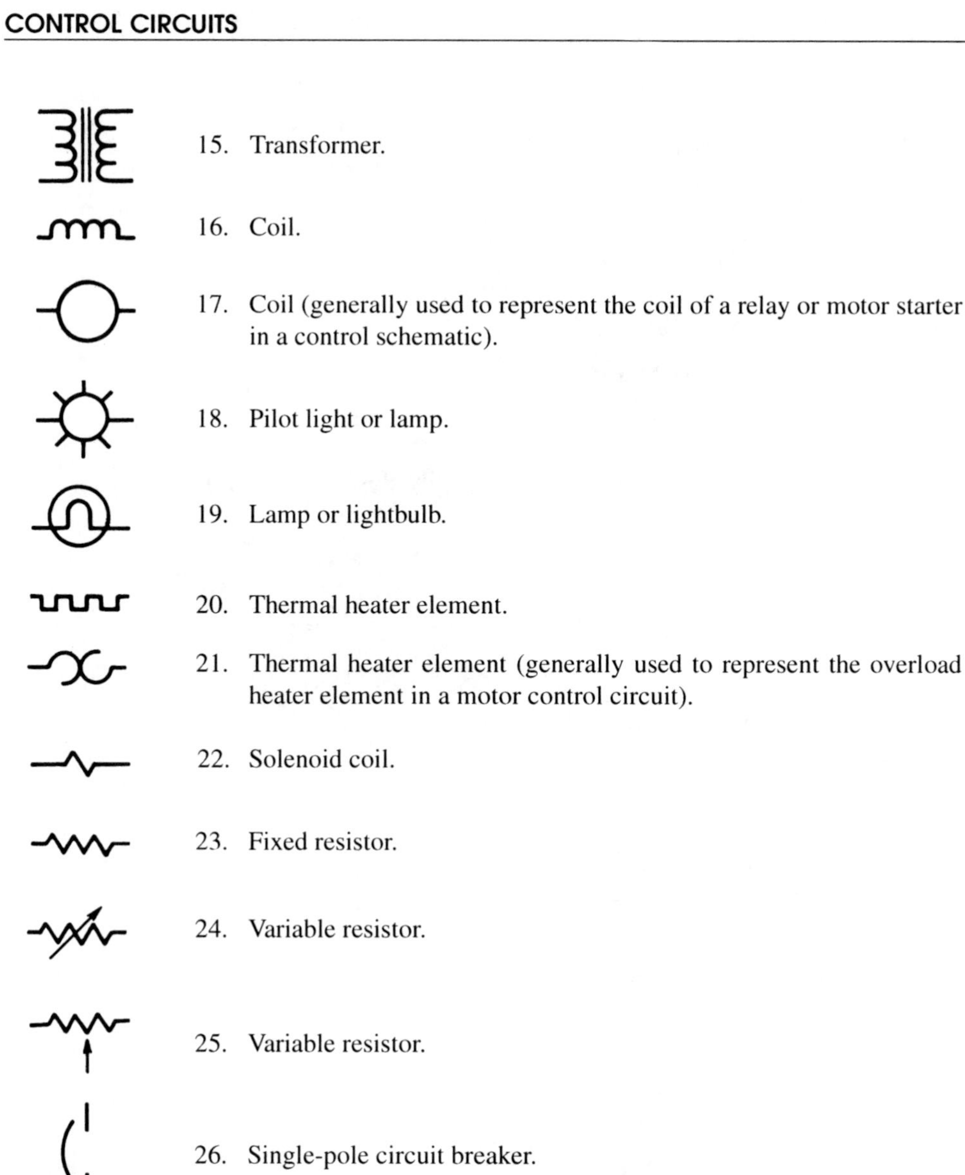

15. Transformer.
16. Coil.
17. Coil (generally used to represent the coil of a relay or motor starter in a control schematic).
18. Pilot light or lamp.
19. Lamp or lightbulb.
20. Thermal heater element.
21. Thermal heater element (generally used to represent the overload heater element in a motor control circuit).
22. Solenoid coil.
23. Fixed resistor.
24. Variable resistor.
25. Variable resistor.

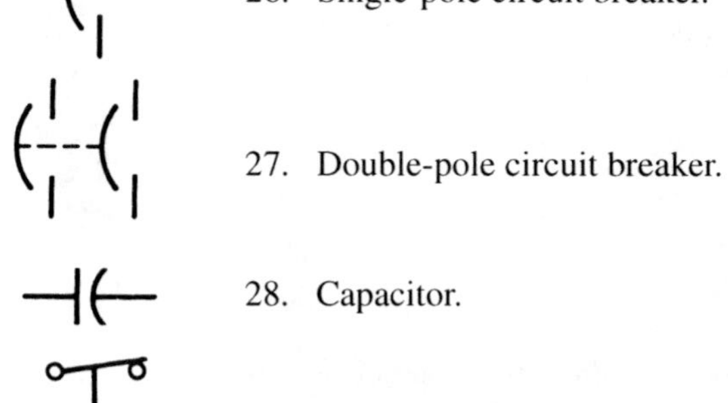

26. Single-pole circuit breaker.
27. Double-pole circuit breaker.
28. Capacitor.
29. Normally closed float switch.

30. Normally open float switch.

31. Normally closed pressure switch.

32. Normally open pressure switch.

33. Normally closed temperature switch (normally closed thermostat).

34. Normally open temperature switch.

35. Normally closed flow switch. This symbol is used to represent both liquid- and air-sensing flow switches.

36. Normally open flow switch.

37. Normally closed limit switch.

38. Normally open limit switch.

39. Normally closed on-delay timer contact. Often shown in schematics as DOE, which stands for delay on energize.

40. Normally on-delay timer contact.

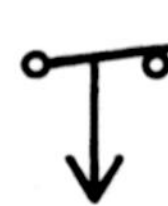

41. Normally closed off-delay timer contact. Often shown on schematics as DODE, which stands for delay on deenergize.

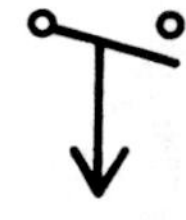

42. Normally open off-delay timer contact.

43. Battery.

44. Electrical ground.

45. Mechanical ground.

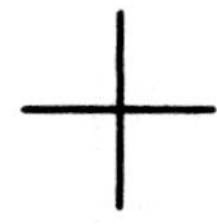

46. Wires crossing without connection.

47. Wires crossing without connection.

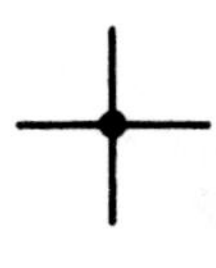

48. Wires connecting. The dot in the center of the cross is known as a *node*. This is used to indicate connection.

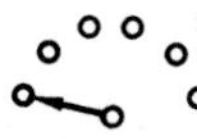

49. Rotary switch.

The contact symbols shown are standard and relatively simple to understand. There can be instances, however, in which symbols can be used to show something that is not at first apparent. For example, the symbol for a normally closed pressure switch is shown in Figure 19–2. Notice that this symbol shows not only the movable arm making contact with the stationary contact but also the movable arm drawn above the stationary contact. Figure 19–3 shows a normally closed held open pressure switch. This symbol is not a normally open contact symbol even though the contact arm is shown not making connection with the stationary contact because pressure is used to keep the contact open. If pressure decreases to a certain point, the switch contact will close.

FIGURE 19–2 Normally closed pressure switch (*Delmar/Cengage Learning*)

FIGURE 19–3 Normally closed held open pressure switch (*Delmar/Cengage Learning*)

FIGURE 19–4 Normally open pressure switch (*Delmar/ Cengage Learning*)

FIGURE 19–5 Normally open held closed pressure switch (*Delmar/Cengage Learning*)

Figure 19–4 shows a normally open pressure switch. Notice that the contact arm is drawn below the stationary contact. Figure 19–5 shows the same symbol except that the movable arm is making connection with the stationary contact. This symbol represents a normally open held closed pressure switch, that is, a pressure switch that is wired normally open but has its contact held closed by pressure. If the pressure decreases to a certain level, the switch will open and break connection to the rest of the circuit.

SCHEMATIC AND WIRING DIAGRAMS

Schematic diagrams show components in their electrical sequence without respect to physical location. Schematic diagrams are used to troubleshoot and install control circuits. Schematics are generally easier to read and understand than are wiring diagrams.

Wiring diagrams show components mounted in their general location with connecting wires. A wiring diagram is used to represent how the circuit generally appears. To help illustrate the differences between wiring diagrams and schematics, a basic control circuit will first be explained as a schematic and then shown as a wiring diagram.

Reading Schematic Diagrams

To read a schematic diagram, a few rules must first be learned. Commit the following rules to memory:

1. Reading a schematic diagram is similar to reading a book in English. It is read from left to right and from top to bottom.
2. Electrical symbols are always shown in their off or deenergized position.
3. Relay contact symbols are shown with the same numbers or letters that are used to designate the relay coil. All contact symbols that have the same number or letter as a coil are controlled by that coil regardless of where in the circuit they are located.
4. When a relay is energized, or turned on, all of its contacts change position. If a contact is shown as normally open, it will close when the coil is energized. If the contact is shown normally closed, it will open when the coil is turned on.
5. There must be a complete circuit before current can flow through a component.
6. Components used to provide a function of stop are generally wired normally closed and connected in series. Figure 19–6 illustrates this concept. Both switches A and B are normally closed and connected in series. If either switch is opened, connection to the lamp will be broken and current will stop flowing in the circuit.
7. Components used to provide the function of start are generally wired normally open and connected in parallel. In Figure 19–7, switches A and B are normally open and connected in parallel with each other. If either switch is closed, a current path will be provided for the lamp and it will turn on.

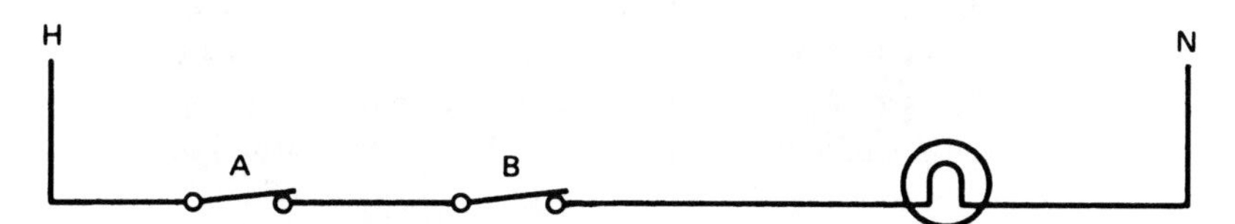

FIGURE 19–6 Components used to perform the function of stop are normally closed and connected in series. (*Delmar/Cengage Learning*)

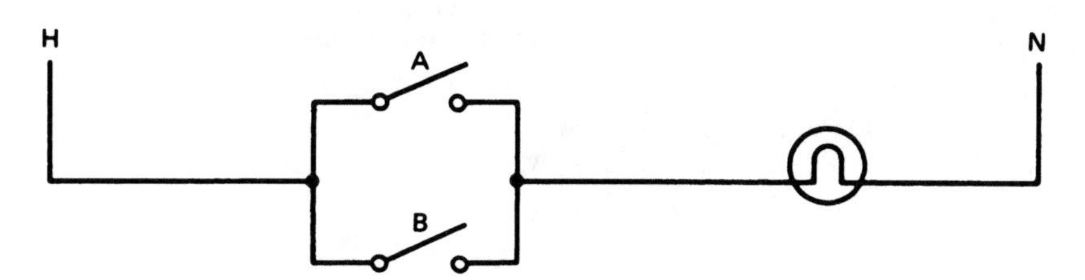

FIGURE 19–7 Components used to perform the function of start are normally open and connected in parallel. (*Delmar/Cengage Learning*)

The circuit to be discussed is a basic control circuit used throughout industry. Figure 19–8 shows a start–stop push button circuit. This schematic shows both the control circuit and the motor circuit. Schematic diagrams do not always show both control and motor connections. Many schematic diagrams show only the control circuit.

Notice in this schematic that there is no complete circuit to M motor starter coil because of the open-start push button and open M auxiliary contacts. There is also no connection to

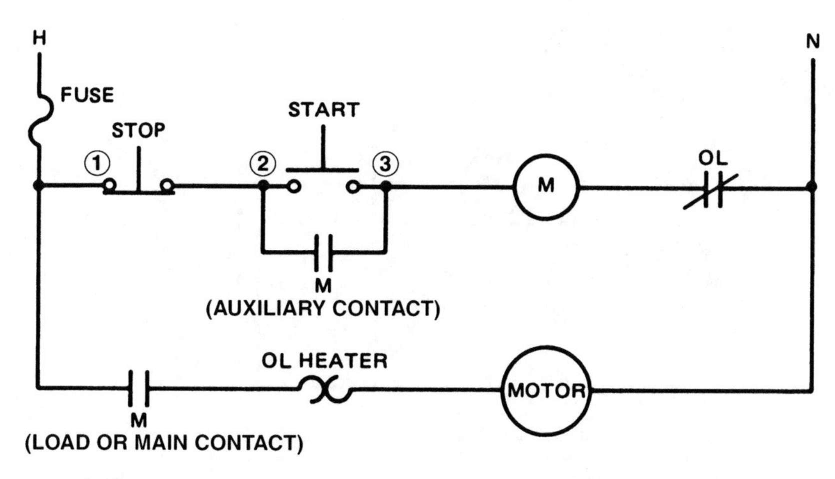

FIGURE 19–8 Start–stop push button control circuit (*Delmar/Cengage Learning*)

the motor because of the open M load contacts. The open M contacts connected in parallel with the start button are small contacts intended to be used as part of the control circuit. This set of contacts is generally referred to as the *holding, sealing,* or *maintaining* contacts. These contacts are used to provide a continued circuit to M coil when the start button is released.

The second set of M contacts is connected in series with the overload heater element and the motor, and the contacts are known as *load* contacts. These contacts are large and designed to carry the current needed to operate the load. Notice that these contacts are normally open and there is no current path to the motor.

When the start button is pushed, a path for current flow is provided to M motor starter coil. When M coil energizes, both M contacts close (Figure 19–9). The small auxiliary contact provides a continued current path to the motor starter coil when the start button is released and returns to its open position. The large M load contact closes and provides a complete circuit to the motor and the motor begins to run. The motor will continue to operate in this manner as long as M coil remains energized.

If the stop button is pushed (Figure 19–10), the current path to M coil is broken and the coil deenergized. This causes both M contacts to return to their normally open position. When M holding contacts open, there is no longer a complete circuit provided to the coil when the stop button is returned to its normal position. The circuit remains in the off position until the start button is again pushed.

Notice that the overload contact is connected in series with the motor starter coil. If the overload contact opens, it has the same effect as pressing the stop button. The fuse is connected in series with both the control circuit and the motor. If the fuse should open, it has the effect of disconnecting power from the line.

A wiring diagram for the start–stop push button circuit is shown in Figure 19–11. Although this diagram looks completely different, it is the same electrically as the schematic diagram. Notice that the push button symbols indicate double-acting push buttons. The stop

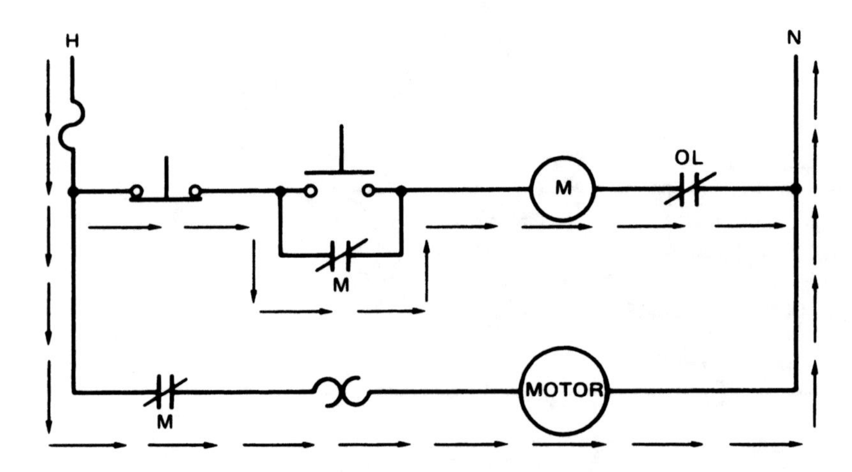

FIGURE 19–9 Current path through the circuit (*Delmar/Cengage Learning*)

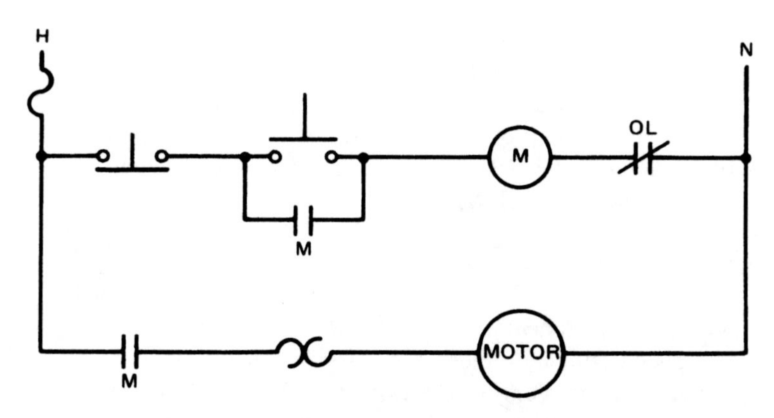

FIGURE 19–10 The stop button breaks the circuit. (*Delmar/Cengage Learning*)

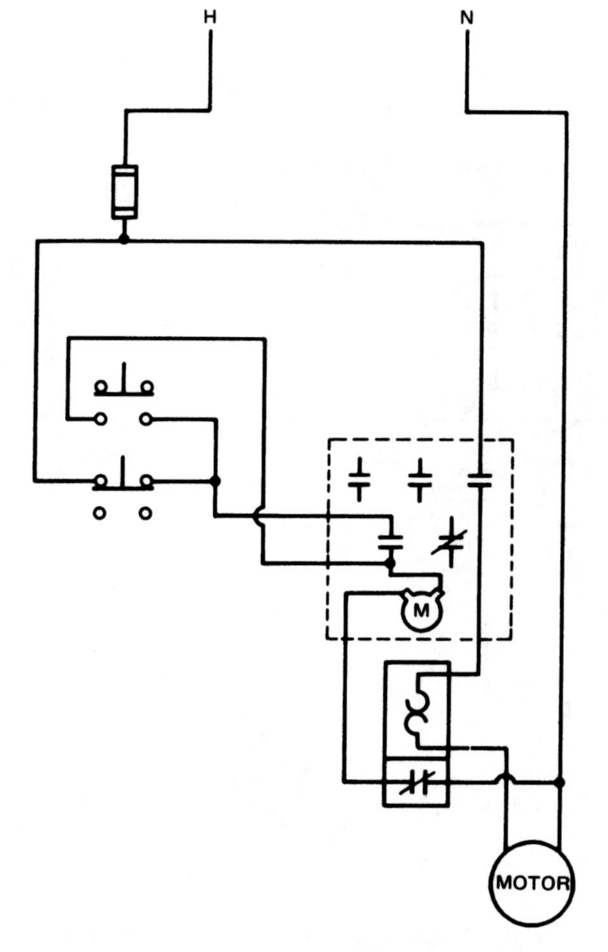

FIGURE 19–11 Wiring diagram of start–stop push button control circuit (*Delmar/Cengage Learning*)

button, however, uses only the normally closed section, and the start button uses only the normally open section. The motor starter shows three load contacts and two auxiliary contacts. One auxiliary contact is open and one is closed, but only the open contact is used.

The overload relay shows two different sections. One section contains the thermal heater element connected in series with the motor. The other section contains the normally closed contact connected in series with the coil of M starter.

INTERLOCKING

Interlocking is used to prevent some function from happening until some other function or action has occurred. A good example of interlocking is shown in the circuit in Figure 19–12. This circuit is a forward–reverse control. The two normally closed contacts labeled F and R are used to interlock the system. In this circuit, the motor is reversed by changing two of the input lines to the stator. If both motor starters F and R are energized

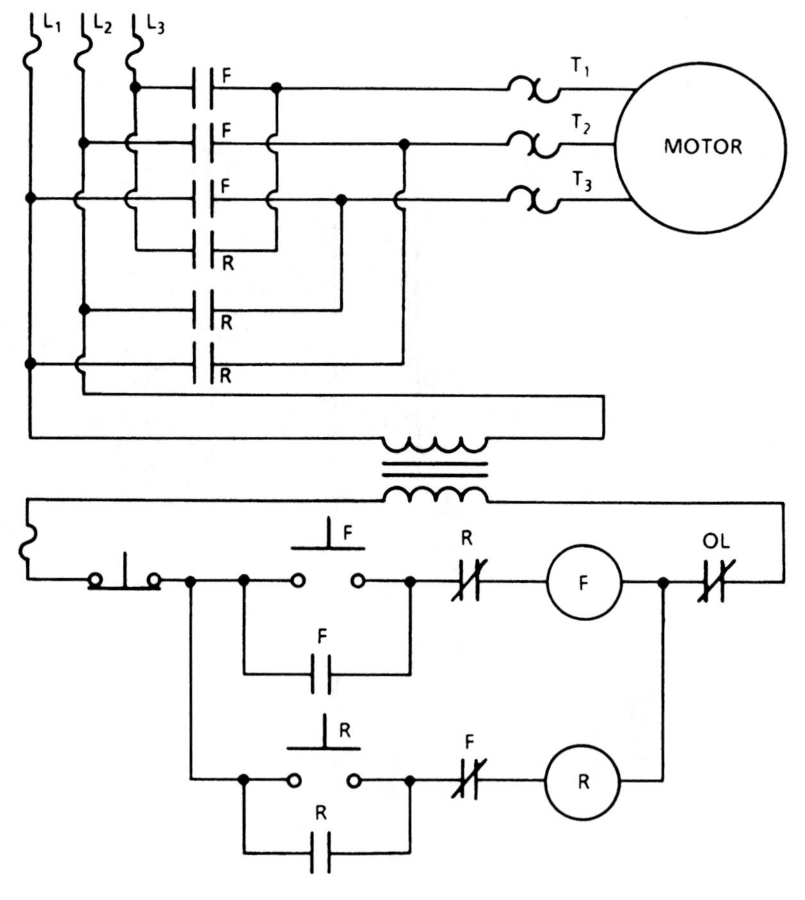

FIGURE 19–12 Forward–reverse control with interlocking (*Delmar/Cengage Learning*)

at the same time, there would be a direct short circuit across two of the phases. Forward–reverse starters are generally interlocked mechanically as well as electrically, but this example shows only electrical interlock.

It will first be assumed that the motor is to be operated in the forward direction. When the forward push button is pressed, a circuit is completed to F coil. This causes all F contacts to change position. The normally open F load contacts close and connect the motor to the line. The normally open auxiliary contact closes to complete the circuit around the normally open F push button when it is released, and the normally closed F contact opens. When the normally open F contact opens, it breaks the circuit to R coil. If R push button were to be pressed, R coil could not energize because of the now open contact connected in series with it.

Before the motor can be reversed, the stop button must be pushed and F coil deenergized. This permits all F contacts to return to their normal position. If the reverse push button is now pressed, current can flow through the normally closed F contact and energize R coil. When R coil energizes, all R contacts change position. The R load contacts connect the motor to the line, the normally open R contact closes to complete a circuit around R push button, and the normally closed R contact connected in series with F coil opens. This prevents F coil from being energized if F push button should be pressed.

JOGGING

Jogging in this context is moving a machine with short jabs of power. It is used to bring a machine into some particular position, generally for loading or unloading. Another term, *inching,* is very similar to jogging. The difference is that jogging applies full voltage to the motor and inching applies reduced voltage to the motor. Regardless of which method is used, the basic control requirement for either is for the motor starter to energize when the jog button is pushed and deenergize when the jog button is released. This control function can be accomplished by several methods. One method is shown in Figure 19–13. In this circuit, a double-acting push button is used as the jog button. When the jog button is pressed, the normally open section of the button closes to complete a circuit to M coil. The normally closed section opens to prevent the normally open M contact from holding the coil in after the button is released. This permits M coil to deenergize and disconnect the motor from the line. If the start button is pressed, M coil will energize and close all M contacts. The normally open auxiliary contact can now provide a current path through the closed push button to keep the coil energized.

TIMERS

There are two basic types of time-delay relays: the *on* delay and the *off* delay. The on-delay timer is often referred to on a schematic as *DOE,* which stands for delay on energize. Off-delay timers are often referred to as *DODE,* which stands for delay on deenergize. Timers may contain only time-operated contacts or a combination of both time-operated and instantaneous contacts. Time-operated contacts are operated by the timer mechanism, and instantaneous contacts operate like normal relay contacts.

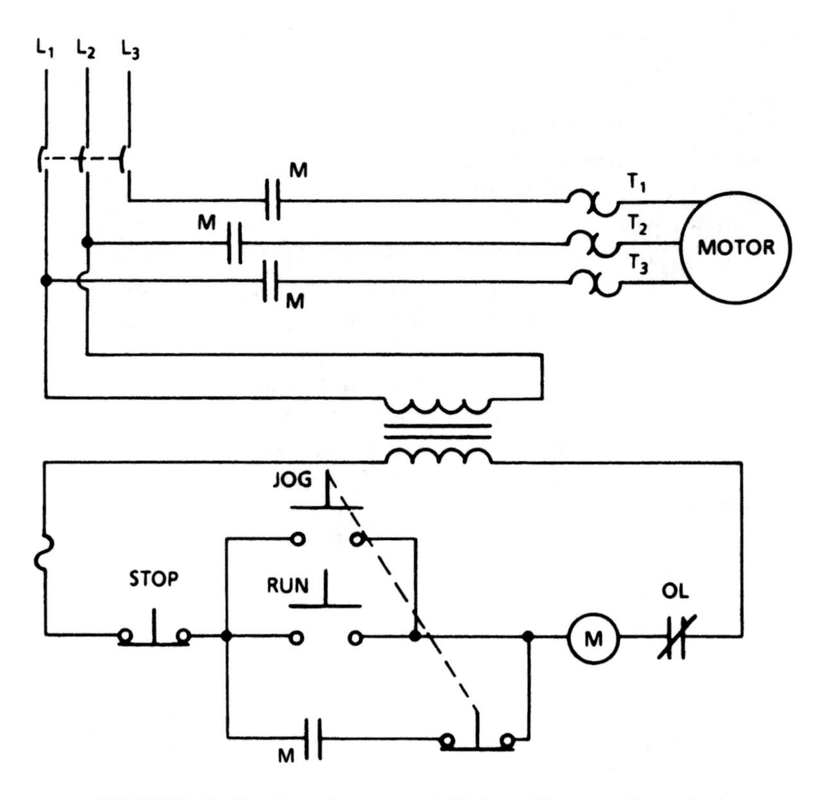

FIGURE 19–13 Run-jog control (*Delmar/Cengage Learning*)

On-Delay Timer

The operation of an on-delay timer is as follows. When the coil is energized, the contacts delay changing from their normal position. When the coil is deenergized, they change back to their normal position immediately.

The circuit shown in Figure 19–14 illustrates the operation of an on-delay timer. It is assumed that the timer has been set for a delay of 10 seconds. When the normally open push button is pressed, a circuit is completed to TR coil. The normally open contacts connected around the start push button are instantaneous contacts. These contacts close immediately and act as holding contacts to keep the relay energized. A normally open time-operated contact is connected in series with a pilot light. This contact will close 10 seconds after TR coil has been energized. When the stop button is pressed, the circuit to TR coil is opened and the relay deenergizes. All TR contacts open immediately. This disconnects the pilot light from the line.

Off-Delay Timer

The off-delay timer functions as follows. When the relay coil is energized, the contacts change position immediately. When the coil is deenergized, the contacts delay changing back to their normal position.

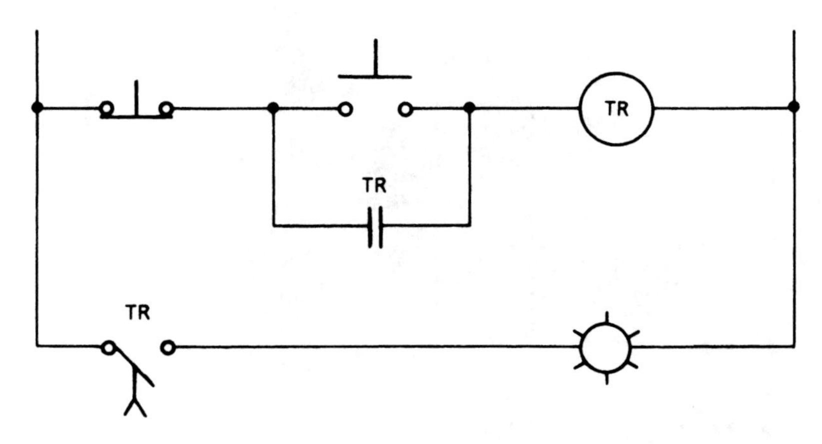

FIGURE 19–14 On-delay relay (*Delmar/Cengage Learning*)

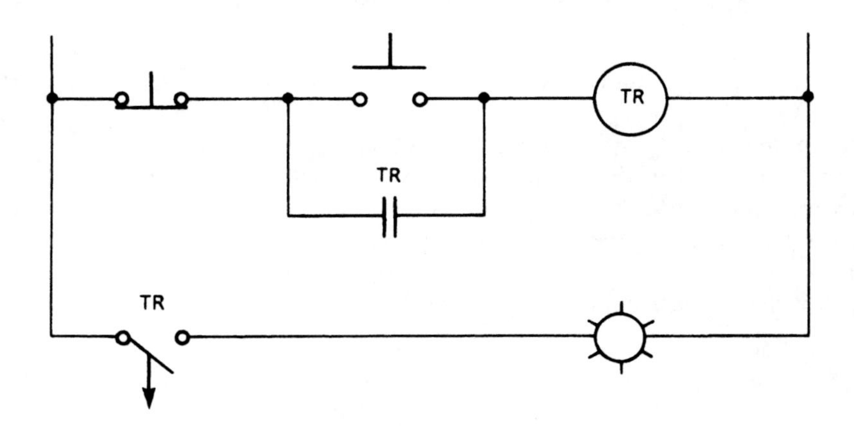

FIGURE 19–15 Off-delay relay (*Delmar/Cengage Learning*)

The circuit in Figure 19–15 illustrates the operation of an off-delay timer. A normally open instantaneous contact is used as the holding contact around the start button. A normally open time-operated contact is connected in series with a pilot light. When the start button is pressed, TR coil energizes and both TR contacts close immediately. When the stop button is pressed, TR coil is deenergized, but the normally open time-operated contact remains closed for 10 seconds before it reopens.

Types of Timers

There are several methods that can be used to achieve a time delay. One type of timer is known as a *pneumatic timer* because it uses air to provide a time delay. Figure 19–16 illustrates this type of timer. When rod A pushes against the bellows, air is forced out the check valve. Retraction of the bellows causes TR contact to close. When rod A is pulled back, the spring pushes outward on the bellows. Before the bellows can expand, however, air must enter through the air inlet. The needle valve controls the rate at which air can

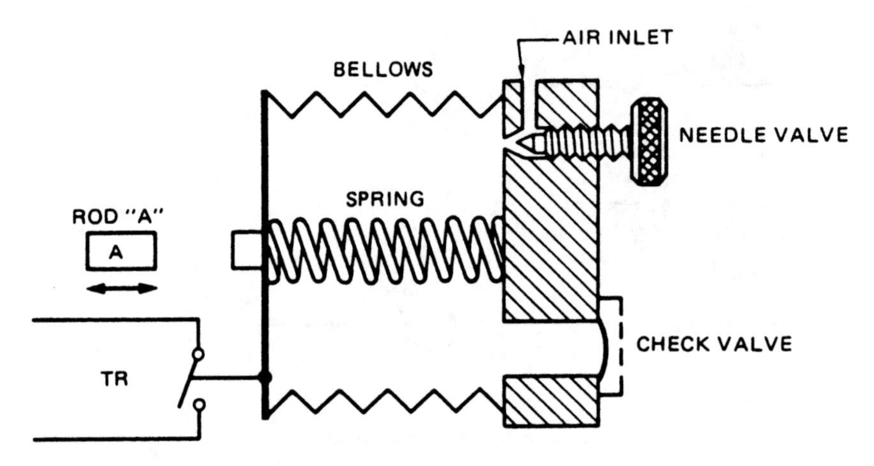

FIGURE 19–16 Bellows-operated pneumatic timer (*Delmar/Cengage Learning*)

enter the bellows and therefore the time required to expand the bellows and reopen the TR contact.

Dashpot timers use a piston moving through oil to create a time delay. Some timers use electric clocks to provide time delays. Electronic timers have become very popular because they are inexpensive and have good reliability and repeat accuracy.

The circuit shown in Figure 19–17 illustrates an electronic circuit that can be used as an on-delay timer. When switch S_1 closes, capacitor C_1 begins to charge through resistor RT. When C_1 reaches about 10 V, the unijunction transistor turns on. This provides gate current

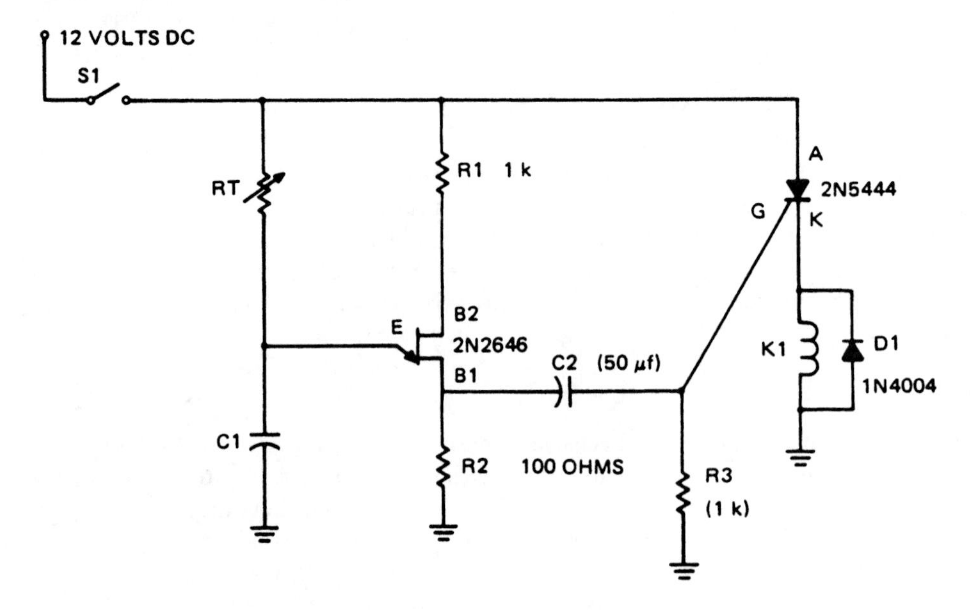

FIGURE 19–17 Schematic of electronic on-delay timer (*Delmar/Cengage Learning*)

to the silicon controlled rectifier (SCR). When the SCR turns on, relay coil K_1 energizes. The SCR remains turned on until switch S_1 is again opened. Two types of electronic timers are shown in Figures 19–18 and 19–19. One is adjustable and the other is not.

FIGURE 19–18 Nonadjustable electronic timer (*Courtesy of Allen-Bradley Corp.*)

FIGURE 19–19 Adjustable electronic timer (*Courtesy of Eagle Signal Industrial Controls*)

WOUND-ROTOR MOTOR CONTROL

An automatic time-operated speed control for a wound-rotor motor is shown in Figure 19–20. In this circuit, the stator of the motor is connected to the line when M load contacts close. Resistors are connected to the rotor of the motor during starting. When the motor is first started, all resistors are connected in series with the rotor. It will be assumed that all timers have been set for a delay of 10 seconds.

When the start button is pushed, a circuit is completed through M coil and TR_1 coil. When M contacts close, the stator of the motor is connected to the line, and M auxiliary contact is used as a holding contact. Ten seconds after TR_1 coil is energized, TR_1 contact closes. This energizes S_1 coil and TR_2 coil. When S_1 coil energizes, S_1 contacts close and short out the first set of resistors. This permits the motor to increase to the second step of speed.

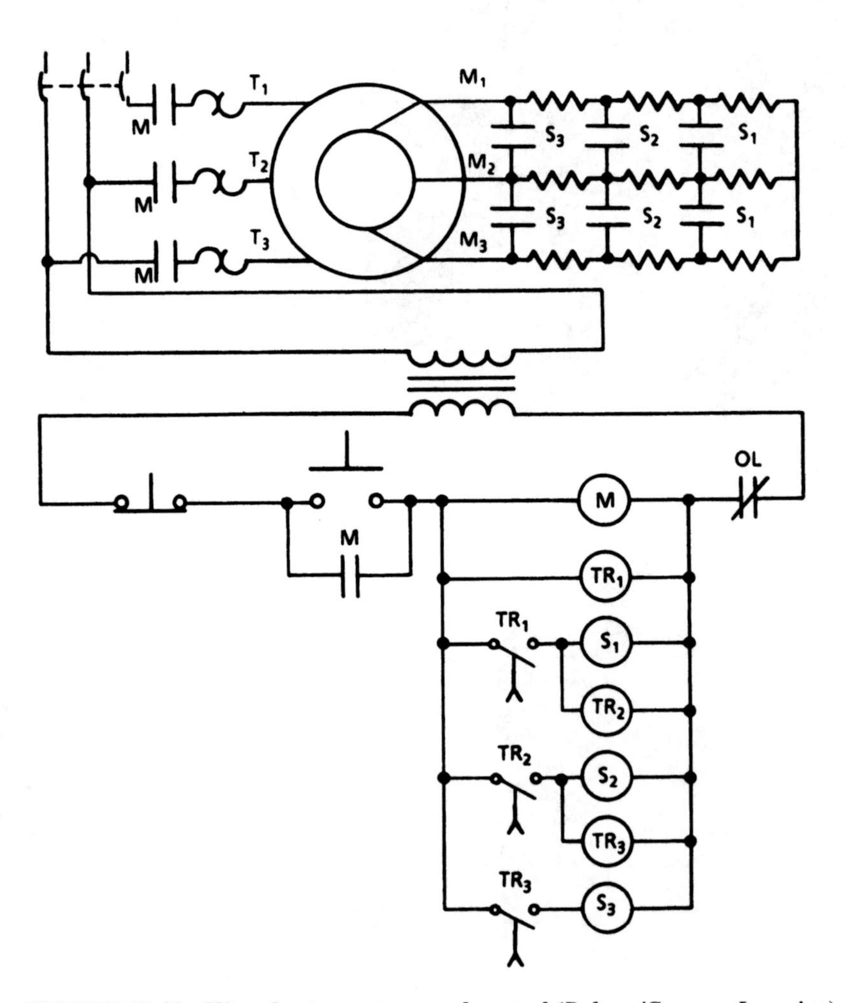

FIGURE 19–20 Wound-rotor motor speed control (*Delmar/Cengage Learning*)

Ten seconds after coil TR_2 is energized, contact TR_2 closes and energizes coils S_2 and TR_3. When S_2 energizes, S_2 contacts close and short out the second set of resistors. This permits the motor to increase to the third step of speed.

Ten seconds after coil TR_3 energizes, contact TR_3 closes and energizes coil S_3. This causes S_3 contacts to close and short out the remaining resistance and the motor operates at its full speed.

SUMMARY

- The two basic types of control circuits are two-wire and three-wire.
- A schematic diagram shows components in their electrical sequence without respect to physical location.
- A wiring diagram shows components in their physical location with connecting wires.
- Control diagrams are read like a book, from top to bottom and from left to right.
- Control components are always shown in their normal deenergized position.
- Interlocking is used to keep some function or action from happening until some other function or action has taken place.
- Jogging is used to position a machine in some position by starting it with short jabs of power.
- On-delay relays delay changing their contacts when the coil is energized, but change back immediately when the coil is deenergized.
- Off-delay relays change their contacts immediately when they are energized but delay changing them back when they are deenergized.

Achievement Review

1. What are the two basic types of motor control?
2. Define a schematic diagram.
3. Define a wiring diagram.
4. Determine how components for the function of stop are generally wired and connected with the stop button.
5. Determine how components used for the function of start are generally wired and connected with the start button.
6. When reading a schematic diagram, are the components shown in their energized or deenergized position?

7. What does a dashed line drawn between components represent?
8. What is an auxiliary contact?
9. Explain the operation of an on-delay timer.
10. Explain the operation of an off-delay timer.

Appendices

APPENDIX 1
TRIGONOMETRY—SIMPLE FUNCTIONS

Sine (sin) of an angle $= \dfrac{\text{opposite side}}{\text{hypotenuse}}$ $\qquad \sin A = \dfrac{a}{c}$

Cosine (cos) of an angle $= \dfrac{\text{adjacent side}}{\text{hypotenuse}}$ $\qquad \cos A = \dfrac{b}{c}$

Tangent (tan) of an angle $= \dfrac{\text{opposite side}}{\text{adjacent side}}$ $\qquad \tan A = \dfrac{a}{b}$

Cotangent (cot) $= \dfrac{1}{\tan} = \dfrac{\text{adjacent side}}{\text{opposite side}}$ $\qquad \cot A = \dfrac{b}{a} = \dfrac{1}{\tan A}$

Secant (sec) $= \dfrac{1}{\cos} = \dfrac{\text{hypotenuse}}{\text{adjacent side}}$ $\qquad \sec A = \dfrac{c}{b} = \dfrac{1}{\cos A}$

Cosecant (cosec) $= \dfrac{1}{\sin} = \dfrac{\text{hypotenuse}}{\text{opposite side}}$ $\qquad \operatorname{cosec} A = \dfrac{c}{a} = \dfrac{1}{\sin A}$

$$\sin B = \frac{b}{c} = \cos A = \cos(90° - B), \text{ since } A = 90° - B.$$

$$\cos B = \frac{a}{c} = \sin A = \sin(90° - B).$$

$$\frac{\sin A}{\cos A} = \frac{a/c}{b/c} = \frac{a}{b} = \tan A.$$

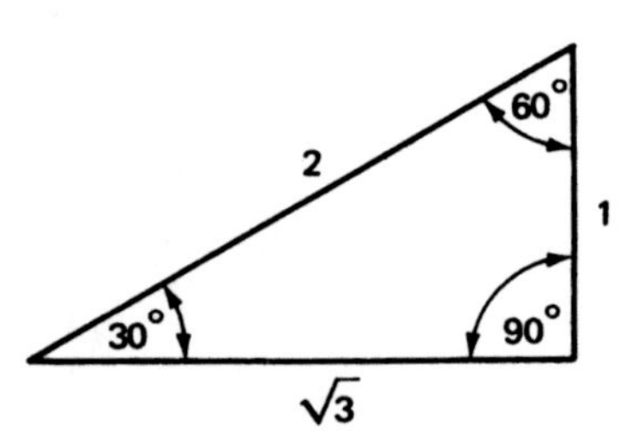

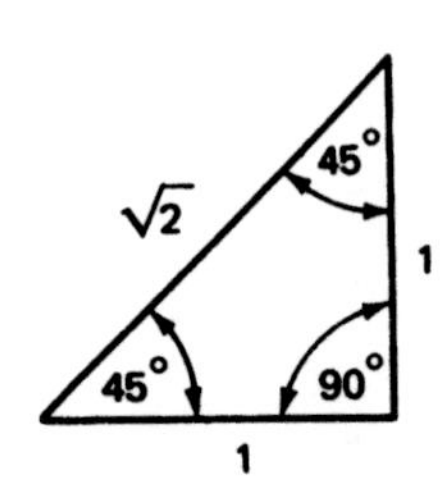

Angle	sin	cos	tan
30°	0.5	$\sqrt{3}/2 = 0.866$	$1/\sqrt{3} = 0.577$
60°	$\sqrt{3}/2 = 0.866$	0.5	$\sqrt{3} = 1.732$
45°	$1/\sqrt{2} = 0.707$	$1/\sqrt{2} = 0.707$	1.0

APPENDIX 2
FUNCTIONS OF ANGLES GREATER THAN 90°

Note: The radius vector, ob, is always positive.

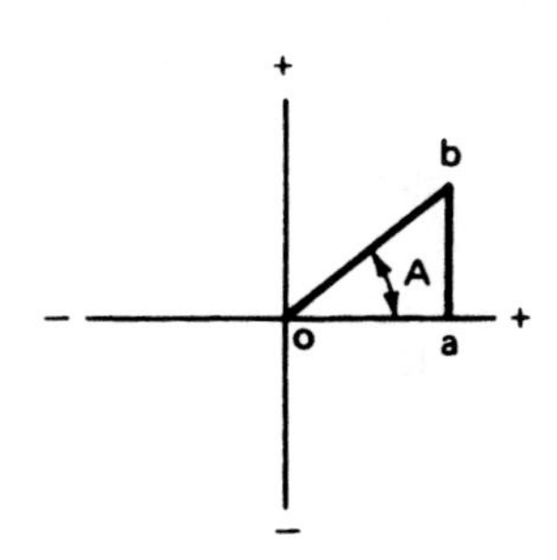

FIRST QUADRANT

$\sin A = \frac{+ab}{+ob}$ sin is (+)

$\cos A = \frac{+oa}{+ob}$ cos is (+)

$\tan A = \frac{+ab}{+oa}$ tan is (+)

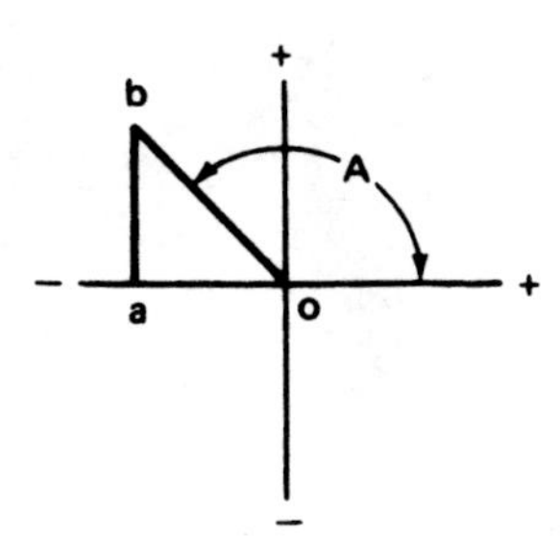

SECOND QUADRANT

$\sin A = \frac{+ab}{+ob}$ sin is (+)

$\cos A = \frac{-oa}{+ob}$ cos is (−)

$\tan A = \frac{+ab}{-oa}$ tan is (−)

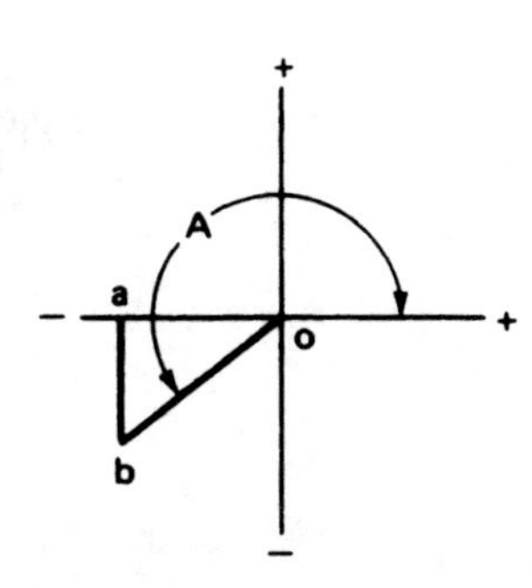

THIRD QUADRANT

$\sin A = \frac{-ab}{+ob}$ sin is (−)

$\cos A = \frac{-oa}{+ob}$ cos is (−)

$\tan A = \frac{-ab}{-oa}$ tan is (+)

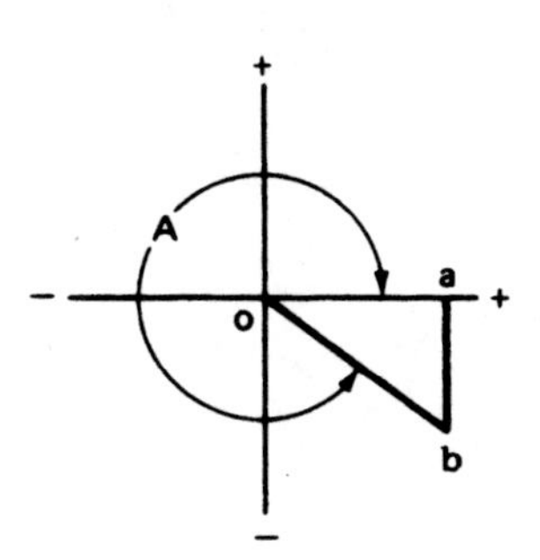

FOURTH QUADRANT

$\sin A = \frac{-ab}{+ob}$ sin is (−)

$\cos A = \frac{+oa}{+ob}$ cos is (+)

$\tan A = \frac{-ab}{+oa}$ tan is (−)

APPENDIX 3
TRIGONOMETRIC FORMULAS

$a^2 + b^2 = c^2$

$\frac{a^2}{c^2} + \frac{b^2}{c^2} = \frac{c^2}{c^2} = 1$

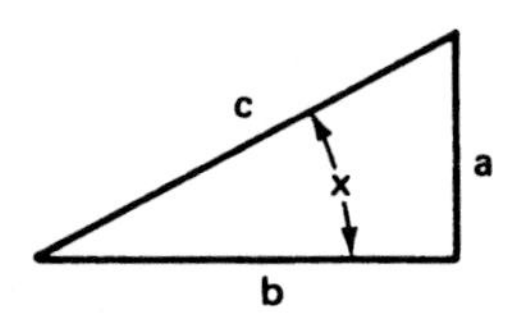

Because $\sin x = \frac{a}{c}$; $\cos x = \frac{b}{c}$, it follows that

$\sin^2 x + \cos^2 x = 1$

$\sec^2 x = 1 + \tan^2 x$

$\sin(90° + x) = \cos x$

$\cos(90° + x) = -\sin x$

$\tan(90° + x) = -\cot x$

$\sin(180° - x) = \sin x$

$\cos(180° - x) = -\cos x$

$\tan(180° - x) = -\tan x$

$\sin(x + y) = \sin x \cos y + \cos x \sin y$

$\sin(x - y) = \sin x \cos y - \cos x \sin y$

$\cos(x + y) = \cos x \cos y - \sin x \sin y$

$\cos(x - y) = \cos x \cos y + \sin x \sin y$

$\tan(x + y) = \frac{\tan x + \tan y}{1 - \tan x \tan y}$

$\tan(x - y) = \frac{\tan x - \tan y}{1 + \tan x \tan y}$

$\sin 2x = 2 \sin x \cos x$

$\cos 2x = \cos^2 x - \sin^2 x$

$\tan 2x = \frac{2 \tan x}{1 - \tan^2 x}$

Law of sines — In any triangle, the sides are proportional to the sines of the opposite angles:

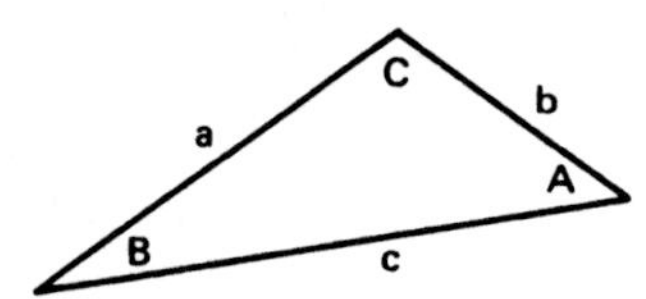

$$\frac{a}{\sin A} = \frac{b}{\sin B} = \frac{c}{\sin C}$$

Law of cosines — In any triangle, the square of any side is equal to the sum of the squares of the other two sides minus twice the product of these two sides and the cosine of their included angle:

$a^2 = b^2 + c^2 - 2bc \cos A$

$\cos A = \frac{b^2 + c^2 - a^2}{2bc}$

$\cos B = \frac{c^2 + a^2 - b^2}{2ca}$

$\cos C = \frac{a^2 + b^2 - c^2}{2ab}$

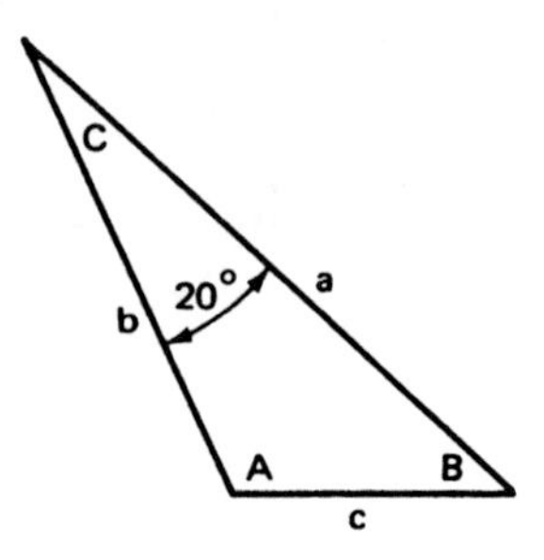

APPENDIX 4
TRIGONOMETRIC FUNCTIONS—NATURAL SINES AND COSINES

Note: For cosines, use right-hand column of degrees and lower line of tenths.

Deg.	°0.0	°0.1	°0.2	°0.3	°0.4	°0.5	°0.6	°0.7	°0.8	°0.9	°1.0	
0°	0.0000	0.0017	0.0035	0.0052	0.0070	0.0087	0.0105	0.0122	0.0140	0.0157	0.0175	89
1	0.0175	0.0192	0.0209	0.0227	0.0244	0.0262	0.0279	0.0297	0.0314	0.0332	0.0349	88
2	0.0349	0.0366	0.0384	0.0401	0.0419	0.0436	0.0454	0.0471	0.0488	0.0506	0.0523	87
3	0.0523	0.0541	0.0558	0.0576	0.0593	0.0610	0.0628	0.0645	0.0663	0.0680	0.0698	86
4	0.0698	0.0715	0.0732	0.0750	0.0767	0.0785	0.0802	0.0819	0.0837	0.0854	0.0872	85
5	0.0872	0.0889	0.0906	0.0924	0.0941	0.0958	0.0976	0.0993	0.1011	0.1028	0.1045	84
6	0.1045	0.1063	0.1080	0.1097	0.1115	0.1132	0.1149	0.1167	0.1184	0.1201	0.1219	83
7	0.1219	0.1236	0.1253	0.1271	0.1288	0.1305	0.1323	0.1340	0.1357	0.1374	0.1392	82
8	0.1392	0.1409	0.1426	0.1444	0.1461	0.1478	0.1495	0.1513	0.1530	0.1547	0.1564	81
9	0.1564	0.1582	0.1599	0.1616	0.1633	0.1650	0.1668	0.1685	0.1702	0.1719	0.1736	80°
10°	0.1736	0.1754	0.1771	0.1788	0.1805	0.1822	0.1840	0.1857	0.1874	0.1891	0.1908	79
11	0.1908	0.1925	0.1942	0.1959	0.1977	0.1994	0.2011	0.2028	0.2045	0.2062	0.2079	78
12	0.2079	0.2096	0.2113	0.2130	0.2147	0.2164	0.2181	0.2198	0.2215	0.2232	0.2250	77
13	0.2250	0.2267	0.2284	0.2300	0.2317	0.2334	0.2351	0.2368	0.2385	0.2402	0.2419	76
14	0.2419	0.2436	0.2453	0.2470	0.2487	0.2504	0.2521	0.2538	0.2554	0.2571	0.2588	75
15	0.2588	0.2605	0.2622	0.2639	0.2656	0.2672	0.2689	0.2706	0.2723	0.2740	0.2756	74
16	0.2756	0.2773	0.2790	0.2807	0.2823	0.2840	0.2857	0.2874	0.2890	0.2907	0.2924	73
17	0.2924	0.2940	0.2957	0.2974	0.2990	0.3007	0.3024	0.3040	0.3057	0.3074	0.3090	72
18	0.3090	0.3107	0.3123	0.3140	0.3156	0.3173	0.3190	0.3206	0.3223	0.3239	0.3256	71
19	0.3256	0.3272	0.3289	0.3305	0.3322	0.3338	0.3355	0.3371	0.3387	0.3404	0.3420	70°
20°	0.3420	0.3437	0.3453	0.3469	0.3486	0.3502	0.3518	0.3535	0.3551	0.3567	0.3584	69
21	0.3584	0.3600	0.3616	0.3633	0.3649	0.3665	0.3681	0.3697	0.3714	0.3730	0.3746	68
22	0.3746	0.3762	0.3778	0.3795	0.3811	0.3827	0.3843	0.3859	0.3875	0.3891	0.3907	67
23	0.3907	0.3923	0.3939	0.3955	0.3971	0.3987	0.4003	0.4019	0.4035	0.4051	0.4067	66
24	0.4067	0.4083	0.4099	0.4115	0.4131	0.4147	0.4163	0.4179	0.4195	0.4210	0.4226	65
25	0.4226	0.4242	0.4258	0.4274	0.4289	0.4305	0.4321	0.4337	0.4352	0.4368	0.4384	64
26	0.4384	0.4399	0.4415	0.4431	0.4446	0.4462	0.4478	0.4493	0.4509	0.4524	0.4540	63
27	0.4540	0.4555	0.4571	0.4586	0.4602	0.4617	0.4633	0.4648	0.4664	0.4679	0.4695	62
28	0.4695	0.4710	0.4726	0.4741	0.4756	0.4772	0.4787	0.4802	0.4818	0.4833	0.4848	61
29	0.4848	0.4863	0.4879	0.4894	0.4909	0.4924	0.4939	0.4955	0.4970	0.4985	0.5000	60°
30°	0.5000	0.5015	0.5030	0.5045	0.5060	0.5075	0.5090	0.5105	0.5120	0.5135	0.5150	59
31	0.5150	0.5165	0.5180	0.5195	0.5210	0.5225	0.5240	0.5255	0.5270	0.5284	0.5299	58
32	0.5299	0.5314	0.5329	0.5344	0.5358	0.5373	0.5388	0.5402	0.5417	0.5432	0.5446	57
33	0.5446	0.5461	0.5476	0.5490	0.5505	0.5519	0.5534	0.5548	0.5563	0.5577	0.5592	56
34	0.5592	0.5606	0.5621	0.5635	0.5650	0.5664	0.5678	0.5693	0.5707	0.5721	0.5736	55
35	0.5736	0.5750	0.5764	0.5779	0.5793	0.5807	0.5821	0.5835	0.5850	0.5864	0.5878	54
36	0.5878	0.5892	0.5906	0.5920	0.5934	0.5948	0.5962	0.5976	0.5990	0.6004	0.6018	53
37	0.6018	0.6032	0.6046	0.6060	0.6074	0.6088	0.6101	0.6115	0.6129	0.6143	0.6157	52
38	0.6157	0.6170	0.6184	0.6198	0.6211	0.6225	0.6239	0.6252	0.6266	0.6280	0.6293	51
39	0.6293	0.6307	0.6320	0.6334	0.6347	0.6361	0.6374	0.6388	0.6401	0.6414	0.6428	50°
40°	0.6428	0.6441	0.6455	0.6468	0.6481	0.6494	0.6508	0.6521	0.6534	0.6547	0.6561	49
41	0.6561	0.6574	0.6587	0.6600	0.6613	0.6626	0.6639	0.6652	0.6665	0.6678	0.6691	48
42	0.6691	0.6704	0.6717	0.6730	0.6743	0.6756	0.6769	0.6782	0.6794	0.6807	0.6820	47
43	0.6820	0.6833	0.6845	0.6858	0.6871	0.6884	0.6896	0.6909	0.6921	0.6934	0.6947	46
44	0.6947	0.6959	0.6972	0.6984	0.6997	0.7009	0.7022	0.7034	0.7046	0.7059	0.7071	45
	°1.0	°0.9	°0.8	°0.7	°0.6	°0.5	°0.4	°0.3	°0.2	°0.1	°0.0	Deg.

Note: For cosines, use right-hand column of degrees and lower line of tenths.

Deg.	°0.0	°0.1	°0.2	°0.3	°0.4	°0.5	°0.6	°0.7	°0.8	°0.9	°1.0	
45	0.7071	0.7083	0.7096	0.7108	0.7120	0.7133	0.7145	0.7157	0.7169	0.7181	0.7193	44
46	0.7193	0.7206	0.7218	0.7230	0.7242	0.7254	0.7266	0.7278	0.7290	0.7302	0.7314	43
47	0.7314	0.7325	0.7337	0.7349	0.7361	0.7373	0.7385	0.7396	0.7408	0.7420	0.7431	42
48	0.7431	0.7443	0.7455	0.7466	0.7478	0.7490	0.7501	0.7513	0.7524	0.7536	0.7547	41
49	0.7547	0.7559	0.7570	0.7581	0.7593	0.7604	0.7615	0.7627	0.7638	0.7649	0.7660	40°
50°	0.7660	0.7672	0.7683	0.7694	0.7705	0.7716	9.7727	0.7738	0.7749	0.7760	0.7771	39
51	0.7771	0.7782	0.7793	0.7804	0.7815	0.7826	0.7837	0.7848	0.7859	0.7869	0.7880	38
52	0.7880	0.7891	0.7902	0.7912	0.7923	0.7934	0.7944	0.7955	0.7965	0.7976	0.7986	37
53	0.7986	0.7997	0.8007	0.8018	0.8028	0.8039	0.8049	0.8059	0.8070	0.8080	0.8090	36
54	0.8090	0.8100	0.8111	0.8121	0.8131	0.8141	0.8151	0.8161	0.8171	0.8181	0.8192	35
55	0.8192	0.8202	0.8211	0.8221	0.8231	0.8241	0.8251	0.8261	0.8271	0.8281	0.8290	34
56	0.8290	0.8300	0.8310	0.8320	0.8329	0.8339	0.8348	0.8358	0.8368	0.8377	0.8387	33
57	0.8387	0.8396	0.8406	0.8415	0.8425	0.8434	0.8443	0.8453	0.8462	0.8471	0.8480	32
58	0.8480	0.8490	0.8499	0.8508	0.8517	0.8526	0.8536	0.8545	0.8554	0.8563	0.8572	31
59	0.8572	0.8581	0.8590	0.8599	0.8607	0.8616	0.8625	0.8634	0.8643	0.8652	0.8660	30°
60°	0.8660	0.8669	0.8678	0.8686	0.8695	0.8704	0.8712	0.8721	0.8729	0.8738	0.8746	29
61	0.8746	0.8755	0.8763	0.8771	0.8780	0.8788	0.8796	0.8805	0.8813	0.8821	0.8829	28
62	0.8829	0.8838	0.8846	0.8854	0.8862	0.8870	0.8878	0.8886	0.8894	0.8902	0.8910	27
63	0.8910	0.8918	0.8926	0.8934	0.8942	0.8949	0.8957	0.8965	0.8973	0.8980	0.8988	26
64	0.8988	0.8996	0.9003	0.9011	0.9018	0.9026	0.9033	0.9041	0.9048	0.9056	0.9063	25
65	0.9063	0.9070	0.9078	0.9085	0.9092	0.9100	0.9107	0.9114	0.9121	0.9128	0.9135	24
66	0.9135	0.9143	0.9150	0.9157	0.9164	0.9171	0.9178	0.9184	0.9191	0.9198	0.9205	23
67	0.9205	0.9212	0.9219	0.9225	0.9232	0.9239	0.9245	0.9252	0.9259	0.9265	0.9272	22
68	0.9272	0.9278	0.9285	0.9291	0.9298	0.9304	0.9311	0.9317	0.9323	0.9330	0.9336	21
69	0.9336	0.9342	0.9348	0.9354	0.9361	0.9367	0.9373	0.9379	0.9385	0.9391	0.9397	20°
70°	0.9397	0.9403	0.9409	0.9415	0.9421	0.9426	0.9432	0.9438	0.9444	0.9449	0.9455	19
71	0.9455	0.9461	0.9466	0.9472	0.9478	0.9483	0.9489	0.9494	0.9500	0.9505	0.9511	18
72	0.9511	0.9516	0.9521	0.9527	0.9532	0.9537	0.9542	0.9548	0.9553	0.9558	0.9563	17
73	0.9563	0.9568	0.9573	0.9578	0.9583	0.9588	0.9593	0.9598	0.9603	0.9608	0.9613	16
74	0.9613	0.9617	0.9622	0.9627	0.9632	0.9636	0.9641	0.9646	0.9650	0.9655	0.9659	15
75	0.9659	0.9664	0.9668	0.9673	0.9677	0.9681	0.9686	0.9690	0.9694	0.9699	0.9703	14
76	0.9703	0.9707	0.9711	0.9715	0.9720	0.9724	0.9728	0.9732	0.9736	0.9740	0.9744	13
77	0.9744	0.9748	0.9751	0.9755	0.9759	0.9763	0.9767	0.9770	0.9774	0.9778	0.9781	12
78	0.9781	0.9785	0.9789	0.9792	0.9796	0.9799	0.9803	0.9806	0.9810	0.9813	0.9816	11
79	0.9816	0.9820	0.9823	0.9826	0.9829	0.9833	0.9836	0.9839	0.9842	0.9845	0.9848	10°
80°	0.9848	0.9851	0.9854	0.9857	0.9860	0.9863	0.9866	0.9869	0.9871	0.9874	0.9877	9
81	0.9877	0.9880	0.9882	0.9885	0.9888	0.9890	0.9893	0.9895	0.9898	0.9900	0.9903	8
82	0.9903	0.9905	0.9907	0.9910	0.9912	0.9914	0.9917	0.9919	0.9921	0.9923	0.9925	7
83	0.9925	0.9928	0.9930	0.9932	0.9934	0.9936	0.9938	0.9940	0.9942	0.9943	0.9945	6
84	0.9945	0.9947	0.9949	0.9951	0.9952	0.9954	0.9956	0.9957	0.9959	0.9960	0.9962	5
85	0.9962	0.9963	0.9965	0.9966	0.9968	0.9969	0.9971	0.9972	0.9973	0.9974	0.9976	4
86	0.9976	0.9977	0.9978	0.9979	0.9980	0.9981	0.9982	0.9983	0.9984	0.9985	0.9986	3
87	0.9986	0.9987	0.9988	0.9989	0.9990	0.9990	0.9991	0.9992	0.9993	0.9993	0.9994	2
88	0.9994	0.9995	0.9995	0.9996	0.9996	0.9997	0.9997	0.9997	0.9998	0.9998	0.9998	1
89	0.9998	0.9999	0.9999	0.9999	0.9999	1.0000	1.0000	1.0000	1.0000	1.0000	1.0000	0°
	°1.0	°0.9	°0.8	°0.7	°0.6	°0.5	°0.4	°0.3	°0.2	°0.1	°0.0	Deg.

APPENDIX 5
TRIGONOMETRIC FUNCTIONS—NATURAL TANGENTS AND COTANGENTS

Note: For cotangents, use right-hand column of degrees and lower line of tenths.

Deg.	°0.0	°0.1	°0.2	°0.3	°0.4	°0.5	°0.6	°0.7	°0.8	°0.9	°1.0	
0°	0.0000	0.0017	0.0035	0.0052	0.0070	0.0087	0.0105	0.0122	0.0140	0.0157	0.0175	89
1	0.0175	0.0192	0.0209	0.0227	0.0244	0.0262	0.0279	0.0297	0.0314	0.0332	0.0349	88
2	0.0349	0.0367	0.0384	0.0402	0.0419	0.0437	0.0454	0.0472	0.0489	0.0507	0.0524	87
3	0.0524	0.0542	0.0559	0.0577	0.0594	0.0612	0.0629	0.0647	0.0664	0.0682	0.0699	86
4	0.0699	0.0717	0.0734	0.0752	0.0769	0.0787	0.0805	0.0822	0.0840	0.0857	0.0875	85
5	0.0875	0.0892	0.0910	0.0928	0.0945	0.0963	0.0981	0.0998	0.1016	0.1033	0.1051	84
6	0.1051	0.1069	0.1086	0.1104	0.1122	0.1139	0.1157	0.1175	0.1192	0.1210	0.1228	83
7	0.1228	0.1246	0.1263	0.1281	0.1299	0.1317	0.1334	0.1352	0.1370	0.1388	0.1405	82
8	0.1405	0.1423	0.1441	0.1459	0.1477	0.1495	0.1512	0.1530	0.1548	0.1566	0.1584	81
9	0.1584	0.1602	0.1620	0.1638	0.1655	0.1673	0.1691	0.1709	0.1727	0.1745	0.1763	80°
10°	0.1763	0.1781	0.1799	0.1817	0.1835	0.1853	0.1871	0.1890	0.1908	0.1926	0.1944	79
11	0.1944	0.1962	0.1980	0.1998	0.2016	0.2035	0.2053	0.2071	0.2089	0.2107	0.2126	78
12	0.2126	0.2144	0.2162	0.2180	0.2199	0.2217	0.2235	0.2254	0.2272	0.2290	0.2309	77
13	0.2309	0.2327	0.2345	0.2364	0.2382	0.2401	0.2419	0.2438	0.2456	0.2475	0.2493	76
14	0.2493	0.2512	0.2530	0.2549	0.2568	0.2586	0.2605	0.2623	0.2642	0.2661	0.2679	75
15	0.2679	0.2698	0.2717	0.2736	0.2754	0.2773	0.2792	0.2811	0.2830	0.2849	0.2867	74
16	0.2867	0.2886	0.2905	0.2924	0.2943	0.2962	0.2981	0.3000	0.3019	0.3038	0.3057	73
17	0.3057	0.3076	0.3096	0.3115	0.3134	0.3153	0.3172	0.3191	0.3211	0.3230	0.3249	72
18	0.3249	0.3269	0.3288	0.3307	0.3327	0.3346	0.3365	0.3385	0.3404	0.3424	0.3443	71
19	0.3443	0.3463	0.3482	0.3502	0.3522	0.3541	0.3561	0.3581	0.3600	0.3620	0.3640	70°
20°	0.3640	0.3659	0.3679	0.3699	0.3719	0.3739	0.3759	0.3779	0.3799	0.3819	0.3839	69
21	0.3839	0.3859	0.3879	0.3899	0.3919	0.3939	0.3959	0.3979	0.4000	0.4020	0.4040	68
22	0.4040	0.4061	0.4081	0.4101	0.4122	0.4142	0.4163	0.4183	0.4204	0.4224	0.4245	67
23	0.4245	0.4265	0.4286	0.4307	0.4327	0.4348	0.4369	0.4390	0.4411	0.4431	0.4452	66
24	0.4452	0.4473	0.4494	0.4515	0.4536	0.4557	0.4578	0.4599	0.4621	0.4642	0.4663	65
25	0.4663	0.4684	0.4706	0.4727	0.4748	0.4770	0.4791	0.4813	0.4834	0.4856	0.4877	64
26	0.4877	0.4899	0.4921	0.4942	0.4964	0.4986	0.5008	0.5029	0.5051	0.5073	0.5095	63
27	0.5095	0.5117	0.5139	0.5161	0.5184	0.5206	0.5228	0.5250	0.5272	0.5295	0.5317	62
28	0.5317	0.5340	0.5362	0.5384	0.5407	0.5430	0.5452	0.5475	0.5498	0.5520	0.5543	61
29	0.5543	0.5566	0.5589	0.5612	0.5635	0.5658	0.5681	0.5704	0.5727	0.5750	0.5774	60°
30°	0.5774	0.5797	0.5820	0.5844	0.5867	0.5890	0.5914	0.5938	0.5961	0.5985	0.6009	59
31	0.6009	0.6032	0.6056	0.6080	0.6104	0.6128	0.6152	0.6176	0.6200	0.6224	0.6249	58
32	0.6249	0.6273	0.6297	0.6322	0.6346	0.6371	0.6395	0.6420	0.6445	0.6469	0.6494	57
33	0.6494	0.6519	0.6544	0.6569	0.6594	0.6619	0.6644	0.6669	0.6694	0.6720	0.6745	56
34	0.6745	0.6771	0.6796	0.6822	0.6847	0.6873	0.6899	0.6924	0.6950	0.6976	0.7002	55
35	0.7002	0.7028	0.7054	0.7080	0.7107	0.7133	0.7159	0.7186	0.7212	0.7239	0.7265	54
36	0.7265	0.7292	0.7319	0.7346	0.7373	0.7400	0.7427	0.7454	0.7481	0.7508	0.7536	53
37	0.7536	0.7563	0.7590	0.7618	0.7646	0.7673	0.7701	0.7729	0.7757	0.7785	0.7813	52
38	0.7813	0.7841	0.7869	0.7898	0.7926	0.7954	0.7983	0.8012	0.8040	0.8069	0.8098	51
39	0.8098	0.8127	0.8156	0.8185	0.8214	0.8243	0.8273	0.8302	0.8332	0.8361	0.8391	50°
40°	0.8391	0.8421	0.8451	0.8481	0.8511	0.8541	0.8571	0.8601	0.8632	0.8662	0.8693	49
41	0.8693	0.8724	0.8754	0.8785	0.8816	0.8847	0.8878	0.8910	0.8941	0.8972	0.9004	48
42	0.9004	0.9036	0.9067	0.9099	0.9131	0.9163	0.9195	0.9228	0.9260	0.9293	0.9325	47
43	0.9325	0.9358	0.9391	0.9424	0.9457	0.9490	0.9523	0.9556	0.9590	0.9623	0.9657	46
44	0.9657	0.9691	0.9725	0.9759	0.9793	0.9827	0.9861	0.9896	0.9930	0.9965	1.0000	45
	°1.0	°0.9	°0.8	°0.7	°0.6	°0.5	°0.4	°0.3	°0.2	°0.1	°0.0	Deg.

Note: For cotangents, use right-hand column of degrees and lower line of tenths.

Deg.	°0.0	°0.1	°0.2	°0.3	°0.4	°0.5	°0.6	°0.7	°0.8	°0.9	°1.0	
45	1.0000	1.0035	1.0070	1.0105	1.0141	1.0176	1.0212	1.0247	1.0283	1.0319	1.0355	44
46	1.0355	1.0392	1.0428	1.0464	1.0501	1.0538	1.0575	1.0612	1.0649	1.0686	1.0724	43
47	1.0724	1.0761	1.0799	1.0837	1.0875	1.0913	1.0951	1.0990	1.1028	1.1067	1.1106	42
48	1.1106	1.1145	1.1184	1.1224	1.1263	1.1303	1.1343	1.1383	1.1423	1.1463	1.1504	41
49	1.1504	1.1544	1.1585	1.1626	1.1667	1.1708	1.1750	1.1792	1.1833	1.1875	1.1918	40°
50°	1.1918	1.1960	1.2002	1.2045	1.2088	1.2131	1.2174	1.2218	1.2261	1.2305	1.2349	39
51	1.2349	1.2393	1.2437	1.2482	1.2527	1.2572	1.2617	1.2662	1.2708	1.2753	1.2799	38
52	1.2799	1.2846	1.2892	1.2938	1.2985	1.3032	1.3079	1.3127	1.3175	1.3222	1.3270	37
53	1.3270	1.3319	1.3367	1.3416	1.3465	1.3514	1.3564	1.3613	1.3663	1.3713	1.3764	36
54	1.3764	1.3814	1.3865	1.3916	1.3968	1.4019	1.4071	1.4124	1.4176	1.4229	1.4281	35
55	1.4281	1.4335	1.4388	1.4442	1.4496	1.4550	1.4605	1.4659	1.4715	1.4770	1.4826	34
56	1.4826	1.4882	1.4938	1.4994	1.5051	1.5108	1.5166	1.5224	1.5282	1.5340	1.5399	33
57	1.5399	1.5458	1.5517	1.5577	1.5637	1.5697	1.5757	1.5818	1.5880	1.5941	1.6003	32
58	1.6003	1.6066	1.6128	1.6191	1.6255	1.6319	1.6383	1.6447	1.6512	1.6577	1.6643	31
59	1.6643	1.6709	1.6775	1.6842	1.6909	1.6977	1.7045	1.7113	1.7182	1.7251	1.7321	30°
60°	1.7321	1.7391	1.7461	1.7532	1.7603	1.7675	1.7747	1.7820	1.7893	1.7966	1.8040	29
61	1.8040	1.8115	1.8190	1.8265	1.8341	1.8418	1.8495	1.8572	1.8650	1.8728	1.8807	28
62	1.8807	1.8887	1.8967	1.9047	1.9128	1.9210	1.9292	1.9375	1.9458	1.9542	1.9626	27
63	1.9626	1.9711	1.9797	1.9883	1.9970	2.0057	2.0145	2.0233	2.0323	2.0413	2.0503	26
64	2.0503	2.0594	2.0686	2.0778	2.0872	2.0965	2.1060	2.1155	2.1251	2.1348	2.1445	25
65	2.1445	2.1543	2.1642	2.1742	2.1842	2.1943	2.2045	2.2148	2.2251	2.2355	2.2460	24
66	2.2460	2.2566	2.2673	2.2781	2.2889	2.2998	2.3109	2.3220	2.3332	2.3445	2.3559	23
67	2.3559	2.3673	2.3789	2.3906	2.4023	2.4142	2.4262	2.4383	2.4504	2.4627	2.4751	22
68	2.4751	2.4876	2.5002	2.5129	2.5257	2.5386	2.5517	2.5649	2.5782	2.5916	2.6051	21
69	2.6051	2.6187	2.6325	2.6464	2.6605	2.6746	2.6889	2.7034	2.7179	2.7326	2.7475	20°
70°	2.7475	2.7625	2.7776	2.7929	2.8083	2.8239	2.8397	2.8556	2.8716	2.8878	2.9042	19
71	2.9042	2.9208	2.9375	2.9544	2.9714	2.9887	3.0061	3.0237	3.0415	3.0595	3.0777	18
72	3.0777	3.0961	3.1146	3.1334	3.1524	3.1716	3.1910	3.2106	3.2305	3.2506	3.2709	17
73	3.2709	3.2914	3.3122	3.3332	3.3544	3.3759	3.3977	3.4197	3.4420	3.4646	3.4874	16
74	3.4874	3.5105	3.5339	3.5576	3.5816	3.6059	3.6305	3.6554	3.6806	3.7062	3.7321	15
75	3.7321	3.7583	3.7848	3.8118	3.8391	3.8667	3.8947	3.9232	3.9520	3.9812	4.0108	14
76	4.0108	4.0408	4.0713	4.1022	4.1335	4.1653	4.1976	4.2303	4.2635	4.2972	4.3315	13
77	4.3315	4.3662	4.4015	4.4374	4.4737	4.5107	4.5483	4.5864	4.6252	4.6646	4.7046	12
78	4.7046	4.7453	4.7867	4.8288	4.8716	4.9152	4.9594	5.0045	5.0504	5.0970	5.1446	11
79	5.1446	5.1929	5.2422	5.2924	5.3435	5.3955	5.4486	5.5026	5.5578	5.6140	5.6713	10°
80°	5.6713	5.7297	5.7894	5.8502	5.9124	5.9758	6.0405	6.1066	6.1742	6.2432	6.3138	9
81	6.3138	6.3859	6.4596	6.5350	6.6122	6.6912	6.7720	6.8548	6.9395	7.0264	7.1154	8
82	7.1154	7.2066	7.3002	7.3962	7.4947	7.5958	7.6996	7.8062	7.9158	8.0285	8.1443	7
83	8.1443	8.2636	8.3863	8.5126	8.6427	8.7769	8.9152	9.0579	9.2052	9.3572	9.5144	6
84	9.5144	9.677	9.845	10.02	10.20	10.39	10.58	10.78	10.99	11.20	11.43	5
85	11.43	11.66	11.91	12.16	12.43	12.71	13.00	13.30	13.62	13.95	14.30	4
86	14.30	14.67	15.06	15.46	15.89	16.35	16.83	17.34	17.89	18.46	19.08	3
87	19.08	19.74	20.45	21.20	22.02	22.90	23.86	24.90	26.03	27.27	28.64	2
88	28.64	30.14	31.82	33.69	35.80	38.19	40.92	44.07	47.74	52.08	57.29	1
89	57.29	63.66	71.62	81.85	95.49	114.6	143.2	191.0	286.5	573.0	∞	0°
	°1.0	°0.9	°0.8	°0.7	°0.6	°0.5	°0.4	°0.3	°0.2	°0.1	°0.0	Deg.

APPENDIX 6
LOGARITHMS

N	0	1	2	3	4	5	6	7	8	9
10	0000	0043	0086	0128	0170	0212	0253	0294	0334	0374
11	0414	0453	0492	0531	0569	0607	0645	0682	0719	0755
12	0792	0828	0864	0899	0934	0969	1004	1038	1072	1106
13	1139	1173	1206	1239	1271	1303	1335	1367	1399	1430
14	1461	1492	1523	1553	1584	1614	1644	1673	1703	1732
15	1761	1790	1818	1847	1875	1903	1931	1959	1987	2014
16	2041	2068	2095	2122	2148	2175	2201	2227	2253	2279
17	2304	2330	2355	2380	2405	2430	2455	2480	2504	2529
18	2553	2577	2601	2625	2648	2672	2695	2718	2742	2765
19	2788	2810	2833	2856	2878	2900	2923	2945	2967	2989
20	3010	3032	3054	3075	3096	3118	3139	3160	3181	3201
21	3222	3243	3263	3284	3304	3324	3345	3365	3385	3404
22	3424	3444	3464	3483	3502	3522	3541	3560	3579	3598
23	3617	3636	3655	3674	3692	3711	3729	3747	3766	3784
24	3802	3820	3838	3856	3874	3892	3909	3927	3945	3962
25	3979	3997	4014	4031	4048	4065	4082	4099	4116	4133
26	4150	4166	4183	4200	4216	4232	4249	4265	4281	4298
27	4314	4330	4346	4362	4378	4393	4409	4425	4440	4456
28	4472	4487	4502	4518	4533	4548	4564	4579	4594	4609
29	4624	4639	4654	4669	4683	4698	4713	4728	4742	4757
30	4771	4786	4800	4814	4829	4843	4857	4871	4886	4900
31	4914	4928	4942	4955	4969	4983	4997	5011	5024	5038
32	5051	5065	5079	5092	5105	5119	5132	5145	5159	5172
33	5185	5198	5211	5224	5237	5250	5263	5276	5289	5302
34	5315	5328	5340	5353	5366	5378	5391	5403	5416	5428
35	5441	5453	5465	5478	5490	5502	5514	5527	5539	5551
36	5563	5575	5587	5599	5611	5623	5635	5647	5658	5670
37	5682	5694	5705	5717	5729	5740	5752	5763	5775	5786
38	5798	5809	5821	5832	5843	5855	5866	5877	5888	5899
39	5911	5922	5933	5944	5955	5966	5977	5988	5999	6010
40	6021	6031	6042	6053	6064	6075	6085	6096	6107	6117
41	6128	6138	6149	6160	6170	6180	6191	6201	6212	6222
42	6232	6243	6253	6263	6274	6284	6294	6304	6314	6325
43	6335	6345	6355	6365	6375	6385	6395	6405	6415	6425
44	6435	6444	6454	6464	6474	6484	6493	6503	6513	6522
45	6532	6542	6551	6561	6571	6580	6590	6599	6609	6618
46	6628	6637	6646	6656	6665	6675	6684	6693	6702	6712
47	6721	6730	6739	6749	6758	6767	6776	6785	6794	6803
48	6812	6821	6830	6839	6848	6857	6866	6875	6884	6893
49	6902	6911	6920	6928	6937	6946	6955	6964	6972	6981
50	6990	6998	7007	7016	7024	7033	7042	7050	7059	7067
51	7076	7084	7093	7101	7110	7118	7126	7135	7143	7152
52	7160	7168	7177	7185	7193	7202	7210	7218	7226	7235
53	7243	7251	7259	7267	7275	7284	7292	7300	7308	7316
54	7324	7332	7340	7348	7356	7364	7372	7380	7388	7396

N	0	1	2	3	4	5	6	7	8	9
55	7404	7412	7419	7427	7435	7443	7451	7459	7466	7474
56	7482	7490	7497	7505	7513	7520	7528	7536	7543	7551
57	7559	7566	7574	7582	7589	7597	7604	7612	7619	7627
58	7634	7642	7649	7657	7664	7672	7679	7686	7694	7701
59	7709	7716	7723	7731	7738	7745	7752	7760	7767	7774
60	7782	7789	7796	7803	7810	7818	7825	7832	7839	7846
61	7853	7860	7868	7875	7882	7889	7896	7903	7910	7917
62	7924	7931	7938	7945	7952	7959	7966	7973	7980	7987
63	7993	8000	8007	8014	8021	8028	8035	8041	8048	8055
64	8062	8069	8075	8082	8089	8096	8102	8109	8116	8122
65	8129	8136	8142	8149	8156	8162	8169	8176	8182	8189
66	8195	8202	8209	8215	8222	8228	8235	8241	8248	8254
67	8261	8267	8274	8280	8287	8293	8299	8306	8312	8319
68	8325	8331	8338	8344	8351	8357	8363	8370	8376	8382
69	8388	8395	8401	8407	8414	8420	8426	8432	8439	8445
70	8451	8457	8463	8470	8476	8482	8488	8494	8500	8506
71	8513	8519	8525	8531	8537	8543	8549	8555	8561	8567
72	8573	8579	8585	8591	8597	8603	8609	8615	8621	8627
73	8633	8639	8645	8651	8657	8663	8669	8675	8681	8686
74	8692	8698	8704	8710	8716	8722	8727	8733	8739	8745
75	8751	8756	8762	8768	8774	8779	8785	8791	8797	8802
76	8808	8814	8820	8825	8831	8837	8842	8848	8854	8859
77	8865	8871	8876	8882	8887	8893	8899	8904	8910	8915
78	8921	8927	8932	8938	8943	8949	8954	8960	8965	8971
79	8976	8982	8987	8993	8998	9004	9009	9015	9020	9025
80	9031	9036	9042	9047	9053	9058	9063	9069	9074	9079
81	9085	9090	9096	9101	9106	9112	9117	9122	9128	9133
82	9138	9143	9149	9154	9159	9165	9170	9175	9180	9186
83	9191	9196	9201	9206	9212	9217	9222	9227	9232	9238
84	9243	9248	9253	9258	9263	9269	9274	9279	9284	9289
85	9294	9299	9304	9309	9315	9320	9325	9330	9335	9340
86	9345	9350	9355	9360	9365	9370	9375	9380	9385	9390
87	9395	9400	9405	9410	9415	9420	9425	9430	9435	9440
88	9445	9450	9455	9460	9465	9469	9474	9479	9484	9489
89	9494	9499	9504	9509	9513	9518	9523	9528	9533	9538
90	9542	9547	9552	9557	9562	9566	9571	9576	9581	9586
91	9590	9595	9600	9605	9609	9614	9619	9624	9628	9633
92	9638	9643	9647	9652	9657	9661	9666	9671	9675	9680
93	9685	9689	9694	9699	9703	9708	9713	9717	9722	9727
94	9731	9736	9741	9745	9750	9754	9759	9763	9768	9773
95	9777	9782	9786	9791	9795	9800	9805	9809	9814	9818
96	9823	9827	9832	9836	9841	9845	9850	9854	9859	9863
97	9868	9872	9877	9881	9886	9890	9894	9899	9903	9908
98	9912	9917	9921	9926	9930	9934	9939	9943	9948	9952
99	9956	9961	9965	9969	9974	9978	9983	9987	9991	9996

APPENDIX 7
ELECTRICAL PROPERTIES OF METALS AND ALLOYS

Metals	Resistivity in microhms-cm at 20°C	Resistivity in ohms per mil-foot at 20°C (K)	Temp. Coeff. or Resistance	Melting Point (°C)	Specific Gravity
Aluminum	2.83	17	0.004	659	2.67
Alumel	33.3	200.3	0.0012		
Aluminum Bronze (90% CR, 10% AL)	12.6	75.8	0.003	1050	7.5
Brass (various comp.)	6.2–8.3	37.3–49.9	0.002	880	8.4
Bronze (88% Cu, 12% tin)	18	108.3	0.005	1000	8.8
Bronze (commercial) (Cu, Zn)	4.2	25.3	0.002	1050	8.8
Carbon, amorphous	3500–4100	21,052.5–24,661.5	–0.0005	3500	1.85
Carbon graphite (for furnace electrodes at 2500°C)	720–1000	4330.8–6015	–0.0005	3500	1.85
Chromel (Ni, Cr)	70–110	421–661.6	0.0001	1350	8.3–8.5
Constatan (60% Cu, 40% Ni)	44–49	264.7–294.7	0.000002	1190	8.9
Copper, annealed	1.724–1.73	10.4	0.004	1083	8.89
Copper, hard-drawn	1.77	10.6	0.0039	1083	8.89
German silver (Cu, Zn, 18% Ni)	33	198.5	0.004	1100	8.4
Gold	2.44	14.7	0.0034	1063	19.3
Invar (65% Fe, 35% Ni)	81	487.2		1495	8
Iron & Steel					
Iron, pure	10	60.15	.005	1530	7.8
Iron, cast	60	360.9			
Soft steel	15.9	95.6	0.0016	1510	7.8
Steel, glass hard	45.7	274.9			
Steel 4% Si	51	306.8			
Steel, Transformer	11.09	66.7			
Lead	22	132.3	0.00387	327	11.4
Magnesium	4.6	27.7	0.004	651	1.74
Manganin (84% Cu, 12% Mn, 4% Ni)	44–48	264.7–288.7	0.000000	1020	8.4
Mercury	95.7	575.6	0.0008	–39	13.6
Monel (Ni, Cu)	42.5	255.6	0.00019	1300	8.9
Nichrome (Ni, Fe, Cr)	90–112	541.3–673.7	0.00017	1350–1500	8.25
Nichrome V (Ni, Cr)	108	649.6	0.00017	1150	8.41
Phosphor Bronze	9.39	56.5	0.0039		8.66
Platinum	11.5	69.2	0.003	1750	21.4
Silver	1.628	9.79	0.0038	960	10.5
Sodium	4.4	26.5	0.004		
Tin	11.5	69.2	0.004	232	7.3
Tungsten	5.51	33.1	0.0045	3410	18.8
at 1730°C	60	360.9			
at 2760°C	100	601.5			
Zinc	6	36.1	0.004	419	7.1

Resistivity in microhm-centimeter is the resistance of a centimeter-cube in millionths of an ohm.
Resistivity in ohms-per-milfoot is the resistance of a conductor one mil in diameter (0.001 inch) and one foot long.
The conversion factor from microhms-cm and ohms-per-milfoot is 6.015.

APPENDIX 8
AMERICAN WIRE GAUGE TABLE

B & S Gauge Number	Diameter in Mils	Area in Circular Mils	Ohms per 1000 ft (ohms per 100 m)			Pounds per 1000 ft (kg per 100 m)	
			Copper* 68°F (20°C)	Copper* 167°F (75°C)	Aluminum 68°F (20°C)	Copper	Aluminum
0000	460	211 600	.049 (.016)	.0596 (.0195)	.0804 (.0263)	640 (95.2)	195 (29.0)
000	410	167 800	.0618 (.020)	.0752 (.0246)	.101 (.033)	508 (75.5)	154 (22.9)
00	365	133 100	.078 (.026)	.0948 (.031)	.128 (.042)	403 (59.9)	122 (18.1)
0	325	105 500	.0983 (.032)	.1195 (.0392)	.161 (.053)	320 (47.6)	97 (14.4)
1	289	83 690	.1239 (.0406)	.151 (.049)	.203 (.066)	253 (37.6)	76.9 (11.4)
2	258	66 370	.1563 (.0512)	.190 (.062)	.526 (.084)	201 (29.9)	61.0 (9.07)
3	229	52 640	.1970 (.0646)	.240 (.079)	.323 (.106)	159 (23.6)	48.4 (7.20)
4	204	41 740	.2485 (.0815)	.302 (.099)	.408 (.134)	126 (18.7)	38.4 (5.71)
5	182	33 100	.3133 (.1027)	.381 (.125)	.514 (.168)	100 (14.9)	30.4 (4.52)
6	162	26 250	.395 (1.29)	.481 (.158)	.648 (.212)	79.5 (11.8)	24.1 (3.58)
7	144	20 820	.498 (.163)	.606 (.199)	.817 (.268)	63.0 (9.37)	19.1 (2.84)
8	128	16 510	.628 (.206)	.764 (.250)	1.03 (.338)	50.0 (7.43)	15.2 (2.26)
9	114	13 090	.792 (.260)	.963 (.316)	1.30 (.426)	39.6 (5.89)	12.0 (1.78)
10	102	10 380	.999 (.327)	1.215 (.398)	1.64 (.538)	31.4 (4.67)	9.55 (1.42)
11	91	8 234	1.260 (.413)	1.532 (.502)	2.07 (.678)	24.9 (3.70)	7.57 (1.13)
12	81	6 530	1.588 (.520)	1.931 (.633)	2.61 (.856)	19.8 (2.94)	6.00 (.89)
13	72	5 178	2.003 (.657)	2.44 (.80)	3.29 (1.08)	15.7 (2.33)	4.8 (.71)
14	64	4 107	2.525 (.828)	3.07 (1.01)	4.14 (1.36)	12.4 (1.84)	3.8 (.56)
15	57	3 257	3.184 (1.043)	3.98 (1.27)	5.22 (1.71)	9.86 (1.47)	3.0 (.45)
16	51	2 583	4.016 (1.316)	4.88 (1.60)	6.59 (2.16)	7.82 (1.16)	2.4 (.36)
17	45.3	2 048	5.06 (1.66)	6.16 (2.02)	8.31 (2.72)	6.20 (.922)	1.9 (.28)
18	40.3	1 624	6.39 (2.09)	7.77 (2.55)	10.5 (3.44)	4.92 (.731)	1.5 (.22)
19	35.9	1 288	8.05 (2.64)	9.97 (3.21)	13.2 (4.33)	3.90 (.580)	1.2 (.18)
20	32.0	1 022	10.15 (3.33)	12.35 (4.05)	16.7 (5.47)	3.09 (.459)	0.94 (.14)
21	28.5	810	12.8 (4.2)	15.6 (5.11)	21.0 (6.88)	2.45 (.364)	.745 (.110)
22	25.4	642	16.1 (5.3)	19.6 (6.42)	26.5 (8.69)	1.95 (.290)	.591 (.09)
23	22.6	510	20.4 (6.7)	24.8 (8.13)	33.4 (10.9)	1.54 (.229)	.468 (.07)
24	20.1	404	25.7 (8.4)	31.2 (10.2)	42.1 (13.8)	1.22 (.181)	.371 (.05)
25	17.9	320	32.4 (10.6)	39.4 (12.9)	53.1 (17.4)	0.970 (.14)	.295 (.04)
26	15.9	254	40.8 (13.4)	49.6 (16.3)	67.0 (22.0)	.769 (.11)	.234 (.03)
27	14.2	202	51.5 (16.9)	62.6 (20.5)	84.4 (27.7)	.610 (.09)	.185 (.03)
28	12.6	160	64.9 (21.3)	78.9 (25.9)	106 (34.7)	.484 (.07)	.147 (.02)
29	11.3	126.7	81.8 (26.8)	99.5 (32.6)	134 (43.9)	.384 (.06)	.117 (.02)
30	10.0	100.5	103.2 (33.8)	125.5 (41.1)	169 (55.4)	.304 (.04)	.092 (.01)
31	8.93	79.7	130.1 (42.6)	158.5 (51.9)	213 (69.8)	.241 (.04)	.073 (.01)
32	7.95	63.2	164.1 (53.8)	199.5 (65.4)	269 (88.2)	.191 (.03)	.058 (.01)
33	7.08	50.1	207 (68)	252 (82.6)	339 (111)	.152 (.02)	.046 (.01)
34	6.31	39.8	261 (86)	317 (104)	428 (140)	.120 (.02)	.037 (.01)
35	5.62	31.5	329 (108)	400 (131)	540 (177)	.095 (.01)	.029
36	5.00	25.0	415 (136)	505 (165)	681 (223)	.076 (.01)	.023
37	4.45	19.8	523 (171)	636 (208)	858 (281)	.0600 (.01)	.0182
38	3.96	15.7	660 (216)	802 (263)	1080 (354)	.0476 (.01)	.0145
39	3.53	12.5	832 (273)	1012 (332)	1360 (446)	.0377 (.01)	.0115
40	3.15	9.9	1049 (344)	1276 (418)	1720 (564)	.0299 (.01)	.0091
41							
42	2.50	6.3					
43							
44	1.97	3.9					

*Resistance figures for standard annealed copper. For hard-drawn, add 2%.

APPENDIX 9
ALLOWABLE AMPACITIES OF SINGLE-INSULATED CONDUCTORS

Rated 0 through 2000 volts in free air based on an ambient air temperature of 30°C (86°F).

Table 310.15(B)(17) (formerly Table 310.17) Allowable Ampacities of Single-Insulated Conductors Rated Up to and Including 2000 Volts in Free Air, Based on Ambient Temperature of 30°C (86°F)*

	Temperature Rating of Conductor [See Table 310.104(A).]						
	60°C (140°F)	75°C (167°F)	90°C (194°F)	60°C (140°F)	75°C (167°F)	90°C (194°F)	
	Types TW, UF	Types RHW, THHW, THW, THWN, XHHW, ZW	Types TBS, SA, SIS, FEP, FEPB, MI, RHH, RHW-2, THHN, THHW, THW-2, THWN-2, USE-2, XHH, XHHW, XHHW-2, ZW-2	Types TW, UF	Types RHW, THHW, THW, THWN, XHHW	Types TBS, SA, SIS, THHN, THHW, THW-2, THWN-2, RHH, RHW-2, USE-2, XHH, XHHW, XHHW-2, ZW-2	
Size AWG or kcmil	COPPER			ALUMINUM OR COPPER-CLAD ALUMINUM			Size AWG or kcmil
18	—	—	18	—	—	—	—
16	—	—	24	—	—	—	—
14**	25	30	35	—	—	—	—
12**	30	35	40	25	30	35	12**
10**	40	50	55	35	40	45	10**
8	60	70	80	45	55	60	8
6	80	95	105	60	75	85	6
4	105	125	140	80	100	115	4
3	120	145	165	95	115	130	3
2	140	170	190	110	135	150	2
1	165	195	220	130	155	175	1
1/0	195	230	260	150	180	205	1/0
2/0	225	265	300	175	210	235	2/0
3/0	260	310	350	200	240	270	3/0
4/0	300	360	405	235	280	315	4/0
250	340	405	455	265	315	355	250
300	375	445	500	290	350	395	300
350	420	505	570	330	395	445	350
400	455	545	615	355	425	480	400
500	515	620	700	405	485	545	500
600	575	690	780	455	545	615	600
700	630	755	850	500	595	670	700
750	655	785	885	515	620	700	750
800	680	815	920	535	645	725	800
900	730	870	980	580	700	790	900
1000	780	935	1055	625	750	845	1000
1250	890	1065	1200	710	855	965	1250
1500	980	1175	1325	795	950	1070	1500
1750	1070	1280	1445	875	1050	1185	1750
2000	1155	1385	1560	960	1150	1295	2000

*Refer to 310.15(B)(2) for the ampacity correction factors where the ambient temperature is other than 30°C (86°F).

**Refer to 240.4(D) for conductor overcurrent protection limitations.

APPENDIX 10
ALLOWABLE AMPACITIES OF SINGLE-INSULATED CONDUCTORS

Rated 0 through 2000 volts, 60°C through 90°C (140°F through 194°F) not more than three current-carrying conductors in a raceway, cable, or earth (directly buried), based on an ambient temperature of 30°C (86°F).

Table 310.15(B)(16) (formerly Table 310.16) Allowable Ampacities of Insulated Conductors Rated Up to and Including 2000 Volts, 60°C Through 90°C (140°F Through 194°F), Not More Than Three Current-Carrying Conductors in Raceway, Cable, or Earth (Directly Buried), Based on Ambient Temperature of 30°C (86°F)*

	Temperature Rating of Conductor [See Table 310.104(A).]						
	60°C (140°F)	75°C (167°F)	90°C (194°F)	60°C (140°F)	75°C (167°F)	90°C (194°F)	
Size AWG or kcmil	Types TW, UF	Types RHW, THHW, THW, THWN, XHHW, USE, ZW	Types TBS, SA, SIS, FEP, FEPB, MI, RHH, RHW-2, THHN, THHW, THW-2, THWN-2, USE-2, XHH, XHHW, XHHW-2, ZW-2	Types TW, UF	Types RHW, THHW, THW, THWN, XHHW, USE	Types TBS, SA, SIS, THHN, THHW, THW-2, THWN-2, RHH, RHW-2, USE-2, XHH, XHHW, XHHW-2, ZW-2	
	COPPER			ALUMINUM OR COPPER-CLAD ALUMINUM			Size AWG or kcmil
18	—	—	14	—	—	—	—
16	—	—	18	—	—	—	—
14**	15	20	25	—	—	—	—
12**	20	25	30	15	20	25	12**
10**	30	35	40	25	30	35	10**
8	40	50	55	35	40	45	8
6	55	65	75	40	50	55	6
4	70	85	95	55	65	75	4
3	85	100	115	65	75	85	3
2	95	115	130	75	90	100	2
1	110	130	145	85	100	115	1
1/0	125	150	170	100	120	135	1/0
2/0	145	175	195	115	135	150	2/0
3/0	165	200	225	130	155	175	3/0
4/0	195	230	260	150	180	205	4/0
250	215	255	290	170	205	230	250
300	240	285	320	195	230	260	300
350	260	310	350	210	250	280	350
400	280	335	380	225	270	305	400
500	320	380	430	260	310	350	500
600	350	420	475	285	340	385	600
700	385	460	520	315	375	425	700
750	400	475	535	320	385	435	750
800	410	490	555	330	395	445	800
900	435	520	585	355	425	480	900
1000	455	545	615	375	445	500	1000
1250	495	590	665	405	485	545	1250
1500	525	625	705	435	520	585	1500
1750	545	650	735	455	545	615	1750
2000	555	665	750	470	560	630	2000

*Refer to 310.15(B)(2) for the ampacity correction factors where the ambient temperature is other than 30°C (86°F).

**Refer to 240.4(D) for conductor overcurrent protection limitations.

APPENDIX 11
AMBIENT TEMPERATURE CORRECTION FACTORS

Table 310.15(B)(2)(a) Ambient Temperature Correction Factors Based on 30°C (86°F)

For ambient temperatures other than 30°C (86°F) multiply the allowable ampacities specified in the ampacity tables by the appropriate correction factor shown below.				
Ambient Temperature (°C)	**Temperature Rating of Conductor**			**Ambient Temperature (°F)**
	60°C	**75°C**	**90°C**	
10 or less	1.29	1.20	1.15	50 or less
11–15	1.22	1.15	1.12	51–59
16–20	1.15	1.11	1.08	60–68
21–25	1.08	1.05	1.04	69–77
26–30	1.00	1.00	1.00	78–86
31–35	0.91	0.94	0.96	87–95
36–40	0.82	0.88	0.91	96–104
41–45	0.71	0.82	0.87	105–113
46–50	0.58	0.75	0.82	114–122
51–55	0.41	0.67	0.76	123–131
56–60	—	0.58	0.71	132–140
61–65	—	0.47	0.65	141–149
66–70	—	0.33	0.58	150–158
71–75	—	—	0.50	159–167
76–80	—	—	0.41	168–176
81–85	—	—	0.29	177–185

APPENDIX 12
FULL-LOAD CURRENTS IN AMPERES, SINGLE-PHASE ALTERNATING-CURRENT MOTORS

Table 430.248 Full-Load Currents in Amperes, Single-Phase Alternating-Current Motors

The following values of full-load currents are for motors running at usual speeds and motors with normal torque characteristics. The voltages listed are rated motor voltages. The currents listed shall be permitted for system voltage ranges of 110 to 120 and 220 to 240 volts.

Horsepower	115 Volts	200 Volts	208 Volts	230 Volts
1/6	4.4	2.5	2.4	2.2
1/4	5.8	3.3	3.2	2.9
1/3	7.2	4.1	4.0	3.6
1/2	9.8	5.6	5.4	4.9
3/4	13.8	7.9	7.6	6.9
1	16	9.2	8.8	8.0
1 1/2	20	11.5	11.0	10
2	24	13.8	13.2	12
3	34	19.6	18.7	17
5	56	32.2	30.8	28
7 1/2	80	46.0	44.0	40
10	100	57.5	55.0	50

APPENDIX 13
FULL-LOAD CURRENT, THREE-PHASE AC MOTORS

Table 430.250 Full-Load Current, Three-Phase Alternating-Current Motors

The following values of full-load currents are typical for motors running at speeds usual for belted motors and motors with normal torque characteristics.

The voltages listed are rated motor voltages. The currents listed shall be permitted for system voltage ranges of 110 to 120, 220 to 240, 440 to 480, and 550 to 600 volts.

	Induction-Type Squirrel Cage and Wound Rotor (Amperes)							Synchronous-Type Unity Power Factor* (Amperes)			
Horsepower	**115 Volts**	**200 Volts**	**208 Volts**	**230 Volts**	**460 Volts**	**575 Volts**	**2300 Volts**	**230 Volts**	**460 Volts**	**575 Volts**	**2300 Volts**
½	4.4	2.5	2.4	2.2	1.1	0.9	—	—	—	—	—
¾	6.4	3.7	3.5	3.2	1.6	1.3	—	—	—	—	—
1	8.4	4.8	4.6	4.2	2.1	1.7	—	—	—	—	—
1½	12.0	6.9	6.6	6.0	3.0	2.4	—	—	—	—	—
2	13.6	7.8	7.5	6.8	3.4	2.7	—	—	—	—	—
3	—	11.0	10.6	9.6	4.8	3.9	—	—	—	—	—
5	—	17.5	16.7	15.2	7.6	6.1	—	—	—	—	—
7½	—	25.3	24.2	22	11	9	—	—	—	—	—
10	—	32.2	30.8	28	14	11	—	—	—	—	—
15	—	48.3	46.2	42	21	17	—	—	—	—	—
20	—	62.1	59.4	54	27	22	—	—	—	—	—
25	—	78.2	74.8	68	34	27	—	53	26	21	—
30	—	92	88	80	40	32	—	63	32	26	—
40	—	120	114	104	52	41	—	83	41	33	—
50	—	150	143	130	65	52	—	104	52	42	—
60	—	177	169	154	77	62	16	123	61	49	12
75	—	221	211	192	96	77	20	155	78	62	15
100	—	285	273	248	124	99	26	202	101	81	20
125	—	359	343	312	156	125	31	253	126	101	25
150	—	414	396	360	180	144	37	302	151	121	30
200		552	528	480	240	192	49	400	201	161	40
250	—	—	—	—	302	242	60	—	—	—	—
300	—	—	—	—	361	289	72	—	—	—	—
350	—	—	—	—	414	336	83	—	—	—	—
400	—	—	—	—	477	382	95	—	—	—	—
450	—	—	—	—	515	412	103	—	—	—	—
500	—	—	—	—	590	472	118	—	—	—	—

*For 90 and 80 percent power factor, the figures shall be multiplied by 1.1 and 1.25, respectively.

APPENDIX 14
COMMON PREFIXES FOR METRIC UNITS

Metric Unit Prefix	Symbol	Power of Ten	Example			
pico-	p	10^{-12}	1 picofarad	1 pf	=	10^{-12} farad
nano-	n	10^{-9}	1 nanosecond	1 ns	=	10^{-9} second
micro-	μ	10^{-6}	1 microvolt	1 μV	=	10^{-6} volt
milli-	m	10^{-3}	1 milligram	1 mg	=	10^{-3} gram
centi-	c	10^{-2}	1 centimeter	1 cm	=	10^{-2} meter
kilo-	k	10^{3}	1 kilowatt	1 kW	=	10^{3} watts
mega-	M	10^{6}	1 megajoule	1 MJ	=	10^{6} joules
giga-	G	10^{9}	1 gigaelectronvolt	1 GeV	=	10^{9} electron volts

APPENDIX 15
SYMBOLS AND DERIVED UNITS OF PHYSICAL QUANTITIES—METRIC SYSTEM

Quantity	Symbol	Derived Units*
Acceleration	a	meters/second^2
Angular acceleration	α	radians/second^2
Angular displacement	θ	radian
Angular momentum	L	kilogram-meters^2/second
Angular velocity	ω	radians/second
Area	A	meter^2
Capacitance	C	farad
Charge	q	coulomb
Conductivity	σ	ohm-meter^{-1}
Current (electrons)	i	ampere
Current density	j	amperes/meter^2
Density, mass	ρ	kilograms/meter^3
Displacement	r,d	meter
Electric dipole moment	p	coulomb-meter
Electric field intensity	E	volts/meter
Electric flux	ϕ_E	volt-meter
Electric potential	V	volt
Electromotive force	emf	volt
Energy: heat	Q	joule
internal	U	joule
kinetic	K	joule
potential	U	joule
total	E	joule
Force	F	newton
Frequency	f	hertz (cycles/second)
Gravitational field intensity	γ	newtons/kilogram
Inductance	L	henry
Length	L, 1	meter
Linear momentum	p	kilogram-meters/second
Magnetic dipole moment	μ	ampere-meter^2
Magnetic flux	ϕ_B	weber
Magnetic induction	B	webers/meter^2
Mass	m	kilogram
Period	T	second
Power	W	watt
Pressure	P	newtons/meter^2
Resistance	R	ohm
Resistivity	ρ	ohm-meter
Rotational inertia	I	kilogram-meter^2
Temperature	T	Kelvin
Time	t	second
Torque	τ	newton-meter
Velocity	v	meters/second
Voltage	V	volt
Volume	V	meter^3
Wavelength	λ	meter
Work	W	joule

*The derived units are stated in MKS system (meter-kilogram-second-coulomb)

APPENDIX 16
ALTERNATING-CURRENT FORMULAS

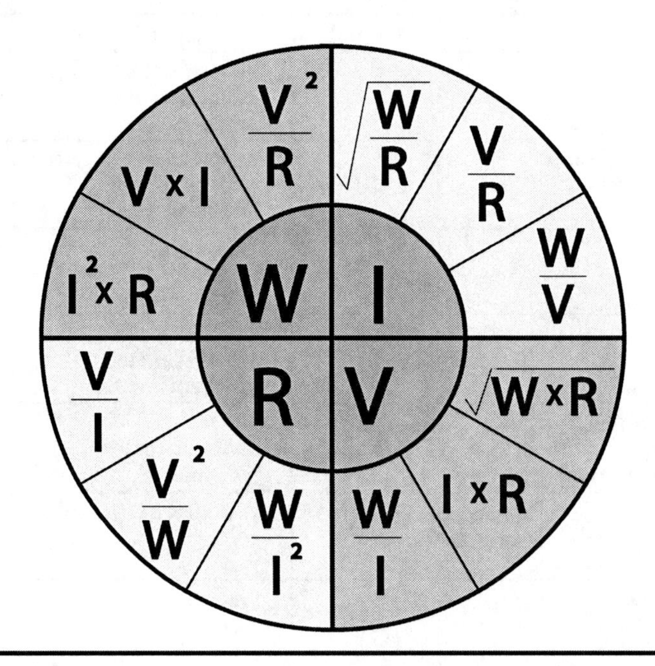

INSTANTANEOUS AND MAXIMUM VALUES

The instantaneous value of voltage and current for a sine wave is equal to the peak or maximum value of the waveform times the sine of the angle. For example, a waveform has a peak value of 300 V. What is the voltage at 22°?

$$V(inst) = V(max) \times \sin\angle$$
$$V(inst) = 300 \times 0.3746 \text{ (Sin of } 22^\circ)$$
$$V(inst) = 112.328 \text{ V}$$

$$V(max) = \frac{V(inst)}{\sin\angle}$$

$$\sin\angle = \frac{V(inst)}{V(max)}$$

CHANGING PEAK, RMS, AND AVERAGE VALUES

TO CHANGE	TO	MULTIPLY BY
Peak	RMS	0.707
Peak	Average	0.637
Peak	Peak to Peak	2
RMS	Peak	1.414
Average	Peak	1.567
RMS	Average	0.9
Average	RMS	1.111

PURE RESISTIVE CIRCUIT

$V = I \times R$ $\qquad I = \frac{V}{R}$ $\qquad R = \frac{V}{I}$ $\qquad W = V \times I$

$V = \frac{W}{I}$ $\qquad I = \frac{W}{V}$ $\qquad R = \frac{V^2}{W}$ $\qquad W = I^2 \times R$

$V = \sqrt{W \times R}$ $\qquad I = \sqrt{\frac{W}{R}}$ $\qquad R = \frac{W}{I^2}$ $\qquad W = \frac{V^2}{R}$

SERIES RESISTIVE CIRCUITS

$R_T = R_1 + R_2 + R_3$

$I_T = I_1 = I_2 = I_3$

$V_T = V_1 + V_2 + V_3$

$W_T = W_1 + W_2 + W_3$

PARALLEL RESISTIVE CIRCUITS

$$R_T = \frac{1}{\frac{1}{R_1} + \frac{1}{R_2} + \frac{1}{R_3}}$$

$$R_T = \frac{R_1 \times R_2}{R_1 + R_2}$$

$I_T = I_1 + I_2 + I_3$

$V_T = V_1 = V_2 = V_3$

$W_T = W_1 + W_2 + W_3$

PURE INDUCTIVE CIRCUITS

NOTE: In a pure inductive circuit, the current lags the voltage by 90°. Therefore, there is no true power or watts and the power factor is 0. VARs is the inductive equivalent of watts.

$L = \frac{X_L}{2 \times \pi \times f}$ $\qquad X_L = 2 \times \pi \times f \times L$

$V_L = I_L \times X_L$ $\qquad I_L = \frac{V_L}{X_L}$ $\qquad X_L = \frac{V_L}{I_L}$ $\qquad VARs_L = \quad \times I_L$

$V_L = \sqrt{VARs_L \times X_L}$ $\qquad I_L = \frac{VARs_L}{V_L}$ $\qquad X_L = \frac{V_L^2}{VARs_L}$ $\qquad VARs_L = I_L^2 \times X_L$

$V_L = \frac{VARs_L}{I_L}$ $\qquad I_L = \sqrt{\frac{VARs_L}{X_L}}$ $\qquad X_L = \frac{VARs_L}{I_L^2}$ $\qquad VARs_L = \frac{V_L^2}{X_L}$

SERIES INDUCTIVE CIRCUITS

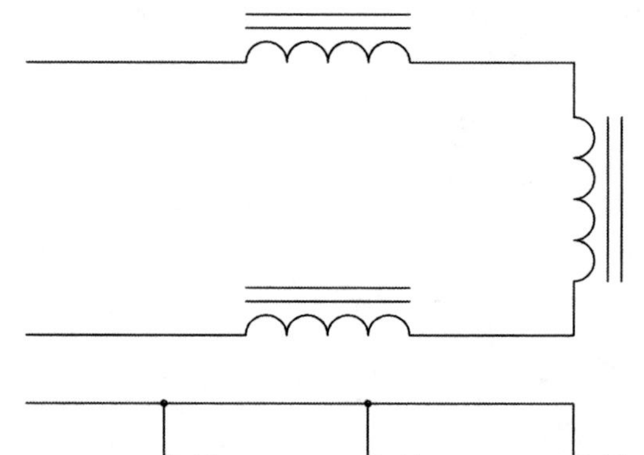

$V_{LT} = V_{L1} + V_{L2} + V_{L3}$

$X_{LT} = X_{L1} + X_{L2} + X_{L3}$

$I_{LT} = I_{L1} = I_{L2} = I_{L3}$

$VARs_{LT} = VARs_{L1} + VARs_{L2} + VARs_{L3}$

$L_T = L_1 + L_2 + L_3$

PARALLEL INDUCTIVE CIRCUITS

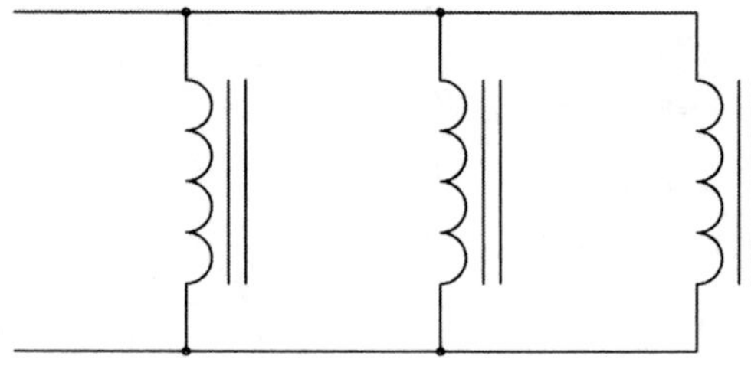

$V_{LT} = V_{L1} = V_{L2} = V_{L3}$

$I_{LT} = I_{L1} + I_{L2} + I_{L3}$

$$X_{LT} = \frac{X_{L1} \times X_{L2}}{X_{L1} + X_{L2}}$$

$$X_{LT} = \frac{1}{\frac{1}{X_{L1}} + \frac{1}{X_{L2}} + \frac{1}{X_{L3}}} \qquad L_T = \frac{1}{\frac{1}{L_1} + \frac{1}{L_2} + \frac{1}{L_3}} \qquad L_T = \frac{L_1 \times L_2}{L_1 + L_2}$$

$VARs_{LT} = VARs_{L1} + VARs_{L2} + VARs_{L3}$

PURE CAPACITIVE CIRCUITS

NOTE: In a pure capacitive circuit, the current leads the voltage by 90°. For this reason there is no true power or watts, and no power factor. VARs are equivalent of watts in a pure capacitive circuit.
NOTE: The value of C in the formula for finding capacitance is in farads and must be changed into the capacitive units being used.

$$C = \frac{1}{2 \times \pi \times f \times X_C} \qquad X_C = \frac{1}{2 \times \pi \times f \times C}$$

$$V_C = I_C \times X_C \qquad I_C = \frac{V_C}{X_C} \qquad X_C = \frac{V_C}{I_C} \qquad VARs_C = V_C \times I_C$$

$$V_C = \sqrt{VARs_C \times X_C} \qquad I_C = \frac{VARs_C}{V_C} \qquad X_C = \frac{V_C^2}{VARs_C} \qquad VARs_C = I_C^2 \times X_C$$

$$V_C = \frac{VARs_C}{I_C} \qquad I_C = \sqrt{\frac{VARs_C}{X_C}} \qquad X_C = \frac{VARs_C}{I_C^2} \qquad VARs_C = \frac{V_C^2}{X_C}$$

SERIES CAPACITIVE CIRCUITS

$V_{CT} = V_{C1} + V_{C2} + V_{C3}$

$I_{CT} = I_{C1} = I_{C2} = I_{C3}$

$X_{CT} = X_{C1} + X_{C2} + X_{C3}$

$VARs_{CT} = VARs_{C1} + VARs_{C2} + VARs_{C3}$

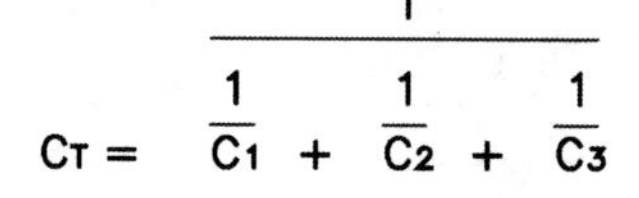

$$C_T = \frac{1}{\frac{1}{C_1} + \frac{1}{C_2} + \frac{1}{C_3}} \qquad C_T = \frac{C_1 \times C_2}{C_1 + C_2}$$

PARALLEL CAPACITIVE CIRCUITS

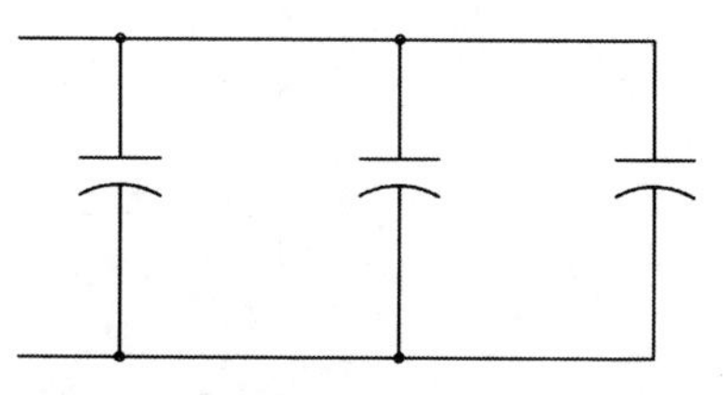

$V_{CT} = V_{C1} = V_{C2} = V_{C3}$

$I_{CT} = I_{C1} + I_{C2} + I_{C3}$

$$X_{CT} = \frac{1}{\frac{1}{X_{C1}} + \frac{1}{X_{C2}} + \frac{1}{X_{C3}}} \qquad X_{CT} = \frac{X_{C1} \times X_{C2}}{X_{C1} + X_{C2}}$$

$C_T = C_1 + C_2 + C_3 \qquad VARs_{CT} = VARs_{C1} + VARs_{C2} + VARs_{C3}$

RESISTIVE–INDUCTIVE SERIES CIRCUITS

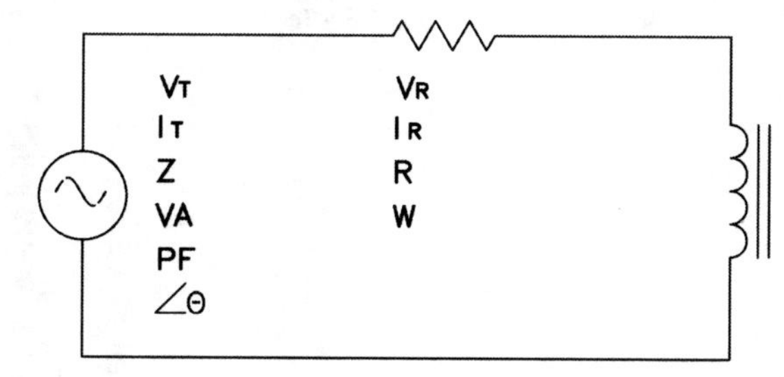

L
V_L
I_L
X_L
$VARs_L$

NOTE: To find values for the reisistor, use the formulas in the PURE RESISTIVE section.

NOTE: To find values for the inductor, use the formulas in the PURE INDUCTIVE section.

$VA = V_T \times I_T \qquad L = \frac{X_L}{2 \times \pi \times f}$

$V_T = \sqrt{V_R^2 + V_L^2} \qquad Z = \sqrt{R^2 + X_L^2} \qquad VA = I_T^2 \times Z \qquad I_T = I_R = I_L$

$V_T = I_T \times Z \qquad Z = \frac{V_T}{I_T} \qquad VA = \frac{V_T^2}{Z} \qquad I_T = \frac{V_T}{Z}$

$V_T = \frac{VA}{I_T} \qquad Z = \frac{VA}{I_T^2} \qquad VA = \sqrt{W^2 + VARs_L^2} \qquad I_T = \frac{VA}{V_T}$

$V_T = \frac{V_R}{PF} \qquad Z = \frac{R}{PF} \qquad Z = \frac{V_T^2}{VA} \qquad VA = \frac{W}{PF}$

RESISTIVE–INDUCTIVE SERIES CIRCUITS (CONTINUED)

$PF = \frac{R}{Z}$ $\quad W = V_R \times I_R$ $\quad V_R = I_R \times R$ $\quad I_R = I_T = I_L$

$PF = \frac{W}{VA}$ $\quad W = \sqrt{VA^2 - VARs_L^2}$ $\quad V_R = \sqrt{W \times R}$ $\quad I_R = \frac{V_R}{R}$

$PF = \frac{V_R}{V_T}$ $\quad W = \frac{V_R^2}{R}$ $\quad V_R = \frac{W}{I_R}$ $\quad I_R = \frac{W}{V_R}$

$PF = \cos\angle\Theta$ $\quad W = I_R^2 \times R$ $\quad V_R = \sqrt{V_T^2 - V_L^2}$ $\quad I_R = \sqrt{\frac{W}{R}}$

$W = VA \times PF$ $\quad V_R = V_T \times PF$

$R = \sqrt{Z^2 - X_L^2}$ $\quad V_L = I_L \times X_L$ $\quad I_L = I_R = I_T$ $\quad X_L = \sqrt{Z^2 - R^2}$

$R = \frac{V_R}{I_R}$ $\quad V_L = \sqrt{V_T^2 - V_R^2}$ $\quad I_L = \frac{V_L}{X_L}$ $\quad X_L = \frac{V_L}{I_L}$

$R = \frac{V_R^2}{W}$ $\quad V_L = \sqrt{VARs_L \times X_L}$ $\quad I_L = \frac{VARs_L}{V_L}$ $\quad X_L = \frac{V_L^2}{VARs_L}$

$R = \frac{W}{I_R^2}$ $\quad V_L = \frac{VARs_L}{I_L}$ $\quad I_L = \sqrt{\frac{VARs_L}{X_L}}$ $\quad X_L = \frac{VARs_L}{I_L^2}$

$R = Z \times PF$ $\quad X_L = 2 \times \pi \times f \times L$

$VARs_L = \sqrt{VA^2 - P^2}$ $\quad VARs_L = V_L \times I_L$ $\quad VARs_L = \frac{V_L^2}{X_L}$ $\quad VARs_L = I_L^2 \times X_L$

RESISTIVE–INDUCTIVE PARALLEL CIRCUITS

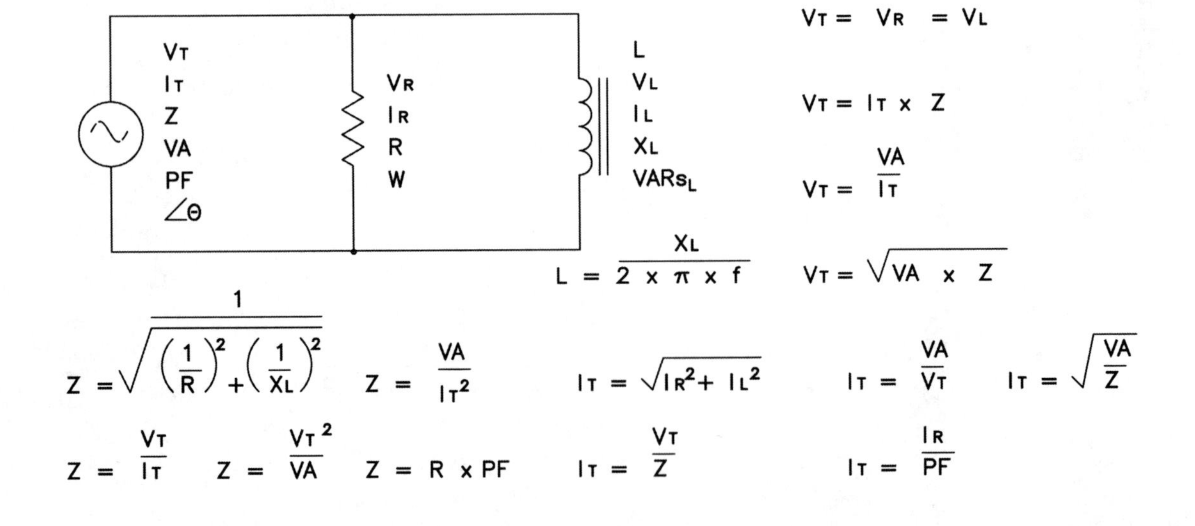

$V_T = V_R = V_L$

$V_T = I_T \times Z$

$V_T = \frac{VA}{I_T}$

$L = \frac{X_L}{2 \times \pi \times f}$ $\quad V_T = \sqrt{VA \times Z}$

$Z = \frac{1}{\sqrt{\left(\frac{1}{R}\right)^2 + \left(\frac{1}{X_L}\right)^2}}$ $\quad Z = \frac{VA}{I_T^2}$ $\quad I_T = \sqrt{I_R^2 + I_L^2}$ $\quad I_T = \frac{VA}{V_T}$ $\quad I_T = \sqrt{\frac{VA}{Z}}$

$Z = \frac{V_T}{I_T}$ $\quad Z = \frac{V_T^2}{VA}$ $\quad Z = R \times PF$ $\quad I_T = \frac{V_T}{Z}$ $\quad I_T = \frac{I_R}{PF}$

RESISTIVE–INDUCTIVE PARALLEL CIRCUITS (CONTINUED)

$VA = V_T \times I_T$

$VA = I_T^2 \times Z$

$VA = \frac{V_T^2}{Z}$

$VA = \sqrt{W^2 + VARs_L^2}$

$VA = \frac{W}{PF}$

$PF = \frac{Z}{R}$

$PF = \frac{W}{VA}$

$PF = \frac{I_R}{I_T}$

$PF = \cos \angle\theta$

$V_L = I_L \times X_L$

$V_L = V_T = V_R$

$V_L = \sqrt{VARs_L \times X_L}$

$E_L = \frac{VARs_L}{I_L}$

$I_L = \sqrt{I_T^2 - I_R^2}$

$I_L = \frac{V_L}{X_L}$

$I_L = \frac{VARs_L}{V_L}$

$I_L = \sqrt{\frac{VARs_L}{X_L}}$

$VARs_L = \sqrt{VA^2 - W^2}$

$VARs_L = V_L \times I_L$

$VARs_L = \frac{V_L^2}{X_L}$

$VARs_L = I_L^2 \times X_L$

$X_L = \frac{V_L}{I_L}$

$X_L = \frac{V_L^2}{VARs_L}$

$X_L = \frac{VARs_L}{I_L^2}$

$X_L = \frac{1}{\sqrt{\left(\frac{1}{Z}\right)^2 - \left(\frac{1}{R}\right)^2}}$

$X_L = 2 \times \pi \times f \times L$

$R = \frac{V_R}{I_R}$

$V_R = I_R \times R$

$V_R = \sqrt{W \times R}$

$V_R = \frac{W}{I_R}$

$V_R = V_T = V_L$

$I_R = \sqrt{I_T^2 - I_L^2}$

$I_R = \frac{V_R}{R}$

$I_R = \frac{W}{V_R}$

$I_R = \sqrt{\frac{W}{R}}$

$R = \frac{W}{I_R^2}$

$I_R = I_T \times PF$

$R = \frac{1}{\sqrt{\left(\frac{1}{Z}\right)^2 - \left(\frac{1}{X_L}\right)^2}}$

$R = \frac{V_R^2}{W}$

$R = \frac{Z}{PF}$

$W = \sqrt{VA^2 - VARs_L^2}$

$W = V_R \times I_R$

$W = \frac{V_R^2}{R}$

$W = I_R^2 \times R$

$W = VA \times PF$

RESISTIVE–CAPACITIVE SERIES CIRCUITS

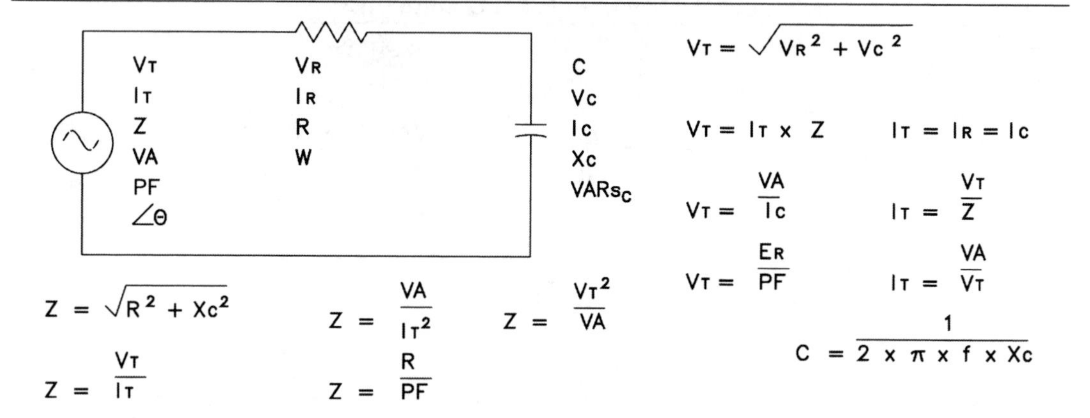

$V_T = \sqrt{V_R^2 + V_C^2}$

$V_T = I_T \times Z$

$I_T = I_R = I_C$

$V_T = \frac{VA}{I_C}$

$I_T = \frac{V_T}{Z}$

$V_T = \frac{E_R}{PF}$

$I_T = \frac{VA}{V_T}$

$C = \frac{1}{2 \times \pi \times f \times X_C}$

$Z = \sqrt{R^2 + X_C^2}$

$Z = \frac{VA}{I_T^2}$

$Z = \frac{V_T^2}{VA}$

$Z = \frac{V_T}{I_T}$

$Z = \frac{R}{PF}$

RESISTIVE–CAPACITIVE SERIES CIRCUITS (CONTINUED)

$VA = V_T \times I_T$ $\quad PF = \dfrac{R}{Z}$ $\quad W = V_R \times I_R$ $\quad V_R = I_R \times R$

$VA = I_T^2 \times Z$ $\quad PF = \dfrac{W}{VA}$ $\quad W = \sqrt{VA^2 - VARs_C^2}$ $\quad V_R = \sqrt{W \times R}$

$VA = \dfrac{V_T^2}{Z}$ $\quad PF = \dfrac{V_R}{V_T}$ $\quad W = \dfrac{V_R^2}{R}$ $\quad V_R = \dfrac{W}{I_R}$

$VA = \sqrt{W^2 + VARs_C^2}$ $\quad PF = \cos\angle\Theta$ $\quad W = I_R^2 \times R$ $\quad V_R = \sqrt{V_T^2 - V_C^2}$

$VA = \dfrac{W}{PF}$ $\quad R = \sqrt{Z^2 - X_C^2}$ $\quad W = VA \times PF$ $\quad V_R = V_T \times PF$

$I_R = I_T = I_C$ $\quad R = \dfrac{V_R}{I_R}$ $\quad V_C = I_C \times X_C$ $\quad I_C = I_R = I_T$

$I_R = \dfrac{V_R}{R}$ $\quad R = \dfrac{V_R^2}{W}$ $\quad V_C = \sqrt{V_T^2 - V_R^2}$ $\quad I_C = \dfrac{V_C}{X_C}$

$I_R = \dfrac{W}{E_R}$ $\quad R = \dfrac{W}{I_R^2}$ $\quad V_C = \sqrt{VARs_C \times X_C}$ $\quad I_C = \dfrac{VARs_C}{V_C}$

$I_R = \sqrt{\dfrac{W}{R}}$ $\quad R = Z \times PF$ $\quad V_C = \dfrac{VARs_C}{I_C}$ $\quad I_C = \sqrt{\dfrac{VARs_C}{X_C}}$

$X_C = \sqrt{Z^2 - R^2}$ $\quad X_C = \dfrac{V_C^2}{VARs_C}$ $\quad X_C = \dfrac{1}{2 \times \pi \times f \times C}$ $\quad VARs_C = I_C^2 \times X_C$

$X_C = \dfrac{V_C}{I_C}$ $\quad X_C = \dfrac{VARs_C}{I_C^2}$ $\quad VARs_C = V_C \times I_C$ $\quad VARs_C = \dfrac{V_C^2}{X_C}$

$Vars\,c = \sqrt{VA^2 - W^2}$

RESISTIVE–CAPACITIVE PARALLEL CIRCUITS

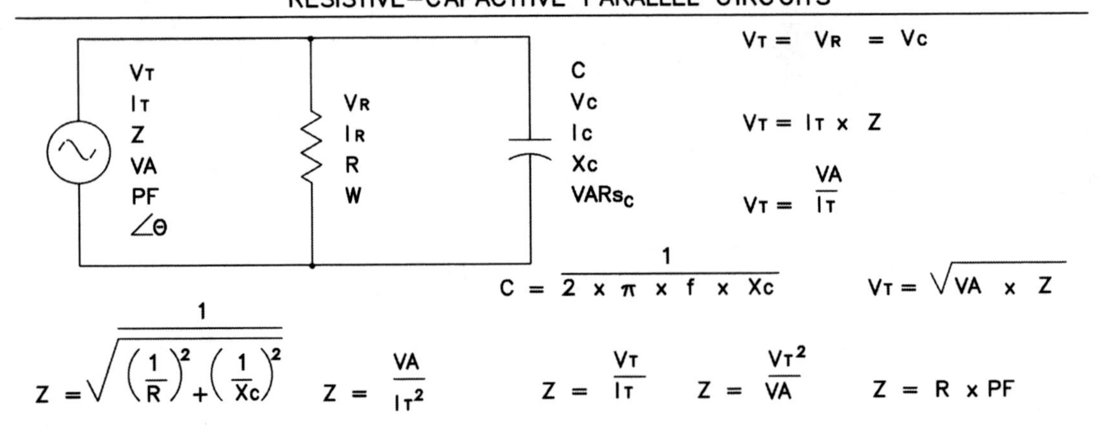

$V_T = V_R = V_C$

$V_T = I_T \times Z$

$V_T = \dfrac{VA}{I_T}$

$C = \dfrac{1}{2 \times \pi \times f \times X_C}$ $\quad V_T = \sqrt{VA \times Z}$

$Z = \dfrac{1}{\sqrt{\left(\dfrac{1}{R}\right)^2 + \left(\dfrac{1}{X_C}\right)^2}}$ $\quad Z = \dfrac{VA}{I_T^2}$ $\quad Z = \dfrac{V_T}{I_T}$ $\quad Z = \dfrac{V_T^2}{VA}$ $\quad Z = R \times PF$

RESISTIVE–CAPACITIVE PARALLEL CIRCUITS (CONTINUED)

$I_T = \sqrt{I_R^2 + I_C^2}$ $\quad VA = V_T \times I_T$ $\quad PF = \frac{R}{Z}$ $\quad W = V_R \times I_R$

$I_T = \frac{V_T}{Z}$ $\quad VA = I_T^2 \times Z$ $\quad PF = \frac{W}{VA}$ $\quad W = \sqrt{VA^2 - VARs_C^2}$

$I_T = \frac{VA}{V_T}$ $\quad VA = \frac{V_T^2}{Z}$ $\quad PF = \frac{V_R}{V_T}$ $\quad W = \frac{V_R^2}{R}$

$I_T = \frac{I_R}{PF}$ $\quad VA = \sqrt{W^2 + VARs_C^2}$ $\quad PF = \cos \angle\theta$ $\quad W = I_R^2 \times R$

$VA = \frac{W}{PF}$ $\quad W = VA \times PF$

$I_T = \sqrt{\frac{VA}{Z}}$

$V_R = I_R \times R$ $\quad I_R = \sqrt{I_T^2 - I_C^2}$ $\quad R = \frac{1}{\sqrt{\left(\frac{1}{Z}\right)^2 - \left(\frac{1}{X_C}\right)^2}}$ $\quad V_C = I_C \times X_C$

$V_R = \sqrt{W \times R}$ $\quad I_R = \frac{V_R}{R}$ $\quad R = \frac{V_R}{I_R}$ $\quad V_C = V_T = V_R$

$V_R = \frac{W}{I_R}$ $\quad I_R = \frac{W}{V_R}$ $\quad R = \frac{V_R^2}{W}$ $\quad V_C = \sqrt{VARs_C \times X_C}$

$V_R = V_T = V_C$ $\quad I_R = \sqrt{\frac{W}{R}}$ $\quad R = \frac{W}{I_R^2}$ $\quad V_C = \frac{VARs_C}{I_C}$

$V_R = V_T \times PF$

$R = Z \times PF$ $\quad VARs_C = I_C^2 \times X_C$

$I_C = \sqrt{I_T^2 - I_R^2}$ $\quad X_C = \frac{1}{\sqrt{\left(\frac{1}{Z}\right)^2 - \left(\frac{1}{R}\right)^2}}$ $\quad X_C = \frac{V_C^2}{VARs_C}$ $\quad VARs_C = \frac{V_C^2}{X_C}$

$I_C = \frac{V_C}{X_C}$ $\quad X_C = \frac{VARs_C}{I_C^2}$ $\quad VARs_C = V_C \times I_C$

$I_C = \frac{VARs_C}{V_C}$ $\quad X_C = \frac{V_C}{I_C}$ $\quad X_C = \frac{1}{2 \times \pi \times f \times C}$ $\quad VARs_C = \sqrt{VA^2 - W^2}$

$I_C = \sqrt{\frac{VARs_C}{X_C}}$

RESISTIVE–INDUCTIVE–CAPACITIVE SERIES CIRCUITS

$V_T = \sqrt{V_R^2 + (V_L - V_C)^2}$ $\quad V_T = \frac{VA}{I_T}$ $\quad Z = \sqrt{R^2 + (X_L - X_C)^2}$ $\quad Z = \frac{VA}{I_T^2}$

$V_T = I_T \times Z$ $\quad V_T = \frac{V_R}{PF}$ $\quad Z = \frac{V_T}{I_T}$ $\quad Z = \frac{R}{PF}$

RESISTIVE–INDUCTIVE–CAPACITIVE SERIES CIRCUITS (CONTINUED)

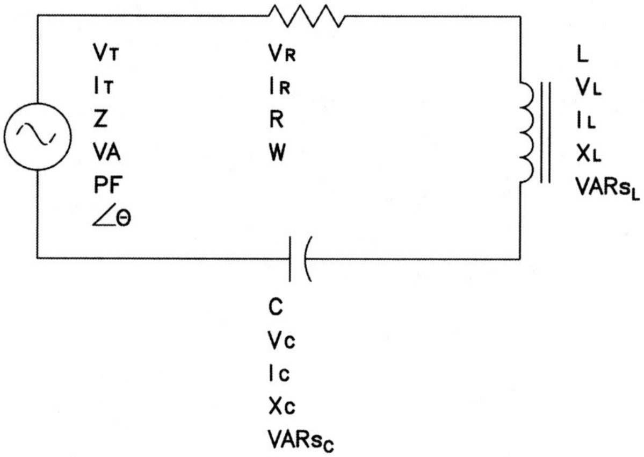

Source: V_T, I_T, Z, VA, PF, $\angle\theta$

Resistor: V_R, I_R, R, W

Inductor: L, V_L, I_L, X_L, $VARs_L$

Capacitor: C, V_C, I_C, X_C, $VARs_C$

$I_T = I_R = I_L = I_C$ $\quad I_T = \frac{VA}{V_T}$

$I_T = \frac{V_T}{Z}$ $\quad I_T = \sqrt{\frac{VA}{Z}}$

$VA = V_T \times I_T$ $\quad VA = \frac{W}{PF}$

$VA = I_T^2 \times Z$ $\quad VA = \frac{V_T^2}{Z}$

$VA = \sqrt{W^2 + (VARs_L - VARs_C)^2}$

$PF = \frac{R}{Z}$ $\quad W = V_R \times I_R$ $\quad W = VA \times PF$ $\quad V_R = I_R \times R$

$PF = \frac{W}{VA}$ $\quad W = \sqrt{VA^2 - (VARs_L - VARs_C)^2}$ $\quad V_R = \frac{W}{I_R}$

$PF = \frac{V_R}{V_T}$ $\quad W = \frac{V_R^2}{R}$ $\quad V_R = \sqrt{W \times R}$ $\quad V_R = \sqrt{V_T^2 - (V_L - V_C)^2}$

$PF = \cos\angle\theta$ $\quad W = I_R^2 \times R$ $\quad V_R = V_T \times PF$

$I_R = I_T = I_C = I_L$ $\quad R = \sqrt{Z^2 - (X_L - X_C)^2}$ $\quad V_C = I_C \times X_C$

$I_R = \frac{V_R}{R}$ $\quad R = \frac{V_R}{I_R}$ $\quad R = Z \times PF$ $\quad V_C = \sqrt{VARs_C \times X_C}$

$I_R = \frac{W}{V_R}$ $\quad R = \frac{V_R^2}{W}$ $\quad R = \frac{W}{I_R^2}$ $\quad V_C = \frac{VARs_C}{I_C}$

$I_R = \sqrt{\frac{W}{R}}$ $\quad X_C = \frac{1}{2 \times \pi \times f \times C}$ $\quad C = \frac{1}{2 \times \pi \times f \times X_C}$

$I_C = I_R = I_T = I_L$ $\quad X_C = \frac{V_C^2}{VARs_C}$ $\quad X_C = \frac{VARs_C}{I_C^2}$ $\quad VARs_C = V_C \times I_C$

$I_C = \frac{V_C}{X_C}$ $\quad X_C = \frac{V_C}{I_C}$ $\quad L = \frac{X_L}{2 \times \pi \times f}$ $\quad X_L = 2 \times \pi \times f \times L$

$I_C = \frac{VARs_C}{V_C}$ $\quad VARs_C = I_C^2 \times X_C$ $\quad X_L = \frac{V_L}{I_L}$ $\quad X_L = \frac{VARs_L}{I_L^2}$

$I_C = \sqrt{\frac{VARs_C}{X_C}}$ $\quad VARs_C = \frac{V_C^2}{X_C}$ $\quad X_L = \frac{V_L^2}{VARs_L}$

RESISTIVE–INDUCTIVE–CAPACITIVE SERIES CIRCUITS (CONTINUED)

$V_L = I_L \times X_L$ | $I_L = I_R = I_T = I_C$ | $VARs_L = V_L \times I_L$

$V_L = \sqrt{VARs_L \times X_L}$ | $I_L = \frac{V_L}{X_L}$ | $VARs_L = \frac{V_L^2}{X_L}$

$V_L = \frac{VARs_L}{I_L}$ | $I_L = \frac{VARs_L}{V_L}$ | $VARs_L = I_L^2 \times X_L$

$I_L = \sqrt{\frac{VARs_L}{X_L}}$

RESISTIVE–INDUCTIVE–CAPACITIVE PARALLEL CIRCUITS

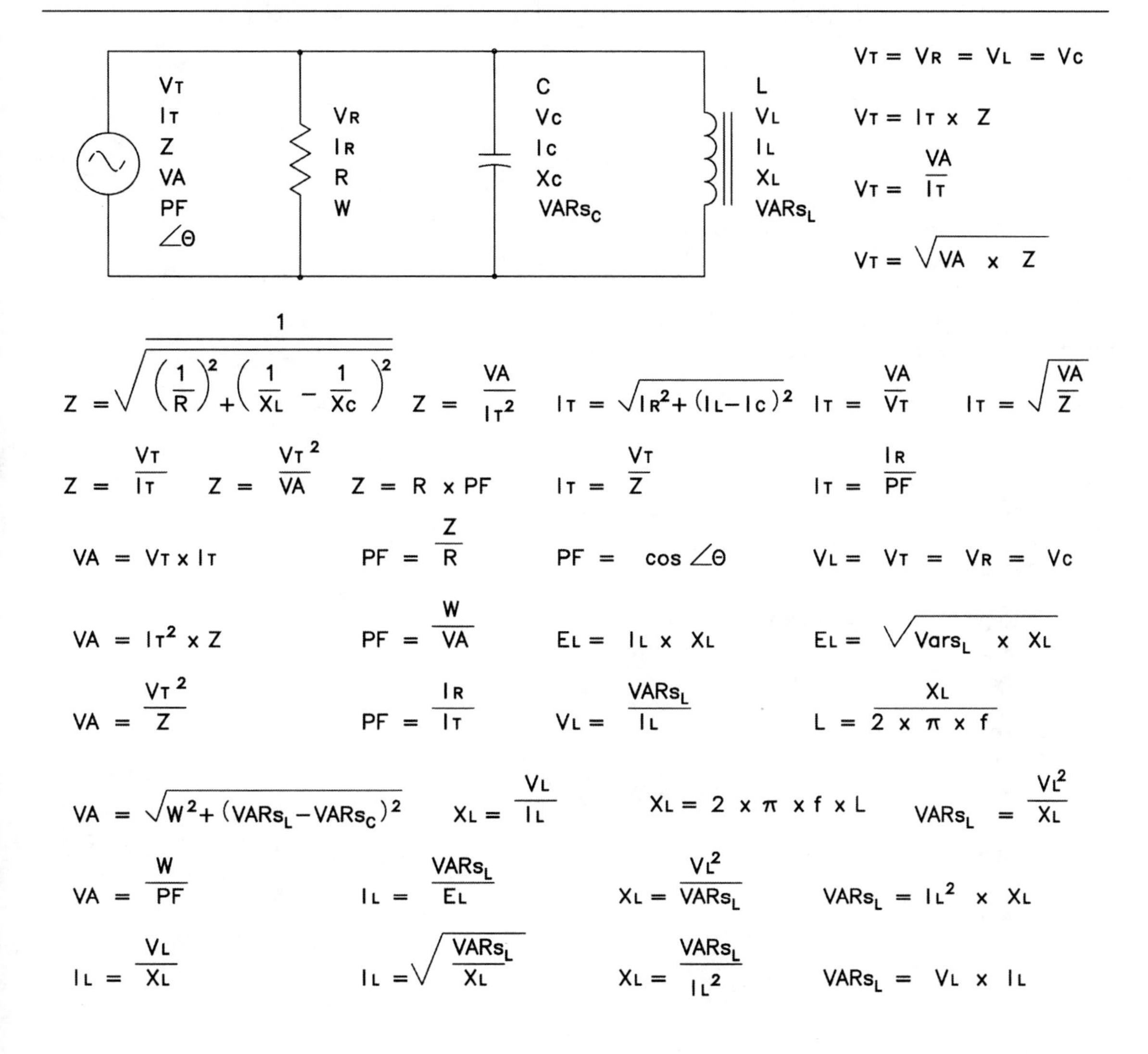

$V_T = V_R = V_L = V_C$

$V_T = I_T \times Z$

$V_T = \frac{VA}{I_T}$

$V_T = \sqrt{VA \times Z}$

$$Z = \frac{1}{\sqrt{\left(\frac{1}{R}\right)^2 + \left(\frac{1}{X_L} - \frac{1}{X_C}\right)^2}} \quad Z = \frac{VA}{I_T^2} \quad I_T = \sqrt{I_R^2 + (I_L - I_C)^2} \quad I_T = \frac{VA}{V_T} \quad I_T = \sqrt{\frac{VA}{Z}}$$

$$Z = \frac{V_T}{I_T} \quad Z = \frac{V_T^{\,2}}{VA} \quad Z = R \times PF \quad I_T = \frac{V_T}{Z} \quad I_T = \frac{I_R}{PF}$$

$$VA = V_T \times I_T \quad PF = \frac{Z}{R} \quad PF = \cos \angle\Theta \quad V_L = V_T = V_R = V_C$$

$$VA = I_T^2 \times Z \quad PF = \frac{W}{VA} \quad E_L = I_L \times X_L \quad E_L = \sqrt{Vars_L \times X_L}$$

$$VA = \frac{V_T^{\,2}}{Z} \quad PF = \frac{I_R}{I_T} \quad V_L = \frac{VARs_L}{I_L} \quad L = \frac{X_L}{2 \times \pi \times f}$$

$$VA = \sqrt{W^2 + (VARs_L - VARs_C)^2} \quad X_L = \frac{V_L}{I_L} \quad X_L = 2 \times \pi \times f \times L \quad VARs_L = \frac{V_L^2}{X_L}$$

$$VA = \frac{W}{PF} \quad I_L = \frac{VARs_L}{E_L} \quad X_L = \frac{V_L^2}{VARs_L} \quad VARs_L = I_L^2 \times X_L$$

$$I_L = \frac{V_L}{X_L} \quad I_L = \sqrt{\frac{VARs_L}{X_L}} \quad X_L = \frac{VARs_L}{I_L^2} \quad VARs_L = V_L \times I_L$$

RESISTIVE–INDUCTIVE–CAPACITIVE PARALLEL CIRCUITS (CONTINUED)

$V_R = I_R \times R$ $\quad I_R = \sqrt{I_T^2 - (I_L - I_C)^2}$ $\quad I_R = I_T \times PF$

$V_R = \sqrt{W \times R}$ $\quad I_R = \dfrac{V_R}{R}$ $\quad R = \dfrac{V_R}{I_R}$ $\quad R = \dfrac{1}{\sqrt{\left(\dfrac{1}{Z}\right)^2 - \left(\dfrac{1}{X_L} - \dfrac{1}{X_C}\right)^2}}$

$E_R = \dfrac{W}{I_R}$ $\quad I_R = \dfrac{W}{E_R}$

$V_R = V_T = V_L = V_C$ $\quad I_R = \sqrt{\dfrac{W}{R}}$ $\quad R = \dfrac{W}{I_R^{\,2}}$ $\quad R = \dfrac{Z}{PF}$ $\quad R = \dfrac{V_R^{\,2}}{W}$

$W = \sqrt{VA^2 - (VARs_L - VARs_C)^2}$ $\quad W = V_R \times I_R$ $\quad W = I_R^{\,2} \times R$

$V_C = \dfrac{VARs_C}{I_C}$ $\quad W = VA \times PF$ $\quad W = \dfrac{V_R^{\,2}}{R}$ $\quad X_C = \dfrac{1}{2 \times \pi \times f \times C}$

$V_C = I_C \times X_C$ $\quad I_C = \dfrac{V_C}{X_C}$ $\quad X_C = \dfrac{V_C}{I_C}$ $\quad VARs_C = V_C \times I_C$

$V_C = V_T = V_R = V_L$ $\quad I_C = \dfrac{VARs_C}{V_C}$ $\quad X_C = \dfrac{V_C^{\,2}}{VARs_C}$ $\quad VARs_C = I_C^2 \times X_C$

$V_C = \sqrt{VARs_C \times X_C}$ $\quad I_C = \sqrt{\dfrac{VARs_C}{X_C}}$ $\quad X_C = \dfrac{VARs_C}{I_C^2}$ $\quad VARs_C = \dfrac{V_C^{\,2}}{X_C}$

TRANSFORMERS

$\dfrac{V_P}{V_S} = \dfrac{N_P}{N_S}$ $\quad \dfrac{V_P}{V_S} = \dfrac{I_S}{I_P}$ $\quad \dfrac{N_P}{N_S} = \dfrac{I_S}{I_P}$ $\quad Z_P = Z_S \left(\dfrac{N_P}{N_S}\right)^2$

$Z_S = Z_P \left(\dfrac{N_P}{N_S}\right)^2$

V_P (Voltage of the primary)

V_S (Voltage of the secondary)

I_P (Current of the primary)

I_S (Current of the secondary)

N_P (Number of turns of wire in the primary)

N_S (Number of turns of wire in the secondary)

Z_P (Impedance of the primary)

Z_S (Impedance of the secondary)

THREE PHASE CONNECTIONS

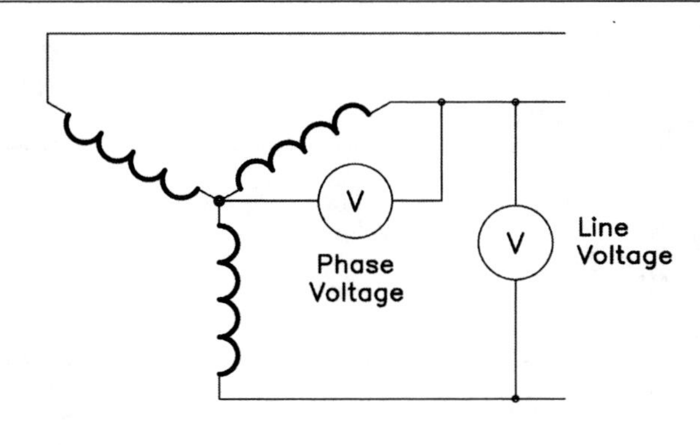

In a WYE connection, the line current and phase current are the same.

$$I_{(Line)} = I_{(Phase)}$$

In a WYE connection, the line voltage is higher than the phase voltage by a factor of the square root of three.

$$V_{(Line)} = V_{(Phase)} \times \sqrt{3}$$

$$V_{(Phase)} = \frac{V_{(Line)}}{\sqrt{3}}$$

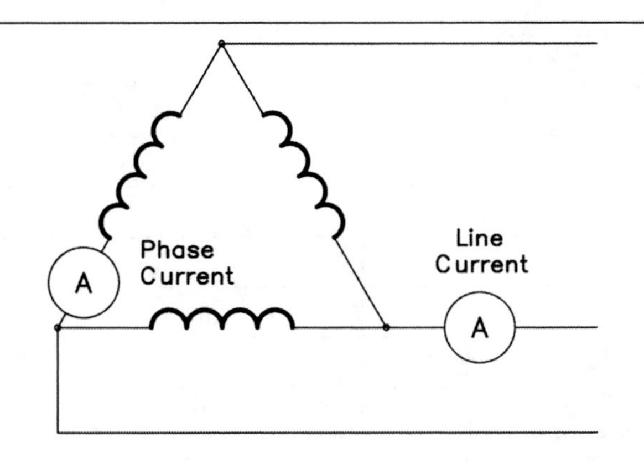

In a DELTA connected system, the line voltage and phase voltage are the same.

$$V_{(Line)} = V_{(Phase)}$$

In a DELTA connected system, the line current is higher than the phase current by a factor of the square root of three.

$$I_{(Line)} = I_{(Phase)} \times \sqrt{3}$$

$$I_{(Phase)} = \frac{I_{(Line)}}{\sqrt{3}}$$

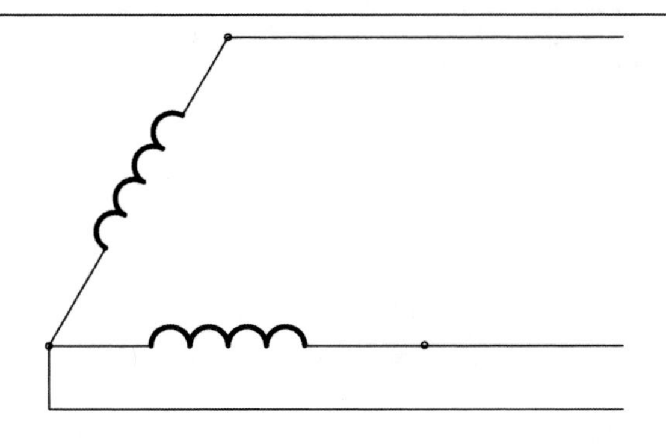

OPEN DELTA connections provide 86.6% of the sum of the power rating of the two transformers. EXAMPLE: The two transformers shown are rated at 60 kVA each. the total power rating for this connection would be:

60 + 60 = 120 kVA

120 kVA x 0.866 = 103.92 kVA

All values of voltage and current are computed in the same manner as a closed delta connection.

$$VA = \sqrt{3} \times V_{(Line)} \times I_{(Line)}$$

$$W = \sqrt{3} \times V_{(Line)} \times I_{(Line)} \times PF$$

$$VA = 3 \times V_{(Phase)} \times I_{(Phase)}$$

$$W = 3 \times V_{(Phase)} \times I_{(Phase)} \times PF$$

Answers to Practice Problems

ANSWERS TO PRACTICE PROBLEMS FOR UNIT 1

PEAK VOLTS	INSTANTANEOUS VOLTS	PHASE ANGLE
347	208	36.8°
780	536.9	43.5°
80.36	24.3	17.6°
224	5.65	1.44°
48.7	43.99	64.6°
339.4	240	45°
87.2	23.7	15.77°
156.9	155.48	82.3°
110.4	62.7	34.6°
1,256	400	18.57°
15,720	3268.37	12°
126.86	72.4	34.8°

ANSWERS TO PRACTICE PROBLEMS FOR UNIT 2

PEAK	RMS	AVERAGE
12.7	8.979	8.09
76.073	53.8	48.42
257.3	182.262	164.2
1235	873.145	786.695
339.36	240	216
26.012	18.443	16.6
339.7	240.168	216.389
17.816	12.6	11.34
14.103	9.99	9
123.7	87.456	78.797
105.767	74.8	67.32
169.236	119.88	108

ANSWERS TO PRACTICE PROBLEMS FOR UNIT 3

INDUCTANCE	FREQUENCY	INDUCTIVE REACTANCE
1.2 H	60 Hz	452.4 Ω
0.085 H	400 Hz	213.628 Ω
0.75 H	1000 Hz	4712.389 Ω
0.65 H	600 Hz	2450.442 Ω
3.6 H	30 Hz	678.584 Ω
2.62 H	25 Hz	411.459 Ω
0.5 H	60 Hz	188.5 Ω
0.85 H	1200 Hz	6408.849 Ω
1.6 H	20 Hz	201.062 Ω
0.45 H	400 Hz	1130.973 Ω
4.8 H	80 Hz	2412.743 Ω
0.0065 H	1000 Hz	40.841 Ω

ANSWERS TO PRACTICE PROBLEMS FOR UNIT 4

1.

V_T = 480 V | V_R = 247.136 V | V_L = 411.509 V

I_T = 20.595 A | I_R = 20.595 A | I_L = 20.595 A

Z = 23.307 Ω | R = 12 Ω | X_L = 19.981 Ω

VA = 9885.442 VA (volt-amperes) | P = 5089.686 W | $VARs_L$ = 8474.751 VARs

PF = 51.5% | $\angle\theta$ = 59° | L = 0.053 H

2.

V_T = 130 V | V_R = 78 V | V_L = 104 V

I_T = 6.5 A | I_R = 6.5 A | I_L = 6.5 A

Z = 20 Ω | R = 12 Ω | X_L = 16 Ω

VA = 845 VA | P = 507 W | $VARs_L$ = 676 VARs

PF = 60% | $\angle\theta$ = 53.13° | L = 0.0424 H

3.

V_T = 120 V | V_R = 96 V | V_L = 72 V

I_T = 1.2 A | I_R = 1.2 A | I_L = 1.2 A

Z = 100 Ω | R = 80 Ω | X_L = 60 Ω

VA = 144 VA | P = 115.2 W | $VARs_L$ = 86.4 VARs

PF = 80% | $\angle\theta$ = 36.87° | L = 0.15915 H

4.

V_T = 240 V | V_R = 187.408 V | V_L = 149.927 V

I_T = 1.562 A | I_R = 1.562 A | I_L = 1.562 A

Z = 153.675 Ω | R = 120 Ω | X_L = 96 Ω

VA = 374.817 VA | P = 292.683 W | $VARs_L$ = 234.146 VARs

PF = 78% | $\angle\theta$ = 38.66° | L = 0.0382 H

ANSWERS TO PRACTICE PROBLEMS FOR UNIT 5

RESISTANCE	CAPACITANCE	TIME CONSTANT	TOTAL TIME
150 kΩ	100 μF	15 s	75 s
350 kΩ	20 μF	7 s	35 s
142,857,142.9 Ω	350 pF	0.05 s	0.25 s
40 MΩ	0.05 μF	2 s	10 s
1.2 MΩ	0.47 μF	0.564 s	2.82 s
4166.67 Ω	12 μF	0.05 s	0.25 s
86 kΩ	3.488 μF	0.3 s	1.5 s
120 kΩ	470 pF	56.4 μs	282 μs
80 kΩ	250 nF	20 ms	100 ms
3.75 Ω	8 μF	30 μs	150 μs
100 kΩ	1.5 μF	150 ms	750 ms
33 kΩ	4 μF	32 ms	660 ms

ANSWERS TO PRACTICE PROBLEMS FOR UNIT 6

1.

V_T = 480 V
I_T = 24 A
Z = 20 Ω
VA = 11,520 VA (volt-amperes)
PF = 60%

V_R = 288 V
I_R = 24 A
R = 12 Ω
P = 6912 W
$\angle\theta$ = 53.13°

V_C = 384 V
I_C = 24 A
X_C = 16 Ω
$VARs_C$ = 9216 VARs
C = 165.782 μF

2.

V_T = 130 V
I_T = 6.5 A
Z = 20 Ω
VA = 845 VA
PF = 60%

V_R = 78 V
I_R = 6.5 A
R = 12 Ω
P = 507 W
$\angle\theta$ = 53.13°

V_C = 104 V
I_C = 6.5 A
X_C = 16 Ω
$VARs_C$ = 676 VARs
C = 165.782 μF

3.

V_T = 318.584 V
I_T = 0.452 A
Z = 704.832 Ω
VA = 144 VA (volt-amperes)
PF = 80%

V_R = 254.867 V
I_R = 0.452 A
R = 563.865 Ω
P = 115.2 W
$\angle\theta$ = 36.87°

V_C = 191.15 V
I_C = 0.452 A
X_C = 422.154 Ω
$VARs_C$ = 86.4 VARs
C = 6.2883 μF

4.

V_T = 185.874 V
I_T = 1.614 A
Z = 115.163 Ω
VA = 300 VA
PF = 68%

V_R = 126.394 V
I_R = 1.614 A
R = 78.311 Ω
P = 204 W
$\angle\theta$ = 47.156°

V_C = 136.285 V
I_C = 1.614 A
X_C = 84.432 Ω
$VARs_C$ = 219.964 VARs
C = 4.7125 μF

ANSWERS TO PRACTICE PROBLEMS FOR UNIT 7

1.

V_T = 120 V
I_T = 2 A
Z = 60 Ω
VA = 240 VA (volt-amperes)
PF = 60%

V_R = 72 V
I_R = 2 A
R = 36 Ω
P = 144 W
$\angle\theta$ = 53.13°

V_L = 200 V
I_L = 2 A
X_L = 100 Ω
$VARs_L$ = 400 VARs
L = 0.265 H

V_C = 104 V
I_C = 2 A
X_C = 52 Ω
$VARs_C$ = 208 VARs
C = 51 μF

2.

$V_T = 35.678$ V
$I_T = 0.6$ A
$Z = 59.464\ \Omega$
$VA = 21.407$ VA
$PF = 67.3\%$

$V_R = 24$ V
$I_R = 0.6$ A
$R = 40\ \Omega$
$P = 14.4$ W
$\angle\theta = 47.7°$

$V_L = 21.6$ V
$I_L = 0.6$ A
$X_L = 36\ \Omega$
$VARs_L = 12.96$ VARs
$L = 0.0143$ H

$V_C = 48$ V
$I_C = 0.6$ A
$X_C = 80\ \Omega$
$VARs_C = 28.8$ VARs
$C = 5\ \mu F$

3.

$V_T = 24$ V
$I_T = 1.25$ A
$Z = 19.2\ \Omega$
$VA = 29.985$ VA
$PF = 62.5\%$

$V_R = 14.993$ V
$I_R = 1.25$ A
$R = 12\ \Omega$
$P = 18.74$ W
$\angle\theta = 51.32°$

$V_L = 74.963$ V
$I_L = 1.25$ A
$X_L = 60\ \Omega$
$VARs_L = 93.75$ VARs
$L = 0.159$ H

$V_C = 56.222$ V
$I_C = 1.25$ A
$X_C = 45\ \Omega$
$VARs_C = 70.312$ VARs
$C = 58.9\ \mu F$

4.

$V_T = 349.057$ V
$I_T = 0.148$ A
$Z = 2358.493\ \Omega$
$VA = 51.8$ VA
$PF = 52.86\%$

$V_R = 185$ V
$I_R = 0.148$ A
$R = 1250\ \Omega$
$P = 27.38$ W
$\angle\theta = 58.1°$

$V_L = 740$ V
$I_L = 0.148$ A
$X_L = 5000\ \Omega$
$VARs_L = 109.52$ VARs
$L = 0.796$ H

$V_C = 444$ V
$I_C = 0.148$ A
$X_C = 3000\ \Omega$
$VARs_C = 65.712$ VARs
$C = 0.053\ \mu F$

ANSWERS TO PRACTICE PROBLEMS FOR UNIT 8

1.

$V_T = 240$ V
$I_T = 34.553$ A
$Z = 6.946\ \Omega$
VA = 8292.72 VA (volt-amperes)
PF = 49.6%

$V_R = 240$ V
$I_R = 17.143$ A
$R = 14\ \Omega$
P = 4114.286 W
$\angle\theta = 60.25°$

$V_L = 240$ V
$I_L = 30$ A
$X_L = 8\ \Omega$
$VARs_L = 7200$ VARs
L = 0.02122 H

2.

$V_T = 54$ V
$I_T = 2.5$ A
$Z = 21.6\ \Omega$
VA = 135 VA
PF = 60%

$V_R = 54$ V
$I_R = 1.5$ A
$R = 36\ \Omega$
P = 81 W
$\angle\theta = 53.13°$

$V_L = 54$ V
$I_L = 2$ A
$X_L = 27\ \Omega$
$VARs_L = 108$ VARs
L = 0.07162 H

3.

$V_T = 480$ V
$I_T = 176.238$ A
$Z = 2.7236\ \Omega$
VA = 84.594.24 VA
PF = 46.5%

$V_R = 480$ V
$I_R = 82$ A
$R = 5.854\ \Omega$
P = 39,360 W
$\angle\theta = 62.3°$

$V_L = 480$ V
$I_L = 156$ A
$X_L = 3.077\ \Omega$
$VARs_L = 74{,}880$ VARs
L = 0.00816 H

4.

V_T = 67.314 V	V_R = 67.314 V	V_L = 67.314 V
I_T = 6 A	I_R = 5.571 A	I_L = 2.228 A
Z = 11.219 Ω	R = 12.083 Ω	X_L = 30.208 Ω
VA = 403.887 VA	P = 375 W	$VARs_L$ = 150 VARs
PF = 92.8%	$\angle\theta$ = 21.8°	L = 0.0801 H

5.

V_T = 120 V	V_R = 120 V	V_C = 120 V
I_T = 10.463 A	I_R = 8.571 A	I_C = 6 A
Z = 11.469 Ω	R = 14 Ω	X_C = 20 Ω
VA = 1255.56 VA	P = 1028.52 W	$VARs_C$ = 720 VARs
PF = 81.9%	$\angle\theta$ = 35°	C = 132.626 μF

6.

V_T = 162 V	V_R = 162 V	V_C = 162 V
I_T = 2.5 A	I_R = 1.5 A	I_C = 2 A
Z = 64.8 Ω	R = 108 Ω	X_C = 81 Ω
VA = 405 VA	P = 243 W	$VARs_C$ = 324 VARs
PF = 60%	$\angle\theta$ = 53.13°	C = 4.9 μF

7.

V_T = 307.202 V	V_R = 307.202 V	V_C = 307.202 V
I_T = 140.99 A	I_R = 65.6 A	I_C = 124.8 A
Z = 2.17888 Ω	R = 4.693 Ω	X_C = 2.461 Ω
VA = 43.312 kVA (kilovolt-amperes)	P = 20.125 kW	$VARs_C$ = 38.339 kilovars
PF = 46.5%	$\angle\theta$ = 62.27°	C = 107.785 μF

8.

$V_T = 69.549$ V
$I_T = 7.5$ A
$Z = 9.273\ \Omega$
VA = 521.615 VA
PF = 94.9%

$V_R = 69.549$ V
$I_R = 7$ A
$R = 9.935\ \Omega$
P = 486.75 W
$\angle\theta = 18.28°$

$V_C = 69.549$ V
$I_C = 2.296$ A
$X_C = 30.291\ \Omega$
$VARs_C = 187.5$ VARs
C = 5.254 μF

9.

$V_T = 120$ V
$I_T = 3.387$ A
$Z = 35.43\ \Omega$
VA = 406.44 VA
PF = 98.4%

$V_R = 120$ V
$I_R = 3.333$ A
$R = 36\ \Omega$
P = 400 W
$\angle\theta = 10.2°$

$V_L = 120$ V
$I_L = 3$ A
$X_L = 40\ \Omega$
$VARs_L = 360$ VARs
L = 0.106 H

$V_C = 120$ V
$I_C = 2.4$ A
$X_C = 50\ \Omega$
$VARs_C = 288$ VARs
C = 53.05 μF

10.

$V_T = 240$ V
$I_T = 22.267$ A
$Z = 10.607\ \Omega$
VA = 5430.58 VA
PF =70.7%

$V_R = 240$ V
$I_R = 16$ A
$R = 15\ \Omega$
P = 3840 W
$\angle\theta = 45°$

$V_L = 240$ V
$I_L = 8$ A
$X_L = 30\ \Omega$
$VARs_L = 1920$ VARs
L = 0.0119 H

V_C = 240 V
I_C = 24 A
X_C = 10 Ω
$VARs_C$ = 5760 VARs
C = 39.79 μF

11.

V_T = 24 V
I_T = 2.004 A
Z = 11.973 Ω
VA = 48.106 VA
PF = 99.8%

V_R = 24 V
I_R = 2 A
R = 12 Ω
P = 48 W
$\angle\theta$ = 3.81°

V_L = 24 V
I_L = 0.4 A
X_L = 60 Ω
$VARs_L$ = 9.6 VARs
L = 0.159 H

V_C = 24 V
I_C = 0.533 A
X_C = 45 Ω
$VARs_C$ = 12.8 VARs
C = 58.94 μF

12.

V_T = 480 V
I_T = 100 A
Z = 4.8 Ω
VA = 48 kVA
PF = 60%

V_R = 480 V
I_R = 60 A
R = 8 Ω
P = 28.8 kW
$\angle\theta$ = 53.14°

V_L = 480 V
I_L = 150 A
X_L = 3.2 Ω
$VARs_L$ = 72 kilovars
L = 0.509 mH

V_C = 480 V
I_C = 70 A
X_C = 6.857 Ω
$VARs_C$ = 33.6 kilovars
C = 23.2 μF

ANSWERS TO PRACTICE PROBLEMS FOR UNIT 13

1.

V_P = 13,800 V	V_S = 480 V
I_P = 1.391 V	I_S = 40 A
N_P = 5750	N_S = 200
	R = 12 Ω

2.

V_P = 120 V	V_S = 8500 V
I_P = 17.708 A	I_S = 0.25 A
N_P = 162	N_S = 11,475
	R = 34 kΩ

3.

V_P = 120 V	V_S = 24 V
I_P = 0.8 A	I_S = 4 A
N_P = 300	N_S = 60
	R = 6 Ω

4.

V_P = 480 V	V_S = 277 V
I_P = 39.963 V	I_S = 69.25 A
N_P = 338	N_S = 195
	R = 4 Ω

5.

$V_P = 240\ V$	$V_{S_1} = 360\ V$	$V_{S_2} = 48\ V$	$V_{S_3} = 24\ V$
$I_P = 2.525\ A$	$I_{S_1} = 0.75\ A$	$I_{S_2} = 3\ A$	$I_{S_3} = 8\ A$
$N_P = 400$	$N_{S_1} = 600$	$N_{S_2} = 80$	$N_{S_3} = 40$
	$R_1 = 480\ \Omega$	$R_2 = 16\ \Omega$	$R_3 = 3\ \Omega$

6.

$V_P = 208\ V$	$V_{S_1} = 120\ V$	$V_{S_2} = 60\ V$	$V_{S_3} = 12\ V$
$I_P = 1.24\ A$	$I_{S_1} = 0.6\ A$	$I_{S_2} = 2.5\ A$	$I_{S_3} = 3\ A$
$N_P = 520$	$N_{S_1} = 300$	$N_{S_2} = 150$	$N_{S_3} = 30$
	$R_1 = 200\ \Omega$	$R_2 = 24\ \Omega$	$R_3 = 4\ \Omega$

7.

$V_P = 480\ V$	$V_{S_1} = 240\ V$	$V_{S_2} = 120\ V$	$V_{S_3} = 24\ V$
$I_P = 1.45\ A$	$I_{S_1} = 0.5\ A$	$I_{S_2} = 4\ A$	$I_{S_3} = 4\ A$
$N_P = 400$	$N_{S_1} = 200$	$N_{S_2} = 100$	$N_{S_3} = 20$
	$R_1 = 480\ \Omega$	$R_2 = 30\ \Omega$	$R_3 = 6\ \Omega$

ANSWERS TO PRACTICE PROBLEMS FOR UNIT 14

1.

$V_{P_P} = 120\ V$	$V_{P_S} = 60\ V$	$V_{P_L} = 34.6\ V$
$I_{P_P} = 2.5\ A$	$I_{P_S} = 5\ A$	$I_{P_L} = 8.66\ A$
$V_{L_P} = 208\ V$	$V_{L_S} = 60\ V$	Ratio = 2:1
$I_{L_P} = 2.5\ A$	$I_{L_S} = 8.66\ A$	$Z = 4\ \Omega$ per phase

2.

$V_{P_P} = 120\ V$

$I_{P_P} = 22.5\ A$

$V_{L_P} = 208\ V$

$I_{L_P} = 22.5\ A$

$V_{P_S} = 60\ V$

$I_{P_S} = 45\ A$

$V_{L_S} = 104\ V$

$I_{L_S} = 45\ A$

$V_{P_L} = 104\ V$

$I_{P_L} = 26\ A$

Ratio = 2:1

$Z = 4\ \Omega$ per phase

3.

$V_{P_P} = 208\ V$

$I_{P_P} = 39\ A$

$V_{L_P} = 208\ V$

$I_{L_P} = 67.5\ A$

$V_{P_S} = 104\ V$

$I_{P_S} = 78\ A$

$V_{L_S} = 180\ V$

$I_{L_S} = 78\ A$

$V_{P_L} = 180\ V$

$I_{P_L} = 45\ A$

Ratio = 2:1

$Z = 4\ \Omega$ per phase

4.

$V_{P_P} = 208\ V$

$I_{P_P} = 4.333\ A$

$V_{L_P} = 208\ V$

$I_{L_P} = 7.506\ A$

$V_{P_S} = 104\ V$

$I_{P_S} = 8.667\ A$

$V_{L_S} = 104\ V$

$I_{L_S} = 15.011\ A$

$V_{P_L} = 60.046\ V$

$I_{P_L} = 15.011\ A$

Ratio = 2:1

$Z = 4\ \Omega$ per phase

Glossary

AC (alternating current) Current that reverses its direction of flow periodically. Reversals generally occur at regular intervals.

Across-the-Line A method of motor starting that connects the motor directly to the supply line on starting or running (also known as *full-voltage starting*).

Alternator A machine used to generate alternating current by rotating conductors through a magnetic field.

Ambient Temperature The temperature surrounding a device.

American Wire Gauge (AWG) A measurement of the diameter of wire. The gauge scale was formally known as the Brown & Sharp scale. There is a fixed constant of 1.123 between gauge sizes.

Ammeter A device used to measure current.

Amortisseur Winding The squirrel-cage winding on the rotor of a synchronous motor used for starting.

Ampacity The maximum current rating of a wire or device.

Ampere A unit of measure for the rate of current flow.

Amplifier A device used to increase signal strength.

Amplitude The highest value reached by a signal, voltage, or current.

Analog Voltmeter A voltmeter that uses a meter movement to indicate the voltage value.

Anode The positive terminal of an electrical device.

Apparent Power The value found by multiplying together the applied voltage and current of an ac circuit. Apparent power, or volt-amperes, is not the true power of the circuit because it does not take power factor into consideration.

Applied Voltage The amount of voltage connected to a circuit or device.

ASA American Standards Association.

Atom The smallest part of an element that contains all the properties of that element.

Attenuator A device that decreases the amount of signal voltage or current.

Automatic Self-acting; operation by its own mechanical or electrical mechanism.

Autotransformer A transformer that uses one winding for both the primary and secondary.

Auxiliary Contacts Small contacts located on relays and motor starters used for the purpose of operating other control components.

Branch Circuit That portion of a wiring system that extends beyond the circuit protective device such as a fuse or circuit breaker.

Breakdown Torque The maximum amount of torque that can be developed by a motor at rated voltage and frequency before an abrupt change in speed occurs.

Bridge Circuit A circuit that consists of four sections connected in series to form a closed loop.

Busway An enclosed system used for power transmission that is voltage and current rated.

Capacitance The electrical size of a capacitor.

Capacitive Reactance The force of a capacitor that acts like resistance to limit the flow of current. Capacitive reactance is inversely proportional to the capacitance of the capacitor and the frequency.

Capacitor A device made with two conductive plates separated by an insulator or dielectric.

Capacitor-Start Motor A single-phase induction motor that uses a capacitor connected in series with the start winding to increase starting torque.

Cathode The negative terminal of an electrical device.

Center-Tapped Transformer A transformer that has a wire connected to the electrical midpoint of its winding. Generally, the secondary winding is tapped.

Choke An inductor designed to present an impedance to ac current or to be used as the current filter of a dc power supply.

Circuit Breaker A device designed to open under an abnormal amount of current flow. The device is not damaged and may be used repeatedly. It is rated by voltage, current, and horsepower.

Clock Timer A time-delay device that uses an electric clock to measure the delay time.

Collapse (of a magnetic field) Occurs when a magnetic field suddenly changes from its maximum value to a zero value.

Conduction Level The point at which an amount of voltage or current will cause a device to conduct.

Conductor A device or material that permits current to flow through it easily.

Contact A conducting part of a relay that acts as a switch to connect or disconnect a circuit or component.

Continuity A complete path for current flow.

Coulomb A quantity measurement of electrons. One coulomb equals 6,250,000,000,000,000,000 (6.25×10^{18}) electrons.

Current The rate of flow of electrons.

Current Rating The amount of current flow a device is designed to withstand.

Current Relay A relay that is operated by a predetermined amount of current flow. Current relays are often used as one type of starting relay for air-conditioning and refrigeration equipment.

DC (direct current) Current that does not reverse its direction of flow.

Delta Connection A circuit formed by connecting three electrical devices in series to form a closed loop. It is used most often in three-phase connections.

Dielectric An electrical insulator.

Digital Device A device that has only two states of operation: on or off.

Digital Logic Circuit elements connected in such a manner as to solve problems using components that have only two states of operation.

Digital Voltmeter A voltmeter that uses direct reading numerical display as opposed to a meter movement.

Disconnecting Means (disconnect) A device or group of devices used to disconnect a circuit or device from its source of supply.

Dynamic Braking (1) Using a dc motor as a generator to produce counter torque and thereby produce a braking action. (2) Applying direct current to the stator winding of an ac induction motor to cause a magnetic braking action.

Eddy Current Circular induced current contrary to the main currents. Eddy currents are a source of heat and power loss in magnetically operated devices.

Effective Value Also called the RMS value. It is the alternating current value that will produce the same heating effect through a resistive load as like amount of direct current.

Electrical Interlock When the contacts of one device or circuit prevent the operation of some other device or circuit.

Electric Controller A device or group of devices used to govern in some predetermined manner the operation of a circuit or piece of electrical apparatus.

Electromotive Force Abbreviated emf; the force described as voltage.

Electron One of the three major parts of an atom. The electron carries a negative charge.

Electronic Control A control circuit that uses solid-state devices as control components.

Enclosure Mechanical, electrical, or environmental protection for components used in a system.

Eutectic Alloy A metal with a low and sharp melting point used in thermal overload relays.

Feeder The circuit conductor between the service equipment or the generator switchboard of an isolated plant and the branch circuit overcurrent protective device.

Filter A device used to remove the ripple produced by a rectifier.

Frequency The number of complete cycles of ac voltage that occur in one second.

Full-Load Torque The amount of torque necessary to produce the full horsepower of a motor at rated speed.

Fuse A device used to protect a circuit or electrical device from excessive current. Fuses operate by melting a metal link when current becomes excessive.

Gain The increase in signal power produced by an amplifier.

Gate (1) A device that has multiple inputs and a single output. There are five basic types of gates: the *and, or, nand, nor,* and *inverter*. (2) One terminal of some electronic devices such as silicon controlled rectifiers (SCRs), triacs, and field effect transistors.

Heat Sink A metallic device designed to increase the surface area of an electronic component for the purpose of removing heat at a faster rate.

Hertz The international unit of frequency.

Holding Contacts Contacts used for the purpose of maintaining current flow to the coil of a relay.

Holding Current The amount of current needed to keep an SCR or triac turned on.

Horsepower A measure of power for electrical and mechanical devices.

Hysteresis Loop A graphical curve that shows the value of magnetizing force for a particular type of material.

Hysteresis Loss A loss that occurs as a result of molecular friction caused by molecules being rubbed together. Hysteresis loss increases with an increase of frequency.

Impedance The total opposition to current flow in an electrical circuit.

Induced Current Current produced in a conductor by the cutting action of a magnetic field.

Inductive Reactance The force that acts like resistance in an inductor to limit the flow of current. Inductive reactance is proportional to the inductance of the coil in henrys and the frequency in hertz.

Inductor A coil.

Input Voltage The amount of voltage connected to a device or circuit.

Insulator A material used to electrically isolate two conductive surfaces.

Interlock A device used to prevent some action from taking place in a piece of equipment or circuit until some other action has occurred.

Isolation Transformer A transformer whose secondary winding is electrically isolated from its primary winding.

Jumper A short piece of conductor used to make connection between components or a break in a circuit.

Limit Switch A mechanically operated switch that detects the position or movement of an object.

Load Center Generally, the service entrance. A point from which branch circuits originate.

Locked-Rotor Current The amount of current produced when voltage is applied to a motor and the rotor is not turning.

Locked-Rotor Torque The amount of torque produced by a motor at the time of starting.

Lockout A mechanical device used to prevent the operation of some other component.

Low-Voltage Protection A magnetic relay circuit connected so that a drop in voltage causes the motor starter to disconnect the motor from the line.

Magnetic Contactor A contactor operated electromechanically.

Magnetic Field The space in which a magnetic force exists.

Maintaining Contact Also known as a *holding* or *sealing* contact. It is used to maintain the coil circuit in a relay control circuit. The contact is connected in parallel with the start push button.

Manual Controller A controller operated by hand at the location of the controller.

Microfarad A measurement of capacitance.

Microprocessor A small computer. The central processing unit is generally made from a single integrated circuit.

Mode A state of condition.

Motor A device used to convert electrical energy into rotating motion.

Motor Controller A device used to control the operation of a motor.

Multispeed Motor A motor that can be operated at more than one speed.

Negative One polarity of a voltage, current, or charge.

NEMA National Electrical Manufacturers Association.

NEMA Ratings Electrical control device ratings of voltage, current, horsepower, and interrupting capability given by NEMA.

Neutron One of the principal parts of an atom. The neutron has no charge and is part of the nucleus.

Noninductive Load An electrical load that does not have induced voltages caused by a coil. Noninductive loads are generally considered to be resistive but can be capacitive.

Nonreversing A device that can be operated in only one direction.

Normally Closed The contact of a relay that is closed when the coil is deenergized.

Normally Open The contact of a relay that is open when the coil is deenergized.

Off-Delay Timer A timer that delays changing its contacts back to their normal position when the coil is deenergized.

Ohmmeter A device used to measure resistance.

On-Delay Timer A timer that delays changing the position of its contacts when the coil is energized.

Oscillator A device used to change dc voltage into ac voltage.

Oscilloscope A voltmeter that displays a waveform of voltage in proportion to its amplitude with respect to time.

Out-of-Phase The condition in which two components do not reach their positive or negative peaks at the same time.

Overload Relay A relay used to protect a motor from damage due to overloads. The overload relay senses motor current and disconnects the motor from the line if the current is excessive for a certain length of time.

Panelboard A metallic or nonmetallic panel used to mount electrical controls, equipment, or devices.

Parallel Circuit A circuit that contains more than one path for current flow.

Peak-to-Peak Voltage The amplitude of ac voltage measured from its positive peak to its negative peak.

Peak Voltage The amplitude of voltage measured from zero to its highest value.

Permanent Split-Capacitor Motor A single-phase induction motor similar to the capacitor-start motor except that the start windings and the starting capacitor remain connected in the circuit during normal operation.

Phase Shift A change in the phase relationship between two quantities of voltage or current.

Phasor An entity that includes both magnitude and direction.

Pilot Device A control component designed to control small amounts of current. Pilot devices are used to control larger control components.

Pneumatic Timer A device that uses the displacement of air in a bellows or diaphragm to produce a time delay.

Polarity The characteristic of a device that exhibits opposite quantities within itself: positive and negative.

Potentiometer A variable resistor with a sliding contact that is used as a voltage divider.

Power Factor A comparison of the true power (watts) to the apparent power (volt-amperes) in an ac circuit.

Power Rating The rating of a device that indicates the amount of current flow and voltage drop that can be permitted.

Pressure Switch A device that senses the presence or absence of pressure and causes a set of contacts to open or close.

Printed Circuit A board on which a predetermined pattern of printed connections has been made.

Proton One of the three major parts of an atom. The proton has a positive charge.

Push button A pilot control device operated manually by being pushed or pressed.

Reactance The opposition to current flow in an ac circuit offered by pure inductance or pure capacitance.

Rectifier A device or circuit used to change ac voltage into dc voltage.

Regulator A device that maintains a quantity at a predetermined level.

Relay A magnetically operated switch that may have one or more sets of contacts.

Remote Control Controls the functions of an electrical device from a distant location.

Repulsion Motor A type of single-phase ac motor that operates on the principle that like magnetic fields repel each other.

Resistance The opposition to current flow in an ac or dc circuit.

Resistance-Start, Induction-Run Motor One type of split-phase motor that uses the resistance of the start winding to produce a phase shift between the current in the start winding and the current in the run winding.

Resistor A device used to introduce some amount of resistance into an electrical circuit.

Rheostat A variable resistor.

RMS Value The value of ac voltage that will produce as much power when connected across a resistor as a like amount of dc voltage.

Salient Pole A projecting pole piece, generally made of iron or laminated steel and containing its own winding or coil.

Saturation The maximum amount of magnetic flux a material can hold.

Schematic An electrical drawing showing components in their electrical sequence without regard for physical location.
Sealing Contacts Contacts connected in parallel with the start button and used to provide a continued path for current flow to the coil of the contactor when the start button is released. See also *Holding Contacts* and *Maintaining Contact.*
Sensing Device A pilot device that detects some quantity and converts it into an electrical signal.
Series Circuit A circuit that contains only one path for current flow.
Service The conductors and equipment necessary to deliver energy from the electrical supply system to the premises served.
Service Factor An allowable overload for a motor indicated by a multiplier that, when applied to a normal horsepower rating, indicates the permissible loading.
Shaded-Pole Motor An ac induction motor that develops a rotating magnetic field by shading part of the stator windings with a shading loop.
Shading Loop A large copper wire or band connected around part of a magnetic pole piece to oppose a change of magnetic flux.
Short Circuit An electrical circuit that contains no resistance to limit the flow of current.
Short Cycling The starting and stopping of a compressor in rapid succession.
Sine-Wave Voltage A voltage waveform whose value at any point is proportional to the trigonometric sine of the angle of the generator producing it.
Slip The difference in speed between the rotating magnetic field and the speed of the rotor in an induction motor.
Snap-Action The quick opening and closing action of a spring-loaded contact.
Solenoid A magnetic device used to convert electrical energy into linear motion.
Solenoid Valve A valve operated by an electric solenoid.
Solid-State Device An electronic component constructed from semiconductor material.
Split-Phase Motor A type of single-phase motor that uses resistance or capacitance to cause a shift in the phase of the current in the run winding and the current in the start winding. The three primary types of split-phase motors are resistance-start, induction-run; capacitor-start, induction-run; and permanent split-capacitor motor.
Squirrel-Cage Motor A type of ac motor that uses a squirrel-cage rotor. All squirrel-cage motors are induction motors.
Starter A relay used to connect a motor to the power line.
Stator The stationary winding of an ac motor.
Stepdown Transformer A transformer that produces a lower voltage at its secondary than is applied to its primary.
Stepup Transformer A transformer that produces a higher voltage at its secondary than is applied to its primary.
Surge A transient variation in the current or voltage at a point in the circuit. Surges are generally unwanted and temporary.
Switch A mechanical device used to connect or disconnect a component or circuit.
Synchronous Motor A type of ac motor that operates at the speed of the rotating magnetic field. Three-phase synchronous motors can be made to operate with a leading power factor by overexcitation of the rotor field.

Synchronous Speed The speed of the rotating magnetic field of an ac induction motor.

Terminal A fitting attached to a device for the purpose of connecting wires to it.

Thermistor A resistor that changes its resistance with a change of temperature.

Thyristor An electronic component that has only two states of operation: on or off.

Torque The turning force developed by a motor.

Transducer A device that converts one type of energy into another type of energy. For example, a solar cell converts light into electricity.

Transformer An electrical device that changes one value of ac voltage into another value of ac voltage.

Troubleshoot To locate and eliminate problems in a circuit.

Valence Electrons Electrons located in the outer orbit of an atom.

Variable Resistor A resistor whose resistance value can be varied between its minimum and maximum values.

Varistor A resistor that changes its resistance value with a change of voltage.

Vector A quantity that has both magnitude and direction.

Voltage An electrical measurement of potential difference, electrical pressure, or electromotive force (emf).

Voltage Drop The amount of voltage required to cause an amount of current to flow through a certain resistance.

Voltage Rating A rating that indicates the amount of voltage that can be safely connected to a device.

Voltage Regulator A device or circuit that maintains a constant value of voltage.

Voltmeter An instrument used to measure a level of voltage.

Volt-Ohm-Milliammeter (VOM) A test instrument so designed that it can be used to measure voltage, resistance, or milliamperes.

Watt A measure of true power.

Waveform The shape of a wave as obtained by plotting a graph with respect to voltage and time.

Wiring Diagram An electrical diagram used to show components in their approximate physical location with connecting wires.

Wound-Rotor Motor A type of three-phase motor that uses a wound rotor to gain control of the current flow through the rotor.

Wye Connection A connection of three components made in such a manner that one end of each component is connected. This connection is generally used to connect devices to a three-phase power system.

X_C Symbol for capacitive reactance.

X_L Symbol for inductive reactance.

Z Symbol for impedance.

Index

D

E

J

K

L

M

N

Q

R

T

Z

CPSIA information can be obtained
at www.ICGtesting.com
Printed in the USA
FFOW03n1715100816
26696FF